# Analytical Mechanics

# Analytical Mechanics
## An Introduction

### Antonio Fasano
*University of Florence*

### Stefano Marmi
*SNS, Pisa*

Translated by
Beatrice Pelloni
*University of Reading*

**Library
Quest University Canada
3200 University Boulevard
Squamish, BC V8B 0N8**

# OXFORD
## UNIVERSITY PRESS

Great Clarendon Street, Oxford OX2 6DP

Oxford University Press is a department of the University of Oxford.
It furthers the University's objective of excellence in research, scholarship,
and education by publishing worldwide in

Oxford  New York

Auckland  Cape Town  Dar es Salaam  Hong Kong  Karachi
Kuala Lumpur  Madrid  Melbourne  Mexico City  Nairobi
New Delhi  Shanghai  Taipei  Toronto

With offices in

Argentina  Austria  Brazil  Chile  Czech Republic  France  Greece
Guatemala  Hungary  Italy  Japan  Poland  Portugal  Singapore
South Korea  Switzerland  Thailand  Turkey  Ukraine  Vietnam

Oxford is a registered trade mark of Oxford University Press
in the UK and in certain other countries

Published in the United States
by Oxford University Press Inc., New York

© 2002, Bollati Boringhieri editore, Torino

English translation © Oxford University Press 2006

Translation of Meccanica Analitica by Antonio Fasano and
Stefano Marmi originally published in
Italian by Bollati Boringhieri editore, Torino 2002

The moral rights of the authors have been asserted
Database right Oxford University Press (maker)

First published in English 2006

All rights reserved. No part of this publication may be reproduced,
stored in a retrieval system, or transmitted, in any form or by any means,
without the prior permission in writing of Oxford University Press,
or as expressly permitted by law, or under terms agreed with the appropriate
reprographics rights organization. Enquiries concerning reproduction
outside the scope of the above should be sent to the Rights Department,
Oxford University Press, at the address above

You must not circulate this book in any other binding or cover
and you must impose the same condition on any acquirer

British Library Cataloguing in Publication Data
Data available

Library of Congress Cataloging in Publication Data

Fasano, A. (Antonio)
Analytical mechanics : an introduction / Antonio Fasano, Stefano Marmi;
translated by Beatrice Pelloni.
   p. cm.
Includes bibliographical references and index.
ISBN-13: 978–0–19–850802–1
ISBN-10: 0–19–850802–6
1. Mechanics, Analytic.   I. Marmi, S. (Stefano), 1963-   II. Title.
QA805.2.F29 2002
531'.01—dc22                                                            2005028822

Typeset by Newgen Imaging Systems (P) Ltd., Chennai, India
Printed in Great Britain
on acid-free paper by
Antony Rowe Ltd., Chippenham, Wiltshire

ISBN 0–19–850802–6   978–0–19–850802–1

1 3 5 7 9 10 8 6 4 2

# Preface to the English Translation

The proposal of translating this book into English came from Dr. Sonke Adlung of OUP, to whom we express our gratitude. The translation was preceded by hard work to produce a new version of the Italian text incorporating some modifications we had agreed upon with Dr. Adlung (for instance the inclusion of worked out problems at the end of each chapter). The result was the second Italian edition (Bollati-Boringhieri, 2002), which was the original source for the translation. However, thanks to the kind collaboration of the translator, Dr. Beatrice Pelloni, in the course of the translation we introduced some further improvements with the aim of better fulfilling the original aim of this book: to explain analytical mechanics (which includes some very complex topics) with mathematical rigour using nothing more than the notions of plain calculus. For this reason the book should be readable by undergraduate students, although it contains some rather advanced material which makes it suitable also for courses of higher level mathematics and physics.

Despite the size of the book, or rather because of it, conciseness has been a constant concern of the authors. The book is large because it deals not only with the basic notions of analytical mechanics, but also with some of its main applications: astronomy, statistical mechanics, continuum mechanics and (very briefly) field theory.

The book has been conceived in such a way that it can be used at different levels: for instance the two chapters on statistical mechanics can be read, skipping the chapter on ergodic theory, etc. The book has been used in various Italian universities for more than ten years and we have been very pleased by the reactions of colleagues and students. Therefore we are confident that the translation can prove to be useful.

Antonio Fasano<br>
Stefano Marmi

# Contents

## 1 Geometric and kinematic foundations of Lagrangian mechanics ... 1
1.1 Curves in the plane ... 1
1.2 Length of a curve and natural parametrisation ... 3
1.3 Tangent vector, normal vector and curvature of plane curves ... 7
1.4 Curves in $\mathbf{R}^3$ ... 12
1.5 Vector fields and integral curves ... 15
1.6 Surfaces ... 16
1.7 Differentiable Riemannian manifolds ... 33
1.8 Actions of groups and tori ... 46
1.9 Constrained systems and Lagrangian coordinates ... 49
1.10 Holonomic systems ... 52
1.11 Phase space ... 54
1.12 Accelerations of a holonomic system ... 57
1.13 Problems ... 58
1.14 Additional remarks and bibliographical notes ... 61
1.15 Additional solved problems ... 62

## 2 Dynamics: general laws and the dynamics of a point particle ... 69
2.1 Revision and comments on the axioms of classical mechanics ... 69
2.2 The Galilean relativity principle and interaction forces ... 71
2.3 Work and conservative fields ... 75
2.4 The dynamics of a point constrained by smooth holonomic constraints ... 77
2.5 Constraints with friction ... 80
2.6 Point particle subject to unilateral constraints ... 81
2.7 Additional remarks and bibliographical notes ... 83
2.8 Additional solved problems ... 83

## 3 One-dimensional motion ... 91
3.1 Introduction ... 91
3.2 Analysis of motion due to a positional force ... 92
3.3 The simple pendulum ... 96
3.4 Phase plane and equilibrium ... 98
3.5 Damped oscillations, forced oscillations. Resonance ... 103
3.6 Beats ... 107
3.7 Problems ... 108
3.8 Additional remarks and bibliographical notes ... 112
3.9 Additional solved problems ... 113

## 4 The dynamics of discrete systems. Lagrangian formalism .... 125
- 4.1 Cardinal equations .... 125
- 4.2 Holonomic systems with smooth constraints .... 127
- 4.3 Lagrange's equations .... 128
- 4.4 Determination of constraint reactions. Constraints with friction .... 136
- 4.5 Conservative systems. Lagrangian function .... 138
- 4.6 The equilibrium of holonomic systems with smooth constraints .... 141
- 4.7 Generalised potentials. Lagrangian of an electric charge in an electromagnetic field .... 142
- 4.8 Motion of a charge in a constant electric or magnetic field .... 144
- 4.9 Symmetries and conservation laws. Noether's theorem .... 147
- 4.10 Equilibrium, stability and small oscillations .... 150
- 4.11 Lyapunov functions .... 159
- 4.12 Problems .... 162
- 4.13 Additional remarks and bibliographical notes .... 165
- 4.14 Additional solved problems .... 165

## 5 Motion in a central field .... 179
- 5.1 Orbits in a central field .... 179
- 5.2 Kepler's problem .... 185
- 5.3 Potentials admitting closed orbits .... 187
- 5.4 Kepler's equation .... 193
- 5.5 The Lagrange formula .... 197
- 5.6 The two-body problem .... 200
- 5.7 The $n$-body problem .... 201
- 5.8 Problems .... 205
- 5.9 Additional remarks and bibliographical notes .... 207
- 5.10 Additional solved problems .... 208

## 6 Rigid bodies: geometry and kinematics .... 213
- 6.1 Geometric properties. The Euler angles .... 213
- 6.2 The kinematics of rigid bodies. The fundamental formula .... 216
- 6.3 Instantaneous axis of motion .... 219
- 6.4 Phase space of precessions .... 221
- 6.5 Relative kinematics .... 223
- 6.6 Relative dynamics .... 226
- 6.7 Ruled surfaces in a rigid motion .... 228
- 6.8 Problems .... 230
- 6.9 Additional solved problems .... 231

## 7 The mechanics of rigid bodies: dynamics .... 235
- 7.1 Preliminaries: the geometry of masses .... 235
- 7.2 Ellipsoid and principal axes of inertia .... 236

|       |                                              |     |
|-------|----------------------------------------------|-----|
| 7.3   | Homography of inertia                        | 239 |
| 7.4   | Relevant quantities in the dynamics of rigid bodies | 242 |
| 7.5   | Dynamics of free systems                     | 244 |
| 7.6   | The dynamics of constrained rigid bodies     | 245 |
| 7.7   | The Euler equations for precessions          | 250 |
| 7.8   | Precessions by inertia                       | 251 |
| 7.9   | Permanent rotations                          | 254 |
| 7.10  | Integration of Euler equations               | 256 |
| 7.11  | Gyroscopic precessions                       | 259 |
| 7.12  | Precessions of a heavy gyroscope (spinning top) | 261 |
| 7.13  | Rotations                                    | 263 |
| 7.14  | Problems                                     | 265 |
| 7.15  | Additional solved problems                   | 266 |

## 8 Analytical mechanics: Hamiltonian formalism ... 279
| 8.1 | Legendre transformations | 279 |
| 8.2 | The Hamiltonian | 282 |
| 8.3 | Hamilton's equations | 284 |
| 8.4 | Liouville's theorem | 285 |
| 8.5 | Poincaré recursion theorem | 287 |
| 8.6 | Problems | 288 |
| 8.7 | Additional remarks and bibliographical notes | 291 |
| 8.8 | Additional solved problems | 291 |

## 9 Analytical mechanics: variational principles ... 301
| 9.1 | Introduction to the variational problems of mechanics | 301 |
| 9.2 | The Euler equations for stationary functionals | 302 |
| 9.3 | Hamilton's variational principle: Lagrangian form | 312 |
| 9.4 | Hamilton's variational principle: Hamiltonian form | 314 |
| 9.5 | Principle of the stationary action | 316 |
| 9.6 | The Jacobi metric | 318 |
| 9.7 | Problems | 323 |
| 9.8 | Additional remarks and bibliographical notes | 324 |
| 9.9 | Additional solved problems | 324 |

## 10 Analytical mechanics: canonical formalism ... 331
| 10.1 | Symplectic structure of the Hamiltonian phase space | 331 |
| 10.2 | Canonical and completely canonical transformations | 340 |
| 10.3 | The Poincaré–Cartan integral invariant. The Lie condition | 352 |
| 10.4 | Generating functions | 364 |
| 10.5 | Poisson brackets | 371 |
| 10.6 | Lie derivatives and commutators | 374 |
| 10.7 | Symplectic rectification | 380 |

|  |  |  |
|---|---|---|
| 10.8 | Infinitesimal and near-to-identity canonical transformations. Lie series | 384 |
| 10.9 | Symmetries and first integrals | 393 |
| 10.10 | Integral invariants | 395 |
| 10.11 | Symplectic manifolds and Hamiltonian dynamical systems | 397 |
| 10.12 | Problems | 399 |
| 10.13 | Additional remarks and bibliographical notes | 404 |
| 10.14 | Additional solved problems | 405 |

## 11 Analytic mechanics: Hamilton–Jacobi theory and integrability .......... 413

|  |  |  |
|---|---|---|
| 11.1 | The Hamilton–Jacobi equation | 413 |
| 11.2 | Separation of variables for the Hamilton–Jacobi equation | 421 |
| 11.3 | Integrable systems with one degree of freedom: action-angle variables | 431 |
| 11.4 | Integrability by quadratures. Liouville's theorem | 439 |
| 11.5 | Invariant $l$-dimensional tori. The theorem of Arnol'd | 446 |
| 11.6 | Integrable systems with several degrees of freedom: action-angle variables | 453 |
| 11.7 | Quasi-periodic motions and functions | 458 |
| 11.8 | Action-angle variables for the Kepler problem. Canonical elements, Delaunay and Poincaré variables | 466 |
| 11.9 | Wave interpretation of mechanics | 471 |
| 11.10 | Problems | 477 |
| 11.11 | Additional remarks and bibliographical notes | 480 |
| 11.12 | Additional solved problems | 481 |

## 12 Analytical mechanics: canonical perturbation theory .......... 487

|  |  |  |
|---|---|---|
| 12.1 | Introduction to canonical perturbation theory | 487 |
| 12.2 | Time periodic perturbations of one-dimensional uniform motions | 499 |
| 12.3 | The equation $D_\omega u = v$. Conclusion of the previous analysis | 502 |
| 12.4 | Discussion of the fundamental equation of canonical perturbation theory. Theorem of Poincaré on the non-existence of first integrals of the motion | 507 |
| 12.5 | Birkhoff series: perturbations of harmonic oscillators | 516 |
| 12.6 | The Kolmogorov–Arnol'd–Moser theorem | 522 |
| 12.7 | Adiabatic invariants | 529 |
| 12.8 | Problems | 532 |

12.9   Additional remarks and bibliographical notes . . . . . . . . . . . 534
  12.10  Additional solved problems . . . . . . . . . . . . . . . . . . . . . . 535

# 13 Analytical mechanics: an introduction to ergodic theory and to chaotic motion . . . . . . . . . . . . . . . . . 545
  13.1   The concept of measure . . . . . . . . . . . . . . . . . . . . . . . . . 545
  13.2   Measurable functions. Integrability . . . . . . . . . . . . . . . . 548
  13.3   Measurable dynamical systems . . . . . . . . . . . . . . . . . . . 550
  13.4   Ergodicity and frequency of visits . . . . . . . . . . . . . . . . . 554
  13.5   Mixing . . . . . . . . . . . . . . . . . . . . . . . . . . . . . . . . . . . . . 563
  13.6   Entropy . . . . . . . . . . . . . . . . . . . . . . . . . . . . . . . . . . . . 565
  13.7   Computation of the entropy. Bernoulli schemes.
         Isomorphism of dynamical systems . . . . . . . . . . . . . . . . 571
  13.8   Dispersive billiards . . . . . . . . . . . . . . . . . . . . . . . . . . . . 575
  13.9   Characteristic exponents of Lyapunov.
         The theorem of Oseledec . . . . . . . . . . . . . . . . . . . . . . . . 578
  13.10  Characteristic exponents and entropy . . . . . . . . . . . . . . . 581
  13.11  Chaotic behaviour of the orbits of planets
         in the Solar System . . . . . . . . . . . . . . . . . . . . . . . . . . . 582
  13.12  Problems . . . . . . . . . . . . . . . . . . . . . . . . . . . . . . . . . . . 584
  13.13  Additional solved problems . . . . . . . . . . . . . . . . . . . . . . 586
  13.14  Additional remarks and bibliographical notes . . . . . . . . . . . 590

# 14 Statistical mechanics: kinetic theory . . . . . . . . . . . . . . . . . . 591
  14.1   Distribution functions . . . . . . . . . . . . . . . . . . . . . . . . . . 591
  14.2   The Boltzmann equation . . . . . . . . . . . . . . . . . . . . . . . . 592
  14.3   The hard spheres model . . . . . . . . . . . . . . . . . . . . . . . . 596
  14.4   The Maxwell–Boltzmann distribution . . . . . . . . . . . . . . . 599
  14.5   Absolute pressure and absolute temperature
         in an ideal monatomic gas . . . . . . . . . . . . . . . . . . . . . . . 601
  14.6   Mean free path . . . . . . . . . . . . . . . . . . . . . . . . . . . . . . 604
  14.7   The 'H theorem' of Boltzmann. Entropy . . . . . . . . . . . . . 605
  14.8   Problems . . . . . . . . . . . . . . . . . . . . . . . . . . . . . . . . . . . 609
  14.9   Additional solved problems . . . . . . . . . . . . . . . . . . . . . . 610
  14.10  Additional remarks and bibliographical notes . . . . . . . . . . . 611

# 15 Statistical mechanics: Gibbs sets . . . . . . . . . . . . . . . . . . . . . 613
  15.1   The concept of a statistical set . . . . . . . . . . . . . . . . . . . 613
  15.2   The ergodic hypothesis: averages and
         measurements of observable quantities . . . . . . . . . . . . . . 616
  15.3   Fluctuations around the average . . . . . . . . . . . . . . . . . . 620
  15.4   The ergodic problem and the existence of first integrals . . . . 621
  15.5   Closed isolated systems (prescribed energy).
         Microcanonical set . . . . . . . . . . . . . . . . . . . . . . . . . . . . 624

| | | |
|---|---|---|
| 15.6 | Maxwell–Boltzmann distribution and fluctuations in the microcanonical set | 627 |
| 15.7 | Gibbs' paradox | 631 |
| 15.8 | Equipartition of the energy (prescribed total energy) | 634 |
| 15.9 | Closed systems with prescribed temperature. Canonical set | 636 |
| 15.10 | Equipartition of the energy (prescribed temperature) | 640 |
| 15.11 | Helmholtz free energy and orthodicity of the canonical set | 645 |
| 15.12 | Canonical set and energy fluctuations | 646 |
| 15.13 | Open systems with fixed temperature. Grand canonical set | 647 |
| 15.14 | Thermodynamical limit. Fluctuations in the grand canonical set | 651 |
| 15.15 | Phase transitions | 654 |
| 15.16 | Problems | 656 |
| 15.17 | Additional remarks and bibliographical notes | 659 |
| 15.18 | Additional solved problems | 662 |

**16 Lagrangian formalism in continuum mechanics** . . . . . . . . . . . . . 671
    16.1  Brief summary of the fundamental laws of continuum mechanics . . . . . . . . . . . . . . . . . . . . . . . . . . . 671
    16.2  The passage from the discrete to the continuous model. The Lagrangian function . . . . . . . . . . . . . . . . . . . . . . . . . . 676
    16.3  Lagrangian formulation of continuum mechanics . . . . . . . . . 678
    16.4  Applications of the Lagrangian formalism to continuum mechanics . . . . . . . . . . . . . . . . . . . . . . . . . . . . . . . 680
    16.5  Hamiltonian formalism . . . . . . . . . . . . . . . . . . . . . . . 684
    16.6  The equilibrium of continua as a variational problem. Suspended cables . . . . . . . . . . . . . . . . . . . . . . . . . . 685
    16.7  Problems . . . . . . . . . . . . . . . . . . . . . . . . . . . . . . 690
    16.8  Additional solved problems . . . . . . . . . . . . . . . . . . . . 691

**Appendices**
    Appendix 1: Some basic results on ordinary differential equations . . . . . . . . . . . . . . . . . . . . . . . . . . . . . 695
        A1.1  General results . . . . . . . . . . . . . . . . . . . . . . . . 695
        A1.2  Systems of equations with constant coefficients . . . . . . . . 697
        A1.3  Dynamical systems on manifolds . . . . . . . . . . . . . . . 701
    Appendix 2: Elliptic integrals and elliptic functions . . . . . . . . . . . 705
    Appendix 3: Second fundamental form of a surface . . . . . . . . . . . 709
    Appendix 4: Algebraic forms, differential forms, tensors . . . . . . . . 715
        A4.1  Algebraic forms . . . . . . . . . . . . . . . . . . . . . . . . 715
        A4.2  Differential forms . . . . . . . . . . . . . . . . . . . . . . . 719
        A4.3  Stokes' theorem . . . . . . . . . . . . . . . . . . . . . . . . 724
        A4.4  Tensors . . . . . . . . . . . . . . . . . . . . . . . . . . . . 726

Appendix 5: Physical realisation of constraints ............. 729
Appendix 6: Kepler's problem, linear oscillators
and geodesic flows ................................. 733
Appendix 7: Fourier series expansions ................... 741
Appendix 8: Moments of the Gaussian distribution
and the Euler $\Gamma$ function ............................ 745

**Bibliography** ........................................... 749

**Index** ................................................. 759

# 1 GEOMETRIC AND KINEMATIC FOUNDATIONS OF LAGRANGIAN MECHANICS

Geometry is the art of deriving good reasoning from badly drawn pictures[1]

The first step in the construction of a mathematical model for studying the motion of a system consisting of a certain number of points is necessarily the investigation of its geometrical properties. Such properties depend on the possible presence of limitations (constraints) imposed on the position of each single point with respect to a given reference frame. For a one-point system, it is intuitively clear what it means for the system to be constrained to lie on a curve or on a surface, and how this constraint limits the possible motions of the point. The geometric and hence the kinematic description of the system becomes much more complicated when the system contains two or more points, mutually constrained; an example is the case when the distance between each pair of points in the system is fixed. The correct set-up of the framework for studying this problem requires that one first considers some fundamental geometrical properties; the study of these properties is the subject of this chapter.

## 1.1 Curves in the plane

Curves in the plane can be thought of as *level sets* of functions $F : U \to \mathbf{R}$ (for our purposes, it is sufficient for $F$ to be of class $\mathcal{C}^2$), where $U$ is an open connected subset of $\mathbf{R}^2$. The curve $C$ is defined as the set

$$C = \{(x_1, x_2) \in U | F(x_1, x_2) = 0\}. \tag{1.1}$$

We assume that this set is non-empty.

DEFINITION 1.1  *A point $P$ on the curve (hence such that $F(x_1, x_2) = 0$) is called* non-singular *if the gradient of $F$ computed at $P$ is non-zero:*

$$\nabla F(x_1, x_2) \neq 0. \tag{1.2}$$

*A curve $C$ whose points are all non-singular is called a regular curve.* ∎

By the implicit function theorem, if $P$ is non-singular, in a neighbourhood of $P$ the curve is representable as the graph of a function $x_2 = f(x_1)$, if $(\partial F/\partial x_2)_P \neq 0$,

---

[1] Anonymous quotation, in Felix Klein, *Vorlesungen über die Entwicklung der Mathematik im 19. Jahrhundert*, Springer-Verlag, Berlin 1926.

or of a function $x_1 = f(x_2)$, if $(\partial F/\partial x_1)_P \neq 0$. The function $f$ is differentiable in the same neighbourhood. If $x_2$ is the dependent variable, for $x_1$ in a suitable open interval $I$,

$$C = \text{graph}(f) = \{(x_1, x_2) \in \mathbf{R}^2 | x_1 \in I, x_2 = f(x_1)\}, \tag{1.3}$$

and

$$f'(x_1) = -\frac{\partial F/\partial x_1}{\partial F/\partial x_2}.$$

Equation (1.3) implies that, at least locally, the points of the curve are in one-to-one correspondence with the values of one of the Cartesian coordinates.

The *tangent line* at a non-singular point $\mathbf{x}_0 = \mathbf{x}(t_0)$ can be defined as the first-order term in the series expansion of the difference $\mathbf{x}(t) - \mathbf{x}_0 \sim (t-t_0)\dot{\mathbf{x}}(t_0)$, i.e. as the best linear approximation to the curve in the neighbourhood of $\mathbf{x}_0$. Since $\dot{\mathbf{x}} \cdot \nabla F(\mathbf{x}(t)) = 0$, the vector $\dot{\mathbf{x}}(t_0)$, which characterises the tangent line and can be called the *velocity* on the curve, is orthogonal to $\nabla F(\mathbf{x}_0)$ (Fig. 1.1).

More generally, it is possible to use a *parametric representation* (of class $\mathcal{C}^2$) $\mathbf{x} : (a, b) \to \mathbf{R}^2$, where $(a, b)$ is an open interval in $\mathbf{R}$:

$$C = \mathbf{x}((a,b)) = \{(x_1, x_2) \in \mathbf{R}^2 | \text{ there exists } t \in (a,b), (x_1, x_2) = \mathbf{x}(t)\}. \tag{1.4}$$

Note that the graph (1.3) can be interpreted as the parametrisation $\mathbf{x}(t) = (t, f(t))$, and that it is possible to go from (1.3) to (1.4) introducing a function $x_1 = x_1(t)$ of class $\mathcal{C}^2$ and such that $\dot{x}_1(t) \neq 0$.

It follows that Definition 1.1 is equivalent to the following.

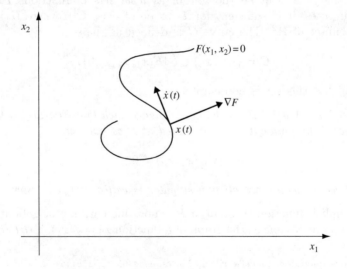

**Fig. 1.1**

DEFINITION 1.2 *If the curve $C$ is given in the parametric form $\mathbf{x} = \mathbf{x}(t)$, a point $\mathbf{x}(t_0)$ is called* non-singular *if $\dot{\mathbf{x}}(t_0) \neq 0$.* ∎

*Example* 1.1
A circle $x_1^2 + x_2^2 - R^2 = 0$ centred at the origin and of radius $R$ is a regular curve, and can be represented parametrically as $x_1 = R \cos t$, $x_2 = R \sin t$; alternatively, if one restricts to the half-plane $x_2 > 0$, it can be represented as the graph $x_2 = \sqrt{1 - x_1^2}$. The circle of radius 1 is usually denoted $\mathbf{S}^1$ or $\mathbf{T}^1$. ∎

*Example* 1.2
Conic sections are the level sets of the second-order polynomials $F(x_1, x_2)$. The ellipse (with reference to the principal axes) is defined by

$$\frac{x_1^2}{a^2} + \frac{x_2^2}{b^2} - 1 = 0,$$

where $a > b > 0$ denote the lengths of the semi-axes. One easily verifies that such a level set is a regular curve and that a parametric representation is given by $x_1 = a \sin t$, $x_2 = b \cos t$. Similarly, the hyperbola is given by

$$\frac{x_1^2}{a^2} - \frac{x_2^2}{b^2} - 1 = 0$$

and admits the parametric representation $x_1 = a \cosh t$, $x_2 = b \sinh t$. The parabola $x_2 - ax_1^2 - bx_1 - c = 0$ is already given in the form of a graph. ∎

*Remark* 1.1
In an analogous way one can define the curves in $\mathbf{R}^n$ (cf. Giusti 1989) as maps $\mathbf{x} : (a, b) \to \mathbf{R}^n$ of class $\mathcal{C}^2$, where $(a, b)$ is an open interval in $\mathbf{R}$. The vector $\dot{\mathbf{x}}(t) = (\dot{x}_1(t), \ldots, \dot{x}_n(t))$ can be interpreted as the velocity of a point moving in space according to $\mathbf{x} = \mathbf{x}(t)$ (i.e. along the parametrised curve).

The concept of curve can be generalised in various ways; as an example, when considering the kinematics of rigid bodies, we shall introduce 'curves' defined in the space of matrices, see Examples 1.27 and 1.28 in this chapter. ∎

## 1.2 Length of a curve and natural parametrisation

Let $C$ be a regular curve, described by the parametric representation $\mathbf{x} = \mathbf{x}(t)$.

DEFINITION 1.3 *The length $l$ of the curve $\mathbf{x} = \mathbf{x}(t)$, $t \in (a, b)$, is given by the integral*

$$l = \int_a^b \sqrt{\dot{\mathbf{x}}(t) \cdot \dot{\mathbf{x}}(t)} \, \mathrm{d}t = \int_a^b |\dot{\mathbf{x}}(t)| \, \mathrm{d}t. \tag{1.5}$$

∎

In the particular case of a graph $x_2 = f(x_1)$, equation (1.5) becomes

$$l = \int_a^b \sqrt{1 + (f'(t))^2}\, dt. \tag{1.6}$$

*Example 1.3*
Consider a circle of radius $r$. Since $|\dot{\mathbf{x}}(t)| = |(-r\sin t, r\cos t)| = r$, we have $l = \int_0^{2\pi} r\, dt = 2\pi r$. ∎

*Example 1.4*
The length of an ellipse with semi-axes $a \geq b$ is given by

$$l = \int_0^{2\pi} \sqrt{a^2 \cos^2 t + b^2 \sin^2 t}\, dt = 4a \int_0^{\pi/2} \sqrt{1 - \frac{a^2 - b^2}{a^2} \sin^2 t}\, dt$$

$$= 4a\mathbf{E}\left(\sqrt{\frac{a^2 - b^2}{a^2}}\right) = 4a\mathbf{E}(e),$$

where $\mathbf{E}$ is the complete elliptic integral of the second kind (cf. Appendix 2) and $e$ is the ellipse eccentricity. ∎

*Remark 1.2*
The length of a curve does not depend on the particular choice of parametrisation. Indeed, let $\tau$ be a new parameter; $t = t(\tau)$ is a $\mathcal{C}^2$ function such that $dt/d\tau \neq 0$, and hence invertible. The curve $\mathbf{x}(t)$ can thus be represented by

$$\mathbf{x}(t(\tau)) = \mathbf{y}(\tau),$$

with $t \in (a, b)$, $\tau \in (a', b')$, and $t(a') = a$, $t(b') = b$ (if $t'(\tau) > 0$; the opposite case is completely analogous). It follows that

$$l = \int_a^b |\dot{\mathbf{x}}(t)|\, dt = \int_{a'}^{b'} \left|\frac{d\mathbf{x}}{dt}(t(\tau))\right| \left|\frac{dt}{d\tau}\right| d\tau = \int_{a'}^{b'} \left|\frac{d\mathbf{y}}{d\tau}(\tau)\right| d\tau. \qquad ∎$$

Any differentiable, non-singular curve admits a *natural parametrisation* with respect to a parameter $s$ (called the *arc length*, or *natural parameter*). Indeed, it is sufficient to endow the curve with a positive orientation, to fix an origin $O$ on it, and to use for every point $P$ on the curve the length $s$ of the arc $OP$ (measured with the appropriate sign and with respect to a fixed unit measure) as a coordinate of the point on the curve:

$$s(t) = \pm \int_0^t |\dot{\mathbf{x}}(\tau)|\, d\tau \tag{1.7}$$

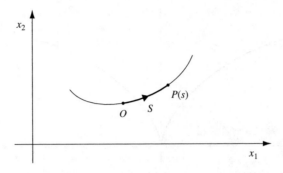

**Fig. 1.2**

(the choice of sign depends on the orientation given to the curve, see Fig. 1.2). Note that $|\dot{s}(t)| = |\dot{\mathbf{x}}(t)| \neq 0$.

Considering the natural parametrisation, we deduce from the previous remark the identity

$$s = \int_0^s \left|\frac{d\mathbf{x}}{d\sigma}\right| d\sigma,$$

which yields

$$\left|\frac{d\mathbf{x}}{ds}(s)\right| = 1 \quad \text{for all } s. \tag{1.8}$$

*Example 1.5*
For an ellipse of semi-axes $a \geq b$, the natural parameter is given by

$$s(t) = \int_0^t \sqrt{a^2 \cos^2 \tau + b^2 \sin^2 \tau}\, d\tau = 4a\mathbf{E}\left(t, \sqrt{\frac{a^2 - b^2}{a^2}}\right)$$

(cf. Appendix 2 for the definition of $\mathbf{E}(t,e)$). ∎

*Remark 1.3*
If the curve is of class $\mathcal{C}^1$, but the velocity $\dot{\mathbf{x}}$ is zero somewhere, it is possible that there exist singular points, i.e. points in whose neighbourhoods the curve cannot be expressed as the graph of a function $x_2 = f(x_1)$ (or $x_1 = g(x_2)$) of class $\mathcal{C}^1$, or else for which the tangent direction is not uniquely defined. ∎

*Example 1.6*
Let $\mathbf{x}(t) = (x_1(t), x_2(t))$ be the curve

$$x_1(t) = \begin{cases} -t^4, & \text{if } t \leq 0, \\ t^4, & \text{if } t > 0, \end{cases}$$

$$x_2(t) = t^2,$$

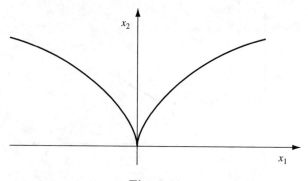

Fig. 1.3

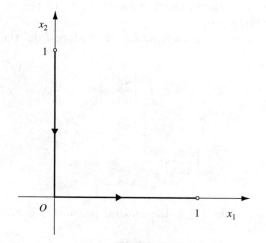

Fig. 1.4

given by the graph of the function $x_2 = \sqrt{|x_1|}$ (Fig. 1.3). The function $x_1(t)$ is of class $\mathcal{C}^3$, but the curve has a cusp at $t = 0$, where the velocity is zero. ∎

*Example* 1.7
Consider the curve

$$x_1(t) = \begin{cases} 0, & \text{if } t \leq 0, \\ e^{-1/t}, & \text{if } t > 0, \end{cases} \qquad x_2(t) = \begin{cases} e^{1/t}, & \text{if } t < 0, \\ 0, & \text{if } t \geq 0. \end{cases}$$

Both $x_1(t)$ and $x_2(t)$ are of class $\mathcal{C}^\infty$ but the curve has a corner corresponding to $t = 0$ (Fig. 1.4). ∎

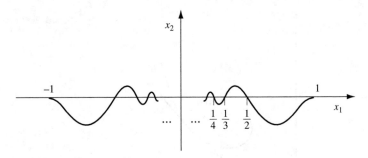

Fig. 1.5

*Example* 1.8
For the plane curve defined by

$$x_1(t) = \begin{cases} e^{1/t}, & \text{if } t < 0, \\ 0, & \text{if } t = 0, \\ -e^{-1/t}, & \text{if } t > 0, \end{cases}$$

$$x_2(t) = \begin{cases} e^{1/t}\sin(\pi e^{-1/t}), & \text{if } t < 0, \\ 0, & \text{if } t = 0, \\ e^{-1/t}\sin(\pi e^{1/t}), & \text{if } t > 0, \end{cases}$$

the tangent direction is not defined at $t = 0$ in spite of the fact that both functions $x_1(t)$ and $x_2(t)$ are in $\mathcal{C}^\infty$.

Such a curve is the graph of the function

$$x_2 = x_1 \sin \frac{\pi}{x_1}$$

with the origin added (Fig. 1.5). ∎

For more details on singular curves we recommend the book by Arnol'd (1991).

## 1.3 Tangent vector, normal vector and curvature of plane curves

Consider a plane regular curve $C$ defined by equation (1.1). It is well known that $\nabla F$, computed at the points of $C$, is orthogonal to the curve. If one considers any parametric representation, $\mathbf{x} = \mathbf{x}(t)$, then the vector $d\mathbf{x}/dt$ is tangent to the curve. Using the natural parametrisation, it follows from (1.8) that the vector $d\mathbf{x}/ds$ is of unit norm. In addition,

$$\frac{d^2\mathbf{x}}{ds^2} \cdot \frac{d\mathbf{x}}{ds} = 0,$$

which is valid for any vector of constant norm. These facts justify the following definitions.

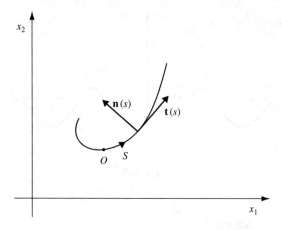

Fig. 1.6

DEFINITION 1.4  *The unit vector*

$$\mathbf{t}(s) = \frac{\mathrm{d}\mathbf{x}(s)}{\mathrm{d}s} \qquad (1.9)$$

*is called the* unit tangent vector *to the curve.* ∎

DEFINITION 1.5  *At any point at which* $\mathrm{d}^2\mathbf{x}/\mathrm{d}s^2 \neq 0$ *it is possible to define the unit vector*

$$\mathbf{n}(s) = \frac{1}{k(s)} \frac{\mathrm{d}^2\mathbf{x}}{\mathrm{d}s^2}, \qquad (1.10)$$

*called the* principal unit normal vector *(Fig. 1.6), where* $k(s) = |\mathrm{d}^2\mathbf{x}/\mathrm{d}s^2|$ *is the* curvature *of the plane curve.* $R(s) = 1/k(s)$ *is the* radius of curvature. ∎

It easily follows from the definition that straight lines have zero curvature (hence their radius of curvature is infinite) and that the circle of radius $R$ has curvature $1/R$.

*Remark 1.4*
Given a point on the curve, it follows from the definition that $\mathbf{n}(s)$ lies in the half-plane bounded by the tangent $\mathbf{t}(s)$ and containing the curve in a neighbourhood of the given point. The orientation of $\mathbf{t}(s)$ is determined by the positive orientation of the curve. ∎

*Remark 1.5*
Consider a point of unit mass, constrained to move along the curve with a time dependence given by $s = s(t)$. We shall see that in this case the curvature determines the strength of the constraining reaction at each point. ∎

The radius of curvature has an interesting geometric interpretation. Consider the family of circles that are tangent to the curve at a point $P$. Then the circle

## 1.3  Geometric and kinematic foundations of Lagrangian mechanics

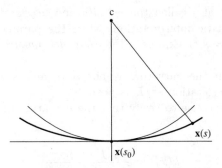

Fig. 1.7

that best approximates the curve in a neighbourhood of $P$ has radius equal to the radius of curvature at the point $P$. Indeed, choosing a circle of radius $r$ and centred in a point $\mathbf{c} = (c_1, c_2)$ lying on the normal line to the curve at a point $\mathbf{x}(s_0)$, we can measure the difference between the circle and the curve (Fig. 1.7) by the function

$$g(s) = |\mathbf{x}(s) - \mathbf{c}| - r,$$

with $s$ a variable in a neighbourhood of $s_0$. Since

$$g'(s_0) = \frac{1}{r}(\mathbf{x}(s_0) - \mathbf{c}) \cdot \mathbf{t}(s_0) = 0,$$

$$g''(s_0) = \frac{1}{r}(1 - kr),$$

it follows that $g(s)$ is an infinitesimal of order greater than $(s-s_0)^2$ if $g''(s_0) = 0$, and hence if $\mathbf{c} - \mathbf{x}(s_0) = R(s_0)\mathbf{n}(s_0)$.

DEFINITION 1.6  *The circle tangent to the given curve, with radius equal to the radius of curvature and centre belonging to the half-plane containing the unit vector* $\mathbf{n}$ *is called the* osculating circle. ■

Considering a generic parametrisation $\mathbf{x} = \mathbf{x}(t)$, one obtains the following relations:

$$\dot{\mathbf{x}}(t) = \mathbf{v}(t) = \dot{s}\mathbf{t} \tag{1.11}$$

and

$$\ddot{\mathbf{x}}(t) = \mathbf{a}(t) = \ddot{s}\mathbf{t} + \frac{\dot{s}^2}{R}\mathbf{n}, \tag{1.12}$$

which implies for the curvature

$$k(t) = \frac{1}{|\mathbf{v}(t)|^2}\left|\mathbf{a}(t) - \frac{\mathbf{v}(t) \cdot \mathbf{a}(t)}{|\mathbf{v}(t)|^2}\mathbf{v}(t)\right|. \tag{1.13}$$

The vectors **v**, **a** are also called the *velocity* and *acceleration*, respectively; this refers to their kinematic interpretation, when the parameter $t$ represents time and the function $s = s(t)$ expresses the *time dependence* of the point moving along the curve.

We remark that, if the curvature is non-zero, and $\dot{s} \neq 0$, then the normal component of the acceleration $\dot{s}^2/R$ is positive.

We leave it as an exercise to verify that the curvature of the graph $x_2 = f(x_1)$ is given by

$$k(x_1) = \frac{|f''(x_1)|}{[1 + f'^2(x_1)]^{3/2}}, \tag{1.14}$$

while, if the curve is expressed in polar coordinates and $r = r(\varphi)$, then the curvature is given by

$$k(\varphi) = \frac{|2r'^2(\varphi) - r(\varphi)r''(\varphi) + r^2(\varphi)|}{[r'^2(\varphi) + r^2(\varphi)]^{3/2}}. \tag{1.15}$$

*Example 1.9*
Consider an ellipse

$$x_1(t) = a \cos t, \quad x_2(t) = b \sin t.$$

In this case, the natural parameter $s$ cannot be expressed in terms of $t$ through elementary functions (indeed, $s(t)$ is given by an elliptic integral). The velocity and acceleration are:

$$\mathbf{v}(t) = (-a \sin t, b \cos t) = \dot{s}\mathbf{t}, \quad \mathbf{a}(t) = (-a \cos t, -b \sin t) = \ddot{s}\mathbf{t} + \frac{\dot{s}^2}{R}\mathbf{n},$$

and using equation (1.13) it is easy to derive the expression for the curvature. Note that $\mathbf{v}(t) \cdot \mathbf{a}(t) = \dot{s}\ddot{s} \neq 0$ because the parametrisation is not the natural one. ∎

THEOREM 1.1 (Frenet) *Let* $s \to \mathbf{x}(s) = (x_1(s), x_2(s))$ *be a plane curve of class at least* $\mathcal{C}^3$, *parametrised with respect to the natural parameter* $s$. *Then*

$$\begin{aligned} \frac{d\mathbf{t}}{ds} &= k(s)\mathbf{n}, \\ \frac{d\mathbf{n}}{ds} &= -k(s)\mathbf{t}. \end{aligned} \tag{1.16}$$

*Proof*
The first formula is simply equation (1.10). The second can be trivially derived from

$$\frac{d}{ds}(\mathbf{n} \cdot \mathbf{n}) = 0, \quad \frac{d}{ds}(\mathbf{n} \cdot \mathbf{t}) = 0.$$

∎

## 1.3 Geometric and kinematic foundations of Lagrangian mechanics

We end the analysis of plane curves by remarking that the curvature function $k(s)$ completely defines the curve up to plane congruences. Namely, ignoring the trivial case of zero curvature, we have the following.

**THEOREM 1.2** *Given a regular function $k : (a,b) \to \mathbf{R}$ such that $k(s) > 0$ for every $s \in (a,b)$, there exists a unique plane regular curve, defined up to translations and rotations, such that $k(s)$ is its curvature, and $s$ its natural parameter.*

*Proof*
The proof of this theorem depends on Frenet's formulae and on the existence and uniqueness theorem for solutions of ordinary differential equations. Indeed, from (1.16) it follows that

$$\frac{d^2\mathbf{t}}{ds^2} - \frac{k'(s)}{k(s)}\frac{d\mathbf{t}}{ds} + k^2(s)\mathbf{t} = 0; \qquad (1.17)$$

after integration this yields $\mathbf{t} = d\mathbf{x}/ds$, up to a constant vector (i.e. a rotation of the curve). One subsequent integration yields $\mathbf{x}(s)$ up to a second constant vector (i.e. a translation of the curve). ∎

*Remark 1.6*
Uniqueness can only be guaranteed if the curvature is not zero. As a counterexample, consider the two curves of class $\mathcal{C}^2$ (Fig. 1.8)

$$\mathbf{x}(t) = (t, t^3);$$

$$\mathbf{y}(t) = \begin{cases} (t, t^3), & \text{if } t < 0, \\ (t, -t^3), & \text{if } t \geq 0. \end{cases}$$

These curves are evidently distinct for $t > 0$, but their curvatures are equal for every $t$ and vanish for $t = 0$. ∎

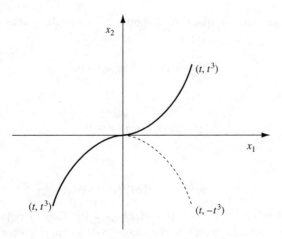

**Fig. 1.8**

## 1.4 Curves in $\mathbf{R}^3$

We have already remarked how it is possible to define regular curves in $\mathbf{R}^3$ in analogy with (1.4): such curves are maps $\mathbf{x} : (a,b) \to \mathbf{R}^3$ of class $\mathcal{C}^2$, with $\dot{\mathbf{x}} \neq 0$. Consider now a curve $t \to \mathbf{x}(t) = (x_1(t), x_2(t), x_3(t)) \in \mathbf{R}^3$; the equation defining the natural parameter is

$$\frac{ds}{dt} = \sqrt{\dot{x}_1^2 + \dot{x}_2^2 + \dot{x}_3^2}.$$

Suppose that the curve is parametrised through the natural parameter $s$. As for the case of a plane curve, we can introduce the unit tangent vector $\mathbf{t}$, the unit normal vector $\mathbf{n}$, and the curvature $k(s)$ according to Definitions 1.4 and 1.5. However, contrary to the plane case, these quantities are not sufficient to fully characterise a curve in three-dimensional space.

DEFINITION 1.7 *The unit vector*

$$\mathbf{b} = \mathbf{t} \times \mathbf{n} \tag{1.18}$$

*is called a* binormal unit vector. *The triple of vectors* $(\mathbf{t}, \mathbf{n}, \mathbf{b})$ *is orthonormal.* ∎

In the case of a plane curve, it is easy to verify that $d\mathbf{b}/ds = 0$, and hence that the binormal unit vector is constant and points in the direction orthogonal to the plane containing the curve. Hence the derivative $d\mathbf{b}/ds$ quantifies how far the curve is from being a plane curve. To be more precise, consider a point $\mathbf{x}(s_0)$ on the curve, and the pencil of planes whose axis is given by the line tangent to the curve at $\mathbf{x}(s_0)$. The equation of the plane of the pencil with unit normal vector $\nu$ is

$$(\mathbf{x} - \mathbf{x}(s_0)) \cdot \nu = 0.$$

The distance from such a plane of a point $\mathbf{x}(s)$ on the curve is given (up to sign) by

$$g(s) = [\mathbf{x}(s) - \mathbf{x}(s_0)] \cdot \nu,$$

and hence

$$g'(s_0) = \mathbf{t}(s_0) \cdot \nu = 0;$$

in addition,

$$g''(s_0) = k(s_0)\mathbf{n}(s_0) \cdot \nu.$$

It follows that if $\mathbf{n}(s_0)$ is defined (i.e. if $k(s_0) \neq 0$), there exists a unique plane such that $g''(s_0) = 0$; this plane is the one whose normal vector is precisely the unit vector $\mathbf{b}(s_0)$.

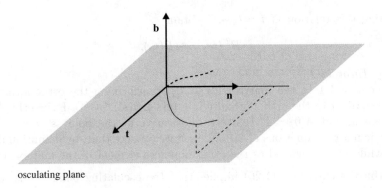

**Fig. 1.9**

DEFINITION 1.8 *The plane normal to* $\mathbf{b}(s_0)$ *is called the* osculating plane *to the curve at the point* $\mathbf{x}(s_0)$ *(Fig. 1.9).* ∎

Hence the osculating plane has parametric equation

$$\mathbf{y} = \mathbf{x}(s_0) + \lambda \mathbf{t}(s_0) + \mu k(s_0)\mathbf{n}(s_0). \tag{1.19}$$

In the case of curves in space as well, we have the following.

THEOREM 1.3 (Frenet) *Let* $s \to \mathbf{x}(s) = (x_1(s), x_2(s), x_3(s))$ *be a curve in* $\mathbf{R}^3$ *endowed with the natural parametrisation. Then the following equations hold:*

$$\begin{aligned} \frac{d\mathbf{t}}{ds} &= &+k(s)\mathbf{n}(s), & \\ \frac{d\mathbf{n}}{ds} &= -k(s)\mathbf{t}(s) & &-\chi(s)\mathbf{b}(s), \\ \frac{d\mathbf{b}}{ds} &= &+\chi(s)\mathbf{n}(s), & \end{aligned} \tag{1.20}$$

*where* $\chi(s)$ *is called the* torsion *(or second curvature) of the curve.* ∎

The proof of Frenet's theorem is based on the following lemma, of interest in its own right.

LEMMA 1.1 *Let* $A : (t_1, t_2) \to O(l)$ *be a function of class* $\mathcal{C}^1$, *taking values in the group of orthogonal matrices* $l \times l$, *such that* $A(t_0) = 1$. *Then* $\dot{A}(t_0)$ *is a skew-symmetric matrix.*

*Proof*
By differentiation of the orthogonality relation

$$A^T(t)A(t) = 1$$

for all $t \in (t_1, t_2)$, if $B(t) = dA/dt\,(t)$, one obtains

$$B^T(t)A(t) + A^T(t)B(t) = 0.$$

Evaluating this relation at $t = t_0$, we obtain

$$B^T(t_0) = -B(t_0).$$ ∎

*Proof of Theorem 1.3*
Apply Lemma 1.1 to the matrix $A(s'-s)$, transforming the orthonormal triple $(\mathbf{t}(s), \mathbf{n}(s), \mathbf{b}(s))$ to the orthonormal triple $(\mathbf{t}(s'), \mathbf{n}(s'), \mathbf{b}(s'))$. Evidently $A(s'-s)$ is orthogonal and $A(0) = 1$. Hence its derivative at the point $s' = s$ is a skew-symmetric matrix; equations (1.20) follow if we observe that, by definition $d\mathbf{t}/ds = k(s)\mathbf{n}$, while $\chi(s)$ is defined as the other non-zero element of the matrix $A'(0)$. ∎

The third of equations (1.20) implies that the osculating plane tends to rotate around the tangent line with velocity equal to the torsion $\chi(s)$. The second of equations (1.20) shows what causes variation in $\mathbf{n}$: under the effect of curvature, the normal vector tends to rotate in the osculating plane, while under the effect of torsion it tends to follow the rotation of the osculating plane. Moreover, if $\chi(s) \neq 0$, the curve crosses the osculating plane. This follows from the fact that

$$\frac{d^3\mathbf{x}}{ds^3} = \frac{d^2\mathbf{t}}{ds^2} = \frac{dk}{ds}\mathbf{n} - k^2\mathbf{t} - k\chi\mathbf{b};$$

hence for $s \simeq s_0$ one has $\mathbf{x}(s) - \mathbf{x}(s_0) \simeq (s-s_0)\mathbf{t} + \frac{1}{2}\cdot(s-s_0)^2 k\mathbf{n} + \frac{1}{6}(s-s_0)^3(k'\mathbf{n} - k^2\mathbf{t} - k\chi\mathbf{b})$, and thus $(\mathbf{x}(s) - \mathbf{x}(s_0))\cdot\mathbf{b} \simeq -\frac{1}{6}k\chi(s-s_0)^3$.

*Example 1.10*
Consider the cylindrical circular helix

$$x_1 = R\cos\varphi, \quad x_2 = R\sin\varphi, \quad x_3 = \lambda\varphi.$$

If the origin of the arcs is at $A$ (Fig. 1.10), we have $s(\varphi) = \sqrt{R^2 + \lambda^2}\,\varphi$; hence

$$\mathbf{t} = \frac{d\mathbf{x}}{d\varphi}\frac{d\varphi}{ds} = \frac{1}{\sqrt{R^2+\lambda^2}}(-R\sin\varphi, R\cos\varphi, \lambda),$$

$$\frac{d\mathbf{t}}{ds} = \frac{d\mathbf{t}}{d\varphi}\frac{d\varphi}{ds} = -\frac{R}{R^2+\lambda^2}(\cos\varphi, \sin\varphi, 0),$$

from which it follows that

$$\mathbf{n} = (-\cos\varphi, -\sin\varphi, 0), \quad k(s) = \frac{R}{R^2+\lambda^2},$$

and finally

$$\mathbf{b} = \frac{1}{\sqrt{R^2+\lambda^2}}(\lambda\sin\varphi, -\lambda\cos\varphi, R).$$

It is easy to compute that

$$\frac{d\mathbf{b}}{ds} = -\frac{\lambda}{R^2+\lambda^2}\mathbf{n},$$

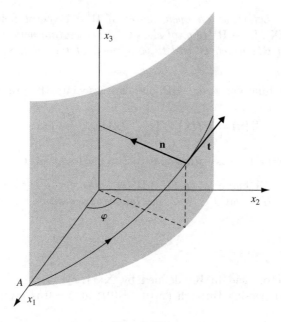

Fig. 1.10

yielding for the torsion

$$\chi = -\frac{\lambda}{R^2 + \lambda^2}.$$

∎

Curvature and torsion are the only two geometric invariants of a curve in space. Namely we have the following.

THEOREM 1.4   Let $k(s) > 0$ and $\chi(s)$ be two given regular functions. There exists a unique curve in space, up to congruences (rotations and translations), which has $s$ as natural parameter, and $k$ and $\chi$ as curvature and torsion, respectively.

∎

The proof is similar to the proof of Theorem 1.2 and is based on the fact that $\mathbf{t}(s)$ solves the differential equation

$$\frac{d^2\mathbf{t}}{ds^2} - \frac{k'}{k}\frac{d\mathbf{t}}{ds} + k^2\mathbf{t} + \chi\mathbf{t} \times \frac{d\mathbf{t}}{ds} = 0. \tag{1.21}$$

## 1.5  Vector fields and integral curves

In complete analogy with (1.4), a regular curve in $\mathbf{R}^l$ is a map $\mathbf{x} : (a, b) \to \mathbf{R}^l$ of class $\mathcal{C}^1$ such that $\dot{\mathbf{x}} \neq 0$.

In this section we shall investigate the relation between curves and vector fields.

DEFINITION 1.9  *Let $U$ be an open subset of $\mathbf{R}^l$. A vector field $\mathbf{X}$ on $U$ is a regular function $\mathbf{X} : U \to \mathbf{R}^l$ (e.g. of class $\mathcal{C}^\infty$) associating with every point $\mathbf{x} \in U$ a vector $\mathbf{X}(\mathbf{x})$ of $\mathbf{R}^l$, which is said to be* applied at the point $\mathbf{x}$. ∎

*Example* 1.11
To every regular function $f : U \to \mathbf{R}$ one can associate the gradient vector field

$$\mathbf{X}(\mathbf{x}) = \nabla f(\mathbf{x}) = \left( \frac{\partial f}{\partial x_1}(\mathbf{x}), \dots, \frac{\partial f}{\partial x_l}(\mathbf{x}) \right).$$

The gradient vector field is orthogonal to the level sets of $f$. ∎

DEFINITION 1.10  *A curve $\mathbf{x} : (a,b) \to \mathbf{R}^l$ is called an* integral curve *of a vector field $\mathbf{X} : U \to \mathbf{R}^l$ if for all $t \in (a,b)$ the following conditions hold:*

(a) $\mathbf{x}(t) \in U$;
(b) $\dot{\mathbf{x}}(t) = \mathbf{X}(\mathbf{x}(t))$. ∎

*Example* 1.12
Consider the vector field in $\mathbf{R}^2$ defined by $\mathbf{X}(x_1, x_2) = (x_2, -x_1)$. The integral curve of the field passing through $(x_1(0), x_2(0))$ at $t = 0$ is given by

$$x_1(t) = x_1(0) \cos t + x_2(0) \sin t,$$
$$x_2(t) = -x_1(0) \sin t + x_2(0) \cos t.$$

Note that, if $(x_1(0), x_2(0)) = (0,0)$, the integral curve is degenerate at the point $(0,0)$. This is possible because at the point $(0,0)$ the vector field vanishes, i.e. it has a *singular point*. ∎

It evidently follows from Definition 1.10 that the existence and uniqueness theorem for ordinary differential equations ensures the existence of a unique integral curve of a vector field passing through a given point. The question of the continuation of solutions of differential equations (hence of the existence of a maximal integral curve) yields the following definition.

DEFINITION 1.11  *A vector field is called* complete *if for every point $\mathbf{x}$ the maximal integral curve (cf. Appendix 1) passing through $\mathbf{x}$ is defined over all of $\mathbf{R}$.* ∎

*Example* 1.13
The vector field given in Example 1.12 is complete. The field $X : \mathbf{R} \to \mathbf{R}$, $X(x) = 1 + x^2$ is not complete. ∎

When not otherwise stated, we shall implicitly assume that the vector fields considered are complete.

## 1.6  Surfaces

The study of the local properties of plane curves, which we considered in the first three sections of this chapter, is rather simple: one invariant—curvature (as a function of arc length)—is sufficient to characterise the curve. Matters are not much more complicated in the case of curves in $\mathbf{R}^3$. The essential reason for this is that the *intrinsic* geometry of curves is 'trivial', in the sense that for all curves there exists a natural parametrisation, i.e. a map $\mathbf{x}(s)$ from an interval $(a,b)$ of $\mathbf{R}$ to the curve, such that the distance between any two points $\mathbf{x}(s_1)$ and $\mathbf{x}(s_2)$ of the curve, measured along the curve, is equal to $|s_2 - s_1|$. Hence the *metric* (i.e. the notion of distance) defined by means of the arc length coincides with that of $\mathbf{R}$.

The situation is much more complicated for the case of surfaces in $\mathbf{R}^3$. We shall see that the intrinsic geometry of surfaces is *non-trivial* due to the fact that, in general, there is no isometry property between surfaces and subsets of $\mathbf{R}^2$ analogous to that of the previous case, and it is not possible to define a metric using just one scalar function.

In analogy with the definition of a curve in the plane (as the level set of a function of two variables), surfaces in $\mathbf{R}^3$ can be obtained by considering the level sets of a function $F : U \to \mathbf{R}$ (for simplicity, we assume that this function is of class $\mathcal{C}^\infty$, but it would be sufficient for the function to be of class $\mathcal{C}^2$), where $U$ is an open subset of $\mathbf{R}^3$. The surface $S$ is hence defined by

$$S = \{(x_1, x_2, x_3) \in U | F(x_1, x_2, x_3) = 0\}, \qquad (1.22)$$

assuming that such a set is non-empty.

DEFINITION 1.12   *A point $(x_1, x_2, x_3)$ of the surface $F(x_1, x_2, x_3) = 0$ is called* non-singular *if the gradient of $F$ computed at the point is non-vanishing:*

$$\nabla F(x_1, x_2, x_3) \neq 0. \qquad (1.23)$$

*A surface $S$ whose points are all non-singular points is called* regular. ∎

By the implicit function theorem, if $P$ is non-singular, in a neighbourhood of $P$ the surface can be written as the graph of a function. For example, if $(\partial F/\partial x_3)_P \neq 0$ there exists a regular function $f : U \to \mathbf{R}$ (where $U$ is an open neighbourhood of the projection of $P$ onto the $(x_1, x_2)$ plane) such that

$$S = \mathrm{graph}\,(f) = \{(x_1, x_2, x_3) \in \mathbf{R}^3 | (x_1, x_2) \in U,\, x_3 = f(x_1, x_2)\}. \qquad (1.24)$$

In addition, from $F(x_1, x_2, x_3) = 0$ it follows that

$$\frac{\partial F}{\partial x_1} \mathrm{d}x_1 + \frac{\partial F}{\partial x_2} \mathrm{d}x_2 + \frac{\partial F}{\partial x_3} \mathrm{d}x_3 = 0;$$

hence from $F(x_1, x_2, f(x_1, x_2)) = 0$ it follows that

$$\frac{\partial f}{\partial x_1} = -\frac{\partial F/\partial x_1}{\partial F/\partial x_3}, \quad \frac{\partial f}{\partial x_2} = -\frac{\partial F/\partial x_2}{\partial F/\partial x_3}.$$

The analogous analysis can be performed if $(\partial F/\partial x_2)_P \neq 0$, or $(\partial F/\partial x_1)_P \neq 0$. Equation (1.24) highlights the fact that the points of a regular surface are, at least locally, in bijective and continuous correspondence with an open subset of $\mathbf{R}^2$.

It is an easy observation that at a non-singular point $\mathbf{x}_0$ there exists the tangent plane, whose equation is

$$(\mathbf{x} - \mathbf{x}_0) \cdot \nabla F = 0.$$

More generally, it is possible to consider a parametric representation of the form $\mathbf{x} : U \to \mathbf{R}^3$, $\mathbf{x} = \mathbf{x}(u,v)$, where $U$ is an open subset of $\mathbf{R}^2$:

$$S = \mathbf{x}(U) = \{(x_1, x_2, x_3) \in \mathbf{R}^3 | \text{there exist } (u,v) \in U, (x_1, x_2, x_3) = \mathbf{x}(u,v)\}. \tag{1.25}$$

Note that the graph of (1.24) is a particular case of the expression (1.25), in which the parametrisation is given by $\mathbf{x}(u,v) = (u, v, f(u,v))$. It is always possible to transform (1.24) into (1.25) by the change of variables on the open set $U$ of $\mathbf{R}^2$, $x_1 = x_1(u,v)$, $x_2 = x_2(u,v)$, provided the invertibility condition $\det[\partial(x_1, x_2)/\partial(u,v)] \neq 0$ holds.

The latter condition expresses the fact that the *coordinate lines* $u = $ constant and $v = $ constant in the $(x_1, x_2)$ plane are not tangent to each other (Fig. 1.11).

It follows that Definition 1.12 is equivalent to the following.

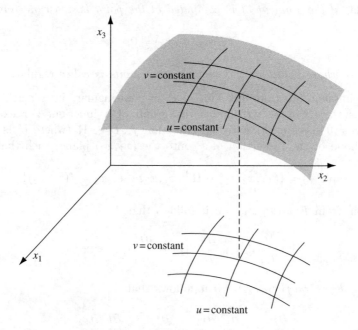

Fig. 1.11

DEFINITION 1.13 *If the surface $S$ is given in parametric form as $\mathbf{x} = \mathbf{x}(u,v)$, a point $P$ is called* non-singular *if*

$$\operatorname{rank} \begin{pmatrix} \dfrac{\partial x_1}{\partial u} & \dfrac{\partial x_2}{\partial u} & \dfrac{\partial x_3}{\partial u} \\ \dfrac{\partial x_1}{\partial v} & \dfrac{\partial x_2}{\partial v} & \dfrac{\partial x_3}{\partial v} \end{pmatrix}_P = 2. \tag{1.26}$$

∎

Equation (1.26) is equivalent to requiring that the vectors $\mathbf{x}_u, \mathbf{x}_v$ are linearly independent.

*Example* 1.14

The sphere of radius $R > 0$ is a regular surface; it is the level set of

$$F(x_1, x_2, x_3) = x_1^2 + x_2^2 + x_3^2 - R^2.$$

A parametrisation of the sphere is given by

$$\mathbf{x}(u,v) = R(\cos v \sin u, \sin v \sin u, \cos u),$$

where $(u,v) \in [0,\pi] \times [0, 2\pi]$. Here $v$ is also called the *longitude*, and $u$ the *colatitude*, as it is equal to $\pi/2$ minus the latitude (Fig. 1.12). This parametrisation of the sphere is regular everywhere except at the two poles $(0, 0, \pm 1)$. The sphere of radius 1 is usually denoted $\mathbf{S}^2$. ∎

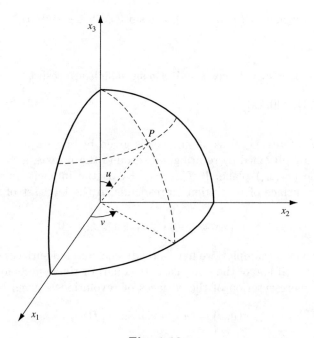

Fig. 1.12

*Example 1.15*
The *ellipsoid* is a regular surface; it is the level set of

$$F(x_1, x_2, x_3) = \frac{x_1^2}{a^2} + \frac{x_2^2}{b^2} + \frac{x_3^2}{c^2} - 1,$$

where $a > b > c > 0$ are the semi-axes of the ellipsoid. A parametrisation is given by

$$\mathbf{x}(u, v) = (a \cos v \sin u, b \sin v \sin u, c \cos u),$$

with $(u, v) \in [0, \pi] \times [0, 2\pi]$. Note that this parametrisation is not regular at the points $(0, 0, \pm c)$; however at these points the surface is regular. ■

*Example 1.16*
The *one-sheeted hyperboloid*, level set $S = F^{-1}(0)$ of

$$F(x_1, x_2, x_3) = \frac{x_1^2}{a^2} + \frac{x_2^2}{b^2} - \frac{x_3^2}{c^2} - 1,$$

or the *two-sheeted hyperboloid* with

$$F(x_1, x_2, x_3) = -\frac{x_1^2}{a^2} - \frac{x_2^2}{b^2} + \frac{x_3^2}{c^2} - 1,$$

are regular surfaces. A parametric representation is given, respectively, by

$$\mathbf{x}(u, v) = (a \cos v \cosh u, b \sin v \cosh u, c \sinh u),$$

and

$$\mathbf{x}(u, v) = (a \cos v \sinh u, b \sin v \sinh u, c \cosh u),$$

where $(u, v) \in \mathbf{R} \times [0, 2\pi]$. ■

*Example 1.17*
A particularly interesting class of surfaces is given by the *surfaces of revolution*; these surfaces are obtained by rotating, e.g. around the $x_3$-axis, a curve (implicitly defined) in the $(x_1, x_3)$ plane. If $f(x_1, x_3) = 0$ is the implicit representation of the curve, the surface of revolution corresponds to the level set of the function

$$F(x_1, x_2, x_3) = f(\sqrt{x_1^2 + x_2^2}, x_3) = 0.$$

Among the previous examples, we have already encountered surfaces of revolution, e.g. the ellipsoids (if two of the semi-axes are equal) or the hyperboloids (if $a = b$). A parametric representation of the surfaces of revolution is given by

$$\mathbf{x}(u, v) = (u \cos v, u \sin v, f(u)),$$

if the generating curve has equation $x_3 = f(x_1)$. ■

## 1.6 Geometric and kinematic foundations of Lagrangian mechanics

*Example* 1.18
The *elliptic paraboloid* is the graph of

$$x_3 = \frac{x_1^2}{a^2} + \frac{x_2^2}{b^2}, \quad a > b > 0, \quad (x_1, x_2) \in \mathbf{R}^2,$$

while the *hyperbolic paraboloid* is the graph of

$$x_3 = \frac{x_1^2}{a^2} - \frac{x_2^2}{b^2}, \quad a > b > 0, \quad (x_1, x_2) \in \mathbf{R}^2. \qquad \blacksquare$$

*Remark* 1.7
In analogy with the definition of surfaces in $\mathbf{R}^3$ one can introduce *(hyper)surfaces* in $\mathbf{R}^l$, as:

(1) level sets of functions from (subsets of) $\mathbf{R}^l$ into $\mathbf{R}$;
(2) graphs of functions defined in an open subset of $\mathbf{R}^{l-1}$ and taking values in $\mathbf{R}$;
(3) through a parametric representation, with $l-1$ parameters $\mathbf{x}(u_1, \ldots, u_{l-1})$.
$\blacksquare$

In this section we will focus primarily on studying surfaces in $\mathbf{R}^3$, while in the next section we shall define the notion of a differentiable manifold, of which surfaces and hypersurfaces are special cases.

Let $F : U \to \mathbf{R}$ be a $\mathcal{C}^\infty$ function, $U$ an open subset of $\mathbf{R}^3$, and denote by $S$ the surface $S = F^{-1}(0)$. It is important to remark that, in general, it is not possible to find a natural parametrisation that is *globally* non-singular for the whole of a regular surface.

*Example* 1.19
The *bidimensional torus* $\mathbf{T}^2$ is the surface of revolution around the $x_3$-axis obtained from the circle in the $(x_1, x_3)$ plane, given by the equation

$$x_3^2 + (x_1 - a)^2 = b^2,$$

thus with centre $x_1 = a$, $x_3 = 0$ and radius $b$, such that $0 < b < a$. Hence its implicit equation is

$$F(x_1, x_2, x_3) = x_3^2 + (\sqrt{x_1^2 + x_2^2} - a)^2 - b^2 = 0.$$

It is easy to verify that a parametrisation of $\mathbf{T}^2$ is given by

$$x_1 = \cos v(a + b \cos u),$$
$$x_2 = \sin v(a + b \cos u),$$
$$x_3 = b \sin u,$$

where $(u, v) \in [0, 2\pi] \times [0, 2\pi]$ (Fig. 1.13). The torus $\mathbf{T}^2$ is a regular surface. Indeed,

$$\nabla F(x_1, x_2, x_3) = \left( 2x_1 - \frac{4ax_1}{\sqrt{x_1^2 + x_2^2}}, 2x_2 - \frac{4ax_2}{\sqrt{x_1^2 + x_2^2}}, 2x_3 \right) \neq 0 \text{ on } \mathbf{T}^2,$$

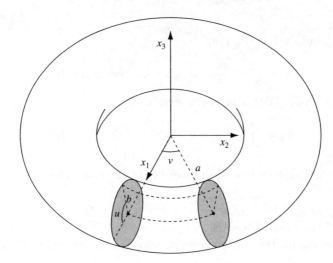

**Fig. 1.13**

and correspondingly

$$\frac{\partial(x_1, x_2, x_3)}{\partial(u,v)} = \begin{pmatrix} -b\sin u \cos v & -b\sin u \sin v & b\cos u \\ -(a+b\cos u)\sin v & (a+b\cos u)\cos v & 0 \end{pmatrix}$$

has rank 2 on $\mathbf{T}^2$. ∎

*Example* 1.20
The sphere $\mathbf{S}^2$

$$x_1^2 + x_2^2 + x_3^2 - 1 = 0$$

is a regular surface; the parametrisation

$$\mathbf{x}(u,v) = (\cos v \cos u, \sin v \cos u, \sin u)$$

in non-singular everywhere except at the points $u = \pm\pi/2$ (corresponding to the north pole $\mathbf{x} = (0,0,1)$ and the south pole $\mathbf{x} = (0,0,-1)$ of the sphere) where the parametrisation is singular (this is intuitively evident by observing that the parallels degenerate to a point at the poles, and hence that the longitude is not defined at these points). However, the parametrisation

$$\mathbf{x}(u,v) = (\sin u, \cos v \cos u, \sin v \cos u)$$

is regular at the poles, while it is singular at $\mathbf{x} = (\pm 1, 0, 0)$. The stereographic projection from one of the poles of the sphere (cf. Example 1.29) is an example of a parametrisation that is regular over the whole sphere minus one point. There is no global regular parametrisation of the whole sphere. ∎

### 1.6  Geometric and kinematic foundations of Lagrangian mechanics

*Example* 1.21
The *cone*

$$\frac{x_1^2}{a^2} + \frac{x_2^2}{b^2} - \frac{x_3^2}{c^2} = 0$$

is *not* a regular surface: the origin $x_1 = x_2 = x_3 = 0$ belongs to the cone but it is a singular point. Excluding this point, the surface becomes regular (but it is no longer connected), and $\mathbf{x}(u,v) = (au \cos v, bu \sin v, cu)$ is a global non-singular parametrisation. ∎

Consider a surface $S = F^{-1}(0)$, and a regular point $P \in S$. At such a point it is possible to define the *tangent space* $T_P S$ to the surface $S$ at the point $P$.

DEFINITION 1.14  *A vector* $\mathbf{w} \in \mathbf{R}^3$ *at the point* $P$ *is said to be* tangent to the surface $S$ at the point $P$, *or* $\mathbf{w} \in T_P S$ *(tangent space to the surface at the point* $P$*) if and only if there exists a curve* $\mathbf{x}(t)$ *on the surface, i.e. such that* $F(x_1(t), x_2(t), x_3(t)) = 0$ *for all* $t$, *passing through the point* $P$ *for some time* $t_0$, $\mathbf{x}(t_0) = P$, *with velocity* $\dot{\mathbf{x}}(t_0) = \mathbf{w}$. ∎

In the expression for the tangent vector at a point $\mathbf{x}(u_0, v_0)$

$$\dot{\mathbf{x}} = \mathbf{x}_u \dot{u} + \mathbf{x}_v \dot{v} \tag{1.27}$$

we can consider $\dot{u}, \dot{v}$ as real parameters, in the sense that, given two numbers $\alpha, \beta$, it is always possible to find two functions $u(t), v(t)$ such that $u(t_0) = u_0$, $v(t_0) = v_0$, $\dot{u}(t_0) = \alpha$, $\dot{v}(t_0) = \beta$. Hence we can identify $T_p S$ with the vector space, of dimension 2, generated by the vectors $\mathbf{x}_u, \mathbf{x}_v$ (Fig. 1.14).

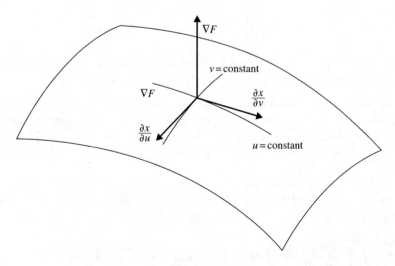

Fig. 1.14

DEFINITION 1.15 *A vector field* $\mathbf{X}$ *over a surface* $S$ *is a function assigning to every point* $P$ *of the surface, a vector* $\mathbf{X}(P) \in \mathbf{R}^3$ *applied at the point* $P$. *The field* $\mathbf{X}$ *is called a* tangent field *if* $\mathbf{X}(P) \in T_P S$ *for every* $P \in S$; *the field is a* normal field *if* $\mathbf{X}(P) \in (T_P S)^\perp$ *for every point* $P \in S$. ∎

*Remark* 1.8
Since a vector field tangent to $S$ is expressed by $\mathbf{X} = X_1(u,v)\mathbf{x}_u + X_2(u,v)\mathbf{x}_v$, the equations of its integral curves are $\dot{u} = X_1(u,v)$, $\dot{v} = X_2(u,v)$ and the curves lie on $S$. ∎

THEOREM 1.5 *Let* $P$ *be a non-singular point of the surface* $F(x_0) = 0$. *Then the tangent space to the surface at* $P$ *coincides with the orthogonal space to the gradient of* $F$ *at* $P$:

$$T_P S = (\nabla F(P))^\perp. \tag{1.28}$$

*Proof*
Differentiating the expression $F(\mathbf{x}(u,v)) = 0$ we obtain $\nabla F \cdot \mathbf{x}_u = \nabla F \cdot \mathbf{x}_v = 0$. Hence $\nabla F$ is orthogonal to every vector of $T_p S$. Conversely, if $\mathbf{w}$ is orthogonal to $\nabla F$ at $P \in S$, it must necessarily belong to the plane generated by $\mathbf{x}_u, \mathbf{x}_v$. ∎

DEFINITION 1.16 *A connected surface* $S$ *is said to be* oriented *when a unitary normal vector field is uniquely assigned on the surface.* ∎

*Remark* 1.9
The regular surfaces we have defined (as level sets $S = F^{-1}(0)$) are always orientable, with two possible orientations corresponding to the two unitary normal vector fields

$$\mathbf{n}_1(P) = \frac{\nabla F(P)}{|\nabla F(P)|}, \quad \mathbf{n}_2(P) = -\frac{\nabla F(P)}{|\nabla F(P)|}. \tag{1.29}$$

However, it is possible in general to extend the definition of surface to also admit non-orientable cases, such as the *Möbius strip* (Fig. 1.15). ∎

For applications in mechanics, it is very important to be able to endow the surface with a *distance* or *metric*, inherited from the natural immersion in three-dimensional Euclidean space. To this end, one can use the notion of length of a curve in space, using the same definition as for curves lying on a surface.

If $S = F^{-1}(0)$ is a regular surface, $\mathbf{x} = \mathbf{x}(u,v)$ is a parametric representation for it, and $t \to (u(t), v(t))$, $t \in (a,b)$ is a curve on $S$, the length of the curve is given by (cf. (1.5))

$$l = \int_a^b \left| \frac{d\mathbf{x}(u(t), v(t))}{dt} \right| dt = \int_a^b \sqrt{(\mathbf{x}_u \dot{u} + \mathbf{x}_v \dot{v}) \cdot (\mathbf{x}_u \dot{u} + \mathbf{x}_v \dot{v})} \, dt. \tag{1.30}$$

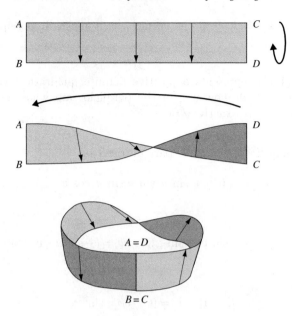

**Fig. 1.15** Möbius strip.

If we define

$$E(u,v) = \mathbf{x}_u \cdot \mathbf{x}_u = \left(\frac{\partial x_1}{\partial u}\right)^2 + \left(\frac{\partial x_2}{\partial u}\right)^2 + \left(\frac{\partial x_3}{\partial u}\right)^2,$$

$$F(u,v) = \mathbf{x}_u \cdot \mathbf{x}_v = \frac{\partial x_1}{\partial u}\frac{\partial x_1}{\partial v} + \frac{\partial x_2}{\partial u}\frac{\partial x_2}{\partial v} + \frac{\partial x_3}{\partial u}\frac{\partial x_3}{\partial v}, \quad (1.31)$$

$$G(u,v) = \mathbf{x}_v \cdot \mathbf{x}_v = \left(\frac{\partial x_1}{\partial v}\right)^2 + \left(\frac{\partial x_2}{\partial v}\right)^2 + \left(\frac{\partial x_3}{\partial v}\right)^2,$$

equation (1.30) can be rewritten as

$$l = \int_a^b \sqrt{E(u(t),v(t))\dot{u}^2 + 2F(u(t),v(t))\dot{u}\dot{v} + G(u(t),v(t))\dot{v}^2}\, dt. \quad (1.32)$$

Setting

$$(ds)^2 = d\mathbf{x} \cdot d\mathbf{x}, \quad (1.33)$$

we obtain for $(ds)^2$ the expression

$$(ds)^2 = E(u,v)(du)^2 + 2F(u,v)(du)(dv) + G(u,v)(dv)^2. \quad (1.34)$$

DEFINITION 1.17 *The quadratic form* (1.34) *is called the* first fundamental form *of the surface*. ∎

This form fixes the *metric* on the surface, as it makes it possible to compute lengths.

*Remark* 1.10
The expression (1.34) represents a positive definite quadratic form: this means $E > 0$ and $EG - F^2 > 0$. The area of the parallelogram whose sides are $\mathbf{x}_u, \mathbf{x}_v$ (linearly independent) is exactly $\sqrt{EG - F^2}$. ∎

*Example* 1.22
Consider the sphere $\mathbf{S}^2$ of radius 1, parametrised by

$$\mathbf{x} = (\cos v \sin u, \sin v \sin u, \cos u).$$

Then

$$\mathbf{x}_u = (\cos u \cos v, \cos u \sin v, -\sin u), \quad \mathbf{x}_v = (-\sin u \sin v, \sin u \cos v, 0),$$

and hence

$$E = 1, \quad F = 0, \quad G = \sin^2 u,$$

from which it follows that

$$(ds)^2 = (du)^2 + \sin^2 u (dv)^2.$$

For example, the length of a parallel at colatitude $u_0$ is given by

$$l = \int_0^{2\pi} \sqrt{\dot{u}^2 + (\sin u_0)^2 \dot{v}^2} \, dt = 2\pi \sin u_0,$$

since the curve has parametric equations $u = u_0$, $v = t$. ∎

Having defined the first fundamental form, it is possible to compute not only the lengths of curves lying on the surface, but also the *angle* $\varphi$ between two intersecting curves: if their parametric representation is

$$u = u_1(t), \ v = v_1(t) \quad \text{and} \quad u = u_2(t), \ v = v_2(t) \tag{1.35}$$

and the intersection point is denoted by $P$, corresponding to the value $t = t_0$, the velocity vectors of the two curves in $P$

$$\mathbf{w}_1 = \dot{u}_1(t_0)\mathbf{x}_u(u_1(t_0), v_1(t_0)) + \dot{v}_1(t_0)\mathbf{x}_v(u_1(t_0), v_1(t_0)),$$
$$\mathbf{w}_2 = \dot{u}_2(t_0)\mathbf{x}_u(u_2(t_0), v_2(t_0)) + \dot{v}_2(t_0)\mathbf{x}_v(u_2(t_0), v_2(t_0))$$

are both tangent to the surface at the point $P$. The angle between the two vectors is given by

$$\cos \varphi = \frac{\mathbf{w}_1 \cdot \mathbf{w}_2}{|\mathbf{w}_1||\mathbf{w}_2|} = \frac{E\dot{u}_1\dot{u}_2 + F(\dot{u}_1\dot{v}_2 + \dot{v}_1\dot{u}_2) + G\dot{v}_1\dot{v}_2}{\sqrt{E\dot{u}_1^2 + 2F\dot{u}_1\dot{v}_1 + G\dot{v}_1^2}\sqrt{E\dot{u}_2^2 + 2F\dot{u}_2\dot{v}_2 + G\dot{v}_2^2}}. \tag{1.36}$$

## 1.6  Geometric and kinematic foundations of Lagrangian mechanics

*Remark* 1.11
The parametrisation of a surface is called *orthogonal* if $F = 0$:
$$(ds)^2 = E(u,v)(du)^2 + G(u,v)(dv)^2.$$
In this case the curves $\mathbf{x}(u,v_0)$, $\mathbf{x}(u_0,v)$ on the surface, obtained by fixing one of the two parameters, are mutually orthogonal. If in addition $E = G = g(u,v)$, and hence
$$(ds)^2 = g(u,v)((du)^2 + (dv)^2),$$
the parametrisation is called *conformal*, since the angle in (1.36) between the two curves on the surface is equal to the angle between the two curves (1.35) in the $(u,v)$ plane. It can be proved (cf. Dubrovin *et al.* 1991a,b) that given a regular surface, there always exist orthogonal as well as conformal coordinates. ∎

Moreover, the first fundamental form allows one to compute the *area* of the surface. Consider the tangent parallelogram defined by the vectors $\mathbf{x}_u \Delta u$ and $\mathbf{x}_v \Delta v$. The total area of this parallelogram is given by
$$|\mathbf{x}_u \Delta u \times \mathbf{x}_v \Delta v| = |\mathbf{x}_u \times \mathbf{x}_v||\Delta u \Delta v| = \sqrt{EG - F^2}|\Delta u \Delta v|.$$
The area of the part $S_D$ of the surface corresponding to the parameters $(u,v)$ varying within a bounded domain $D$ is
$$\text{area}(S_D) = \int_D \sqrt{EG - F^2}\, du\, dv. \tag{1.37}$$

A very important feature of the first fundamental form of a surface is how it behaves under coordinate transformations.

THEOREM 1.6  *The first fundamental form is a covariant tensor of rank 2 (cf. Appendix 4).*

*Proof*
Let $(u', v')$ be a new parametrisation of the surface. From the identities
$$du = \frac{\partial u}{\partial u'} du' + \frac{\partial u}{\partial v'} dv',$$
$$dv = \frac{\partial v}{\partial u'} du' + \frac{\partial v}{\partial v'} dv'$$
it follows immediately that
$$(ds)^2 = (du\; dv) \begin{pmatrix} E & F \\ F & G \end{pmatrix} \begin{pmatrix} du \\ dv \end{pmatrix} = (du'\; dv') J^T \begin{pmatrix} E & F \\ F & G \end{pmatrix} J \begin{pmatrix} du' \\ dv' \end{pmatrix}, \tag{1.38}$$
where
$$J = \begin{pmatrix} \frac{\partial u}{\partial u'} & \frac{\partial u}{\partial v'} \\ \frac{\partial v}{\partial u'} & \frac{\partial v}{\partial v'} \end{pmatrix}, \quad \begin{pmatrix} E' & F' \\ F' & G' \end{pmatrix} = J^T \begin{pmatrix} E & F \\ F & G \end{pmatrix} J,$$
and $E', F', G'$ are expressed in terms of the new parameters. ∎

Among all the possible curves on a surface, the class of *geodesics* deserves special attention. Indeed, we shall see that geodesic curves play a very important role in mechanics.

Let $S$ be a regular surface, and $\mathbf{x}(u,v)$ its parametric representation. Consider a curve on the surface parametrised with respect to the natural parameter $s$:

$$s \to (u(s), v(s)) \to \mathbf{x}(u(s), v(s)). \tag{1.39}$$

The unit vector $\mathbf{t}$ tangent to the curve is given by

$$\mathbf{t}(s) = \frac{\mathrm{d}\mathbf{x}}{\mathrm{d}s}(s) = u'(s)\mathbf{x}_u(u(s), v(s)) + v'(s)\mathbf{x}_v(u(s), v(s)) \in T_{\mathbf{x}(u(s),v(s))}S$$

and the normal unit vector $\mathbf{n}$ is given by

$$\mathbf{n}(s) = \frac{1}{k(s)}\frac{\mathrm{d}^2\mathbf{x}}{\mathrm{d}s^2} = \frac{1}{k(s)}\left(u''\mathbf{x}_u + (u')^2\mathbf{x}_{uu} + 2u'v'\mathbf{x}_{uv} + (v')^2\mathbf{x}_{vv} + v''\mathbf{x}_v\right), \tag{1.40}$$

where $k(s)$ is the curvature,

$$\mathbf{x}_{uu} = \frac{\partial^2 \mathbf{x}}{\partial u^2}, \quad \mathbf{x}_{uv} = \frac{\partial^2 \mathbf{x}}{\partial u \partial v}, \quad \mathbf{x}_{vv} = \frac{\partial^2 \mathbf{x}}{\partial v^2}.$$

DEFINITION 1.18 *The curve (1.39) is called a* geodesic *if at every point of the curve the unit vector* $\mathbf{n}$ *normal to the curve belongs to the space normal to the surface, i.e. if*

$$\mathbf{n}(s) \in (T_{\mathbf{x}(u(s),v(s))}S)^\perp \tag{1.41}$$

*for all $s$, and hence if and only if*

$$\mathbf{n}(s) \cdot \mathbf{x}_u(u(s), v(s)) = 0,$$
$$\mathbf{n}(s) \cdot \mathbf{x}_v(u(s), v(s)) = 0. \tag{1.42}$$

∎

*Remark* 1.12
Given a curve with an arbitrary parametrisation, denoting by $s = s(t)$ the time dependence, its acceleration $\mathbf{a}$ is given by the expression (1.12), and the condition for this curve to be a geodesic consists in this case of imposing the condition that the acceleration be orthogonal to the surface.

The condition for a curve in the Euclidean space $\mathbf{R}^3$ to be a geodesic is satisfied by straight lines, for which $\mathrm{d}^2\mathbf{x}/\mathrm{d}s^2 = 0$.
∎

*Example* 1.23
It is easy to convince oneself that the maximal circles are geodesics on the sphere, while on a cylinder with circular normal section, the geodesics are the generating lines and helices (cf. Example 1.10), including the ones that degenerate to circles.
∎

## 1.6  Geometric and kinematic foundations of Lagrangian mechanics

From equations (1.40) and (1.42) it is easy to derive a system of ordinary differential equations which the geodesics must satisfy:

$$(u''\mathbf{x}_u + (u')^2\mathbf{x}_{uu} + 2u'v'\mathbf{x}_{uv} + (v')^2\mathbf{x}_{vv} + v''\mathbf{x}_v) \cdot \mathbf{x}_u = 0,$$
$$(u''\mathbf{x}_u + (u')^2\mathbf{x}_{uu} + 2u'v'\mathbf{x}_{uv} + (v')^2\mathbf{x}_{vv} + v''\mathbf{x}_v) \cdot \mathbf{x}_v = 0. \qquad (1.43)$$

Recall that $E = \mathbf{x}_u \cdot \mathbf{x}_u$, $F = \mathbf{x}_u \cdot \mathbf{x}_v$ and $G = \mathbf{x}_v \cdot \mathbf{x}_v$, and note that

$$\frac{\partial E}{\partial u} = 2\mathbf{x}_{uu} \cdot \mathbf{x}_u, \qquad \frac{\partial E}{\partial v} = 2\mathbf{x}_{uv} \cdot \mathbf{x}_u,$$
$$\frac{\partial F}{\partial u} = \mathbf{x}_{uv} \cdot \mathbf{x}_u + \mathbf{x}_{uu} \cdot \mathbf{x}_v, \qquad \frac{\partial F}{\partial v} = \mathbf{x}_{uv} \cdot \mathbf{x}_v + \mathbf{x}_{vv} \cdot \mathbf{x}_u,$$
$$\frac{\partial G}{\partial u} = 2\mathbf{x}_{uv} \cdot \mathbf{x}_v, \qquad \frac{\partial G}{\partial v} = 2\mathbf{x}_{vv} \cdot \mathbf{x}_v;$$

hence equations (1.43) become

$$Eu'' + Fv'' + \frac{1}{2}\frac{\partial E}{\partial u}(u')^2 + \frac{\partial E}{\partial v}u'v' + \left(\frac{\partial F}{\partial v} - \frac{1}{2}\frac{\partial G}{\partial u}\right)(v')^2 = 0,$$
$$Fu'' + Gv'' + \frac{1}{2}\frac{\partial G}{\partial v}(v')^2 + \frac{\partial G}{\partial u}u'v' + \left(\frac{\partial F}{\partial u} - \frac{1}{2}\frac{\partial E}{\partial v}\right)(u')^2 = 0. \qquad (1.44)$$

Denoting the matrix representing the first fundamental form by

$$(g_{ij}) = \begin{pmatrix} E & F \\ F & G \end{pmatrix}, \qquad (1.45)$$

and its inverse by

$$(g^{kl}) = \frac{1}{EG - F^2}\begin{pmatrix} G & -F \\ -F & E \end{pmatrix}, \qquad (1.46)$$

we can introduce the so-called *Christoffel symbols*

$$\Gamma_{ij}^k = \frac{1}{2}\sum_{l=1}^{2} g^{kl}\left(\frac{\partial g_{lj}}{\partial u^i} + \frac{\partial g_{il}}{\partial u^j} - \frac{\partial g_{ij}}{\partial u^l}\right), \qquad (1.47)$$

where $u^1 = u$, $u^2 = v$. Using Christoffel symbols, one finds that the system of differential equations (1.44) for the geodesics can be written in the form

$$\frac{d^2 u^k}{ds^2} + \sum_{i,j=1}^{2} \Gamma_{ij}^k \frac{du^i}{ds}\frac{du^j}{ds} = 0, \quad k = 1, 2. \qquad (1.48)$$

*Example* 1.24
For a cylinder with generic section $x_1 = f_1(v)$, $x_2 = f_2(v)$, $x_3 = u$ and $(f_1')^2 + (f_2')^2 = 1$, one obtains $E = G = 1$, $F = 0$ and equations (1.44) yield $u'' = v'' = 0$, i.e. $u = as + b$, $v = cs + d$, with $a$, $b$, $c$, $d$ arbitrary constants. When $c = 0$ one

obtains the generating lines; $a = 0$ yields the normal sections; in all other cases $v - d = c/a\,(u - b)$, and hence one finds helices. Since $du/ds = a$, the geodesics intersect the generating lines at a constant angle. ∎

*Example 1.25*
The first fundamental form of a surface of revolution with the parametrisation $\mathbf{x} = (u\cos v, u\sin v, f(u))$ can be written as

$$(ds)^2 = [1 + (f'(u))^2](du)^2 + u^2(dv)^2, \tag{1.49}$$

and hence the Christoffel symbols have the values

$$\Gamma^1_{11} = \frac{f'(u)f''(u)}{1 + (f'(u))^2}, \quad \Gamma^1_{22} = -\frac{u}{1 + (f'(u))^2}, \quad \Gamma^2_{12} = \Gamma^2_{21} = \frac{1}{u},$$

while $\Gamma^1_{12} = \Gamma^1_{21} = \Gamma^2_{11} = \Gamma^2_{22} = 0$. The geodesic equation (1.48) on the surface is thus equivalent to the system

$$\frac{d^2 u}{ds^2} + \frac{f'(u)f''(u)}{1 + (f'(u))^2}\left(\frac{du}{ds}\right)^2 - \frac{u}{1 + (f'(u))^2}\left(\frac{dv}{ds}\right)^2 = 0,$$
$$\frac{d^2 v}{ds^2} + \frac{2}{u}\left(\frac{du}{ds}\right)\left(\frac{dv}{ds}\right) = 0. \tag{1.50}$$

The second of equations (1.50) can be rewritten as

$$\frac{1}{u^2}\frac{d}{ds}\left[u^2\left(\frac{dv}{ds}\right)\right] = 0,$$

from which it follows that there exists a constant $c \in \mathbf{R}$ such that for every $s$

$$u^2 \frac{dv}{ds} = c, \tag{1.51}$$

and hence, if $c \neq 0$,

$$ds = \frac{1}{c} u^2 dv.$$

Substituting the latter expression into the first fundamental form (1.49) one obtains the relation

$$u^4 (dv)^2 = c^2[1 + (f'(u))^2](du)^2 + c^2 u^2 (dv)^2; \tag{1.52}$$

this leads to the elimination of $ds$ and one can hence consider $v$ as a function of $u$. The geodesics on a surface of revolution thus have the implicit form

$$v - v_0 = \pm c \int_{u_0}^{u} \frac{\sqrt{1 + (f'(\xi))^2}}{\xi \sqrt{\xi^2 - c^2}}\, d\xi. \tag{1.53}$$

If $c = 0$, from equation (1.51) it follows that $u^2\,(dv/ds) = 0$, i.e. that $v$ is constant: the *meridians* are geodesic curves. On the other hand, the *parallels* (the curves corresponding to $u = $ constant) are geodesics only if

$$\frac{u}{1+(f'(u))^2}\left(\frac{dv}{ds}\right)^2 = 0,$$

$$\frac{d^2v}{ds^2} = 0,$$

i.e. only if $dv/ds$ is in turn constant, and if $dx_3/du = f'(u) = \infty$, which implies that along the given parallel, the planes tangent to the surface envelop a cylinder whose generator lines are parallel to the $x_3$-axis. The relation (1.51) has an interesting consequence. Let $\alpha$ be the angle between the geodesic $(u(s), v(s))$ at $s = s_0$ and the meridian $v = v(s_0)$ (Fig. 1.16). It is immediate to verify that

$$u(s_0)\frac{dv}{ds}(s_0) = \sin \alpha,$$

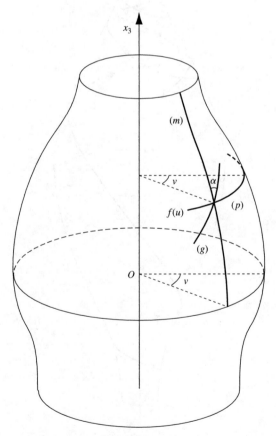

($p$): parallel, ($m$): meridian, ($g$): geodesic

**Fig. 1.16** Geodesics on a surface of revolution.

since the unit vector tangent to the parallel is simply $(-\sin v, \cos v, 0)$; hence substituting in the expression (1.51) we obtain *Clairaut's theorem*:

$$u(s) \sin \alpha(s) = c. \tag{1.54}$$

Hence the geodesic must lie in the region $u(s) \geq |c|$.

In the case of a surface of revolution, with a cusp at infinity, i.e. such that $\lim_{u \to 0} f(u) = \infty$ (Fig. 1.17), every geodesic, after attaining the minimum value of $u$ allowed by equation (1.54), reverses the motion (along the $x_3$-axis) and comes back into the region corresponding to values of $u$ satisfying $|u| > |c|$.

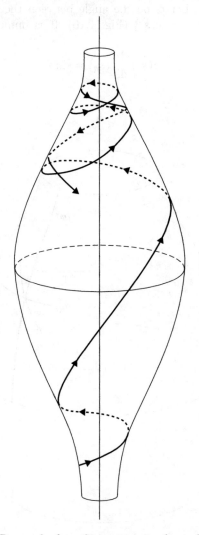

**Fig. 1.17** Reversal of geodesics on a surface of revolution.

It is possible to prove that geodesics on a surface of revolution which are neither meridians nor closed curves are dense in the region $u \geq |c|$. ∎

## 1.7 Differentiable Riemannian manifolds

Let $\mathbf{x}$ be a point in Euclidean $n$-dimensional space $\mathbf{R}^n$, and let $f_1, \ldots, f_m$ be $m$ regular real-valued functions defined on the same connected open subset $A \subset \mathbf{R}^n$. Just as the level set of a real function of three real variables identifies a surface in Euclidean three-dimensional space, the level sets of any of the functions $f_j$ identify a (hyper)surface in $\mathbf{R}^n$. With the requirement that $\mathbf{x}$ lies in the intersection (supposed non-empty) of the level sets of all the functions $f_j$, one identifies a *submanifold* of $\mathbf{R}^n$. In analogy with the notion of a regular surface introduced in the previous section, as a surface endowed with a tangent plane to all of its points, we can introduce the notion of a *regular* submanifold of $\mathbf{R}^n$ by imposing the condition that at each of its points there is defined a tangent plane (and a normal space). The *dimension* of the submanifold is then defined as the dimension of its tangent space. These sketchy introductory remarks justify the following definition.

DEFINITION 1.19   *Let $A$ be an open connected subset of $\mathbf{R}^n$, $n > 1$, and $\mathbf{f} : A \to \mathbf{R}^{n-l}$, $1 \leq l < n$, a map of class $C^k$, $k \geq 2$. The zero level set $V = \{\mathbf{x} \in A | \mathbf{f}(\mathbf{x}) = 0\}$ of $\mathbf{f}$, assumed non-empty, is called a* regular submanifold of $\mathbf{R}^n$ of class $C^k$ *and of dimension $l$ if the Jacobian matrix of the map $\mathbf{f}$ is of maximal rank (hence if its rank is equal to $n - l$) at every point of $V$.* ∎

Remark 1.13
Evidently the condition that the Jacobian matrix of $\mathbf{f} = (f_1, \ldots, f_{n-l})$ be of rank $n - l$ at every point of $V$ is equivalent to requiring that the gradient vectors $\nabla_{\mathbf{x}} f_1, \ldots, \nabla_{\mathbf{x}} f_{n-l}$ be an $(n-l)$-tuple of vectors in $\mathbf{R}^n$ which are linearly independent on $V$. ∎

Consider as an example the case shown in Fig. 1.18, for which $n = 3$, $l = 1$, $\mathbf{f} = (f_1, f_2)$, where

$$f_1(x_1, x_2, x_3) = x_3 - \sqrt{x_1^2 + x_2^2}, \quad f_2(x_1, x_2, x_3) = x_1^2 + x_2^2 + x_3^2 - 1.$$

The set $V$ is a circle. Note that the vectors

$$\nabla f_1 = \left( \frac{-x_1}{\sqrt{x_1^2 + x_2^2}}, \frac{-x_2}{\sqrt{x_1^2 + x_2^2}}, 1 \right), \quad \nabla f_2 = 2(x_1, x_2, x_3)$$

are linearly independent on $V$.

This definition includes in particular plane regular curves ($n = 2$, $l = 1$), regular curves in $\mathbf{R}^3$ ($n = 3$, $l = 1$), considered as the intersection of two non-tangential surfaces, and regular surfaces in $\mathbf{R}^3$ ($n = 3$, $l = 2$).

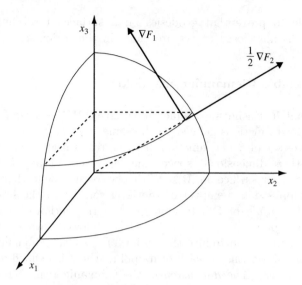

**Fig. 1.18**

DEFINITION 1.20   *The* tangent space $T_P V$ *to a regular submanifold $V$ at the point $P$ is the $l$-dimensional vector space of the velocities $\dot{\mathbf{x}}(t_0)$ along the curves $\mathbf{x}(t)$ belonging to $V$ (hence such that $\mathbf{f}(\mathbf{x}(t)) = 0$ for every $t$) and passing through $P$ for $t = t_0$.* ∎

*Remark* 1.14
It is easy to verify that $T_P V$ coincides with the vector space generated by the vectors which are orthogonal to the gradients $\nabla_{\mathbf{x}} f_1(P), \ldots, \nabla_{\mathbf{x}} f_{n-l}(P)$ (cf. Theorem 1.5). The latter will be called a basis of the *normal space* to $V$ in $P$. ∎

Having chosen a local parametrisation $\mathbf{x} = \mathbf{x}(u_1, \ldots, u_l)$ of $V$, whose existence is guaranteed by the implicit function theorem, the tangent space at a point $P$ of $V$ has as a basis the vectors $\mathbf{x}_{u_1}, \ldots, \mathbf{x}_{u_l}$, where

$$\mathbf{x}_{u_i} = \frac{\partial \mathbf{x}}{\partial u_i} \tag{1.55}$$

and derivatives are computed at the point $P$.

*Example* 1.26
The sphere $\mathbf{S}^l$ of unit radius is the regular submanifold of $\mathbf{R}^{l+1}$ defined by

$$f(x_1, \ldots, x_{l+1}) = x_1^2 + \cdots + x_{l+1}^2 - 1 = 0.$$

The tangent space at one of its points $P$, with coordinates $(\hat{x}_1, \ldots, \hat{x}_{l+1})$, is the hyperplane of $\mathbf{R}^{l+1}$ described by the equation

$$\mathbf{x} \cdot \hat{\mathbf{x}} = 0.$$

∎

*Example* 1.27
The group of real $n \times n$ matrices $A$ with unit determinant, denoted by $\mathrm{SL}(n, \mathbf{R})$, is a regular submanifold of $\mathbf{R}^{n^2}$ of dimension $n^2 - 1$, defined by the equation

$$\det(A) = 1.$$

Its tangent space at the point corresponding to the identity matrix can be identified with the space of $n \times n$ matrices of zero trace. Indeed, if $A(t)$ is any curve in $\mathrm{SL}(n, \mathbf{R})$ passing through the identity at $t = 0$, and thus such that $A(0) = 1$, we have that

$$0 = \frac{\mathrm{d}}{\mathrm{d}t} \det A(t)|_{t=0} = \mathrm{Tr}\, \dot{A}(0).$$

Indeed, if we set $X = \dot{A}(0)$ we have that $\det A(t) = \det(1 + tX) + \mathcal{O}(t) = 1 + t\, \mathrm{Tr}\, X + \mathcal{O}(t)$. ∎

*Example* 1.28
The group of real orthogonal $n \times n$ matrices $A$, denoted by $\mathrm{O}(n)$, is a regular submanifold of $\mathbf{R}^{n^2}$ of dimension $n(n-1)/2$ defined by the system of equations

$$AA^T = 1.$$

Its tangent space at the point corresponding to the identity matrix can be identified with the vector space of $n \times n$ skew-symmetric matrices (cf. Lemma 1.1). The connected component of $\mathrm{O}(n)$ containing the identity matrix coincides with the group $\mathrm{SO}(n)$ of orthogonal matrices of unit determinant. ∎

We now turn to the problem of parametrising regular submanifolds.

We have already remarked that for surfaces in $\mathbf{R}^3$ it is not possible in general to give a *global* parametric representation. For example, the sphere $\mathbf{S}^2$ is a regular submanifold of $\mathbf{R}^3$, but the parametrisation given by the spherical coordinates $\mathbf{x}_1 = (\sin u_1 \cos u_2, \sin u_1 \sin u_2, \cos u_1)$ is singular at the points $(0, 0, 1)$ and $(0, 0, -1)$. A regular parametrisation at those points is given instead by $\mathbf{x}_2 = (\cos \bar{u}_1, \sin \bar{u}_1 \cos \bar{u}_2, \sin \bar{u}_1 \sin \bar{u}_2)$, which however is singular at $(1, 0, 0)$ and $(-1, 0, 0)$.

Hence there exist two regular injective maps $\mathbf{x}_1, \mathbf{x}_2$ defined on $R = (0, \pi) \times [0, 2\pi)$ such that $\mathbf{S}^2 = \mathbf{x}_1(R) \cup \mathbf{x}_2(R)$. Moreover, if we consider the intersection $W = \mathbf{x}_1(R) \cap \mathbf{x}_2(R) = \mathbf{S}^2 \setminus \{(0,0,1), (0,0,-1), (1,0,0), (-1,0,0)\}$, the preimages $\mathbf{x}_1^{-1}(W) = R \setminus \{(\pi/2, 0), (\pi/2, \pi)\}$ and $\mathbf{x}_2^{-1}(W) = R \setminus \{(\pi/2, \pi/2), (\pi/2, 3\pi/2)\}$ are set in one-to-one correspondence by the map $\mathbf{x}_2^{-1} \circ \mathbf{x}_1$, which expresses $\bar{u}_1, \bar{u}_2$ as functions of $u_1, u_2$, and by its inverse $\mathbf{x}_1^{-1} \circ \mathbf{x}_2$.

In summary, these are the properties of any 'good' parametrisation of a regular submanifold. We can now consider the problem of parametric representation in a more general context, by referring to a set $M$ which is not necessarily endowed with a metric structure, as in the case of regular submanifolds of $\mathbf{R}^n$.

DEFINITION 1.21 *A differentiable manifold of dimension $l$ and class $\mathcal{C}^k$ consists of a non-empty set $M$ and of a family of injective maps $\mathbf{x}_\alpha : U_\alpha \subset \mathbf{R}^l \to M$, with*

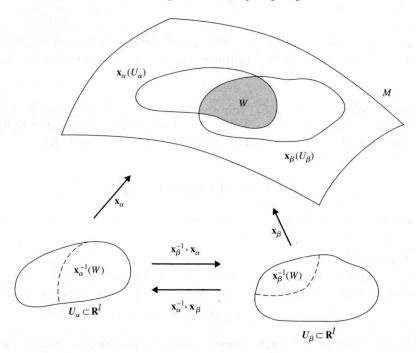

Fig. 1.19

$U_\alpha$ open and connected and $\alpha \in \mathcal{A} \neq \emptyset$ such that:

(a) $\bigcup_{\alpha \in \mathcal{A}} \mathbf{x}_\alpha(U_\alpha) = M$;

(b) for any $\alpha$ and $\beta$ in $\mathcal{A}$, if $\mathbf{x}_\alpha(U_\alpha) \cap \mathbf{x}_\beta(U_\beta) = W \neq \emptyset$ the sets $\mathbf{x}_\alpha^{-1}(W)$ and $\mathbf{x}_\beta^{-1}(W)$ are open subsets of $\mathbf{R}^l$ and the maps $\mathbf{x}_\beta^{-1} \circ \mathbf{x}_\alpha$ and $\mathbf{x}_\alpha^{-1} \circ \mathbf{x}_\beta$ (inverses of each other) are differentiable maps of class $\mathcal{C}^k$.

The pair $(U_\alpha, \mathbf{x}_\alpha)$ (or the map $\mathbf{x}_\alpha$) is called a local parametrisation or a chart of $M$, while a family $\{(U_\alpha, \mathbf{x}_\alpha)\}_{\alpha \in \mathcal{A}}$ with the properties listed in the definition is called a differentiable structure on $M$ or an atlas of $M$ (Fig. 1.19). ∎

In the example of the sphere in $\mathbf{R}^3$, $\mathcal{A}$ is the set of indices $\{1, 2\}$.

The set $\mathcal{A}$ may have only one element if the representation of $M$ is global.

Evidently the Euclidean space $\mathbf{R}^l$ endowed with the differential structure induced by the identity map is a differentiable manifold of dimension $l$.

*Example* 1.29

Consider the $l$-dimensional sphere

$$\mathbf{S}^l = \{(x_1, \ldots, x_l, x_{l+1}) \in \mathbf{R}^{l+1} | x_1^2 + \cdots + x_{l+1}^2 = 1\}$$

with the atlas given by the stereographic projections $\pi_1 : \mathbf{S}^l \backslash \{N\} \to \mathbf{R}^l$ and $\pi_2 : \mathbf{S}^l \backslash \{S\} \to \mathbf{R}^l$ from the north pole $N = (0, \ldots, 0, 1)$ and from the south pole

## 1.7 Geometric and kinematic foundations of Lagrangian mechanics

$S = (0, \ldots, 0, -1)$, respectively:

$$\pi_1(x_1, \ldots, x_l, x_{l+1}) = \left( \frac{x_1}{1 - x_{l+1}}, \ldots, \frac{x_l}{1 - x_{l+1}} \right),$$

$$\pi_2(x_1, \ldots, x_l, x_{l+1}) = \left( \frac{x_1}{1 + x_{l+1}}, \ldots, \frac{x_l}{1 + x_{l+1}} \right).$$

It is immediate to verify that the parametrisations $(\mathbf{R}^l, \pi_1^{-1})$, $(\mathbf{R}^l, \pi_2^{-1})$ define the structure of a differentiable manifold. ∎

Comparing this with the definition of a regular submanifold of $\mathbf{R}^n$, we note that the common feature of both definitions is the existence of local regular parametrisations (i.e. parametrisations without singular points). Indeed, we have the following.

THEOREM 1.7 *Every regular $l$-dimensional submanifold $V$ of $\mathbf{R}^n$ is a differentiable manifold.*

*Proof*
It follows from the implicit function theorem that to every point $p$ of $V$ one can associate an open neighbourhood $A \subset \mathbf{R}^n$, a point $\mathbf{u}$ of $\mathbf{R}^l$, an open neighbourhood $U$ of $\mathbf{u}$ and a differentiable, invertible map $\mathbf{x}_p : U \to V$ such that $\mathbf{x}_p(\mathbf{u}) = p$ and $\mathbf{x}_p(U) = V \cap A$, and hence a local parametrisation of $V$ (Fig. 1.20).

Consider now the pairs $(U_p, \mathbf{x}_p)$ as $p$ varies in $V$; clearly the conditions of Definition 1.21 are satisfied, and thus $\{(U_p, \mathbf{x}_p)\}_{p \in V}$ is an atlas for $V$. ∎

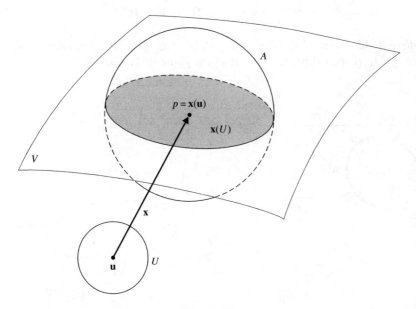

Fig. 1.20

*Remark* 1.15
The definition of a differentiable manifold naturally yields a topological space structure: we will say that a subset $A$ of $M$ is *open* if $\mathbf{x}_\alpha^{-1}(A \cap \mathbf{x}_\alpha(U_\alpha))$ is an open subset of $\mathbf{R}^l$ for every $\alpha \in \mathcal{A}$. Hence a subset $K$ of $M$ is *compact* if every covering of $K$ with open sets $A$ has a finite subcovering. The manifold $M$ is *connected* if for any two points $P_1, P_2 \in M$ there exists a finite sequence of charts $\{(U_j, \mathbf{x}_j)\}_{j=1,\ldots,N}$ such that $P_1 \in \mathbf{x}_1(U_1)$, $P_N \in \mathbf{x}_N(U_N)$, the open sets $U_j$ are connected and $U_j \cap U_{j+1} \neq \emptyset$ for every $j = 1, \ldots, N-1$. ∎

*Remark* 1.16
With the topology induced by the differentiable structure, the manifold $M$ is *separable* (i.e. every pair of points $m_1$, $m_2$ in $M$ has two open disjoint neighbourhoods $A_1$ and $A_2$, $m_1 \in A_1$ and $m_2 \in A_2$) and the topology has a *countable base* (there is no loss of generality in assuming that $\mathcal{A}$ is countable). ∎

DEFINITION 1.22 *A differentiable manifold $M$ is* orientable *if it admits a differentiable structure $\{(U_\alpha, \mathbf{x}_\alpha)\}_{\alpha \in \mathcal{A}}$ such that for every pair $\alpha, \beta \in \mathcal{A}$ with $\mathbf{x}_\alpha(U_\alpha) \cap \mathbf{x}_\beta(U_\beta) \neq \emptyset$ the Jacobian of the change of coordinates $\mathbf{x}_\alpha^{-1} \circ \mathbf{x}_\beta$ is positive. Otherwise the manifold is called* non-orientable. ∎

DEFINITION 1.23 *Let $M_1$ and $M_2$ be two differentiable manifolds of dimension $l$ and $m$, respectively. A map $g : M_1 \to M_2$ is differentiable at a point $p \in M_1$ if given an arbitrary parametrisation $\mathbf{y} : V \subset \mathbf{R}^m \to M_2$ with $\mathbf{y}(V) \ni g(p)$, there exists a parametrisation $\mathbf{x} : U \subset \mathbf{R}^l \to M_1$ with $\mathbf{x}(U) \ni p$, such that $g(\mathbf{x}(U)) \subset \mathbf{y}(V)$ and the function*

$$\mathbf{y}^{-1} \circ g \circ \mathbf{x} : U \subset \mathbf{R}^l \to V \subset \mathbf{R}^m \tag{1.56}$$

*is differentiable in $\mathbf{x}^{-1}(p)$ (Fig. 1.21). The map $g$ is differentiable in an open subset of $M_1$ if it is differentiable at every point of the subset.* ∎

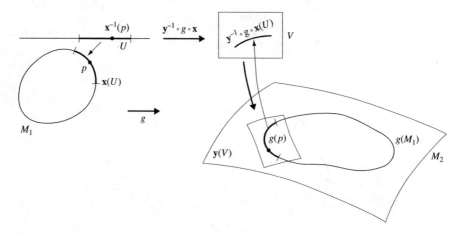

Fig. 1.21

Note that by choosing $M_2 = \mathbf{R}$ this defines the notion of a differentiable map (in an obvious way we can also define the notion of a map of class $\mathcal{C}^k$ or $\mathcal{C}^\infty$) from $M$ to $\mathbf{R}$.

If we denote by $\mathbf{f} = (f_1, \ldots, f_m)$ the map (1.56), we have $v_i = f_i(u_1, \ldots, u_l)$, $i = 1, \ldots, m$, where $f_i$ are differentiable functions.

DEFINITION 1.24  A curve on a manifold $M$ is a differentiable map $\gamma : (a, b) \to M$. ∎

If $(U, \mathbf{x})$ is a local parametrisation of $M$ in a neighbourhood of a point $p = \mathbf{x}(0)$, we can express a curve $\gamma : (-\varepsilon, \varepsilon) \to M$ using the parametrisation

$$(\mathbf{x}^{-1} \circ \gamma)(t) = (u_1(t), \ldots, u_l(t)) \in U. \tag{1.57}$$

In spite of the fact that $M$ has no metric structure, we can define at every point $p$ of the curve the velocity vector through the $l$-tuple $(\dot{u}_1, \ldots, \dot{u}_l)$. It is then natural to consider the velocity vectors corresponding to the $l$-tuples $(1, 0, \ldots, 0), (0, 1, \ldots, 0), \ldots, (0, 0, \ldots, 1)$. We denote these vectors by the symbols

$$\frac{\partial}{\partial u_1}, \ldots, \frac{\partial}{\partial u_l};$$

the generic velocity vector is expressed in the form of a linear combination

$$\dot{\mathbf{x}} = \sum_{i=1}^{l} \dot{u}_i \frac{\partial \mathbf{x}}{\partial u_i}, \tag{1.58}$$

exactly as in the case of a regular $l$-dimensional submanifold.

It is now easy to show that for $p \in M$ and $\mathbf{v} \in T_pM$, it is possible to find a curve $\gamma : (-\varepsilon, \varepsilon) \to M$ such that $\gamma(0) = p$ and $\dot{\gamma}(0) = \mathbf{v}$. Indeed, it is enough to consider the decomposition

$$\mathbf{v} = \sum_{i=1}^{l} v_i \frac{\partial \mathbf{x}}{\partial u_i}(0)$$

for some local parametrisation $(U, \mathbf{x})$, and to construct a map $\mu : (-\varepsilon, \varepsilon) \to U$ such that its components $u_i(t)$ have derivatives $u'_i(0) = v_i$. The composite map $\mathbf{x} \circ \mu$ hence defines the required function $\gamma$ (Fig. 1.22).

DEFINITION 1.25  The tangent space $T_pM$ to a differentiable manifold $M$ at a point $p$ is the space of vectors tangent to the curves on $M$ passing through $p$. ∎

The notion of a tangent space allows us to define the differential of a differentiable map $g$ between two differentiable manifolds $M_1, M_2$. Given a point $p \in M_1$, we define a linear map between $T_pM_1$ and $T_{g(p)}M_2$. Consider a curve $\gamma : (-\epsilon, \epsilon) \to M_1$, such that $\gamma(0) = p$ and $\dot{\gamma}(0) = \mathbf{v}$, the given element of $T_pM_1$. The map $g$ defines a curve on $M_2$ through $\beta = g \circ \gamma$. It is natural to associate with $\mathbf{v} \in T_pM_1$ the vector $\mathbf{w} = \dot{\beta}(0) \in T_{g(p)}M_2$.

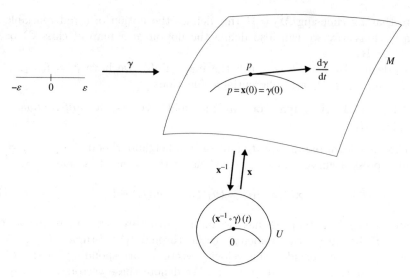

**Fig. 1.22**

The construction of the vector **w** is easy after remarking that, if the curve $\gamma(t)$ on $M_1$ possesses the local parametrisation $(u_1(t), \ldots, u_l(t))$, then the curve $\beta(t)$ on $M_2$ has the parametrisation $(v_1(t), \ldots, v_m(t))$, where $v_i = f_i(u_1, \ldots, u_l)$, $i = 1, \ldots, m$ (cf. (1.56)). Hence if the vector $\mathbf{v} = \dot{\gamma}(0)$ is characterised with respect to the basis

$$\frac{\partial}{\partial u_1}, \ldots, \frac{\partial}{\partial u_l}$$

by having components $(\dot{u}_1(0), \ldots, \dot{u}_l(0))$, the vector $\mathbf{w} = \dot{\beta}(0)$ with respect to the basis

$$\frac{\partial}{\partial v_1}, \ldots, \frac{\partial}{\partial v_m}$$

has components $(\dot{v}_1(0), \ldots, \dot{v}_m(0))$, where

$$\dot{v}_i(0) = \sum_{j=1}^{l} \frac{\partial f_i}{\partial u_j}(u_1(0), \ldots, u_l(0))\dot{u}_j(0).$$

We can thus give the following definition.

**DEFINITION 1.26** *Let $g : M_1 \to M_2$ be a differentiable map between the differentiable manifolds $M_1, M_2$ of dimension $l, m$, respectively. The linear map which with every $\mathbf{v} \in T_pM_1$, defined by $\mathbf{v} = \dot{\gamma}(0)$, associates $\mathbf{w} \in T_{g(p)}M_2$, defined by $\mathbf{w} = \dot{\beta}(0)$, with $\beta = g \circ \gamma$, is the differential $\mathrm{d}g_p : T_pM_1 \to T_{g(p)}M_2$.* ∎

We showed that the map $\mathrm{d}g_p$ acts on the components of the vectors in $T_pM_1$ as the row-by-column product with the Jacobian matrix $\partial(f_1, \ldots, f_m)/\partial(u_1, \ldots, u_l)$.

## 1.7 Geometric and kinematic foundations of Lagrangian mechanics

This happens in particular when the map is the change of parametrisation on a manifold (the Jacobian is in this case a square matrix).

**DEFINITION 1.27** *Let $M_1$ and $M_2$ be two differentiable manifolds, both of dimension $l$. A map $g : M_1 \to M_2$ is a* diffeomorphism *if it is differentiable, bijective and its inverse $g^{-1}$ is differentiable; $g$ is a* local diffeomorphism *at $p \in M_1$ if there exist two neighbourhoods, $A$ of $p$ and $B$ of $g(p)$, such that $g : A \to B$ is a diffeomorphism.* ∎

Applying the theorem of local invertibility, it is not difficult to prove the following.

**THEOREM 1.8** *Let $g : M_1 \to M_2$ be a differentiable map, and let $p \in M_1$ be such that $\mathrm{d}g_p : T_p M_1 \to T_{g(p)} M_2$ is an isomorphism. Then $g$ is a local diffeomorphism.* ∎

Given a differentiable manifold $M$ of dimension $\ell$, the set of its tangent spaces $T_p M$ when $p$ varies inside $M$ has a natural structure as a differentiable manifold. Indeed, if $\{(U_\alpha, \mathbf{x}_\alpha)\}_{\alpha \in A}$ is an atlas for $M$ and we indicate by $(u_1^{(\alpha)}, \ldots, u_\ell^{(\alpha)})$ the local coordinates of $U_\alpha$, at every point of $U_\alpha$ the vectors $e_i^{(\alpha)} = \partial/\partial u_i^{(\alpha)}$ when $i = 1, \ldots, \ell$ are a basis for the tangent space of $M$, and every tangent vector $\mathbf{v} \in T_p M$ can be written as

$$\mathbf{v} = \sum_{i=1}^{\ell} v_i^{(\alpha)} \left. \frac{\partial}{\partial u_i^{(\alpha)}} \right|_p.$$

**DEFINITION 1.28** *We call the* tangent bundle *of $M$, denoted by $TM$, the differentiable manifold of dimension $2\ell$:*

$$TM = \bigcup_{p \in M} \{p\} \times T_p M \tag{1.59}$$

*with the differentiable structure $\{(U_\alpha \times \mathbf{R}^\ell, \mathbf{y}_\alpha)\}_{\alpha \in A}$, where $\mathbf{y}_\alpha(\mathbf{u}^{(\alpha)}, \mathbf{v}^{(\alpha)}) = (\mathbf{x}_\alpha(\mathbf{u}^{(\alpha)}), \mathbf{v}^{(\alpha)})$, with $\mathbf{u}^{(\alpha)} \in U_\alpha$ being the vector of local coordinates in $U_\alpha$ and $\mathbf{v}^{(\alpha)}$ is a vector in the tangent space at a point $\mathbf{x}_\alpha(\mathbf{u}^{(\alpha)})$. The manifold $M$ is called the* base space *of the tangent bundle.* ∎

The map $\pi : TM \to M$ which associates with every point $(p, \mathbf{v}) \in TM$ the point $p$ itself (at which $\mathbf{v}$ is tangent to $M$: $\mathbf{v} \in T_p M$) is called the *projection onto the base*. Clearly

$$T_p M = \pi^{-1}(p), \tag{1.60}$$

and $T_p M$ is also called the *fibre corresponding to the point $p$* of the tangent bundle.

The notion of a tangent bundle of a manifold is important as it allows one to extend to manifolds the notions of a vector field and a differential equation.

**DEFINITION 1.29** *A (tangent)* vector field *on $M$ is a map $X : M \to TM$ which associates with every point $p \in M$ a vector $\mathbf{v}_p \in T_p M$ in a differentiable way, i.e. it is a differentiable map $X$ such that $\pi(X(p)) = p$, $\forall p \in M$.*

For a given vector field, the integral curves are the curves $\gamma : (a,b) \to M$ such that

$$\dot\gamma(t) = X(\gamma(t)). \tag{1.61}$$

∎

It is now natural to consider the problem of integrating differential equations on a manifold.

Recalling equation (1.58), equation (1.61) can be written as a system of first-order differential equations: namely, if $X$ is given in the form

$$X(p) = \sum_{i=1}^{\ell} \alpha_i(u_1, \ldots, u_\ell) \frac{\partial \mathbf{x}}{\partial u_i},$$

with $p = \mathbf{x}(\mathbf{u})$, then equation (1.61) is simply

$$\dot u_i(t) = \alpha_i(u_1(t), \ldots, u_\ell(t)), \quad i = 1, \ldots, \ell.$$

*Example* 1.30
Let $M$ be the unit sphere; consider the parametrisation

$$\mathbf{x} = (\sin u_1 \cos u_2, \sin u_1 \sin u_2, \cos u_1),$$

with the tangent vectors

$$\frac{\partial \mathbf{x}}{\partial u_1} = (\cos u_1 \cos u_2, \cos u_1 \sin u_2, -\sin u_1),$$

$$\frac{\partial \mathbf{x}}{\partial u_2} = (-\sin u_1 \sin u_2, \sin u_1 \cos u_2, 0).$$

A vector field tangent over $M$ takes the form

$$\alpha_1(u_1, u_2) \frac{\partial \mathbf{x}}{\partial u_1} + \alpha_2(u_1, u_2) \frac{\partial \mathbf{x}}{\partial u_2}.$$

For example, if $\alpha_1 = $ constant, $\alpha_2 = $ constant the integral curves are given by $u_1(t) = \alpha_1 t + u_1^{(0)}$, $u_2(t) = \alpha_2 t + u_2^{(0)}$.

∎

We now extend the fundamental notion of a metric to differentiable manifolds.

DEFINITION 1.30   A Riemannian metric *on a differentiable manifold $M$ of dimension $\ell$ is a symmetric, positive definite bilinear form $(\,,\,)_p$ defined in the tangent space $T_p M$, which has differentiable dependence on $p$. A differentiable manifold with a given Riemannian metric is called a* Riemannian manifold.

∎

*Example* 1.31
The first fundamental form (1.34) is a Riemannian metric for any regular surface in $\mathbf{R}^3$.

∎

## 1.7  Geometric and kinematic foundations of Lagrangian mechanics

Let $\mathbf{x} : U \to M$ be a local parametrisation in $p \in M$ with local coordinates $(u_1, \ldots, u_\ell)$. We saw that at every point $q \in \mathbf{x}(U)$, $q = \mathbf{x}(u_1, \ldots, u_\ell)$, the vectors

$$e_i(q) = \left.\frac{\partial}{\partial u_i}\right|_q, \quad i = 1, \ldots, \ell,$$

are a basis for $T_q M$. If $(\,,\,)_p$ is a Riemannian metric on $M$ the functions

$$g_{ij}(u_1, \ldots, u_\ell) = (e_i(q), e_j(q))_q \tag{1.62}$$

are differentiable in $U$ for every $i, j = 1, \ldots, \ell$. Evidently $g_{ij} = g_{ji}$ and if $(u'_1, \ldots, u'_\ell)$ is a new local parametrisation, compatible with the former one, setting $g'_{ij} = (e'_i(q), e'_j(q))_q$ we have

$$g'_{ij} = \sum_{m,n=1}^{\ell} J_{mi} g_{mn} J_{nj}, \tag{1.63}$$

where $J_{mi} = \partial u_m / \partial u'_i$. Hence a Riemannian metric defines a *symmetric covariant tensor of order* 2 on the manifold (cf. Appendix 4). In analogy with the case of surfaces, we write

$$(\mathrm{d}s)^2 = \sum_{i,j=1}^{\ell} g_{ij}(u_1, \ldots, u_\ell)\, \mathrm{d}u_i\, \mathrm{d}u_j. \tag{1.64}$$

It is possible to prove that every differentiable manifold can be endowed with a Riemannian metric. Using this metric, one can define—in analogy with equation (1.32)—the notion of the length of a curve over $M$ and of the arc length parameter $s$.

We can also say that the metric tensor $g_{ij}(\mathbf{u})$ defines the *scalar product* in $T_p M$ and hence the norm of a vector in $T_p M$. In particular, on the curve $(u_1(s), \ldots, u_\ell(s))$ written with respect to the natural parametrisation, the tangent vector has unit norm.

*Example* 1.32
The *Lobačevskij half-plane* is the Riemannian manifold given by $\{(x_1, x_2) \in \mathbf{R}^2 | x_2 > 0\}$ with the usual differentiable structures (**H** is an open set of $\mathbf{R}^2$) and the metric

$$(\mathrm{d}s)^2 = \frac{(\mathrm{d}x_1)^2 + (\mathrm{d}x_2)^2}{x_2^2},$$

i.e. $g_{11} = g_{22} = 1/x_2^2$, $g_{12} = g_{21} = 0$. A curve $\gamma : (a, b) \to \mathbf{H}$, $\gamma(t) = (x_1(t), x_2(t))$ has length

$$\ell = \int_a^b \frac{1}{x_2(t)} \sqrt{\dot{x}_1^2(t) + \dot{x}_2^2(t)}\, \mathrm{d}t.$$

For example, if $\gamma(t) = (c, t)$ we have

$$\ell = \int_a^b \frac{dt}{t} = \log \frac{b}{a}.$$
■

**DEFINITION 1.31** *Let $M$ and $N$ be two Riemannian manifolds. A diffeomorphism $g : M \to N$ is an* isometry *if*

$$(\mathbf{v}_1, \mathbf{v}_2)_p = (\mathrm{d}g_p(\mathbf{v}_1), \mathrm{d}g_p(\mathbf{v}_2))_{g(p)} \tag{1.65}$$

*for every $p \in M$ and $\mathbf{v}_1, \mathbf{v}_2 \in T_p M$. If $N = M$, $g$ is called an* isometry of $M$. ■

It is not difficult to prove that the isometries of a Riemannian manifold form a group, denoted $\mathrm{Isom}(M)$.

*Example 1.33*
Let $M = \mathbf{R}^\ell$ be endowed with the Euclidean metric. The isometry group of $\mathbf{R}^\ell$ contains translations, rotations and reflections. ■

*Example 1.34*
Consider the sphere $\mathbf{S}^\ell$ as immersed in $\mathbf{R}^{\ell+1}$, with the Riemannian metric induced by the Euclidean structure of $\mathbf{R}^{\ell+1}$. It is not difficult to prove that $\mathrm{Isom}(\mathbf{S}^\ell) = O(\ell+1)$, the group of $(\ell+1) \times (\ell+1)$ orthogonal matrices. ■

*Example 1.35*
Consider the Lobačevskij plane $\mathbf{H}$. Setting $z = x_1 + ix_2$ (where $i = \sqrt{-1}$) the mappings

$$w = \frac{az+b}{cz+d}, \tag{1.66}$$

with $a, b, c, d \in \mathbf{R}$, $ad - bc = 1$, are isometries of $\mathbf{H}$. Indeed,

$$(\mathrm{d}s)^2 = \frac{(\mathrm{d}x_1)^2 + (\mathrm{d}x_2)^2}{x_2^2} = -4 \frac{\mathrm{d}z\, \mathrm{d}\bar{z}}{(z - \bar{z})^2}.$$

To prove that (1.66) is an isometry, we compute

$$4 \frac{\mathrm{d}w\, \mathrm{d}\bar{w}}{(w - \bar{w})^2} = 4 \frac{\mathrm{d}w}{\mathrm{d}z} \overline{\left(\frac{\mathrm{d}w}{\mathrm{d}z}\right)} \frac{\mathrm{d}z\, \mathrm{d}\bar{z}}{\left(\frac{az+b}{cz+d} - \frac{a\bar{z}+b}{c\bar{z}+d}\right)^2}. \tag{1.67}$$

Immediately one can verify that

$$\frac{\mathrm{d}w}{\mathrm{d}z} = \frac{1}{(cz+d)^2}, \quad \overline{\left(\frac{\mathrm{d}w}{\mathrm{d}z}\right)} = \frac{1}{(c\bar{z}+d)^2},$$

and that

$$\frac{az+b}{cz+d} - \frac{a\bar{z}+b}{c\bar{z}+d} = \frac{z - \bar{z}}{(cz+d)(c\bar{z}+d)}.$$

Substituting these relations into (1.67) yields

$$4\frac{dw\,d\overline{w}}{(w-\overline{w})^2} = 4\frac{dz\,d\overline{z}}{(cz+d)^2(c\overline{z}+d)^2}\frac{(cz+d)^2(c\overline{z}+d)^2}{(z-\overline{z})^2} = 4\frac{dz\,d\overline{z}}{(z-\overline{z})^2}.\quad\blacksquare$$

Among all curves on a Riemannian manifold $M$ we now consider the particular case of the *geodesics*.

DEFINITION 1.32  Given a local parametrisation $(u_1,\ldots,u_\ell)$ of $M$, and denoting by $s$ the natural parameter along the curve, a geodesic $s \to (u_1(s),\ldots,u_\ell(s))$ is a solution of the system of equations

$$\frac{d^2 u_k}{ds^2} + \sum_{i,j=1}^{\ell} \Gamma_{ij}^k \frac{du_i}{ds}\frac{du_j}{ds} = 0, \quad k=1,\ldots,\ell, \tag{1.68}$$

where the Christoffel symbols $\Gamma_{ij}^k$ are given by

$$\Gamma_{ij}^k = \frac{1}{2}\sum_{n=1}^{\ell} g^{kn}\left(\frac{\partial g_{ni}}{\partial u_j} + \frac{\partial g_{nj}}{\partial u_i} - \frac{\partial g_{ij}}{\partial u_n}\right) \tag{1.69}$$

and $(g^{kn})$ is the matrix inverse to $(g_{ij})$, which defined the metric (1.64).  $\blacksquare$

We shall consider in Chapter 9 the geometric interpretation of these equations, which are obviously an extension of equations (1.47), (1.48).

*Example 1.36*
The Christoffel symbols corresponding to the Riemannian metric of the Lobačevskij half-plane are

$$\Gamma_{12}^1 = \Gamma_{21}^1 = -\frac{1}{x_2}, \quad \Gamma_{11}^2 = \frac{1}{x_2}, \quad \Gamma_{22}^2 = -\frac{1}{x_2},$$

while $\Gamma_{11}^1 = \Gamma_{22}^1 = \Gamma_{12}^2 = \Gamma_{21}^2 = 0$. The geodesic equations are then given by the system

$$\frac{d^2 x_1}{ds^2} - \frac{2}{x_2}\frac{dx_1}{ds}\frac{dx_2}{ds} = 0,$$

$$\frac{d^2 x_2}{ds^2} + \frac{1}{x_2}\left(\frac{dx_1}{ds}\right)^2 - \frac{1}{x_2}\left(\frac{dx_2}{ds}\right)^2 = 0.$$

The first equation can be written as

$$x_2^2 \frac{d}{ds}\left[\frac{1}{x_2^2}\left(\frac{dx_1}{ds}\right)\right] = 0;$$

it follows that there exists a constant $c \in \mathbf{R}$ such that

$$\frac{dx_1}{ds} = cx_2^2.$$

If $c=0$ it follows that $x_1 = $ constant, and hence vertical lines are geodesics.

Otherwise, substituting

$$\frac{d}{ds} = cx_2^2 \frac{d}{dx_1}$$

into the second geodesic equation yields

$$x_2 \frac{d^2 x_2}{dx_1^2} + \left(\frac{dx_2}{dx_1}\right)^2 + 1 = 0.$$

The general integral of this equation is given by $x_2 = \sqrt{R^2 - (x_1 - A)^2}$, and hence the geodesics corresponding to the values of $c \neq 0$ are semicircles with the centre on the $x_1$-axis (i.e. on $\partial \mathbf{H}$). ∎

*Remark 1.17*
Geodesics are invariant under any isometry of a Riemannian manifold. Indeed, thanks to (1.65) the Christoffel symbols (1.69) do not change. More generally, if $g : M \to N$ is an isometry, the geodesics on $N$ are the images, through the isometry $g$, of geodesics on $M$ and vice versa (cf. Problem 13.29). ∎

## 1.8 Actions of groups and tori

One way of constructing a differentiable manifold $M$ from another manifold $\widetilde{M}$ is to consider the quotient of $\widetilde{M}$ with respect to an equivalence relation. This situation occurs frequently in mechanics.

DEFINITION 1.33  *A group $G$ acts (to the left) on a differentiable manifold $\widetilde{M}$ if there exists a map $\varphi : G \times \widetilde{M} \to \widetilde{M}$ such that:*

(a) *for every $g \in G$ the map $\varphi_g : \widetilde{M} \to \widetilde{M}$, $\varphi_g(p) = \varphi(g, p)$, where $p \in \widetilde{M}$, is a diffeomorphism;*
(b) *if $e$ denotes the unit element in $G$, $\varphi_e = $ identity;*
(c) *for any choice of $g_1, g_2 \in G$, $\varphi_{g_1 g_2} = \varphi_{g_1} \varphi_{g_2}$.*

The action of $G$ on $\widetilde{M}$ is *free* if for every $p \in \widetilde{M}$ the unit element $e \in G$ is the only element of $G$ such that $\varphi_e(p) = p$. The action is *discontinuous if every point $p \in \widetilde{M}$ has a neighbourhood $A \subset \widetilde{M}$ such that $A \cap \varphi_g(A) = \emptyset$ for every $g \in G$, $g \neq e$.* ∎

The action of a group on a manifold determines an equivalence relation on the manifold.

DEFINITION 1.34  *Two points $p_1, p_2 \in \widetilde{M}$ are equivalent (denoted $p_1 \sim p_2$) if and only if there exists an element $g \in G$ such that $p_2 = \varphi_g(p_1)$.* ∎

Two points of the manifold are equivalent if they belong to the same *orbit* $Gp = \{\varphi_g(p) | g \in G\}$. The orbits of the points of $\widetilde{M}$ under the action of the group $G$ are the equivalence classes $[p] = Gp = \{p' \in \widetilde{M} | p' \sim p\}$.

## 1.8 Geometric and kinematic foundations of Lagrangian mechanics

The *quotient space*

$$\widetilde{M}/G = \{[p] | p \in \widetilde{M}\}, \tag{1.70}$$

with respect to the equivalence relation introduced, is a topological space, with the topology induced by the requirement that the *projection*

$$\pi : \widetilde{M} \to \widetilde{M}/G, \quad \pi(p) = [p] \tag{1.71}$$

is continuous and open (hence the open subsets of $\widetilde{M}/G$ are the projections of the open subsets of $\widetilde{M}$).

It is not difficult to prove (cf. Do Carmo 1979) the following.

THEOREM 1.9 *Let $\widetilde{M}$ be a differentiable manifold and let $\varphi : G \times \widetilde{M} \to \widetilde{M}$ be the free discontinuous action of a group $G$ on $\widetilde{M}$. The quotient $M = \widetilde{M}/G$ is a differentiable manifold and the projection $\pi : \widetilde{M} \to M$ is a local diffeomorphism.*

*Proof*
A local parametrisation of $\widetilde{M}/G$ is obtained by considering the restrictions of the local parametrisations $\widetilde{\mathbf{x}} : \widetilde{U} \to \widetilde{M}$ to open neighbourhoods $U \subset \mathbf{R}^l$ of $\widetilde{\mathbf{x}}^{-1}(\widetilde{p})$, where $\widetilde{p} \in \widetilde{\mathbf{x}}(\widetilde{U})$, such that $\widetilde{\mathbf{x}}(\widetilde{U}) \cap \varphi_g(\widetilde{\mathbf{x}}(\widetilde{U})) = \emptyset$ for every $g \in G$, $g \neq e$. We can then define the atlas of $\widetilde{M}/G$ through the charts $(U, \mathbf{x})$, where $\mathbf{x} = \pi \circ \widetilde{\mathbf{x}} : U \to \widetilde{M}/G$ (notice that, by the choice of $U$, $\pi|_{\widetilde{\mathbf{x}}(U)}$ is injective). We leave it as a problem for the reader to verify that these charts define an atlas. ∎

*Example 1.37*
The group $2\pi \mathbf{Z}$ acts on $\mathbf{R}^2$ as a group of translations: $\varphi_k(x_1, x_2) = (x_1 + 2\pi k, x_2)$. The action is free and discontinuous, and the quotient is diffeomorphic to the cylinder $\mathbf{S}^1 \times \mathbf{R}$. ∎

*Example 1.38*
The group $(2\pi \mathbf{Z})^l$ (whose elements are the vectors of $\mathbf{R}^l$ of the form $2\pi \mathbf{m}$, where $\mathbf{m} \in \mathbf{Z}^l$) acts on $\mathbf{R}^l$ as the translation group: $\varphi(\mathbf{x}) = \mathbf{x} + 2\pi \mathbf{m}$. It is easy to verify that the action is free and discontinuous, and that the quotient $\mathbf{R}^l/(2\pi \mathbf{Z})^l$ is a compact and connected differentiable manifold of dimension $l$ called the *$l$-dimensional torus* $\mathbf{T}^l$. Its elements are the equivalence classes $[\mathbf{x}]$ of $l$-tuples of real numbers $\mathbf{x} = (x_1, \ldots, x_l)$ with respect to the equivalence relation $\mathbf{x} \sim \mathbf{y} \Leftrightarrow \mathbf{x} - \mathbf{y} \in (2\pi \mathbf{Z})^l$, and hence if and only if $(x_j - y_j)/2\pi$ is an integer for every $j = 1, \ldots, l$. A geometric representation of $\mathbf{T}^l$ is obtained by considering the cube of side $2\pi$ in $\mathbf{R}^l$, identifying opposites sides (Fig. 1.23). ∎

An alternative way to construct a manifold is to start from two manifolds $M_1$ and $M_2$ (of dimension $l_1$ and $l_2$, respectively) and consider their Cartesian product, endowed with the product topology.

THEOREM 1.10 *The Cartesian product $M_1 \times M_2$ is a differentiable manifold of dimension $l_1 + l_2$ called the* product manifold *of $M_1$ and $M_2$.*

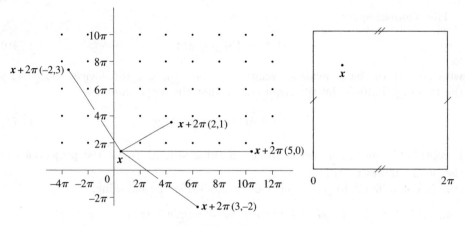

Fig. 1.23

*Proof*
It is immediate to verify that, if $\{(U_\alpha^{(1)}, \mathbf{x}_\alpha^{(1)})\}_{\alpha \in \mathcal{A}^{(1)}}$, $\{(U_\alpha^{(2)}, \mathbf{x}_\alpha^{(2)})\}_{\alpha \in \mathcal{A}^{(2)}}$ are atlases of $M_1$ and $M_2$, then $\{(U_\alpha^{(1)} \times U_\beta^{(2)}, \mathbf{y}_{\alpha\beta})\}_{(\alpha,\beta) \in \mathcal{A}^{(1)} \times \mathcal{A}^{(2)}}$ is an atlas of $M_1 \times M_2$, where we set

$$\mathbf{y}_{\alpha\beta}(\mathbf{u}_1, \mathbf{u}_2) = (\mathbf{x}_\alpha^{(1)}(\mathbf{u}_1), \mathbf{x}_\beta^{(2)}(\mathbf{u}_2))$$

with $\mathbf{u}_1 \in U_\alpha^{(1)}$, $\mathbf{u}_2 \in U_\beta^{(2)}$. Moreover, the projections $\pi_1 : M_1 \times M_2 \to M_1$ and $\pi_2 : M_1 \times M_2 \to M_2$, easily defined as $(\pi_1(\mathbf{u}_1, \mathbf{u}_2) = \mathbf{u}_1, \pi_2(\mathbf{u}_1, \mathbf{u}_2) = \mathbf{u}_2)$, are differentiable maps. ∎

*Example* 1.39
The torus $\mathbf{T}^l$ is diffeomorphic to the manifold obtained as the product of $l$ circles

$$\mathbf{T}^l \simeq \mathbf{S}^1 \times \ldots \times \mathbf{S}^1 \ (l \text{ times}). \tag{1.72}$$

This manifold is also called an $l$-dimensional torus. Indeed, considering $\mathbf{S}^1 \times \ldots \times \mathbf{S}^1$ as the regular submanifold of $\mathbf{R}^{2l}$ defined by

$$\mathbf{S}^1 \times \ldots \times \mathbf{S}^1 = \{(x_1, \ldots, x_{2l}) \in \mathbf{R}^{2l} | x_{2j-1}^2 + x_{2j}^2 = 1 \text{ for all } j = 1, \ldots, l\}, \tag{1.73}$$

the differentiable map $\widetilde{f} : \mathbf{R}^l \to \mathbf{R}^{2l}$ given by

$$\widetilde{f}(t_1, \ldots, t_l) = (\cos t_1, \sin t_1, \cos t_2, \sin t_2, \ldots, \cos t_l, \sin t_l)$$

has as image $\mathbf{S}^1 \times \ldots \times \mathbf{S}^1$ and satisfies $\widetilde{f}(\mathbf{t}+2\pi\mathbf{m}) = \widetilde{f}(\mathbf{t})$ for every $\mathbf{t} = (t_1, \ldots, t_l) \in \mathbf{R}^l$ and for every $\mathbf{m} \in \mathbf{Z}^l$. Hence it induces a diffeomorphism $f : \mathbf{T}^l \to \mathbf{S}^1 \times \ldots \times \mathbf{S}^1$, $f([\mathbf{t}]) = \widetilde{f}(\mathbf{t})$. Note that in general, everyfunction $\widetilde{g} : \mathbf{R}^l \to \mathbf{R}$, $2\pi$-periodic with respect to all its arguments, induces a function $g : \mathbf{T}^l \to \mathbf{R}$, and vice versa (every function on the torus can be identified with a single $2\pi$-periodic function of $\mathbf{R}^l$).

One can also allow the periods with respect to different arguments $t$ to be different, as it is easy to show that the torus $\mathbf{T}^l$ is diffeomorphic to the quotient of $\mathbf{R}^l$ with respect to the action of the translation group $\mathbf{x} \to \mathbf{x} + \mathbf{a} \cdot \mathbf{m}$, where $\mathbf{m} \in \mathbf{Z}^l$ and $\mathbf{a}$ is a given vector in $\mathbf{R}^l$ whose components $a_i$ are all different from zero. ∎

The torus $\mathbf{T}^l$ inherits the Riemannian metric from passing to the quotient of $\mathbf{R}^l$ on $(2\pi\mathbf{Z})^l$:

$$(\mathrm{d}s)^2 = (\mathrm{d}x_1)^2 + \cdots + (\mathrm{d}x_l)^2. \tag{1.74}$$

The resulting manifold is called a *flat torus*. Geodesics on $\mathbf{T}^l$ are clearly the projection of lines on $\mathbf{R}^l$, and hence they take the form

$$s \to (\alpha_1 s + \beta_1, \ldots, \alpha_l s + \beta_l) \pmod{(2\pi\mathbf{Z})^l}, \tag{1.75}$$

where $\alpha_1^2 + \cdots + \alpha_l^2 = 1$ and $s$ is the natural parameter. It is not difficult to prove that a geodesic is closed if and only if there exist $l$ rational numbers $m_1/n_1, \ldots, m_l/n_l$ and one real number $\alpha$ such that $\alpha_i = (m_i/n_i)\alpha$ for every $i$.

*Remark* 1.18

The flat torus $\mathbf{T}^2$ is *not* isometric to the 'doughnut', i.e. to the two-dimensional torus immersed in $\mathbf{R}^3$ (cf. Example 1.9) with the metric defined by the first fundamental form, although these two manifolds are diffeomorphic. Indeed, the geodesics on the latter are not obtained by setting $u = \alpha_1 s + \beta_1$, $v = \alpha_2 s + \beta_2$ in the parametrisation, because the two-dimensional torus immersed in $\mathbf{R}^3$ is a surface of revolution and its geodesics verify Clairaut's theorem (1.54); it is enough to note that among all curves obtained by setting $u = \alpha_1 s + \beta_1$, $v = \alpha_2 s + \beta_2$ are also the parallels ($\alpha_1 = 0$), which are not geodesics. ∎

## 1.9 Constrained systems and Lagrangian coordinates

We now start the study of dynamical systems consisting of a finite number of points, without taking into account that these points might be interacting with other objects. The background space is the physical space, i.e. $\mathbf{R}^3$, where we suppose that we have fixed a reference frame, and hence an origin $O$ and an orthonormal basis $\mathbf{e}_1, \mathbf{e}_2, \mathbf{e}_3$.

If $P_1, \ldots, P_n$ are the points defining the system, to assign the *configuration* of the system in the chosen reference frame means to give the Cartesian coordinates of all the $P_i$s. If all configurations are possible, the system is *free* (or *unconstrained*). If however there are limitations imposed on the allowed configurations (called *constraints*) the system is said to be *constrained*.

For example, we can require that some or all of the points of the system belong to a given curve or surface, which we will always assume to be regular.

The simplest is the case of a single point $P(x_1, x_2, x_3)$ constrained to be on the surface

$$F(x_1, x_2, x_3) = 0 \tag{1.76}$$

(simple constraint), or on the curve obtained as the intersection of two surfaces

$$F_1(x_1, x_2, x_3) = 0, \quad F_2(x_1, x_2, x_3) = 0 \tag{1.77}$$

(double constraint).

The analysis carried out in the previous sections shows that it is possible in the case (1.76) to introduce a local parametrisation of the surface, of the form

$$x_1 = x_1(q_1, q_2), \quad x_2 = x_2(q_1, q_2), \quad x_3 = x_3(q_1, q_2) \tag{1.78}$$

with the property (cf. (1.26)) that the Jacobian matrix has maximum rank

$$\operatorname{rank} \begin{pmatrix} \dfrac{\partial x_1}{\partial q_1} & \dfrac{\partial x_1}{\partial q_2} \\ \dfrac{\partial x_2}{\partial q_2} & \dfrac{\partial x_2}{\partial q_2} \\ \dfrac{\partial x_3}{\partial q_1} & \dfrac{\partial x_3}{\partial q_2} \end{pmatrix} = 2, \tag{1.79}$$

where $(q_1, q_2)$ vary in an appropriate open subset of $\mathbf{R}^2$. The vectors $\partial \mathbf{x}/\partial q_1, \partial \mathbf{x}/\partial q_2$ are then linearly independent and form a basis in the tangent space, while $\nabla F$ forms a basis in the normal space (Fig. 1.24). The vectors $\partial \mathbf{x}/\partial q_1, \partial \mathbf{x}/\partial q_2$ are tangent to the curves obtained by setting $q_2 = $ constant and $q_1 = $ constant, respectively, in equations (1.78).

One can use for the curves (1.77) the (local) parametrisation

$$x_1 = x_1(q), \quad x_2 = x_2(q), \quad x_3 = x_3(q), \tag{1.80}$$

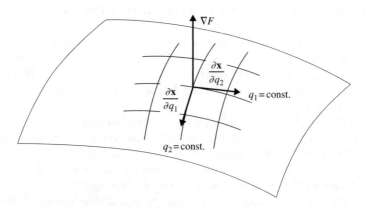

Fig. 1.24

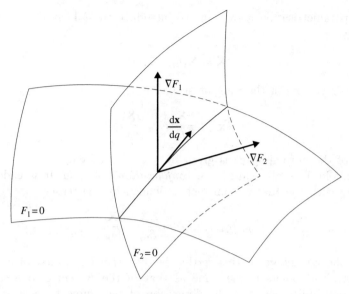

Fig. 1.25

where $d\mathbf{x}/dq \neq 0$, which is a basis for the tangent space, while the normal space has basis $\nabla F_1, \nabla F_2$ (Fig. 1.25).

The dimension of the tangent space gives the number of *degrees of freedom* of the point (2 and 1, respectively). The coordinates $(q_1, q_2)$ and the coordinate $q$ in the two cases are called *Lagrangian coordinates* of the point.

We now consider a system of several points $P_1, P_2, \ldots, P_n$; we can then impose constraints of the form $f(P_1, P_2, \ldots, P_n) = 0$. It appears natural to describe the system in the space $\mathbf{R}^{3n}$, by establishing a bijective correspondence between the configurations of the system and the vectors $\mathbf{X} = \oplus_{i=1,\ldots,n} \mathbf{x}_i$.

Thus imposing $m < 3n$ independent constraints is equivalent to the condition that the representative vector $\mathbf{X}$ belongs to a submanifold $\mathcal{V}$ of dimension $l = 3n - m$ (cf. Definition 1.19), and hence that the equations

$$f_j(\mathbf{X}) = 0, \quad j = 1, 2, \ldots, m, \tag{1.81}$$

are satisfied, with the vectors $\nabla_{\mathbf{X}} f_1, \ldots, \nabla_{\mathbf{X}} f_m$ being linearly independent on $\mathcal{V}$, or equivalently, with the Jacobian matrix

$$\begin{pmatrix} \dfrac{\partial f_1}{\partial X_1} & \dfrac{\partial f_1}{\partial X_2} & \cdots & \dfrac{\partial f_1}{\partial X_{3n}} \\ \cdots \cdots \\ \dfrac{\partial f_m}{\partial X_1} & \dfrac{\partial f_m}{\partial X_2} & \cdots & \dfrac{\partial f_m}{\partial X_{3n}} \end{pmatrix} \tag{1.82}$$

being of rank $m$ on $\mathcal{V}$. Hence $\mathcal{V}$ is a submanifold of $\mathbf{R}^{3n}$ of dimension $l$ having the same regularity as the functions $f_i$; in particular, $\mathcal{V}$ is also a differentiable manifold (Definition 1.21 and Theorem 1.7). The system has $l$ degrees of freedom.

A local parametrisation allows one to introduce the $l$ Lagrangian coordinates $q_1, q_2, \ldots, q_l$:

$$\mathbf{X} = \mathbf{X}(q_1, \ldots, q_l) \tag{1.83}$$

and the basis vectors of the tangent space

$$T_{\mathbf{X}}\mathcal{V} : \frac{\partial \mathbf{X}}{\partial q_1}, \ldots, \frac{\partial \mathbf{X}}{\partial q_l}.$$

The basis of the normal space is given by $\nabla_{\mathbf{X}} f_1, \ldots, \nabla_{\mathbf{X}} f_m$.

The manifold $\mathcal{V}$ is also called the *configuration manifold*. It is endowed in a natural way with the Riemannian metric defined by the tensor

$$g_{ij}(q_1, \ldots, q_l) = \frac{\partial \mathbf{X}}{\partial q_i} \cdot \frac{\partial \mathbf{X}}{\partial q_j}.$$

Note that the advantage of this setting is that the description of a system of many constrained points is the same as that of the system of one constrained point; the only difference is in the dimension of the ambient space. In the next paragraph we shall study the motion of these systems.

*Example* 1.40
The system of two points $P_1$, $P_2$ with the rigidity constraint

$$\sum_{i=1}^{3} [x_i^{(1)} - x_i^{(2)}]^2 - R^2 = 0$$

has five degrees of freedom and admits, e.g. the parametrisation

$$\begin{aligned}
x_1^{(1)} &= \xi_1, \quad x_2^{(1)} = \xi_2, \quad x_3^{(1)} = \xi_3, \\
x_1^{(2)} &= \xi_1 + R \cos \varphi \cos \theta, \\
x_2^{(2)} &= \xi_2 + R \sin \varphi \cos \theta, \\
x_3^{(2)} &= \xi_3 + R \sin \theta.
\end{aligned}$$
∎

## 1.10 Holonomic systems

A further step in the construction of a mathematical model of the mechanics of discrete systems is to introduce a temporal variable, and correspondingly the concept of motion with respect to an observer, i.e. to a triple $(O, \mathbf{e}_1, \mathbf{e}_2, \mathbf{e}_3)$ and a temporal scale.[2]

We assume that the fundamental notions of the mechanics of a single point are known, and we stress that when passing from a purely geometrical description to the more complex notion of kinematics, the concept of constraint needs to

---

[2] We will remain within the scope of the well-known axioms of classic kinematics.

be considerably extended. It is, for example, possible to impose constraints on the velocity of a point, or on the minimal radius of curvature of a trajectory, and so on.

The most natural extension of the concept of constraint from geometry to kinematics consists of imposing the validity of the constraint equations (1.81), which we considered in the previous section, in a certain time interval; we shall say that the system is subject to *fixed constraints* in the given time interval.

More generally, we can consider a system of constraint equations of the form

$$f_j(\mathbf{X}, t) = 0, \quad j = 1, \ldots, m < 3n, \quad \mathbf{X} \in \mathbf{R}^{3n}, \quad t \in I, \qquad (1.84)$$

where we assume that in the given time interval $I$, the usual regularity and compatibility conditions, as well as the linear independence of the vectors $\nabla_{\mathbf{X}} f_j$, are satisfied. The configuration space can be considered to be a *moving* differentiable manifold $\mathcal{V}(t)$.

Thus we can make use of the local representation of the manifold $\mathcal{V}(t)$ described by equation (1.84) through a vector $\mathbf{q}$ of Lagrangian coordinates

$$\mathbf{X} = \mathbf{X}(\mathbf{q}, t), \quad \mathbf{q} \in \mathbf{R}^l, \quad l = 3n - m, \qquad (1.85)$$

with the important property that the vectors $\partial \mathbf{X}/\partial q_k$, $k = 1, \ldots, l$, are linearly independent for every $t$ in the given interval, and they form a basis of the tangent space $T_{\mathbf{X}} \mathcal{V}(t)$, for every fixed $t$.

DEFINITION 1.35 *The constraints (1.84) satisfying the properties described above are called* holonomic[3] *(the systems subject to such constraints are themselves called holonomic). If* $\partial f_j / \partial t \not\equiv 0$ *for some* $j$, *the constraints are said to be* moving constraints. *The constraints (or systems) that are not holonomic are called* non-holonomic. ∎

*Example* 1.41
Consider a system consisting of a single point $P$ moving in space, and impose the condition that the velocity of the point not be external to a certain given cone $\Phi(P)$ with vertex in $P$ (if, e.g. $\Phi$ is a circular right-angle cone, this is equivalent to a limitation imposed on the angle between $\mathbf{v}$ and the cone axis). This is typically a non-holonomic constraint, as it is expressed exclusively on the velocity of the point $P$ and does not affect its position. To understand the effect of this constraint, imagine moving $P$ from a position $P'$ to a position $P'' \notin \Phi(P')$. Clearly not all the trajectories are allowed, because the velocity direction must constantly belong to $\Phi(P)$. If, for example, $\Phi(P)$ varies with $P$ only by translation, the point can follow a straight line connecting $P'$ with a point $P^*$ such that $P'' \in \Phi(P^*)$ and then follow the segment between $P^*$ and $P''$.

A similar situation is found for the problem of parking a car (the condition for the wheels not to slip and the minimal radius of the turn are typically non-holonomic constraints). ∎

---

[3] The etymology of the name (literally, 'integer law') refers to the absence of derivatives in (1.84).

We shall encounter another example of a non-holonomic constraint in Chapter 6 (Example 6.2).

*Remark 1.19*
It may happen that constraints imposed on the velocities are in fact holonomic. The typical case is the case of a *plane rigid system* (see Chapter 6), bounded by (or consisting of) a regular curve $\gamma$, constrained to roll without sliding on another given regular curve $\Gamma$. This constraint is only apparently a kinematic constraint (vanishing velocity at the contact point between $\gamma$ and $\Gamma$). Indeed, choosing a configuration $\gamma_0$ of $\gamma$ where $P_0$ represents the contact point with $\Gamma$, the coordinates of all the points of the system in a generic configuration of the contact point $P$ between $\gamma$ and $\Gamma$ are known functions of the length $s$ of the arc $\widehat{P_0 P}$ on $\Gamma$. Hence the system is holonomic with a single degree of freedom and $s$ can be chosen as the Lagrangian coordinate. ∎

In the generic case, the basis of the normal space $\nabla_\mathbf{X} f_j$, $j = 1,\ldots,m$, and that of the tangent space $\partial \mathbf{X}/\partial q_k$, $k = 1,\ldots,l$, at every point $\mathbf{X}$ of the manifold (1.84), depend on time.

An important class of holonomic system consists of the so-called *rigid systems*; these are treated in Chapters 6 and 7.

## 1.11 Phase space

We start by observing that given a particular motion of the system $\{P_1,\ldots,P_n\}$, one has $\dot{\mathbf{X}} = \oplus_{i=1}^n \dot{P}_i$, and hence the vector $\dot{\mathbf{X}} \in \mathbf{R}^{3n}$ represents the velocities of the points of the system. Clearly this is the velocity of the representative vector $\mathbf{X}$.

There are two ways of describing the effects of the constraints (1.84) upon the vector $\dot{\mathbf{X}}$, by projecting it either onto the normal space or onto the tangent space. Suppose that a motion of the system, compatible with the constraints, is known. By differentiating with respect to time equations (1.84) we find

$$\dot{\mathbf{X}} \cdot \nabla_\mathbf{X} f_j(\mathbf{X}, t) + \frac{\partial f_j}{\partial t} = 0, \quad j = 1,\ldots,m, \tag{1.86}$$

which provides information on the projection of $\dot{\mathbf{X}}$ onto the normal space.

By assigning the motion through equations (1.85), choosing $\mathbf{q} = \mathbf{q}(t) \in \mathcal{C}^1$, by differentiating (1.85) we obtain the representation

$$\dot{\mathbf{X}} = \sum_{k=1}^l \frac{\partial \mathbf{X}}{\partial q_k} \dot{q}_k + \frac{\partial \mathbf{X}}{\partial t}. \tag{1.87}$$

Both equation (1.86) and equation (1.87) imply, e.g. that *for the case of fixed constraints*, $\dot{\mathbf{X}}$ *belongs to the tangent space*.

Equation (1.87) suggests the decomposition

$$\dot{\mathbf{X}} = \widehat{\mathbf{V}} + \mathbf{V}^*, \tag{1.88}$$

## 1.11  Geometric and kinematic foundations of Lagrangian mechanics

where

$$\widehat{\mathbf{V}} = \sum_{k=1}^{l} \frac{\partial \mathbf{X}}{\partial q_k} \dot{q}_k \tag{1.89}$$

is called the *virtual velocity* of the representative point $\mathbf{X}$, while the interpretation of

$$\mathbf{V}^* = \frac{\partial \mathbf{X}}{\partial t} \tag{1.90}$$

is that of the velocity of the point $\mathbf{X}^* \in \mathcal{V}(t)$ for constant values of the Lagrangian coordinates.

Both $\widehat{\mathbf{V}}$ and $\mathbf{V}^*$ depend on the Lagrangian coordinate system and are clearly transformed by a time-dependent transformation of Lagrangian coordinates. It is interesting, however, to note the following.

PROPOSITION 1.1  *The projection of $\mathbf{V}^*$ onto the normal space is independent of the system of Lagrangian coordinates.*

*Proof*
Let

$$\mathbf{Q} = \mathbf{Q}(\mathbf{q}, t) \tag{1.91}$$

be a Lagrangian coordinate transformation, and

$$\mathbf{q} = \mathbf{q}(\mathbf{Q}, t) \tag{1.92}$$

its inverse. Defining

$$\overline{\mathbf{X}}(\mathbf{Q}, t) = \mathbf{X}[\mathbf{q}(\mathbf{Q}, t), t] \tag{1.93}$$

one can compute

$$\frac{\partial \overline{\mathbf{X}}}{\partial t} - \frac{\partial \mathbf{X}}{\partial t} = \sum_{k=1}^{l} \frac{\partial \mathbf{X}}{\partial q_k} \frac{\partial q_k}{\partial t}, \tag{1.94}$$

which yields the result. ∎

Again, as suggested by equations (1.86), we find that, fixing the Cartesian coordinate system, the vector $\dot{\mathbf{X}}$ can be intrinsically decomposed into its tangential and normal components; the latter is due to the motion of the constraints, and can be called the *drag velocity* of the constraints.

*Example 1.42*
Consider the point $P$ subject to the moving constraint

$$x_1 = R\cos(\varphi + \alpha(t)), \quad x_2 = R\sin(\varphi + \alpha(t)), \quad x_3 = \lambda\varphi,$$

where $R, \lambda$ are positive constants. A computation yields

$$\hat{\mathbf{v}} = \frac{\partial \mathbf{x}}{\partial \varphi} \dot{\varphi}, \quad \mathbf{v}^* = \frac{\partial \mathbf{x}}{\partial t},$$

$$\hat{\mathbf{v}} = \dot{\varphi} \begin{pmatrix} -R\sin(\varphi + \alpha) \\ R\cos(\varphi + \alpha) \\ \lambda \end{pmatrix}, \quad \mathbf{v}^* = \begin{pmatrix} -R\dot{\alpha}\sin(\varphi + \alpha) \\ R\dot{\alpha}\cos(\varphi + \alpha) \\ 0 \end{pmatrix}.$$

The projection of $\mathbf{v}^*$ on the space normal to the constraint is characterised by

$$\mathbf{v}^* \cdot \mathbf{n} = 0, \quad \mathbf{v}^* \cdot \mathbf{b} = -\frac{\lambda R \dot{\alpha}}{\sqrt{R^2 + \lambda^2}}.$$

In this reference system the helix spirals around the $x_3$-axis. However, making the change of coordinates $\varphi' = \varphi + \alpha(t)$, one has

$$x_1' = R\cos\varphi', \quad x_2' = R\sin\varphi', \quad x_3' = \lambda(\varphi' - \alpha(t)).$$

The new decomposition $\mathbf{v} = \hat{\mathbf{v}}' + \mathbf{v}^{*'}$ of the velocity is given by

$$\hat{\mathbf{v}}' = \dot{\varphi}' \begin{pmatrix} -R\sin\varphi' \\ R\cos\varphi' \\ \lambda \end{pmatrix}, \quad \mathbf{v}^{*'} = \begin{pmatrix} 0 \\ 0 \\ -\lambda\dot{\alpha} \end{pmatrix},$$

and now

$$\mathbf{v}^{*'} \cdot \mathbf{n} = 0, \quad \mathbf{v}^{*'} \cdot \mathbf{b} = -\frac{\lambda R \dot{\alpha}}{\sqrt{R^2 + \lambda^2}} = \mathbf{v}^* \cdot \mathbf{b}.$$

Note that in this example $\mathbf{v}^*$ and $\mathbf{v}^{*'}$ are orthogonal to each other. ∎

For a fixed time $t$ consider a point $\mathbf{X} \in \mathcal{V}(t)$. In a chosen system of Lagrangian coordinates, equation (1.87) describes all the velocities $\dot{\mathbf{X}}$ compatible with the constraints, as long as the coefficients $\dot{q}_k$ are considered to be variable parameters in $\mathbf{R}$. Thus the components of the vector $\dot{\mathbf{q}} \in \mathbf{R}^l$ take on the role of *kinetic coordinates*.

DEFINITION 1.36 *The space in which the pair* $(\mathbf{q}, \dot{\mathbf{q}})$ *varies is called the* phase space *of the system. This space parametrises the vector bundle* $T\mathcal{V}(t)$ *of the configuration manifold* $\mathcal{V}(t)$. ∎

At every time $t$ the pairs $(\mathbf{q}, \dot{\mathbf{q}})$ are in bijective correspondence with the pairs $(\mathbf{X}, \dot{\mathbf{X}})$ that are compatible with the constraints; we call these pairs the *kinematic states* of the system. It is useful to recall that equation (1.87) summarises the information on the velocity of the single points of the system:

$$\dot{P}_i = \sum_{k=1}^{\ell} \frac{\partial P_i}{\partial q_k} \dot{q}_k + \frac{\partial P_i}{\partial t}. \qquad (1.95)$$

## 1.12 Accelerations of a holonomic system

The results of the previous section yield information about the vector $\ddot{\mathbf{X}}$ for a holonomic system.

Differentiation of equation (1.86) with respect to time, for a given motion (assuming the $f_j$ are sufficiently regular), yields

$$\ddot{\mathbf{X}} \cdot \nabla_{\mathbf{X}} f_j + \dot{\mathbf{X}} \cdot \mathbf{H}_j \dot{\mathbf{X}} + 2\left(\frac{\partial}{\partial t} \nabla_{\mathbf{X}} f_j\right) \cdot \dot{\mathbf{X}} + \frac{\partial^2 f_j}{\partial t^2} = 0, \quad j = 1, \ldots, m, \qquad (1.96)$$

where $\mathbf{H}_j$ is the Hessian matrix of $f_j$. What is interesting about equations (1.96) is summarised in the following.

PROPOSITION 1.2 *For every time the projection $\ddot{\mathbf{X}}$ onto the normal space is determined by the pair $(\mathbf{X}, \dot{\mathbf{X}})$.* ∎

In the case of fixed constraints, equations (1.96) reduce to

$$\ddot{\mathbf{X}} \cdot \nabla_{\mathbf{X}} f_j = -\dot{\mathbf{X}} \cdot \mathbf{H}_j \dot{\mathbf{X}}. \qquad (1.97)$$

In particular, for a point constrained to a fixed surface, given by the equation $F(\mathbf{x}) = 0$, we have

$$\ddot{\mathbf{x}} \cdot \nabla F = -\dot{\mathbf{x}} \cdot \mathbf{H} \dot{\mathbf{x}}, \qquad (1.98)$$

and if $\mathbf{x} = \mathbf{x}(s)$ is the natural parametrisation of the trajectory, then also

$$\ddot{\mathbf{x}} = \frac{d^2 \mathbf{x}}{ds^2} \dot{s}^2 + \frac{d\mathbf{x}}{ds} \ddot{s} \qquad (1.99)$$

and hence, if $\mathbf{N}$ denotes the normal vector to the surface at the point $\mathbf{x}(s)$,

$$\ddot{\mathbf{x}} \cdot \mathbf{N} = \frac{d\mathbf{t}}{ds} \cdot \mathbf{N} \dot{s}^2 = k_n \dot{s}^2, \qquad (1.100)$$

where $k_n = k\mathbf{n} \cdot \mathbf{N}$ is the *normal curvature*. Setting $\mathbf{N} = \nabla F/|\nabla F|$, a comparison between equations (1.100) and (1.98) yields an expression for $k_n$:

$$|k_n| = \left|\frac{\mathbf{t} \cdot \mathbf{H} \mathbf{t}}{|\nabla F|}\right|. \qquad (1.101)$$

*Example* 1.43
Given any point on the sphere $x_1^2 + x_2^2 + x_3^2 = R^2$, the normal curvature of a curve on the sphere at any one of its points is equal to $1/R$. ∎

Reverting to equation (1.99), we note how it indicates that the acceleration of the point belongs to the osculating plane to the trajectory, on which it has the decomposition

$$\ddot{\mathbf{x}} = k(s)\dot{s}^2 \mathbf{n} + \ddot{s} \mathbf{t} \qquad (1.102)$$

($\mathbf{n}$ is the principal normal vector, and $k(s)$ is the curvature of the trajectory).

For a point in the plane constrained to belong to a curve $f(x_1, x_2) = 0$, the same computation yielding equation (1.101) easily yields that the same formula gives the expression for the curvature, with the only difference that in this case one can set $\mathbf{t} = \mathbf{e}_3 \times \nabla f / |\nabla f|$, and obtain

$$k(s) = \frac{\left| \frac{\partial^2 f}{\partial x_1^2} \left(\frac{\partial f}{\partial x_2}\right)^2 - 2 \frac{\partial^2 f}{\partial x_1 \partial x_2} \frac{\partial f}{\partial x_1} \frac{\partial f}{\partial x_2} + \frac{\partial^2 f}{\partial x_2^2} \left(\frac{\partial f}{\partial x_1}\right)^2 \right|}{\left[ \left(\frac{\partial f}{\partial x_1}\right)^2 + \left(\frac{\partial f}{\partial x_2}\right)^2 \right]^{3/2}}. \tag{1.103}$$

*Example 1.44*
For a generic point of the cylinder given by the equation $F(x_1, x_2) = 0$, varying $\mathbf{t} = \cos \vartheta \mathbf{e}_3 + \sin \vartheta \mathbf{e}_3 \times \nabla F / |\nabla F|$ the normal curvature is obtained by using equation (1.101); this yields $|k_n| = k \sin^2 \vartheta$, where $k$ is the curvature of the normal section, given by equation (1.103). ∎

## 1.13 Problems

1. Compute the length and the natural parametrisation of the following plane curves:
    (a) $x_1(t) = t$, $x_2(t) = \log t$;
    (b) $x_1(t) = t$, $x_2(t) = t^2$;
    (c) $x_1(t) = a(1 + \cos t) \cos t$, $x_2 = a(1 + \cos t) \sin t$
       (hint: change to polar coordinates);
    (d) $x_1(t) = t$, $x_2(t) = e^t$.
2. Compute the velocity of the following plane curve, and sketch its graph:
$$x_1(t) = 2 \cos\left(t - \frac{\pi}{2}\right), \quad x_2(t) = \sin\left(2\left(t - \frac{\pi}{2}\right)\right).$$
3. Consider the *spiral of Archimedes*
$$x_1(t) = rt \cos t, \quad x_2(t) = rt \sin t$$
and compute the velocity, acceleration, natural parametrisation, unit normal and tangent vectors, and curvature.
4. Determine the curve described by a point in uniform motion along a line through the origin, rotating uniformly (answer: *spiral of Archimedes*).
5. Determine the curve described by a point in motion with velocity proportional to the distance from the origin along a line, through the origin, rotating uniformly (answer: $x_1(t) = ce^{kt} \cos t$, $x_2(t) = ce^{kt} \sin t$, a *logarithmic spiral*, with $c$ and $k$ constant).
6. Prove that the curvature $k(t)$ of the plane curve $t \to (x_1(t), x_2(t))$ is
$$k(t) = \frac{|\dot{x}_1 \ddot{x}_2 - \ddot{x}_1 \dot{x}_2|}{(\dot{x}_1^2 + \dot{x}_2^2)^{3/2}}.$$

## 1.13 Geometric and kinematic foundations of Lagrangian mechanics

**7.** Find a global parametrisation and compute the curvature of the following plane curves:
   (a) $x_2 - ax_1^2 = c$, with $a \neq 0$;
   (b) $x_1^2 - x_2^2 = 1$, $x_1 > 0$.

**8.** Compute the natural parametrisation and the tangent, normal and binormal unit vectors, as well as the curvature and torsion, of the following curves:
   (a) $t \to (rt \cos t, rt \sin t, bt)$;
   (b) $t \to (re^t \cos t, re^t \sin t, bt)$;
   (c) $t \to (t^2, 1-t, t^3)$;
   (d) $t \to (\cosh t, \sinh t, t)$, where $b \in \mathbf{R}$ is a given constant.

**9.** Verify that the curve given by $t \to (a \sin^2 t, a \sin t \cos t, a \cos t)$, where $a \in \mathbf{R}$ is a given constant, lies inside a sphere, and that all its normal planes pass through the origin. Prove that the curve is of order 4.

**10.** Prove that the curve $t \to (at + b, ct + d, t^2)$ where $a, b, c, d \in \mathbf{R}$ are given constants, $c \neq 0$, has the same osculating plane in all points. What can you conclude? Compute the torsion.

**11.** Prove that the solutions of the vector differential equations (1.17) and (1.21) with natural initial conditions, for $\mathbf{t}$ and $d\mathbf{t}/ds$, have the following properties: $|\mathbf{t}| = 1, |d\mathbf{t}/ds| = k(s)$.
*Sketch.* Setting $\theta = |\mathbf{t}|^2$, $\Xi = |d\mathbf{t}/ds|^2$, from equation (1.17) one obtains the system

$$\frac{1}{2}\theta'' - \frac{1}{2}\frac{k'}{k}\theta' + k^2\theta - \Xi = 0,$$

$$\frac{1}{2}\Xi' - \frac{k'}{k}\Xi + \frac{1}{2}k^2\theta' = 0$$

(multiply, respectively, by $\mathbf{t}$ and $d\mathbf{t}/ds$). With the natural initial conditions (i.e. $\mathbf{t}(0)$ an arbitrary unit vector, $\mathbf{t}'(0)$ orthogonal to $\mathbf{t}(0)$ with absolute value $k(0)$), this system admits the unique solution $\theta = 1$, $\Xi = k^2(s)$. By the same manipulation one can derive from equation (1.21) exactly the same system.

**12.** Find the level sets and sketch the graph of $f(x_1, x_2) = x_2^2 - 3x_1^2 x_2$.

**13.** Given any hypersurface in $\mathbf{R}^n$, $S = F^{-1}(0)$, where $F : U \to \mathbf{R}$, $U \subset \mathbf{R}^n$ is open, the *cylinder* $C$ over $S$ is the hypersurface in $\mathbf{R}^{n+1}$ defined by $C = G^{-1}(0)$, where $G : U \times \mathbf{R} \to \mathbf{R}$, $G(x_1, \ldots, x_n, x_{n+1}) = F(x_1, \ldots, x_n)$. Draw the cylinders on the following hypersurfaces $S = F^{-1}(0)$:
   (a) $F(x_1) = x_1^2 - 1$;
   (b) $F(x_1) = x_1$;
   (c) $F(x_1, x_2) = x_1^2 + x_2^2 - 1$;
   (d) $F(x_1, x_2) = x_1 - x_2^2$;
   (e) $F(x_1, x_2) = x_1^2/4 + x_2^2/9 - 1$.

Find parametric representations, and verify that these cylinders are regular surfaces.

**14.** Prove that the cylinder over a regular surface (see Problem 13) is a regular surface.

**15.** Find the equation of the tangent plane in an arbitrary point of a sphere, a cylinder, a cone and an ellipsoid.

**16.** Compute the first fundamental form of an ellipsoid, of a one- and a two-sheeted hyperboloid, and of the elliptic paraboloid.

**17.** Determine the curves on the unit sphere which intersect the meridians at a constant angle $\alpha$, and compute their length (these curves are called *loxodromes*).

**18.** Prove that the area of a geodesic triangle $A$ on the sphere of radius 1 is given by

$$A = \alpha + \beta + \gamma - \pi,$$

where $\alpha$, $\beta$ and $\gamma$ are the internal angles of the triangle (a geodesic triangle is a triangle which has as sides geodesic arcs, in this case arcs of maximal circles). How does the formula change if the sphere has radius $r$?

**19.** The sphere of radius 1 and centre $(0,0,1)$ can be parametrised, except at the north pole $(0,0,2)$, by a stereographic projection. Find the first fundamental form of the sphere using this parametrisation. Find the image of the meridians, parallels, and loxodromes under the stereographic projection.

**20.** Prove that if a surface contains a line segment, then this segment is a geodesic curve on the surface.

**21.** Prove that the curve $t \to (t \cos \alpha, t \sin \alpha, t^2)$, where $\alpha \in \mathbf{R}$ is given, is a geodesic curve on the circular paraboloid $x_1^2 + x_2^2 - x_3 = 0$.

**22.** Prove that the plane, cylinder and cone are isometric surfaces.

**23.** Prove that the geodesics on a surface whose first fundamental form is given by $(ds)^2 = v((du)^2 + (dv)^2)$, $v > 0$, are straight lines parallel to the axis $v$ or else they are parabolas with axes parallel to the axis $v$.

**24.** Determine the geodesics on a surface whose first fundamental form is given by $(ds)^2 = (du)^2 + e^{2u}(dv)^2$.

**25.** The unit disc $\mathbf{D} = \{(\xi, \eta) \in \mathbf{R}^2 | \xi^2 + \eta^2 < 1\}$ has a metric with constant curvature equal to $-1$:

$$(ds)^2 = 4 \frac{(d\xi)^2 + (d\eta)^2}{(1 - \xi^2 - \eta^2)^2} \quad (Poincaré \; disc).$$

Prove that the geodesics are the diameters and the arcs of circles that intersect orthogonally the boundary of the disc $\partial \mathbf{D} = \{\xi^2 + \eta^2 = 1\}$.

**26.** Consider $\mathbf{R}^2$ as identified with $\mathbf{C}$. Setting $z = x + iy$ and $w = \xi + i\eta$, prove that the transformation

$$w = T(z) = \frac{z - i}{z + i}$$

from the Lobačevskij half-plane **H** to the Poincaré disc **D** is an isometry. Determine $T^{-1}$.

**27.** Compute the area of the disc centred at the origin and with radius $r < 1$ in the Poincaré disc. Compute the limit for $r \to 1^-$.

**28.** Prove that the geodesics on the bidimensional torus immersed in $\mathbf{R}^3$ (the 'doughnut', cf. Example 1.19) are obtained by integrating the relation

$$\mathrm{d}v = C \frac{b\,\mathrm{d}r}{r\sqrt{r^2 - C^2}\sqrt{b^2 - (r-a)^2}},$$

where $C$ is any integration constant, $r = a + b \cos u$.

**29.** Prove that if two Riemannian manifolds $M$ and $N$ are isometric, then the geodesics of $M$ are the image through the isometry of the geodesics of $N$ (and vice versa).

## 1.14 Additional remarks and bibliographical notes

In this chapter we have introduced some elementary notions of differential geometry, of fundamental importance for the study of analytical mechanics.

The study of local properties of curves and surfaces was the object of intense research by several mathematicians of the eighteenth century (Clairaut, Euler, Monge, Serret, Frenet, among the most famous). This was motivated by the development of the calculus of variations (cf. Chapter 9) and by the mechanics of a constrained point. Riemannian geometry, the natural development of the work of these mathematicians, was founded by Gauss and Riemann during the nineteenth century (it is curious that the notion of a differentiable manifold, while necessary for the rigorous development of their results, was introduced for the first time by Hermann Weyl in 1913). These two mathematicians, together with Lobačevskij, Bolyai and Beltrami, developed 'non-Euclidean geometry'. An excellent historical discussion of the beginnings of differential geometry is given by Paulette Libermann (in Dieudonné 1978, Chapter 9).

Weeks' book (1985) is an example of 'high level popularisation', containing an intuitive introduction to the concept of a manifold. We recommend it for its clarity and readability. However, we must warn the reader that this clarity of exposition may give a misleading impression of simplicity; it is necessary to read this book carefully, considering the proposed (often humorous) problems, in order to develop a good geometric intuition and familiarity with the subject. We recommend in particular the reading of the beautiful section on the Gauss–Bonnet formula and its consequences.

For a particularly accessible introduction to the concepts developed in the first six sections, along with a discussion of much additional material (covariant derivative, Gauss map, second fundamental form, principal and Gaussian curvatures, etc.) which we could not include in our exposition (cf. Appendix 3 for some of it) we recommend Thorpe's textbook (1978). More advanced texts, for the further analysis of the notions of a manifold and a Riemannian metric, are Do Carmo (1979) and Singer and Thorpe (1980). The first two volumes of Dubrovin et al. (1991a,b) contain a very clear and profound exposition of the basic notions of differential geometry, nowadays indispensable for the study of theoretical physics (to which the authors devote a lot of attention in the exposition) and of dynamical systems. The first volume in particular should be accessible to any student familiar with the concepts introduced in the basic analysis and geometry courses in the first two years of university studies. The same can be said for the book of Arnol'd (1978b), which contains in Chapter 5 a very good introduction to differentiable manifolds and to the study of differential equations on a manifold, including an introduction to topological methods and to the index theorem.

## 1.15 Additional solved problems

*Problem 1*
Consider the family of plane curves $\varphi(x_1, x_2, l) = 0$ with $\nabla_\mathbf{x}\varphi \neq 0$ and $\partial\varphi/\partial l > 0$, $l \in (a, b)$. Construct the family of curves intersecting the given curves orthogonally.

*Solution*
Since $\varphi$ is strictly monotonic as a function of $l$, the curves belonging to the given family do not intersect. A field of directions orthogonal to the curves is defined in the region of the plane containing these curves. The flux lines of this field (i.e. the orthogonal trajectories) have equation

$$\dot{\mathbf{x}} = \nabla_\mathbf{x}\varphi(\mathbf{x}, l) \tag{1.104}$$

and the condition for intersection $\mathbf{x}(0) = \mathbf{x}_0$ determines $l$. Indeed, thanks to the hypothesis $\partial\varphi/\partial l > 0$, we can write $l_0 = \Lambda(\mathbf{x}_0)$.

This is in fact the general procedure, but it is interesting to examine a few explicit cases.

(i) $\varphi(x_1, x_2, l) = f_1(x_1, l) + f_2(x_2, l)$

with the obvious hypotheses on $f_1, f_2$. In this case, equation (1.104) becomes

$$\dot{x}_1 = \frac{\partial f_1}{\partial x_1}, \quad \dot{x}_2 = \frac{\partial f_2}{\partial x_2}$$

and both equations are separately integrable. Setting

$$F_i(x_i, l) = \int \left(\frac{\partial f_i}{\partial x_i}\right)^{-1} dx_i, \quad i = 1, 2,$$

we can find the parametric solution

$$F_1(x_1, l_0) - F_1(x_1^0, l_0) = t, \quad F_2(x_2, l_0) - F_2(x_2^0, l_0) = t,$$

with $l_0$ determined by $(x_1^0, x_2^0)$. As an example, consider the family of parabolas

$$\varphi(x_1, x_2, l) = lx_1^2 - x_2 + l = 0,$$

satisfying the conditions $\nabla_\mathbf{x}\varphi = (2lx_1, -1) \neq 0$ and $\partial\varphi/\partial l = 1 + x_1^2 > 0$.

## 1.15 Geometric and kinematic foundations of Lagrangian mechanics

The equations for the orthogonal trajectories are

$$\dot{x}_1 = 2lx_1, \quad \dot{x}_2 = -1,$$

to be integrated subject to the conditions $x_i(0) = x_i^0$, $l_0 = x_2^0/(1+x_1^{0^2})$.
One finds $x_1 = x_1^0 e^{2l_0 t}$, $x_2 - x_2^0 = -t$.
Hence the trajectory, orthogonal to the family of parabolas, and passing through $(x_1^0, x_2^0)$, can be written in the form of a graph:

$$x_1 = x_1^0 \exp\left[\frac{2x_2^0}{1+x_1^{0^2}}(x_2^0 - x_2)\right].$$

(ii) $\varphi(x_1, x_2, l) = \xi(x_1, x_2) + l$
with $\nabla_\mathbf{x}\xi \neq 0$. The parameter $l$ does not appear in the field equations

$$\dot{\mathbf{x}} = \nabla \xi(\mathbf{x}), \tag{1.105}$$

but only in the intersection conditions.

### Problem 2
Consider the cone projecting, from the point $(0,0,1)$ into the $(x,y)$ plane, the curve of equation $x = f_1(\sigma)$, $y = f_2(\sigma)$, where $\sigma$ is the arc length parameter of the curve.

(i) Write the parametric equations, using the coordinates $\sigma, z$.
(ii) Find the first fundamental form.
(iii) In the case that $f_1(\sigma) = R\cos\sigma$, $f_2(\sigma) = R\sin\sigma$ study the set of geodesics (for $z < 1$).

### Solution
(i) The parametric equations of the cone are

$$x = (1-z)f_1(\sigma), \quad y = (1-z)f_2(\sigma), \quad z = z. \tag{1.106}$$

(ii) In the representation considered, the vectors forming the basis of the tangent space are

$$\mathbf{x}_\sigma = (1-z)\begin{pmatrix} f_1' \\ f_2' \\ 0 \end{pmatrix}, \quad \mathbf{x}_z = \begin{pmatrix} -f_1 \\ -f_2 \\ 1 \end{pmatrix}.$$

Hence we have

$$E = \mathbf{x}_\sigma^2 = (1-z)^2, \quad F = \mathbf{x}_\sigma \cdot \mathbf{x}_z = -(1-z)(f_1 f_1' + f_2 f_2'),$$
$$G = \mathbf{x}_z^2 = 1 + f_1^2 + f_2^2.$$

Note that we used the fact that $f_1'^2 + f_2'^2 = 1$.

(iii) If the cone is a right circular cone, $f_1^2 + f_2^2 = R^2$ and then $F = 0$, $G = 1+R^2$. In this case it is easy to compute the Christoffel symbols. The only non-zero ones are

$$\Gamma_{12}^1 = \Gamma_{21}^1 = -\frac{1}{1-z}, \qquad \Gamma_{11}^2 = \frac{1-z}{1+R^2}.$$

If the independent variable is the arc length parameter $s$ on the geodesic, we obtain the equations

$$\sigma'' - \frac{2}{1-z}\sigma' z' = 0, \tag{1.107}$$

$$z'' + \frac{1-z}{1+R^2}\sigma'^2 = 0. \tag{1.108}$$

The first equation can be written as $\sigma''/\sigma' = 2z'/(1-z)$ and by integrating one obtains

$$\sigma' = \frac{c}{(1-z)^2}, \qquad c = \text{constant}; \tag{1.109}$$

hence from equation (1.109) we can derive an equation for $z$ only:

$$z'' + \frac{c^2}{1+R^2}\frac{1}{(1-z)^3} = 0. \tag{1.110}$$

Multiplying equation (1.110) by $z'$ and integrating, it is easy to obtain a first integral. This can also be obtained through a different procedure, highlighting the geometrical meaning. Compute the unit vector $\tau$ tangent to the geodesic

$$\tau = \begin{pmatrix} (1-z)f_1'(\sigma)\sigma'(s) - z'(s)f_1(\sigma) \\ (1-z)f_2'(\sigma)\sigma'(s) - z'(s)f_2(\sigma) \\ z'(s) \end{pmatrix}$$

and write explicitly that its absolute value is 1:

$$(1-z)^2\sigma'^2 + (1+R^2)z'^2 = 1,$$

and owing to equation (1.109) this yields the first integral of (1.110):

$$\frac{c^2}{(1-z)^2} + (1+R^2)z'^2 = 1. \tag{1.111}$$

The two terms on the left-hand side of equation (1.111) are the squares, respectively, of

$$\sin\varphi = \tau \cdot \frac{\mathbf{x}_\sigma}{1-z}, \qquad \cos\varphi = \tau \cdot \frac{\mathbf{x}_z}{(1+R^2)^{1/2}},$$

where $\varphi$ is the angle between the geodesic and the cone generatrix. By requiring that the curve passes through the point of coordinates $(\sigma_0, z_0)$

## 1.15  Geometric and kinematic foundations of Lagrangian mechanics

forming an angle $\varphi_0$, one can determine the constant $c = (1-z_0)\sin\varphi_0$. The sign of $c$ determines the orientation. Equation (1.111) is easily integrated and yields the solution (with $z<1$)

$$1 - z = \left[c^2 + \frac{(s-c_1)^2}{1+R^2}\right]^{1/2}. \tag{1.112}$$

The constant $c_1$ is determined by the condition $z(s_0) = z_0$. For $c = 0$ (the condition of tangency to the generatrix) equation (1.109) implies $\sigma = $ constant, and hence the geodesic corresponds to the generatrix $\sigma = \sigma_0$. As we know, equation (1.112) implies that the parallels are not geodesics. Clairaut's theorem has a clear interpretation. From the relation $\sin\varphi = c/(1-z)$ it follows that once the constant $c$ is fixed, one must have $1 - z > |c|$. Hence the only geodesics passing through the vertex are the generating straight lines. The maximum value of $z$ on a non-linear geodesic is

$$z_{\max} = -|c| + 1,$$

where the geodesic is tangent to a parallel ($z' = 0$). Notice that from equation (1.110) it follows that $z'' < 0$ for $c \neq 0$. This implies that after attaining the maximum height, $z'$ decreases. In particular it implies that no geodesic can be closed. For $z \to -\infty$ the geodesic tends to a generatrix. To find which one, we need to integrate equation (1.109):

$$\sigma(s) - \sigma(s_0) = \int_{s_0}^{s} \frac{c}{c^2 + \frac{(s'-c_1)^2}{1+R^2}}\, ds'. \tag{1.113}$$

Choosing $s_0 = 0$, $\sigma(s_0) = 0$, $z(s_0) = z_{\max} = 1 - c$ ($c > 0$), from equation (1.112) we find that $c_1 = 0$ and equation (1.113) implies

$$\sigma(s) = \sqrt{1+R^2}\arctan\frac{s}{c\sqrt{1+R^2}}. \tag{1.114}$$

Hence for $s \to \pm\infty$, $\sigma \to \pm\frac{\pi}{2}\sqrt{1+R^2}$. The equation

$$1 - z = \left(c^2 + \frac{s^2}{1+R^2}\right)^{1/2} \tag{1.115}$$

together with (1.114) describes the maximum height geodesic $z_{\max} = 1-c$, positively oriented ($c > 0$) with $s = 0$ at the highest point. The arc between $z_{\max}$ and $z$ has length $s = \sqrt{1+R^2(z_{\max}-z)[2-(z_{\max}+z)]}$. We can now proceed to compute $d\tau/ds$, recalling that $f_1 = R\cos(\sigma/R)$, $f_2 = R\sin(\sigma/R)$:

$$\frac{d\tau}{ds} = \begin{pmatrix} 2z'\sin\frac{\sigma}{R}\sigma'(s) - \frac{1-z}{R}\cos\frac{\sigma}{R}\sigma'^2 - (1-z)\sin\frac{\sigma}{R}\sigma'' - z''R\cos\frac{\sigma}{R} \\ -2z'\cos\frac{\sigma}{R}\sigma'(s) - \frac{1-z}{R}\sin\frac{\sigma}{R}\sigma'^2 + (1-z)\cos\frac{\sigma}{R}\sigma'' - z''R\sin\frac{\sigma}{R} \\ z'' \end{pmatrix},$$

whose absolute value gives the curvature. Exploiting equations (1.107)–(1.110) one finds

$$\frac{d\boldsymbol{\tau}}{ds} = -\frac{c^2}{1+R^2}\frac{1}{(1-z)^2}\begin{pmatrix}\frac{1}{R}\cos\frac{\sigma}{R}\\ \frac{1}{R}\sin\frac{\sigma}{R}\\ 1\end{pmatrix}.$$

Hence

$$k(s) = \frac{c^2}{R\sqrt{1+R^2}}\frac{1}{(1-z)^3} \qquad (1.116)$$

at the point of maximum height $k(0) = 1/cR\sqrt{1+R^2}$. Note that the unit normal vector

$$\mathbf{n}(s) = -\frac{1}{\sqrt{1+R^2}}\begin{pmatrix}\cos\frac{\sigma}{R}\\ \sin\frac{\sigma}{R}\\ R\end{pmatrix} \qquad (1.117)$$

has constant component along the cone axis, as expected. Finally, we have

$$\mathbf{b} = \boldsymbol{\tau}\times\mathbf{n} = -\frac{1}{\sqrt{1+R^2}}\begin{pmatrix}\frac{cR}{1-z}\cos\frac{\sigma}{R} - \sqrt{1+R^2}\sqrt{1-\frac{c^2}{(1-z)^2}}\sin\frac{\sigma}{R}\\ \frac{cR}{1-z}\sin\frac{\sigma}{R} + \sqrt{1+R^2}\sqrt{1-\frac{c^2}{(1-z)^2}}\cos\frac{\sigma}{R}\\ \frac{-c}{1-z}\end{pmatrix}. \qquad (1.118)$$

Hence, excluding the case of the generating straight lines ($c = 0$), **b** is not constant and the geodesics are not plane curves (hence they are not conic sections).

*Problem 3*
In the right circular cone of Problem 2 consider the two elicoidal curves obtained by setting, respectively,

(a) $z = \sigma/2\pi R$, $\sigma \in (0, 2\pi R)$,
(b) $z = \sin(\sigma/4R)$, $\sigma \in (0, 2\pi R)$.

Prove that these curves are not geodesics and compute their length.

*Solution*
The curves are not geodesics since they pass through the vertex of the cone ($\sigma = 2\pi R$). Recall that in the representation of the parameters $z, \sigma$ one has $E = (1-z)^2$, $F = 0$, $G = 1+R^2$; hence the formula for the length of a curve

expressed as $z = z(\sigma)$ for $\sigma \in (0, 2\pi R)$ is

$$l = \int_0^{2\pi R} [(1 - z(\sigma))^2 + (1+R)^2 z'^2(\sigma)]^{1/2} \, d\sigma.$$

Thus in the two cases we have

(a) $\quad l = \displaystyle\int_0^{2\pi R} \left[\left(1 - \frac{\sigma}{2\pi R}\right)^2 + \frac{1+R^2}{4\pi^2 R^2}\right]^{1/2} d\sigma = 2\pi R \int_0^1 \left(\xi^2 + \frac{1+R^2}{4\pi^2 R^2}\right)^{1/2} d\xi,$

(b) $\quad l = \displaystyle\int_0^{2\pi R} \left[\left(1 - \sin\frac{\sigma}{4R}\right)^2 + \frac{1+R^2}{16R^2} \cos^2 \frac{\sigma}{4R}\right]^{1/2} d\sigma.$

Setting $\sin(\sigma/4R) = x$, the latter integral is transformed to

$$\int_0^1 \left(\frac{1-x}{1+x} + \frac{1+R^2}{16R^2}\right)^{1/2} dx,$$

which can be easily computed.

### Problem 4
On a surface of revolution $(u \cos v, u \sin v, f(u))$, $u =$ radius, $v =$ angle, find the curves that intersect the meridians at a constant angle. Under what conditions are these curves geodesics?

### Solution
Let us start by answering the last question. We know that for the natural parametrisation $u = u(s)$, $v = v(s)$ of a geodesic, Clairaut's theorem (1.54) holds: $u(s) \sin \alpha(s) = c$, where $\alpha(s)$ is the angle between the geodesic and the meridian. Hence $\alpha =$ constant $(\neq 0)$ is equivalent to $u =$ constant, which corresponds to the case of a cylinder with a circular section, or else $\alpha = \pi/2$ which is the exceptional case of a geodesic parallel. On a cylinder with circular section, the helices are the only geodesics with the property that we are considering here (with $\alpha \neq 0$). We need to include in this class the meridians, corresponding to the case $\alpha = 0$, $c = 0$. Consider now the problem of finding the curves that form a given angle $\alpha$ with the meridians. We seek such curves in the parametric form $u = g(v)$. The vector tangent to the curve sought is given by

$$\tau = [g^2 + g'^2(1 + f'^2)]^{-1/2} \begin{pmatrix} -g \sin v + g' \cos v \\ g \cos v + g' \sin v \\ f'(g) g' \end{pmatrix}.$$

The vector tangent to the parallel is

$$\tau_p = \begin{pmatrix} -\sin v \\ \cos v \\ 0 \end{pmatrix},$$

and hence the condition we need to impose is

$$\sin \alpha = g[g^2 + g'^2(1 + f'^2)]^{-1/2}, \tag{1.119}$$

i.e.

$$\int_{g_0}^{g} \gamma^{-1}\sqrt{1 + f'^2(\gamma)}\, d\gamma = (v - v_0)\cos\alpha. \tag{1.120}$$

Obviously, for any surface, equation (1.120) includes the parallels ($\cos\alpha = 0$, $u = g(v) = g_0$, constant).

# 2 DYNAMICS: GENERAL LAWS AND THE DYNAMICS OF A POINT PARTICLE

## 2.1 Revision and comments on the axioms of classical mechanics

The discussion of the phenomenological aspects of classical mechanics is beyond the scope of this book. We shall restrict ourselves to a summary of the fundamental concepts following Mach (1883), without any historical introduction, and hence overlooking the work of Galileo and Newton, who laid the foundations of mechanics; for this we refer the reader to Truesdell (1968).

Up to this point we have modelled physical bodies by a finite number of points, without any reference to their dimensions or internal structure. By physical bodies we mean bodies that can interact with each other; hence this interaction must be precisely quantified. To be able to express this quantitatively, we need to select a class of observers with respect to whom to formulate the laws governing such an interaction.

To define a suitable class of observers, we start with the simpler case of an *isolated point particle*, assuming that any other system that might interact with the given particle is at infinity.

DEFINITION 2.1 *An* inertial *observer is any observer for whom, at every time and for any kinematic state, an isolated point particle has zero acceleration.* ∎

The existence of such inertial observers is an axiom.

**Axiom I** *There exists an inertial observer.*

To proceed further, we must make use of the basic notions of relative kinematics, which we assume known; we shall however review them in the context of the kinematics of rigid bodies, see Chapter 6. Recall that systems whose relative motion is a uniform translation (preserving the direction of the axes) will measure the same acceleration; moreover, a translation of the time-scale will similarly leave the measurement of accelerations unchanged.

Axiom I is equivalent to the assumption that there exists a class of inertial observers, which can be identified up to translation along the time-scale and/or because they move relative to each other with a rectilinear, uniform translation.

It is easy to point out the intrinsic weakness of Definition 2.1: the concept of an isolated point particle is in direct contrast with the possibility of performing measurements of its acceleration, and these alone can establish if the observer is indeed inertial. However, we shall accept the existence of inertial observers, and let us proceed by assuming that one of them measures the accelerations of two point particles, corresponding to various kinematic states; in addition, we assume that the two-point system is isolated.

It is possible to use these measurements to give a quantitative definition of the concept of interaction.[1]

**Axiom II**  *Consider an isolated system comprising two point particles $\{P_1, P_2\}$ and let $a(P_1), a(P_2)$ be the magnitudes of their accelerations, measured by an inertial observer. The quotient $m_{1,2} = a(P_2)/a(P_1)$ is independent of the kinematic state of the system, and of the instant at which the measurement is taken. In addition, the quotients $m_{1,0}$ and $m_{2,0}$ obtained by considering the interaction of $P_1$ and $P_2$, respectively, with a third point $P_0$ satisfy the relation*

$$m_{1,2} = \frac{m_{1,0}}{m_{2,0}}. \tag{2.1}$$

The point $P_0$ can then be taken as a *reference point particle*; in order to obtain the interaction constant $m_{1,2}$ between two point particles, it is sufficient to know the interaction constants of these points with the reference point. This allows us to define the concept of *inertial mass*.

DEFINITION 2.2  *Associate with the reference point particle the unit mass $m_0$. The interaction constant $m$ of a point particle $P$ with respect to $P_0$ is assumed to be the measure of the inertial mass of $P$ with respect to the unit of measure $m_0$.*  ∎

From now on we use the notation $(P, m)$ to indicate a point particle and its mass.

We still need information on the direction of the interaction accelerations. This is provided by a third axiom.

**Axiom III**  *For an inertial observer, the accelerations $\mathbf{a}(P_1), \mathbf{a}(P_2)$ considered in Axiom II are directed as the vector $P_1 - P_2$ and have opposite orientation.*

In order to be able to study systems of higher complexity, we must make the following further assumption on the mutual interactions within the system.

**Axiom IV**  *The acceleration of a point particle $(P, m)$ due to the interaction with a system of other point particles is the sum of the accelerations due to the interaction of $(P, m)$ with each one of the other particles, taken separately.*

The reference to acceleration is a way to express the fundamental axioms (and the definition of mass) so as to be invariant with respect to the class of inertial observers.

If we now define the *force* applied to the point particle $(P, m)$ by the equation

$$m\mathbf{a} = \mathbf{F}, \tag{2.2}$$

this quantity will have the same invariance property.

---

[1] In the context of classical mechanics, this interaction is instantaneous, and hence the propagation time is taken to be zero.

Equation (2.2) and Axiom III are known jointly as the 'action and reaction principle'.

When $\mathbf{F}$ is specified as a function of $P$, of the velocity $\mathbf{v}$ and of time, equation (2.2) is the well-known fundamental equation of the dynamics of a point particle. This equation can be integrated once initial conditions are prescribed:

$$P(0) = P_0, \qquad \mathbf{v}(0) = \mathbf{v}_0. \tag{2.3}$$

This approach to the dynamics of a point particle must be justified; indeed, this is evident when one considers the so-called Galilean relativity principle, one of the most profound intuitions of classical mechanics.

## 2.2 The Galilean relativity principle and interaction forces

In a celebrated passage of his *Dialogue on the two chief world systems* (1632), Galileo states very clearly the principle according to which two observers who are moving relative to each other in uniform translation will give identical descriptions of mechanical phenomena.

More precisely, we define a *Galilean space* to be a space of the form $\mathbf{R} \times \mathbf{R}^3$. The natural coordinates $(t, x_1, x_2, x_3)$ parametrising this space are called the Galilean coordinates. The space component is endowed with a Euclidean structure: two simultaneous events $(t, x_1, x_2, x_3)$ and $(t, y_1, y_2, y_3)$ are separated by a distance

$$\sqrt{(x_1 - y_1)^2 + (x_2 - y_2)^2 + (x_3 - y_3)^2}.$$

The *Galilean group* is the group of all transformations of the Galilean space which preserve its structure. Each transformation in this group can be uniquely written as the composition of:

(1) a rotation in the subspace $\mathbf{R}^3$ of the space coordinates:

$$\mathbf{x} = A\mathbf{y}, \qquad A \in O(3, \mathbf{R})$$

(where $O(3, \mathbf{R})$ indicates the group of $3 \times 3$ orthogonal matrices);
(2) a translation of the origin:

$$(t, \mathbf{x}) = (t' + s, \mathbf{y} + \mathbf{b}), \quad \text{where } (s, \mathbf{b}) \in \mathbf{R} \times \mathbf{R}^3;$$

(3) a linear uniform motion with velocity $\mathbf{v}$:

$$(t, \mathbf{x}) = (t, \mathbf{y} + \mathbf{v}t).$$

With this notation, the Galilean relativity principle can be expressed as follows: *The trajectories of an isolated mechanical system are mapped by any Galilean transformation into trajectories of the same system.*

Let us illustrate this basic principle by means of a simple example. Consider an isolated system of $n$ free point particles $\{(P_1, m_1), \ldots, (P_n, m_n)\}$ and specify the following:

(a) a time $t_0$;
(b) $3n$ Cartesian coordinates to be assigned sequentially to the points in the system;
(c) $n$ velocity vectors for each one of the points $P_1, \ldots, P_n$.

Consider now two inertial observers, and suppose they are given the data (a)–(c); let us imagine that they use these data to construct a kinematic state, relative to their respective coordinate axes, and at a time $t_0$ of the respective time-scales. The Galilean relativity principle states that by integrating the system of equations

$$m_i \mathbf{a}_i = \mathbf{F}_i(P_1, \ldots, P_n, \mathbf{v}_1, \ldots, \mathbf{v}_n, t) \tag{2.4}$$

with initial time $t = t_0$, and prescribing the above conditions, the two observers will obtain two identical solutions $P_i = P_i(t)$, $i = 1, \ldots, n$.

This means that simply observing mechanical phenomena due only to the interaction between point particles, the two observers will not be able to detect if:

($\alpha$) the respective temporal scales are not synchronised;
($\beta$) their coordinate axes have different orientation;
($\gamma$) they move relative to each other.[2]

These facts clearly have three consequences for the structure of interaction forces:

(a) they cannot depend explicitly on time (since ($\alpha$) implies that such forces are invariant under a translation of the temporal axis);
(b) they can only depend on the differences $P_i - P_j, \mathbf{v}_i - \mathbf{v}_j$;
(c) if all the vectors $P_i - P_j, \mathbf{v}_i - \mathbf{v}_j$ are rotated by the same angle, then all the vectors $\mathbf{F}_i$ will be subject to the same rotation.

It is therefore evident that there cannot exist privileged instants or points or directions, where privileged means that they can be singled out purely by the experience of a mechanical phenomenon.

The question then is how to reconcile this necessity with the well-known equation

$$m\mathbf{a} = \mathbf{F}(P, \mathbf{v}, t) \tag{2.5}$$

and in particular, with the existence of force fields $\mathbf{F} = \mathbf{F}(P)$. Consider, as an example, a central field, in which the presence of a centre destroys the spatial homogeneity, and allows two inertial observers to discover that they are indeed moving with respect to one another.

---

[2] The inclusion in the relativity principle of electromagnetic phenomena (in particular the invariance of the speed of light) will yield the special relativity theory of Einstein.

To answer this question, it is convenient to consider more carefully the dynamics of a point particle $(P, m)$, subject to the action of other point particles $(P_i, m_i)$, $i = 1, \ldots, n$. The correct way to consider this problem is to integrate the system of equations $m\mathbf{a} = \mathbf{F}, m_1 \mathbf{a}_1 = \mathbf{F}_1, \ldots, m_n \mathbf{a}_n = \mathbf{F}_n$, taking into account that the interaction forces $\mathbf{F}, \mathbf{F}_1, \ldots, \mathbf{F}_n$ depend on the kinematic state of the whole system.

However, when we write equation (2.5) we assume a priori the knowledge of the motion of the point particles $(P_i, m_i)$ generating the force $\mathbf{F}$. By doing this we necessarily introduce an approximation: we neglect the influence of the point particle $(P, m)$ on the other points of the system.

For example, consider a system consisting of a pair of point particles $(P, m), (O, M)$, attracting each other with an elastic force with constant $k$ (Fig. 2.1). In an inertial system, the equation

$$m\ddot{P} = -k(P - O) \tag{2.6}$$

is to be considered jointly with

$$M\ddot{O} = -k(O - P). \tag{2.7}$$

As a consequence, the centre of mass $P_0$ (defined by the requirement that $m(P - P_0) + M(O - P_0) = 0$) must have zero acceleration. We can hence introduce

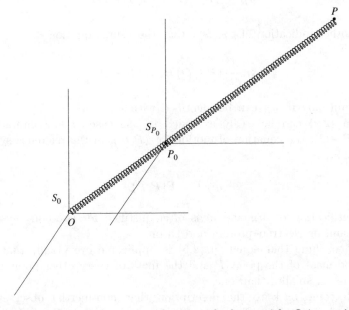

**Fig. 2.1** The reference frame translating with the particle $O$ is not inertial.

an inertial system $S_{P_0}$, where $P_0$ has zero velocity. Since

$$P - O = \left(1 + \frac{m}{M}\right)(P - P_0),$$

we can write the equation of motion of the point particle $(P, m)$ with respect to $S_{P_0}$ as

$$\frac{mM}{m+M}\frac{d^2}{dt^2}(P - P_0) = -k(P - P_0). \tag{2.8}$$

We conclude that the centre of elastic attraction of $P$ in $S_{P_0}$ appears to be $P_0$ and the mass $m$ must be replaced by the 'reduced mass'

$$m_R = \frac{mM}{m+M} < m.$$

However, when $M/m \gg 1$ it is justified to identify $P_0$ with $O$ and $m_R$ with $m$. Notice that to be entirely rigorous, a system $S_O$ where $O$ has null velocity and whose coordinate axes are in uniform linear motion with respect to the above system $S_{P_0}$ is not inertial, because its points have acceleration $\ddot{O} \neq 0$ with respect to any inertial system.

To write the equation of motion of $P$ with respect to $S_O$, we compute $(d^2/dt^2)(P - O)$ from equations (2.6), (2.7), and we find

$$m_R \frac{d^2}{dt^2}(P - O) = -k(P - O). \tag{2.9}$$

This is another indication of the fact that the usual equation

$$m \frac{d^2}{dt^2}(P - O) = -k(P - O)$$

is meaningful only if $m_R$ can be identified with $m$.

Equation (2.9) can be easily extended to the case of any interaction force $\mathbf{F}(P - O, \dot{P} - \dot{O})$; the equation of motion for $(P, m)$, in the reference system used to write (2.9), is

$$m_R \ddot{P} = \mathbf{F}(P, \dot{P}). \tag{2.10}$$

The identification of $m_R$ and $m$ is often justified for two-body systems such as planet–sun, or electron–proton, and so on.

We can conclude that equation (2.5) is applicable every time that the ratio between the mass of the point $P$ and the mass of every other point interacting with $P$ is much smaller than one.

We shall come back to the description that non-inertial observers give of mechanical phenomena in Chapter 6 (Section 6.6).

## 2.3 Work and conservative fields

Let $(P, m)$ be a point particle in motion under the action of a force $\mathbf{F}(P, \mathbf{v}, t)$. During its motion, at every time $t$ we can define the *power*

$$W(t) = \mathbf{F}(P(t), \mathbf{v}(t), t) \cdot \mathbf{v}(t) \tag{2.11}$$

and the *work*

$$\mathcal{L}(t) = \int_{t_0}^{t} W(\tau) \, d\tau \tag{2.12}$$

done by the force $\mathbf{F}$ in the time interval $(t_0, t)$.

Note that the derivative of the kinetic energy $T = \frac{1}{2} m v^2$ along the path of the motion is given by $dT/dt = m\mathbf{v} \cdot \mathbf{a} = W$; it is therefore easy to compute the *energy integral*

$$T(t) - T(t_0) = \mathcal{L}(t). \tag{2.13}$$

In practice, to compute the work $\mathcal{L}(t)$ one must know the motion (hence the complete integral of equation (2.5)). However, when $\mathbf{F}$ depends only on $P$, i.e. if the point is moving in a *positional force field*, $\mathbf{F}(P)$, the work can be expressed as a line integral in the form

$$\mathcal{L}_\gamma = \int_\gamma \mathbf{F} \cdot d P = \int_\gamma \sum_{i=1}^{3} F_i \, dx_i, \tag{2.14}$$

where $\gamma$ is the arc of the trajectory travelled in the time interval $(t_0, t)$.

On the other hand, the integral (2.14) can be computed not only along the trajectory of $P$, but along any rectifiable path. Hence we can distinguish the *dynamic* notion of work, expressed by equation (2.12), from the purely *geometrical* one, expressed, for positional force fields, by equation (2.14).

When the structure of the force field is such that the value of the integral (2.14) is independent of the curve joining the endpoints, one can establish a deep connection between geometry and dynamics: the energy integral fixes a scalar field of the kinetic energy.

It is well known that the independence of work on the integration path is a characteristic property of *conservative fields*; such fields are of the form

$$\mathbf{F} = \nabla U(\mathbf{x}), \tag{2.15}$$

where $U(\mathbf{x})$ is the field *potential*. Since

$$\int_{\widehat{AB}} \mathbf{F} \cdot dP = \int_{\widehat{AB}} dU = U(B) - U(A),$$

independent of the arc $\widehat{AB}$, it follows that
$$T(\mathbf{x}) - T(\mathbf{x}_0) = U(\mathbf{x}) - U(\mathbf{x}_0). \tag{2.16}$$
This is the form of the energy integral which defines the function $T(\mathbf{x})$, and that can be interpreted as the *conservation of the total energy*
$$E = T - U = T + V, \tag{2.17}$$
where $V = -U$ is identified with the *potential energy*. This is the reason these fields are called conservative.

Recall that a conservative field is also irrotational:
$$\operatorname{rot} \mathbf{F} = 0. \tag{2.18}$$
Conversely, in every simply connected region where it applies, equation (2.18) guarantees the existence of a potential.

Recall also that the fact that work is independent of the integration path is equivalent to the statement that work is zero along any closed path.

*Example* 2.1
The Biot–Savart field in $\mathbf{R}^3 \setminus \{x_1 = x_2 = 0\}$, given by
$$F(x_1, x_2, x_3) = c \frac{\mathbf{e}_3 \times \mathbf{x}}{|\mathbf{e}_3 \times \mathbf{x}|^2} = c \frac{(-x_2, x_1, 0)}{x_1^2 + x_2^2}, \tag{2.19}$$
where $c \in \mathbf{R}$, is irrotational but it is not conservative. ∎

*Example* 2.2
The force field in $\mathbf{R}^3$ given by
$$F(x_1, x_2, x_3) = (ax_1 x_2, ax_1 x_2, 0), \quad a \neq 0$$
is not conservative, despite the fact that the work along any path symmetric with respect to the $x_3$-axis is zero. ∎

*Example* 2.3
The force fields in $\mathbf{R}^3$ of the form
$$\mathbf{F} = f(r, \theta, \varphi) \widehat{e}_r,$$
where $(r, \theta, \varphi)$ are spherical coordinates, are conservative if and only if $\partial f / \partial \theta = \partial f / \partial \varphi = 0$, and hence if $f$ depends only on $r$. Such fields are called central force fields, and will be studied in detail in Chapter 5. ∎

## 2.4 The dynamics of a point constrained by smooth holonomic constraints

It is useful to consider the problem of the dynamics of a constrained point. Indeed, this will indicate the way in which to consider the more general problem of the dynamics of holonomic systems.

Let $(P, m)$ be a point particle subject to a holonomic constraint; suppose a force $\mathbf{F} = \mathbf{F}(P, \mathbf{v}, t)$, due to the interaction with objects other than the constraint, is applied to the point. First of all, by integrating the equation $m\mathbf{a} = \mathbf{F}$ with initial conditions compatible with the constraint, one obtains in general a motion which does not satisfy the constraint equations.

Hence it is necessary to modify the equation of motion, adding to the right-hand side a force term $\phi(t)$, expressing the dynamic action of the constraint, and called the *constraint reaction*:

$$m\ddot{\mathbf{x}}(t) = F(\mathbf{x}(t), \dot{\mathbf{x}}(t), t) + \phi(t). \tag{2.20}$$

The force $\phi(t)$ is unknown, and it is evidently impossible to determine the two vectors $\mathbf{x}(t)$ and $\phi(t)$ only from equation (2.20) and the constraint equations (which are one or two scalar equations).

It is therefore necessary to find additional information concerning the *mechanics of the constraints*. The simplest hypothesis is to assume that the constraint is *smooth*, in the following sense.

DEFINITION 2.3   *A holonomic constraint acting on a point particle $(P, m)$ is called* smooth *(or* idealised *or* frictionless*) if the constraint reaction is orthogonal to the constraint configuration, at every instant and for every kinematic state of the point on the constraint.* ■

Hence a simple constraint (Section 1.10)

$$f(\mathbf{x}, t) = 0 \tag{2.21}$$

is smooth if and only if

$$\phi(t) = \lambda(t) \nabla f(\mathbf{x}(t), t), \tag{2.22}$$

whereas for a double constraint

$$f_1(\mathbf{x}, t) = 0, \qquad f_2(\mathbf{x}, t) = 0 \tag{2.23}$$

the analogous condition is

$$\phi(t) = \lambda_1(t) \nabla f_1 + \lambda_2(t) \nabla f_2. \tag{2.24}$$

Equations (2.22), (2.24) must hold for every $t$, with $\mathbf{x}(t)$ the solution of (2.20). The coefficients $\lambda(t)$, $\lambda_1(t)$, $\lambda_2(t)$ are unknown.

Equation (2.20) is then supplemented by equations (2.21), (2.22) or with (2.23), (2.24); notice that formally we now have the same number of equations and unknowns.

From the point of view of energy balance, it is important to note that for a smooth constraint the only contribution to the power of the constraint reaction comes from the component of the velocity orthogonal to the constraint, which must be attributed exclusively to the motion of the constraint itself. We can therefore state the following.

PROPOSITION 2.1  *When a point particle moves along a smooth, fixed constraint, the work done by the constraint reaction is zero.* ∎

COROLLARY 2.1  *For a point particle in a conservative force field, constrained by a smooth fixed constraint, the conservation of energy (2.7) holds.* ∎

*Example 2.4: a single point particle constrained along a smooth, fixed curve*
It is convenient to decompose equation (2.20) with respect to the principal reference frame (Fig. 2.2):

$$m\ddot{s} = \mathbf{F}(s, \dot{s}, t) \cdot \mathbf{t}(s), \quad \dot{s}(0) = v_0, \quad s(0) = s_0, \qquad (2.25)$$

$$m\frac{\dot{s}^2}{R(s)} = \mathbf{F}(s, \dot{s}, t) \cdot \mathbf{n}(s) + \boldsymbol{\phi}(t) \cdot \mathbf{n}(s), \qquad (2.26)$$

$$0 = \mathbf{F}(s, \dot{s}, t) \cdot \mathbf{b}(s) + \boldsymbol{\phi}(t) \cdot \mathbf{b}(s), \qquad (2.27)$$

where $R(s)$ is the radius of curvature.

The unknowns are the function $s = s(t)$ and the two components $\boldsymbol{\phi} \cdot \mathbf{n}, \boldsymbol{\phi} \cdot \mathbf{b}$. Equation (2.25) is the differential equation governing the motion along the

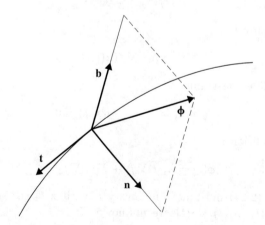

**Fig. 2.2** Decomposition of the constraint reaction.

constraint; after integration of equation (2.25), equations (2.26) and (2.27) determine $\phi$.

If **F** is conservative we can write

$$T(s) - \mathcal{U}(s) = E, \tag{2.28}$$

where $\mathcal{U}(s)$ denotes the restriction of the potential to the constraint. Using equations (2.26), (2.27), this yields $\phi$ as a function of $s$. ∎

We discuss the equation of motion (2.25) in the next chapter.

*Example 2.5: a single point particle constrained on a smooth, fixed surface $f(\mathbf{x}) = 0$*
We fix a parametrisation $\mathbf{x} = \mathbf{x}(u, v)$ of the surface; hence the equations of motion are obtained by projecting equation (2.20) onto the tangent vectors $\mathbf{x}_u, \mathbf{x}_v$:

$$E(u,v)\ddot{u} + F(u,v)\ddot{v} = \frac{1}{m}\mathbf{F}(u,v,\dot{u},\dot{v},t) \cdot \mathbf{x}_u - (\mathbf{x}_{uu}\dot{u}^2 + 2\mathbf{x}_{uv}\dot{u}\dot{v} + \mathbf{x}_{vv}\dot{v}^2) \cdot \mathbf{x}_u, \tag{2.29}$$

$$F(u,v)\ddot{u} + G(u,v)\ddot{v} = \frac{1}{m}\mathbf{F}(u,v,\dot{u},\dot{v},t) \cdot \mathbf{x}_v - (\mathbf{x}_{uu}\dot{u}^2 + 2\mathbf{x}_{uv}\dot{u}\dot{v} + \mathbf{x}_{vv}\dot{v}^2) \cdot \mathbf{x}_v. \tag{2.30}$$

Equations (2.29) and (2.30) yield a system which must be integrated, after assigning initial conditions for $u, v, \dot{u}, \dot{v}$. Once the solutions $u(t), v(t)$ have been determined, one can compute the constraint reaction $\phi = \lambda(t)\nabla f(\mathbf{x})$ by writing

$$-\dot{\mathbf{x}} \cdot H\dot{\mathbf{x}} = \frac{1}{m}\mathbf{F}(u,v,\dot{u},\dot{v},t) \cdot \nabla f + \frac{1}{m}\lambda(t)|\nabla f|^2, \tag{2.31}$$

where $H(\mathbf{x})$ is the Hessian matrix of $f$; this equation is obtained by multiplying both sides of equation (2.20) by $\nabla f$ and using (1.98). ∎

We end this section by proving an interesting property of the motion of a point particle on an equipotential surface.

PROPOSITION 2.2 *Let $(P, m)$ be a point particle subject to a conservative force and constrained on an equipotential surface. The possible trajectories of the point are the geodesics of the surface.*

*Proof*
Consider the generic motion of the point on the constraint; it is enough to prove that the principal unit vector orthogonal to the trajectory is parallel to $\nabla U$ (if the trajectory is a straight line, the problem is trivial). Suppose this is not the case; we then have $\mathbf{b} \cdot \nabla U \neq 0$, because any vector normal to the surface lies in the plane $(\mathbf{n}, \mathbf{b})$.

However, since $\phi$ is parallel to $\nabla U$, equation (2.27) implies that $\phi + \nabla U = 0$, which contradicts (2.26) (recall that we are considering $\dot{s} \neq 0, 1/R \neq 0$). ∎

**PROPOSITION 2.3** *The same conclusion holds true for the so-called* spontaneous motion *on the constraint* $(\mathbf{F} = 0)$. ∎

The proof is even easier in this case, as equation (2.27) implies that $\phi \cdot \mathbf{b} = 0$, and hence that $\phi$ is parallel to $\mathbf{n}$. This is equivalent to the orthogonality of $\mathbf{n}$ to the surface constraint.

On the other hand, it is easy to identify equations (2.29), (2.30) with the geodesic equations (1.25), when $\mathbf{F} \cdot \mathbf{x}_u = \mathbf{F} \cdot \mathbf{x}_v = 0$ (using that $\ddot{s} = 0$).

## 2.5 Constraints with friction

When the hypothesis that the constraint is frictionless is not justifiable, it is necessary to introduce a criterion to define the tangential component of the constraint reaction. For this we must distinguish between the static and dynamic cases. We only consider here fixed constraints.

In case of equilibrium we assume, on the basis of experimental observations, that the following inequality must hold:

$$|\phi_\theta| \leq f_s |\phi_N|, \tag{2.32}$$

where $\phi_\theta$ and $\phi_N$ represent the tangential and normal components of the reaction, respectively, and the number $f_s > 0$ is called the *static friction coefficient*. This implies that the reaction $\phi$ must belong to the so-called *static friction cone* (Fig. 2.3).

Note that in the case of a simple constraint, the static friction cone contains the axis (which corresponds in this instance with the normal to the constraint), while for a double constraint the axis of the cone is tangent to the constraint and the static friction cone coincides with the region containing the normal plane.

The static equation, given by

$$\mathbf{F} + \phi = 0, \tag{2.33}$$

yields the following.

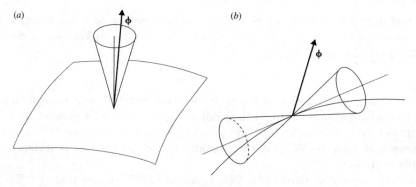

**Fig. 2.3** Static friction cone: (*a*) simple constraint; (*b*) double constraint.

PROPOSITION 2.4   *The equilibrium positions are only those for which* **F** *belongs to the static friction cone.*   ∎

We now consider the dynamics. The absolute value of the tangential reaction is defined by the identity

$$|\phi_\theta| = f_d |\phi_N|; \qquad (2.34)$$

its direction is that of the velocity **v**, with opposite orientation.

The coefficient $f_d$ in equation (2.34) is called the *coefficient of dynamic friction*; in physical situations, $0 < f_d < f_s$. The condition imposed on $\phi$ implies that $\phi \cdot \mathbf{v} < 0$; hence mechanical energy is dissipated by friction. Equation (2.34) defines a conical surface, the *dynamic friction cone*, that must contain $\phi$.

We now reconsider the solution of the equations of motion.

*Example 2.6: single particle constrained with friction on a fixed curve*
The equations of motion are

$$m\ddot{s} = \mathbf{F}(s, \dot{s}, t) \cdot \mathbf{t}(s) + \phi(t) \cdot \mathbf{t}(s), \qquad (2.35)$$

replacing equation (2.25), while equations (2.26), (2.27) are unchanged. We start from the two latter ones to determine

$$|\phi_N| = [(\phi \cdot \mathbf{n})^2 + (\phi \cdot \mathbf{b})^2]^{1/2} \qquad (2.36)$$

as a function of $s, \dot{s}, t$. We can use equation (2.34) to obtain $\phi \cdot \mathbf{t} = -|\phi_\theta|\dot{s}/|\dot{s}|$, a known function of $s, \dot{s}, t$. In principle, it is thus possible to integrate equation (2.35), starting from prescribed initial conditions. Equations (2.26), (2.27) and (2.36) yield the determination of the unknown $\phi(t)$.   ∎

*Example 2.7: motion of a single particle constrained with friction on a fixed surface*
Consider the projection of equation $m\mathbf{a} = \mathbf{F} + \phi$ onto the vector normal to the surface; by using equation (1.98) it is possible to determine the expression for $\phi_N$ as a function of the particle's position and velocity. Finally, using the conditions defining the vector $\phi_\theta$ we arrive at a well-determined problem for the motion of the particle on the constraint.   ∎

## 2.6   Point particle subject to unilateral constraints

We now consider the case of a point particle $(P, m)$ subject to the constraint

$$f(\mathbf{x}) \leq 0, \qquad (2.37)$$

where $f$ is a function in the usual class.

As long as the particle is moving inside the region $f(\mathbf{x}) < 0$ the constraint exerts no force. If for some time interval the motion evolves on the surface $f(\mathbf{x}) = 0$ then the previous analysis applies.

We still need to consider the case that the particle only comes into contact with the constraint instantaneously; in this case we need to make a physical assumption. The contact may happen according to an idealised law of reflection, i.e. with a simple inversion of the component of the velocity orthogonal to the constraint (elastic shock); else it may happen with partial (or even total) absorption of the kinetic energy.

We consider only the case of pure reflection; obviously, this provides the 'initial' conditions to integrate the equations of motion until the next contact between the particle and the constraint.

It is interesting to note how, if the particle moves in a conservative field, it is possible to incorporate the effect of the constraint in the potential. To this end, we define the constraint as an improper function: if $V(\mathbf{x})$ is the potential energy of the field acting on the point, we set $V(\mathbf{x}) = +\infty$ in the region $f(\mathbf{x}) > 0$. Since the particle has a finite energy, which remains constant along the motion, this results in creating artificially a region in space that is inaccessible to the particle. This point of view will be useful in other contexts—in statistical mechanics, one often considers systems of particles confined inside a container with reflecting walls. It is possible to justify this approach by a limiting argument; for simplicity, we illustrate this for the case that the constraint is given by

$$x_3 \leq 0$$

(since the impact is purely local, we can consider the plane tangent to the constraint at the point of contact). For every $\varepsilon > 0$ we introduce in the region $0 < x_3 < \varepsilon$ a potential energy field $V_\varepsilon(x_3)$, with $V'_\varepsilon(x_3) > 0$ and $\lim_{x_3 \to \varepsilon} V(x_3) = +\infty$. If the point $(P, m)$ enters this region with a velocity whose normal component is $v_3^0 > 0$, during the motion inside the region the components $v_1$, $v_2$ of the velocity remain unchanged, while $v_3$ vanishes when $x_3$ reaches the value $x_3^*$; this value

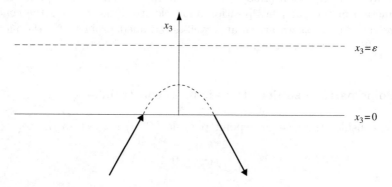

**Fig. 2.4** Mollifying the shock on a rigid wall.

is uniquely defined by $V_\varepsilon(x_3^*) = \frac{1}{2}m(v_3^0)^2$ (we set $V_\varepsilon(0) = 0$). Subsequently, $v_3$ changes sign and eventually the point exits the region with a velocity which is obtained by reflection from the entry velocity. Hence the motion on exit from the region is symmetric to the motion on entry, see Fig. 2.4. If we let $\varepsilon$ tend to zero, the exit point converges to the entry point and we can deduce that the effect of an infinity potential barrier is an elastic reflection.

## 2.7 Additional remarks and bibliographical notes

For the historical discussion of the development of classical mechanics, besides the cited book of Truesdell, the most important sources are: Galileo Galilei (*Dialogo sui due massimi sistemi del mondo*, 1632), Isaac Newton (*Principia Mathematica Philosophiae Naturalis*, 1686, 1687), Giuseppe Luigi Lagrange (*Mécanique Analitique*, 1788), Carl Jacobi (*Vorlesungen über Dynamik*, 1866), and Henri Poincaré (*Les Méthodes Nouvelles de la Mécanique Céleste*, 1892–1899).

## 2.8 Additional solved problems

*Problem 1*
A point particle $(P, m)$ is at one end of a perfectly flexible and inextensible string, of zero mass. The string is turned around a circumference of radius $R$. At time $t = 0$, the free part of the string has length $l$ and the point's velocity is $v_0$.

(i) Study the trajectory of the particle.
(ii) Assuming that the only force acting on the point is the tension of the string, study the motion of the particle and compute the tension.
(iii) If the motion is confined to a vertical plane and the particle is subject to gravity, find the conditions necessary for the string to remain under tension.

*Solution*

(i) Let $\varphi$ be the angle describing how much the string turns around the circumference starting from the initial configuration $AP_0$. Then the free part of the string has length $l - R\varphi$ ($R\varphi < l$). In the system in which $P_0$ has coordinates $(R, l)$, the coordinates of the point particle $P$ are (see Fig. 2.5)

$$x = R\cos\varphi - (l - R\varphi)\sin\varphi, \quad y = R\sin\varphi + (l - R\varphi)\cos\varphi,$$

which give the parametric equations describing the trajectory. Obviously

$$\frac{dx}{d\varphi} = -(l - R\varphi)\cos\varphi, \quad \frac{dy}{d\varphi} = -(l - R\varphi)\sin\varphi.$$

Hence the unit tangent vector is given by $\mathbf{t} = -(\cos\varphi, \sin\varphi)$ and $\mathbf{n} = (\sin\varphi, -\cos\varphi)$. The relation between $s$ and $\varphi$ is given by $s = \int_0^\varphi (l - R\psi)\, d\psi$,

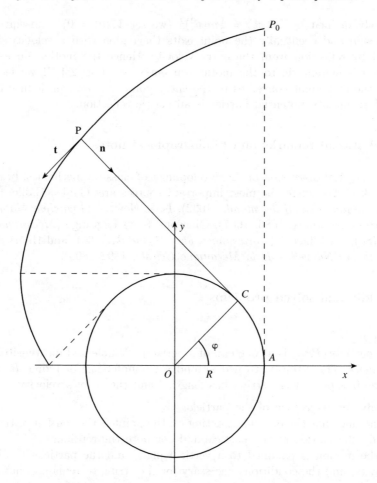

**Fig. 2.5** The motion of a point attached to a winding string.

and hence the curvature is

$$k(s) = \left|\frac{d\mathbf{t}}{ds}\right| = \frac{d\varphi}{ds} = \frac{1}{l - R\varphi(s)},$$

where $\varphi(s)$ can be found by inverting $s = l\varphi - \frac{1}{2}R\varphi^2$, or $R\varphi = l - \sqrt{l^2 - 2sR}$ (the other solution $R\varphi = l + \sqrt{l^2 - 2sR}$ corresponds to the string unravelling).

(ii) The string's tension does not do any work because it is orthogonal to the velocity. It follows that the kinetic energy is constant, and hence $\dot{s} = v_0$ and the tension is given by $\tau = mkv_0^2 = mv_0^2/(l - R\varphi)$.

(iii) If the point is subject to weight, the motion depends on the initial conditions. If the $y$-axis is vertical and we wish to start from a generic configuration, the equations need to be written in a different way.

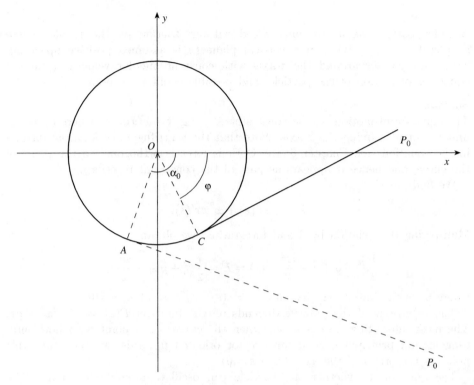

**Fig. 2.6** Selecting the initial condition in the presence of gravity.

Let $\alpha_0$ be the angle that $OA$ makes with the $x$-axis (negative in Fig. 2.6). If $\varphi$ denotes as before the angle between $OC$ and the $x$-axis, it is sufficient to replace $l$ with $l + R\alpha_0$ in the parametric equations of the trajectory. The conservation of energy is now expressed by

$$\frac{1}{2}\dot{s}^2 + gy = \frac{1}{2}v_0^2 + gy_0, \quad y_0 = R\sin\alpha_0 + l\cos\alpha_0.$$

The maximum value of $y$, when admissible, is $y_{\max} = y(0) = l + R\alpha_0$; hence if $gy_{\max} < \frac{1}{2}v_0^2 + gy_0$ the motion does not change direction. Otherwise, the motion is oscillatory, as long as the string's tension remains positive.

The tension can be deduced from

$$m k \dot{s}^2 = mg\cos\varphi + \tau \Rightarrow \tau = m\left[\frac{1}{l + R\varphi_0 - R\varphi}(v_0^2 + 2gy_0 - 2gy(\varphi)) - g\cos\varphi\right].$$

If $\tau$ vanishes for a certain value of $\varphi$, from that time on we need to solve the unconstrained problem, until the point in the new trajectory intersects the previous constraint.

*Problem 2*
A point particle $(P, m)$ is constrained without friction on the regular curve $x = x(s)$, $z = z(s)$ lying in a vertical plane ($z$ is assumed positive upwards). The plane rotates around the $z$-axis with constant angular velocity. Find the equations of motion of the particle, and possible equilibrium points.

*Solution*
The equation of motion on the constraint is $\ddot{s} = \mathbf{g} \cdot \mathbf{t} + \omega^2 x \mathbf{e}_1 \cdot \mathbf{t}$, where $\mathbf{e}_1$ is the unit vector identifying the $x$-axis. Note that the centrifugal acceleration appears in the equation (see Chapter 6; the Coriolis force is orthogonal to the plane of the curve, and hence it appears as part of the constraint reaction).
We find
$$\ddot{s} = -g z'(s) + \omega^2 x x'(s).$$
Multiplying this relation by $\dot{s}$ and integrating, we obtain
$$\frac{1}{2}\dot{s}^2 + g z(s) - \frac{\omega^2}{2} x^2(s) = E = \frac{1}{2} v_0^2 + g z_0 - \frac{\omega^2}{2} x_0^2,$$
where $v_0$ is the initial velocity and $z_0 = z(s_0)$, $x_0 = x(s_0)$, $s_0 = s(0)$.

The behaviour of the particle depends on the function $F(s) = -\frac{1}{2}\omega^2 x^2 + g z$. The most interesting case is found when the curve has a point with horizontal tangent and principal normal unit vector oriented upwards; we then take this point as the origin of the axes (also $s = 0$).

The question is whether the particle can oscillate around this point. The equation $\dot{s}^2 = 2(E - F(s))$ implies an oscillatory motion in any interval $(s_1, s_2) \ni 0$ as long as $E$ is such that there exist two simple zeros $s_1, s_2$ of $F(s_1) = F(s_2) = E$, and $F(s) < E$ for $s \in (s_1, s_2)$.

Consider the case when $z$ is the graph of the function $z = \lambda |x|^n$, $n > 1$. We can then study $F(x) = -\frac{1}{2}\omega^2 x^2 + \lambda g |x|^n$. The derivative $F'(x) = x(-\omega^2 + \text{sign}(x) n \lambda g |x|^{n-2})$ vanishes for $x = 0$ and for
$$|\bar{x}| = \left(\frac{\omega^2}{n \lambda g}\right)^{1/(n-2)},$$
where
$$F(|\bar{x}|) = -\left(\frac{\omega^2}{n \lambda g}\right)^{2/(n-2)} \omega^2 \left(\frac{1}{2} - \frac{1}{n}\right).$$

For $n > 2$ and $1 < n < 2$ we find the following graphs of the function $F$ (Fig. 2.7); for $n > 2$ there exist oscillatory motions around $\bar{x}$ (or $-\bar{x}$) if $E < 0$ and oscillatory motions around the origin if $E > 0$ (the curve $E = 0$ is a separatrix in the phase plane).

For $1 < n < 2$ there exist oscillatory motions around the origin only if
$$0 < E < F(\bar{x}) = \left(\frac{m \lambda g}{\omega^2}\right)^{2/(2-n)} \frac{2-n}{2n} \omega^2,$$

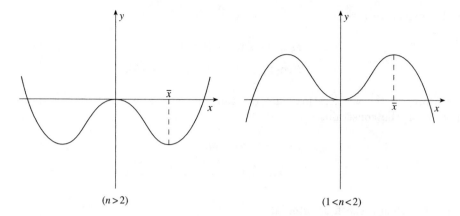

$(n>2)$ $\qquad$ $(1<n<2)$

**Fig. 2.7** Graphs of the potential energy.

otherwise the kinetic energy grows indefinitely while the particle escapes to infinity (as centrifugal acceleration prevails).

In the limiting case $n=2$ there are three possibilities:

$\lambda g > \omega^2/2$, oscillatory motion for any value of the energy $E$;

$\lambda g < \omega^2/2$, the particle always escapes to infinity with velocity growing to infinity;

$\lambda g = \omega^2/2$, uniform motion for any initial condition.

Regarding equilibrium, we find the following cases:

$n > 2$: $x = 0$ (unstable), $x = \pm \bar{x}$ (stable);

$1 < n < 2$: $x = 0$ (stable), $x = \pm \bar{x}$ (unstable);

$n = 2$: $x = 0$ (stable for $\lambda g > \omega^2/2$, unstable for $\lambda g \leq \omega^2/2$), all points are equilibrium points if $\lambda g = \omega^2/2$.

Equilibrium can be attained at the points where the sum of $\mathbf{g}$ and $\omega^2 x \mathbf{e}_1$ is orthogonal to the constraint.

*Remark* 2.1

An equilibrium configuration is called *stable* if the system can oscillate around it. For more details on stability, see Chapter 4. ∎

*Problem 3*

Describe the motion of a point particle subject to its own weight and constrained on a smooth sphere.

*Solution*

Consider the parametrisation

$$x_1 = R\sin\theta\cos\varphi, \quad x_2 = R\sin\theta\sin\varphi, \quad x_3 = R\cos\theta,$$

with tangent vectors

$$\mathbf{x}_\varphi = R\sin\theta(-\sin\varphi, \cos\varphi, 0),$$
$$\mathbf{x}_\theta = R(\cos\theta\cos\varphi, \cos\theta\sin\varphi, -\sin\theta).$$

If we project the acceleration onto the tangent vectors, we find the equations of motion on the constraint:

$$\sin^2\theta\,\ddot\varphi + 2\sin\theta\cos\theta\,\dot\theta\dot\varphi = 0,$$
$$\ddot\theta - \sin\theta\cos\theta\,\dot\varphi^2 = \frac{g}{R}\sin\theta.$$

The first admits the first integral

$$\dot\varphi\sin^2\theta = c \tag{2.38}$$

(the vertical component of the angular momentum); using this, we can rewrite the second in the form

$$\ddot\theta - \frac{\cos\theta}{\sin^3\theta}c^2 = \frac{g}{R}\sin\theta. \tag{2.39}$$

Expressing the constraint reaction as $\lambda\nabla f$, with $f = \frac{1}{2}(x^2+y^2+z^2-R^2)=0$, and recalling equation (1.98), we find after multiplying $m\mathbf{a} = m\mathbf{g}+\lambda\nabla f$ by $\nabla f$ and dividing by $R^2$ that

$$-(\sin^2\theta\,\dot\varphi^2 + \dot\theta^2) = -\frac{gx_3}{R^2} + \frac{\lambda}{m}, \quad \text{with } x_3 = R\cos\theta. \tag{2.40}$$

On the left-hand side we have the projection of $(1/R)\mathbf{a}$ onto the radius of the sphere. It is not easy to integrate equation (2.39). We can naturally find its first integral (multiply by $\dot\theta$ and integrate):

$$\frac{1}{2}\left(\dot\theta^2 + \frac{c^2}{\sin^2\theta}\right) + \frac{g}{R}\cos\theta = \frac{E}{mR^2}. \tag{2.41}$$

In view of (2.38), this is the energy integral $\frac{1}{2}\dot{\mathbf{x}}^2 + gx_3 = E$. If we now combine (2.40) with (2.38) and (2.41) we can determine the scalar field of possible reactions

$$\lambda = -\frac{2E}{R^2} + 3\frac{mg}{R}\cos\theta \tag{2.42}$$

on the sphere. There are two simple cases to examine: $\varphi = $ constant (motion along the meridians) and $\theta = $ constant (motion along the parallels). The motion with $\varphi = \varphi_0$ implies $c = 0$ and the equation of motion (2.39) is reduced to the equation of the pendulum (these are the only trajectories passing through the poles). For the motion with $\theta = \theta_0$ we can deduce the value of $c$ from equation (2.39): $c^2 = -(g/R)\sin^3\theta\tan\theta$, i.e.

$$\dot\varphi^2 = -\frac{g}{R\cos\theta}.$$

Since $\sin\theta > 0$ necessarily $\tan\theta < 0$, and hence the only possible motion is along the parallels, with $\theta_0 \in (\pi/2, \pi)$ (southern hemisphere).

More generally, for $c \neq 0$, the motion is bounded between those values of $\theta$ for which the expression

$$\frac{2E}{mR^2} - \frac{2g}{R}\cos\theta - \frac{c^2}{\sin^2\theta}$$

vanishes, values which are guaranteed to exist because the last term diverges at the poles, and $E$ can be chosen in such a way that $\dot\theta^2 \geq 0$.

### Problem 4
Study the motion of a point mass on a smooth surface of revolution around the vertical axis.

### Solution
Consider the representation

$$x_1 = r(\theta)\sin\theta\cos\varphi, \quad x_2 = r(\theta)\sin\theta\sin\varphi, \quad x_3 = r(\theta)\cos\theta, \quad r(\theta) > 0.$$

The vectors in the basis of the tangent space are

$$\mathbf{x}_\varphi = r(\theta)\sin\theta\begin{pmatrix}-\sin\varphi\\ \cos\varphi\\ 0\end{pmatrix}, \quad \mathbf{x}_\theta = r'(\theta)\begin{pmatrix}\sin\theta\cos\varphi\\ \sin\theta\sin\varphi\\ \cos\theta\end{pmatrix} + r(\theta)\begin{pmatrix}\cos\theta\cos\varphi\\ \cos\theta\sin\varphi\\ -\sin\theta\end{pmatrix},$$

and hence $\mathbf{x}_\varphi^2 = r^2\sin^2\theta$, $\mathbf{x}_\theta^2 = r'^2 + r^2$, $\mathbf{x}_\varphi\cdot\mathbf{x}_\theta = 0$. In addition,

$$\mathbf{x}_{\varphi\varphi} = -r\sin\theta\begin{pmatrix}\cos\varphi\\ \sin\varphi\\ 0\end{pmatrix}, \quad \mathbf{x}_{\varphi\theta} = (r'\sin\theta + r\cos\theta)\begin{pmatrix}-\sin\varphi\\ \cos\varphi\\ 0\end{pmatrix},$$

$$\mathbf{x}_{\theta\theta} = (r''-r)\begin{pmatrix}\sin\theta\cos\varphi\\ \sin\theta\sin\varphi\\ \cos\theta\end{pmatrix} + 2r'\begin{pmatrix}\cos\theta\cos\varphi\\ \cos\theta\sin\varphi\\ -\sin\varphi\end{pmatrix},$$

which implies

$$\mathbf{x}_\varphi\cdot\mathbf{x}_{\varphi\varphi} = 0, \quad \mathbf{x}_\varphi\cdot\mathbf{x}_{\varphi\theta} = r\sin\theta(r'\sin\theta + r\cos\theta), \quad \mathbf{x}_\varphi\cdot\mathbf{x}_{\theta\theta} = 0,$$
$$\mathbf{x}_\theta\cdot\mathbf{x}_{\varphi\varphi} = -r\sin\theta(r'\sin\theta + r\cos\theta), \quad \mathbf{x}_\theta\cdot\mathbf{x}_{\varphi\theta} = 0,$$
$$\mathbf{x}_\theta\cdot\mathbf{x}_{\theta\theta} = (r''+r)r'.$$

In summary, the equations of motion are given by

$$r^2 \sin^2 \theta \ddot\varphi + 2r \sin\theta (r' \sin\theta + r\cos\theta)\dot\theta\dot\varphi = 0,$$

$$(r'^2 + r^2)\ddot\theta - r\sin\theta(r'\sin\theta + r\cos\theta)\dot\varphi^2 + (r'' + r)r'\dot\theta^2 = g(r\sin\theta - r'\cos\theta).$$

As in the spherical case, the first equation has first integral

$$r^2(\theta) \sin^2\theta \, \dot\varphi = c$$

with the same interpretation; we also find the energy integral

$$\frac{1}{2}[r^2 \sin^2\theta \dot\varphi^2 + (r'^2 + r^2)\dot\theta^2] + mgr\cos\theta = E$$

which allows us to eliminate $\dot\varphi$:

$$\frac{1}{2}\left[\frac{c^2}{r^2 \sin^2\theta} + (r'^2 + r^2)\dot\theta^2\right] + mgr\cos\theta = E.$$

Some of the qualitative remarks valid in the spherical case can be extended to the present case, but care must be taken as $r\sin\theta$ does not necessarily tend to zero (for instance it is constant in the cylindrical case, when it is clearly impossible to have motion along the parallels).

# 3 ONE-DIMENSIONAL MOTION

## 3.1 Introduction

In Section 2.4 of the previous chapter, we mentioned the problem of the motion of a point particle $P$ of mass $m$ along a fixed smooth curve. We now want to consider the problem of determining the time dependence $s = s(t)$, and hence of integrating equation (2.25); this equation has the form

$$m\ddot{s} = f(s, \dot{s}, t), \tag{3.1}$$

to which one has to associate initial conditions $s(0) = s_0, \dot{s}(0) = v_0$. In two special cases, the problem is easily solvable: when the force depends only on the position of the particle $f = f(s)$ or only on the velocity of the particle $f = f(\dot{s})$.

The first case is the most interesting, and we will consider it in detail in the following sections. Recall that when the force $f(s)$ is associated with a potential $U(s)$ it is possible to write down the energy integral (2.28). However, in the case we are considering, when the trajectory of the point is prescribed, we can still define a function of $s$:

$$U(s) = \int_0^s f(z)\mathrm{d}z, \tag{3.2}$$

representing the work done along the corresponding arc of the trajectory and that yields the first integral

$$\frac{1}{2}m\dot{s}^2 = E + U(s), \tag{3.3}$$

where $E$ is determined by the initial conditions. Equation (3.3) determines the region where motion is possible, through the inequality $U(s) \geq -E$. It is integrable by separation of variables: for every interval where $U(s) > -E$ we can write

$$\mathrm{d}t = \pm \frac{\mathrm{d}s}{\sqrt{\frac{2}{m}[E + U(s)]}}. \tag{3.4}$$

If the force depends only on the velocity of the particle, $f = f(\dot{s})$, the equation of motion is again solvable by separation of variables: since

$$m\ddot{s} = f(\dot{s}), \tag{3.5}$$

we obtain that, where $f \neq 0$,

$$m \frac{d\dot{s}}{f(\dot{s})} = dt. \tag{3.6}$$

This yields the implicit form $F(\dot{s}) = t+\text{constant}$, which in turn yields, by another integration, the function $s(t)$.

An example is given by motion in a medium dissipating energy by friction, where $f(\dot{s})\dot{s} < 0, f(0) = 0, f' < 0$. The (slow) motion inside viscous fluids belongs to this class; in this case, it is usually assumed that $f(\dot{s}) = -b\dot{s}$, where $b$ is a positive constant depending on the viscosity.

We can summarise what we have just discussed in the following.

THEOREM 3.1   *If in equation (3.1) $f$ is a continuous function depending only on the variable $s$, or a Lipschitz function depending only on the variable $\dot{s}$, the initial value problem is solvable by separation of variables, and hence integrating equation (3.4) or equation (3.6).* ∎

Remark 3.1
It will be evident later that the theory developed here for the case of a constrained point particle can be generalised to the motion of holonomic systems with one degree of freedom (one-dimensional motion). ∎

## 3.2   Analysis of motion due to a positional force

We analyse equation (3.4), where there appears the function

$$\Phi(s) = \frac{2}{m}[E + U(s)];$$

we assume that this function is sufficiently regular. The motion takes place in the intervals defined by the condition $\Phi(s) \geq 0$.

In the plane $(s, \dot{s})$ the equation $\dot{s}^2 = \Phi(s)$ determines a family of curves depending on the parameter $E$. If there exist isolated roots of the function $\Phi(s)$, they separate branches $\dot{s} = \sqrt{\Phi(s)}$ and $\dot{s} = -\sqrt{\Phi(s)}$. Let us consider the case that the initial conditions $s(0) = s_0, \dot{s}(0) = v_0$ determine the branch $\dot{s} > 0$ (i.e. $v_0 > 0$; the other case is analogous). There exist two possibilities: for $s > s_0$ we have $\Phi(s) > 0$, or else there exist roots of $\Phi(s)$ to the right of $s_0$; let us denote the first of these roots by $s_1$.

In the first case

$$t(s) = \int_{s_0}^{s} \frac{d\sigma}{\sqrt{\Phi(\sigma)}} \tag{3.7}$$

is a monotonic function, and hence invertible. If the integral on the right-hand side diverges when $s \to +\infty$, then the function $s(t) \to \infty$ for $t \to \infty$. If, on the other hand, the integral is convergent, then

$$s(t) \to +\infty \quad \text{for } t \to t_\infty = \int_{s_0}^{+\infty} \frac{d\sigma}{\sqrt{\Phi(\sigma)}}.$$

In the other case, the solution attains the value $s_1$ in a finite time

$$t_1 = \int_{s_0}^{s_1} \frac{ds}{\sqrt{\Phi(s)}},$$

provided this integral converges; this is the case if $s_1$ is a simple root (i.e. $\Phi'(s_1) < 0$). Otherwise $s_1$ is an asymptotic value for $s(t)$, but for all times, $s(t) < s_1$.

We must analyse the case that $\Phi(s_0) = 0$. If $(2/m)f(s_0) = \Phi'(s_0) \neq 0$ the sign of this expression determines the initial value of $\ddot{s}$ and the orientation of the motion, and the solution is still expressed by a formula similar to (3.7). In this case, the previous considerations still apply. If, on the other hand, $\Phi(s_0) = \Phi'(s_0) = 0$, and hence if $f(s_0) = 0$, the particle is in an *equilibrium* position and $s(t) = s_0$ is the unique solution of (3.1).

*Remark 3.2*

The motion can never pass through a point parametrised by a value of $s$ which is a multiple root of $\Phi$; the motion can only tend to this position asymptotically, or else remain there indefinitely if this was the initial position. This fact is a consequence of the uniqueness of the solution of the Cauchy problem for the equation $\dot{s} = \sqrt{\Phi(s)}$ when $\sqrt{\Phi(s)}$ is a Lipschitz function. Suppose that a multiple root $s_1$ of $\Phi$ could be reached in a finite time $t_1$. This would imply that for $t < t_1$ the problem $\dot{s} = \sqrt{\Phi(s)}$, $s(t_1) = s_1$ has a solution that is different from the constant solution $s \equiv s_1$. If the regions where motion can take place are bounded, they must lie between two consecutive roots $s_1$ and $s_2$ of $\Phi$. The analysis of such motion in accessible regions lying between two simple roots of $\Phi(s)$ is not difficult. ∎

DEFINITION 3.1 *A simple root $\hat{s}$ of $\Phi$ is called an* inversion point *for the motion.* ∎

THEOREM 3.2 *The motion between two consecutive inversion points $s_1$ and $s_2$ is periodic with period*

$$T(E) = 2\int_{s_1}^{s_2} \frac{ds}{\sqrt{\Phi(s)}} = 2\int_{s_1}^{s_2} \frac{ds}{\sqrt{\frac{2}{m}[E + U(s)]}}. \tag{3.8}$$

*Proof*

Without loss of generality we can assume that $s_1 < s_0 < s_2$. In this interval we can write

$$\Phi(s) = (s - s_1)(s_2 - s)\psi(s),$$

with $\psi(s) > 0$ for $s \in [s_1, s_2]$. Hence

$$\dot{s}^2 = (s - s_1)(s_2 - s)\psi(s),$$

and the sign of the square root is determined by the initial condition. Assume that the sign is positive; the point particle approaches $s_2$, until it reaches it at time $t_1$. At this moment, the velocity is zero and the motion starts again with the orientation of the force acting in $s_2$, given by

$$\left.\frac{d\Phi}{ds}\right|_{s=s_2} = -(s_2 - s_1)\psi(s_2) < 0.$$

Thus the orientation of the motion is inverted in $s_2$; hence this is called an inversion point. For $t > t_1$, $P$ returns to $s_1$ where it arrives at time $t_2$. Again, the velocity is zero and the motion continues with the orientation of the force acting in $s_1$:

$$\left.\frac{d\Phi}{ds}\right|_{s=s_1} = (s_2 - s_1)\psi(s_1) > 0.$$

This implies that the particle passes again through $s_0$ at time $t_3 = t_0 + T$. The motion is periodic: $s(t) = s(t + T)$ for every $t$, and the period $T$ is given by

$$T = t_1 + (t_2 - t_1) + (t_3 - t_2) = \int_{s_0}^{s_2} - \int_{s_2}^{s_1} + \int_{s_1}^{s_0} \frac{ds}{\sqrt{\Phi(s)}} = 2\int_{s_1}^{s_2} \frac{ds}{\sqrt{\Phi(s)}}. \blacksquare$$

*Remark* 3.3
Note that the motion is possible because the Cauchy problem $\dot{s} = \sqrt{\Phi(s)}$, with a simple zero of $\Phi$ (for which the function $\sqrt{\Phi}$ is not Lipschitz) as initial condition, does not have a unique solution. $\blacksquare$

*Example* 3.1
We compute the period of the oscillations of a heavy point particle $(P, m)$ constrained to move on a cycloid, and we show that it is independent of the amplitude.

In the reference frame of Fig. 3.1 the constraint has parametric equations

$$x = R(\psi + \sin \psi),$$
$$z = R(1 - \cos \psi),$$

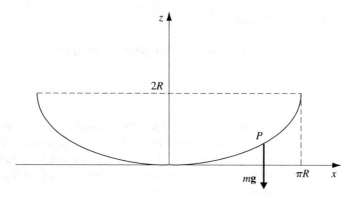

**Fig. 3.1**

and the length of the arc between the origin and the point $P(\psi)$ is

$$s = \sqrt{2}R \int_0^\psi \sqrt{1 + \cos\varphi}\, d\varphi = 4R \sin\frac{\psi}{2}.$$

For the oscillations to be possible we must have $E + U \geq 0$ and $E < 2mgR$. It then follows that $|\psi| \leq \psi_m$, with $\psi_m$ given by

$$\cos\psi_m = 1 - \alpha, \qquad \alpha = \frac{E}{mgR} \in (0, 2).$$

By equation (3.8) the period is

$$T = 4\int_0^{\psi_m} \frac{ds}{d\psi} \left\{\frac{2}{m}[E - mgR(1 - \cos\psi)]\right\}^{-1/2} d\psi.$$

Writing $ds/d\psi = \sqrt{2}R\sqrt{1 + \cos\psi}$ and setting $\cos\psi = \eta$, we arrive at the expression

$$T = 4\sqrt{\frac{R}{g}} \int_{1-\alpha}^1 \frac{d\eta}{\sqrt{(1-\eta)(\eta - 1 + \alpha)}} = 4\pi\sqrt{\frac{R}{g}},$$

showing that the oscillations on the cycloid are isochronous. We shall see that there are no other symmetric curves with this property (Problem 4 in Section 3.9). ■

## 3.3 The simple pendulum

The simple pendulum is a very important model in mechanics. This model has equation

$$\ddot{\vartheta} + \frac{g}{l}\sin\vartheta = 0, \tag{3.8}$$

where $\vartheta$ is the angle measuring the deviation of the pendulum from the vertical direction, $g$ is the acceleration due to gravity and $l$ is the pendulum length.

The phase space of the system is planar, but all angles $\vartheta$ are identified modulo $2\pi$; hence we think of the pendulum phase space as the cylinder $(\vartheta, \dot{\vartheta}) \in \mathbf{S}^1 \times \mathbf{R}$. Let $E = T - U$ be the total mechanical energy

$$E = \frac{1}{2}ml^2\dot{\vartheta}^2 - mgl\cos\vartheta. \tag{3.9}$$

Let $e = E/mgl$ be fixed by the initial conditions. Clearly $e \geq -1$, and

$$\dot{\vartheta}^2 = \frac{2g}{l}(\cos\vartheta + e). \tag{3.10}$$

As $e$ varies we can distinguish two kinds of motion, which differ in the topology of their trajectories in the phase space (Fig. 3.2). The *rotations* correspond to values of $e > 1$ and to trajectories that wind around the cylinder, and hence that cannot be deformed continuously to a point (they are *homotopically non-trivial*).[1] If $|e| < 1$, the motion is *oscillatory*: the trajectories do not wind around the cylinder and are homotopically trivial. The position of stable equilibrium $\vartheta = 0$ of the pendulum corresponds to the value $e = -1$, while to $e = 1$ there correspond both the position of unstable equilibrium $\vartheta = \pi$, and the trajectory asymptotic to it (in the past and in the future), of the equation

$$\frac{1}{2}l^2\dot{\vartheta}^2 - gl(\cos\vartheta + 1) = 0, \tag{3.11}$$

called the *separatrix*, because it separates oscillatory motions from rotations.

By separating variables in the energy equation (3.10) it is possible to compute the time dependence and the period of the pendulum. Setting

$$y = \sin\frac{\vartheta}{2}, \tag{3.12}$$

and substituting this into equation (3.11) we find, after some easy algebraic manipulations,

$$\dot{y}^2 = \frac{g}{l}(1-y^2)\left(\frac{e+1}{2} - y^2\right). \tag{3.13}$$

---

[1] A closed curve $\gamma : [0,1] \to M$ on a manifold is 'homotopically trivial' if there exist a continuous function $F : [0,1] \times [0,1] \to M$ and a point $p \in M$ such that $\forall s \in [0,1]$, $t \to F(s,t)$ is a closed curve which for $s = 0$ coincides with $\gamma$, while $F(1,t) = p$ for every $t \in [0,1]$.

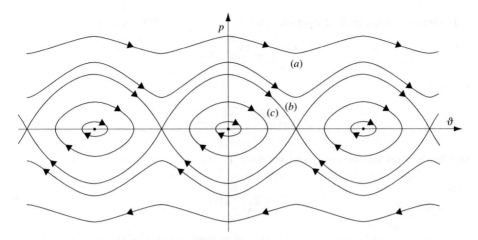

**Fig. 3.2** Pendulum trajectories: (*a*) rotations, (*b*) separatrix, (*c*) oscillations.

If the pendulum oscillates, namely for $0 < e < 1$, we have $(e+1)/2 = k^2$, where $k < 1$. Then equation (3.14) can be written

$$\dot{y}^2 = \frac{gk^2}{l}(1-y^2)\left(1 - \frac{y^2}{k^2}\right), \tag{3.14}$$

yielding

$$\int_0^{y/k} \frac{d\xi}{\sqrt{(1-\xi^2)(1-k^2\xi^2)}} = \sqrt{\frac{g}{l}}(t-t_0), \tag{3.15}$$

where we set $y(t_0) = 0$. This equation can be integrated immediately by using the Jacobi elliptic function (see Appendix 2):

$$y(t) = k \,\operatorname{sn}\left\{\sqrt{\frac{g}{l}}(t-t_0), k\right\}. \tag{3.16}$$

The value of the constant of integration $t_0$ and of $k$ are fixed by the initial conditions. The motion is periodic, with period

$$T = 4\sqrt{\frac{l}{g}}\mathbf{K}(k), \tag{3.17}$$

where **K** is the complete elliptic integral of the first kind. Using the series expansion of **K** (see Appendix 2), we find

$$T = 2\pi\sqrt{\frac{l}{g}}\left\{1 + \sum_{j=1}^{\infty}\left[\frac{(2j-1)!!}{(2j)!!}\right]^2 k^{2j}\right\}, \tag{3.18}$$

which measures the size of the deviations from isochronism.

If the pendulum is in rotation, namely for $e > 1$, after setting

$$\frac{e+1}{2} = \frac{1}{k^2}, \quad k < 1,$$

we find

$$y(t) = \operatorname{sn}\left\{\sqrt{\frac{g}{l}} \frac{t - t_0}{k}, k\right\}, \qquad (3.19)$$

and the period of one complete rotation is expressed by

$$T = 2\sqrt{\frac{l}{g}} k \mathbf{K}(k). \qquad (3.20)$$

Finally for $e = 1$, corresponding to the motion along the separatrix, it is easy to find that the motion is given by

$$y(t) = \tanh\left\{\sqrt{\frac{g}{l}}(t - t_0)\right\}. \qquad (3.21)$$

## 3.4 Phase plane and equilibrium

The equation of motion (3.1) is equivalent to the system of two first-order equations

$$\begin{aligned}\dot{x} &= y, \\ \dot{y} &= \frac{1}{m} f(x, y, t),\end{aligned} \qquad (3.22)$$

where $x$ replaces $s$. Suppose in addition that $f$ is a regular function of all its variables.

DEFINITION 3.2  *The plane $(x, y) \in \mathbf{R}^2$ is called the* phase plane *of equation (3.1); the terms on the left-hand side of the system (3.23) define a vector field whose integral curves are the* phase curves *of the system. The operator $g^t$, associating with every initial point $(x_0, y_0)$ the point $(x(t), y(t))$ on the corresponding phase curve, is called the* flux operator *of system (3.23).*  ∎

The existence and uniqueness theorem for the solutions of the Cauchy problem for ordinary differential equations implies that one and only one phase curve passes through any given point $(x, y)$ in the phase plane.

If the force field is positional, then

$$\begin{aligned}\dot{x} &= y, \\ \dot{y} &= \frac{1}{m} f(x),\end{aligned} \qquad (3.23)$$

the system is *autonomous* (i.e. the terms on the right-hand side of (3.24) do not depend explicitly on the time variable $t$) and the energy is conserved. Along every phase curve, the energy is constant; hence the phase curves belong to the energy level, denoted by $M_e$: for every fixed $e \in \mathbf{R}$,

$$M_e = \{(x,y) \in \mathbf{R}^2 | E(x,y) = \frac{my^2}{2} + V(x) = e\}, \qquad (3.24)$$

where $V(x) = -U(x)$ is the potential energy (recall that $U$ is defined by (3.2)). The level sets can have several connected components, and hence may contain more than one distinct phase curve. In addition, $M_e$ is a regular curve, if $\nabla E \neq (0,0)$, and hence if

$$(my, V'(x)) \neq (0,0). \qquad (3.25)$$

The points where $\nabla E = (0,0)$ are called *critical points*. Note that at a critical point one has $y = 0$ and that every critical point is a stationary point of the potential energy.

DEFINITION 3.3 *A point $(x_0, 0)$ is called an* equilibrium point *of the system (3.24) if any phase curve passing through it reduces to the point itself, and hence if $x(t) \equiv x_0, y(t) \equiv 0$ is a solution of (3.24) with initial condition $(x_0, 0)$.* ∎

Since an equilibrium point for (3.24) has by definition its $y$-coordinate equal to zero, to identify it, it suffices to give $x_0$.

The theorem of the existence and uniqueness of solutions of the Cauchy problem for ordinary differential equations implies that, when the field is conservative, a point $x_0$ is an equilibrium point for (3.24) *if and only if* it is a critical point for the energy.

DEFINITION 3.4 *An equilibrium position $x_0$ is called* Lyapunov stable *if for every neighbourhood $U \subset \mathbf{R}^2$ of $(x_0, 0)$ there exists a neighbourhood $U'$ such that, for every initial condition $(x(0), y(0)) \in U'$, the corresponding solution $(x(t), y(t))$ is in $U$ for every time $t > 0$ (Fig. 3.3). Any point that is not stable is called* unstable. ∎

In other words, the stability condition is the following: for every $\varepsilon > 0$ there exists $\delta > 0$ such that, for every initial condition $(x(0), y(0))$ such that $|x(0) - x_0| < \delta$, $|y(0)| < \delta$, we have $|x(t) - x_0| < \varepsilon$ and $|y(t)| < \varepsilon$ for every time $t > 0$.

DEFINITION 3.5 *A point of stable equilibrium $x_0$ is called* asymptotically stable *if there exists a neighbourhood $U$ of $(x_0, 0)$ such that, for every initial condition $(x(0), y(0)) \in U$, $(x(t), y(t)) \to (x_0, 0)$ for $t \to +\infty$. The maximal neighbourhood $U$ with this property is called the* basin of attraction *of $x_0$.* ∎

If the forces involved are positional, by the theorem of conservation of energy it is impossible to have asymptotically stable equilibrium positions.

PROPOSITION 3.1 *Let $x_0$ be an isolated relative minimum of $V(x)$. Then $x_0$ is a Lyapunov stable equilibrium point.*

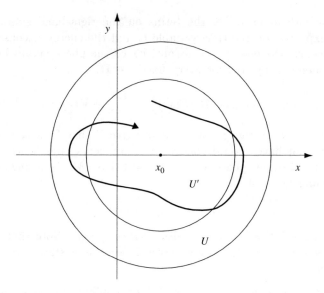

**Fig. 3.3**

*Proof*
We saw that the system preserves the energy, $E(x,y)$. We denote by $e_0 = E(x_0,0) = V(x_0)$ the value of the energy corresponding to the equilibrium position we are considering. Clearly $x_0$ is also an isolated relative minimum for the energy $E$. Let $U$ be any neighbourhood of $(x_0,0)$ and let $\delta > 0$; consider the sublevel set of the energy corresponding to $e_0 + \delta$, namely

$$\{(x,y)|E(x,y) < e_0 + \delta\}.$$

The connected component of this set containing the point $(x_0,0)$ defines, for $\delta$ sufficiently small, a neighbourhood of $(x_0,0)$ in the phase plane. This neighbourhood is contained in $U$ and it is invariant under the flow associated with equations (3.24). ∎

*Remark* 3.4
One could propose that the converse be also true, i.e. that if a point is Lyapunov stable for the system of equations (3.24) then it is a relative minimum for the potential energy. However this is false in the case that the potential energy is not an analytic function but it is only of class $\mathcal{C}^\infty$ (or less regular). In dimensionless coordinates, a counterexample is given by

$$V(x) = \begin{cases} e^{-1/x^2} \sin \dfrac{1}{x}, & \text{if } x \neq 0, \\ 0, & \text{if } x = 0. \end{cases} \qquad (3.26)$$

In this case $x = 0$ is a stable equilibrium point, but not a minimum for the potential energy (see Problem 1 in Section 3.7). ∎

DEFINITION 3.6  *A level set $M_e$ of the energy corresponding to a critical value $e = E(x_0, 0)$ of the energy when $(x_0, 0)$ is an unstable equilibrium point, is called a separatrix.* ∎

A separatrix curve $M_e$ consists in general of several distinct phase curves: the points of equilibrium $x_1, \ldots, x_n \in M_e$ and the connected components $\gamma_1, \ldots, \gamma_k$ of $M_e \setminus \{(x_1, 0), \ldots, (x_n, 0)\}$; see Fig. 3.2 for the example of the pendulum. From Remark 3.2 it follows that the motion along each phase curve $\gamma_i$ tends asymptotically to one of the equilibrium points, an endpoint for the curve under consideration.

*Example 3.2*
Consider the case corresponding to the elastic force $f(x) = -kx$. The only point of equilibrium is $x = 0$, and the level sets $M_e$ of the energy $E = \frac{1}{2}my^2 + \frac{1}{2}kx^2$, with $e > 0$ are ellipses centred at the origin. This system does not admit any separatrix. ∎

*Example 3.3*
Consider a one-dimensional system subject to a conservative force with the potential energy $V(x)$ whose graph is shown in Fig. 3.4b.

The corresponding phase curves are shown in Fig. 3.4a. The separatrix curves are the level sets $M_{e_4}$ and $M_{e_3}$. ∎

In a neighbourhood of any equilibrium point it is possible to approximate equations (3.24) by a system of linear equations. Indeed, if $\xi = x - x_0$ is a new coordinate measuring the displacement from the equilibrium position, setting $\eta = y$, equations (3.24) can be written as

$$\dot{\xi} = \eta,$$
$$\dot{\eta} = -\frac{1}{m} V'(x_0 + \xi). \tag{3.27}$$

Considering the Taylor expansion of the potential energy, this yields

$$\dot{\eta} = -\frac{1}{m}\left(V''(x_0)\xi + \frac{1}{2}V'''(x_0)\xi^2 + \ldots\right),$$

where the dots stand for terms of order higher than two in $\xi$. Note that the term $V'(x_0)$ is missing; it vanishes because of the hypothesis that $x_0$ is an equilibrium position. The linear equations are obtained by considering the term $V''(x_0)\xi$ and neglecting all others. The linearised motion is then governed by the equation

$$m\ddot{\xi} + \overline{V}\xi = 0, \tag{3.28}$$

where $\overline{V} = V''(x_0)$. Equation (3.29) describes a harmonic oscillator if $\overline{V} > 0$, so that the equilibrium position of the system is stable. In this case we call

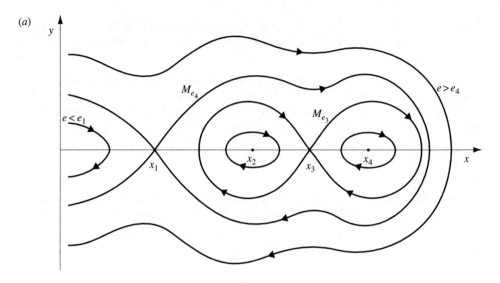

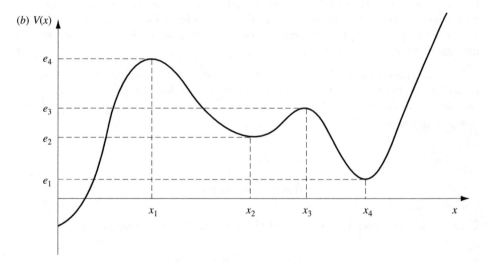

**Fig. 3.4**

(3.29) the equation of *small oscillations* around the stable equilibrium position. If $\omega = \sqrt{|\overline{V}|/m}$, the solutions of (3.29) are given by

$$\xi(t) = \begin{cases} \xi(0)\cos(\omega t) + \dfrac{\eta(0)}{\omega}\sin(\omega t), & \text{if } \overline{V} > 0, \\ \xi(0)\cosh(\omega t) + \dfrac{\eta(0)}{\omega}\sinh(\omega t), & \text{if } \overline{V} < 0. \end{cases} \quad (3.29)$$

The corresponding phase trajectories $t \mapsto (\xi(t), \eta(t))$ are ellipses and branches of hyperbolas, respectively. In the latter case, equation (3.29) is only valid in a sufficiently small time interval.

If $\overline{V} > 0$ it is easy to verify that the periods of the solutions of (3.24)—with initial conditions $(x(0), y(0))$ close to $(x_0, 0)$—tend to $T = 2\pi/\omega$ when $(x(0), y(0)) \to (x_0, 0)$.

In the case that $\overline{V} > 0$ it is possible to study the behaviour of the period $T(E)$ near $E = 0$ (see Problem 4, Section 3.9).

*Example 3.4*
Consider the motion of a point particle with mass $m$ under the action of gravity, and constrained to move along a prescribed curve in a vertical plane. If the natural parametrisation of the curve is given by $s \to (x(s), y(s))$ the energy of the point is

$$E(s, \dot{s}) = \frac{1}{2}m\dot{s}^2 + mgy(s), \tag{3.30}$$

where $g$ denotes the acceleration due to gravity. The equilibrium positions correspond to the critical points $y'(\overline{s}) = 0$, and a position $\overline{s}$ of relative minimum of $y$ with $y''(\overline{s}) > 0$, is Lyapunov stable. Denoting by $\sigma = s - \overline{s}$ the distance along the curve of the equilibrium position, the energy corresponding to the linearised equation can be written as

$$E(\sigma, \dot{\sigma}) = \frac{1}{2}(m\dot{\sigma}^2 + mgy''(\overline{s})\sigma^2). \tag{3.31}$$

This implies the equation of motion

$$\ddot{\sigma} + [gy''(\overline{s})]\sigma = 0, \tag{3.32}$$

corresponding to a harmonic oscillator of frequency $\omega^2 = gy''(\overline{s})$. Note that the curvature at the equilibrium position is $k(\overline{s}) = y''(\overline{s})$, and hence that

$$k(\overline{s}) = \frac{\omega^2}{g}. \tag{3.33}$$

Namely, the frequency of the harmonic oscillations around the equilibrium position is proportional to the square root of the curvature. ∎

## 3.5 Damped oscillations, forced oscillations. Resonance

Consider the one-dimensional motion of a point particle with mass $m$ under the action of an elastic force and of a linear dissipative force: $F(x, \dot{x}) = -kx - \alpha\dot{x}$, where $\alpha$ and $k$ are two positive constants. In this case the energy

$$E(x, \dot{x}) = \frac{m}{2}\dot{x}^2 + \frac{k}{2}x^2$$

is strictly decreasing in time, unless the point is not in motion; indeed, from the equation of motion it follows that $dE/dt = -\alpha \dot{x}^2$, and consequently the point of equilibrium (0,0) in the phase plane is asymptotically stable. Its basin of attraction is the whole of $\mathbf{R}^2$.

Setting $\omega^2 = k/m, \beta = \alpha/2m$, the equation of motion can be written as

$$\ddot{x} + 2\beta \dot{x} + \omega^2 x = 0. \tag{3.34}$$

To find the solutions of equation (3.35), substitute $x(t) = e^{\lambda t}$ into (3.35); $\lambda$ must be a root of the characteristic polynomial

$$\lambda^2 + 2\beta\lambda + \omega^2 = 0. \tag{3.35}$$

If $\Delta = \beta^2 - \omega^2 \neq 0$, the two roots are $\lambda_{\pm} = -\beta \pm \sqrt{\Delta}$ and the solutions are

$$x(t) = A_1 e^{-(\beta+\sqrt{\beta^2-\omega^2})t} + A_2 e^{-(\beta-\sqrt{\beta^2-\omega^2})t}, \tag{3.36}$$

where $A_1$ and $A_2$ are determined by the initial conditions $x(0) = x_0, \dot{x}(0) = v_0$.

It is immediate to verify that if $\beta > \omega$ the motion has at most one inversion point and $x(t) \to 0$ for $t \to \infty$ (Fig. 3.5a). If $\omega > \beta$, $\lambda_{\pm} = -\beta \pm i\sqrt{\omega^2 - \beta^2}$ and equation (3.37) can be rewritten as

$$x(t) = Be^{-\beta t} \cos(\sqrt{\omega^2 - \beta^2}\, t + C), \tag{3.37}$$

where the constants $B, C$ depend on $A_1, A_2$ and the initial conditions through the relations

$$\begin{aligned} x_0 &= B\cos C = A_1 + A_2, \\ v_0 &= -\beta B \cos C - \sqrt{\omega^2 - \beta^2}\, B \sin C \\ &= -(\beta + i\sqrt{\omega^2 - \beta^2})A_1 - (\beta - i\sqrt{\omega^2 - \beta^2})A_2. \end{aligned} \tag{3.38}$$

Once again $x(t) \to 0$ for $t \to \infty$, but the function $x(t)e^{\beta t}$ is now periodic, of period $2\pi/\sqrt{\omega^2 - \beta^2}$ (Fig. 3.5b).

Finally, if $\Delta = 0$ the solution is critically damped:

$$x(t) = e^{-\beta t}(A_1 + A_2 t). \tag{3.39}$$

If in addition to the elastic and dissipative forces the point particle is under the action of an external periodic force $F(t) = F(t+T)$, the equation of motion becomes

$$\ddot{x} + 2\beta \dot{x} + \omega^2 x = \frac{F(t)}{m}. \tag{3.40}$$

Suppose that $F(t) = F_0 \cos(\Omega t + \gamma)$, where $\Omega = 2\pi/T$. The general solution of the non-homogeneous linear equation (3.41) is given by the sum of the general

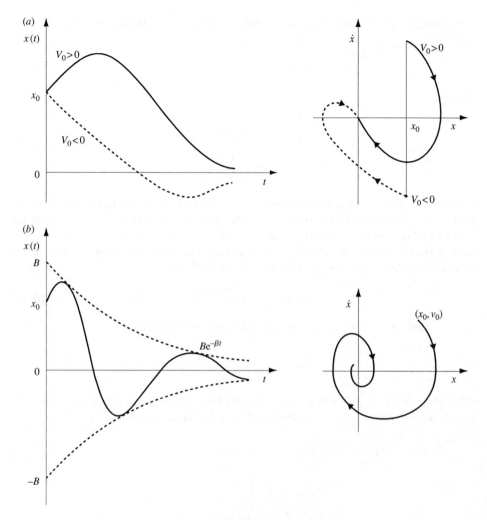

**Fig. 3.5**

solution of equation (3.35) and of one particular solution of equation (3.41). To determine the latter, observe that (3.41) is the real part of

$$\ddot{z} + 2\beta\dot{z} + \omega^2 z = \frac{F_0}{m} e^{i(\Omega t + \gamma)}. \tag{3.41}$$

This equation admits the particular solution $z_p(t) = b e^{i\Omega t}$, where $b \in \mathbf{C}$ can be determined by requiring that $z_p(t)$ solves equation (3.42):

$$b = \frac{F_0}{m} e^{i\gamma} (\omega^2 - \Omega^2 + 2i\beta\Omega)^{-1}. \tag{3.42}$$

Setting $b = Be^{iC}$, with $B$ and $C$ real, we find the particular solution $x_p(t) = z_p(t)$:

$$x_p(t) = B\cos(\Omega t + C), \tag{3.43}$$

where

$$B = \frac{F_0}{m}\frac{1}{\sqrt{(\omega^2 - \Omega^2)^2 + 4\beta^2\Omega^2}},$$
$$C = \gamma + \arctan\frac{2\beta\Omega}{\Omega^2 - \omega^2}. \tag{3.44}$$

We showed that the general solution of equation (3.35) is damped; hence, if the time $t$ is sufficiently large, relative to the damping constant $1/\beta$, the solution $x(t)$ of (3.41) is approximately equal to $x_p(t)$. This is a periodic function of time, with period equal to the period of the forcing term, and amplitude $B$. The latter depends on the frequency $\Omega$ (Fig. 3.6). When $\omega^2 > 2\beta^2$, for

$$\Omega = \omega_R = \sqrt{\omega^2 - 2\beta^2}, \tag{3.45}$$

the so-called *resonance frequency*, $B$, takes the maximum value

$$B_{\max} = \frac{F}{m}\frac{1}{2\beta\sqrt{\omega^2 - \beta^2}};$$

otherwise, $B(\Omega)$ is decreasing. Note that in the case of weak dissipation, namely if $\beta \ll \omega$, we obtain $\omega_R = \omega + \mathcal{O}((\beta/\omega)^2)$, and $B(\omega_R) \to +\infty$ for $\beta \to 0$.

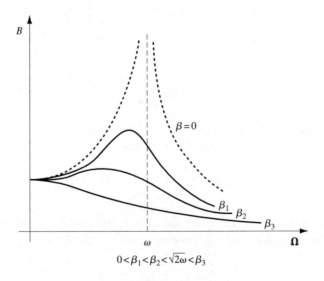

Fig. 3.6

In the general case, the function $F(t)$ can be expanded in Fourier series (see Appendix 7):

$$F(t) = \sum_{n=0}^{\infty} \widehat{F}_n \cos(n\Omega t + \gamma_n). \tag{3.46}$$

From the linearity of equation (3.41) it follows that the corresponding particular solution $x_p(t)$ is also given by the Fourier series

$$x_p(t) = \sum_{n=0}^{\infty} B_n \cos(n\Omega t + C_n), \tag{3.47}$$

where $B_n$ and $C_n$ can be found by replacing $\gamma$ by $\gamma_n$ and $\Omega$ by $\Omega_n$ in equation (3.45).

## 3.6 Beats

A particular phenomenon known as *beats* is due to the superposition of harmonic oscillations of frequencies that while different, are close together. More precisely, if $w_1, w_2$ are such frequencies ($w_1 > w_2$), then $(w_1 - w_2)/w_1 \ll 1$. This can happen in various circumstances, but the most important cases occur in acoustics: it can easily be heard by playing the same note on two different instruments, not perfectly tuned to the same pitch. Mathematically, it reduces to the study of sums of the following kind:

$$x(t) = A_1 \cos(w_1 t + \alpha_1) + A_2 \cos(w_2 t + \alpha_2), \tag{3.48}$$

where we can assume $A_1 = A_2 = A$, and isolating the excessive amplitude in one of the two vibrations, which henceforth does not contribute to the occurrence of this particular phenomenon. Under this assumption, equation (3.49) is equivalent to

$$x(t) = 2A \cos(\overline{w} t + \overline{\alpha}) \cos(\varepsilon t + \beta), \tag{3.49}$$

where

$$\overline{w} = \frac{w_1 + w_2}{2}, \quad \varepsilon = \frac{w_1 - w_2}{2}, \quad \overline{\alpha} = \frac{\alpha_1 + \alpha_2}{2}, \quad \beta = \frac{\alpha_1 - \alpha_2}{2}.$$

The term $\cos(\overline{w} t + \overline{\alpha})$ produces an oscillation with a frequency very close to the frequencies of the single component motions. The amplitude of this oscillation is *modulated* in a periodic motion by the factor $\cos(\varepsilon t + \beta)$, whose frequency is much smaller than the previous one. To be able to physically perceive the phenomenon of beats, the base and modulating frequency must be very different. In this case, in a time interval $\tau$ much larger than the period $2\pi/\varepsilon$ there can be found many oscillations of pulse $\overline{w}$ and nearly-constant amplitude; one has the impression of a sound of frequency $\overline{w}$ with amplitude slowly varying in time.

## 3.7 Problems

**0.** Draw the graph of the function $T(e)$ showing the period of the pendulum when $e = E/mgl$ varies in $[-1, +\infty]$.

**1.** Prove that $x = 0$ is a point of stable equilibrium for the potential (3.27).

*Solution*

Write the energy (in dimensionless coordinates) in the form

$$E = \frac{1}{2}y^2 + V(x).$$

Clearly $V(x) \leq E$. We now distinguish between the cases where $E > 0$ or $E \leq 0$ (Fig. 3.7).

If $E > 0$, we can define $x_E > 0$ such that, if $x(0)$ is initially in the interval $(-x_E, x_E)$, then $x(t)$ must remain in the same interval. Since in the same interval we must have $V(x) > -E$, it follows that $|y| < 2\sqrt{E}$. Hence the trajectory is confined inside a rectangle; this rectangle is interior to any neighbourhood of the origin if $E$ is chosen sufficiently small.

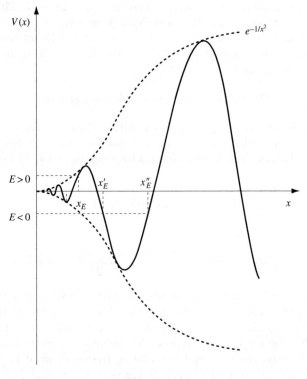

Fig. 3.7

If $E \leq 0$, we can choose $x(0)$ between two consecutive roots $x'_E, x''_E$ of the equation $V(x) = E$. Once again $x(0) \in (x'_E, x''_E) \Rightarrow x(t) \in (x'_E, x''_E)$, and the interval $(x'_E, x''_E)$ can be chosen as near to the origin as desired by appropriately selecting $|E|$. In this interval we have $V(x) > -e^{-1/x_M^2}$, where $x_M = \max(|x'_E|, |x''_E|)$. It follows that $\frac{1}{2}y^2 < E + e^{-1/x_M^2}$ and therefore the trajectory can be confined in an arbitrarily small neighbourhood.

**2.** A point in a horizontal plane is constrained without friction to the curve $y = a \cos(2\pi(x/\lambda))$, with $a$, $\lambda$ positive constants. The point is attracted to the origin by an elastic force. Discuss the dependence of the equilibrium of the point on the parameters $a, \lambda$.

**3.** Study the motion of a point particle subject to gravity on a cylindrical smooth helix in the following cases:
   (a) the helix has a vertical axis;
   (b) the helix has a horizontal axis. Find also the constraint reaction.

**4.** For a point constrained on a helix as in the previous problem, substitute the gravity force with an elastic force with centre on the axis of the helix. Study the equilibrium of the system.

**5.** Study the motion of a point particle constrained on a conic section, subject to an attractive elastic force with centre in a focus of the curve.

**6.** A cylinder of radius $R$ and height $h$ contains a gas subject to the following law: pressure × volume = constant. An airtight disc slides without friction inside the cylinder. The system is in equilibrium when the disc is in the middle position. Study the motion of the disc if its initial position at time $t = 0$ is not the equilibrium position, and the initial velocity is zero.

**7.** A point particle is constrained to move along a line with friction proportional to $v^p$ ( $v$ = absolute value of the velocity, $p > 0$ real). Suppose that no other force acts on the point. Find for which values of $p$ the point comes to a stop in finite time, and what is the stopping time as a function of the initial velocity.

**8.** A point particle of unit mass is constrained to move along the $x$-axis, under the action of a conservative force field, with potential

$$V(x) = \frac{x^2}{2} + \frac{x^3}{3}.$$

Determine the equilibrium positions and discuss their stability. Find the equation of the separatrix in the phase plane and draw the phase curves corresponding to the energy values $0$, $\frac{1}{8}$, $\frac{1}{6}$, $\frac{1}{5}$, $\frac{1}{2}$. Compute to first order the variation of the frequency of the motion as a function of amplitude for orbits close to the position of stable equilibrium.

**9.** A point particle of mass $m$ is moving along the $x$-axis under the action of a conservative force field with potential

$$V(x) = V_0 \left(\frac{x}{d}\right)^{2n},$$

where $V_0$ and $d$ are positive real constants, and $n$ is an integer, $n \geq 1$. Prove that the period of the motion, corresponding to a fixed value $E > 0$ of the system energy, is

$$T = 2d\sqrt{\frac{2m}{E}} \left(\frac{E}{V_0}\right)^{1/2n} \int_0^1 \frac{dy}{\sqrt{1 - y^{2n}}}.$$

**10.** A point particle of mass $m$ is moving along the $x$-axis under the action of a conservative force field with potential $V(x) = V_0 \tan^2(x/d)$, where $V_0$ and $d$ are positive real constants. Prove that the period $T$ of the motions corresponding to a fixed value $E > 0$ of the system energy is

$$T = \frac{2\pi m d}{\sqrt{2m(E + V_0)}}.$$

**11.** A point particle of unit mass is moving along the $x$-axis under the action of a conservative force with potential

$$V(x) = \begin{cases} (x+1)^2, & \text{if } x \leq -1, \\ 0, & \text{if } -1 < x < 1, \\ (x-1)^2, & \text{if } x \geq 1. \end{cases}$$

Draw the phase curves corresponding to the values $E = 0, \frac{1}{2}, 1$. Prove that the period $T$ of the motion corresponding to a fixed value $E > 0$ of the system energy is

$$T = 2\pi \left(\frac{1}{\sqrt{2}} + \frac{1}{\pi}\sqrt{\frac{2}{E}}\right).$$

**12.** A point particle of unit mass is moving along the $x$-axis under the action of a conservative force field with potential $V(x)$ periodic of period $2\pi$ in $x$ and such that

$$V(x) = \frac{V_0}{\pi}|x| \quad \text{if } x \in [-\pi, \pi],$$

where $V_0$ is a fixed constant. Draw the phase curves of the system corresponding to the values $E = V_0/2, V_0, 2V_0$ of the energy. (Be careful! The potential energy is a function which is continuous but not of class $\mathcal{C}^1$, therefore ...) Compute the period of the motion as a function of the energy corresponding to oscillatory motions.

**13.** A point particle of mass $m$ is moving along the $x$-axis under the action of a conservative force field with potential

$$V(x) = -\frac{V_0}{\cosh^2(x/d)},$$

where $V_0$ and $d$ are positive real constants. Determine the equilibrium position, discuss its stability and linearise the equation of motion around it. Draw the phase curves of the system corresponding to the energy values $E = -V_0, -V_0/2, 0, V_0$. Compute the explicit value of $T$ for $E = -V_0/2$.

**14.** A point particle of mass $m$ is moving along the $x$-axis under the action of a conservative force field with potential $V(x) = V_0(e^{-2x/d} - 2e^{-x/d})$, where $V_0$ and $d$ are two real positive constants. Prove that the motion is bounded only for the values of $E$ in the interval $[-V_0, 0)$, and that in this case the period $T$ is given by

$$T = 2\pi d\sqrt{\frac{m}{-2E}}.$$

**15.** A point particle of unit mass is moving along the $x$-axis under the action of a conservative force field with potential

$$V(x) = \begin{cases} (x+1)^2, & \text{for } x < -\frac{1}{2}, \\ -x^2 + \frac{1}{2}, & \text{for } |x| \leq \frac{1}{2}, \\ (x-1)^2, & \text{for } x > \frac{1}{2}. \end{cases}$$

Write the equation of the separatrix, draw the phase curves corresponding to values of $E = 0, \frac{1}{5}, \frac{1}{4}, \frac{1}{2}, 1, \frac{3}{2}$ (hint: the potential energy is of class $\mathcal{C}^1$ but not $\mathcal{C}^2$, therefore ...) and compute the period $T$ of the motion as a function of the energy.

**16.** A point particle of unit mass is moving along the $x$-axis according to the following equation of motion:

$$\ddot{x} = -\Omega^2(t)x,$$

where

$$\Omega(t) = \begin{cases} \omega + \varepsilon, & \text{if } 0 < t < \pi, \\ \omega - \varepsilon, & \text{if } \pi \leq t < 2\pi, \end{cases}$$

Here $\omega > 0$ is a fixed constant, $0 < \varepsilon \ll \omega$ and $\Omega(t) = \Omega(t + 2\pi)$ for every $t$. Prove that, if $(x_n, \dot{x}_n)$ denotes the position and the velocity of the particle at time $t = 2\pi n$, then

$$\begin{pmatrix} x_n \\ \dot{x}_n \end{pmatrix} = A^n \begin{pmatrix} x_0 \\ \dot{x}_0 \end{pmatrix},$$

where $A = A_+ A_-$ is a $2 \times 2$ real matrix and

$$A_\pm = \begin{pmatrix} \cos(\omega \pm \varepsilon) & \dfrac{1}{\omega \pm \varepsilon} \sin(\omega \pm \varepsilon) \\ -(\omega \pm \varepsilon)\sin(\omega \pm \varepsilon) & \cos(\omega \pm \varepsilon) \end{pmatrix}.$$

Prove that if $\omega$ and $\varepsilon$ satisfy the inequality

$$|\omega - k| < \frac{\varepsilon^2}{k^2} + \mathcal{O}(\varepsilon^2),$$

or

$$\left|\omega - k - \frac{1}{2}\right| < \frac{\varepsilon}{\pi\left(k + \frac{1}{2}\right)} + \mathcal{O}(\varepsilon),$$

where $k$ is any integer, $k \geq 1$; it follows that $|\operatorname{Tr} A| > 2$. Deduce that the matrix $A$ has two real distinct eigenvalues $\lambda_1 = 1/\lambda_2$ and prove that in this case the equilibrium position $x_0 = \dot{x}_0 = 0$ is unstable. This instability phenomenon, due to a periodic variation in the frequency of a harmonic motion synchronised with the period of the motion, is called *parametric resonance*. See the books by Arnol'd (1978a, §25) and Landau and Lifschitz (1976, §27) for a more detailed discussion and applications (such as the swing).

## 3.8 Additional remarks and bibliographical notes

Whittaker's book (1936, Chapter IV) contains a good discussion of the simple pendulum. More specifically, one can find there the derivation of the double periodicity of the elliptic functions, using the following general result (§34):

in a mechanical system subject only to fixed holonomic constraints and positional forces, the solutions of the equations of motion are still real if the time $t$ is replaced by $\sqrt{-1}t$ and the initial velocities $(v_1, \ldots, v_n)$ are replaced by $(-\sqrt{-1}v_1, \ldots, -\sqrt{-1}v_n)$. The expressions obtained represent the motion of the same system, with the same initial conditions, but with forces acting in the opposite orientation.

Struik (1988, Chapter 1) gives a very detailed description of curves, which one can use as a starting-point for a deeper understanding of the topics considered in this chapter.

## 3.9 Additional solved problems

*Problem 1*
A point particle of unit mass is moving along a line under the action of a conservative force field with potential energy

$$V(x) = \frac{x^4}{\alpha + \lambda x^2},$$

where $\alpha, \lambda$ are two given real parameters, not both zero (all variables are dimensionless).

(a) Determine for which values of $\alpha$ and of $\lambda$ the origin $x = 0$ is a position of stable equilibrium. Linearise the equations around this point and determine the frequency of small oscillations.
(b) Consider the motion, with initial conditions $x(0) = 0, \dot{x}(0) = 1$. For which values of $\alpha$ and $\lambda$ is the motion periodic? For which values of $\alpha, \lambda$ does the particle go to infinity in finite time?
(c) Determine all the periodic motions and the integral of the period.
(d) Draw the phase portrait of the system in the case $\alpha > 0, \lambda < 0$.

*Solution*
(a) It is immediate to verify that $V'(0) = 0$, for every choice of $\alpha$ and $\lambda$. In addition, if $\alpha = 0$ ($\lambda = 0$, respectively) the origin is stable if and only if $\lambda > 0$ ($\alpha > 0$, respectively); in this case, it is also an absolute minimum of the potential energy. If $\alpha\lambda \neq 0$ we need to distinguish the case $\alpha\lambda > 0$, when the potential energy $V$ is defined on the entire line **R**, and the case $\alpha\lambda < 0$ when the two lines $x = \pm\sqrt{-\alpha/\lambda}$ are vertical asymptotes of $V$. In both cases, $x = 0$ is stable if and only if $\alpha > 0$. Notice, however, that $V(x) = x^4/\alpha + \mathcal{O}(x^6)$ if $\alpha \neq 0$ and $V(x) = x^2/\lambda$ if $\alpha = 0$. This implies that, if $\alpha \neq 0$, the linearised equation is simply $\ddot{x} = 0$ (the system is not linearly stable), while if $\alpha = 0$ one has $\ddot{x} + (2/\lambda)x = 0$ and the frequency of small oscillations is $\sqrt{2/\lambda}$.
(b) To ensure that the motion corresponding to the initial condition $x(0) = 0, \dot{x}(0) = 1$, is periodic it is necessary for it to take place between two inversion points $x_+ = -x_- > 0$, that are solutions of $V(x_\pm) = E = \frac{1}{2}$. This is possible in the following cases: $\alpha > 0$ for any $\lambda$; $\alpha = 0$ and $\lambda > 0$; $\alpha < 0$, $\lambda > 0$ and $|\alpha| < \frac{1}{8}\lambda^2$.

To ensure that the motion corresponding to the initial condition $x(0) = 0, \dot{x}(0) = 1$ reaches infinity in finite time, $T_\infty$, we must have $\lambda = 0, \alpha < 0$; indeed, in this case we obtain the integral

$$T_\infty = \int_0^{+\infty} \frac{\mathrm{d}x}{\sqrt{1 - 2x^4/\alpha}} < +\infty.$$

(c) For arbitrary initial conditions, the motions are periodic of period $T$ in the following cases.

- For $\alpha = 0, \lambda > 0$
  In this case the period does not depend on the initial conditions $(x(0), \dot{x}(0))$ and is given by $T = 2\pi\sqrt{\lambda/2}$.

  In all other cases, setting $E = \frac{1}{2}\dot{x}(0)^2 + V(x(0))$, we have the following situations.

- For $\lambda = 0, \alpha > 0$
  All motions are periodic, $E \geq 0$, $x_+ = (\alpha E)^{1/4} = -x_-$,
  $$T = 4 \int_0^{(\alpha E)^{1/4}} \frac{dx}{\sqrt{2[E - \frac{x^4}{\alpha}]}}.$$

- For $\lambda > 0, \alpha > 0$
  All motions are periodic, $E \geq 0$,
  $$x_+ = \left[\frac{\lambda E}{2}\left(1 + \sqrt{1 + \frac{4\alpha}{\lambda^2 E}}\right)\right]^{1/2} = -x_-, \quad T = 4\int_0^{x_+} \frac{dx}{\sqrt{2[E - \frac{x^4}{\alpha + \lambda x^2}]}}.$$

- For $\lambda > 0, \alpha < 0$
  Only motions corresponding to $E \geq -4\alpha/\lambda^2 > 0$ are periodic, and take place in the interval $(-x_-, -x_+)$, or $(x_-, x_+)$, where $x_-$ and $x_+$ are the positive roots of $V(x_\pm) = E$:
  $$x_+ = \left[\frac{\lambda E}{2}\left(1 + \sqrt{1 + \frac{4\alpha}{\lambda^2 E}}\right)\right]^{1/2}, \quad x_- = \left[\frac{\lambda E}{2}\left(1 - \sqrt{1 + \frac{4\alpha}{\lambda^2 E}}\right)\right]^{1/2},$$
  the period is given by
  $$T = 2\int_{x_-}^{x_+} \frac{dx}{\sqrt{2[E - \frac{x^4}{\alpha + \lambda x^2}]}}.$$

- Finally, for $\lambda < 0, \alpha > 0$
  Only motions corresponding to $E > 0$ and the initial condition $x(0) \in \left(-\sqrt{-\alpha/\lambda}, \sqrt{-\alpha/\lambda}\right)$ are periodic. The inversion points are
  $$x_+ = \left[-\frac{\lambda E}{2}\left(\sqrt{1 + \frac{4\alpha}{\lambda^2 E}} - 1\right)\right]^{1/2} = -x_-$$
  and the period is given by
  $$T = 4\int_0^{x_+} \frac{dx}{\sqrt{2[E - \frac{x^4}{\alpha + \lambda x^2}]}}.$$

(d) The phase portrait in the case $\alpha > 0, \lambda < 0$ is shown in Fig. 3.8.

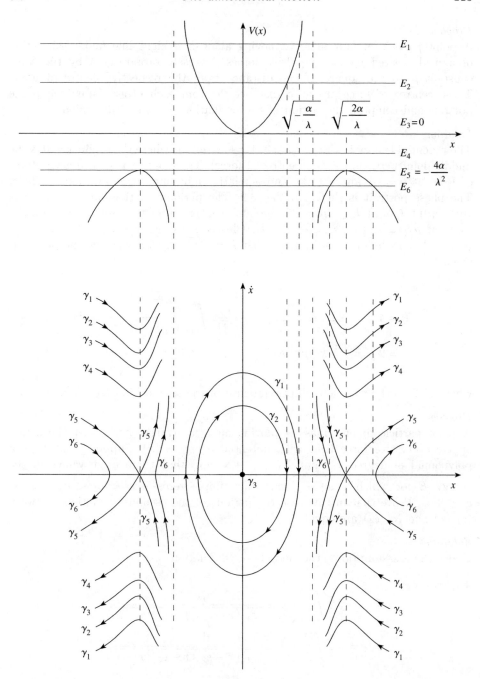

Fig. 3.8

*Problem 2*
A point particle of unit mass is moving along a straight line under the action of a field created by two repulsive forces, inversely proportional by the same constant $\mu$ to the square of the distance from the respective centre of force. These centres of force are at a distance $2c$ from each other. Draw the phase portrait and compute the period of the oscillations of bounded motions.

*Solution*
The potential energy is $V(x) = \mu(|x+c|^{-1}+|x-c|^{-1})$ (where the origin $x = 0$ is the middle point between the two centres of force). There are two vertical asymptotes of $V(x)$ at $x = -c$ and $x = +c$ and a relative minimum at $x = 0$: $V(0) = 2\mu/c$. The phase portrait is shown in Fig. 3.9. The periodic motions are the motions with $x(0) \in (-c, c)$, $E \geq 2\mu/c$. The points of inversion of the motion, $x_\pm$, are the roots of $\mu/|x_\pm + c| + \mu/|x_\pm - c| = E$. Clearly $-c < x_- = -x_+ \leq 0 \leq x_+ < c$. Setting $k = x_+/c$, one readily finds that $k = \sqrt{1 - (2\mu/cE)}$. Hence the period is given by

$$T = 4\int_0^{kc} \frac{dx}{\sqrt{2[E - V(x)]}} = \frac{4}{k\sqrt{2E}} \int_0^{kc} \sqrt{\frac{1 - x^2 c^{-2}}{1 - x^2 k^{-2} c^{-2}}}\, dx$$
$$= 2\sqrt{c^3(1 - k^2)\mu^{-1}} E(k),$$

where $E(k)$ is the complete elliptic integral of the second kind (see Appendix 2).

*Problem 3*
A point particle of unit mass is moving along a straight line under the action of a conservative force field with potential energy $V(x)$. Suppose that $V$ is a polynomial of degree 4, $\lim_{x \to \pm\infty} V(x) = +\infty$ and that there exist values of the energy $E$ for which $V(x) - E$ has four simple zeros $-\infty < e_1 < e_2 < e_3 < e_4 < +\infty$. Prove that in this case the periods of the oscillatory motions between $(e_1, e_2)$ and $(e_3, e_4)$ are equal.

*Solution*
Under these assumptions, the periods of the motions are

$$T_{12} = \sqrt{2} \int_{e_1}^{e_2} \frac{dx}{\sqrt{(x - e_1)(x - e_2)(x - e_3)(x - e_4)}},$$

$$T_{34} = \sqrt{2} \int_{e_3}^{e_4} \frac{dx}{\sqrt{(x - e_1)(x - e_2)(x - e_3)(x - e_4)}},$$

respectively. Since the four points $(e_1, e_2, e_3, e_4)$ and $(e_3, e_4, e_1, e_2)$ have the same *cross-ratio* (see Sernesi 1989, p. 325) $((e_1 - e_2)/(e_2 - e_3)) \cdot ((e_3 - e_4)/(e_4 - e_1))$, there exists a rational transformation $\xi = g \cdot x = (Ax + B)/(Cx + D)$,

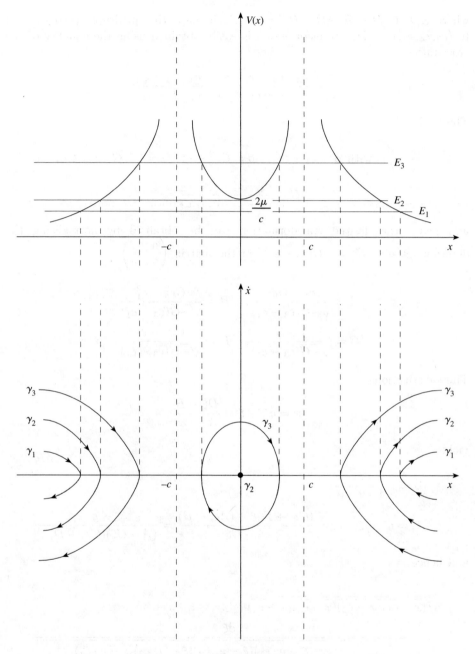

Fig. 3.9

where $A, B, C, D \in \mathbf{R}, AD - BC = 1$, which maps the quadruple $(e_1, e_2, e_3, e_4)$ to $(e_3, e_4, e_1, e_2)$. This transformation is easily obtained using the equality of the cross-ratios:

$$\frac{e_1 - e_2}{e_2 - e_3} \cdot \frac{e_3 - x}{x - e_1} = \frac{e_3 - e_4}{e_4 - e_1} \cdot \frac{e_1 - \xi}{\xi - e_3}.$$

This yields

$$\xi = \frac{\alpha x + \beta}{\gamma x + \delta} \text{ with } \alpha = e_1 - Ge_3,\ \beta = Ge_3^2 - e_1^2,\ \gamma = 1 - G,\ \delta = Ge_3 - e_1,$$

$$G = \frac{e_4 - e_1}{e_2 - e_3} \cdot \frac{e_1 - e_2}{e_3 - e_4},$$

and hence the desired transformation can be obtained by normalising the determinant $\alpha\delta - \beta\gamma = -G(e_3 - e_1)^2$ of the matrix $\begin{pmatrix} \alpha & \beta \\ \gamma & \delta \end{pmatrix}$:

$$A = \frac{e_1 - Ge_3}{\sqrt{-G}(e_3 - e_1)}, \quad B = \frac{Ge_3^2 - e_1^2}{\sqrt{-G}(e_3 - e_1)},$$

$$C = \frac{1 - G}{\sqrt{-G}(e_3 - e_1)}, \quad D = \frac{Ge_3 - e_1}{\sqrt{-G}(e_3 - e_1)}.$$

The substitution

$$x = g^{-1} \cdot \xi = \frac{D\xi - B}{A - C\xi}$$

yields

$$dx = \frac{d\xi}{(A - C\xi)^2},$$

$$x - e_i = g^{-1} \cdot \xi - e_i = \frac{(Ce_i + D)\xi - (Ae_i + B)}{A - C\xi} = \frac{\xi - g \cdot e_i}{(A - C\xi)(Ce_i + D)^{-1}},$$

and hence

$$\frac{dx}{\sqrt{(x - e_1)(x - e_2)(x - e_3)(x - e_4)}}$$

$$= \frac{d\xi}{(A - C\xi)^2 \sqrt{\dfrac{(\xi - g \cdot e_1)(\xi - g \cdot e_2)(\xi - g \cdot e_3)(\xi - g \cdot e_4)}{(A - C\xi)^4 [(Ce_1 + D)(Ce_2 + D)(Ce_3 + D)(Ce_4 + D)]^{-1}}}}$$

$$= \frac{\sqrt{(Ce_1 + D)(Ce_2 + D)(Ce_3 + D)(Ce_4 + D)}}{\sqrt{(\xi - e_3)(\xi - e_4)(\xi - e_1)(\xi - e_2)}} d\xi,$$

where we have used the fact that $ge_i = e_{i+2}$ for $i = 1, 2$ and $ge_i = e_{i-2}$ for $i = 3, 4$. Computing the product $(Ce_1 + D) \cdots (Ce_4 + D)$, we find:

$$(Ce_1 + D) \cdots (Ce_4 + D) = \frac{(1-G)e_1 + Ge_3 - e_1}{\sqrt{-G}(e_3 - e_1)} \cdots \frac{(1-G)e_4 + Ge_3 - e_1}{\sqrt{-G}(e_3 - e_1)}$$

$$= \frac{1}{G^2(e_3 - e_1)^4} G(e_3 - e_1)[e_2 - e_1 + G(e_3 - e_2)]$$

$$\times (e_3 - e_1)[e_4 - e_1 + G(e_3 - e_4)].$$

Finally, since

$$e_2 - e_1 + G(e_3 - e_2) = e_2 - e_1 + \frac{e_4 - e_1}{e_3 - e_4}(e_2 - e_1) = (e_2 - e_1)\frac{e_3 - e_1}{e_3 - e_4},$$

$$e_4 - e_1 + G(e_3 - e_4) = e_4 - e_1 + \frac{e_1 - e_2}{e_2 - e_3}(e_4 - e_1) = (e_4 - e_1)\frac{e_3 - e_1}{e_3 - e_2},$$

we arrive at

$$(Ce_1 + D)(Ce_2 + D)(Ce_3 + D)(Ce_4 + D) = \frac{1}{G(e_3 - e_1)^2} \frac{e_2 - e_1}{e_3 - e_4} \frac{e_4 - e_1}{e_3 - e_2}(e_3 - e_1)^2$$

$$= 1$$

and the substitution

$$x = \frac{A\xi + B}{C\xi + D}$$

transforms the integral

$$\int_{e_1}^{e_2} \frac{\mathrm{d}x}{\sqrt{(x - e_1)(x - e_2)(x - e_3)(x - e_4)}}$$

into

$$\int_{e_3}^{e_4} \frac{\mathrm{d}\xi}{\sqrt{(\xi - e_3)(\xi - e_4)(\xi - e_1)(\xi - e_2)}},$$

yielding $T_{12} = T_{34}$. It is possible to prove in an analogous way that if $V$ is a polynomial of degree 3 and $V(x) - E$ has three simple roots $-\infty < e_1 < e_2 < e_3 < +\infty$, the period of oscillation in the interval $[e_1, e_2]$ is equal to twice the (finite) time needed for the point with energy $E$ to travel the distance $[e_3, +\infty)$:

$$T_3 = \int_{e_1}^{e_2} \frac{\mathrm{d}x}{\sqrt{(x - e_1)(x - e_2)(x - e_3)}} = \int_{e_3}^{+\infty} \frac{\mathrm{d}x}{\sqrt{(x - e_1)(x - e_2)(x - e_3)}}.$$

The basic idea is to construct the rational transformation of the projective line mapping the quadruple $(e_1, e_2, e_3, \infty)$ to $(e_3, \infty, e_1, e_2)$:

$$\frac{e_1 - e_2}{e_2 - e_3} \cdot \frac{e_3 - x}{x - e_1} = -\frac{e_1 - \xi}{\xi - e_3}.$$

These relations have a more general interpretation in the theory of elliptic curves and their periods. This is the natural geometric formulation highlighting the properties of elliptic integrals (see e.g. McKean and Moll 1999). Indeed, both $T_{12}$ and $T_3$ can be reduced to a complete elliptic integral of the first kind by the transformation that maps the quadruples $(e_1, e_2, e_3, e_4)$ and $(e_1, e_2, e_3, \infty)$ to $(1, -1, 1/k, -1/k)$, with $k$ determined by the equality of the cross-ratios.

*Problem 4*

Consider the motion with potential energy $V \in \mathcal{C}^\infty$ such that $V(0) = 0, V'(0) = 0, V''(0) > 0$. The motion around $x = 0$ is periodic, with period $T(E)$ given by (3.8) with $s_1, s_2$ roots of $V(x) = E$, for $E \in (0, E_0)$, for an appropriate $E_0$.

(i) What are the conditions on $V$ that ensure that $T(E)$ is constant?
(ii) Study in the general case the behaviour of $T(E)$ for $E \to 0$.
(iii) Using the result of (i) consider the problem of finding a (smooth) curve $z = f(x), f \in \mathcal{C}^\infty, f(0) = 0, f'(0) = 0, f''(0) > 0, f(-x) = f(x)$, such that the motion on it due to gravity is isochronous.

*Solution*

(i) To answer this we follow Gallavotti (1980, §2.10).
Start from the case that $V(-x) = V(x)$, when equation (3.8) becomes

$$T(E) = 4\sqrt{\frac{m}{2}} \int_0^{x(E)} \frac{dx}{\sqrt{E - V(x)}}. \tag{3.50}$$

Introduce the inverse function of $V$ in the interval $(0, x(E))$: $x = \xi(V)$ and use it as a change of variable in (3.51), observing that $(0, x(E)) \to (0, E)$:

$$T(E) = 4\sqrt{\frac{m}{2}} \int_0^E \frac{\xi'(V)}{\sqrt{E - V}} dV. \tag{3.51}$$

It is well known (and easily verified) that *Abel's integral equation*

$$\phi(t) = \int_0^t \frac{\psi(r) dr}{\sqrt{t - r}} \tag{3.52}$$

($\phi$ known, $\phi(0) = 0$, $\psi$ unknown) has the unique solution

$$\psi(t) = \frac{1}{\pi} \int_0^t \frac{\phi'(r)}{\sqrt{t - r}} dr. \tag{3.53}$$

Comparing this with equation (3.52) we conclude

$$\frac{1}{\pi}\phi' = 4\sqrt{\frac{m}{2}}\xi'$$

and $T(E)$ satisfies the equation

$$4\pi\sqrt{\frac{m}{2}}\xi(V) = \int_0^V \frac{T(z)}{\sqrt{V-z}}dz. \qquad (3.54)$$

Hence, if we want $T(z) = T_0$, we find that

$$2\pi\sqrt{\frac{m}{2}}\xi(V) = \sqrt{V}T_0, \qquad (3.55)$$

or

$$V = \frac{m}{2}\omega^2 x^2, \quad \omega = \frac{2\pi}{T_0}. \qquad (3.56)$$

It follows that *the only symmetric potential that generates isochronous motions is the elastic potential.*

Considering a generic potential, we need to introduce the right inverse function $\xi_+(V) > 0$ and the left inverse function $\xi_-(V) < 0$. In equation (3.55), $2\xi(V)$ is replaced by $\xi_+(V) - \xi_-(V)$; a similar modification appears in (3.56), characterising those perturbations of the elastic potential that preserve isochronicity.

(ii) Consider the fourth-order expansion of $V$:

$$V(x) \simeq \frac{m\omega^2 x^2}{2}(1 + c_1 x + c_2 x^2), \qquad (3.57)$$

assuming that $c \neq 0$. The expansion of $\xi_\pm(V)$ to order $V^2$ is

$$\xi_\pm(V) \simeq \pm\sqrt{\frac{2V}{m\omega}} + k_\pm V + k'_\pm V^{3/2} + k''_\pm V^2, \qquad (3.58)$$

where the coefficients need to be determined.

Substituting (3.59) into (3.58) and imposing an identity to order $V^{3/2}$, we find

$$k_\pm = -\frac{c_1}{m\omega^2}, \quad k'_\pm = \pm\frac{1}{\sqrt{2m^3\omega^3}}\left(\frac{5}{2}c_1^2 - 2c_2\right), \quad k''_\pm = \frac{6c_1 c_2 - 4c_1^3}{m^2\omega^4}. \qquad (3.59)$$

In the light of the result of (i) (see (3.55) with $2\xi$ replaced by $\xi_+ - \xi_-$) the linear term in $V$ in the expansion of $\xi$ does not contribute to the difference $\xi_+ - \xi_-$. Hence the first correction of $\xi_+ - \xi_-$ is of order $V^{3/2}$:

$$\xi_+ - \xi_- = 2\sqrt{\frac{2V}{m\omega^2}} + \sqrt{\frac{2}{m^3\omega^6}}\left(\frac{5}{2}c_1^2 - 2c_2\right)V^{3/2}. \qquad (3.60)$$

Replacing in equation (3.55) $\xi(V)$ by $\frac{1}{2}(\xi_+ - \xi_-)/2$, and writing $T(z) = T_0 + T_1(z)$ with $T_0 = 2\pi/\omega$ yields for $T_1(z)$ Abel's equation

$$2\pi \left(\frac{5}{2}c_1^2 - 2c_2\right) \frac{V^{3/2}}{m\omega^3} = \int_0^V \frac{T_1(z)}{\sqrt{V-z}} dz, \tag{3.61}$$

with solution

$$T_1(E) = 3\left(\frac{5}{2}c_1^2 - 2c_2\right) \frac{1}{m\omega^3} \int_0^E \sqrt{\frac{V}{E-V}} dV = \frac{3}{2}\pi \frac{E}{m\omega^3} \left(\frac{5}{2}c_1^2 - 2c_2\right). \tag{3.62}$$

We conclude that $T(E)$ can be differentiated for $E = 0$, with

$$T'(0) = \frac{3\pi}{2m\omega^3} \left(\frac{5}{2}c_1^2 - 2c_2\right). \tag{3.63}$$

(iii) We already know one solution: the cycloid (Example 3.1). We want here to start the problem from the equation

$$\frac{1}{2}\dot{s}^2 + gf(x(s)) = E, \tag{3.64}$$

where $s = \int_0^x \sqrt{1 + f'^2(\xi)}\, d\xi$. Recalling (i), we impose on the function $f$ the condition

$$gf(x) = \frac{1}{2}\omega^2 s^2, \tag{3.65}$$

which produces a harmonic motion of period $2\pi/\omega$; it follows that $s = A\sin\omega t$. To construct the curve corresponding to (3.66) it is convenient to set $\omega t = \gamma/2$ and remark that

$$\left(\frac{dx}{d\gamma}\right)^2 = \left(\frac{ds}{d\gamma}\right)^2 - \left(\frac{dz}{d\gamma}\right)^2,$$

where

$$z(\gamma) = \frac{\omega^2}{2g} A^2 \sin^2 \frac{\gamma}{2}.$$

Also set $A = \alpha g/\omega^2$, which yields

$$\left(\frac{dx}{d\gamma}\right)^2 = \alpha^2 \left(\frac{g}{2\omega^2}\right)^2 \cos^2 \frac{\gamma}{2} \left(1 - \alpha^2 \sin^2 \frac{\gamma}{2}\right). \tag{3.66}$$

Let us consider first the case that $\alpha = 1$ (note that $0 < \alpha \leq 1$), when clearly

$$\frac{dx}{d\gamma} = \frac{g}{2\omega^2}\cos^2\frac{\gamma}{2} = R(1 + \cos\gamma), \quad R = \frac{g}{4\omega^2}; \qquad (3.67)$$

together with the condition $x(0) = 0$ this yields

$$x = R(\gamma + \sin\gamma). \qquad (3.68)$$

Since $A = g/\omega^2$ we find, for $z(\gamma)$,

$$z = R(1 - \cos\gamma), \qquad (3.69)$$

the cycloid. The choice $\alpha = 1$ corresponds to the value of the energy allowing the motion to reach the highest possible points ($z = 2R$) of the cycloid. For $\alpha < 1$ consider the cycloid

$$x = R(\psi + \sin\psi), \quad z = R(1 - \cos\psi) \qquad (3.70)$$

and let us verify if the arc defined by (3.67) and by $z(\gamma) = \alpha^2 R(1 - \cos\gamma)$ lies on it.
Write the relation between $\gamma$ and $\psi$, expressed by $z(\psi) = z(\gamma)$, namely

$$\alpha^2(1 - \cos\gamma) = 1 - \cos\psi$$

and compute

$$\frac{dx}{d\gamma} = \frac{dx}{d\psi}\frac{d\psi}{d\gamma}.$$

It is evident that

$$\frac{d\psi}{d\gamma} = \alpha^2 \frac{\sin\gamma}{\sin\psi}$$

and that (3.71) yields

$$\frac{dx}{d\gamma} = R(1 + \cos\psi)\alpha^2\frac{\sin\gamma}{\sin\psi}.$$

Expressing the right-hand side as a function of $\gamma$ we find that $(dx/d\gamma)^2$ coincides with (3.67).
Hence we have proved that *the cycloid is the only symmetric curve producing isochronous oscillations under the gravity field.*

# 4 THE DYNAMICS OF DISCRETE SYSTEMS. LAGRANGIAN FORMALISM

## 4.1 Cardinal equations

The mathematical modelling of the dynamics of a constrained system of point particles $(P_1, m_1), \ldots, (P_n, m_n)$ is based on the equations of motions for the single points:

$$m_i \mathbf{a}_i = \mathbf{R}_i, \quad i = 1, \ldots, n, \qquad (4.1)$$

where $\mathbf{R}_i$ denotes the sum of all forces acting on the point $(P_i, m_i)$. In addition to equations (4.1), one has to consider the constraint equations.

The forces acting on a single particle can be classified in two different ways: either distinguishing between *internal forces* and *external forces*, or using instead the distinction between *constraint reactions* and the so-called *active forces* which we used in the study of the dynamics of a single point particle. These two different classifications yield two different mathematical schemes describing the dynamics of systems.

In this section we consider the former possibility, distinguishing between internal forces, i.e. the forces due to the interaction of the points of the system among themselves, and external forces, due to the interaction between the points of the system and points outside the system.[1] We note two important facts:

(a) internal forces are in equilibrium;
(b) the (unknown) constraint reactions may appear among the external as well as among the internal forces.

As a consequence of (a) we obtain the cardinal equations of dynamics:

$$\dot{\mathbf{Q}} = \mathbf{R}^{(e)}, \qquad (4.2)$$

$$\dot{\mathbf{L}}(O) + \mathbf{v}(O) \times \mathbf{Q} = \mathbf{M}^{(e)}(O), \qquad (4.3)$$

using the standard notation for the linear momentum $\mathbf{Q} = \sum_{i=1}^{n} m_i \mathbf{v}_i$ and the angular momentum $\mathbf{L}(O) = \sum_{i=1}^{n} m_i (P_i - O) \times \mathbf{v}_i$ (the first can be derived by adding each side of equations (4.1), while the second is obtained by taking the vector product of the two sides of (4.1) with $P_i - O$, with $O$ an arbitrary point, and then adding). Here $\mathbf{R}^{(e)}$ and $\mathbf{M}^{(e)}$ are the resultant and the resultant moment, respectively, of the external force system.

---

[1] To non-inertial observers the so-called *apparent forces* will also seem external (see Section 6.6 or Chapter 6).

Define the centre of mass $P_0$ by

$$m(P_0 - O) = \sum_{i=1}^{n} m_i(P_i - O) \qquad (4.4)$$

($m = \sum_{i=1}^{n} m_i$, $O$ an arbitrary point). Since

$$\mathbf{Q} = m\dot{P}_0, \qquad (4.5)$$

equation (4.2) can be interpreted as the *equation of motion of the centre of mass* (i.e. of the point particle $(P_0, m)$):

$$m\ddot{P}_0 = \mathbf{R}^{(e)}.$$

Equation (4.3) can be reduced to the form

$$\dot{\mathbf{L}}(O) = \mathbf{M}^{(e)}(O) \qquad (4.6)$$

if $\mathbf{v}(O) \times \mathbf{Q} = 0$, and hence in particular when $O$ is fixed or coincides with the centre of mass.

The cardinal equations are valid for any system. On the other hand, in general they contain too many unknowns to yield the solution of the problem.

The most fruitful application of such equations is the dynamics of rigid bodies (see Chapters 6 and 7). This is because all reactions due to rigid constraints are internal and hence do not appear in the cardinal equations. In the relevant chapters we will discuss the use of such equations. Here, we only consider the energy balance of the system, which has the form

$$\frac{dT}{dt} = W, \qquad (4.7)$$

where

$$T = \frac{1}{2} \sum_{i=1}^{n} m_i v_i^2, \quad W = \sum_{i=1}^{n} \mathbf{R}_i \cdot \mathbf{v}_i.$$

Equation (4.7) can be deduced by differentiating $T$ with respect to time and from equations (4.1). In correspondence with the two proposed subdivisions of the forces $\mathbf{R}_i$ we can isolate the following contributions to the power $W$:

$$W = W^{(e)} + W^{(i)} \qquad (4.8)$$

($W^{(e)}$ is the power of the external forces, $W^{(i)}$ is the power of the internal forces), or else

$$W = W^{(a)} + W^{(r)} \qquad (4.9)$$

($W^{(a)}$ is the power of the active forces, $W^{(r)}$ is the power of the constraint reactions).

*Remark* 4.1
Equation (4.7) is not in general a consequence of the cardinal equations (4.2) and (4.3). Indeed, in view of equation (4.8) we can state that it is independent of the cardinal equations whenever $W^{(i)} \neq 0$. It follows that when the internal forces perform non-vanishing work, the cardinal equations cannot contain all the information on the dynamics of the system. ∎

Equation (4.9) suggests that we can expect a considerable simplification of the problem when the constraint reactions have vanishing resulting power. We will examine this case in detail in the next section.

## 4.2 Holonomic systems with smooth constraints

Holonomic systems have been introduced in Chapter 1 (Section 1.10). In analogy with the theory discussed in Chapter 2 (Section 2.4) on the dynamics of the constrained point, we say that a holonomic system has *smooth constraints* if the only contribution of the constraint reactions to the resulting power $W^{(r)}$ is due to the possible motion of the constraints.

Let
$$\Phi = \bigoplus_{i=1,\ldots,n} \phi_i \tag{4.10}$$
be the vector representing all constraint reactions. Then the power $W^{(r)}$ is expressed as
$$W^{(r)} = \Phi \cdot \dot{\mathbf{X}} \tag{4.11}$$
and in view of the decomposition (1.88) for the velocity $\dot{\mathbf{X}}$ of a representative point, we can write
$$W^{(r)} = \widehat{W}^{(r)} + W^{(r)*}, \tag{4.12}$$
with
$$\widehat{W}^{(r)} = \Phi \cdot \widehat{\mathbf{V}}, \quad W^{(r)*} = \Phi \cdot \mathbf{V}^*. \tag{4.13}$$

We call the quantity $\widehat{W}^{(r)}$ the *virtual power of the system of constraint reactions*. We can now give the precise definition of a holonomic system with smooth constraints.

DEFINITION 4.1  *A holonomic system has* smooth constraints *if the virtual power of the constraint reaction is zero at every time and for any kinematic state of the system.* ∎

Equivalently we can say that a holonomic system has smooth constraints if and only if $\Phi$ is orthogonal to the configuration space:
$$\Phi \in (T_{\mathbf{X}} \mathcal{V}(t))^{\perp} \tag{4.14}$$

at every time and for any kinematic state. The latter definition is analogous to that of the orthogonality property for a smooth constraint in the context of the dynamics of a point particle (Definition 2.3).

The characterisation of the vector $\Phi$ yields the possibility of a unique decomposition in terms of the basis of the orthogonal space:

$$\Phi = \sum_{j=1}^{3n-\ell} \lambda_j(t) \nabla_{\mathbf{X}} f_j(\mathbf{X}, t), \qquad (4.15)$$

where $\lambda_j(t)$ are (unknown) multipliers.

The property (4.14) yields a system of differential equations characterising the motion of any holonomic system with smooth constraints, from which all constraint reactions can be eliminated. To find this system, consider the vector representing the active forces

$$\mathbf{F}^{(a)} = \bigoplus_{i=1,\ldots,n} \mathbf{F}_i^a \qquad (4.16)$$

and the vector representing the momenta

$$\mathfrak{Q} = \bigoplus_{i=1,\ldots,n} m_i \mathbf{v}_i. \qquad (4.17)$$

Then from equations (4.1) it follows that

$$\dot{\mathfrak{Q}} = \mathbf{F}^{(a)} + \Phi. \qquad (4.18)$$

Imposing the property (4.14), the two vectors $\dot{\mathfrak{Q}}$ and $\mathbf{F}^{(a)}$ must have the same projection onto the tangent space $(T_{\mathbf{X}}\mathcal{V}(t))$.

Since the vectors $(\partial \mathbf{X}/\partial q_k)_{k=1,\ldots,\ell}$ are a basis for the tangent space $(T_{\mathbf{X}}\mathcal{V}(t))$ in a fixed system of Lagrangian coordinates $(q_1,\ldots,q_\ell)$, the projection onto $(T_{\mathbf{X}}\mathcal{V}(t))$ of a vector $\mathbf{Z} = (\mathbf{Z}_1,\ldots,\mathbf{Z}_n) \in \mathbf{R}^{3n}$ is uniquely determined by the components

$$Z_{\Theta,k} = \mathbf{Z} \cdot \frac{\partial \mathbf{X}}{\partial q_k} = \sum_{i=1}^n \mathbf{Z}_i \cdot \frac{\partial P_i}{\partial q_k}.$$

It follows that

$$F^{(a)}_{\Theta,k} = \sum_{i=1}^n \mathbf{F}^{(a)}_i \cdot \frac{\partial P_i}{\partial q_k}, \quad k=1,\ldots,\ell, \qquad (4.19)$$

and the equation of motion can be written as

$$(\dot{\mathfrak{Q}})_{\Theta,k} = F^{(a)}_{\Theta,k}, \quad k=1,\ldots,\ell. \qquad (4.20)$$

## 4.3 Lagrange's equations

The kinematic term (i.e. the left-hand side) in (4.20) has an interesting connection with the kinetic energy $T$. To show this connection, we first deduce the expression for $T$ through the Lagrangian coordinates of the phase space.

## 4.3     The dynamics of discrete systems. Lagrangian formalism

By the definition of $T$ and equations (1.95) we easily find

$$T = \frac{1}{2} \sum_{h,k=1}^{\ell} a_{hk} \dot{q}_h \dot{q}_k + \sum_{k=1}^{\ell} b_k \dot{q}_k + c, \qquad (4.21)$$

where

$$a_{hk}(\mathbf{q},t) = \sum_{i=1}^{n} m_i \frac{\partial P_i}{\partial q_h} \cdot \frac{\partial P_i}{\partial q_k} = a_{kh}(\mathbf{q},t), \qquad (4.22)$$

$$b_k(\mathbf{q},t) = \sum_{i=1}^{n} m_i \frac{\partial P_i}{\partial q_k} \cdot \frac{\partial P_i}{\partial t}, \qquad (4.23)$$

$$c(\mathbf{q},t) = \frac{1}{2} \sum_{i=1}^{n} m_i \left(\frac{\partial P_i}{\partial t}\right)^2. \qquad (4.24)$$

In the case of fixed constraints, it is possible to choose a Lagrangian coordinate system with respect to which time does not explicitly appear in the expression for $\mathbf{X} = \mathbf{X}(\mathbf{q})$, and hence in the equations $P_i = P_i(\mathbf{q})$. For holonomic systems with fixed constraints we henceforth assume that such a coordinate system has been chosen.

In this system all terms in equations (4.21) which are not quadratic vanish.

PROPOSITION 4.1   *For a holonomic system with fixed constraints, the kinetic energy is a homogeneous quadratic form in the components of the vector* $\dot{\mathbf{q}}$.   ∎

In the general case, note that

$$T = \frac{1}{2} \mathfrak{Q} \cdot \mathbf{V} \qquad (4.25)$$

(the vector $\mathfrak{Q}$ is defined by (4.17)) and that the quadratic term in equations (4.25), i.e.

$$\widehat{T} = \frac{1}{2} \sum_{h,k=1}^{\ell} a_{hk} \dot{q}_h \dot{q}_k, \qquad (4.26)$$

is the kinetic energy due to the virtual component of the velocity.

The expression (4.26) possesses a fundamental property.

THEOREM 4.1   $\widehat{T}$ *is a positive definite quadratic form.*   ∎

Note that the matrix $(a_{hk})_{h,k=1,\ldots,\ell}$ is the Hessian of $T$ with respect to the variables $\dot{q}_k$:

$$a_{hk} = \frac{\partial^2 T}{\partial \dot{q}_h \partial \dot{q}_k}. \qquad (4.27)$$

Denoting it by $\mathbf{H}_T$ we can write (4.26) as

$$\widehat{T} = \frac{1}{2} \dot{\mathbf{q}} \cdot \mathbf{H}_T \dot{\mathbf{q}}. \qquad (4.28)$$

As a consequence of Theorem 4.1 we have the following.

COROLLARY 4.1  *The matrix $\mathbf{H}_T$ is positive definite.*  ∎

*Proof of Theorem 4.1*
Given a holonomic system, and a Lagrangian coordinate system, the coefficients $a_{hk}$, $b_k$, $c$ are uniquely determined. We need to show that

$$\widehat{T} > 0, \quad \forall \dot{\mathbf{q}} \neq 0. \tag{4.29}$$

To this end we express (4.26) in terms of the virtual velocities of the single point particles:

$$\widehat{T} = \frac{1}{2} \sum_{i=1}^{n} m_i \widehat{v}_i^2 \tag{4.30}$$

and observe that $\dot{\mathbf{q}} \neq 0 \Leftrightarrow \widehat{\mathbf{V}} \neq 0$, and hence not all $\widehat{\mathbf{v}}_i$ can be zero.  ∎

We have already remarked how the components of the vector $\dot{\mathbf{q}}$, for every fixed $\mathbf{q}$ and $t$, can be viewed as parameters, playing the role of kinetic coordinates. We consider the derivatives of $T$ with respect to such parameters:

$$p_k = \frac{\partial T}{\partial \dot{q}_k}, \quad k = 1, \ldots, \ell, \tag{4.31}$$

and hence

$$p_k = \sum_{h=1}^{\ell} a_{hk} \dot{q}_h + b_k, \quad k = 1, \ldots, \ell. \tag{4.32}$$

The system (4.32) is linear in $p_k, \dot{q}_k$ and, because of Theorem 4.1, it is invertible.

It is easy to recognise that the $p_k$ are the Lagrangian components of the vector $\mathcal{Q}$:

$$p_k = \mathcal{Q} \cdot \frac{\partial \mathbf{X}}{\partial q_k}, \quad k = 1, \ldots, \ell. \tag{4.33}$$

For this reason, these are called the *kinetic momenta* conjugate to the corresponding $q_k$. The variables $p_k$ have great importance in mechanics.

*Remark 4.2*
Let us set

$$\widehat{p}_k = \sum_{h=1}^{\ell} a_{hk} \dot{q}_k.$$

It is useful to note that $\widehat{T} = \frac{1}{2} \widehat{\mathbf{p}} \cdot \dot{\mathbf{q}}$. In the case of fixed constraints, this implies

$$T = \frac{1}{2} \mathbf{p} \cdot \dot{\mathbf{q}}. \tag{4.34}$$

∎

## 4.3 The dynamics of discrete systems. Lagrangian formalism

Let us return to our original aim of expressing the left-hand side of equation (4.20) as a function of $T$. Differentiate both sides of (4.33) with respect to time:

$$\dot{p}_k = (\dot{\mathcal{Q}})_{\Theta,k} + \mathcal{Q} \cdot \frac{\partial \dot{\mathbf{X}}}{\partial q_k} \tag{4.35}$$

and note that

$$\mathcal{Q} \cdot \frac{\partial \dot{\mathbf{X}}}{\partial q_k} = \sum_i m_i \mathbf{v}_i \cdot \frac{\partial \mathbf{v}_i}{\partial q_k} = \frac{\partial T}{\partial q_k}, \tag{4.36}$$

which finally yields

$$(\dot{\mathcal{Q}})_{\Theta,k} = \frac{d}{dt}\frac{\partial T}{\partial \dot{q}_k} - \frac{\partial T}{\partial q_k}. \tag{4.37}$$

Equations (4.20) can be written in the form

$$\frac{d}{dt}\frac{\partial T}{\partial \dot{q}_k} - \frac{\partial T}{\partial q_k} = F_{\Theta,k}, \quad k = 1, \ldots, \ell \tag{4.38}$$

and are known as *Lagrange's equations*. The functions $F_{\Theta,k}$ are defined by (4.19). Equations (4.38) are sufficient to find the solution of the motion problem.

THEOREM 4.2 *The Lagrange equations* (4.38) *admit a unique solution satisfying the initial conditions*

$$\mathbf{q}(0) = \mathbf{q}_0, \quad \dot{\mathbf{q}}(0) = \mathbf{w}_0. \tag{4.39}$$

*Proof*
Equations (4.38) are second-order equations, linear with respect to $\ddot{q}_k$. Indeed, denoting by $\dot{\mathbf{q}}$ and $\ddot{\mathbf{q}}$ the $\ell \times 1$ column vectors with components $\dot{q}_k$ and $\ddot{q}_k$ respectively, the system (4.38) can be written as

$$H_T \ddot{\mathbf{q}} + \dot{H}_T \dot{\mathbf{q}} + \dot{\mathbf{b}} - \left(\frac{1}{2}\dot{\mathbf{q}}^T \nabla_\mathbf{q} H_T \dot{\mathbf{q}} + \nabla_\mathbf{q} \mathbf{b}^T \dot{\mathbf{q}} + \nabla_\mathbf{q} c\right) = \mathbf{F}_\Theta, \tag{4.40}$$

where $\mathbf{b}$ and $\mathbf{F}_\Theta$ denote, respectively, the column vectors with components $b_k$ and $F_{\Theta,k}$, and $c$ is given by (4.24). Note that the $k$th component of the column vector $\frac{1}{2}\dot{\mathbf{q}}^T \nabla_q H_T \mathbf{q}$, is given by $\frac{1}{2}\sum_{i,j=1}^{\ell}(\partial a_{ij}/\partial q_k)\dot{q}_i\dot{q}_j$.

Hence Corollary 4.1 yields that the system is solvable with respect to the unknowns $\ddot{q}_k$, i.e. it admits the *normal form*

$$\ddot{q}_k = \chi_k(\mathbf{q}, \dot{\mathbf{q}}, t), \quad k = 1, \ldots, \ell, \tag{4.41}$$

where the functions $\chi_k$ are easily found. Indeed, $\chi_k$ is the $k$th component of the column vector

$$\chi = H_T^{-1}\left(\mathbf{F}_\Theta + \frac{1}{2}\dot{\mathbf{q}}^T\nabla_\mathbf{q}H_T\dot{\mathbf{q}} + \nabla_\mathbf{q}\mathbf{b}^T\dot{\mathbf{q}} + \nabla_\mathbf{q}c - \dot{H}_T\dot{\mathbf{q}} - \mathbf{b}\right). \tag{4.42}$$

The functions $\chi_k$ contain $F_{\Theta,k}$, given functions of $\mathbf{q}$, $\dot{\mathbf{q}}$, $t$; we assume that these functions are regular. The conclusion of the theorem now follows from the existence and uniqueness theorem for the Cauchy problem for a system of ordinary differential equations (cf. Appendix 1, Theorem A1.1): it is sufficient to set $\mathbf{x} = (\mathbf{q}, \dot{\mathbf{q}}) \in \mathbf{R}^{2\ell}$, $\mathbf{x}(0) = (\mathbf{q}_0, \mathbf{w}_0)$ and to write equation (3.21) as $\dot{\mathbf{x}} = \mathbf{v}(\mathbf{x}, t)$ with $\mathbf{v}(\mathbf{x}, t) = (x_{\ell+1}, \ldots, x_{2\ell}, \chi_1(\mathbf{x}, t), \ldots, \chi_\ell(\mathbf{x}, t))$. ∎

*Remark 4.3*
Equations (4.38) imply that the vector $\mathbf{F}^{(a)}$ acts on the motion only through its projection onto the tangent space. ∎

*Remark 4.4*
Consider a point particle of mass $m$ constrained to move on a fixed smooth regular surface $S \subseteq \mathbf{R}^3$ with no active forces. If $\mathbf{x} = \mathbf{x}(q_1, q_2)$ is a local parametrisation of $S$, it follows that the kinetic energy of the particle can be written as

$$T = \frac{m}{2}[E(q_1, q_2)\dot{q}_1^2 + 2F(q_1, q_2)\dot{q}_1\dot{q}_2 + G(q_1, q_2)\dot{q}_2^2], \tag{4.43}$$

where $E$, $F$, $G$ are the entries of the first fundamental form of the surface (1.34). Since there are no active forces, $F_{\Theta 1} = F_{\Theta 2} = 0$ and Lagrange's equations (4.38) take the form

$$E\ddot{q}_1 + F\ddot{q}_2 + \frac{1}{2}\left[\left(\frac{\partial E}{\partial q_1}\right)\dot{q}_1^2 + 2\frac{\partial E}{\partial q_2}\dot{q}_1\dot{q}_2 + \left(2\frac{\partial F}{\partial q_2} - \frac{\partial G}{\partial q_1}\right)\dot{q}_2^2\right] = 0,$$

$$F\ddot{q}_1 + G\ddot{q}_2 + \frac{1}{2}\left[\left(2\frac{\partial F}{\partial q_1} - \frac{\partial E}{\partial q_2}\right)\dot{q}_1^2 + 2\frac{\partial G}{\partial q_1}\dot{q}_1\dot{q}_2 + \frac{\partial G}{\partial q_2}\dot{q}_2^2\right] = 0. \tag{4.44}$$

These can be recognised as the geodesic equations of the surface (1.46).

Once again we find that *the trajectories of a point particle constrained on a fixed smooth regular surface, with no other forces acting on it, are the geodesics of the surface* (Proposition 2.2). Note in addition that this implies that the point acceleration is orthogonal to the surface. ∎

*Example 4.1*
Consider a point particle constrained to move on a surface of rotation without any active force. If $\mathbf{x} = (u\cos v, u\sin v, f(u))$ is a local parametrisation, the kinetic energy of the point is given by

$$T = \frac{m}{2}\left\{[1 + (f'(u))^2]\dot{u}^2 + u^2\dot{v}^2\right\}$$

## 4.3 The dynamics of discrete systems. Lagrangian formalism

and Lagrange's equations are the geodesic's equations discussed and solved in Example 1.25. ∎

A similar conclusion can be reached when considering a holonomic system with fixed smooth constraints, without external forces. In this case the space $\mathcal{V}$ of all possible configurations becomes a Riemannian manifold when endowed with the metric

$$(\mathrm{d}s)^2 = \sum_{i,j=1}^{\ell} a_{ij}(q_1,\ldots,q_\ell)\, \mathrm{d}q_i\, \mathrm{d}q_j, \qquad (4.45)$$

where $a_{ij}$ are given by equation (4.22); Theorem 4.1 ensures that equation (4.45) defines a Riemannian metric (see Definition 1.30).

For this system, Lagrange's equations become

$$\sum_{j=1}^{\ell} a_{ij}\ddot{q}_j + \frac{1}{2}\sum_{j,k=1}^{\ell}\left(\frac{\partial a_{ij}}{\partial q_k} + \frac{\partial a_{ik}}{\partial q_j} - \frac{\partial a_{jk}}{\partial q_i}\right)\dot{q}_j\dot{q}_k = 0, \qquad (4.46)$$

where $i = 1,\ldots,\ell$. Multiplying by $a^{hi}$ and summing over $i$ (where $a^{hi}$ are the components of the inverse matrix of $A = (a_{ij})$), since $\sum_{i=1}^{\ell} a^{hi}a_{ij} = \delta_j^h$, equations (4.46) become

$$\ddot{q}_h + \sum_{j,k=1}^{\ell} \Gamma_{j,k}^h \dot{q}_j\dot{q}_k = 0, \quad h = 1, 2, \ldots, \ell, \qquad (4.47)$$

where $\Gamma_{j,k}^h$ are the Christoffel symbols (1.69) associated with the metric (4.45). Equations (4.47) are the geodesic equations (1.68), (note that $\dot{s}^2 = 2T$ is constant). We have proved the following.

THEOREM 4.3 *The space of configurations of a holonomic system with fixed constraints, endowed with the metric (4.45) induced by the kinetic energy, is a Riemannian manifold. If there are no active forces (and the constraints are smooth) the trajectories of the systems are precisely the geodesics of the Riemannian manifold.* ∎

Systems of this kind are also called *natural Lagrangian systems* (see Arnol'd et al. 1988).

*Example 4.2*
Write down Lagrange's equations for a system of two point particles $(P_1, m_1)$, $(P_2, m_2)$ with $P_1$ constrained to move on a circle of radius $R$ and centre $O$, $P_2$ constrained to move along the line $OP_1$, in the presence of the following forces acting in the plane of the circle:

$\mathbf{F}_1$, applied to $P_1$, of constant norm and tangent to the circle;
$\mathbf{F}_2$, applied to $P_2$, of constant norm, parallel to $F_1$ but with opposite orientation.

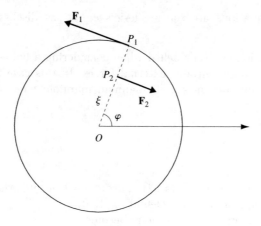

Fig. 4.1

All constraints are smooth.

The system has two degrees of freedom, and we can take as Lagrangian coordinates the angle $\varphi$ between the radius $OP_1$ and a fixed axis, and the abscissa $\xi$ of $P_2$ onto the radius $OP_1$ (Fig. 4.1).

The kinetic energy is

$$T = \frac{1}{2} m_1 R^2 \dot\varphi^2 + \frac{1}{2} m_2(\xi^2 \dot\varphi^2 + \dot\xi^2). \tag{4.48}$$

We need to compute the Lagrangian components $F_\varphi$, $F_\xi$:

$$F_\varphi = \mathbf{F}_1 \cdot \frac{\partial P_1}{\partial \varphi} + \mathbf{F}_2 \cdot \frac{\partial P_2}{\partial \varphi}, \quad F_\xi = \mathbf{F}_1 \cdot \frac{\partial P_1}{\partial \xi} + \mathbf{F}_2 \cdot \frac{\partial P_2}{\partial \xi}. \tag{4.49}$$

Setting $\mathbf{e}_r = (P_1 - O)/|P_1 - O|$, $\mathbf{e}_\varphi = \mathbf{F}_1/|\mathbf{F}_1|$, we find

$$\frac{\partial P_1}{\partial \varphi} = R\mathbf{e}_\varphi, \quad \frac{\partial P_2}{\partial \varphi} = \xi \mathbf{e}_\varphi, \quad \frac{\partial P_1}{\partial \xi} = 0, \quad \frac{\partial P_2}{\partial \xi} = \mathbf{e}_r. \tag{4.50}$$

Hence

$$F_\varphi = RF_1 - \xi F_2, \quad F_\xi = 0. \tag{4.51}$$

Substituting into equations (4.38), we obtain the desired equations

$$(m_1 R^2 + m_2 \xi^2)\ddot\varphi + 2m_2 \xi \dot\xi \dot\psi = RF_1 - \xi F_2, \tag{4.52}$$

$$\ddot\xi - \xi \dot\varphi^2 = 0. \tag{4.53}$$

We easily recognise that equation (4.52) is the second cardinal equation written with reference to the point $O$. Equation (4.53) can instead be interpreted as the equation of motion for $P_2$ in the reference frame rotating with the line $OP_1$, in which $P_2$ is subject to the centrifugal acceleration field $\xi \dot\varphi^2$.

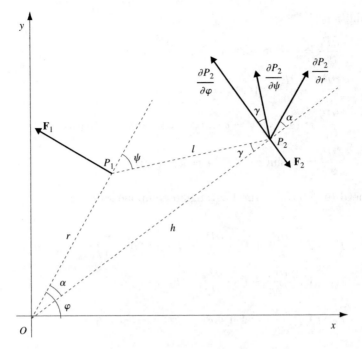

Fig. 4.2

We note finally that the equations do not change if $F_1$ and $F_2$ are given functions of $\xi$. ∎

*Example 4.3*
Two point particles $P_1$, $P_2$ of equal mass $m$ move in a plane in which they are subject to the Biot–Savart field (2.19), with the constant $C$ equal in magnitude for the points but with opposite signs. The two points are constrained to preserve a constant distance $\ell$. Write down the Lagrange equations.

The system has three degrees of freedom. We can choose polar coordinates $(r, \varphi)$ for the point $P_1$ and determine the position of $P_2$ as a function of the angle $\psi$ between the vectors $(P_1 - O)$ and $(P_2 - P_1)$ (Fig. 4.2).

The Cartesian coordinates of the two points $P_1$, $P_2$ are given by

$$\begin{aligned} x_1 &= r \cos \varphi, & y_1 &= r \sin \varphi, \\ x_2 &= r \cos \varphi + \ell \cos(\varphi - \psi), & y_2 &= r \sin \varphi + \ell \sin(\varphi - \psi). \end{aligned} \qquad (4.54)$$

Hence

$$v_1^2 = \dot{r}^2 + r^2 \dot{\varphi}^2, \qquad (4.55)$$

$$v_2^2 = \dot{r}^2 + r^2 \dot{\varphi}^2 + \ell^2 (\dot{\varphi} - \dot{\psi})^2 + 2\ell \left[ \dot{r} \sin \psi + r \dot{\varphi} \cos \psi \right] (\dot{\varphi} - \dot{\psi}), \qquad (4.56)$$

and in addition

$$\frac{\partial P_1}{\partial \varphi} = (-r \sin \varphi, r \cos \varphi), \quad \frac{\partial P_1}{\partial \psi} = 0,$$

$$\frac{\partial P_1}{\partial r} = \frac{\partial P_2}{\partial r} = (\cos \varphi, \sin \varphi),$$

$$\frac{\partial P_2}{\partial \varphi} = (-r \sin \varphi - \ell \sin(\varphi - \psi), r \cos \varphi + \ell \cos(\varphi - \psi)),$$

$$\frac{\partial P_2}{\partial \psi} = (\ell \sin(\varphi - \psi), -\ell \cos(\varphi - \psi)).$$

(4.57)

We still need to determine the Lagrangian components $F_r$, $F_\varphi$, $F_\psi$. As in Fig. 4.2 we have

$$F_r = \mathbf{F}_1 \cdot \frac{\partial P_1}{\partial r} + \mathbf{F}_2 \cdot \frac{\partial P_2}{\partial r} = F_2 \cos\left(\alpha + \frac{\pi}{2}\right) = -F_2 \sin \alpha,$$

$$F_\varphi = \mathbf{F}_1 \cdot \frac{\partial P_1}{\partial \varphi} + \mathbf{F}_2 \cdot \frac{\partial P_2}{\partial \varphi} = F_1 r - F_2 h,$$

$$F_\psi = \mathbf{F}_2 \cdot \frac{\partial P_2}{\partial \psi} = F_2 \ell \cos(\pi - \gamma) = -F_2 \ell \cos \gamma,$$

(4.58)

where

$$h^2 = r^2 + \ell^2 + 2\ell r \cos \psi \qquad (4.59)$$

and

$$\frac{\sin \alpha}{\ell} = \frac{\sin \gamma}{r} = \frac{\sin \psi}{h}. \qquad (4.60)$$

Since $T = \frac{1}{2} m(v_1^2 + v_2^2)$, we can write down Lagrange's equations. This is left as an exercise. ∎

## 4.4 Determination of constraint reactions. Constraints with friction

The solution of the initial value problem for the Lagrange system of equations yields the vector $\mathbf{q} = \mathbf{q}(t)$. Once this vector is known, we can determine the motion of the representative vector $\mathbf{X} = \mathbf{X}(t)$. Hence the kinematic terms in equations (4.18) and the vector $\mathbf{F}(\mathbf{X}(t), \dot{\mathbf{X}}(t), t)$ are known. From this it is easy to find $\Phi = \Phi(t)$ and then $\phi_i = \phi_i(t)$.

As an example, using expression (4.15), where the multipliers $\lambda_j(t)$ are now known, we can write

$$\phi_i = \sum_{j=1}^{3n-\ell} \lambda_j(t) \nabla_{P_i} f_j(\mathbf{X}(t), t), \qquad (4.61)$$

### Example 4.4

Consider again the system studied in Example 4.2. Suppose the integrals $\varphi = \varphi(t)$ and $\xi = \xi(t)$ of the equations of motion are known; find the constraint reactions $\phi_1$ and $\phi_2$ on the two points of the system.

In this particular case, the problem is simple; indeed, once the accelerations $\mathbf{a}_1$, $\mathbf{a}_2$ are known, it is enough to write $\phi_i = m_i \mathbf{a}_i - \mathbf{F}_i$. However, it is useful to illustrate the general procedure. We start by computing the unit base vectors of the normal space, writing the constraint equation in the form (see Fig. 4.1)

$$2f_1 = x_1^2 + y_1^2 - R^2 = 0, \quad f_2 = y_1 x_2 - x_1 y_2 = 0,$$

and hence $\nabla f_1 = (x_1, y_1, 0, 0)$, $\nabla f_2 = (-y_2, x_2, y_1, -x_1)$.

In addition,

$$\dot{\mathcal{Q}} = (m_1 \ddot{x}_1, m_1 \ddot{y}_1, m_2 \ddot{x}_2, m_2 \ddot{y}_2) \quad \text{and} \quad \mathbf{F}^{(a)} = \left( \frac{-F_1 y_1}{R}, \frac{F_1 x_1}{R}, \frac{F_2 y_1}{R}, \frac{-F_2 x_1}{R} \right),$$

implying that the vector $\Phi \in \mathbf{R}^4$ of the constraint reactions can be found by using equation (4.15), and determining the multipliers $\lambda_1$, $\lambda_2$ starting from

$$(\dot{\mathcal{Q}} - \mathbf{F}^{(a)}) \cdot \nabla f_1 = \lambda_1 |\nabla f_1|^2,$$

$$(\dot{\mathcal{Q}} - \mathbf{F}^{(a)}) \cdot \nabla f_2 = \lambda_2 |\nabla f_2|^2.$$

In the present case, we have $\nabla f_1 \cdot \nabla f_2 = 0$. Using the coordinates $\xi$, $\varphi$ of Fig. 4.1 one can easily obtain the equations

$$\lambda_1 = m_1 \dot{\varphi}^2,$$

$$\lambda_2 = \frac{1}{\xi^2 + R^2} \left[ R\xi(m_1 - m_2 \cos^2 \varphi) \ddot{\varphi} - 2m_2 R^2 \dot{\xi} \sin^2 \varphi \dot{\varphi} - F_1 \xi - F_2 R \right].$$

For the case of *constraints with friction* it is necessary to formulate the hypothesis linking the constraint reactions $\phi_i^{(a)}$, due to friction, to the velocities $\mathbf{v}_i$ (strictly speaking, to the virtual velocities $\widehat{\mathbf{v}}_i$) and then include them among the active forces. Using linear links such as[2]

$$\phi_i^{(a)} = -\mu_i \widehat{\mathbf{v}}_i, \tag{4.62}$$

with $\mu_i \geq 0$, one must add the following term to the right-hand side of equation (4.38):

$$\Phi_{\Theta,k}^{(a)} = -\sum_{i=1}^{n} \sum_{h=1}^{\ell} \mu_i \frac{\partial P_i}{\partial q_h} \cdot \frac{\partial P_i}{\partial q_k} \dot{q}_h. \tag{4.63}$$

---

[2] The system of coordinates plays an important role. Consider as an example the case of a rotating sphere. Mathematically, the constraint is fixed, as we can represent it by $|\mathbf{x}| = R$, but in order to take friction into account, the virtual velocity must be computed *relative to the constraint*, and hence in a coordinate system based on the sphere.

These Lagrangian components of the friction forces can be deduced from a kind of *kinetic variables potential*:

$$f_D = -\frac{1}{2} \sum_{i=1}^{n} \sum_{h,k=1}^{\ell} \mu_i \frac{\partial P_i}{\partial q_h} \cdot \frac{\partial P_i}{\partial q_k} \dot{q}_h \dot{q}_k \qquad (4.64)$$

in the sense that

$$\Phi_{\Theta,k}^{(a)} = \frac{\partial f_D}{\partial \dot{q}_k}. \qquad (4.65)$$

The function $f_D$ is called the *Rayleigh dissipation function*; it is equal to half the power dissipated due to the friction

$$W_D = -\sum_{i=1}^{n} \mu_i \hat{\mathbf{v}}_i^2. \qquad (4.66)$$

## 4.5 Conservative systems. Lagrangian function

DEFINITION 4.2 *The system of active forces* $(\mathbf{F}_i^{(a)}, P_i)$, $i = 1, \ldots, n$, *is conservative if there exists a regular function $U$, called the* potential *of the system, such that its representative vector* $\mathbf{F}^{(a)}$ *is given by*

$$\mathbf{F}^{(a)} = \nabla_{\mathbf{X}} U(\mathbf{X}). \qquad (4.67)$$

∎

To determine if a system of forces is conservative, and to determine its potential, it is necessary to consider the subdivision into internal and external forces.

If each of the external forces is a conservative field with potential given by $U_i^{(e)}(P_i)$, the overall potential of the external forces is given by

$$U^{(e)}(\mathbf{X}) = \sum_{i=1}^{n} U_i^{(e)}(P_i). \qquad (4.68)$$

As an example, for the gravity field we find $U = -mgz_G$ ($m = \sum_i m_i$, $z_G$ is the height of the centre of mass, assuming the $z$-axis is vertical and oriented upwards).

The internal forces are given by interaction pairs. As an example, the interaction between the points $P_i$ and $P_j$ is expressed by the pair $(\mathbf{F}_{ij}, P_i), (-\mathbf{F}_{ij}, P_j)$, where

$$\mathbf{F}_{ij} = f_{ij}(P_i - P_j) \frac{(P_i - P_j)}{|P_i - P_j|}. \qquad (4.69)$$

The conservative interaction pairs are characterised as the conservative central force fields (Example 2.3).

PROPOSITION 4.2 *The interaction pair* $(\mathbf{F}_{ij}, P_i), (-\mathbf{F}_{ij}, P_j)$, *where* $\mathbf{F}_{ij}$ *is given by* (4.69), *is conservative if and only if* $f_{ij}$ *depends only on* $r_{ij} = |P_i - P_j|$.

Its potential is given by

$$U_{ij}^{(i)}(r_{ij}) = \int f_{ij}(r_{ij}) \, dr_{ij}. \tag{4.70}$$

*Proof*
The proof is a simple extension of Example 2.3. ∎

Well-known examples of interaction potentials are the elastic potential $U(r) = -\frac{1}{2}kr^2$ and the gravitational potential $U = k/r$.

The overall potential of the internal forces is the sum of the potentials of the interaction pairs:

$$U^{(i)}(\mathbf{X}) = \sum_{1 \leq i < j \leq n} U_{ij}^{(i)}(r_{ij}). \tag{4.71}$$

The potential of the system is given by $U = U^{(e)} + U^{(i)}$.

*Remark 4.5*
The projection of $\mathbf{F}^{(a)}$ onto the tangent space has the structure of a gradient. This means that the Lagrangian components $F_{\Theta,k}^{(a)}$ are given by

$$F_{\Theta,k}^{(a)} = \frac{\partial \mathcal{U}}{\partial q_k}, \quad k = 1, \ldots, \ell, \tag{4.72}$$

where

$$\mathcal{U}(\mathbf{q}, t) = U[\mathbf{X}(\mathbf{q}, t)]. \tag{4.73}$$

The time dependence is introduced only through the constraints' motion (recall that for holonomic systems with fixed constraints the Lagrangian coordinates are by convention such as to yield representations of the form $\mathbf{X} = \mathbf{X}(\mathbf{q})$). ∎

When the forces are conservative, equations (4.38) can be written more concisely, by introducing the *Lagrangian function*

$$L(\mathbf{q}, \dot{\mathbf{q}}, t) = T + \mathcal{U}, \tag{4.74}$$

where $\mathcal{U}(\mathbf{q}, t)$ is the function given by (4.73).

*Remark 4.6*
Recall the Lagrangian expression for $T$, and what we have just mentioned regarding the potential. These facts imply that $\partial L/\partial t \neq 0$ only if the constraints are in motion. ∎

Consider now equations (4.72) and note that all the derivatives $\partial \mathcal{U}/\partial \dot{q}_k$ vanish. We deduce that Lagrange's equations can be written in the form

$$\frac{d}{dt} \frac{\partial L}{\partial \dot{q}_k} - \frac{\partial L}{\partial q_k} = 0, \quad k = 1, \ldots, \ell. \tag{4.75}$$

*Remark 4.7*
If a Lagrangian coordinate $q_k$ does not appear explicitly in the Lagrangian $L$ (in this case we say the coordinate is *cyclic*), from equation (4.75) it follows that there exists the first integral $p_k = \partial L/\partial \dot{q}_k = $ constant. ∎

*Example 4.5*
Write equations (4.75) for a system of two point particles $(P_1, m_1)$, $(P_2, m_2)$ constrained in a vertical plane as in Fig. 4.3 (smooth constraints).

Let $\varphi$ be as shown in the figure and $x_2$ the $x$-coordinate of $P_2$; we take these as the Lagrangian coordinates. Then we have

$$L(\varphi, x_2, \dot{\varphi}, \dot{x}_2) = \frac{1}{2} m_1 R^2 \dot{\varphi}^2 + \frac{1}{2} m_2 \dot{x}_2^2 - m_1 g R(1 - \cos \varphi)$$
$$- \frac{k}{2}(x_2^2 + 2R^2 - 2R x_2 \sin \varphi - 2R^2 \cos \varphi). \tag{4.76}$$

It follows that the required equations are

$$\ddot{\varphi} + \frac{g}{R} \sin \varphi - \frac{k}{m_1}\left(\frac{x_2}{R} \cos \varphi - \sin \varphi\right) = 0, \tag{4.77}$$

$$\ddot{x}_2 + \frac{k}{m_2}(x_2 - R \sin \varphi) = 0. \tag{4.78}$$

∎

It is possible that some of the forces are conservative and some are not. In this case, writing

$$\mathbf{F}^{(a)} = \nabla_{\mathbf{X}} U(\mathbf{X}) + \mathbf{G}(\mathbf{X}, \dot{\mathbf{X}}, t), \tag{4.79}$$

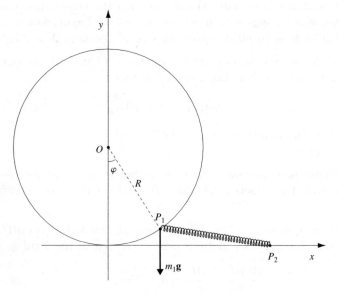

Fig. 4.3

## 4.6 The dynamics of discrete systems. Lagrangian formalism

where $\mathbf{G}$ cannot be derived from any potential, we can define $L = T + U$, which yields a mixed form of Lagrange's equations:

$$\frac{\mathrm{d}}{\mathrm{d}t}\frac{\partial L}{\partial \dot{q}_k} - \frac{\partial L}{\partial q_k} = G_{\Theta,k}. \tag{4.80}$$

**Remark 4.8**
It can be verified immediately that Lagrange's equations (4.75) corresponding to the Lagrangians $L(q,\dot{q},t)$, $L' = cL$, $L'' = L + \mathrm{d}F/\mathrm{d}t$, with $c$ a constant and $F(q,t)$ an arbitrary $\mathcal{C}^2$ function, have the same solutions. To clarify this point, we write explicitly

$$\frac{\mathrm{d}F}{\mathrm{d}t} = \dot{\mathbf{q}} \cdot \nabla_{\mathbf{q}} F(\mathbf{q},t) + \frac{\partial F}{\partial t}. \qquad \blacksquare$$

## 4.6 The equilibrium of holonomic systems with smooth constraints

The equilibrium configurations of a holonomic system with fixed, smooth constraints are given by the constant solutions of equations (4.38). Therefore the equilibrium equations are

$$F_{\Theta,k}^{(a)}(\mathbf{q}) = 0, \quad k = 1, \ldots, \ell \tag{4.81}$$

(in this case, it is natural to consider that forces and constraints are time-independent).

In the case of conservative fields, using equations (4.73), the equilibrium equations can be written as

$$\frac{\partial \mathcal{U}}{\partial q_k} = 0, \quad k = 1, \ldots, \ell \tag{4.82}$$

and express the fact that at equilibrium the restriction of the potential $U(\mathbf{X})$ to the manifold $\mathbf{X} = \mathbf{X}(\mathbf{q})$ is stationary.

Clearly the problem (4.82) can be formulated in $\mathbf{R}^{3n}$ as the *problem of finding the stationary points of the potential on the configuration manifold*, i.e.

$$\nabla_{\mathbf{X}}\left(U(\mathbf{X}) + \sum_{j=1}^{m} \lambda_j f_j(\mathbf{X})\right) = 0, \tag{4.83}$$

$$f_j(\mathbf{X}) = 0, \quad j = 1, \ldots, m. \tag{4.84}$$

The Lagrange multipliers $\lambda_j$ are the same as those that give the constraint reactions

$$\Phi = \sum_{j=1}^{m} \lambda_j \nabla_{\mathbf{X}} f_j(\mathbf{X}), \tag{4.85}$$

since it must be that, at equilibrium,
$$\mathbf{F}^{(a)} + \Phi = 0. \tag{4.86}$$

*Remark 4.9*
When the forces are only due to the presence of mass, the equilibrium configurations are exactly the configurations for which the height of the centre of mass is extremal. ∎

*Example 4.6*
Study the equilibrium of the system considered in Example 4.2.

It is easy to check that of the two equilibrium equations (4.51), $F_\varphi = 0$, $F_\xi = 0$, the latter is an identity while the former admits the unique solution $\xi = RF_1/F_2$. For this value of $\xi$ there exist infinitely many equilibrium configurations, obtained by varying $\varphi$.

If $F_1$ and $F_2$ are functions of $\xi$ the equation $\xi = RF_1/F_2$ admits either one, many, or no solution. ∎

*Example 4.7*
Study the equilibrium of the system considered in Example 4.3.

It is easy to show that there are no equilibrium configurations. Indeed, because of equations (4.58) we should have in particular $\sin \alpha = \cos \gamma = 0$, which is incompatible with equations (4.60), valid for all the possible configurations of the system. ∎

*Example 4.8*
Study the equilibrium of the system considered in Example 4.5.

The equilibrium equations are
$$\frac{\partial \mathcal{U}}{\partial \varphi} = 0, \quad \frac{\partial \mathcal{U}}{\partial x_2} = 0,$$

yielding (set to zero the kinematic terms in (4.77), (4.78)):
$$\frac{m_1 g}{kR} \sin \varphi - \frac{x_2}{R} \cos \varphi + \sin \varphi = 0, \tag{4.87}$$
$$x_2 - R \sin \varphi = 0. \tag{4.88}$$

If $x_2$ is eliminated one obtains
$$\sin \varphi \left[\frac{m_1 g}{kR} - \cos \varphi + 1\right] = 0 \Leftrightarrow \sin \varphi = 0. \tag{4.89}$$

Hence the only solutions are $\varphi = 0$, $x_2 = 0$ and $\varphi = \pi$, $x_2 = 0$. ∎

## 4.7 Generalised potentials. Lagrangian of an electric charge in an electromagnetic field

There are situations when it is possible to define a Lagrangian even if the system of forces depends on velocity. Indeed, note how equations (4.38) imply that, if

## 4.7  The dynamics of discrete systems. Lagrangian formalism

there exists a function $\mathcal{U}(\mathbf{q}, \dot{\mathbf{q}}, t)$ such that

$$F^{(a)}_{\Theta,k} = \frac{\partial \mathcal{U}}{\partial q_k} - \frac{d}{dt}\left(\frac{\partial \mathcal{U}}{\partial \dot{q}_k}\right), \quad k = 1, \ldots, \ell, \tag{4.90}$$

the usual definition of the Lagrangian $L = T + U$ still permits us to write the Lagrange equations in the form (4.75).

The function $\mathcal{U}(\mathbf{q}, \dot{\mathbf{q}}, t)$ is called a *generalised potential*.

An important example when it is possible to define a generalised potential is the case of a force applied to a charge $e$ in an electromagnetic field $(\mathbf{E}, \mathbf{B})$ (the *Lorentz force*):

$$\mathbf{F} = e\left\{\mathbf{E} + \frac{1}{c}\mathbf{v} \times \mathbf{B}\right\}. \tag{4.91}$$

We seek the generalised potential for equation (4.91), starting from Maxwell's equations

$$\operatorname{div} \mathbf{B} = 0, \tag{4.92}$$

$$\operatorname{curl} \mathbf{E} + \frac{1}{c}\frac{\partial \mathbf{B}}{\partial t} = 0. \tag{4.93}$$

From the first it follows that it is possible to express the field $\mathbf{B}$ as

$$\mathbf{B} = \operatorname{curl} \mathbf{A}, \tag{4.94}$$

where $\mathbf{A}(x, t)$ is the so-called *vector potential*, defined up to an irrotational field (which we assume to be independent of time).

Because of equation (4.94), equation (4.93) takes the form

$$\operatorname{curl}\left(\mathbf{E} + \frac{1}{c}\frac{\partial \mathbf{A}}{\partial t}\right) = 0. \tag{4.95}$$

As a consequence of (4.95) there exists a scalar function $\varphi$ (the usual electrostatic potential when $\mathbf{B}$ is independent of time) such that

$$\mathbf{E} + \frac{1}{c}\frac{\partial \mathbf{A}}{\partial t} = -\nabla \varphi. \tag{4.96}$$

Substituting equations (4.94) and (4.96) into (4.91) we obtain

$$\mathbf{F} = e\left\{-\nabla \varphi - \frac{1}{c}\frac{\partial \mathbf{A}}{\partial t} + \frac{1}{c}\mathbf{v} \times \operatorname{curl} \mathbf{A}\right\}. \tag{4.97}$$

We now make a second transformation. Note that

$$(\mathbf{v} \times \operatorname{curl} \mathbf{A})_i = \mathbf{v} \cdot \frac{\partial \mathbf{A}}{\partial x_i} - \mathbf{v} \cdot \nabla A_i, \quad i = 1, 2, 3, \tag{4.98}$$

and that
$$\mathbf{v} \cdot \frac{\partial \mathbf{A}}{\partial x_i} = \frac{\partial}{\partial x_i}(\mathbf{v} \cdot \mathbf{A}), \qquad (4.99)$$

where, as for equation (4.90), we consider $\dot{x}_1$, $\dot{x}_2$, $\dot{x}_3$ as independent variables. Since finally

$$\mathbf{v} \cdot \nabla A_i = \frac{\mathrm{d} A_i}{\mathrm{d} t} - \frac{\partial A_i}{\partial t}, \qquad (4.100)$$

we can write

$$\mathbf{v} \times \operatorname{curl} \mathbf{A} = \nabla(\mathbf{v} \cdot \mathbf{A}) - \frac{\mathrm{d}\mathbf{A}}{\mathrm{d}t} + \frac{\partial \mathbf{A}}{\partial t}, \qquad (4.101)$$

and we arrive at the following expression for $\mathbf{F}$:

$$\mathbf{F} = e \left\{ -\nabla\varphi + \frac{1}{c}\nabla(\mathbf{v} \cdot \mathbf{A}) - \frac{1}{c}\frac{\mathrm{d}\mathbf{A}}{\mathrm{d}t} \right\}. \qquad (4.102)$$

Note that $\mathbf{F}$ can be expressed in the form[3]

$$\mathbf{F} = -\nabla V + \frac{\mathrm{d}}{\mathrm{d}t}\nabla_{\mathbf{v}} V, \qquad (4.103)$$

with

$$V = e\left\{\varphi - \frac{1}{c}\mathbf{v} \cdot \mathbf{A}\right\} \qquad (4.104)$$

and $\nabla_{\mathbf{v}} = \sum_i (\partial/\partial \dot{x}_i)\,\mathbf{e}_i$.

We have finally obtained the Lagrangian of the charge $e$ in the electromagnetic field $(\mathbf{E}, \mathbf{B})$:

$$L = T - e\left\{\varphi - \frac{1}{c}\mathbf{v} \cdot \mathbf{A}\right\}. \qquad (4.105)$$

## 4.8 Motion of a charge in a constant electric or magnetic field

For completeness, we examine the motion of a charge $e$ (positive or negative) of mass $m$ in a constant electric field $\mathbf{E}$ superimposed on a field with constant induction $\mathbf{B}$.

Consider the equation

$$m\mathbf{a} = e\left(\mathbf{E} + \frac{1}{c}\mathbf{v} \times \mathbf{B}\right). \qquad (4.106)$$

---

[3] According to definition (4.90) the generalised potential is $U = -V$. However it is customary to use the potential energy instead of the potential.

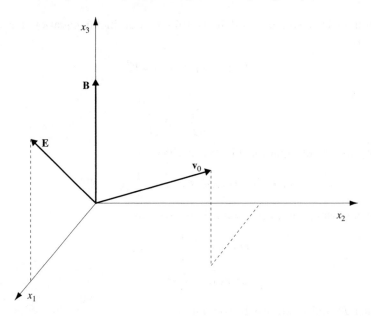

Fig. 4.4

Choose the reference system in such a way that the origin coincides with the initial position of the charge $P(0)$, the axis $x_3$ has the same direction and orientation as $\mathbf{B}$, and the axis $x_2$ is orthogonal to both $\mathbf{E}$ and $\mathbf{B}$, so that $E_1 \geq 0$ (Fig. 4.4).

If $\mathbf{v}(0) = \mathbf{v}_0$, we need to study the following problem:

$$\ddot{x}_1 = \mathcal{E}_1 + \dot{x}_2 \omega, \quad x_1(0) = 0, \quad \dot{x}_1(0) = v_1^0, \tag{4.107}$$

$$\ddot{x}_2 = -\dot{x}_1 \omega, \quad x_2(0) = 0, \quad \dot{x}_1(0) = v_2^0, \tag{4.108}$$

$$\ddot{x}_3 = \mathcal{E}_3, \quad x_3(0) = 0, \quad \dot{x}_3(0) = v_3^0, \tag{4.109}$$

where we set

$$\frac{e}{m} E_i = \mathcal{E}_i, \quad \frac{eB}{mc} = \omega. \tag{4.110}$$

If $B = 0$ the motion is in a uniform electric field, and the generic trajectory is a parabola. Suppose $B \neq 0$ and integrate once; this yields

$$\dot{x}_1 = \mathcal{E}_1 t + \omega x_2 + v_1^0, \tag{4.111}$$

$$\dot{x}_2 = -\omega x_1 + v_2^0, \tag{4.112}$$

$$\dot{x}_3 = \mathcal{E}_3 t + v_3^0. \tag{4.113}$$

Equation (4.112) can be used in (4.107) to obtain an equation for the only variable $x_1$:

$$\ddot{x}_1 + \omega^2 x_1 = \mathcal{E}_1 + \omega v_2^0. \tag{4.114}$$

Set

$$C = \frac{\mathcal{E}_1}{\omega} = c\frac{E_1}{B}. \tag{4.115}$$

Then the integral of equation (4.114) is given by

$$x_1(t) = -D\cos(\omega t + \alpha) + \frac{1}{\omega}(C + v_2^0), \tag{4.116}$$

and after imposing the initial conditions, we find

$$D\cos\alpha = \frac{1}{\omega}(C + v_2^0), \tag{4.117}$$

$$D\sin\alpha = \frac{1}{\omega}v_1^0, \tag{4.118}$$

from which $D$ and $\alpha$ are easily computed.

Equation (4.111) now yields

$$x_2(t) = D\sin(\omega t + \alpha) - Ct - \frac{1}{\omega}v_1^0. \tag{4.119}$$

Therefore the projection of the point onto the plane normal to the magnetic field moves in a circular trajectory, with radius

$$D = \frac{1}{\omega}\left[(v_1^0)^2 + (C + v_2^0)^2\right]^{1/2} \tag{4.120}$$

and frequency $\omega$, around the centre

$$\left(\frac{1}{\omega}(C + v_2^0), -Ct - \frac{1}{\omega}v_1^0\right).$$

The latter moves uniformly according to $\mathbf{E} \times \mathbf{B}$ with velocity $C$ defined by (4.115). The projection motion is circular and uniform if $E_1 = 0$ (implying $C = 0$). In this case equation (4.120) defines the *Larmor radius*.

The motion in the $x_3$-coordinate is due exclusively to the electric field:

$$x_3(t) = \frac{1}{2}\mathcal{E}_3 t^2 + v_3^0 t. \tag{4.121}$$

Note that in correspondence with the zeros of $\dot{x}_1$, given by $\omega t_n + \alpha = (2n+1)\pi/2$, one finds $\dot{x}_2(t_n) = \pm\omega D - C$, and hence if $\omega D > C$, the motion in the $x_2$-direction is periodically inverted and the projection of the trajectory onto the plane $x_3 = 0$ self-intersects (if $\omega D = C$, it forms cusps).

As an exercise, let $\mathcal{E}_3 = 0$, $v_3^0 = 0$ and prove that for $\omega \to 0$ equations (4.116), (4.119) reproduce the motion in a uniform force field.

## 4.9 Symmetries and conservation laws. Noether's theorem

The invariant properties of a system with respect to the action of a group (with one or more parameters) are called symmetries. In a Lagrangian system, to such symmetries there correspond conservation laws, which are first integrals of the motion of the system. We shall see as an example that conservation of momenta corresponds to the invariance of the Lagrangian with respect to coordinate translations, conservation of the angular momentum corresponds to the invariance of the Lagrangian with respect to rotations, and so on. The rigorous mathematical formulation of this relation between symmetries and conservation laws is the content of Noether's theorem.

Consider a Lagrangian system with $l$ degrees of freedom; for simplicity, we assume the system to be independent of the time $t$. Let $L(\mathbf{q}, \dot{\mathbf{q}})$ be its Lagrangian.

DEFINITION 4.3  *An invertible coordinate transformation* $\mathbf{q} = \mathbf{f}(\mathbf{Q})$ *is admissible for a given system if and only if the Lagrangian is invariant under the transformation, and hence if*

$$L(\mathbf{q}, \dot{\mathbf{q}}) = L(\mathbf{Q}, \dot{\mathbf{Q}}). \tag{4.122}$$

■

*Example 4.9*
If a Lagrangian has a cyclic coordinate (see Remark 4.7), it follows that it is invariant under translations in this coordinate. ■

*Example 4.10*
Rotations around the origin

$$q_1 = Q_1 \cos\alpha + Q_2 \sin\alpha,$$
$$q_2 = -Q_1 \sin\alpha + Q_2 \cos\alpha$$

are admissible for the Lagrangian

$$L(q_1, q_2, \dot{q}_1, \dot{q}_2) = \frac{m}{2}(\dot{q}_1^2 + \dot{q}_2^2) - V\left(\sqrt{q_1^2 + q_2^2}\right)$$

corresponding to the plane motion of a point particle of mass $m$ in a central force field. ■

DEFINITION 4.4  *A one-parameter* $s \in \mathbf{R}$ *family of invertible transformations* $\mathbf{q} = \mathbf{f}(\mathbf{Q}, s)$ *is called a* one-parameter group *of transformations if it satisfies the following properties:*

(a) $\mathbf{f}(\mathbf{Q}, 0) = \mathbf{Q}$, *for every* $\mathbf{Q}$;
(b) *for every* $s_1, s_2 \in \mathbf{R}$, $\mathbf{f}(\mathbf{f}(\mathbf{Q}, s_1), s_2) = \mathbf{f}(\mathbf{Q}, s_1 + s_2)$.

*If for every* $s \in \mathbf{R}$ *the transformation* $\mathbf{q} = \mathbf{f}(\mathbf{Q}, s)$ *is admissible, then the group is called* admissible. ■

Note that (a), (b) imply that if $\mathbf{q} = \mathbf{f}(\mathbf{Q}, s)$, then $\mathbf{Q} = \mathbf{f}(\mathbf{q}, -s)$.

THEOREM 4.4 (Noether) *If a Lagrangian $L(\mathbf{q}, \dot{\mathbf{q}})$ admits a one-parameter group of transformations $\mathbf{q} = \mathbf{f}(\mathbf{Q}, s)$, the Lagrange equations associated with $L$ admit the first integral $I(\mathbf{q}, \dot{\mathbf{q}})$ given by*

$$I(\mathbf{q}, \dot{\mathbf{q}}) = \sum_{i=1}^{l} \frac{\partial L}{\partial \dot{q}_i} \frac{\partial f_i}{\partial s}(\mathbf{q}, 0). \tag{4.123}$$

*Proof*
The invariance property of the Lagrangian implies that if $\mathbf{q}(t)$ is a solution of Lagrange's equations (4.75), then $\mathbf{Q}(t, \sigma) = \mathbf{f}(\mathbf{q}(t), \sigma)$ is also a solution, $\forall \sigma \in \mathbf{R}$. This means that

$$\frac{d}{dt} \nabla_{\dot{\mathbf{Q}}} L(\mathbf{Q}, \dot{\mathbf{Q}}) = \nabla_{\mathbf{Q}} L(\mathbf{Q}, \dot{\mathbf{Q}}), \quad \forall \sigma \in \mathbf{R}, \tag{4.124}$$

where $\dot{\mathbf{Q}} = (\partial/\partial t)\mathbf{Q}(t, \sigma)$. In addition, the definition of an admissible transformation yields

$$0 = \frac{\partial}{\partial \sigma} L(\mathbf{Q}, \dot{\mathbf{Q}}) = \nabla_{\mathbf{Q}} L \cdot \frac{\partial \mathbf{Q}}{\partial \sigma} + \nabla_{\dot{\mathbf{Q}}} L \cdot \frac{\partial \dot{\mathbf{Q}}}{\partial \sigma}, \tag{4.125}$$

and using equation (4.125) and multiplying equation (4.124) by $\partial \mathbf{Q}/\partial \sigma$ we find

$$\frac{d}{dt} \left( \nabla_{\dot{\mathbf{Q}}} L \cdot \frac{\partial \mathbf{Q}}{\partial \sigma} \right) = 0. \tag{4.126}$$

For $\sigma = 0$ this is exactly the invariance of $I(\mathbf{q}, \dot{\mathbf{q}})$ along the motion. ∎

*Example 4.11*
If the Lagrangian $L(\mathbf{q}, \dot{\mathbf{q}})$ admits the translations $q_k = Q_k + s$ as transformation group, the coordinate $q_k$ is cyclic and $I(\mathbf{q}, \dot{\mathbf{q}}) = p_k$ is a constant of the motion. ∎

*Example 4.12*
If a Lagrangian $L(\mathbf{q}, \dot{\mathbf{q}})$, where $\mathbf{q} \in \mathbf{R}^3$, admits the rotations around the axis $q_1$:

$$q_1 = Q_1, \quad q_2 = Q_2 \cos s + Q_3 \sin s, \quad q_3 = -Q_2 \sin s + Q_3 \cos s$$

as transformation group, the function

$$I(\mathbf{q}, \dot{\mathbf{q}}) = \frac{\partial L}{\partial \dot{q}_2} q_3 - \frac{\partial L}{\partial \dot{q}_3} q_2 = p_2 q_3 - p_3 q_2$$

is a constant of the motion, coinciding with the component of the angular momentum along the axis $q_1$. ∎

## 4.9 The dynamics of discrete systems. Lagrangian formalism

*Example 4.13*
The Lagrangian of a point particle constrained to move on a surface of revolution around the $z$-axis, with no active forces acting on it, is equal to (see Example 4.1)

$$L(u, v, \dot{u}, \dot{v}) = \frac{m}{2}\left\{\left[1 + (f'(u))^2\right]\dot{u}^2 + u^2\dot{v}^2\right\}.$$

The coordinate $v$ is cyclic, and therefore the conjugate kinetic momentum

$$p_v = mu^2\dot{v}$$

is a constant of the motion. Note that $p_v$ is equal to the component $L_3$ of the angular momentum $p_v = m(x_1\dot{x}_2 - x_2\dot{x}_1)$. ∎

*Remark 4.10*
There are transformations which, although not admissible, leave the equations of motion invariant. In this case there are no associated first integrals, but the study of the equations can help to establish interesting properties of the motion, without explicitly solving them. ∎

*Example 4.14*
Consider a Lagrangian system

$$L(\mathbf{q}, \dot{\mathbf{q}}) = \sum_{i=1}^{l} m\frac{\dot{q}_i^2}{2} - V(\mathbf{q}), \tag{4.127}$$

where $V$ is a homogeneous function of $q_1, \ldots, q_l$ of degree $d$:

$$V(\alpha q_1, \ldots, \alpha q_l) = \alpha^d V(q_1, \ldots, q_l).$$

For every $\alpha > 0$ the transformation of $\mathbf{q}$, $\dot{\mathbf{q}}$ and time $t$:

$$\mathbf{q} = \alpha \mathbf{Q}, \quad t = \beta\tau, \quad \frac{d\mathbf{q}}{dt} = \frac{\alpha}{\beta}\frac{d\mathbf{Q}}{d\tau},$$

where $\beta = \alpha^{1-d/2}$, transforms the Lagrangian (4.127) as follows:

$$\tilde{L}\left(\mathbf{Q}, \frac{d\mathbf{Q}}{d\tau}\right) = \frac{\alpha^2}{\beta^2}\sum_{i=1}^{l}\frac{1}{2}m\left(\frac{dQ_i}{d\tau}\right)^2 - \alpha^d V(\mathbf{Q}) = \alpha^d L\left(\mathbf{Q}, \frac{d\mathbf{Q}}{d\tau}\right). \tag{4.128}$$

Since these two functions are proportional, the equations of motion are invariant. Hence, if $\mathbf{q}(t, \mathbf{q}_0, \dot{\mathbf{q}}_0)$ is a solution of Lagrange's equations associated with (4.127), $\mathbf{Q}(\tau, \mathbf{Q}_0, \dot{\mathbf{Q}}_0)$ is also a solution, and therefore

$$\mathbf{Q}\left(\alpha^{(d/2)-1}t, \frac{1}{\alpha}\mathbf{q}_0, \alpha^{-d/2}\dot{\mathbf{q}}_0\right) = \frac{1}{\alpha}\mathbf{q}(t, \mathbf{q}_0, \dot{\mathbf{q}}_0)$$

is a solution and the two trajectories are called *similar*.

If $d = 2$ one again finds that the period of motion is independent of the amplitude for harmonic oscillators. For the case that $d = -1$, corresponding to a

Newtonian potential, and considering for simplicity circular orbits, we find that if $T$ is the period and $l$ is the orbit length, the ratio $l^{3/2}/T$ is constant for similar trajectories (as stated by the third law of Kepler). ∎

*Example 4.15*
Among the transformations that modify the Lagrangian but not the equations of motion are the *gauge transformations* for the vector and scalar potentials $\mathbf{A}$, $\varphi$ of the electromagnetic field (4.94), (4.96). Let $f(\mathbf{x}, t)$ be an arbitrary regular function, and set

$$\mathbf{A}' = \mathbf{A} + \nabla f,$$
$$\varphi' = \varphi - \frac{1}{c}\frac{\partial f}{\partial t}. \tag{4.129}$$

Then the definitions of the fields $\mathbf{B}$, $\mathbf{E}$ through equations (4.94) and (4.96) are invariant. What changes is the Lagrangian (4.105), which is transformed into

$$L' = T - e\left\{\varphi' - \frac{1}{c}\mathbf{v} \cdot \mathbf{A}'\right\} - \frac{e}{c}\frac{df}{dt}. \tag{4.130}$$

As we remarked (see Remark 4.8), this generates the same motions as the former Lagrangian.

## 4.10 Equilibrium, stability and small oscillations

Consider an autonomous system of differential equations of first order in $\mathbf{R}^n$:

$$\dot{\mathbf{x}} = \mathbf{w}(\mathbf{x}), \tag{4.131}$$

where $\mathbf{w}$ is a regular vector field defined on $\mathbf{R}^n$. Lagrange's equations (4.40) can be written in the form (4.131), where $\mathbf{x}$ represents the vector $(\mathbf{q}, \dot{\mathbf{q}})$ in the phase space. Indeed, after setting the equations in normal form (4.41), $\ddot{\mathbf{q}} = \chi(\mathbf{q}, \dot{\mathbf{q}})$, it is enough to introduce in the phase space the field $\mathbf{w} = (\dot{\mathbf{q}}, \chi(\mathbf{q}, \dot{\mathbf{q}}))$ to obtain equation (4.131).

DEFINITION 4.5 *A point* $\mathbf{x}_0$ *is an* equilibrium point *if the constant function* $\mathbf{x}(t) = \mathbf{x}_0$ *is a solution of the system of differential equations* (4.131). ∎

PROPOSITION 4.3 *A point* $\mathbf{x}_0$ *is an equilibrium point if and only if the vector field* $\mathbf{w}$ *at the point is zero:* $\mathbf{w}(\mathbf{x}_0) = 0$.

*Proof*
It is trivial that at an equilibrium point, the vector field $\mathbf{w}$ is zero: from the definition of an equilibrium point it follows that $\mathbf{w}(\mathbf{x}_0) = \dot{\mathbf{x}}(t) = 0$. Conversely, if $\mathbf{w}(\mathbf{x}_0) = 0$ then $\mathbf{x}(t) = \mathbf{x}_0$ is a solution of the system for all $t$. ∎

The definition of stability of equilibrium is analogous to the definition given in Chapter 3 for a single point.

DEFINITION 4.6  *An equilibrium position* $\mathbf{x}_0$ *is* (Lyapunov) stable *if for every neighbourhood $U$ of the equilibrium point there exists a neighbourhood $U'$ such that, for any initial condition $\mathbf{x}(0)$ in $U'$, the corresponding solution $\mathbf{x}(t)$ is in $U$ for every $t > 0$. If this stability condition does not hold, then the equilibrium is called* unstable. ∎

*Remark* 4.11
Using spherical neighbourhoods, we can equivalently define stability as follows: for every $\varepsilon > 0$ there exists a number $\delta > 0$ such that, for any initial condition $\mathbf{x}(0)$ such that $|\mathbf{x}(0) - \mathbf{x}_0| < \delta$, $|\mathbf{x}(t) - \mathbf{x}_0| < \varepsilon$ for every $t > 0$.

Instability can be characterised by the condition that there exists an $\epsilon > 0$ such that for any fixed $\delta > 0$ there exists an initial condition $\mathbf{x}(0)$ in $|\mathbf{x}(0) - x_0| < \delta$ for which $|\mathbf{x}(t) - \mathbf{x}_0| > \epsilon$ for some $t > 0$.

It is evident from the definitions that we are referring to stability *in the future*, but it is possible to consider the analogous concept *in the past* by inverting the direction of time. ∎

*Example* 4.16
Consider a system of linear equations in $\mathbf{R}^n$

$$\dot{\mathbf{x}} = A\mathbf{x},$$

where $A$ is a real diagonalisable $n \times n$ matrix, with constant coefficients. Suppose that the eigenvalues $\lambda_1, \ldots, \lambda_n$ of $A$ are all distinct and non-zero. Then the general integral of the equation is given by

$$\mathbf{x}(t) = \sum_{j=1}^{n} c_j e^{\lambda_j t} \mathbf{u}_j,$$

where $\mathbf{u}_1, \ldots, \mathbf{u}_n$ are the eigenvectors of $A$. The constants $c_j$ (complex in general) are fixed by the initial conditions. Obviously $\mathbf{x} = 0$ is an equilibrium position, and it is easy to verify that it is stable if the real parts of *all* the eigenvalues are non-positive: Re $\lambda_j \le 0, j = 1, \ldots, n$ (simply use the linear transformation that diagonalises the matrix $A$). ∎

The analysis of the equilibrium stability for systems with one degree of freedom is carried out in Chapter 3 (Section 3.4). We now consider the corresponding problem for autonomous Lagrangian systems with several degrees of freedom.

As we saw (Section 4.6), if $V(q)$ is the potential energy, the equilibrium equations are

$$\frac{\partial V}{\partial q_i} = 0, \quad i = 1, \ldots, l. \tag{4.132}$$

Let $\bar{\mathbf{q}}$ be a solution of equations (4.132). We now prove the following stability criterion, for the case of smooth constraints.

THEOREM 4.5 (Dirichlet)  *If $\bar{\mathbf{q}}$ is an isolated minimum of the potential energy, the corresponding configuration is one of stable equilibrium.*

*Proof*
The hypotheses imply that $\bar{\mathbf{q}}$ solves equations (4.132), and hence that it is an equilibrium configuration. In addition, there exists a neighbourhood $A \subset \mathbf{R}^l$ of $\bar{\mathbf{q}}$ in which $V(\mathbf{q}) > V(\bar{\mathbf{q}})$, $\forall\, \mathbf{q} \neq \bar{\mathbf{q}}$.

We can choose $V(\bar{\mathbf{q}}) = 0$. Consider now any neighbourhood $B \subset \mathbf{R}^{2l}$ of $(\bar{\mathbf{q}}, 0)$ in the phase space, and for any $\varepsilon > 0$, define the energy sublevel set

$$\Omega_\varepsilon = \{(\mathbf{q}, \dot{\mathbf{q}}) | T(\mathbf{q}, \dot{\mathbf{q}}) + V(\mathbf{q}) < \varepsilon\}.$$

Recall that $T(\mathbf{q}, \dot{\mathbf{q}}) \geq a_0 |\dot{\mathbf{q}}^2|$ for some constant $a_0 > 0$ (Theorem 4.1). Consequently $\Omega_\epsilon \subset \Omega'_\epsilon = \{(\mathbf{q}, \dot{\mathbf{q}}) \mid a_0|\dot{\mathbf{q}}^2| + V(\mathbf{q}) < \epsilon\} \subset M_\epsilon \cap N_\epsilon$ where $M_\epsilon = \{(\mathbf{q}, \dot{\mathbf{q}}) \mid |\dot{\mathbf{q}}| < (\epsilon/a_0)^{1/2}\}$, $N_\epsilon = \{(\mathbf{q}, \dot{\mathbf{q}}) \mid V(\mathbf{q}) < \epsilon\}$. Since by hypothesis, the diameter of $M_\epsilon \cap N_\epsilon$ tends to zero when $\epsilon \to 0$, we can find $\varepsilon$ so small that $\Omega_\varepsilon \subset \Omega'_\varepsilon \subset B \cap (A \times \mathbf{R}^l)$. On the other hand, because of conservation of energy, every trajectory originating in $\Omega_\varepsilon$ must remain in $\Omega_\varepsilon$. This yields the stability condition (Definition 4.6). ∎

COROLLARY 4.2 *For any holonomic system with fixed smooth constraints, for which the active forces are only due to gravity, the stable equilibrium configurations occur in correspondence with isolated minima of the height of the centre of mass.* ∎

*Example* 4.17
We refer to Fig. 3.5. The isolated minima of $x_2, x_4$ of $V(x)$ correspond to positions of stable equilibrium. Consider for example the point $(x_2, 0)$ in the phase space, and consider a generic neighbourhood $U$. Define $e_{\max} \in (e_2, e_3)$ in such a way that the trajectory with energy $e_{\max}$ is entirely lying in $U$. The region determined by this trajectory contains all trajectories with energy in the interval $(e_2, e_{\max})$, and hence the definition of stability holds. ∎

We now consider the motion near configurations of stable equilibrium. Rewrite the Lagrangian of the system as

$$L(\mathbf{Q}, \dot{\mathbf{Q}}) = \frac{1}{2} \sum_{i,j=1}^{l} a_{ij}(\mathbf{Q})\dot{Q}_i \dot{Q}_j - V(\mathbf{Q}), \tag{4.133}$$

where there appear the vectors $\mathbf{Q} = \mathbf{q} - \bar{\mathbf{q}}$, $\dot{\mathbf{Q}} = \dot{\mathbf{q}}$, with $\bar{\mathbf{q}}$ an isolated minimum of $V(\mathbf{q})$. As we have seen, it is always possible to choose the initial conditions in such a way that the trajectory in the phase space remains in a fixed neighbourhood of $(\bar{\mathbf{q}}, 0)$. Select now a neighbourhood so small that inside it one can neglect terms of degree greater than two in the expansion of the function $L(\mathbf{Q}, \dot{\mathbf{Q}})$. Hence replace equations (4.133) with the quadratic approximation

$$L(\mathbf{Q}, \dot{\mathbf{Q}}) = \frac{1}{2}\left(\sum_{i,j=1}^{l} \bar{a}_{ij} \dot{Q}_i \dot{Q}_j - \sum_{i,j=1}^{l} \overline{V}_{ij} Q_i Q_j\right), \tag{4.134}$$

## 4.10     The dynamics of discrete systems. Lagrangian formalism

where we set

$$\bar{a}_{ij} = a_{ij}(\bar{\mathbf{q}}),$$
$$\bar{V}_{ij} = \frac{\partial^2 V}{\partial Q_i \partial Q_j}(\bar{\mathbf{q}}). \tag{4.135}$$

Denoting by $\bar{A}$ and $\bar{V}$ the symmetric matrices of the coefficients $\bar{a}_{ij}$ and $\bar{V}_{ij}$, respectively, the Lagrangian (4.133) can be written in matrix notation as

$$L(\mathbf{Q}, \dot{\mathbf{Q}}) = \frac{1}{2}(\dot{\mathbf{Q}}^T \bar{A} \dot{\mathbf{Q}} - \mathbf{Q}^T \bar{V} \mathbf{Q}), \tag{4.136}$$

and the associated Lagrange equations are linear:

$$\bar{A}\ddot{\mathbf{Q}} + \bar{V}\mathbf{Q} = 0. \tag{4.137}$$

Assuming that the matrix $\bar{V}$ is also positive definite, we can prove the following.

THEOREM 4.6   *If $\bar{A}$, $\bar{V}$ are symmetric and positive definite, there exists a linear transformation in $\mathbf{R}^l$ which decouples equations (4.137) into $l$ harmonic oscillations, called* normal modes *of the system and whose frequencies are called* fundamental frequencies *of the system.*

*Proof*
We follow the standard procedure to find the general integral of a system of linear ordinary differential equations with constant coefficients. Hence we seek a solution of (4.137) of the form

$$\mathbf{Q} = \mathbf{w} e^{i\lambda t}, \tag{4.138}$$

where $\mathbf{w}$ is a vector in $\mathbf{R}^l$ to be determined and $\lambda \in \mathbf{C}$. Substituting (4.138) into (4.137) we find

$$e^{i\lambda t}(\bar{V} - \lambda^2 \bar{A})\mathbf{w} = 0$$

and we must therefore study the generalised eigenvalue problem

$$\det(\mu \bar{A} - \bar{V}) = 0. \tag{4.139}$$

Accounting for multiplicity, this system has $l$ solutions $\mu_1, \ldots, \mu_l$ corresponding to the eigenvectors $\mathbf{w}_1, \ldots, \mathbf{w}_l$. We prove that in this case the $l$ roots $\mu_1, \ldots, \mu_l$ are positive. The method consists of reducing (4.137) to diagonal form, by a sequence of linear transformations. The choice of each such transformation must obey the criterion of *symmetry conservation* of the matrices of coefficients.
    Since $\bar{A}$ is a symmetric, positive definite matrix, there exists a unique symmetric, positive definite matrix whose square is equal to $\bar{A}$, which we denote by $\bar{A}^{1/2}$

(the square root of $\overline{A}$). Indeed, since $\overline{A}$ is symmetric, there exists an orthogonal matrix $S$ which diagonalises $\overline{A}$:

$$S\overline{A}S^{-1} = S\overline{A}S^T = \begin{pmatrix} \alpha_1 & 0 & \cdots & 0 \\ 0 & \alpha_2 & \cdots & 0 \\ \vdots & \vdots & \ddots & \vdots \\ 0 & 0 & \cdots & \alpha_l \end{pmatrix}, \tag{4.140}$$

where $\alpha_1, \ldots, \alpha_l$ are precisely the eigenvalues of $\overline{A}$. Since $\overline{A}$ is positive definite, the eigenvalues are all positive, and we can define

$$\overline{A}^{1/2} = S^T \begin{pmatrix} \sqrt{\alpha_1} & 0 & \cdots & 0 \\ 0 & \sqrt{\alpha_2} & \cdots & 0 \\ \vdots & \vdots & \ddots & \vdots \\ 0 & 0 & \cdots & \sqrt{\alpha_l} \end{pmatrix} S. \tag{4.141}$$

It is easily verified that $\overline{A}^{1/2}$ is symmetric and positive definite and that $(\overline{A}^{1/2})^2 = \overline{A}$. Moreover,

$$\overline{A}^{-1/2} = S^T \begin{pmatrix} 1/\sqrt{\alpha_1} & 0 & \cdots & 0 \\ 0 & 1/\sqrt{\alpha_2} & \cdots & 0 \\ \vdots & \vdots & \ddots & \vdots \\ 0 & 0 & \cdots & 1/\sqrt{\alpha_l} \end{pmatrix} S$$

is also symmetric. Through the change of variables

$$\mathbf{Y} = \overline{A}^{1/2} \mathbf{Q} \tag{4.142}$$

equation (4.137) becomes

$$\ddot{\mathbf{Y}} + \overline{A}^{-1/2} \overline{V} \overline{A}^{-1/2} \mathbf{Y} = 0, \tag{4.143}$$

and hence (4.139) is equivalent to

$$\det(\overline{A}^{-1/2} \overline{V} \overline{A}^{-1/2} - \mu) = 0. \tag{4.144}$$

Evidently $\overline{A}^{-1/2} \overline{V} \overline{A}^{-1/2}$ is symmetric and positive definite; it follows that its eigenvalues $\mu_1, \ldots, \mu_l$ are real and positive. We conclude (see Example 4.16) that the configuration $\overline{\mathbf{q}}$ is of stable equilibrium for the linearised system. Setting

$$C = \overline{A}^{-1/2} \overline{V} \overline{A}^{-1/2}, \tag{4.145}$$

## 4.10 The dynamics of discrete systems. Lagrangian formalism

if $W$ is an orthogonal matrix, diagonalising $C$, so that

$$W^T C W = \begin{pmatrix} \mu_1 & 0 & \cdots & 0 \\ 0 & \mu_2 & \cdots & 0 \\ \vdots & \vdots & \ddots & \vdots \\ 0 & 0 & \cdots & \mu_l \end{pmatrix}, \tag{4.146}$$

and if we define

$$\mathbf{Y} = W\mathbf{X}, \tag{4.147}$$

equation (4.143) becomes

$$\ddot{\mathbf{X}} + \begin{pmatrix} \mu_1 & 0 & \cdots & 0 \\ 0 & \mu_2 & \cdots & 0 \\ \vdots & \vdots & \ddots & \vdots \\ 0 & 0 & \cdots & \mu_l \end{pmatrix} \mathbf{X} = 0. \tag{4.148}$$

This equation represents $l$ independent harmonic oscillations with frequency $\omega_i = \sqrt{\mu_i}$, $i = 1,\ldots,l$ (normal modes). The linear transformation yielding the normal modes is hence given by

$$\mathbf{X} = W^T \overline{A}^{1/2} \mathbf{Q}. \tag{4.149}$$

∎

*Remark* 4.12
Recall that if $C$ is a real symmetric $\ell \times \ell$ matrix with eigenvalues $(\mu_1,\ldots,\mu_\ell)$, the orthogonal matrix $W$ diagonalising $C$ can be constructed as follows: $\ell$ orthonormal column vectors $\mathbf{w}^{(1)},\ldots,\mathbf{w}^{(\ell)}$ such that $(C - \mu_j)\mathbf{w}^{(j)} = 0$ can be easily determined. The matrix $W = (\mathbf{w}^{(1)},\ldots,\mathbf{w}^{(\ell)})$ is orthogonal and

$$W^T C W = \begin{pmatrix} \mu_1 & 0 & \cdots & 0 \\ 0 & \mu_2 & \cdots & 0 \\ \vdots & \vdots & \ddots & \vdots \\ 0 & 0 & \cdots & \mu_l \end{pmatrix},$$

As an example, if $C = \begin{pmatrix} 2 & 1 \\ 1 & 2 \end{pmatrix}$, $\mu_1 = 3$, $\mu_2 = 1$,

$$\mathbf{w}^{(1)} = \begin{pmatrix} 1/\sqrt{2} \\ 1/\sqrt{2} \end{pmatrix}, \quad \mathbf{w}^{(2)} = \begin{pmatrix} 1/\sqrt{2} \\ -1/\sqrt{2} \end{pmatrix}, \quad W = \begin{pmatrix} 1/\sqrt{2} & 1/\sqrt{2} \\ 1/\sqrt{2} & -1/\sqrt{2} \end{pmatrix}.$$

∎

*Example* 4.18
Consider a point particle of mass $m$ moving under the action of its weight on a surface of parametric equations

$$\mathbf{x} = (x(q_1,q_2), y(q_1,q_2), z(q_1,q_2)).$$

The Lagrangian of the system is given by

$$L(q_1, q_2, \dot q_1, \dot q_2) = \frac{1}{2} m\Big(E(q_1,q_2)\dot q_1^2 + 2F(q_1,q_2)\dot q_1\dot q_2 + G(q_1,q_2)\dot q_2^2\Big) - mgz(q_1,q_2),$$

where $E$, $F$ and $G$ are the coefficients of the first fundamental form of the surface. A point $(\bar q_1, \bar q_2)$ is an equilibrium point for the system only if it is a critical point of $z = z(q_1, q_2)$. The Lagrangian of the linearised equations is

$$L = \frac{1}{2}\left(m\left(\overline{E}\dot Q_1^2 + 2\overline{F}\dot Q_1\dot Q_2 + \overline{G}\dot Q_2^2\right) - \frac{1}{2} mg\left(\bar z_{11} Q_1^2 + 2\bar z_{12} Q_1 Q_2 + \bar z_{22} Q_2^2\right)\right),$$

where $\mathbf{Q} = \mathbf{q} - \bar{\mathbf{q}}$, $\overline{E}, \overline{F}, \overline{G}$ are the coefficients of first fundamental form evaluated at $\bar{\mathbf{q}}$, and

$$\bar z_{11} = \frac{\partial^2 z}{\partial q_1^2}(\bar q_1, \bar q_2), \quad \bar z_{12} = \frac{\partial^2 z}{\partial q_1 \partial q_2}(\bar q_1, \bar q_2), \quad \bar z_{22} = \frac{\partial^2 z}{\partial q_2^2}(\bar q_1, \bar q_2).$$

The fundamental frequencies of the system, $\omega_1$ and $\omega_2$, are the solutions of the eigenvalue problem with characteristic polynomial

$$\det\left(\omega^2 \begin{pmatrix} E & F \\ F & G \end{pmatrix} - g \begin{pmatrix} \bar z_{11} & \bar z_{12} \\ \bar z_{12} & \bar z_{12} \end{pmatrix}\right) = 0.$$

On the other hand, denoting by $\bar e$, $\bar f$ and $\bar g$ the coefficients of the second fundamental form of the surface (see Appendix 3) evaluated at $(\bar q_1, \bar q_2)$, one verifies that

$$\bar e = \bar z_{11}, \quad \bar f = \bar z_{12}, \quad \bar g = \bar z_{22}.$$

For example,

$$\bar e = \bar z_{11} \left(\frac{\partial x}{\partial q_1}\frac{\partial y}{\partial q_2} - \frac{\partial x}{\partial q_2}\frac{\partial y}{\partial q_1}\right) \Big/ \sqrt{\overline{E}\,\overline{G} - \overline{F}^2},$$

but in $(\bar q_1, \bar q_2)$ we have

$$\frac{\partial z}{\partial q_1} = \frac{\partial z}{\partial q_2} = 0,$$

and therefore

$$\overline{E}\,\overline{G} - \overline{F}^2 = \left[\left(\frac{\partial x}{\partial q_1}\right)^2 + \left(\frac{\partial y}{\partial q_1}\right)^2\right]\left[\left(\frac{\partial x}{\partial q_2}\right)^2 + \left(\frac{\partial y}{\partial q_2}\right)^2\right]$$

$$- \left[\left(\frac{\partial x}{\partial q_1}\right)\left(\frac{\partial x}{\partial q_2}\right) + \left(\frac{\partial y}{\partial q_1}\right)\left(\frac{\partial y}{\partial q_2}\right)\right]^2$$

$$= \frac{\partial x}{\partial q_1}\frac{\partial y}{\partial q_2} - \frac{\partial x}{\partial q_2}\frac{\partial y}{\partial q_1},$$

implying $e = \bar z_{11}$.

## 4.10 The dynamics of discrete systems. Lagrangian formalism

The *principal curvatures* $k_1$ and $k_2$ of the surface (Appendix 3) at the equilibrium point are the solutions of the eigenvalue problem of the first fundamental form with respect to the second, i.e. the roots of the characteristic polynomial

$$\det\left(k\begin{pmatrix}E & F \\ F & G\end{pmatrix} - \begin{pmatrix}e & f \\ f & g\end{pmatrix}\right) = \det\left(k\begin{pmatrix}E & F \\ F & G\end{pmatrix} - \begin{pmatrix}z_{11} & z_{12} \\ z_{12} & z_{22}\end{pmatrix}\right) = 0.$$

It follows that the principal curvatures are directly proportional to the square of the fundamental frequencies of the linearised equations

$$k_1 = \frac{\omega_1^2}{g}, \quad k_2 = \frac{\omega_2^2}{g}. \tag{4.150}$$

∎

We now compute the fundamental frequencies for the case that $l = 2$, and that the matrix $\overline{A}$ is diagonal:

$$\overline{A} = \begin{pmatrix}\alpha_1 & 0 \\ 0 & \alpha_2\end{pmatrix}, \quad \alpha_1, \alpha_2 > 0, \tag{4.151}$$

and of course $\overline{V}$ is symmetric and positive definite. The Lagrangian of the linearised motion is then given by

$$L_2(\mathbf{Q}, \dot{\mathbf{Q}}) = \frac{1}{2}\left(\alpha_1 \dot{Q}_1^2 + \alpha_2 \dot{Q}_2^2 - \overline{V}_{11} Q_1^2 - 2\overline{V}_{12} Q_1 Q_2 - \overline{V}_{22} Q_2^2\right), \tag{4.152}$$

and the matrix (4.145) is

$$C = \begin{pmatrix}\dfrac{\overline{V}_{11}}{\alpha_1} & \dfrac{\overline{V}_{12}}{\sqrt{\alpha_1 \alpha_2}} \\ \dfrac{\overline{V}_{12}}{\sqrt{\alpha_1 \alpha_2}} & \dfrac{\overline{V}_{22}}{\alpha_2}\end{pmatrix}.$$

The eigenvalue equation is

$$\mu^2 - \left(\frac{\overline{V}_{11}}{\alpha_1} + \frac{\overline{V}_{22}}{\alpha_2}\right)\mu + \frac{\overline{V}_{11}\overline{V}_{22} - \overline{V}_{12}^2}{\alpha_1 \alpha_2} = 0.$$

We find the two frequencies

$$\omega_\pm = \left\{\frac{1}{2}\left(\frac{\overline{V}_{11}}{\alpha_1} + \frac{\overline{V}_{22}}{\alpha_2}\right) \pm \frac{1}{2}\left[\left(\frac{\overline{V}_{11}}{\alpha_1} - \frac{\overline{V}_{22}}{\alpha_2}\right)^2 + 4\frac{\overline{V}_{12}^2}{\alpha_1 \alpha_2}\right]^{1/2}\right\}^{1/2}. \tag{4.153}$$

Obviously if $\overline{V}_{12} = 0$ (hence if the original system is in diagonal form) we find $\omega_+ = \sqrt{\overline{V}_{11}/\alpha_1}$, $\omega_- = \sqrt{\overline{V}_{22}/\alpha_2}$.

*Example 4.19*
A cylindrical container of height $h$ is closed at the boundary and is divided into three sections by two pistons of mass $m$, which can slide without friction. Each section contains the same amount of gas, for which we suppose the law $Pv =$ constant is applicable. Write the Lagrange equations describing the motion of the two pistons, find the stable equilibrium configuration and study the small oscillations of the system around it.

Let $x_1, x_2$ indicate the distance of the pistons from one of the two bases. Then $(x_1 < x_2)$, on the first piston there acts the force

$$F_1 = \frac{c}{x_1} - \frac{c}{x_2 - x_1}, \quad c > 0 \text{ constant},$$

and on the second piston the force

$$F_2 = \frac{c}{x_2 - x_1} - \frac{c}{h - x_2}.$$

Use the dimensionless variables $f_i = hF_i/c$, $\xi_i = x_i/h$, $i = 1, 2$, and write

$$f_1 = \frac{1}{\xi_1} - \frac{1}{\xi_2 - \xi_1}, \quad f_2 = \frac{1}{\xi_2 - \xi_1} - \frac{1}{1 - \xi_2}.$$

This is a conservative system of forces, with potential $V(\xi_1, \xi_2) = -\log[\xi_1(\xi_2 - \xi_1)(1 - \xi_2)]$. Recall that $V$ is expressed in dimensionless variables while the corresponding physical quantity is $\widehat{V} = cV$. The Lagrangian in the original variables is $L = \frac{1}{2}m(\dot{x}_1^2 + \dot{x}_2^2) - cV$ and can be replaced by the dimensionless Lagrangian

$$\widehat{L} = \frac{1}{2}\left[\left(\frac{d\xi_1}{d\tau}\right)^2 + \left(\frac{d\xi_2}{d\tau}\right)^2\right] - V(\xi_1, \xi_2),$$

by introducing the change of time-scale $\tau = t/t_0$, with $t_0^2 = mh^2/c$. The equations of motion become

$$\frac{d^2\xi_1}{d\tau^2} = \frac{1}{\xi_1} - \frac{1}{\xi_2 - \xi_1}, \quad \frac{d^2\xi_2}{d\tau^2} = \frac{1}{\xi_2 - \xi_1} - \frac{1}{1 - \xi_2}.$$

It is easily verified that the only equilibrium configuration is given by $\xi_1 = \frac{1}{3}$, $\xi_2 = \frac{2}{3}$. The Hessian matrix of $V(\xi_1, \xi_2)$ is

$$\begin{pmatrix} \dfrac{1}{\xi_1^2} + \dfrac{1}{(\xi_2 - \xi_1)^2} & -\dfrac{1}{(\xi_2 - \xi_1)^2} \\ -\dfrac{1}{(\xi_2 - \xi_1)^2} & \dfrac{1}{(1 - \xi_2)^2} \end{pmatrix}.$$

At the equilibrium, this becomes

$$\overline{V} = 9 \begin{pmatrix} 2 & -1 \\ -1 & 1 \end{pmatrix},$$

which is positive definite, with eigenvalues given as solutions of $\lambda^2 - 27\lambda + 81 = 0$, namely $\lambda_1 = \frac{9}{2}(3 - \sqrt{5})$, $\lambda_2 = \frac{9}{2}(3 + \sqrt{5})$. Hence the equilibrium is stable. The Hessian matrix of the kinetic energy is the identity matrix. Therefore the equations describing small oscillations are

$$\frac{d^2}{d\tau^2}\mathbf{Q} + \overline{V}\mathbf{Q} = 0, \quad \text{with} \quad \mathbf{Q} = \begin{pmatrix} \xi_1 \\ \xi_2 \end{pmatrix},$$

and $\sqrt{\lambda_1}, \sqrt{\lambda_2}$ give the dimensionless frequencies directly (we obtain $\omega_i = \sqrt{\lambda_i}/t_0$, $i = 1, 2$). The normal modes are obtained by setting $\mathbf{X} = W^T\mathbf{Q}$, where $W$ is such that

$$W^T \overline{V} W = \begin{pmatrix} \lambda_1 & 0 \\ 0 & \lambda_2 \end{pmatrix}.$$

We easily find that

$$W = \frac{1}{5^{1/4}} \begin{pmatrix} \left(\dfrac{2}{\sqrt{5}-1}\right)^{1/2} & \left(\dfrac{2}{\sqrt{5}+1}\right)^{1/2} \\ -\left(\dfrac{\sqrt{5}-1}{2}\right)^{1/2} & \left(\dfrac{\sqrt{5}+1}{2}\right)^{1/2} \end{pmatrix}.$$

By writing $\mathbf{Q} = W\mathbf{X}$ we can describe the small motions of the pistons as combinations of the harmonic motions $X_1, X_2$.

## 4.11 Lyapunov functions

In the previous section we have introduced the concept of stability of equilibrium points, for the system of differential equations (4.131). In particular, we have analysed the stability of the equilibrium of holonomic systems, with smooth fixed constraints, and subject to conservative forces. We now discuss some extensions and one additional criterion for stability. We start by observing that the conditions guaranteeing the stability of the equilibrium in the case of conservative forces must still hold if we introduce dissipative forces.

THEOREM 4.7 *Theorem 4.4 is still valid if in addition to forces with potential energy $V(\mathbf{q})$ there exist dissipative forces.*

*Proof*
The proof of Theorem 4.4 is based only on the fact that the trajectories originating within the set $\Omega_\epsilon$ remain there for all subsequent times. This is true if energy is conserved, but also if energy is dissipated. ∎

Dissipation helps stability, and in addition it may have the effect of bringing the system back to the equilibrium configuration, starting from a small enough perturbation, either in finite time or asymptotically for $t \to +\infty$. This is the case of asymptotic stability (see Definition 3.5).

DEFINITION 4.7  *A point $\mathbf{x}_0$ of stable equilibrium for the system (4.131) is asymptotically stable if there exists a $\delta > 0$ such that for every $\mathbf{x}(0)$ in the neighbourhood $|\mathbf{x}(0) - \mathbf{x}_0| < \delta$ one has $|\mathbf{x}(t) - \mathbf{x}_0| \to 0$ for $t \to +\infty$.* ∎

*Example 4.20*
For the harmonic damped motion (3.35) the point $x = 0$ is a point of equilibrium, and it is asymptotically stable (see (3.38)). ∎

Recall the case of the linear system $\dot{\mathbf{x}} = A\mathbf{x}$ (Example 4.14); in this case we can deduce that $\mathbf{x} = 0$ is an equilibrium point which is asymptotically stable if all eigenvalues $\lambda_j$ of the matrix $A$ have negative real part: Re $\lambda_j < 0$, $j = 1, \ldots, n$. The Dirichlet stability criterion (Theorem 4.4) is a special case of a well-known method for analysing stability, based on the so-called *Lyapunov function*. We consider again the system (4.131) and an equilibrium point $\mathbf{x}_0$; with reference to these we give the following definition.

DEFINITION 4.8  *Let $\Omega$ be a neighbourhood of $\mathbf{x}_0$, and let $\Lambda \in C^1(\Omega)$ be a function with an isolated minimum at $\mathbf{x}_0$ (assume $\Lambda(\mathbf{x}_0) = 0$). If for the field $\mathbf{w}(\mathbf{x})$ of system (10.1) we have that*

$$\mathbf{w}(\mathbf{x}) \cdot \nabla \Lambda(\mathbf{x}) \leq 0, \quad \forall\, \mathbf{x} \in \Omega, \tag{4.154}$$

*then $\Lambda$ is a* Lyapunov function *for the system.* ∎

Note that the meaning of (4.154) is that

$$\frac{\mathrm{d}}{\mathrm{d}t} \Lambda(\mathbf{x}(t)) \leq 0$$

along the solutions $\mathbf{x}(t)$ of the system (4.131).

Clearly for any holonomic system the total energy is a Lyapunov function in the phase space, in a neighbourhood of a local isolated minimum of the potential energy.

The following theorem has a proof analogous to the proof of Theorem 4.4.

THEOREM 4.8  *If $\mathbf{x}_0$ is such that there exists a Lyapunov function for the system (4.131) then it is a stable equilibrium point.* ∎

A more specific case is the following.

**THEOREM 4.9** *If*

$$\mathbf{w}(\mathbf{x}) \cdot \Lambda(\mathbf{x}) < 0, \quad \mathbf{x} \neq \mathbf{x}_0, \quad \mathbf{x} \in \Omega, \tag{4.155}$$

*then $\mathbf{x}_0$ is asymptotically stable.*

*Proof*
Consider the sets

$$A_\epsilon = \{\mathbf{x} \in \Omega \mid \Lambda(\mathbf{x}) \leq \epsilon\}.$$

Then $A_{\epsilon'} \subset A_\epsilon$ if $\epsilon' < \epsilon$ and moreover diam $A_\epsilon \to 0$ for $\epsilon \to 0$. Since along the trajectories of (4.131) $\dot{\Lambda} < 0$, any trajectory originating in $\Omega$ must cross the boundary $\partial A_\epsilon$ with $\epsilon$ decreasing. If the point tends to $\partial A_{\epsilon^*}$ for some $\epsilon^* > 0$, we would have $\dot{\Lambda} \leq -\alpha$ for some $\alpha > 0$ and $\forall\, t > 0$, which cannot hold; indeed, this would yield $\Lambda \to -\infty$, contradicting the hypothesis that $\Lambda(\mathbf{x}_0) = 0$ is a minimum. ∎

*Example 4.21*
For the damped harmonic oscillator (3.35), or equivalently for the system

$$\dot{x} = w, \quad \dot{w} = -(2\beta w + \omega^2 x), \quad \beta > 0, \tag{4.156}$$

$\Lambda(x, w) = \frac{1}{2}(w^2 + \omega^2 x^2)$ has an isolated minimum at the equilibrium point and $\dot{\Lambda} = -2\beta w^2 < 0$ away from the origin. We can therefore apply Theorem 4.8. ∎

Lyapunov's method can be invoked to establish instability.

**THEOREM 4.10** *Let $\mathbf{x}_0$ be an equilibrium point for the system (4.131). Suppose that there exist a neighbourhood $\Omega$ of $x_0$ and a function $\Lambda^* \in C^1(\Omega)$ ($\Lambda^*(\mathbf{x}_0) = 0$) such that*

$$\mathbf{w}(\mathbf{x}) \cdot \nabla \Lambda^*(\mathbf{x}) > 0, \quad \mathbf{x} \neq \mathbf{x}_0, \quad \mathbf{x} \in \Omega, \tag{4.157}$$

*and that $\mathbf{x}_0$ is an accumulation point for the positivity set of $\Lambda^*$. Then $\mathbf{x}_0$ is unstable.*

*Proof*
Consider a ball $B_\delta(\mathbf{x}_0)$ of centre $\mathbf{x}_0$ and radius $\delta$ such that $B_\delta \subset \Omega$ and let $\mathbf{x}(0) \in B_\delta(\mathbf{x}_0)$ be such that $\Lambda^*(\mathbf{x}(0)) > 0$. Due to (4.157) the trajectory remains in the set $M_0$, where $\Lambda^*(\mathbf{x}) > \Lambda^*(\mathbf{x}(0))$. In the intersection of this set with $B_\delta(\mathbf{x}_0)$ the scalar product $\mathbf{w} \cdot \nabla \Lambda^*$ has a positive infimum, while in this set $\Lambda^*$ is bounded. It follows that $\mathbf{x}(t)$ must leave $B_\delta$ in a finite time. ∎

*Example 4.22*
Consider the system

$$\dot{x} = w, \quad \dot{w} = \omega^2 x \tag{4.158}$$

for which $(0,0)$ is the (only) equilibrium point. Consider the function $\Lambda^* = xw$. In the plane $(x, w)$ this function is positive in the first and third quadrant and

$$w\frac{\partial \Lambda^*}{\partial x} + \omega^2 x \frac{\partial \Lambda^*}{\partial w} = w^2 + \omega^2 x^2 > 0$$

away from the origin. Instability follows. ∎

Another useful result on instability, whose assumptions are less restrictive than those of Theorem 4.9 is the following.

THEOREM 4.11 (Četaev) *Suppose that there exists an open connected set $\Omega_1$ (possibly unbounded) with $\mathbf{x}_0 \in \partial \Omega_1$, and a function $\Lambda^* \in C^1(\Omega_1)$, such that $\Lambda^* > 0$ in $\Omega_1$ and $\Lambda^*(\mathbf{x}_0) = 0$, for which (4.157) holds inside $\Omega_1$. Then $\mathbf{x}_0$ is unstable.*

*Proof*
This is just an extension of the previous theorem. With $\mathbf{x}(0) \in \Omega_1$, the trajectory cannot reach the boundary of $\Omega_1$ (as $\Lambda^*$ is increasing) and cannot stay indefinitely inside $B_\delta(\mathbf{x}_0) \cap M_0$. ∎

*Example 4.23*
The origin is the only point of equilibrium for the system

$$\dot{x} = w, \quad \dot{w} = \omega^2 |x|. \tag{4.159}$$

The function $\Lambda^* = xw$ is such that $\dot{\Lambda}^* = \omega^2 x|x| + w^2$ and it satisfies the hypotheses of Theorem 4.10, with $\Omega_1$ taken equal to the first quadrant (note that the hypotheses of Theorem 4.9 are not satisfied). ∎

## 4.12 Problems

**1.** Two point particles with mass, $(P_1, m_1)$, $(P_2, m_2)$, are constrained on two vertical lines $r_1$, $r_2$, at a distance $d$. The two points attract each other with an elastic force of constant $k$ and both are attracted by a fixed point $O$, placed at an equal distance from the two lines, with an elastic force of equal constant. Write down Lagrange's equations and show that the motion can be decomposed into two harmonic oscillations around the equilibrium configuration. Determine also the constraint reactions.

**2.** In a horizontal plane, two point particles $(P_1, m_1)$, $(P_2, m_2)$ attract each other with an elastic force of constant $k$ and are constrained on a smooth circle of centre $O$ and radius $R$. They are also attracted by two points $O_1$, $O_2$, respectively, with an elastic force of equal constant. The latter points are at a distance $2R$ from $O$ and such that the radii $O_1 - O$ and $O_2 - O$ form a right angle. Find the equilibrium configurations of the system and study the small oscillations around the stable equilibrium configuration.

**3.** Find the normal modes when the number of degrees of freedom of the system is equal to two, and the matrix $V$ is diagonal.

**4.** In a horizontal plane two point particles $(P_1, m_1)$, $(P_2, m_2)$ are attracted respectively by two fixed points $O_1$, $O_2$ in the plane with elastic forces of equal constant. The two particles are subject to the rigidity constraint $|P_1 - P_2| = |O_1 - O_2|$. Find the normal modes of the system.

**5.** Determine the fundamental frequencies and the normal modes of oscillation of a system of $\ell$ equal point particles constrained to move on a line and sequentially linked by springs with an elastic constant equal to $k$. The first particle is elastically attracted by the origin with a constant $k$ and the last particle is elastically attracted by a fixed point at a distance $a > 0$ from the origin with a constant $k$.

*Solution*

Let $q_i$ be the coordinate of the $i$th particle. Then the equilibrium positions are $\bar{q}_i = ai/(\ell+1)$, $i = 1, \ldots, \ell$, the fundamental frequencies are

$$\omega_i = 2\sqrt{\frac{k}{m}} \sin\left(\frac{\pi}{2} \frac{i}{\ell+1}\right)$$

and the normal modes are

$$q_i = \frac{ai}{\ell+1} + \sum_{j=1}^{\ell} \sqrt{\frac{2}{\ell+1}} \sin\left(\frac{ji\pi}{\ell} + 1\right) X_i.$$

**6.** Consider $l$ equal point particles $P_1, P_2, \ldots, P_l$ ($l > 2$) on a circle of radius $R$ and centre $O$. All particles move without friction and the point $P_i$ is attracted by its neighbouring points $P_{i-1}$, $P_{i+1}$ with an elastic force (set $P_0 = P_l$). Write down the potential of the system and prove that the configurations in which neighbouring rays form equal angles are equilibrium configurations. Study its stability (up to rotations). Compute the fundamental frequencies for $l = 3$. What is the general procedure?

**7.** A point particle of mass $m$ is constrained to move along a curve of equation $\zeta = A\xi^{2n}$, where $A > 0$ and $n \geq 1$ is an integer. The curve rotates in three-dimensional Euclidean space with angular velocity $\omega$ around the $z$-axis and at time $t = 0$ belongs to the vertical $(x, z)$ plane. Prove that, if $\xi$ is chosen as the generalised coordinate, the Lagrangian of the system is equal to

$$L = \frac{m}{2}\left(1 + 4n^2 A^2 \xi^{4n-2}\right)\dot{\xi}^2 - mgA\xi^{2n} - \frac{m}{2}\omega^2\xi^2.$$

Prove that if $n = 1$ the only equilibrium position of the system is $\xi = 0$; the equilibrium is stable if $\omega^2 < 2gA$, and unstable otherwise. If $n > 1$ then

$$\xi = \pm\left(\frac{\omega^2}{2ngA}\right)^{1/(2n-2)}$$

are positions of stable equilibrium, while $\xi = 0$ is unstable. Compute the frequencies of the small oscillations around the stable equilibrium positions.

**8.** A point particle of mass $m$ is constrained to move on an ellipsoid of equation

$$\frac{\xi^2}{a^2} + \frac{\eta^2 + \zeta^2}{b^2} = 1,$$

where $a > b > 0$. The ellipsoid rotates in space around the $y$-axis with angular velocity $\omega$. At the instant $t = 0$ the principal axes $\xi$, $\eta$ and $\zeta$ coincide with the axes $x$, $y$ and $z$. Prove that, after setting

$$\xi = a\cos\theta, \quad \eta = b\sin\theta\sin\varphi, \quad \zeta = b\sin\theta\cos\varphi,$$

the kinetic energy of the point is $T = T_2 + T_1 + T_0$, where

$$T_2 = \frac{m}{2}\left[(a^2\sin^2\theta + b^2\cos^2\theta)\dot\theta^2 + b^2\sin^2\theta\,\dot\varphi^2\right],$$

$$T_1 = abm\omega[\cos\varphi\,\dot\theta - \sin\theta\cos\theta\sin\varphi\,\dot\varphi],$$

$$T_0 = \frac{m}{2}\left[a^2\omega^2\cos^2\theta + b^2\omega^2\sin^2\theta\cos^2\varphi\right].$$

**9.** Two point particles of mass $m$ constrained to the vertical axis mutually interact with an elastic force of constant $k$. The first point is also elastically attracted to the point $z = 0$ by a spring of constant $k$. Let $z_1$ and $z_2$ be the coordinates of the two points. Prove that the Lagrangian of the system is

$$L = \frac{m}{2}(\dot z_1^2 + \dot z_2^2) - \frac{k}{2}\left[z_1^2 + (z_1 - z_2)^2\right] - mgz_1 - mgz_2.$$

Determine the equilibrium positions, discuss their stability and compute the fundamental frequencies of the small oscillations around the equilibrium positions, and the normal modes.

**10.** A point particle of mass $m$ and electric charge $e$ is in motion in space under the action of a central field with potential energy $V$ and of a magnetic field $\mathbf{B} = (0, 0, B)$. Prove that if the initial velocity is $\mathbf{v} = (v_1, v_2, 0)$ the motion takes place in the $(x, y)$ plane. Write the Lagrangian in the plane polar coordinates $(r, \varphi)$, and prove that the coordinate $\varphi$ is cyclic. Use this fact to reduce the problem to one-dimensional motion and find the trajectories in the case $V(r) = \frac{1}{2}\omega^2 r^2$.

**11.** A point particle $P$ of mass $m$ is constrained to move along the parabola $y = a + bx^2$, with $a$, $b$ being given positive constants. A point $Q$ of mass $m$ is constrained to move along the line $y = (\tan\alpha)x$. $P$ and $Q$ interact with an attractive elastic force of constant $k$. Write the expression for the Lagrangian and find the equilibrium positions depending on the parameter $\alpha$. Study the stability and compute the frequency of the small oscillations around the stable equilibrium position.

**12.** A point particle of mass $m$ moves on a torus of equation

$$x^2 + y^2 + z^2 - 2a\sqrt{y^2 + z^2} + a^2 - b^2 = 0,$$

where $0 < b < a$, under the action of the force due to its weight $\mathbf{F} = (0, 0, -mg)$. Write down the Lagrangian, find the equilibrium positions and study their stability. Compute the principal curvature of the torus at the points $(0, 0, -a - b)$, $(0, 0, -a + b)$, $(0, 0, a - b)$, $(0, 0, a + b)$.

**13.** A point particle of unit mass is constrained to move on the sphere $x^2 + y^2 + z^2 = 1$ under the action of the force field $\mathbf{F} = (-ax, ay, -bz)$, where $a$, $b$ are given constants. Write down the Lagrangian and reduce the problem to one-dimensional motion.

## 4.13 Additional remarks and bibliographical notes

The theory of stability is much more extensive than that presented in Section 4.10. The concept of stability is very important when studying all phenomena modelled by systems of differential equations of the same kind as system (4.131). It is not surprising then that the literature on the subject is very extensive, and that research in this field is still very active. The beginning of the theory is in a memoir, published in 1892, by A. Lyapunov (in Russian).

The book of La Salle and Lefschetz (1961) is a particularly simple and concise read. In addition, we note a recent book of Amann (1990), containing a vast bibliography.

Finally, we recall that Definition 4.1 of a holonomic system with smooth constraints is traditionally given by introducing the so-called *virtual (infinitesimal) displacements* instead of the virtual velocities, and hence the definition is known as the *virtual work principle*.

## 4.14 Additional solved problems

*Problem 1*
Consider a rigid plane plate, bounded and with a smooth boundary, lying in a vertical plane. The boundary $\gamma$ (or a part of it) of the plate rolls without sliding on a horizontal line, with respect to which the plate lies in the upper half-plane (the ascending orientation on the vertical is assumed as the positive orientation). In an equilibrium configuration the centre of mass $G$ is on the vertical of the contact point $O$ (Fig. 4.5).

(i) Prove that the stability condition for the equilibrium is that the height $h$ of the centre of mass is less than the curvature radius $k_0^{-1}$ of $\gamma$ at $O$.
(ii) Compute the period of small oscillations under the above hypotheses.

*Solution*
(i) With reference to Fig. 4.5, let us compute the height of the centre of mass in the configuration when the contact point on the supporting line is moved from $O$ to $C$. Equivalently we can compute it in the frame of reference $\mathbf{t}, \mathbf{n}$, the tangent and principal normal unit vectors to $\gamma$ at $C$.

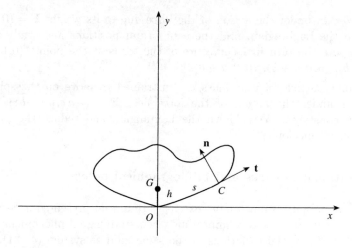

Fig. 4.5

If the parametric equations of $\gamma$ are $x = f(s)$, $y = g(s)$, we are in the conditions ensuring that at the origin $O$

$$f(0) = 0, \quad f'(0) = 1, \quad f''(0) = 0,$$
$$g(0) = 0, \quad g'(0) = 0, \quad g''(0) = k(0) = k_0.$$

In addition along all of the curve, by the orthogonality of $\mathbf{n}$ and $\mathbf{t}$, $f'' = -kg'$, $g'' = kf'$. The coordinates $\xi_0, \eta_0$ of $O$ in the system $(C, \mathbf{t}, \mathbf{n})$ are obtained from $C - O = f\mathbf{e}_1 + g\mathbf{e}_2 = -(\xi_0 \mathbf{t} + \eta_0 \mathbf{n})$, from which it follows that

$$\xi_0 = -(ff' + gg'), \quad \eta_0 = fg' - gf'.$$

The coordinates of $G$ can be found from

$$G - O = h\mathbf{e}_2 = (\xi_G - \xi_0)\mathbf{t} + (\eta_G - \eta_0)\mathbf{n}.$$

We require the height

$$\eta_G = \eta_0 + hf' = fg' - gf' + hf'.$$

By differentiation we find

$$\eta_G' = fg'' - gf'' + hf'' = k(ff' + gg' - hg')$$

(which vanishes at $s = 0$), and

$$\eta_G'' = k'(ff' + gg' - hg') + k[1 + k(g-h)f' - kfg'],$$

which yields

$$\eta_G''(0) = k_0 - k_0^2 h.$$

The stability condition is $\eta_G''(0) > 0$, and hence $h < k_0^{-1}$, proving (i). In the case that $hk_0 = 1$ we can compute $\eta_G'''$ and find $\eta_G'''(0) = -k'(0)$, and hence for stability we must have that $k'(0) = 0$. Computing $\eta_G^{IV}(0) = -k''(0)$ we find the stability condition $k''(0) < 0$, etc.

(ii) The potential energy is

$$V(s) = m\widehat{g}\eta_G(s)$$

($m$ is the mass of the plate, $\widehat{g}$ is the acceleration due to gravity). The kinetic energy is $T = \frac{1}{2}I(s)\dot{\varphi}^2$, where $I(s)$ denotes the moment of inertia with respect to the rotation axis. Let $\varphi$ be the angle that the vector $G - O$ makes with the vertical direction. We then find $\cos\varphi = f'(s)$, $\sin\varphi = g'(s)$. By differentiating the latter with respect to time we obtain $\dot{\varphi} = (g''/f')\dot{s}$. For $I(s)$ we have $I(s) = m(G-C)^2 + I_G$, where $I_G$ is the moment of inertia with respect to the axis normal to the plate for $G$. Since $\xi_G = -(ff' + gg') + hg'$ this yields $(G-C)^2 = \xi_G^2 + \eta_G^2 = f^2 + (h-g)^2$ (see also Fig. 4.5). Therefore, the Lagrangian is

$$L(s,\dot{s}) = \frac{1}{2}\left\{m[f^2 + (h-g)^2] + I_G\right\}k^2\dot{s}^2.$$

Its quadratic approximation is

$$L(s,\dot{s}) = \frac{1}{2}\left(mh^2 + I_G\right)k_0^2\dot{s}^2 - \frac{1}{2}m\widehat{g}\left(k_0 - k_0^2 h\right)s^2,$$

producing harmonic motion $\ddot{s} + \omega^2 s = 0$ of frequency

$$\omega = \left[\frac{\widehat{g}(1 - k_0 h)}{k_0(h^2 + \delta_G^2)}\right]^{1/2}, \quad \delta_G^2 = I_G/m \quad (k_0 h < 1).$$

In the degenerate case $\eta_G(s) = \text{constant}$ (circular profile of radius $h = R$ with $G$ at the centre) we find $f^2 + (h-g)^2 = R^2$, $g''/f' = 1/R^2$ and for the motion $\dot{s} = \text{constant}$. In the case $k_0 h = 1$, $k_0'' < 0$ the coefficient of $\dot{s}^2$ is approximated by $\frac{1}{2}(mh^2 + I_G)k_0^2$ to $\mathcal{O}(s^3)$, and hence the fourth-order approximation of the Lagrangian is

$$L(s,\dot{s}) = \frac{1}{2}\left(mh^2 + I_G\right)k_0^2\dot{s}^2 + \frac{1}{4!}m\widehat{g}k_0''s^4,$$

and the energy integral

$$\frac{1}{2}\left(mh^2 + I_G\right)k_0^2\dot{s}^2 + \frac{1}{4!}m\widehat{g}|k_0''|s^4 = E$$

yields the solution in the form

$$A\int_0^{s(t)} \frac{ds}{\sqrt{E - B^2 s^4}} = t, \quad A^2 = \frac{1}{2}\left(mh^2 + I_G\right)k_0^2, \quad B^2 = \frac{1}{4!}m\widehat{g}|k_0''|$$

if $s(0) = 0$, $\dot{s}(0) = \sqrt{E/A}$, from which we can compute the period

$$\theta = 4A \int_0^{(E/B^2)^{1/4}} \frac{ds}{\sqrt{E - B^2 s^4}}.$$

Apply these results to the following homogeneous systems: an arc of a circle, of an ellipse, and of a cycloid; a half-disc; a disc with a regular circular, but not concentric, hole; and the set bounded by an arc of a parabola and a segment orthogonal to the axis.

### Problem 2
A point particle $(P, m)$ is constrained to move on the smooth paraboloid $z = c(x^2 + y^2)$, $c > 0$, under the action of gravity.

(i) Write down the Lagrangian.
(ii) Prove that the component $L_z$ of the angular momentum is a first integral of the motion.
(iii) Find the value of $L_z$ for which the circle $z = z_0 > 0$ is a trajectory and find the corresponding motion.
(iv) Discuss the stability of circular motions and study the linear perturbations around them.

### Solution
(i) The Lagrangian is

$$L = \frac{1}{2}m\left[\dot{x}^2 + \dot{y}^2 + 4c^2(x\dot{x} + y\dot{y})^2\right] - mgc(x^2 + y^2).$$

It is convenient to express it in polar coordinates $(r, \varphi)$:

$$L = \frac{1}{2}m\left(\dot{r}^2 + r^2\dot{\varphi}^2 + 4c^2 r^2 \dot{r}^2\right) - mgcr^2.$$

(ii) $L_z = x\dot{y} - y\dot{x}$ is a first integral because the quantities

$$\dot{x}^2 + \dot{y}^2, \quad 2(x\dot{x} + y\dot{y}) = \frac{d}{dt}r^2, \quad x^2 + y^2 = r^2$$

are invariant under the action of the group of rotations around the $z$-axis, which is an admissible one-parameter group of symmetries for the Lagrangian.
On the other hand, writing the Lagrange equations in polar coordinates, we obtain:

$$(1 + 2c^2 r^2)\ddot{r} + 2gcr - r\dot{\varphi}^2 = 0,$$

$$\frac{d}{dt}(r^2 \dot{\varphi}) = 0.$$

The second equation expresses the conservation of $L_z = r^2 \dot{\varphi}$. Hence the first equation can be written as

$$(1 + 2c^2 r^2)\ddot{r} + 2gcr - \frac{L_z^2}{r^3} = 0.$$

(iii) From the last equation, imposing the condition that $r^2 = z_0/c$, we find that the required value of $L_z$ is

$$L_z^* = z_0 \sqrt{\frac{2g}{c}}.$$

(iv) Let us first study the perturbations keeping the value $L_z = L_z^*$[4] fixed. Writing

$$r = r_0 + \rho \quad \left(r_0 = \sqrt{\frac{z_0}{c}}\right), \quad \frac{1}{r^3} \simeq \frac{1}{r_0^3}\left(1 - 3\frac{\rho}{r_0}\right),$$

the equation for first-order perturbations is

$$(1 + 2c^2 r_0^2)\ddot{\rho} + 2gc\rho + 3\frac{L_z^{*2}}{r_0^4}\rho = 0.$$

Therefore the perturbations are harmonic oscillations with frequency

$$\omega = (8gc)^{1/2} (1 + 2c^2 r_0^2)^{-1/2}.$$

Allowing also perturbations of $L_z = L_z^*(1 + \epsilon)$ with $\epsilon \ll 1$, to first order we find

$$(1 + 2c^2 r_0^2)\ddot{\rho} + (8gc + 12gc\epsilon)\rho = 0.$$

This equation naturally describes the same oscillations as before, but with respect to the equilibrium orbit corresponding to the perturbed value of $L_z$.

*Problem 3*
A homogeneous circle of mass $M$ and radius $R$ rolls without friction in a vertical plane along a horizontal line. A rod of mass $m$ and length $\ell < 2R$ is constrained in such a way that its ends can slide with no friction on the circle. The centre $O$ of the circle is attracted by a fixed point $C$, at a distance $R$ from the horizontal line, with an elastic force. The system is subject to gravity.

(i) Write down Lagrange's equations.
(ii) Study the equilibrium configurations.
(iii) Study the small oscillations around the configuration of stable equilibrium.

---

[4] After reading Chapter 10, prove that $L_z$ is an integral independent of the Hamiltonian. Therefore $L_z$ and the total energy $E$ can be chosen independently.

*Solution*

(i) For the circle consider the angular coordinate $\varphi$ which a radius forms with the vertical, in such a way that $\varphi = 0$ if $O = C$. For the rod, take the angular coordinate $\psi$ identified by the angle between the vertical and the normal to the rod. Denote by $h = \sqrt{R^2 - \ell^2/4}$ the distance between $O$ and the centre of mass $G$ of the rod. The coordinates of $O$ are $(R\varphi, R)$, the coordinates of $G$ are

$$x_G = R\varphi + h\sin\psi, \quad y_G = R - h\cos\psi.$$

The kinetic energy of the system is

$$T = \frac{1}{2} \cdot 2MR^2\dot\varphi^2 + \frac{1}{2}\frac{1}{12}m\ell^2\dot\psi^2 + \frac{1}{2}m\left(R^2\dot\varphi^2 + h^2\dot\psi^2 + 2Rh\cos\psi\,\dot\varphi\dot\psi\right).$$

The potential energy is

$$V = \frac{1}{2}kR^2\varphi^2 + mg(R - h\cos\psi).$$

Note that the Hessian matrix of $T$ is

$$H_T = \begin{pmatrix} (m+2M)R^2 & mRh\cos\psi \\ mRh\cos\psi & m\left(\dfrac{\ell^2}{12} + h^2\right) \end{pmatrix}.$$

Verify that this matrix is positive definite (since $(m+2M)R^2 > 0$, it is enough to verify that $\det(H_T) > 0$). Lagrange's equations are

$$(2M+m)R^2\ddot\varphi + mRh\cos\psi\,\ddot\psi - mRh\sin\psi\,\dot\psi^2 + kR^2\varphi = 0,$$

$$mRh\cos\psi\,\ddot\varphi + \left(\frac{R^2}{12} + h^2\right)\sin\ddot\psi - mRh\sin\psi\,\dot\psi\dot\varphi + mgh\sin\psi = 0.$$

(ii) It can be easily verified that the equilibrium equations are

$$\varphi = 0, \quad \sin\psi = 0.$$

For $\varphi = 0$, $\psi = 0$ the Hessian matrix of $V$ is

$$H_V(0,0) = \begin{pmatrix} kR^2 & 0 \\ 0 & mgh \end{pmatrix}$$

(stable equilibrium), while for $\varphi = 0$, $\psi = \pi$

$$H_V(0,0) = \begin{pmatrix} kR^2 & 0 \\ 0 & -mgh \end{pmatrix}$$

(unstable equilibrium).

## 4.14 The dynamics of discrete systems. Lagrangian formalism

(iii) The equations of motion linearised around $\varphi = 0$, $\psi = 0$ are

$$\begin{pmatrix} (2M+m)R^2 & mRh \\ mRh & m\left(\dfrac{\ell^2}{12}+h^2\right) \end{pmatrix} \begin{pmatrix} \ddot{\varphi} \\ \ddot{\psi} \end{pmatrix} + \begin{pmatrix} kR^2 & 0 \\ 0 & mgh \end{pmatrix} \begin{pmatrix} \varphi \\ \psi \end{pmatrix} = 0.$$

In the second term there appears a diagonal matrix. After writing the system in the form

$$H_T^0 \ddot{\mathbf{x}} + H_V^0 \mathbf{x} = 0,$$

it is convenient to proceed as in Section 4.10 (but interchanging the procedures applied to the two matrices). Consider

$$(H_V^0)^{1/2} = \begin{pmatrix} \sqrt{kR} & 0 \\ 0 & \sqrt{mgh} \end{pmatrix}$$

and its inverse

$$(H_V^0)^{-1/2} = \begin{pmatrix} 1/\sqrt{kR} & 0 \\ 0 & 1/\sqrt{mgh} \end{pmatrix},$$

and define $\mathbf{y} = (H_V^0)^{1/2}\mathbf{x}$. Then the system is transformed to

$$(H_V^0)^{-1/2} H_T^0 (H_V^0)^{-1/2} \ddot{\mathbf{y}} + \mathbf{y} = 0.$$

Now let $S$ be the orthogonal transformation which diagonalises the matrix $C = (H_V^0)^{-1/2} H_T^0 (H_V^0)^{-1/2}$ and set $\mathbf{y} = S\mathbf{Z}$. The system is now transformed to

$$\begin{pmatrix} \lambda_2 & 0 \\ 0 & \lambda_1 \end{pmatrix} \ddot{\mathbf{Z}} + \mathbf{Z} = 0,$$

where $\lambda_1, \lambda_2$ are the eigenvalues of the matrix $C$. The frequencies of the normal modes are $1/\sqrt{\lambda_1}, 1/\sqrt{\lambda_2}$.

We can solve the problem in general by considering

$$A = \begin{pmatrix} a_{11} & a_{12} \\ a_{12} & a_{22} \end{pmatrix}$$

in place of $H_T^0$ (where $a_{12} \neq 0$), and

$$B = \begin{pmatrix} \gamma_1 & 0 \\ 0 & \gamma_2 \end{pmatrix}$$

instead of $H_V^0$. The matrix $C$ has the form

$$C = \begin{pmatrix} \dfrac{a_{11}}{\gamma_1} & \dfrac{a_{12}}{\sqrt{\gamma_1 \gamma_2}} \\ \dfrac{a_{12}}{\sqrt{\gamma_1 \gamma_2}} & \dfrac{a_{22}}{\gamma_2} \end{pmatrix}$$

and its eigenvalues are

$$\lambda_i = \frac{1}{2}\left\{\frac{a_{11}}{\gamma_1} + \frac{a_{22}}{\gamma_2} + (-1)^{i-1}\left[\left(\frac{a_{11}}{\gamma_1} + \frac{a_{22}}{\gamma_2}\right)^2 - 4\frac{\det(A)}{\gamma_1\gamma_2}\right]^{1/2}\right\}, \quad i = 1, 2.$$

The orthonormal eigenvectors of $C$, $\begin{pmatrix}\alpha_i\\ \beta_i\end{pmatrix}$, $i = 1, 2$, can be found by solving the systems $(i = 1, 2)$

$$\frac{a_{11}}{\gamma_1}\alpha_i + \frac{a_{12}}{\sqrt{\gamma_1\gamma_2}}\beta_i = \lambda_i\alpha_i,$$

$$\alpha_i^2 + \beta_i^2 = 1.$$

Setting

$$\mu_i = \frac{\sqrt{\gamma_1\gamma_2}}{a_{12}}\left(\lambda_i - \frac{a_{11}}{\gamma_1}\right)$$

$$= \frac{1}{2a_{12}}\left\{\xi a_{22} - \frac{1}{\xi}a_{11} + (-1)^{i-1}\left[\left(\xi a_{22} + \frac{1}{\xi}a_{11}\right)^2 - 4\det A\right]^{1/2}\right\}, \quad i = 1, 2,$$

with $\xi = \sqrt{\gamma_1/\gamma_2}$, the eigenvectors are $\begin{pmatrix}1/\sqrt{1+\mu_i^2}\\ \mu_i/\sqrt{1+\mu_i^2}\end{pmatrix}$. The orthogonal matrix $S$ diagonalising $C$ is

$$S = \begin{pmatrix}\frac{1}{\sqrt{1+\mu_1^2}} & \frac{1}{\sqrt{1+\mu_2^2}}\\ \frac{\mu_1}{\sqrt{1+\mu_1^2}} & \frac{\mu_2}{\sqrt{1+\mu_2^2}}\end{pmatrix}$$

and the normal modes are

$$\mathbf{z} = S^T B^{1/2}\mathbf{x} = \begin{pmatrix}\left(\frac{\gamma_1}{1+\mu_1^2}\right)^{1/2} & \mu_1\left(\frac{\gamma_2}{1+\mu_1^2}\right)^{1/2}\\ \left(\frac{\gamma_1}{1+\mu_2^2}\right)^{1/2} & \mu_2\left(\frac{\gamma_2}{1+\mu_2^2}\right)^{1/2}\end{pmatrix}\mathbf{x}.$$

To complete the solution of the problem under consideration it is now sufficient to substitute back.

*Problem 4*
A point particle $(P_1, m)$ moves along the circle

$$x_1 = R\cos\varphi, \quad y_1 = R\sin\varphi$$

## 4.14 The dynamics of discrete systems. Lagrangian formalism

in a horizontal plane. A second point $(P_2, m)$ is constrained on the curve

$$x_2 = R\cos\psi, \quad y_2 = R\sin\psi, \quad z_2 = h\sin\psi.$$

The two points interact with an elastic force of constant $k$; the constraints are smooth. Consider the following three cases:

(i) no gravity, $P_1$ fixed in the position $\varphi = \pi/2$;
(ii) no gravity, $P_1$ free to move on the circle;
(iii) non-zero gravity, $P_1$ free to move on the circle.

Then find what follows.

(a) For case (iii) write down the Lagrangian and Lagrange's equations.
(b) Study the equilibrium in all cases.
(c) How can the fundamental frequencies around the stable equilibrium configuration be found?

*Solution*
(a) Since $|P_1 - P_2|^2 = R^2[2 - 2\cos(\varphi - \psi)] + h^2\sin^2\psi$, the potential energy in case (iii) is

$$V(\varphi, \psi) = \frac{1}{2}kR^2\left(\gamma^2\sin^2\psi - 2\cos(\varphi - \psi)\right) + mgh\sin\psi,$$

with $\gamma = h/R$. For the kinetic energy we have

$$T = \frac{1}{2}mR^2\left[\dot\varphi^2 + (1 + \gamma^2\cos^2\psi)\dot\psi^2\right].$$

It follows that the Lagrangian is given by

$$L = \frac{1}{2}R^2\left[\dot\varphi^2 + (1 + \gamma^2\cos^2\psi)\dot\psi^2\right] - \frac{1}{2}kR^2\left(\gamma^2\sin^2\psi - 2\cos(\varphi - \psi)\right) - mgh\sin\psi$$

and Lagrange's equations are

$$mR^2\ddot\varphi + kR^2\sin(\varphi - \psi) = 0,$$

$$mR^2[(1 + \gamma^2\cos^2\psi)\ddot\psi - \gamma^2\sin\psi\dot\psi^2]$$
$$+ kR^2\left[\frac{1}{2}\gamma^2\sin 2\psi - \sin(\varphi - \psi)\right] + mgh\cos\psi = 0.$$

(b) *Case* (i) $[g = 0, \varphi = \pi/2]$

$$V = \frac{1}{2}kR^2\left(\gamma^2\sin^2\psi - 2\sin\psi\right),$$
$$V' = kR^2(\gamma^2\sin\psi\cos\psi - \cos\psi),$$
$$V'' = kR^2(\gamma^2\cos 2\psi + \sin\psi).$$

The equilibrium corresponds to $\cos\psi = 0$ and also to $\gamma^2 \sin\psi = 1$, if $\gamma > 1$. For $\psi = \pi/2$ we have $V'' = kR^2(1-\gamma^2)$; thus we get a stable equilibrium if $\gamma < 1$, and an unstable equilibrium if $\gamma > 1$. For $\psi = -\pi/2$ we have $V'' = -kR^2(1+\gamma^2)$; thus we get an unstable equilibrium. If $\gamma > 1$, let $\psi^* = \arcsin(1/\gamma^2)$; then

for $\psi = \psi^*$, $V'' = kR^2\left(\gamma^2 - \dfrac{1}{\gamma^2}\right) > 0 \Rightarrow$ stable equilibrium,

for $\psi = \pi - \psi^*$, as above.

Note that if $\gamma = 1$ we have $\psi^* = \pi - \psi^* = \pi/2$, with $V''(\pi/2) = V'''(\pi/2) = 0$, $V^{(IV)}(\pi/2) > 0$, and hence stability follows (even if the oscillations are not harmonic).

Case (ii) [$g = 0$]

$$V = \frac{1}{2}kR^2\left(\gamma^2 \sin^2\psi - 2\cos(\varphi - \psi)\right),$$

$$\frac{\partial V}{\partial \varphi} = kR^2 \sin(\varphi - \psi),$$

$$\frac{\partial V}{\partial \psi} = kR^2\left(\frac{1}{2}\gamma^2 \sin 2\psi - \sin(\varphi - \psi)\right),$$

and therefore the equilibrium equations can be written as

$$\sin(\varphi - \psi) = 0, \quad \sin 2\psi = 0,$$

with solutions

$$(0,0), \quad (0,\pi), \quad (\pi,0), \quad (\pi,\pi), \quad (\pm\pi/2, \pm\pi/2), \quad (\pm\pi/2, \mp\pi/2).$$

We compute the Hessian matrix of $V$ as

$$H_V = kR^2 \begin{pmatrix} \cos(\varphi - \psi) & -\cos(\varphi - \psi) \\ -\cos(\varphi - \psi) & \gamma^2 \cos 2\psi + \cos(\varphi - \psi) \end{pmatrix}.$$

Stability is only possible when $\varphi = \psi$, while all cases when $\varphi - \psi = \pm\pi$ are unstable. In summary:

$(0,0)$ stable, $(\pi,\pi)$ stable,

$(\pm\pi/2, \pm\pi/2) \Rightarrow \det(H_V) < 0 \Rightarrow$ unstable.

Case (iii)

$$\frac{\partial V}{\partial \varphi} = kR^2 \sin(\varphi - \psi),$$

$$\frac{\partial V}{\partial \psi} = kR^2 \left[\frac{1}{2}\gamma^2 \sin 2\psi - \sin(\varphi - \psi)\right] + mgh \cos\psi.$$

The equilibrium equations can be written as

$$\sin(\varphi - \psi) = 0,$$

$$\frac{1}{2}\gamma^2 \sin 2\psi + \alpha \cos \psi = 0, \quad \alpha = \frac{mgh}{kR^2}.$$

We again find the equations $\cos \psi = 0$, $\sin(\varphi - \psi) = 0$, yielding the solutions

$$(\pm \pi/2, \pm \pi/2), \quad (\pm \pi/2, \mp \pi/2).$$

In addition, if $\gamma^2 > \alpha$, there are the solutions of

$$\gamma^2 \sin \psi + \alpha = 0.$$

Setting

$$\chi = \arcsin\left(\frac{\alpha}{\gamma^2}\right) = \arcsin\left(\frac{mg}{kR}\right),$$

the corresponding equilibrium configurations are

$$(-\chi, -\chi), \quad (\chi - \pi, -\chi), \quad (-\chi, \chi - \pi), \quad (\chi - \pi, \chi - \pi).$$

The Hessian matrix of $V$ is

$$H_V = kR^2 \begin{pmatrix} \cos(\varphi - \psi) & -\cos(\varphi - \psi) \\ -\cos(\varphi - \psi) & \gamma^2 \cos 2\psi + \cos(\varphi - \psi) - \alpha \sin \psi \end{pmatrix},$$

and $\det(H_V) = (kR^2)^2 \cos(\varphi - \psi)(\gamma^2 \cos 2\psi - \alpha \sin \psi)$. Stability is possible only when $\varphi = \psi$. We examine these cases as follows.

$(\pi/2, \pi/2)$: $\det(H_V) < 0$, and thus we get an unstable equilibrium.

$(-\pi/2, -\pi/2)$: $\det(H_V)$ has the sign of $-\gamma^2 + \alpha$ and the second diagonal element of $H_V$ is $-\gamma^2 + 1 + \alpha$. It follows that for $\gamma^2 < \alpha$ there is stability, and for $\gamma^2 > \alpha$ there is instability.

$(-\chi, -\chi)$: Note that $\cos(-2\chi) = \cos 2\chi = 1 - 2(\alpha/\gamma^2)^2$. Hence $\det(H_V)$ has the sign of

$$\gamma^2\left(1 - 2\frac{\alpha^2}{\gamma^4}\right) + \frac{\alpha^2}{\gamma^2} = \gamma^2 - \frac{\alpha^2}{\gamma^2};$$

this is positive if $\gamma^2 > \alpha$, which is our assumption. In addition, $(H_V)_{22} = \gamma^2 - \alpha^2/\gamma^2 + 1 > 0$, and thus we get a stable equilibrium.

$(\chi - \pi, \chi - \pi)$: As above.

(c) We only need to note that the Hessian matrix of the kinetic energy

$$H_T = mR^2 \begin{pmatrix} 1 & 0 \\ 0 & 1+\gamma^2 \end{pmatrix}$$

is diagonal. It follows that in all cases examined, the formulae for the fundamental frequencies are as summarised in equations (4.153).

*Problem 5*
Consider the system of two point particles $(P_1, m_1)$, $(P_2, m_2)$ as represented in Fig. 4.6.
Find the stable equilibrium configurations and the frequencies of the normal modes.

*Solution*
Let $k$ be the elastic constant; then the Lagrangian of the system is

$$L = \frac{1}{2} m_1 R^2 \dot\varphi^2 + \frac{1}{2} m_2 \dot\xi^2 + m_1 g R \cos\varphi - \frac{1}{2} k \left[ (R\sin\varphi - \xi)^2 + R^2(2 - \cos\varphi)^2 \right].$$

Dividing this expression by $m_1 R^2$ and setting

$$\eta = \frac{\xi}{R}, \quad \Omega_1^2 = \frac{g}{R}, \quad \Omega_2^2 = \frac{k}{m_1}, \quad \mu = \frac{m_2}{m_1},$$

this can be written as

$$L = \frac{1}{2}\dot\varphi^2 + \frac{1}{2}\mu\dot\eta^2 + \Omega_1^2 \cos\varphi - \frac{1}{2}\Omega_2^2 \left[(\sin\varphi - \eta)^2 + (2 - \cos\varphi)^2\right].$$

The equilibrium equations are

$$\Omega_1^2 \sin\varphi + \Omega_2^2[(\sin\varphi - \eta)\cos\varphi + (2 - \cos\varphi)\sin\varphi] = 0,$$

$$\sin\varphi - \eta = 0,$$

yielding

$$\eta = \sin\varphi = 0.$$

Hence we conclude that there exists a configuration of unstable equilibrium $(\eta = 0, \varphi = \pi)$ and one of stable equilibrium $(\eta = 0, \varphi = 0)$.
The quadratic approximation of the Lagrangian around the latter is

$$L_2 = \frac{1}{2}\dot\varphi^2 + \frac{1}{2}\mu\dot\eta^2 - \frac{1}{2}\Omega_1^2 \varphi^2 - \frac{1}{2}\Omega_2^2 \left(2\varphi^2 - 2\varphi\eta + \eta^2\right).$$

Therefore we identify the two matrices $\overline{A}$ and $\overline{V}$:

$$\overline{A} = \begin{pmatrix} 1 & 0 \\ 0 & \mu \end{pmatrix},$$

$$\overline{V} = \Omega_2^2 \begin{pmatrix} 2 + \left(\frac{\Omega_1}{\Omega_2}\right)^2 & -1 \\ -1 & 1 \end{pmatrix}.$$

## 4.14 The dynamics of discrete systems. Lagrangian formalism

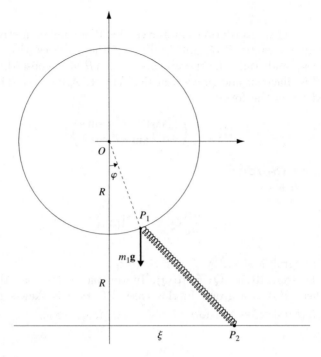

Fig. 4.6

Applying equations (4.153) we then find

$$\omega_{\pm} = \Omega_2 \left\{ \frac{1}{2}\left(1 + \left(\frac{\Omega_1}{\Omega_2}\right)^2 + \frac{1}{\mu}\right) \pm \frac{1}{2}\left[\left(1 + \left(\frac{\Omega_1}{\Omega_2}\right)^2 - \frac{1}{\mu}\right)^2 + \frac{4}{\mu}\right]^{1/2} \right\}^{1/2}.$$

What happens in the limiting cases $\Omega_1/\Omega_2 \gg 1$ and $\Omega_1/\Omega_2 \ll 1$?

*Problem 6*
Consider the holonomic system with smooth fixed constraints, and with $l$ degrees of freedom, associated with the kinetic energy

$$T = \frac{1}{2}\dot{\mathbf{q}}^T S \cdot \mathbf{q},$$

where $S$ is a constant symmetric positive definite matrix. Let $\mathbf{q} = A(s)\mathbf{Q}$ be a group of linear transformations with $A(s)$ an $l \times l$ matrix, such that $A(0) = \mathbf{1}$.

(i) How must $A(s)$ be chosen for $T$ to be invariant?
(ii) If there are no active forces, which is the corresponding first integral (Noether's theorem)?
(iii) If there exists a potential $\mathcal{U}(\mathbf{q})$, what conditions on its structure must be imposed to ensure that the transformation is admissible for the Lagrangian?

*Solution*

(i) $A^T S A = S$. This transformation leaves the Riemannian metric associated with the kinetic energy invariant. Verify that the matrices with this property form a group and study the special case $l = 2$. (*Hint:* Choose the coordinates so that $S$ is diagonal and prove that $\det(A) = 1$, $A_{11} = A_{22}$. Then $A(s)$ can be looked for in the form

$$A(s) = \begin{pmatrix} \cos s & c \sin s \\ -c^{-1} \sin s & \cos s \end{pmatrix},$$

obtaining $c = (S_{22}/S_{11})^{1/2}$.)

(ii) Since $\mathbf{p} = S\dot{\mathbf{q}}$ and

$$\frac{dA}{ds} \mathbf{Q} \bigg|_{s=0} = \dot{A}(0)\mathbf{q},$$

the first integral is given by $I = \dot{\mathbf{q}}^T S \dot{A}(0) \mathbf{q}$.

(iii) It must be that $\mathcal{U}(A(s)\mathbf{Q}) = \mathcal{U}(\mathbf{Q})$. In the particular case that $S = k\mathbf{1}$, $k > 0$, then $A(s)$ is a group In this case $\dot{A}(0) = \Omega$ is skew-symmetric and the first integral takes the form $I = \sum_{i>j} \Omega_{ij}(p_i q_j - p_j q_i)$.

# 5 MOTION IN A CENTRAL FIELD

## 5.1 Orbits in a central field

Consider a point particle of mass $m$ and denote by $\mathbf{r}$ the position vector in the space $\mathbf{R}^3$. Recall that a central field $\mathbf{F}(\mathbf{r})$ of the form

$$\mathbf{F}(\mathbf{r}) = f(r)\frac{\mathbf{r}}{r}, \quad r = |\mathbf{r}| \neq 0, \tag{5.1}$$

where $f : (0, +\infty) \to \mathbf{R}$ is a regular function, is conservative (Example 2.2) with potential energy $V(r) = -\int f(r)\,\mathrm{d}r$. The moment of the field (5.1) with respect to the centre is zero, yielding conservation of the angular momentum $\mathbf{L}$. The motion takes place in the plane passing through the origin and orthogonal to $\mathbf{L}$, namely the plane identified by the initial position vector $\mathbf{r}_0$ and the initial velocity vector $\mathbf{v}_0$ (note that in the case $\mathbf{L} = \mathbf{0}$, the vectors $\mathbf{r}_0$ and $\mathbf{v}_0$ are necessarily parallel and the motion takes place along a line).

We now introduce in the orbit plane (which we assume to be the $(x, y)$ plane, as shown in Fig. 5.1) the polar coordinates

$$x = r\cos\varphi, \quad y = r\sin\varphi. \tag{5.2}$$

The angular momentum of the system, $\mathbf{L}$, can then be identified with the component $L_z$:

$$L_z = m(x\dot{y} - y\dot{x}) = mr^2\dot{\varphi}, \tag{5.3}$$

and the conservation of $\mathbf{L}$ yields that $L_z$ is constant along the motion. The conservation of $L_z$ also yields *Kepler's second law*, about the area swept by the vector $\mathbf{r} = \mathbf{r}(t)$ in the time interval $(0, t)$:

$$S(t) = \frac{1}{2}\int_{\varphi(0)}^{\varphi(t)} r^2(\varphi)\,\mathrm{d}\varphi = \frac{1}{2}\int_0^t r^2\dot{\varphi}\,\mathrm{d}\tau = \frac{L_z t}{2m}. \tag{5.4}$$

THEOREM 5.1 (Kepler's second law)   *The areal velocity*

$$\dot{S}(t) = \frac{L_z}{2m} \tag{5.5}$$

*is a constant, and its value is also known as* the area constant.   ∎

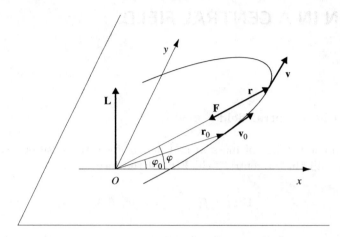

**Fig. 5.1**

Introduce the radial unit vector $\mathbf{e}_r = (\cos\varphi, \sin\varphi)$ and the orthogonal unit vector $\mathbf{e}_\varphi = (-\sin\varphi, \cos\varphi)$. The equation of motion

$$m\ddot{\mathbf{r}} = f(r)\frac{\mathbf{r}}{r} = -V'(r)\frac{\mathbf{r}}{r} \tag{5.6}$$

can then be written componentwise as follows:

$$\frac{1}{r}\frac{d}{dt}(r^2\dot{\varphi}) = 2\dot{r}\dot{\varphi} + r\ddot{\varphi} = 0,$$
$$m\ddot{r} - mr\dot{\varphi}^2 = -\frac{dV}{dr}, \tag{5.7}$$

and the first equation simply expresses the conservation of $L_z$.

DEFINITION 5.1  *The function*

$$V_e(r) = V(r) + \frac{L_z^2}{2mr^2} \tag{5.8}$$

*is called the* effective potential energy. ■

Using $V_e(r)$ in (5.6), and considering equation (5.3), the equation governing the radial motion becomes

$$m\ddot{r} = -\frac{dV_e}{dr}(r). \tag{5.9}$$

The total energy $E$ also takes a simple form, given by

$$E = \frac{1}{2}m\dot{\mathbf{r}}^2 + V(r) = \frac{1}{2}m\dot{r}^2 + V_e(r), \tag{5.10}$$

showing that the problem is equivalent to the one-dimensional motion of a point particle of mass $m$ under the action of a force field with potential energy equal to the effective potential $V_e$. Note that shifting the term $mr\dot\varphi^2$ to the right-hand side of equation (5.7) is equivalent to writing the equation of motion in the non-inertial reference system with an axis coinciding with the direction of the radius $\mathbf{r}$. The effective potential energy is the potential energy computed by such an observer.

*Remark 5.1*
It is possible to reach the same conclusion through the use of the Lagrangian formalism. Indeed, the Lagrangian of a point particle of mass $m$ under the action of a central field can be written as

$$L = \frac{m}{2}(\dot x^2 + \dot y^2 + \dot z^2) - V(\sqrt{x^2 + y^2 + z^2}), \tag{5.11}$$

and is clearly invariant under the action of rotations around the origin. It follows from Noether's theorem (4.4) that the angular momentum $\mathbf{L}$ is conserved. If the motion is in the $(x,y)$ plane and $\dot z \equiv 0$, and after introducing polar coordinates (5.2) the Lagrangian becomes

$$L = \frac{m}{2}(\dot r^2 + r^2 \dot\varphi^2) - V(r). \tag{5.12}$$

The coordinate $\varphi$ is cyclic, and hence $L_z = \partial L/\partial \dot\varphi$ is constant, and the motion is reduced to one-dimensional motion with energy (5.10). ∎

If $L_z = 0$, the motion is along the half-line $\varphi = \varphi(0)$ and can reach the origin. It is a solution of the equation $m\ddot r = f(r)$ which we discussed in Section 3.1. Otherwise the polar angle $\varphi$ is a monotonic function of time (increasing if $L_z > 0$ and decreasing $L_z < 0$). In this case the function $\varphi = \varphi(t)$ is invertible, and hence the trajectory can be parametrised as a function of the angle $\varphi$; we then write

$$\frac{dr}{dt} = \dot\varphi \frac{dr}{d\varphi} = \frac{L_z}{mr^2} \frac{dr}{d\varphi}. \tag{5.13}$$

It follows from the fact that energy is conserved that the equation for the function $r = r(\varphi)$ describing the orbit is

$$\frac{dr}{d\varphi} = \pm \frac{mr^2}{L_z} \sqrt{\frac{2}{m}(E - V_e(r))}. \tag{5.14}$$

This equation is called the *first form of the orbit equation*. The sign in (5.14) is determined by the initial conditions and equation (5.14) can be integrated by separation of variables:

$$\varphi - \varphi_0 = \pm \int_{r_0}^{r} \frac{L_z}{m} \sqrt{\frac{m}{2}} \frac{d\rho}{\rho^2 \sqrt{E - V_e(\rho)}}, \tag{5.15}$$

where $r_0 = r(\varphi_0)$. We find then $\varphi = \varphi_0 + \varphi(r)$, and inverting this expression we obtain $r = r(\varphi)$.

*Remark* 5.2

It is possible to have *circular motion*; by Theorem 5.1 such motion must be uniform, in correspondence with the values of $r$ which annihilate the right-hand side of (5.9), and hence of the stationary points of $V_e(r)$. If $r = r_c$ is one such value, equation (5.10) shows that the energy corresponding to the circular motion is $E_c = V_e(r_c)$. We shall return to this in Section 5.3. ■

*Example* 5.1: *the harmonic potential*
Let

$$V(r) = \frac{1}{2}m\omega^2 r^2 \tag{5.16}$$

(motion in an elastic field).
The effective potential corresponding to it is given by

$$V_e(r) = \frac{L_z^2}{2mr^2} + \frac{1}{2}m\omega^2 r^2. \tag{5.17}$$

It is easily verified (Fig. 5.2) that $V_e(r) \geq E_c = V_e(r_c)$, where (Remark 5.2)

$$r_c = \sqrt{\frac{|L_z|}{m\omega}}, \tag{5.18}$$

$$E_c = \omega|L_z|.$$

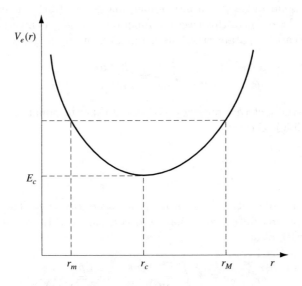

Fig. 5.2

## 5.1 Motion in a central field

For every fixed value of $E > E_c$, the equation $V_e(r) = E$ has two roots:

$$r_m = \sqrt{\frac{E}{m\omega^2}}\sqrt{1 - \sqrt{1 - \frac{E_c^2}{E^2}}},$$

$$r_M = \sqrt{\frac{E}{m\omega^2}}\sqrt{1 + \sqrt{1 - \frac{E_c^2}{E^2}}}.$$

(5.19)

From (5.15) we derive (note that $r_m/r_c = r_c/r_M$)

$$\varphi - \varphi_0 = \int_{r_0}^{r(\varphi)} \frac{dr}{r^2 \sqrt{\frac{2mE}{L_z^2} - \frac{1}{r^2} - \frac{r^2}{r_c^4}}}, \qquad (5.20)$$

from which, setting $w = 1/r^2$,

$$\varphi - \varphi_0 = \int_{1/r(\varphi)^2}^{1/r_0^2} \frac{dw}{2\sqrt{\left(\frac{m^2 E^2}{L_z^4} - \frac{1}{r_c^4}\right) - \left(w - \frac{mE}{L_z^2}\right)^2}}, \qquad (5.21)$$

and by means of the substitution

$$w - \frac{mE}{L_z^2} = \sqrt{\frac{m^2 E^2}{L_z^4} - \frac{1}{r_c^4}} \cos\psi = \frac{mE}{L_z^2}\sqrt{1 - \frac{E_c^2}{E^2}}\cos\psi,$$

we find that the integration yields $\psi/2$. Choosing the polar axis in such a way that $r = r_m$ for $\varphi = 0$, we finally obtain

$$\frac{1}{r(\varphi)^2} = \frac{mE}{L_z^2}\left[1 + \sqrt{1 - \frac{E_c^2}{E^2}}\cos 2\varphi\right]. \qquad (5.22)$$

Equation (5.22) describes an ellipse centred at the origin, whose semi-axes are given by (5.19). Note that the orbit is a circle if $E = E_c$, yielding $r = r_c$. ∎

Another form of the orbit equation can be obtained by the substitution of $u = 1/r$ into the equation of motion (5.9). Since

$$\frac{d}{dt} = \dot\varphi \frac{d}{d\varphi}, \qquad (5.23)$$

we obtain, as in (5.13),

$$\ddot r = \dot\varphi \frac{d}{d\varphi}\dot\varphi \frac{d}{d\varphi}r = \frac{L_z^2 u^2}{m^2}\frac{d}{d\varphi}u^2\frac{du}{d\varphi}\frac{dr}{du} = -\frac{L_z^2 u^2}{m^2}\frac{d^2 u}{d\varphi^2}. \qquad (5.24)$$

On the other hand

$$-\frac{\partial}{\partial r}V_e(r) = u^2 \frac{d}{du}V_e\left(\frac{1}{u}\right), \qquad (5.25)$$

and substituting (5.24) and (5.25) into (5.9) we obtain the equation

$$\frac{d^2u}{d\varphi^2} = -\frac{m}{L_z^2}\frac{d}{du}V_e\left(\frac{1}{u}\right), \qquad (5.26)$$

called *second form of the orbit equation*.

Using the variable $u$ the energy can be written in the form

$$E = \frac{1}{2m}L_z^2\left(\frac{du}{d\varphi}\right)^2 + V_e\left(\frac{1}{u}\right). \qquad (5.27)$$

*Example 5.2*
Consider the motion of a point particle of mass $m = 1$ in the central field $V(r) = -k^2/2r^2$, where $k$ is a real constant. Setting $u = 1/r$, the effective potential is given by $V_e(1/u) = \frac{1}{2}(L_z^2 - k^2)/2u^2$; substituting the latter into (5.26) yields the equation

$$\frac{d^2u}{d\varphi^2} + \left(1 - \frac{k^2}{L_z^2}\right)u = 0. \qquad (5.28)$$

If we set $\omega^2 = |1 - k^2/L_z^2|$, the solution of (5.28) corresponding to the data $u'(0) = -r'(0)/r(0)^2$ is given by

$$u(\varphi) = \begin{cases} u(0)\cos\omega\varphi + \dfrac{u'(0)}{\omega}\sin\omega\varphi, & \text{if } k^2 < L_z^2, \\ u(0) + u'(0)\varphi, & \text{if } k^2 = L_z^2, \\ u(0)\cosh\omega\varphi + \dfrac{u'(0)}{\omega}\sinh\omega\varphi, & \text{if } k^2 > L_z^2. \end{cases}$$

If $k^2 > L_z^2$ and the energy $E = \frac{1}{2}(L_z^2/2)[(u'(0))^2 - \omega^2(u(0))^2]$ is negative, the orbit is bounded (i.e. $u(\varphi)$ does not vanish) and it describes a spiral turning towards the centre of the field if $u'(0) > 0$ (the so-called *Cotes spiral*; see Danby 1988).∎

We now return to the general case and fix a non-zero value of $L_z$; the orbit belongs to

$$A_{E,L_z} = \{(r,\varphi)|V_e(r) \le E\}, \qquad (5.29)$$

consisting of one or more regions bounded by circles. In each region the radius $r$ lies between a minimum value $r_m$ (pericentre) and a maximum $r_M$ (apocentre, see Fig. 5.3), where $r_m$ and $r_M$ are two consecutive roots of $V_e(r) = E$ (except

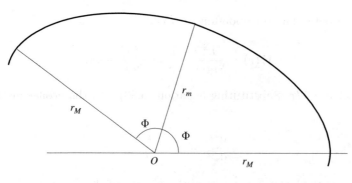

Fig. 5.3

in the case $r_m = 0$ or $r_M = +\infty$). If the point is initially positioned in a region in which

$$0 \leq r_m \leq r_M < +\infty$$

the motion is bounded. If $r_m < r_M$ (otherwise the motion is circular), from equation (5.3) it follows that the polar angle $\varphi$ varies monotonically, while $r$ oscillates periodically between $r_m$ and $r_M$. In general the orbit is not closed. Indeed, from equation (5.15) it follows that the angle $\Phi$ between a pericentre and an apocentre is given by the integral

$$\Phi = \int_{r_m}^{r_M} \frac{L_z}{mr^2} \frac{dr}{\sqrt{\frac{2}{m}[E - V_e(r)]}} \qquad (5.30)$$

(the integral converges provided that $r_m$ and $r_M$ are simple roots of the equation $V_e(r) = E$) and the angle between two consecutive pericentres is given by $2\Phi$. Hence the necessary and sufficient condition that the orbit is closed is that there exist two integers $n_1$ and $n_2$ such that

$$\Phi = 2\pi \frac{n_1}{n_2}, \qquad (5.31)$$

i.e. that the ratio $\Phi/2\pi$ is rational. If, on the other hand, $\Phi/2\pi$ is not rational, one can prove that the orbit is dense in the annulus $r_m < r < r_M$.

## 5.2 Kepler's problem

In this section we study the motion under the action of the Newtonian potential

$$V(r) = -\frac{k}{r}, \qquad k > 0. \qquad (5.32)$$

The effective potential corresponding to (5.32) is

$$V_e(r) = \frac{L_z^2}{2mr^2} - \frac{k}{r} = \frac{L_z^2 u^2}{2m} - ku, \qquad (5.33)$$

where we set $u = 1/r$. Substituting equation (5.33) into the orbit equation (5.26) we find

$$\frac{d^2 u}{d\varphi^2} = -u + \frac{km}{L_z^2}. \qquad (5.34)$$

The solution of the latter is the sum of the integral of the associated homogeneous equation, which we write in the form $u = (e/p)\cos(\varphi - \varphi_0)$, and of a particular solution of the non-homogeneous equation $u = 1/p$, corresponding to the unique circular orbit admissible for the Newtonian potential, of radius

$$r_c = p = \frac{L_z^2}{km}, \qquad (5.35)$$

and corresponding to energy

$$E_c = -\frac{k^2 m}{2L_z^2}. \qquad (5.36)$$

The parametric equation of the orbit is given by

$$u = \frac{1}{p}(1 + e\cos(\varphi - \varphi_0)),$$

and *Kepler's first law* follows:

$$r = \frac{p}{1 + e\cos(\varphi - \varphi_0)}, \qquad (5.37)$$

where $e \geq 0$ is the eccentricity of the orbit. Hence the orbit is a conical section, with one focus at the origin: if $0 \leq e < 1$ the orbit is an ellipse, if $e = 1$ it is a parabola and if $e > 1$ it is a hyperbola. The eccentricity is determined by

$$e = \sqrt{1 + \frac{2L_z^2 E}{k^2 m}} = \sqrt{1 + \frac{E}{|E_c|}}, \qquad E \geq E_c. \qquad (5.38)$$

In the elliptic case $(E < 0)$ the two semi-axes $a$ and $b$ are given by

$$a = \frac{1}{2}(r_m + r_M) = \frac{p}{1 - e^2} = \frac{k}{2|E|},$$

$$b = a\sqrt{1 - e^2} = \frac{p}{\sqrt{1 - e^2}} = \frac{|L_z|}{\sqrt{2|E|m}}, \qquad (5.39)$$

where $r_m$ and $r_M$ denote the distance from the apocentre and pericentre, respectively. The verification of equations (5.39) is immediate, while equation (5.38) is obtained from (5.27):

$$E = V_e(r_m) = -\frac{k}{p}(1+e) + \frac{L_z^2}{2mp^2}(1+e)^2,$$

$$E = V_e(r_M) = -\frac{k}{p}(1-e) + \frac{L_z^2}{2mp^2}(1-e)^2,$$

yielding

$$e^2 = 1 + \frac{2EL_z^2}{k^2 m}. \tag{5.40}$$

From Kepler's second law it follows that the area swept by the radius over a period $T$ of the motion of revolution, and hence the area of the ellipse with semi-axes $a$ and $b$, is proportional to the period $T$, namely (see (5.5))

$$\pi ab = \frac{|L_z|}{2m} T. \tag{5.41}$$

On the other hand, it follows from equations (5.35) and (5.39) that

$$|L_z| = \sqrt{akm(1-e^2)}. \tag{5.42}$$

Substituting this into (5.41) and recalling that $b = a\sqrt{1-e^2}$, we find

$$\pi a^2 \sqrt{1-e^2} = \frac{1}{2}\sqrt{a(1-e^2)}\sqrt{\frac{k}{m}} T,$$

and hence

$$\frac{a^3}{T^2} = \frac{k}{4\pi^2 m}, \tag{5.43}$$

which is the proportionality between the cube of the major semi-axis and the square of the period of revolution (*Kepler's third law*).

## 5.3 Potentials admitting closed orbits

Among all closed orbits, the circular orbits are of particular relevance. These orbits are defined by the parametrisation

$$r(t) = r_c, \qquad \varphi(t) = \varphi(0) + \omega_c t, \tag{5.44}$$

where $\omega_c^2 = f(r_c)/mr_c$ (see (5.7)). By equation (5.9) $r_c$ is necessarily a *critical point of the effective potential* $V_e$, i.e. it is a root of

$$V_e'(r) = 0, \tag{5.45}$$

and in particular it is a regular function of the angular momentum $L_z$. The energy $E_c$ of the orbit can be derived from equation (5.10):

$$E_c = V_e(r_c) = V(r_c) + \frac{L_z^2}{2mr_c^2}. \tag{5.46}$$

This is the *critical value* of $V_e$ corresponding to $r_c$.

Suppose that the point particle moves along a circular orbit and receives a small impulse $\delta \mathbf{p}$. Since the resulting variation $\delta \mathbf{L}$ of the angular momentum satisfies $\delta \mathbf{L} = \mathbf{r} \times \delta \mathbf{p}$, the component of $\delta \mathbf{p}$ which is parallel to the angular momentum $\mathbf{L}$ has the effect of slightly altering $|\mathbf{L}|$ and the plane of the orbit. The component of $\delta \mathbf{p}$ orthogonal to $\mathbf{L}$ can be decomposed into a vector parallel to $\mathbf{e}_r$—which does not change $\mathbf{L}$—and one parallel to $\mathbf{e}_\varphi$, changing the absolute value but not the direction of the angular momentum. The orbit therefore always stays in a plane close to the plane of the initial reference circular orbit, and in studying stability we can as a first approximation neglect the variation of $\mathbf{L}$.

DEFINITION 5.2 *Let $\mathbf{L}$ be fixed. A circular orbit (5.44) is called* (Lagrange) stable *if for every $\varepsilon > 0$ there exists $\delta > 0$ such that, for any initial data $r(0)$, $\dot r(0)$ such that $|r(0) - r_c| < \delta$, $|\dot r(0)| < \delta$, it holds that*

$$|r(t) - r(0)| < \varepsilon, \tag{5.47}$$

*for every $t \in \mathbf{R}$.* ∎

Remark 5.3
Note that $\varphi(0)$ does not influence the computation of the orbit, while the value of $\dot\varphi(0)$ is fixed by the choice of $r(0)$, after we fix the value of $L_z = mr(0)^2\dot\varphi(0)$. ∎

Remark 5.4
'Lagrange' stability of a circular orbit is equivalent to 'Lyapunov' stability of the equilibrium position corresponding to $r = r_c$ for the corresponding one-dimensional motion (5.9). However, it is a weaker notion than that of Lyapunov stability in the phase space $\mathbf{R}^4$ of the original problem. Indeed, we have neglected the change in polar angle, which, in general, differs linearly in time from that of the reference circular orbit. As an example, in Kepler's problem the circular orbit $r = r_c = k/2|E_c|$ is clearly stable. However, if we consider a nearby orbit, of energy $E = E_c + \delta E$ with $\delta E > 0$, by Kepler's third law (5.43) the period of the motion becomes

$$T = \sqrt{\frac{4\pi^2 m}{k}} a^{3/2} = \sqrt{\frac{\pi^2 m k^2}{2}} |E|^{-3/2}, \tag{5.48}$$

by equation (5.39). Hence, if $T_c$ is the period of the circular orbit,

$$T = T_c \frac{1}{(1 - \delta E/|E_c|)^{3/2}} = T_c \left(1 + \frac{3}{2}\frac{\delta E}{|E_c|} + \mathcal{O}((\delta E)^2)\right), \tag{5.49}$$

and thus the difference of the polar angles along the two orbits grows linearly with time. ∎

From Remark 5.4 and applying Proposition 3.1 to the one-dimensional system (5.9) we obtain the following.

THEOREM 5.2  *If the effective potential $V_e$ has an isolated relative minimum at $r_c$, the corresponding circular orbit is stable.* ∎

For the study of orbits near the circular orbits we linearise the equation of motion (5.9) following a procedure analogous to that of Section 3.4.

Setting
$$x = \frac{L_z}{\sqrt{mr}}, \tag{5.50}$$

since
$$\frac{d}{dt} = \frac{x^2}{L_z}\frac{d}{d\varphi} \quad \text{and} \quad \frac{d}{dr} = -\frac{\sqrt{m}}{L_z}x^2\frac{d}{dx},$$

it follows that equation (5.9) can be written as
$$\frac{d^2 x}{d\varphi^2} = -\frac{dW}{dx}, \tag{5.51}$$

where
$$W(x) = V\left(\frac{L_z}{\sqrt{mx}}\right) + \frac{x^2}{2} = V_e\left(\frac{L_z}{\sqrt{mx}}\right). \tag{5.52}$$

Equation (5.51) can be identified with the equation of motion for a one-dimensional mechanical system with potential energy $W(x)$ and total energy
$$E = \frac{x'^2}{2} + W(x), \tag{5.53}$$

where the polar angle $\varphi$ replaces time and $x'$ denotes the derivative of $x$ with respect to $\varphi$. The orbit equation (5.14) becomes
$$\frac{dx}{d\varphi} = x' = \pm\sqrt{2[E - W(x)]}, \tag{5.54}$$

and the angle $\Phi$ between a pericentre and an apocentre is given by
$$\Phi = \int_{x_m}^{x_M} \frac{dx}{\sqrt{2[E - W(x)]}}, \tag{5.55}$$

where $x_m = L_z/\sqrt{mr_M}$, $x_M = L_z/\sqrt{mr_m}$. Equation (5.55) expresses the half-period of the one-dimensional motion (5.51). The circular orbits are obtained in correspondence with the roots $x_c$ of $W'(x_c) = 0$, and if $W''(x_c) > 0$ then the orbit is stable. Consider an orbit near a circular orbit. Setting $x_1 = x - x_c$, the behaviour of the orbit is described by the equation obtained by linearising equation (5.51):

$$\frac{d^2 x_1}{d\varphi^2} = -W'(x_c) - W''(x_c)x_1 + \mathcal{O}(x_1^2). \tag{5.56}$$

From this it follows, by neglecting the nonlinear terms and setting $W_c = \tfrac{1}{2} W''(x_c)$, that

$$\frac{d^2 x_1}{d\varphi^2} + 2W_c x_1 = 0. \tag{5.57}$$

This is the equation of a harmonic oscillator.

The angle $\Phi_c$ between the pericentre (the maximum of $x_1$) and apocentre (minimum of $x_1$) of an orbit close to a circular orbit of radius $r_c$ is to a first approximation equal to the half-period of oscillations of the system (5.57):

$$\Phi_c = \frac{\pi}{\sqrt{2W_c}} = \frac{\pi L_z}{r_c^2 \sqrt{m V_e''(r_c)}} \tag{5.58}$$

because

$$\frac{d}{dx} = -\frac{\sqrt{m} r^2}{L_z} \frac{d}{dr},$$

which yields

$$W_c = \frac{m r_c^4}{2 L_z^2} V_e''(r_c).$$

From the expression (5.8) for the effective potential we find that

$$V_e''(r) = V''(r) + \frac{3 L_z^2}{m r^4}.$$

However $V_e'(r_c) = V'(r_c) - L_z^2/m r_c^3 = 0$, and hence $m r_c^4 / L_z^2 = r_c / V'(r_c)$, and equation (5.58) becomes

$$\Phi_c = \pi \sqrt{\frac{V'(r_c)}{r_c V''(r_c) + 3 V'(r_c)}}. \tag{5.59}$$

We can now give the proof of a theorem due to Bertrand (1873).

THEOREM 5.3 *In a central field with analytic potential energy $V(r)$, all bounded orbits are closed if and only if the potential energy $V(r)$ has one of the following forms:*

$$V(r) = \begin{cases} kr^2, \\ -\dfrac{k}{r}, \end{cases} \tag{5.60}$$

where $k > 0$. ∎

## 5.3 Motion in a central field

Hence the only central potentials for which all bounded orbits are closed are the elastic and Newtonian potentials; Bertrand commented that, 'all attracting laws allow closed orbits but natural law is the only one dictating them.'

Before the proof we consider the following.

LEMMA 5.1 *If in a central field all orbits, close to a circular orbit, are closed, then the potential energy $V(r)$ has the form*

$$V(r) = \begin{cases} ar^b, & b > -2, \; b \neq 0, \\ a \log \dfrac{r}{R}, & \end{cases} \qquad (5.61)$$

*where $a, b$ and $R$ are constants. For these potentials, the angle between a pericentre and an apocentre is given, respectively, by*

$$\Phi_c = \frac{\pi}{\sqrt{b+2}}, \quad \Phi_c = \frac{\pi}{\sqrt{2}}. \qquad (5.62)$$

*Proof*

Since $r_c$ depends continuously on $L_z$, from equation (5.59) it follows that $\Phi_c$ depends continuously on $L_z$. The condition (5.31) that ensures that an orbit is closed is false for a dense set of values of $\Phi_c$. Hence the only way that it can be satisfied for varying $L_z$ is if $\Phi_c$ is a constant, independent of $r_c$. In this case, by imposing

$$\frac{V'}{rV'' + 3V'} = c > 0, \qquad (5.63)$$

where $c$ is a constant, it follows that

$$V'' = \frac{1 - 3c}{rc} V', \qquad (5.64)$$

from which, setting $U = V'$, we get that $U = ar^{(1-3c)/c}$. Integrating the last relation we obtain (5.61). Equation (5.62) follows from (5.59). ∎

*Proof of Theorem 5.3*

By Lemma 5.1 we can assume that $V(r)$ has the form (5.61).

Note that we must have $ab > 0$ (or $a > 0$ in the logarithmic case), otherwise $V_e$ is a monotone function and no closed orbits can exist.

Let $x = L_z/\sqrt{mr}$ as in (5.50). Then we can reduce the study of the variation of the angle between a pericentre and an apocentre (which must be a rational multiple of $2\pi$ for the orbits to be closed) to the study of the variation of the period (5.55) of the one-dimensional motion (5.53), (5.54), with

$$W(x) = \begin{cases} \dfrac{x^2}{2} + \alpha x^{-b}, \; \alpha = a\left(\dfrac{L_z}{\sqrt{m}}\right)^b, & b > -2, \; b \neq 0, \\ \dfrac{x^2}{2} - a \log \dfrac{x}{X}, \; X = \dfrac{L_z}{\sqrt{m}R}. & \end{cases} \qquad (5.65)$$

Each of these potentials has a stable equilibrium point, corresponding to a stable circular orbit, and obtained by imposing $W'(x_c) = 0$:

$$x_c = \begin{cases} (ab)^{1/(2+b)}, & b > -2, \ b \neq 0, \\ \sqrt{a}. \end{cases} \quad (5.66)$$

To conclude the proof we use a result proved in Problem 4(ii) of Section 3.9, and in particular the formula (3.63) expressing the variation of the period of a one-dimensional motion with respect to the period of small oscillations, for orbits near the equilibrium position.

Setting $y = x - x_c$, $\varepsilon = E - E_c$, $\hat{W}_e(y) = W(x_c + y) - W(x_c)$, we have $\hat{W}_e(0) = \hat{W}'_e(0) = 0$,

$$\hat{W}''_e(0) = \begin{cases} b + 2, & b > -2, \ b \neq 0, \\ 2 \end{cases} \quad (5.67)$$

and

$$\Phi = \int_{y_m}^{y_M} \frac{dy}{\sqrt{2[\varepsilon - \hat{W}_e(y)]}}. \quad (5.68)$$

Use of the Taylor series expansion of $\hat{W}_e$ yields

$$\hat{W}_e(y) = \frac{b+2}{2} y^2 \left(1 - \frac{b+1}{3x_c} y + \frac{(b+1)(b+3)}{12 x_c^2} y^2\right) + \mathcal{O}(y^5). \quad (5.69)$$

Note that equation (5.69) for $b = 0$ is precisely this Taylor expansion in the case that

$$W(x) = \frac{x^2}{2} - a \log \frac{x}{X}.$$

It follows then that $\Phi = \Phi_c + \Phi_1(\varepsilon)$, where $\Phi_c$ is given by (5.62) while $\Phi_1$ is obtained from (3.63) (by the substitution $m = 1$, $\omega = 2\pi/\Phi_c$, $c_1 = (b+1)/3x_c$, $c_2 = (b+1)(b+3)/12x_c^2$):

$$\Phi_1(\varepsilon) = \frac{3\pi}{2} \frac{\varepsilon}{\omega^3} \left(\frac{5}{2} c_1^2 - 2c_2\right)$$

$$= \frac{3\varepsilon \, \Phi_c^3}{16\pi^2 x_c^2} \left[\frac{5}{18}(b+1)^2 - \frac{1}{6}(b+1)(b+3)\right]$$

$$= \varepsilon \frac{\Phi_c^3}{48\pi^2 x_c^2} (b+1)(b-2). \quad (5.70)$$

Hence $\Phi_1(\varepsilon)$ is independent of $\varepsilon$ if and only if $(b+1)(b-2) = 0$. In all other cases, the angle between a pericentre and an apocentre varies continuously with $\varepsilon$, and hence not all orbits can be closed. Thus it must be that either $b = -1$ (Newtonian potential, $\Phi_c = \pi$) or $b = 2$ (elastic potential, $\Phi_c = \pi/2$). ∎

The interesting relation between the harmonic oscillator and Kepler's problem is considered in Appendix 6, where we prove the existence of a transformation of coordinates and of time which maps the associated flows into one another.

## 5.4 Kepler's equation

In this section we derive the time dependence in Kepler's problem.

This problem can be addressed directly; this is done in the usual manner for the case of one-dimensional problems, by using the conservation of the total energy

$$E = \frac{m\dot{r}^2}{2} + \frac{L_z^2}{2mr^2} - \frac{k}{r}, \tag{5.71}$$

from which it follows that

$$t = \sqrt{\frac{m}{2}} \int_{r(0)}^{r(t)} \frac{dr}{\sqrt{\frac{k}{r} - \frac{L_z^2}{2mr^2} + E}}. \tag{5.72}$$

Using equations (5.37) and (5.3) we also find

$$t = \frac{mp^2}{L_z} \int_{\varphi(0)}^{\varphi(t)} \frac{d\varphi}{(1 + e\cos(\varphi - \varphi(0)))^2}. \tag{5.73}$$

This integral can be solved in terms of elementary functions, noting that

$$\int \frac{d\varphi}{(1 + e\cos\varphi)^2} = (1-e^2)^{-1} \left[ -\frac{e\sin\varphi}{1 + e\cos\varphi} + \int \frac{d\varphi}{1 + e\cos\varphi} \right]$$

$$= (1-e^2)^{-1} \left[ -\frac{e\sin\varphi}{1 + e\cos\varphi} + \frac{2}{\sqrt{1-e^2}} \arctan\left(\frac{\sqrt{1-e^2}\tan(\varphi/2)}{1+e}\right) \right].$$

However this approach yields a rather complicated form for the time dependence. In addition, it is necessary to invert the relation $t = t(\varphi)$.

A simpler solution, due to Kepler, consists of introducing the so-called *eccentric anomaly* $\xi$ (Fig. 5.4) and the *mean anomaly* (time normalised to an *angle*)

$$l = \frac{2\pi t}{T} = \frac{L_z}{mab} t. \tag{5.74}$$

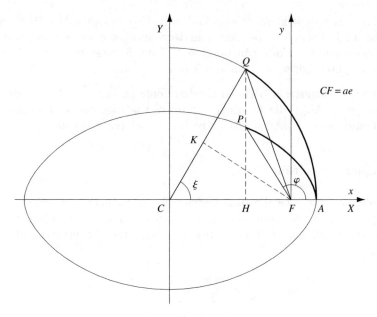

**Fig. 5.4**

The substitution $r - a = -ae \cos \xi$ (see Problem 12) in the integral (5.72) yields Kepler's equation

$$\xi - e \sin \xi = \ell, \tag{5.75}$$

and a parametric representation of the relation between $r$ and $t$:

$$\begin{aligned} r &= a(1 - e \cos \xi), \\ t &= \frac{mab}{L_z}(\xi - e \sin \xi), \end{aligned} \tag{5.76}$$

where we assume that at time $t = 0$ the particle is at the pericentre.

The variable $\xi$, the so-called *eccentric anomaly*, has a remarkable geometrical interpretation. Consider a Cartesian reference system with origin at the centre $C$ of the ellipse traced by the particle along its orbit, and denote by $X$ and $Y$ the coordinates in the standard system, different from $x = r \cos \varphi$ and $y = r \sin \varphi$ which are referred to the system with a focus $F$ at the origin, and axes parallel to $X, Y$. It then follows that

$$\begin{aligned} x &= X - ae, \\ y &= Y, \end{aligned} \tag{5.77}$$

and in the new coordinate system the equation of the ellipse is given by

$$\frac{X^2}{a^2} + \frac{Y^2}{b^2} = 1. \tag{5.78}$$

The eccentric anomaly parametrises the ellipse in the form

$$X = a\cos\xi = r\cos\varphi + ae,$$
$$Y = b\sin\xi = r\sin\varphi, \qquad (5.79)$$

and using this parametrisation the motion is described by the equation $\xi = \xi(l)$.

Trace a circle of centre $C$ and radius equal to the major semi-axis $a$ of the ellipse. If $P$ denotes the point subject to the central force field, moving around the ellipse, denote by $Q$ the point of the circle with the same $x$-coordinate as $P$. Let $H$ be a point on the $X$-axis with the same $x$-coordinate $X_H$ as $P$ and $Q$. It then follows (see Fig. 5.4) that

$$\text{area }(AFP) = -\frac{FH \cdot HP}{2} + \int_{X_H}^{a} \frac{b}{a}\sqrt{a^2 - X^2}\, dX.$$

On the other hand

$$\text{area }(AFQ) = -\frac{FH \cdot HQ}{2} + \int_{X_H}^{a} \sqrt{a^2 - X^2}\, dX,$$

and $HP = (b/a)HQ$. Therefore

$$\text{area }(AFP) = b/a\,\text{area }(AFQ). \qquad (5.80)$$

By Kepler's second law it follows that the area $(AFP) = L_z t/2m$, yielding

$$\frac{L_z t}{2m} = \frac{b}{a}\text{area }(AFQ) = \frac{b}{a}[\text{area }(ACQ) - \text{area }(QFC)]$$
$$= \frac{b}{a}\left[\frac{a^2\xi}{2} - \frac{CF \cdot QH}{2}\right] = \frac{ab}{2}(\xi - e\sin\xi).$$

This finally leads to *Kepler's equation* (5.75).

The solution of Kepler's equation is given by the following theorem (here, to avoid any confusion with number $e = 2.718\ldots$, the eccentricity is denoted by $\epsilon$. See also Problem 6).

THEOREM 5.4 *The eccentric anomaly is an analytic function of the eccentricity $\epsilon$ for $|\epsilon| < 1/e$ and of the mean anomaly $l$. Its series expansion is given by*

$$\xi(\epsilon, l) = \sum_{m=0}^{\infty} c_m(l)\frac{\epsilon^m}{m!}, \qquad (5.81)$$

where
$$c_0(l) = l, \qquad (5.82)$$

$$c_m(l) = \left(\frac{d}{dl}\right)^{m-1} (\sin l)^m \quad \text{for all } m \geq 1. \qquad (5.83)$$

The series (5.81) converges uniformly in $l$ and $\epsilon$ for all values of the eccentricity $|\epsilon| < 1/e$. ∎

*Remark* 5.5

Newton proved that the solution of Kepler's equation, which expresses the coordinates $(x, y)$ of the point particle along the orbit as a function of time, cannot be an algebraic function. His proof can be found in the book by Arnol'd (1990, Chapter 5). ∎

The proof of Theorem 5.4 is a consequence of Lagrange's formula, to be discussed in the next section. As an exercise we now verify that the first two terms in the series expansion (5.71) are correct.

Since when the eccentricity is zero we have that $\xi = l$, it is natural to seek the solution of Kepler's equation in the form of a series in $\epsilon$ with coefficients depending on the mean anomaly $l$ and with the zero-order term equal to $l$:

$$\xi(\epsilon, l) = l + \epsilon \xi_1(l) + \epsilon^2 \xi_2(l) + \mathcal{O}(\epsilon^3). \qquad (5.84)$$

Substituting equation (5.84) into (5.75) we find

$$l + \epsilon \xi_1 + \epsilon^2 \xi_2 - \epsilon[\sin l \cos(\epsilon \xi_1 + \epsilon^2 \xi_2) + \cos l \sin(\epsilon \xi_1 + \epsilon^2 \xi_2)] = l + \mathcal{O}(\epsilon^3).$$

Using the series expansion of the sine and cosine functions, up to second order, this yields

$$\epsilon \xi_1 + \epsilon^2 \xi_2 - \epsilon \sin l - \epsilon^2 \xi_1 \cos l = \mathcal{O}(\epsilon^3).$$

From the latter expression, by equating powers of $\epsilon$, it follows that

$$\xi_1(l) = \sin l,$$
$$\xi_2(l) = \xi_1(l) \cos l = \sin l \cos l,$$

and hence

$$\xi(\epsilon, l) = l + \epsilon \sin l + \epsilon^2 \sin l \cos l + \mathcal{O}(\epsilon^3),$$

which is in agreement with (5.83).

The proof of the uniform convergence, for $|\epsilon| < e^{-1}$, of the series expansion (5.81) of the solution of Kepler equation follows from the formula (see Wintner 1941)

$$c_m(l) = \frac{d^{m-1}}{dl^{m-1}}(\sin l)^m = \sum_{k=0}^{[m/2]} (m-2k)^{m-1} \frac{(-1)^k}{2^{m-1}} \binom{m}{k} \sin[(m-2k)l],$$

where $[m/2]$ denotes the integer part of $m/2$. This formula can be deduced from

$$(\sin l)^m = \sum_{k=0}^{m} \binom{m}{k} \frac{1}{(2\mathrm{i})^m} (-1)^k \mathrm{e}^{\mathrm{i}(m-2k)l}.$$

Applying Stirling's formula (Dieudonné 1968, p. 130)

$$m! \sim \sqrt{2\pi} m^{m+1/2} \mathrm{e}^{-m} \quad (\text{for } m \to +\infty),$$

since $|\sin[(m-2k)l]| \leq 1$ and $(m-2k)^{m-1} \leq m^{m-1}$, we find

$$\max_{0 \leq l \leq 2\pi} |c_m(l)| \leq \frac{1}{2^{m-1}} m^{m-1} \sum_{k=0}^{[m/2]} \binom{m}{k} \leq \frac{1}{2^{m-1}} m^{m-1} 2^m = 2 m^{m-1},$$

yielding

$$\max_{0 \leq l \leq 2\pi} \left| \frac{\epsilon^m}{m!} c_m(l) \right| \sim \frac{2 \epsilon^m m^{m-1}}{\sqrt{2\pi} \mathrm{e}^{-m} m^{m+1/2}} = \sqrt{\frac{2}{\pi}} \mathrm{e}^m m^{-3/2} \epsilon^m.$$

It follows that

$$|\xi(\epsilon, l)| \leq \sqrt{\frac{2}{\pi}} \sum_{m=0}^{\infty} (e\epsilon)^m < +\infty$$

if $0 \leq \epsilon < 1/e$, i.e. the radius of convergence of the power series for the eccentricity

$$\sum_{m=0}^{\infty} \left( \max_{0 \leq l \leq 2\pi} |c_m(l)| \right) \frac{\epsilon^m}{m!}$$

is at least $1/e$.

## 5.5 The Lagrange formula

In the previous section we studied the Kepler equation (5.75) and the convergence of its series solution (5.81), (5.83). The expansion of the eccentric anomaly in power series in terms of the eccentricity is a particular case of a more general formula. This formula was derived by Lagrange in the study of the Kepler equation. In this section we now prove that the series (5.81) is indeed a formal solution of (5.75).

Consider the more general problem of the determination of the solution $x = X(y, \epsilon)$ of an equation

$$x = y + \varepsilon f(x), \tag{5.85}$$

where $\epsilon$ is a parameter and $f$ is an analytic function of $x$ such that $f(0) = 0$. If $|\varepsilon f'(x)| < 1$ the implicit function theorem yields the existence of a unique solution

$x = X(y, \varepsilon)$ in a neighbourhood of $x = y$. We can seek the series expansion of $X$ as a power series in $\varepsilon$ with coefficients depending on $y$:

$$X(y, \varepsilon) = y + \varepsilon X_1(y) + \varepsilon^2 X_2(y) + \cdots .$$

The Lagrange formula (also called the series inversion, as it yields the inversion of the relation between $x$ and $y$ through a series expansion) is a formula for the explicit determination of the coefficients $X_n$ of this expansion, in terms of the function $f$ and of its derivatives.

THEOREM 5.5 (Lagrange) *The solution* $x = X(y, \varepsilon)$ *of (5.85) is given by the series of functions (if it converges)*

$$X(y, \varepsilon) = y + \sum_{n=1}^{\infty} \varepsilon^n X_n(y), \qquad (5.86)$$

*where*

$$X_n(y) = \frac{1}{n!} \frac{d^{n-1}}{dy^{n-1}} [(f(y))^n], \qquad (5.87)$$

*with the convention* $(d^0/dy^0) f(y) = f(y)$. *In addition, if $g$ is an analytic function such that $g(0) = 0$, then*

$$g(X) = g(y) + \sum_{n=1}^{\infty} \varepsilon^n G_n(y), \qquad (5.88)$$

*where*

$$G_n(y) = \frac{1}{n!} \frac{d^{n-1}}{dy^{n-1}} [(f(y))^n g'(y)]. \qquad (5.89)$$

∎

The proof is given below.

*Remark 5.6*
The formulae (5.86) and (5.87) are obtained from equations (5.88) and (5.89) by setting $g(y) = y$. ∎

The previous theorem yields the following corollary.

COROLLARY 5.1 *The series (5.81), (5.83) solves Kepler's equation (5.75).*

*Proof*
It is sufficient to set $x = \xi$, $y = l$, $\varepsilon = e$, $f(\xi) = \sin \xi$ and to apply the result of Theorem 5.5. ∎

*Proof of Theorem 5.5*
Let $x = X(y, \epsilon)$ be the solution; its existence is guaranteed by the implicit function theorem. By differentiating (5.85) we find

$$dX = dy + \epsilon f'(X) \, dX + f(X) \, d\epsilon, \tag{5.90}$$

from which it follows that

$$\frac{\partial X}{\partial \epsilon} = \frac{f(X)}{1 - \epsilon f'(X)} = f(X) \frac{\partial X}{\partial y}. \tag{5.91}$$

Let $F$ and $G$ be any two regular functions. From (5.91) it follows that

$$\frac{\partial}{\partial \epsilon} F(X(y, \epsilon)) = f(X(y, \epsilon)) \frac{\partial}{\partial y} F(X(y, \epsilon)), \tag{5.92}$$

and hence

$$\frac{\partial}{\partial \epsilon} \left[ F(X(y, \epsilon)) \frac{\partial}{\partial y} G(X(y, \epsilon)) \right] = \frac{\partial}{\partial y} \left[ F(X(y, \epsilon)) \frac{\partial}{\partial \epsilon} G(X(y, \epsilon)) \right] \tag{5.93}$$

as

$$\frac{\partial}{\partial \epsilon} \left( F(X) \frac{\partial}{\partial y} G(X) \right) = f(X) \frac{\partial F}{\partial y}(X) \frac{\partial G}{\partial y}(X) + F(X) \frac{\partial^2 G}{\partial \epsilon \partial y}(X),$$

$$\frac{\partial}{\partial y} \left( F(X) \frac{\partial}{\partial \epsilon} G(X) \right) = \frac{\partial F}{\partial y}(X) f(X) \frac{\partial G}{\partial y}(X) + F(X) \frac{\partial^2 G}{\partial y \partial \epsilon}(X).$$

From equations (5.92) and (5.93) we deduce by recurrence that for every integer $n \geq 1$ and for any analytic function $g$ such that $g(0) = 0$ we have

$$\frac{\partial^n g}{\partial \epsilon^n}(X(y, \epsilon)) = \frac{\partial^{n-1}}{\partial y^{n-1}} \left[ (f(X(y, \epsilon)))^n \frac{\partial g}{\partial y}(X(y, \epsilon)) \right].$$

Consequently Taylor's formula yields

$$g(X(y, \epsilon)) = g(X(y, 0)) + \sum_{n=1}^{\infty} \frac{\epsilon^n}{n!} \frac{\partial^{n-1}}{\partial y^{n-1}} \left[ f(X(y, 0))^n \frac{\partial g}{\partial y}(X(y, 0)) \right].$$

Since $X(y, 0) = y$ we find the expression (5.89). ∎

*Example 5.3*
Consider the equation

$$x = y + \varepsilon x^2. \tag{5.94}$$

Applying the Lagrange formula we find the solution

$$X(y,\epsilon) = y + \sum_{n=1}^{\infty} \frac{\epsilon^n}{n!} \frac{(2n)!}{(n+1)!} y^{n+1}.$$

On the other hand, the solution of (5.94) such that $X(0,\epsilon) = 0$ is given by

$$X(y,\epsilon) = \frac{1}{2\epsilon} - \frac{1}{2\epsilon}\sqrt{1 - 4\epsilon y}.$$

Verify as an exercise, using the Taylor series expansion of $\sqrt{1 - 4\epsilon y}$, that the Lagrange series is indeed the correct solution (hint: first show by induction that $2^n(2n-1)!! = (2n)!/n!$). For fixed $y$, what is the radius of convergence of the series? ∎

## 5.6 The two-body problem

Consider two bodies of mass $m_1$ and $m_2$ described by the position vectors $\mathbf{r}_1$ and $\mathbf{r}_2$, respectively, in $\mathbf{R}^3$. Assume the two bodies interact through a central potential $V(|\mathbf{r}_1 - \mathbf{r}_2|)$. We can prove the following.

THEOREM 5.6 *Let*

$$\mathbf{r} = \mathbf{r}_1 - \mathbf{r}_2,$$
$$\mathbf{r}_M = \frac{m_1 \mathbf{r}_1 + m_2 \mathbf{r}_2}{m_1 + m_2} \tag{5.95}$$

*be, respectively, the relative position vector of the two bodies and the position vector of the centre of mass of the system. Then the acceleration of the centre of mass with respect to an inertial system is zero, and in a reference system having the centre of mass as its origin the equations of motion are given by*

$$m\ddot{\mathbf{r}} = -\nabla_{\mathbf{r}} V, \tag{5.96}$$

*where*

$$m = \frac{m_1 m_2}{m_1 + m_2} \tag{5.97}$$

*is the so-called* reduced mass *of the two-body system. In addition, since*

$$\mathbf{r}_1 = \mathbf{r}_M + \frac{m_2}{m_1 + m_2}\mathbf{r},$$
$$\mathbf{r}_2 = \mathbf{r}_M - \frac{m_1}{m_1 + m_2}\mathbf{r}, \tag{5.98}$$

*the trajectories of the two points are planar curves lying in the same plane, and similar to each other, with similarity ratio $m_1/m_2$.*

*Proof*
The verification that $\ddot{\mathbf{r}}_M = 0$ is left to the reader. In addition,

$$m\ddot{\mathbf{r}} = m\ddot{\mathbf{r}}_1 - m\ddot{\mathbf{r}}_2 = \frac{m_1 m_2}{m_1 + m_2}\left(-\frac{1}{m_1}\nabla_{\mathbf{r}_1} V + \frac{1}{m_2}\nabla_{\mathbf{r}_2} V\right) = -\nabla_{\mathbf{r}} V.$$

The remaining claims are of immediate verification. ∎

We saw in Section 4.9 that because of the (rigid) translation invariance of the two-body system, due to the absence of forces external to the system, the centre of mass follows a linear uniform motion. The initial value problem with six degrees of freedom is thus reduced to the problem of the motion of a point particle, with mass equal to the reduced mass of the system, in a central potential field. For this problem, the considerations of the previous sections apply.

## 5.7 The *n*-body problem

The study of the problem of $n$ bodies interacting through a Newtonian potential is central to the study of celestial mechanics (for a very readable introduction, see Saari 1990). In the previous section we saw how the two-body problem is integrable and can be reduced to the motion of a single point in a central potential field. If $n \geq 3$, the resulting motion is much more complicated, and in this short introduction we only list the most classical and elementary results. For a more detailed exposition, we recommend Wintner (1941), Siegel and Moser (1971), Alekseev (1981) and the monographs of Pollard (1966, 1976) which inspired this section.

Consider $n$ bodies with masses $m_i, i = 1, \ldots, n$ and corresponding position vectors $\mathbf{r}_i$ (measured in an inertial reference frame). Let $G$ indicate the gravitational constant. The force of the gravitational attraction between any pair $(i, j)$ has the direction of $\mathbf{r}_j - \mathbf{r}_i$ and intensity equal to $Gm_i m_j / r_{ij}^2$, where $r_{ij} = |\mathbf{r}_i - \mathbf{r}_j|$. The equations describing the motion of the bodies are of the form

$$m_i \ddot{\mathbf{r}}_i = \sum_{j=1, j \neq i}^{n} \frac{Gm_i m_j}{r_{ij}^2} \frac{\mathbf{r}_j - \mathbf{r}_i}{r_{ij}}, \qquad (5.99)$$

where $i = 1, \ldots, n$.

*Remark 5.7*
The problem of the existence of solutions, local in time, of the initial value problem for equation (5.99), for prescribed initial conditions $\mathbf{r}_i(0), \mathbf{v}_i(0)$, and of the possibility of their continuation has been widely studied. Here we only note that from the theorem of the existence and uniqueness for ordinary differential equations there follows the existence of a solution of the system (5.99) for a sufficiently small time interval, assuming that at time $t = 0$ the relative distances $|\mathbf{r}_i(0) - \mathbf{r}_j(0)|$ of the points are bounded from below by a constant $r_0 > 0$. ∎

The system of differential equations (5.99) is a system of order $6n$. Denoting by

$$\mathbf{r}_M = \frac{1}{M} \sum_{i=1}^{n} m_i \mathbf{r}_i \qquad (5.100)$$

(where $M = \sum_{i=1}^{n} m_i$ represents the total mass of the system) the position vector of the centre of mass, it is easy to verify, by summing over all $i$ in equations (5.99), that

$$\ddot{\mathbf{r}}_M = \mathbf{0}. \qquad (5.101)$$

From this, it follows that *the centre of mass moves with a linear uniform motion in the chosen inertial frame of reference*. Hence the coordinates and the velocity of the centre of mass constitute a set of six first integrals of the motion for the system. In what follows we suppose that the chosen frame of reference has origin coinciding with the centre of mass, and axes parallel to those of the initial inertial system, so that

$$\mathbf{r}_M = \dot{\mathbf{r}}_M = \mathbf{0}. \qquad (5.102)$$

In addition it is easy to verify that in this frame of reference the energy integral can be written as

$$E = T + V = \sum_{i=1}^{n} \frac{1}{2} m_i |\dot{\mathbf{r}}_i|^2 - \sum_{1 \le i < j \le n} \frac{G m_i m_j}{r_{ij}}, \qquad (5.103)$$

where

$$T = \sum_{i=1}^{n} \frac{1}{2} m_i |\dot{\mathbf{r}}_i|^2 \quad \text{and} \quad V = - \sum_{1 \le i < j \le n} \frac{G m_i m_j}{r_{ij}}$$

represent the kinetic and potential energy of the system.

An equivalent formulation of the conservation of $E$ can be obtained when considering the *polar moment of inertia* of the system

$$\mathcal{I} = \frac{1}{2} \sum_{i=1}^{n} m_i r_i^2, \qquad (5.104)$$

where $r_i = |\mathbf{r}_i|$. Indeed, by differentiating twice with respect to time, we find

$$\ddot{\mathcal{I}} = \sum_{i=1}^{n} m_i |\dot{\mathbf{r}}_i|^2 + \sum_{i=1}^{n} \mathbf{r}_i \cdot m_i \ddot{\mathbf{r}} = 2T + \sum_{i=1}^{n} \mathbf{r}_i \cdot m_i \ddot{\mathbf{r}}_i,$$

and, recalling (5.99), it follows that

$$\ddot{J} = 2T + \sum_{i=1}^{n} \sum_{j=1, j \neq i}^{n} \frac{Gm_i m_j}{r_{ij}^3} [\mathbf{r}_i \cdot \mathbf{r}_j - r_i^2]$$

$$= 2T + \frac{1}{2} \sum_{i=1}^{n} \sum_{j=1, j \neq i}^{n} \frac{Gm_i m_j}{r_{ij}^3} [r_j^2 - r_i^2 - r_{ij}^2] \qquad (5.105)$$

$$= 2T - \frac{1}{2} \sum_{i=1}^{n} \sum_{j=1, j \neq i}^{n} \frac{Gm_i m_j}{r_{ij}} = 2T + V = E + T.$$

The identity $\ddot{J} = E + T$ is called the *Lagrange–Jacobi identity*.

The coordinates, the velocity of the centre of mass of the system and the total energy $E$ constitute a set of seven independent constants of the motion for the $n$-body problem. Three more constants are given by the *conservation of the total angular momentum* $\mathbf{L}$:

$$\mathbf{L} = \sum_{i=1}^{n} m_i \mathbf{r}_i \times \dot{\mathbf{r}}_i. \qquad (5.106)$$

Indeed, since $\dot{\mathbf{r}}_i \times \dot{\mathbf{r}}_i = 0$, we find

$$\dot{\mathbf{L}} = \sum_{i=1}^{n} m_i \mathbf{r}_i \times \ddot{\mathbf{r}}_i = \sum_{i=1}^{n} \sum_{j=1, j \neq i}^{n} \frac{Gm_i m_j}{r_{ij}^3} \mathbf{r}_i \times \mathbf{r}_j = 0,$$

because in the last summation there appear both $\mathbf{r}_i \times \mathbf{r}_j$ and $\mathbf{r}_j \times \mathbf{r}_i = -\mathbf{r}_i \times \mathbf{r}_j$ with equal coefficients. This yields the proof of the following theorem.

THEOREM 5.7 *The system of equations (5.99) admits ten first integrals of the motion.* ∎

We end this brief introduction with an important result (see Sundman 1907) regarding the possibility that the system of $n$ points undergoes *total collapse*, i.e. that all $n$ particles are found in the same position at the same time, colliding with each other (in this case, of course equation (5.99) becomes singular).

THEOREM 5.8 (Sundman) *A necessary condition that the system undergoes total collapse is that the total angular momentum vanishes.* ∎

Before proving Theorem 5.8, we consider a few lemmas.

LEMMA 5.2 (Sundman inequality) *If $L$ is the magnitude of the total angular momentum (5.106) then*

$$L^2 \leq 4J(\ddot{J} - E). \qquad (5.107)$$

*Proof*
From definition (5.106) it follows that

$$L \leq \sum_{i=1}^{n} m_i |\mathbf{r}_i \times \dot{\mathbf{r}}_i| \leq \sum_{i=1}^{n} m_i r_i v_i,$$

where $v_i = |\dot{\mathbf{r}}_i|$. Applying the Cauchy–Schwarz inequality we find

$$L^2 \leq \left(\sum_{i=1}^{n} m_i r_i^2\right)\left(\sum_{i=1}^{n} m_i v_i^2\right) = 4\mathcal{J}T.$$

Hence the result follows from the Lagrange–Jacobi identity (5.105). ∎

**LEMMA 5.3** *Let $f : [a,b] \to \mathbf{R}$ be a function of class $\mathcal{C}^2$ such that $f(x) \geq 0$, $f''(x) \geq 0$ for every $x \in [a,b]$. If $f(b) = 0$ then $f'(x) \leq 0$ for every $x \in [a,b]$.* ∎

The proof of Lemma 5.3 is left to the reader as an exercise.

**LEMMA 5.4** *The polar moment of inertia (5.104) is given by*

$$\mathcal{J} = \frac{1}{2M} \sum_{1 \leq i < j \leq n} m_i m_j r_{ij}^2. \tag{5.108}$$

*Proof*
We have that

$$\sum_{i=1}^{n} m_i (\mathbf{r}_i - \mathbf{r}_j)^2 = \sum_{i=1}^{n} m_i r_i^2 - 2\mathbf{r}_j \cdot \sum_{i=1}^{n} m_i \mathbf{r}_i + \left(\sum_{i=1}^{n} m_i\right) r_j^2.$$

Hence from (5.102) it follows that

$$\sum_{i=1}^{n} m_i (\mathbf{r}_i - \mathbf{r}_j)^2 = 2\mathcal{J} + M r_j^2.$$

Multiplying both sides by $m_j$, summing over $j$ and using that $(\mathbf{r}_i - \mathbf{r}_j)^2 = r_{ij}^2$ we find

$$\sum_{i=1}^{n}\sum_{j=1}^{n} m_i m_j r_{ij}^2 = 2\mathcal{J} \sum_{j=1}^{n} m_j + M \sum_{j=1}^{n} m_j r_j^2 = 4\mathcal{J}M.$$

From this relation, since $r_{ii} = 0$, we deduce equation (5.108). ∎

Equation (5.108) shows that total collapse implies the vanishing of $\mathcal{J}$ (all particles collide at the origin).

*Proof of Theorem 5.8*
We first show that any total collapse must necessarily happen in finite time, i.e. it is impossible that $\mathcal{J}(t) \to 0$ for $t \to +\infty$. Indeed, if for $t \to +\infty$ we find $r_{ij} \to 0$ for every $i$ and $j$, then $V \to -\infty$. From the Lagrange–Jacobi identity (5.104) it follows that $\ddot{\mathcal{J}} \to +\infty$. There then exists a time $\hat{t}$ such that for every $t \geq \hat{t}$

we have $\ddot{\mathfrak{I}}(t) \geq 2$, and hence $\mathfrak{I}(t) \geq t^2 + At + B$ for $t \to \infty$, contradicting the hypothesis $\mathfrak{I} \to 0$.

Hence any total collapse must happen at some finite time $t_1$. We have just showed that $V \to -\infty$, $\mathfrak{I} \to 0$ and $\ddot{\mathfrak{I}} \to +\infty$ as $t \to t_1$, and thus an application of Lemma 5.3 to $\mathfrak{I}(t)$ yields that $\dot{\mathfrak{I}}(t) \leq 0$ for $t_2 < t < t_1$. Multiplying both sides of the Sundman inequality (5.107) by $-\dot{\mathfrak{I}}/\mathfrak{I}$ we find

$$-\frac{1}{4}L^2 \frac{\dot{\mathfrak{I}}}{\mathfrak{I}} \leq E\dot{\mathfrak{I}} - \dot{\mathfrak{I}}\ddot{\mathfrak{I}},$$

and integrating both sides of the latter with respect to time, we find for $t \in (t_2, t_1)$ that

$$\frac{1}{4}L^2 \log \frac{\mathfrak{I}(t_2)}{\mathfrak{I}(t)} \leq E[\mathfrak{I}(t) - \mathfrak{I}(t_2)] - \frac{1}{2}[\dot{\mathfrak{I}}^2(t) - \dot{\mathfrak{I}}^2(t_2)] \leq E\mathfrak{I}(t) + C,$$

where $C$ is a constant. Hence

$$\frac{1}{4}L^2 \leq \frac{E\mathfrak{I}(t) + C}{\log(\mathfrak{I}(t_2)/\mathfrak{I}(t))},$$

which tends to 0 for $t \to t_1$. Since $L$ is constant, it follows that $L = 0$. ∎

## 5.8 Problems

**1.** Study the existence and stability of circular orbits for the following central potentials:

$$V(r) = ar^{-3/2} + br^{-1},$$
$$V(r) = ae^{br},$$
$$V(r) = ar\sin(br),$$

for varying real parameters $a$ and $b$.

**2.** Find a central potential for which the polar angle varies with time as $\varphi(t) = \arctan(\omega t)$, with $\omega \in \mathbf{R}$ fixed by the initial conditions.

**3.** Solve the orbit equation for the potential $V(r) = -kr^{-1} + ar^{-2}$, where $k > 0$ and $a$ are prescribed constants. (Answer: $r = p/(1 + e\cos(\omega\varphi))$ with $\omega^2 = 1 + 2am/L_z^2$.)

**4.** A spherical galaxy has approximately constant density near its centre, while the density tends to zero as the radius increases. The gravitational potential it generates is then proportional to $r^2$, a constant for small values of $r$, and proportional to $r^{-1}$ for large values of $r$. An example of such a potential is given by the so-called *isochronous potential* (see Binney and Tremaine 1987, p. 38):

$$V(r) = -\frac{k}{b + \sqrt{b^2 + r^2}}.$$

Introduce an auxiliary variable $s = 1 + \sqrt{r^2/b^2 + 1}$. Prove that, if $s_1$ and $s_2$ correspond, respectively, to the distance of the pericentre and of the apocentre, the radial period is given by

$$T_r = \frac{2\pi b}{\sqrt{-2E}} \left[1 - \frac{1}{2}(s_1 + s_2)\right].$$

In addition, prove that $s_1 + s_2 = 2(1 - k/2Eb)$, and hence that

$$T_r = \frac{2\pi k}{(-2E)^{3/2}};$$

thus $T_r$ depends only on the energy $E$, and not on the angular momentum $L$ (this is the reason $V(r)$ is called isochronous). Prove also that the increment $\Phi$ of the azimuthal angle between two consecutive passages at pericentre and apocentre is given by

$$\Phi = \frac{\pi}{2}\left[1 + \frac{L}{\sqrt{L^2 + 4kb}}\right].$$

Note that $\Phi \to \pi$ when $b \to 0$.

5. Find the series expansion of the solution of the equation

$$\xi - e \cos \xi = l$$

and prove that it converges uniformly in $l$ for $|e|$ small enough.

6. Let $y = \sin x = x - x^3/3! + x^5/5! + \mathcal{O}(x^7)$. Compute the expansion of $x = X(y)$ up to terms of order $\mathcal{O}(y^7)$. Verify the accuracy by comparing with the Taylor series expansion of $x = \arcsin y$.

7. Let $y = x - x^2 + x^3 - x^4 + \cdots$. Compute $x = x(y)$ up to fourth order. (Answer: $x = y + y^2 + y^3 + y^4 + \cdots$.)

8. Let $y = x - 1/4x^2 + 1/8x^3 - 15/192x^4 + \cdots$. Compute $x = X(y)$ up to fourth order. (Answer: $x = y + 1/4y^2 + \mathcal{O}(y^5)$.)

9. Solve, using the Lagrange formula, the equation

$$x = y + \varepsilon \sin(hx).$$

For fixed $y$, for which values of $\varepsilon$ does the series converge?

10. Solve, using the Lagrange formula, the equation

$$x = y + \varepsilon x^3$$

and discuss the convergence of the series.

**11.** In the Kepler problem, express the polar angle $\varphi$ as a function of the average anomaly $l$ and of the eccentricity $e$ up to terms of order $\mathcal{O}(e^3)$. (Answer: $\varphi = l - 2e\sin l + 5/4 e^2 \sin(2l) + \mathcal{O}(e^3)$.)

**12.** With the help of Fig. 5.4 show that $r = a(1 - e\cos\xi)$ (remember that $CF = ae$).

## 5.9 Additional remarks and bibliographical notes

In this chapter we studied central motions and we have seen the most elementary results on the $n$-body problem. What we proved constitutes a brief and elementary introduction to the study of celestial mechanics. According to Poincaré, the final aim of this is 'to determine if Newton's law is sufficient to explain all the astronomical phenomena'.[1] Laskar (1992) wrote an excellent introduction to the history of research on the stability of the solar system, from the first studies of Newton and Laplace up to the most recent developments of the numerical simulations which seem to indicate that the motion of the planets, on long time-scales, is better analysed with the tools of the theory of chaotic dynamical systems. This is at odds with the previous firm (but unproven) belief that planets are unchangeable, stable systems.[2]

For a deeper study of these topics, we suggest the book of Pollard (1976), of exceptional clarity and depth. In it, after a summary of the fundamental notions of mechanics and analytical mechanics, the reader can find the most important results regarding the $n$-body problem. The problem with $n = 3$ is treated in depth, with a careful analysis of all solutions discovered by Lagrange and Euler and of the equilibrium positions of the reduced circular plane problem (where it is assumed that the three bodies belong to the same plane and that two of them—the *primary* bodies—move with a uniform circular motion around the centre of mass). Another advanced work on the same subject is the book by Meyer and Hall (1992). These authors choose from the first pages a 'dynamical systems' approach to celestial mechanics, which requires a greater mathematical background (the present book should contain all necessary prerequisites to such a reading).

The monograph by Wintner (1941) is of fundamental importance, but it is not easy to read.

In Wintner's book one can find a discussion of Bertrand's theorem (1873) (see also Arnol'd 1978a and Albouy 2000).

The Kepler equation, and the various analytical and numerical methods for its solution, are discussed in detail by Danby (1988) and by Giorgilli (1990). In

---

[1] '*Le but final de la mécanique céleste est de résoudre cette grand question de savoir si la loi de Newton explique à elle seule tous les phénomènes astronomiques*' (Poincaré 1892, p. 1).

[2] However, Newton had already expressed doubts in the *Principia Mathematica* as he considered that it was necessary to allow for the intervention of a superior being in order to maintain the planets near their Keplerian orbits for very long times.

the former, one can also find the listings of some easy BASIC programs to solve Kepler's equation and determine the time dependence in Kepler's problem. In addition, the book contains a detailed discussion of the problem of determining all elements of a Keplerian orbit (eccentricity, semi-axes, inclination, etc., see Section 9.8) starting from astronomical data.

In this book it was impossible, due to space constraints, to go into a more detailed study of the geometric and topological aspects of the two-body and $n$-body problems (although Appendix 6 partially fills this gap). The articles by Smale (1970a,b) and the discussion by Abraham and Marsden (1978) are excellent but very difficult; the work of Alekseev (1981) is more accessible but less complete.

Finally, a curious observation: the Ptolemaic theory of epicycles has recently been interpreted through Fourier series expansions and the theory of quasi-periodic functions; we recommend the first volume of Sternberg's book (1969).

## 5.10 Additional solved problems

*Problem 1*
Determine a central force field in which a particle of mass $m$ is allowed to describe the orbit $r = r_0 e^{c\varphi}$, where $r_0 > 0$ is fixed, $c$ is a non-zero constant and $\varphi$ is the polar angle. Compute $\varphi = \varphi(t)$ and $r = r(t)$.

*Solution*
Setting $u = 1/r$ we have $u = u_0 e^{-c\varphi}$, where $u_0 = 1/r_0$. From equation (5.26) it follows that

$$\frac{d}{du} V_e\left(\frac{1}{u}\right) = -\frac{c^2 L_z^2}{m} u,$$

from which by integration we obtain

$$V(r) = -\frac{L_z^2}{2mr^2}(1 + c^2).$$

We find $\varphi(t)$ starting from the conservation of the angular momentum: $mr^2 \dot\varphi = L_z$, from which it follows that $mr_0^2 e^{2c\varphi(t)} \dot\varphi(t) = L_z$. The last relation can be integrated by separation of variables.

*Problem 2*
Prove that in a central force field with potential energy $V(r) = -\alpha e^{-kr}/r$, where $\alpha$ and $k$ are two positive constants, for sufficiently small values of the angular momentum there can exist a stable circular orbit.

*Solution*
The effective potential energy is given by

$$V_e(r) = \frac{L_z^2}{2mr^2} - \alpha \frac{e^{-kr}}{r},$$

and hence $\lim_{r \to \infty} V_e(r) = 0$, $\lim_{r \to 0+} V_e(r) = +\infty$. Differentiating once we find

$$V_e'(r) = -\frac{L_z^2}{mr^3} + \alpha k \frac{e^{-kr}}{r} + \alpha \frac{e^{-kr}}{r^2}.$$

Circular orbits correspond to critical points of the effective potential energy. Thus we must study the equation

$$re^{-kr}(1+kr) = \frac{L_z^2}{m\alpha}. \tag{5.109}$$

The function $f(x) = xe^{-x}(1+x)$, when $x \in \mathbf{R}^+$ varies, has a unique critical point (an absolute maximum) for $x = (\sqrt{5}+1)/2$. If we set $M = f\left((\sqrt{5}+1)/2\right)$, we find that (5.109) has a solution if and only if $L_z^2 \leq m\alpha M/k$. In addition, if $L_z^2 < m\alpha M/k$ there are two solutions corresponding to a relative minimum and maximum of $V_e$, and hence there exists a stable circular orbit, whose radius tends to zero if $L_z^2 \to 0$.

*Problem 3*
(From Milnor 1983, pp. 353–65.)
Consider Kepler's problem and suppose that the angular momentum is non-zero. Prove that, as time $t$ varies, the velocity vector $\mathbf{v} = \dot{\mathbf{r}}(t)$ moves along a circle $C$ lying in a plane $P$ passing through the origin. This circle, and its orientation, uniquely determines the orbit $\mathbf{r} = \mathbf{r}(t)$. The orbit is elliptic, hyperbolic or parabolic according to whether the origin is at the interior, at the exterior, or exactly on the circle $C$.

*Solution*
Let $R = k/L_z$. From equation (5.6) (with $V(r) = -k/r$) it follows that $d\mathbf{v}/d\varphi = -R\mathbf{r}/r$ which yields, by integration, $\mathbf{v} = R\mathbf{e}_\varphi + \mathbf{c}$, with $\mathbf{c} = (c_1, c_2)$ an integration constant. This shows that $\mathbf{v}$ moves along a circle $C$ with centre $\mathbf{c}$, radius $R$, lying in the same plane containing the orbit. Let $\epsilon = |\mathbf{c}|/R$ be the distance of $\mathbf{c}$ from the origin, divided by the radius of the circle $C$. If we choose the orientation of the axes $x, y$ of the plane so that $\mathbf{c}$ lies on the $y$-axis, we find

$$\mathbf{v} = R(-\sin\varphi, \epsilon + \cos\varphi). \tag{5.110}$$

It follows that $L_z = mR(1 + \epsilon\cos\varphi)$ and we find again that $r = L_z/[mR(1+\epsilon\cos\varphi)]$, and hence equation (5.37) in which the eccentricity is given by $\epsilon$. This shows how $\mathbf{c}$ lies at the interior or exterior, or it belongs to the circle $C$ according to whether the orbit is elliptic, hyperbolic or parabolic.

*Problem 4*
Prove that there are no equilibrium points for the $n$-body problem.

*Solution*
An equilibrium point is a solution of the system of equations

$$-\nabla_{\mathbf{r}_i} V(\mathbf{r}_1,\ldots,\mathbf{r}_n) = 0, \quad i = 1,\ldots,n,$$

where $V$ is the potential energy (see (5.103)). Since $V$ is homogeneous of degree $-1$ we have

$$-\sum_{i=1}^n \mathbf{r}_i \cdot \nabla_{\mathbf{r}_i} V = V. \qquad (5.111)$$

However, since $V$ is a sum of negative terms, it follows that $V < 0$, contrary to the requirement that, at an equilibrium point, the left-hand side of (5.111) vanishes.

*Problem 5*
A *central configuration* in the $n$-body problem is a solution of (5.99) of the form $\mathbf{r}_i(t) = \psi(t)\mathbf{a}_i$, where $\psi$ is a real function and the vectors $\mathbf{a}_1,\ldots,\mathbf{a}_n$ are constant. Prove that if $n=3$, for any values of the masses there exists a central configuration in which the three particles are at the vertices of an equilateral triangle (Lagrange's solutions).

*Solution*
Without loss of generality we can assume that the centre of mass of the system is fixed at the origin. Substituting $\mathbf{r}_i = \psi(t)\mathbf{a}_i$ into equation (5.98) we find

$$|\psi|^3 \psi^{-1} \ddot{\psi} m_i \mathbf{a}_i = \sum_{j \neq i} \frac{G m_i m_j}{|\mathbf{a}_j - \mathbf{a}_i|^3}(\mathbf{a}_j - \mathbf{a}_i). \qquad (5.112)$$

Since the right-hand side is constant, by separation of variables we find

$$\ddot{\psi} = -\frac{\lambda \psi}{|\psi|^3}, \qquad (5.113)$$

$$-\lambda m_i \mathbf{a}_i = \sum_{j \neq i} \frac{G m_i m_j}{|\mathbf{a}_j - \mathbf{a}_i|^3}(\mathbf{a}_j - \mathbf{a}_i). \qquad (5.114)$$

Equation (5.113) has infinitely many solutions while (5.114) can be written as

$$\nabla_{\mathbf{r}_i} V(\mathbf{a}_1,\ldots,\mathbf{a}_n) + \lambda \nabla_{\mathbf{r}_i} \mathfrak{I}(\mathbf{a}_1,\ldots,\mathbf{a}_n) = 0. \qquad (5.115)$$

By the theorem of Lagrange multipliers, the system of vectors $\mathbf{a}_1,\ldots,\mathbf{a}_n$ yielding a central configuration corresponds to an extremal point of the potential energy, under the constraint that the polar moment of inertia (see (5.104)) $\mathfrak{I}$ is fixed. Let $a_{ij} = |\mathbf{a}_i - \mathbf{a}_j|$. Then $1 \leq i < j \leq 3$, $M = \sum_{i=1}^3 m_i$ and since

$$4M\mathfrak{I} = \sum_{1 \leq i < j \leq 3} m_i m_j a_{ij}^2,$$

equation (5.115) can be written as

$$-\frac{Gm_i m_j}{a_{ij}^2} + \frac{\lambda m_i m_j a_{ij}}{2M} = 0 \qquad (5.116)$$

and admits the unique solution $a_{ij} = (2MG/\lambda)^{1/3}$.

## Problem 6
A limiting case of the three-body problem is the so-called reduced plane circular three-body problem, when one considers the motion of a point particle (of mass $m$) under the action of two masses moving along circular orbits. Suppose that the two masses are $\alpha$ and $1-\alpha$, with centre of mass at the origin. Show that in the frame of reference moving with the two points, there exist three equilibrium configurations when the three bodies are aligned.

## Solution
The chosen reference system rotates with angular velocity $\omega$. Choose the axes in such a way that the coordinates of the points of mass $\alpha$ and $1-\alpha$ are $(1-\alpha, 0)$ and $(-\alpha, 0)$, respectively. The kinetic energy of the point particle is

$$T(x,y,\dot{x},\dot{y}) = \frac{1}{2} m[\dot{x}^2 + \dot{y}^2 + 2\omega y \dot{x} - 2\omega x \dot{y} + \omega^2 x^2 + \omega^2 y^2], \qquad (5.117)$$

while the potential energy is

$$V(x,y) = -\frac{Gm\alpha}{\sqrt{(x-1+\alpha)^2 + y^2}} - \frac{Gm(1-\alpha)}{\sqrt{(x+\alpha)^2 + y^2}}. \qquad (5.118)$$

It follows that the equations of motion are

$$m\ddot{x} - 2m\omega\dot{y} = \frac{\partial V_e}{\partial x}(x,y),$$
$$m\ddot{y} - 2m\omega\dot{x} = \frac{\partial V_e}{\partial y}(x,y), \qquad (5.119)$$

where $V_e(x,y) = V(x,y) - \frac{1}{2}m\omega^2(x^2+y^2)$. Hence the equilibrium positions are given by $\dot{x} = \dot{y} = 0$ and solutions $(x,y)$ of $\nabla V_e(x,y) = 0$. Along the axis $y=0$ we find

$$V_e(x,0) = \frac{1}{2}m\omega^2 x^2 \pm \frac{Gm\alpha}{x-1+\alpha} \pm \frac{Gm(1-\alpha)}{x+\alpha}, \qquad (5.120)$$

where the signs are chosen in such a way that each term is positive: if $x < -\alpha$ the signs are $(-,-)$, if $-\alpha < x < 1-\alpha$ we have $(-,+)$, if $x > 1-\alpha$ we have $(+,+)$. Since $V_e(x,0) \to +\infty$ when $x \to \pm\infty$, $x \to -\alpha$, $x \to 1-\alpha$ and $V_e(x,0)$ is a convex function (verify that

$$\frac{\partial^2 V_e}{\partial x^2}(x,0) > 0$$

for all $x$), $V_e$ has exactly one critical point in each of these three intervals.

*Problem 7*

An alternative way to solve Kepler's equation (5.75) is obtained by computing the Fourier series expansion (see Appendix 7) of the eccentric anomaly as a function of the mean anomaly. Using the definition

$$J_n(x) = \frac{1}{2\pi} \int_0^{2\pi} \cos(ny - x \sin y) \, dy \tag{5.121}$$

of the $n$th *Bessel function* prove that

$$\xi = l + \sum_{n=1}^{\infty} \frac{2}{n} J_n(ne) \sin(nl). \tag{5.122}$$

*Solution*

Let $e \sin \xi = \sum_{n=1}^{\infty} A_n \sin(nl)$, where the coefficients $A_n$ are given by $A_n = (2/\pi) \int_0^\pi e \sin \xi \sin(nl) \, dl$. Integrating by parts and using the substitution $l = \xi - e \sin \xi$, we find

$$A_n = \frac{2}{n\pi} \int_0^\pi \cos(nl) \frac{d}{dl}(e \sin \xi) \, dl = \frac{2}{n\pi} \int_0^\pi \cos(nl) \frac{e \cos \xi}{1 - e \cos \xi} \, dl$$

$$= \frac{2}{n\pi} \int_0^\pi \cos(nl) \left( \frac{d\xi}{dl} - 1 \right) dl = \frac{2}{n\pi} \int_0^\pi \cos[n(\xi - e \sin \xi)] \, d\xi = \frac{2}{n} J_n(ne).$$

Recalling the power series expansion of the Bessel functions

$$J_n(x) = \sum_{k=0}^{\infty} \frac{(-1)^k x^{n+2k}}{2^{n+2k} k!(n+k)!}, \tag{5.123}$$

it is possible to verify that (5.122) leads to (5.81)–(5.83) and vice versa. It is, however, necessary to appeal to equation (5.121) to get a better version of Theorem 5.4 and to prove the convergence of the series expansions (5.81)–(5.83) for all values of the eccentricity $|e| < 0.6627434\ldots$. The series (5.122) converges for all values of $e \in [0, 1]$ (see Watson 1980, in particular Chapter XVII '*Kapteyn series*').

# 6 RIGID BODIES: GEOMETRY AND KINEMATICS

## 6.1 Geometric properties. The Euler angles

DEFINITION 6.1  *A rigid body is a set of points* $\{P_1, \ldots, P_n\}$, $n \geq 2$ *which satisfy the rigidity constraints*

$$|P_i - P_j| = r_{ij} > 0, \qquad 1 \leq i < j \leq n, \tag{6.1}$$

*where $r_{ij}$ are prescribed lengths.* ■

We note that:

(a) the constraints (6.1) must satisfy certain compatibility conditions (e.g. the triangle inequalities $r_{ij} < r_{ik} + r_{kj}, 1 \leq i < k < j \leq n$);
(b) the $n(n-1)/2$ equations (6.1) are not all independent when $n > 3$.

Regarding (b), if we consider three points $P_1, P_2, P_3$ in the system, not lying on the same line, for a chosen reference frame $\Sigma = (\Omega, \xi_1, \xi_2, \xi_3)$ we can assign to $P_1$ three coordinates, to the point $P_2$ two coordinates, and to the point $P_3$ one coordinate.

It is clear that having fixed the configuration of the triangle $P_1 P_2 P_3$ with respect to $\Sigma$, the coordinates of every other point in the system are automatically determined as functions of the six prescribed parameters. From this follows the well-known property stating that *a rigid system containing at least three points, not all lying on the same line, has six degrees of freedom.*

In the case of a rigid body with all points lying on the same line, the number of degrees of freedom is five (it is sufficient to determine the configuration of the segment $P_1 P_2$ with respect to $\Sigma$), while for a rigid body in the plane the degrees of freedom are three.

We now introduce the concept of a *body reference frame*. Since the configuration of a rigid body is determined by that of any triangle $P_1 P_2 P_3$ formed by three of its points, we can associate with it a frame $S \equiv (O, x_1, x_2, x_3)$, called a body frame, for example assuming that the origin $O$ coincides with $P_1$, the $x_1$-axis contains the side $P_1 P_2$, and the plane $(x_1, x_2)$ contains the triangle $P_1 P_2 P_3$ and the $x_3$-axis, with the orientation of the unit vector $\mathbf{e}_3 = \mathbf{e}_1 \times \mathbf{e}_2$, where $\mathbf{e}_1$ and $\mathbf{e}_2$ are the unit vectors of the $x_1$- and $x_2$-axes (Fig. 6.1).

Hence the problem of the determination of the configuration of a rigid body with respect to a frame $\Sigma$ (which we call 'fixed') is equivalent to the problem of determining the configuration with respect to $\Sigma$ of one of its body frames $S$.

A body frame is determined by choosing an element $g$ of the group of orientation-preserving isometries of three-dimensional Euclidean space (we exclude reflections). We denote by SO(3) the group of real $3 \times 3$ orientation-preserving

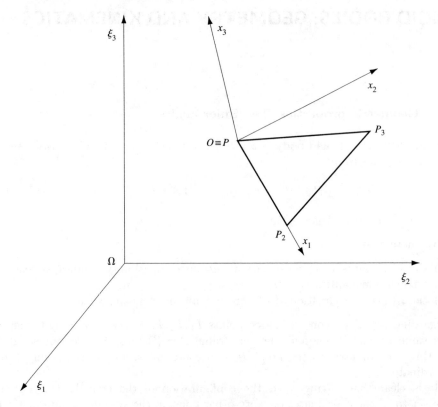

**Fig. 6.1** Body reference frame.

orthogonal matrices $A$, i.e. such that $AA^T = 1$ and $\det A = +1$. Then $g \in \mathbf{R}^3 \times \mathrm{SO}(3)$ is the composition of a translation $\Sigma \to \Sigma'$, determined by the three coordinates of $O$ in $\Sigma$, and of a rotation $\Sigma' \to S$ around $O$.

The column vectors of the rotation matrix $A$ are the direction cosines of the axes of $S$ with respect to $\Sigma'$ (nine parameters related by the six orthonormality conditions). This proves the following important property.

PROPOSITION 6.1 *The configuration space of a rigid body with at least three non-collinear points in three-dimensional Euclidean space is* $\mathbf{R}^3 \times \mathrm{SO}(3)$. *If the system has a fixed point it must have three degrees of freedom and its configuration space is* $\mathrm{SO}(3)$. ∎

A more direct representation of the transformation $\Sigma' \to S$ is given by the so-called *Euler angles* (Fig. 6.2). It is easy to verify that the transformation $A : \Sigma' \to S$ can be obtained by composing three rotations. Let $N$ be the line of nodes, i.e. the intersection between the planes $x_3 = 0$ and $\xi_3' = 0$ and denote by $A_a(\alpha)$ the rotation by an angle $\alpha$ around the axis $a$. We then have

$$A = A_{x_3}(\varphi) A_N(\theta) A_{\xi_3'}(\psi), \tag{6.2}$$

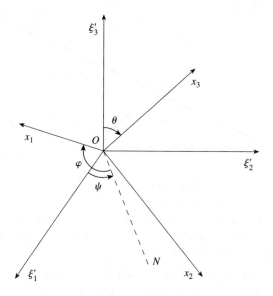

**Fig. 6.2**

since

(a) the rotation $A_{\xi_3'}(\psi)$ by an angle $\psi$ (*precession*) around the axis $\xi_3'$ makes the axis $\xi_1'$ coincide with the line of nodes $N$;
(b) the rotation $A_N(\theta)$ by an angle $\theta$ (*nutation*) around the line of nodes $N$ brings the axis $\xi_3'$ to coincide with the axis $x_3$;
(c) the rotation $A_{x_3}(\varphi)$ by an angle $\varphi$ around the axis $x_3$ makes the axes $\xi_1', \xi_3'$ coincide with the axes $x_1, x_2$.

*Remark 6.1*
In order to determine the configurations of a rigid body it is not important to know the actual geometric structure of the system. We only need to know if the system has three non-collinear points. ∎

Often one considers rigid bodies subject to additional constraints.
In general, given a set of $N$ rigid bodies each with an associated vector $\mathbf{q}^{(i)}$ with six components (three in the planar case), we can impose constraints of the following kind:

$$f_j(\mathbf{q}^{(1)}, \ldots, \mathbf{q}^{(N)}) = 0, \qquad j = 1, \ldots, m, \tag{6.3}$$

subject to the same criteria as seen in Chapter 1.

*Example 6.1*
We study the planar system made up of two rods $AB$, $CD$ of length $\ell_1, \ell_2$, respectively ($\ell_1 \geq \ell_2$), with the endpoints $A, C$ fixed at a distance $\ell$ and with the endpoint $D$ constrained to lie on the segment $AB$ (Fig. 6.3).

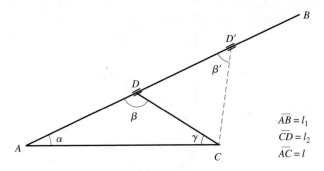

Fig. 6.3

The constraint imposed on $D$ makes sense only if $\ell < \ell_1 + \ell_2$. In this case the system has only one degree of freedom. We can take the angle $\alpha$ as a Lagrangian coordinate. We distinguish the following cases.

(1) $\ell_2 > \ell$. The angle $\alpha$ is variable, without limitations. Setting $\ell/\ell_2 = \xi$ we have $\sin \beta = \xi \sin \alpha$, $\gamma = \pi - [\alpha + \beta(\alpha)]$, from which we can obtain the values of $\sin \gamma, \cos \gamma$, if we note that $\cos \beta = \sqrt{1 - \xi^2 \sin^2 \alpha}$ ($\beta$ is always acute, as its opposite side is not the longest of the three sides, in any configuration). Since

$$\frac{d\gamma}{d\alpha} = -1 - \frac{d\beta}{d\alpha} \quad \text{and} \quad \frac{d\beta}{d\alpha} = \xi \frac{\cos \alpha}{\cos \beta},$$

it is easy to check that $d\gamma/d\alpha$ is bounded and does not vanish, and hence the constraint is non-singular.

(2) $\ell_2 < \ell$. Since now $\xi > 1$ we have the restriction $|\sin \alpha| < 1/\xi$. The extreme values correspond to $\beta = \pm \pi/2$ and give a singularity of $d\gamma/d\alpha$; this is the configuration which lies between those characterised by $\cos \beta = \sqrt{1 - \xi^2 \sin^2 \alpha}$ and those for which $\cos \beta = -\sqrt{1 - \xi^2 \sin^2 \alpha}$.

(3) $\ell_2 = \ell$. The configuration with $D \equiv A$ is degenerate; it can be obtained as a limiting case for $\alpha \to \pi/2$, but if the point $D$ remains fixed then the rod $AB$ can rotate around $A$. ∎

## 6.2 The kinematics of rigid bodies. The fundamental formula

Since the configuration of a rigid system with respect to a frame $\Sigma$ is determined by the configuration of any of its body frames $S$, the study of the kinematics of rigid systems consists of the description of the motion of $S$ with respect to $\Sigma$.

A fundamental property of the velocity field of $S$ with respect to $\Sigma$ is the following.

THEOREM 6.1 *The velocity field of a rigid motion is expressed by the formula*

$$\mathbf{v}(P) = \mathbf{v}(O) + \boldsymbol{\omega} \times (P - O), \tag{6.4}$$

## 6.2    Rigid bodies: geometry and kinematics

where $\mathbf{v}(O)$ is the velocity of a prescribed point $O$ and $\boldsymbol{\omega}$ is called the *angular velocity*. ∎

Equation (6.4) is a consequence of the following result.

**THEOREM 6.2**  *In every rigid motion there exists one and only one vector $\boldsymbol{\omega}$, which is a function of time, through which we can express the variation of any unit vector $\mathbf{e}$ in the body by means of the formula*

$$\frac{d\mathbf{e}}{dt} = \boldsymbol{\omega} \times \mathbf{e} \tag{6.5}$$

*(Poisson's formula, also known as the attitude equation).*

*Proof*
The variation of a unit vector $\mathbf{e}$ in the time interval $(t, t + \Delta t)$ can be expressed as

$$\mathbf{e}(t + \Delta t) - \mathbf{e}(t) = A(\Delta t)\mathbf{e}(t) - \mathbf{e}(t) = (A(\Delta t) - \mathbf{1})\mathbf{e}(t),$$

where $A(\Delta t)$ is an orthogonal matrix such that $A(0) = \mathbf{1}$. It follows that

$$\frac{d\mathbf{e}}{dt} = \dot{A}(0)\mathbf{e}(t). \tag{6.6}$$

Recalling Lemma 1.1, which states that $\dot{A}(0)$ is a skew-symmetric matrix, if we write it in the form

$$\dot{A}(0) = \Omega = \begin{pmatrix} 0 & -\omega_3 & \omega_2 \\ \omega_3 & 0 & -\omega_1 \\ -\omega_2 & \omega_1 & 0 \end{pmatrix}, \tag{6.7}$$

where $\omega_1, \omega_2, \omega_3$ define the vector $\boldsymbol{\omega}(t)$, we see that (6.6) can be written exactly in the form (6.5), as $\Omega \mathbf{e} = \boldsymbol{\omega} \times \mathbf{e}$.

The uniqueness of the vector $\boldsymbol{\omega}$ follows by noting that if there existed a different vector $\boldsymbol{\omega}'$ with the same characterisation, the difference would satisfy

$$(\boldsymbol{\omega} - \boldsymbol{\omega}') \times \mathbf{e} = 0 \tag{6.8}$$

for every unit vector $\mathbf{e}$ of $S$, implying $\boldsymbol{\omega} = \boldsymbol{\omega}'$. ∎

It is interesting to note that if a unit vector $\mathbf{e}(t)$ satisfies equation (6.5) then it must be fixed in the body frame, as these are the only unit vectors satisfying the transformation law (6.6).

The Poisson formula is clearly valid for any fixed vector in the body frame $\mathbf{W}$:

$$\frac{d\mathbf{W}}{dt} = \boldsymbol{\omega} \times \mathbf{W}. \tag{6.9}$$

For any two points $P, O$ in $S$, by applying equation (6.9) to the vector $\mathbf{W} = P - O$, one finds the formula (6.4).

Clearly the condition $\boldsymbol{\omega} \neq 0$ characterises any motion which is not purely a translation.

PROPOSITION 6.2 *A necessary and sufficient condition for a rigid motion to be a translation is that* $\boldsymbol{\omega} = 0$. ∎

PROPOSITION 6.3 *If the direction of* $\boldsymbol{\omega}$ *is constant in the reference frame* $\Sigma$ *then it is constant in the body frame* $S$, *and vice versa*.

*Proof*
Suppose that $\mathbf{e} = \boldsymbol{\omega}/|\boldsymbol{\omega}|$ is constant in $S$. We can then apply to it equation (6.5) and conclude that $d\mathbf{e}/dt = 0$, i.e. that $\mathbf{e}$ does not vary in $\Sigma$. Conversely, if $d\mathbf{e}/dt = 0$ in $\Sigma$, the unit vector $\mathbf{e}$ satisfies (6.5), since trivially $\boldsymbol{\omega} \times \mathbf{e} = 0$. It follows that $\mathbf{e}$ is constant in $S$. ∎

Rigid systems can be subject to non-holonomic constraints. This is illustrated in the following example.

*Example 6.2*
Show that a disc that rolls without sliding on a plane $\pi$, and always orthogonal to it, is not a holonomic system (Fig. 6.4).

If we ignored the kinematic condition which requires that the velocity at the contact point is zero, we would have a holonomic system with four degrees of freedom. The corresponding Lagrangian coordinates can be taken as:

- two coordinates $x_1, x_2$ which determine the position of the centre $O$;
- the angle $\psi$ formed by the plane of the disc with a fixed plane, normal to the base plane;
- the angle $\varphi$ formed by a radius fixed on the disc with the normal to the base plane.

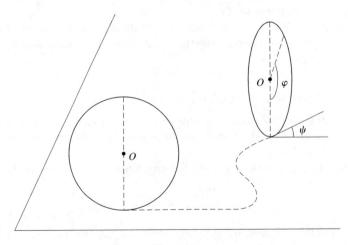

Fig. 6.4

If the addition of the kinematic constraint preserved the holonomic property of the system, then it would be possible to determine a relation of the kind $f(x_1, x_2, \varphi) = 0$; indeed this constraint cannot have any effect on the determination of the angle $\psi$. Hence to prove that the system is not holonomic, it is enough to prove that the coordinates $x_1$, $x_2$, $\varphi$ remain independent.

Hence we show that, for a given configuration of the system (for example, one associated with zero values for $x_1, x_2, \varphi, \psi$), we can move the disc in such a way that we can reach any other configuration, characterised by arbitrary values of $x_1, x_2, \varphi, \psi$. Ignoring the coordinate $\psi$, which can be chosen once the other three parameters are fixed, there exist infinitely many motions achieving this aim; if $R$ is the radius of the given disc, and if $x_1^{(0)}, x_2^{(0)}, \varphi^{(0)}$ are the coordinates of the final configuration, it suffices to connect the points having coordinates $(0,0)$ and $(x_1^{(0)}, x_2^{(0)})$ in the plane $\pi$ with an arc of any regular curve of length $R(2k\pi + \varphi^{(0)})$, with $k \geq 0$ any integer, allowing the disc to roll along this arc. ∎

This example shows that constraints that are not holonomic do not prevent certain configurations, but limit to certain classes the motions that connect two prescribed configurations (a familiar example is the parking of a car, when some manoeuvring is necessary due to the fact that the wheels cannot slide on the road).

*Remark* 6.2
If one prescribes a curve $\gamma$ along which one constrains the disc of Example 6.2 to roll $(x_1 = x_1(q), x_2 = x_2(q))$ the system is again holonomic. Indeed, this gives a relation

$$R\varphi = \int_0^q (x_1'^2(\eta) + x_2'^2(\eta))^{1/2} \, d\eta + \text{constant}.$$

If in addition the disc is constrained in the plane containing the unit vector tangent to $\gamma$ then the system has only one degree of freedom, corresponding to the Lagrangian coordinate $q$. ∎

## 6.3 Instantaneous axis of motion

THEOREM 6.3  *If $\omega \neq 0$ at a given time instant, then there exists at that instant a straight line parallel to $\omega$, whose points have velocity parallel to $\omega$ or zero.*

*Proof*
First of all, note that by taking the scalar product of (6.4) with $\omega$ we find

$$\mathbf{v}(P) \cdot \boldsymbol{\omega} = \mathbf{v}(O) \cdot \boldsymbol{\omega}, \tag{6.10}$$

i.e. *the product $\mathbf{v} \cdot \boldsymbol{\omega}$ is invariant in the velocity field of a rigid system.* We want to show that there exists a line, parallel to $\omega$, along which the velocity reduces only to the component parallel to $\boldsymbol{\omega}$ (common to all the points in the field). To this aim we consider the plane $\pi$ normal to $\boldsymbol{\omega}$ and passing through $O$ and we look for $P^* \in \pi$ with the property that $\mathbf{v}(P^*) \times \boldsymbol{\omega} = 0$. Evaluating equation (6.4) at $P \in \pi$, and taking the vector product of both sides with $\boldsymbol{\omega}$, in view of the fact that $\boldsymbol{\omega} \cdot (P - O) = 0$, we find

$$\mathbf{v} \times \boldsymbol{\omega} = (P - O)\omega^2 + \mathbf{v}(O) \times \boldsymbol{\omega}.$$

This yields the sought solution $P^*$ as:

$$P^* - O = \boldsymbol{\omega} \times \frac{\mathbf{v}(O)}{\omega^2}. \tag{6.11}$$

In addition, from (6.4) we immediately deduce that *all points belonging to the same line parallel to $\boldsymbol{\omega}$ have the same velocity.*

This completes the proof of the theorem. ∎

DEFINITION 6.2 *The straight line of Theorem 6.3 is called the* instantaneous axis of motion *(*instantaneous axis of rotation, *if the invariant $\mathbf{v} \cdot \boldsymbol{\omega}$ vanishes).* ∎

For the previous considerations it follows that *the velocity field in a rigid motion has rotational symmetry with respect to the instantaneous axis of motion.* Figure 6.5 justifies the fact that the generic rigid motion is called *helical*.

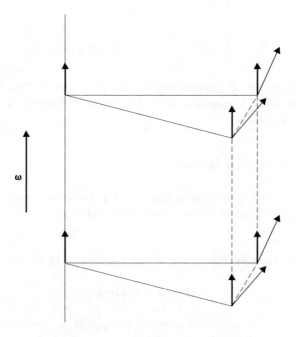

**Fig. 6.5** Axial symmetry of the velocity field around the instantaneous axis of motion.

The kinematics of rigid systems has far-reaching developments, which we do not consider here.[1] We recall the following special kinds of rigid motion.

(a) *Plane rigid motions*: these are the motions defined by the presence of a plane in the system superimposed on a fixed plane. They can be characterised by the condition that

$$\mathbf{v} \cdot \boldsymbol{\omega} = 0, \quad \boldsymbol{\omega} \text{ has constant direction,} \quad (6.12)$$

and can be studied in a *representative plane*. The intersection of the latter with the instantaneous axis of rotation is called the *instantaneous centre of rotation*.

(b) *Precessions*: rigid motions with a fixed point $O : \mathbf{v}(O) \equiv 0$. Clearly $\mathbf{v} \cdot \boldsymbol{\omega} = 0$ and the point $O$, called the *pole* of the precession, belongs to the instantaneous axis of rotation for all times.

(c) *Rotations*: rigid motions with a fixed axis (a particular case of Proposition 6.3). The rotation if said to be uniform if $\boldsymbol{\omega} = $ constant.

## 6.4 Phase space of precessions

We state a result that we have already discussed.

**THEOREM 6.4** *The tangent space to the space of orthogonal $3 \times 3$ matrices computed at the identity matrix is the space of $3 \times 3$ skew-symmetric matrices (6.7), denoted by* so(3). ∎

We should add that when $\mathbf{v}(0) = 0$, equation (6.4) reduces to $\dot{\mathbf{x}} = \Omega \mathbf{x}$, where $\Omega$ is the matrix given by (6.7). Setting $\mathbf{x}(0) = \boldsymbol{\xi}$, we can describe the precession by the equation $\mathbf{x} = A(t)\boldsymbol{\xi}$, with $A(0) = \mathbf{1}$ and $AA^T = \mathbf{1}$. It follows that $\dot{\mathbf{x}} = \dot{A}(t)\boldsymbol{\xi}$ and finally, from (6.4) $\dot{A}\boldsymbol{\xi} = \Omega A \boldsymbol{\xi}$, $\forall \boldsymbol{\xi} \in \mathbf{R}^3$. Hence the matrix $A$ satisfies the Cauchy problem

$$\dot{A} = \Omega A, \quad A(0) = \mathbf{1} \quad (6.13)$$

whose solution, if $\Omega = $ constant, is given by

$$A(t) = \exp(t\Omega) = \sum_{n=0}^{\infty} \frac{t^n}{n!} \Omega^n. \quad (6.14)$$

It is easy to compute the powers of the matrix $\Omega$. The elements of $\Omega^2$ are $(\Omega^2)_{ij} = \omega_i \omega_j$ for $i \neq j$, $(\Omega^2)_{ii} = -(\omega_j^2 + \omega_k^2)$, with $i, j, k$ all distinct, $\Omega^3 = -\omega^2 \Omega$ and

$$\Omega^{2n} = (-1)^{n-1} \omega^{2(n-1)} \Omega^2, \quad \Omega^{2n+1} = (-1)^n \omega^{2n} \Omega, \quad n \geq 1.$$

---

[1] See for instance Fasano *et al.* (2001).

This yields the explicit expression for the series (6.14):

$$\exp(t\Omega) = 1 + \frac{\Omega^2}{\omega^2}(1 - \cos \omega t) + \frac{\Omega}{\omega} \sin \omega t. \quad (6.15)$$

In the particular case of a rotation around the axis $x_3$ the matrix $\Omega$ reduces to

$$\Omega_3 = \omega \begin{pmatrix} 0 & -1 & 0 \\ 1 & 0 & 0 \\ 0 & 0 & 0 \end{pmatrix}, \quad \frac{\Omega_3^2}{\omega^2} = \begin{pmatrix} -1 & 0 & 0 \\ 0 & -1 & 0 \\ 0 & 0 & 0 \end{pmatrix},$$

and we find the well-known result that

$$A(t) = \omega \begin{pmatrix} \cos \omega t & -\sin \omega t & 0 \\ \sin \omega t & \cos \omega t & 0 \\ 0 & 0 & 0 \end{pmatrix}.$$

If $\Omega$ is not constant, equation (6.13) is equivalent to the integral equation

$$A(t) = 1 + \int_0^t \Omega(t')A(t') \, dt', \quad (6.16)$$

which can be solved by iteration, and yields the formula

$$A(t) = 1 + \sum_{n=1}^{\infty} \int_0^t dt_1 \, \Omega(t_1) \int_0^{t_1} dt_2 \, \Omega(t_2) \ldots \int_0^{t_{n-1}} dt_n \, \Omega(t_n). \quad (6.17)$$

We now give an equivalent, more concise formulation of the latter expression. Note that the order of multiplication in (6.17) is relevant, as in general $\Omega(t_i)\Omega(t_j) \neq \Omega(t_j)\Omega(t_i)$ (obviously if $\Omega$ is constant, the series (6.17) reduces to (6.14)). Let

$$\theta(t) = \begin{cases} 0, & t < 0, \\ 1, & t > 0. \end{cases} \quad (6.18)$$

The general term of the series (6.17) can be written as

$$\int_0^t dt_1 \ldots \int_0^t dt_n \, \theta(t_1 - t_2) \cdots \theta(t_{n-1} - t_n) \, \Omega(t_1) \cdots \Omega(t_n)$$

$$= \frac{1}{n!} \int_0^t dt_1 \ldots \int_0^t dt_n \sum_{\sigma \in \mathcal{P}(n)} \theta(t_{\sigma(1)} - t_{\sigma(2)}) \cdots \theta(t_{\sigma(n-1)} - t_{\sigma(n)})$$

$$\times \Omega(t_{\sigma(1)}) \cdots \Omega(t_{\sigma(n)}), \quad (6.19)$$

where $\mathcal{P}(n)$ denotes the group of permutations of the indices $\{1,\ldots,n\}$; it is clear that the number of elements $\sigma$ of $\mathcal{P}(n)$ is $n!$. We then define the *T-product* (time-ordered product) of the matrices $\Omega(t_1),\ldots,\Omega(t_n)$ as

$$T[\Omega(t_1)\ldots\Omega(t_n)] = \sum_{\sigma \in \mathcal{P}(n)} \theta(t_{\sigma(1)} - t_{\sigma(2)}) \cdots \theta(t_{\sigma(n-1)} - t_{\sigma(n)})$$
$$\times \Omega(t_{\sigma(1)}) \cdots \Omega(t_{\sigma(n)}) \quad (6.20)$$

in such a way that if $\bar{\sigma}$ denotes the permutation of $\{1,\ldots,n\}$ such that $t_{\bar{\sigma}(1)} > t_{\bar{\sigma}(2)} > \cdots > t_{\bar{\sigma}(n-1)} > t_{\bar{\sigma}(n)}$ we have

$$T[\Omega(t_1)\ldots\Omega(t_n)] = \Omega(t_{\bar{\sigma}(1)})\cdots\Omega(t_{\bar{\sigma}(n)}).$$

Recalling (6.19) and the definition (6.20) we find that the solution (6.17) of equation (6.13) can be written

$$\Omega(t) = 1 + \sum_{n=1}^{\infty} \frac{1}{n!} \int_0^t dt_1 \cdots \int_0^t dt_n\, T[\Omega(t_1)\cdots\Omega(t_n)]. \tag{6.21}$$

This solution is also known as the *T-exponential* (or time-ordered exponential)

$$\Omega(t) = T - \exp\left[\int_0^t \Omega(t')\,dt'\right]$$
$$= 1 + \sum_{n=1}^{\infty} \frac{1}{n!} \int_0^t dt_1 \cdots \int_0^t dt_n\, T[\Omega(t_1)\cdots\Omega(t_n)]. \tag{6.22}$$

Again we note that if $\Omega$ is constant then equation (6.22) becomes (6.14). In the special case that, for any $t',t'' \in [0,t]$ one has $\Omega(t')\Omega(t'') = \Omega(t'')\Omega(t')$, equation (6.22) simplifies to

$$\exp\left(\int_0^t \Omega(t')\,dt'\right),$$

where the order of multiplication of the matrices $\Omega(t_i)$ is not important.

## 6.5 Relative kinematics

After studying the motion of a frame $S$ with respect to a frame $\Sigma$, we summarise the main results of relative kinematics. This is concerned with the mutual relations between kinematic quantities of a point in motion as observed by $S$ (called *relative*) and as observed by $\Sigma$ (called *absolute*).

We use the subscript $R$ for relative quantities, the subscript $T$ for quantities corresponding to the rigid motion of $S$ with respect to $\Sigma$, and no subscript for absolute quantities. The following relations hold:

$$\mathbf{v} = \mathbf{v}_R + \mathbf{v}_T, \tag{6.23}$$
$$\mathbf{a} = \mathbf{a}_R + \mathbf{a}_T + \mathbf{a}_C, \tag{6.24}$$

where $\boldsymbol{\omega}$ is the rotational velocity of $S$ with respect to $\Sigma$ and the term $\mathbf{a}_C = 2\boldsymbol{\omega} \times \mathbf{v}_R$ is called the *Coriolis acceleration*.

The former can be found by differentiating the vector $P - O = \sum_i x_i \mathbf{e}_i$ in the $\Sigma$ reference frame and using equation (6.5):

$$\frac{d(P-O)}{dt} = \sum_i \dot{x}_i \mathbf{e}_i + \boldsymbol{\omega} \times \sum_i x_i \mathbf{e}_i.$$

Equation (6.23) expresses the relation between the absolute derivative (in $\Sigma$) and the relative derivative (in $S$):

$$\frac{d\mathbf{W}}{dt} = \left(\frac{d\mathbf{W}}{dt}\right)_R + \boldsymbol{\omega} \times \mathbf{W}, \tag{6.25}$$

where $\mathbf{W}$ is any vector, variable in both $S$ and $\Sigma$.

Applying (6.24) to $\mathbf{v}_R$ we find

$$\frac{d\mathbf{v}_R}{dt} = \mathbf{a}_R + \boldsymbol{\omega} \times \mathbf{v}_R. \tag{6.26}$$

For the derivative of $\mathbf{v}_T = \mathbf{v}(0) + \boldsymbol{\omega} \times (P - O)$ we then obtain

$$\frac{d\mathbf{v}_T}{dt} = \mathbf{a}_T + \boldsymbol{\omega} \times \mathbf{v}_R, \tag{6.27}$$

where the last term has its origin in the fact that the relative motion in general produces a variation of $\mathbf{v}_T$, because the point in question moves in the velocity field of the motion $S$ with respect to $\Sigma$ (note that this effect vanishes if the relative motion is in the direction of $\boldsymbol{\omega}$, since $\mathbf{v}_T$ does not vary in this direction).

We can now consider a triple $S' = (O', x'_1, x'_2, x'_3)$, in motion with respect to both $S$ and $\Sigma$ (Fig. 6.6), and find the relations between the characteristic vectors of the relative motion and of the absolute motion.

We start by expressing that for every $P \in S'$

$$\mathbf{v}_R(P) = \mathbf{v}_R(O') + \boldsymbol{\omega}_R \times (P - O'), \tag{6.28}$$

where $\boldsymbol{\omega}_R$ is the angular velocity of $S'$ with respect to $S$. Let $\boldsymbol{\omega}_T$ be the angular velocity of $S$ with respect to $\Sigma$. Then we have

$$\mathbf{v}_T(P) = \mathbf{v}_T(O) + \boldsymbol{\omega}_T \times (P - O). \tag{6.29}$$

From this it follows by adding (6.28) and (6.29) that

$$\mathbf{v}(P) = \mathbf{v}_R(O') + \mathbf{v}_T(O) + \boldsymbol{\omega}_T \times (O' - O) + (\boldsymbol{\omega}_T + \boldsymbol{\omega}_R) \times (P - O'), \tag{6.30}$$

which contains $\mathbf{v}_T(O) + \boldsymbol{\omega}_T \times (O' - O) = \mathbf{v}_T(O')$.

Hence we find the expression for the absolute velocity field:

$$\mathbf{v}(P) = \mathbf{v}(O') + \boldsymbol{\omega} \times (P - O'), \tag{6.31}$$

with

$$\mathbf{v}(O') = \mathbf{v}_R(O') + \mathbf{v}_T(O') \tag{6.32}$$

## 6.5 Rigid bodies: geometry and kinematics

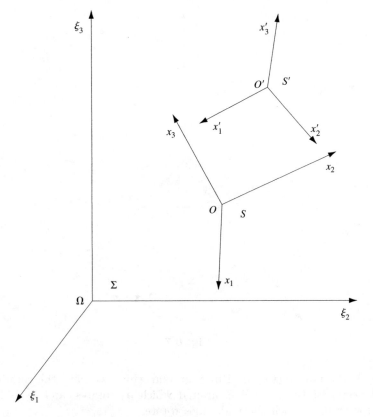

**Fig. 6.6** Composition of rigid motions.

and
$$\omega = \omega_R + \omega_T. \tag{6.33}$$

The absolute rigid motion is commonly called *compound rigid motion*.

*Example 6.3: composition of precessions with the same pole*
We immediately find that the compound motion is a precession whose angular velocity is the sum of the angular velocities of the component motions. When two uniform rotations are composed, the vector $\omega = \omega_R + \omega_T$ forms constant angles with $\omega_T$ fixed in $\Sigma$ (known as the *precession axis*) and with $\omega_R$ fixed in $S$ (*spin axis*). The resulting precession is called *regular*. ∎

Clearly $\omega$ can be thought of as the sum of the rotations driven by the variations of the associated Euler angles:
$$\omega = \dot{\varphi}\mathbf{e}_3 + \dot{\theta}\mathbf{N} + \dot{\psi}\boldsymbol{\varepsilon}_3, \tag{6.34}$$

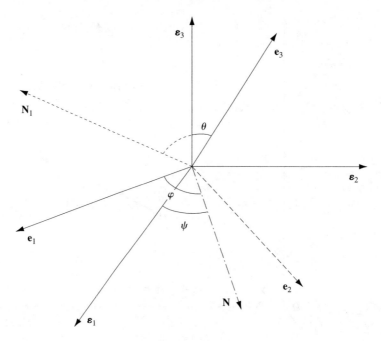

**Fig. 6.7**

where $e_3$ is the unit vector fixed in $S$ around which we have the rotation, $\mathbf{N}$ is the unit vector of the node line, around which $e_3$ rotates, and $\varepsilon_3$ is the fixed unit vector around which the node lines rotates.

We can find the decomposition of $\boldsymbol{\omega}$ in the body frame by appealing to Fig. 6.7, where the unit vectors $\mathbf{N}_1, \mathbf{N}, \mathbf{e}_3$ form an orthogonal frame, so that $\varepsilon_3$ (normal to $\mathbf{N}$) is in the plane of $\mathbf{e}_3$ and $\mathbf{N}_1$:

$$\varepsilon_3 = \sin\theta \mathbf{N}_1 + \cos\theta \mathbf{e}_3, \tag{6.35}$$

$$\mathbf{N} = \cos\varphi \mathbf{e}_1 - \sin\varphi \mathbf{e}_2, \tag{6.36}$$

$$\mathbf{N}_1 = \sin\varphi \mathbf{e}_1 + \cos\varphi \mathbf{e}_2. \tag{6.37}$$

It follows that by projecting $\boldsymbol{\omega}$ onto the vectors of the body frame, we find

$$\begin{aligned}\boldsymbol{\omega} &= (\dot\theta \cos\varphi + \dot\psi \sin\theta \sin\varphi)\mathbf{e}_1 + (-\dot\theta \sin\varphi + \dot\psi \sin\theta \cos\varphi)\mathbf{e}_2 \\ &\quad + (\dot\varphi + \dot\psi \cos\theta)\mathbf{e}_3.\end{aligned} \tag{6.38}$$

## 6.6 Relative dynamics

As an appendix to the study of relative kinematics we now present some observations on the relative dynamical picture. What we have just presented allows us to complete the list of fundamental facts of the mechanics of a point particle

(Chapter 2), by answering the following question: how can we write the equation of motion of a point with respect to a non-inertial observer?

Consider the motion of a point particle $(P, m)$ under the action of a force $\mathbf{F}$ measured by a given inertial observer $\Sigma$, for whom the equation $m\mathbf{a} = \mathbf{F}$ is valid. Given a second observer $S$ in motion with respect to the first one, using the notation of (6.24) we can write

$$m\mathbf{a}_R = \mathbf{F} - m\mathbf{a}_T - m\mathbf{a}_C, \tag{6.39}$$

where $\mathbf{a}_R$ is the acceleration as measured by the non-inertial observer and $\mathbf{a}_T$ and $\mathbf{a}_C$ are computed with respect to the motion of this observer relative to the fixed inertial observer.

Ignoring for the moment the force $\mathbf{F}$, the non-inertial observer measures some *apparent forces*: the force connected to the acceleration of $S$ relative to $\Sigma$, $\mathbf{F}_T = -m\mathbf{a}_T$, and the *Coriolis force* $\mathbf{F}_C = -2m\boldsymbol{\omega} \times \mathbf{v}_R$.

Recall that

$$\mathbf{a}_T(P, t) = \mathbf{a}(O) + \dot{\boldsymbol{\omega}} \times (P - O) + \boldsymbol{\omega} \times [\boldsymbol{\omega} \times (P - O)]. \tag{6.40}$$

In particular, in the case that the reference frame of the observer $S$ is in uniform rotation with respect to that of $\Sigma$, the acceleration $\mathbf{a}_T$ is reduced to the last term of (6.40) (*centripetal* acceleration) and correspondingly, $\mathbf{F}_T$ is a *centrifugal force*.

If the reference frame $S$ translates with respect to $\Sigma$ ($\boldsymbol{\omega} = 0$) we have $\mathbf{a}_T = \mathbf{a}(O)$. If $S$ is chosen in such a way that $P$ has zero relative velocity (let $S_P$ denote such an observer) then $\mathbf{a}(O) = \mathbf{a}$ and the apparent force observed by $S_P$ acting on $P$ is the so-called *inertial force* $-m\mathbf{a}$.

In this case, the equation

$$0 = \mathbf{F} - m\mathbf{a} \tag{6.41}$$

is interpreted by $S_P$ as the equilibrium equation.

If the point is constrained and $\phi$ is the constraint reaction, instead of equation (6.41) we have

$$0 = \mathbf{F} - m\mathbf{a} + \phi, \tag{6.42}$$

which again is interpreted by $S_P$ as the balance of forces in the equilibrium position.

*Remark 6.3*
Despite their name, apparent forces are measurable in practice by a non-inertial observer. This is true in particular for the inertial force; imagine carrying an object in your hand and giving it an upwards or downwards acceleration (less than the acceleration due to gravity, so the object does not leave your hand): in the two cases one perceives that the force exerted on the hand (corresponding to $-\phi$) is either increased or decreased. These variations correspond to the inertial force. ∎

The interpretation of (6.42) as an equilibrium equation in $S_P$, with the introduction of inertial forces, is often referred to as *d'Alembert's principle*.

An important property in the study of relative dynamics is that the Galilean relativity principle, stated in Section 2.2 of Chapter 2 for the case of inertial observers, is valid for any class of observers whose relative motion is a uniform translation.

PROPOSITION 6.4 *The equation of motion (6.39) for a point particle $(P, m)$ is invariant in the class of observers that move with respect to one another with zero acceleration.*

*Proof*
If $S$ and $S'$ are two observers whose relative motion has zero acceleration, the sum $\mathbf{a}_T + \mathbf{a}_C$ in $S$ with respect to an inertial observer $\Sigma$ is equal to the sum $\mathbf{a}'_T + \mathbf{a}'_C$ in $S'$. Indeed, the relative accelerations $\mathbf{a}_R$ (with respect to $S$) and $\mathbf{a}'_R$ (with respect to $S'$) are equal.

Note also that $\mathbf{a}'_T = \mathbf{a}_T + 2\boldsymbol{\omega} \times \mathbf{v}_0$, where $\mathbf{v}_0$ is the translation velocity of $S'$ with respect to $S$. Similarly $\mathbf{a}'_C = 2\boldsymbol{\omega} \times \mathbf{v}'_R = 2\boldsymbol{\omega} \times (\mathbf{v}_R - \mathbf{v}_0)$. Hence the variation $\mathbf{a}'_T - \mathbf{a}_T = 2\boldsymbol{\omega} \times \mathbf{v}_0$ is balanced by the variation of $\mathbf{a}'_C - \mathbf{a}_C$. Since the observers $S$ and $S'$ cannot distinguish the individual sources of apparent forces, but can only measure their sum, it follows that it is impossible for them to be aware of their relative motion solely on the evidence of mechanical observations. ∎

## 6.7 Ruled surfaces in a rigid motion

DEFINITION 6.3 *A fixed ruled surface (respectively body ruled surface) of a given rigid motion is the locus of the lines which take the role of instantaneous axes of motion in the fixed (respectively, in the body) reference frame.* ∎

A trivial example is given by the purely rolling motion of a cylinder on a plane; the fixed ruled surface is the plane; the body ruled surface is the cylinder.

These surfaces are important because they can generate a prescribed rigid motion when the moving ruled surface rolls onto the fixed one (this fact is exploited in the theory of gears).

THEOREM 6.5 *In a generic rigid motion the body ruled surface rolls onto the fixed one. Along the contact line, the sliding velocity is equal to the invariant component of the velocity field along the angular velocity.*

*Proof*
We only need to prove that the two ruled surfaces at every instant are tangent to each other (this fact characterises the rolling motion); indeed, the last claim of the theorem is evidently true, as the contact line coincides at every instant with the axis of instantaneous motion.

Recalling equation (6.11), which gives the intersection point $P^*$ between the axis of motion and the plane orthogonal to it and passing through $O$, we can write the equation for the axis of motion $r(t)$ as

$$P - O = \lambda \boldsymbol{\omega} + \boldsymbol{\omega} \times \mathbf{v}(0)/\omega^2, \qquad \lambda \in (-\infty, +\infty). \tag{6.43}$$

The two ruled surfaces are generated by the absolute and relative motion of the line $\bar{r}(t)$ which moves with respect to the two reference frames so as to coincide at every instant with $r(t)$. To understand more clearly the different role played by $r(t)$ and by $\bar{r}(t)$, consider the case of the cylinder rolling on a fixed plane: the points of the line $r(t)$, considered in either $\Sigma$ or $S$, have zero velocity, while the points of the line $\bar{r}(t)$ move, sweeping the plane (absolute motion) or the cylinder (relative motion).

We must then study the absolute and relative motion of the point $\overline{P}(\lambda, t) \in \bar{r}(t)$, defined by (6.43): the absolute velocity is tangent to the fixed ruled surface, while the relative velocity is tangent to the body ruled surface. The velocity of the rigid motion on the instantaneous axis of motion is $\mathbf{v}_\omega = \mathbf{v}(O) \cdot \boldsymbol{\omega}/|\boldsymbol{\omega}|$. It follows that

$$\mathbf{v}(\overline{P}) - \mathbf{v}_R(\overline{P}) = \mathbf{v}_\omega. \tag{6.44}$$

If we ignore the degenerate case when the instantaneous axis of motion is stationary, we have that neither $\mathbf{v}(\overline{P})$ nor $\mathbf{v}_R(\overline{P})$ are parallel to $\boldsymbol{\omega}$. Equation (6.44) then implies that their difference must be either zero $(\mathbf{v}(O) \cdot \boldsymbol{\omega} = 0)$ or parallel to $\boldsymbol{\omega}$, and hence that the three vectors $\boldsymbol{\omega}$, $\mathbf{v}(\overline{P})$, $\mathbf{v}_R(\overline{P})$ lie in the same plane. Since the plane determined by the axis of motion and by $\mathbf{v}(\overline{P})$ is tangent to the fixed ruled surface, and the plane determined by the axis of motion and by $\mathbf{v}_R(\overline{P})$ is tangent to the body surface, this proves that the two planes coincide and that the two ruled surfaces are tangent. ∎

COROLLARY 6.1 *For the motions with $\boldsymbol{\omega}$ of constant direction (in particular for plane rigid motions) the ruled surfaces are cylinders. For plane motions the intersections of the ruled surfaces with the representative plane are called* conjugate profiles. ∎

COROLLARY 6.2 *In any precession, the ruled surfaces are cones with their vertex in the precession pole* (Poinsot cones). ∎

The Poinsot cones are circular in the case of regular precession (Example 6.3) and they both degenerate to a line in the case of rotations.

*Example 6.4*
We determine the conjugate profiles for the motion of a rod with endpoints sliding along two orthogonal lines (Fig. 6.8).

The instantaneous centre of motion $C$ can be trivially determined by using Chasles' theorem (see Problem 2), and considering the vectors normal to the direction of the velocities of the two extremes $A$, $B$ of the rod. We deduce that in every configuration of the pair of reference axes $(O, x_1, x_2)$ the point $C$ is characterised as follows:

(a) in the fixed reference frame $(\Omega, \xi_1, \xi_2)$ it is the point at a distance $2\ell = \overline{AB}$ from the point $\Omega$;

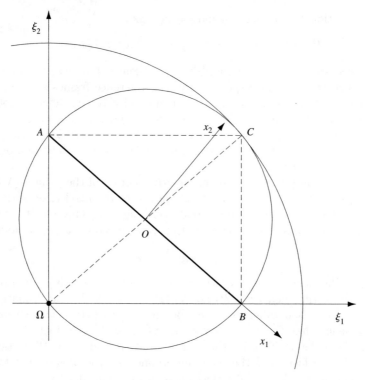

**Fig. 6.8**

(b) in the reference frame $(O, x_1, x_2)$ it is the point at a distance $\ell$ from the point $O$.

Hence the conjugate profiles in the fixed and in the body plane are the circle of centre $\Omega$ and radius $2\ell$ and the circle of centre $O$ and radius $\ell$, respectively. The motion of the pole can be generated by laying the rod on the diameter of a circle of radius $\ell$ and making it rotate without sliding on a fixed circle of twice the radius, as in Fig. 6.8 ∎

## 6.8 Problems

**1.** Compute the direction cosines of a moving reference frame with respect to a fixed reference frame as a function of the Euler angles.

**2.** Prove that in a plane rigid motion the instantaneous centre of rotation lies on the normal to the velocity of each point which is distinct from it (*Chasles' theorem*).

**3.** A disc of radius $R$ moves on a plane, rolling and sliding along a straight line. Assuming that the velocity of the centre and the sliding velocity are known, find the instantaneous centre of motion. Find also the conjugate profiles when the above velocities are constant.

**4.** A rod $AB$ moves on a plane while being tangent to a given circle of radius $R$. The rod has one of its endpoints constrained to slide on a line tangent to the circle. Prove that the conjugate profiles are two parabolas.

**5.** Compose two uniform rotations around distinct parallel axes (consider both the case of rotations with the same or opposite direction).

**6.** Compose two uniform rotations around incident axes (the result is a regular precession).

**7.** Compose two uniform rotations around skew lines, proving that the two ruled surfaces of the compound motion are one-sheeted hyperboloids.

**8.** A disc of radius $r$ rolls without sliding on a circle of radius $R$, while being orthogonal to the circle plane and moving in such a way that its axis intersects at every instant the normal to the circle plane passing through its centre. Prove that the motion is a precession and that the Poinsot cones are circular.

**9.** Determine the generalised potentials of the centrifugal force and of the Coriolis force.

## 6.9 Additional solved problems

*Problem 1*
It is well known that if $A, B$ are two $n \times n$ matrices ($n \geq 2$) then $e^A e^B = e^{A+B} = e^B e^A$ if and only if $[A, B] = AB - BA = 0$. More generally, show that if $[C, A] = [C, B] = 0$, where $C = [A, B]$, then

$$e^A e^B = e^{A+B+[A,B]/2}. \tag{6.45}$$

*Solution*
We start by showing that if $F$ is a function analytic in $x \in \mathbf{R}$ and entire (i.e. its radius of convergence is $+\infty$) then

$$[A, F(B)] = [A, B]\frac{\mathrm{d}F}{\mathrm{d}x}(B).$$

Indeed it can be seen that if $[C, B] = 0$, then $[C, B^n] = 0$ and, by induction on $n \geq 0$, that

$$[A, B^n] = [A, B]nB^{n-1}.$$

From this it follows that

$$[A, F(B)] = \sum_{n=1}^{\infty} f_n[A, B^n] = [A, B]\sum_{n=1}^{\infty} f_n n B^{n-1} = [A, B]\frac{\mathrm{d}F}{\mathrm{d}x}(B),$$

where $F(x) = \sum_{n=0}^{\infty} f_n x^n$. Now let $G(t) = e^{tA}e^{tB}$; it is immediate to verify that

$$\frac{dG}{dt}(t) = (A + e^{tA}Be^{-tA})G(t).$$

Since $[B, e^{tA}] = [B, A]te^{tA}$ we have

$$\frac{dG}{dt} = (A + B + t[A, B])G(t).$$

Finally, since $A + B$ under our assumptions commutes with $[A, B]$ the latter differential equation admits the solution (note that $G(0) = 1$)

$$G(t) = \exp\left((A + B)t + [A, B]\frac{t^2}{2}\right),$$

from which equation (6.45) follows by choosing $t = 1$.

*Problem 2*

Let $A$ and $B$ be two $n \times n$ matrices ($n \geq 2$) and let $t \in \mathbf{R}$. Prove that

$$e^{tA}e^{tB} = e^{t(A+B)+(t^2/2)[A,B]+(t^3/12)([A,[A,B]]+[B,[B,A]])+\mathcal{O}(t^4)}. \tag{6.46}$$

*Solution*

By definition $e^C = 1 + \sum_{n=1}^{\infty} C^n/n!$, from which it follows that

$$e^{tA}e^{tB} = 1 + t(A + B) + \frac{t^2}{2}(A^2 + B^2 + 2AB)$$

$$+ \frac{t^3}{6}(A^3 + B^3 + 3AB^2 + 3A^2B) + \mathcal{O}(t^4), \tag{6.47}$$

$$\exp\left\{t(A+B) + \frac{t^2}{2}[A, B] + \frac{t^3}{12}([A,[A,B]] + [B,[B,A]]) + \mathcal{O}(t^4)\right\}$$

$$= 1 + t(A + B) + \frac{t^2}{2}[A, B] + \frac{t^3}{12}([A,[A,B]] + [B,[B,A]])$$

$$+ \frac{1}{2}\left\{t^2(A+B)^2 + \frac{t^3}{2}[(A+B)[A,B] + [A,B](A+B)]\right\} \tag{6.48}$$

$$+ \frac{t^3}{6}(A+B)^3 + \mathcal{O}(t^4),$$

and to obtain (6.46) one only needs to identify the coefficients of $t^0, t^1, t^2$ and $t^3$. In the case of $t^0$ and $t^1$ this is obvious, while for $t^2/2$ we find

$$A^2 + B^2 + 2AB = (A + B)^2 + [A, B] = A^2 + AB + BA + B^2 + AB - BA.$$

## 6.9 Rigid bodies: geometry and kinematics

For $t^3/12$ the computation is more tedious. The coefficient of $t^3/12$ in (6.48) is

$$[A, [A, B]] + [B, [B, A]] + 3(A+B)[A, B] + 3[A, B](A+B) + 2(A+B)^3. \quad (6.49)$$

By iterating the identity $BA = AB - [A, B]$ we find

$$\begin{aligned}(A+B)^3 &= (A+B)(A^2 + AB + BA + B^2) = (A+B)(A^2 + 2AB + B^2 - [A, B]) \\ &= A^3 + B^3 + 2A^2B + AB^2 + BA^2 + 2BAB - (A+B)[A, B] \\ &= A^3 + B^3 + 3A^2B + 3AB^2 - (2A+B)[A, B] - [A, B](A+2B).\end{aligned} \quad (6.50)$$

Comparing (6.49) with the coefficient of $t^3/6$ in (6.47), recalling also (6.50), it follows that the proof of (6.46) is complete if we can show that

$$[A, [A, B]] + [B, [B, A]] + 3(A+B)[A, B] + 3[A, B](A+B) \\ - 2(2A+B)[A, B] - 2[A, B](A+2B) = 0.$$

This reduces to

$$[A, [A, B]] + [B, [B, A]] + (B-A)[A, B] + [A, B](A-B) = 0$$

with immediate verification.

### Problem 3
Consider a plane rigid motion and suppose that the conjugate profiles, as well as the angular velocity $\omega(t)$, are known. Determine at every instant $t$:

(i) the locus of points $P$ for which $\mathbf{a}(P) \parallel \mathbf{v}(P)$;
(ii) the locus of points $P$ for which $\mathbf{a}(P) \perp \mathbf{v}(P)$.

### Solution
Let $C(t)$ be the instantaneous centre of rotation. This is the point at time $t$ where the conjugate profiles are tangent. By definition, the velocity $\mathbf{v}(C) = 0$ in the rigid motion (while in general the acceleration is not zero). Introduce the point $\overline{C}(t)$ (moving with respect to both the fixed and the body reference frame), which at every instant coincides with $C(t)$. In its absolute motion, $\overline{C}(t)$ travels over the fixed conjugate profile, while its trajectory in the relative motion is the body conjugate profile. It is easily seen that $\overline{C}(t)$ has equal absolute and relative velocities. Writing the absolute velocity field at every instant $t$ in the form

$$\mathbf{v}(P) = \omega \times (P - \overline{C}) \quad (6.51)$$

we obtain, by differentiating, the acceleration field

$$a(P) = \dot{\omega} \times (P - \overline{C}) + \omega \times [\omega \times (P - \overline{C}) - \overline{\mathbf{v}}], \quad (6.52)$$

where $\overline{\mathbf{v}} = d\overline{C}/dt$. The condition for a point to belong to the locus (i) is $\mathbf{a}(P) \cdot (P - \overline{C}) = 0$, which reduces to

$$\omega^2 (P - \overline{C})^2 + \omega \times \overline{\mathbf{v}} \cdot (P - \overline{C}) = 0. \quad (6.53)$$

Setting $P = C = \overline{C}$ in expression (6.52) we find the acceleration of the instantaneous centre $C$, i.e. $\mathbf{a}(C) = -\omega \times \overline{\mathbf{v}}$, orthogonal to $\overline{\mathbf{v}}$. Setting $(P-\overline{C})/|P-\overline{C}| = \mathbf{e}$ and $2\mathbf{R} = \mathbf{a}(C)/\omega^2$, expression (6.53) can be written

$$|P - \overline{C}| = 2\mathbf{R} \cdot \mathbf{e},$$

clarifying the following structure of the locus (i).

*The locus of points $P$ for which $\mathbf{a}(P) \parallel \mathbf{v}(P)$ at time $t$ is a circle tangent to the conjugate profiles, and of radius $R = \overline{v}/2\omega$.*

The points having such a property find themselves at a point of the trajectory with *vanishing curvature*. For this reason, the locus (i) is also called the *circle of inflection points*. Analogously, imposing the condition $\mathbf{a}(P) \times (P - \overline{C}) = 0$ one arrives at the equation for the locus (ii):

$$\dot{\omega}(P - \overline{C})^2 + \omega[\overline{\mathbf{v}} \cdot (P - \overline{C})] = 0.$$

If $\dot{\omega} = 0$ this locus reduces to the line passing through $\overline{C}$ and orthogonal to $\overline{\mathbf{v}}$. If $\dot{\omega} \neq 0$, setting $2\widehat{R} = |\omega/\dot{\omega}|\overline{v}$, we conclude that *the locus of points $P$ for which $\mathbf{a}(P) \perp \mathbf{v}(P)$ at time $t$, if $\dot{\omega} \neq 0$, is a circle through $C$ and orthogonal to the conjugate profiles of radius $\widehat{R} = \frac{1}{2}|\omega/\dot{\omega}|\overline{v}$.*

*If $\dot{\omega} = 0$ it degenerates to a line normal to these profiles.* The kinematic meaning of the locus (ii) lies in the fact that the magnitude of the velocity of its points has at that time *zero time derivative*, and hence the name *stationary circle*. We note finally that the intersection $H$ of the two circles, different from $\overline{C}$, has zero acceleration. Hence the acceleration field of the body frame can be written in the form

$$\mathbf{a}(P) = \frac{\mathrm{d}}{\mathrm{d}t}[\omega \times (P - H)].$$

For this reason $H$ is called the *pole of accelerations*.

To complete the problem, we now find the relation between $\overline{v}$ and $\omega$. Consider the case that the principal normal vectors to the conjugate profiles at the point of contact have opposite orientation. Let $k_f, k_b$ be their curvatures. Considering the osculating circles, it is easy to find that if the point $\overline{C}$ undergoes a displacement $ds = \overline{v}\,dt$, the angular displacement of the respective normal vectors to the two curves at the contact point is $d\varphi_f = k_f\,ds$, $d\varphi_b = k_b\,ds$, in the fixed and body frames, respectively. Consequently the variation of the angle that a fixed direction forms with a body direction is $\omega\,dt = d\varphi_f + d\varphi_b = (k_f + k_b)\overline{v}\,dt$, yielding $\omega = (k_f + k_b)\overline{v}$. Hence we conclude that the radius of the circle of inflection points is $R = \frac{1}{2}(k_f + k_b)$, while the radius of the stationariness circle can be written as $\widehat{R} = \frac{1}{2}(\omega^2(k_f + k_b))/|\dot{\omega}|$. In the case that the two principal normal vectors have the same orientation, we substitute $k_f + k_b$ with $|k_f - k_b|$ (obviously we must have $k_f \neq k_b$).

# 7 THE MECHANICS OF RIGID BODIES: DYNAMICS

## 7.1 Preliminaries: the geometry of masses

In contrast with kinematics, the dynamics of rigid bodies depends on the specific distribution of masses. Hence it is necessary to review some results on the geometry and kinematics of masses; we shall limit ourselves to the essential facts.[1]

Rigid bodies can be treated equally well by a discrete or by a continuum model. The latter consists of defining a *mass density function* in the region occupied by the system. This is due to the fact that for rigid continua it makes sense to consider forces applied to single points, or rather, to substitute a force field (such as weight) with equivalent systems of forces applied to various points of the system, or rigidly connected to it. This should be contrasted with the case of deformable continua, and for which one must define force densities.

To simplify notation, we consider in this chapter discrete rigid systems, but the results can be easily extended to continua: it is enough to substitute any expression of the type $\sum_{i=1}^{n} m_i f(P_i)$, where the sum extends to all points $(P_1, m_1), \ldots, (P_n, m_n)$ of the system, with an integral of the form $\int_{\mathcal{R}} \rho(P) f(P) \, dV$, where $\rho$ is the density and $\mathcal{R}$ is the region occupied by the body.

We start by recalling the notion of *centre of mass*:[2]

$$m(P_0 - O) = \sum_{i=1}^{n} m_i(P_i - O) \tag{7.1}$$

($O$ is an arbitrary point in $\mathbf{R}^3$, $m = \sum_{i=1}^{n} m_i$). The *moment of inertia* with respect to a line $r$ is given by

$$I_r = \sum_{i=1}^{n} m_i [(P_i - O) \times \mathbf{e}]^2 \tag{7.2}$$

($O \in r$, $\mathbf{e}$ is a unit vector of $r$). The *centrifugal moment* or *product of inertia* with respect to a pair of non-parallel planes $\pi$, $\pi'$ with normal vectors $\mathbf{n}$, $\mathbf{n}'$ is

$$I_{\pi\pi'} = -\sum_{i=1}^{n} m_i [(P_i - O) \cdot \mathbf{n}][(P_i - O) \cdot \mathbf{n}'] \tag{7.3}$$

---

[1] A more detailed description can be found in Fasano *et al.* (2001).
[2] Recall that *in the gravity field the centre of mass coincides with the baricentre*.

($O \in \pi \cap \pi'$). Given a reference frame $(O, x_1, x_2, x_3)$, we denote by $I_{11}, I_{22}, I_{33}$ the moments of inertia with respect to the axes and by $I_{ij}$, $i \neq j$, the products of inertia with respect to the pairs of coordinate planes $x_i = 0$, $x_j = 0$. Note that the matrix $(I_{ij})$ is symmetric.

It is also possible to define other quadratic moments, but we will not consider them here. For each of them *Huygens' theorem* holds, which in the cases of interest here, takes the form

$$I_r = I_{r_0} + m[(P_0 - O) \times \mathbf{e}]^2, \tag{7.4}$$

$$I_{\pi\pi'} = I_{\pi_0\pi'_0} - m[(P_0 - O) \cdot \mathbf{n}][(P_0 - O) \cdot \mathbf{n}'], \tag{7.5}$$

where $r_0$ is the line parallel to $r$ and passing through the centre of mass $P_0$, and $\pi_0$, $\pi'_0$ are planes parallel to $\pi$, $\pi'$, respectively, and passing through $P_0$. The other symbols in (7.4), (7.5) keep the same meaning as in formulas (7.1)–(7.3). It is customary to also define the *radius of gyration* with respect to a line:

$$\delta_r = (I_r/m)^{1/2}, \tag{7.6}$$

and hence equation (7.4) can also be written as

$$\delta_r^2 = \delta_{r_0}^2 + \delta^2, \tag{7.7}$$

where $\delta$ is the distance between $P_0$ and $r$.

*Remark* 7.1

Equation (7.4) implies that among all lines in a given direction, the one for which the moment of inertia has a minimum is the one passing through $P_0$. ∎

## 7.2 Ellipsoid and principal axes of inertia

The distribution of moments of inertia with respect to the set of lines through a given point $O$ plays an important role in the dynamics of rigid bodies.

Given a reference frame $(O, x_1, x_2, x_3)$, we now compute the moment of inertia with respect to the line passing through $O$ with direction cosines $\alpha_1, \alpha_2, \alpha_3$. Applying the definition (7.2) we find

$$I(\alpha_1, \alpha_2, \alpha_3) = \sum_{i,j=1}^{3} \alpha_i \alpha_j I_{ij}. \tag{7.8}$$

Since in equation (7.2) we have that $I_r > 0$ for any unit vector $\mathbf{e}$, the quadratic form (7.8) is positive definite (at least excluding the degenerate case of a rigid system with mass distributed along a straight line passing through $O$, in which the matrix is positive semidefinite). It follows that the level sets are the ellipsoid of centre $O$ (*ellipsoid of inertia*) given by the equation

$$\sum_{ij} I_{ij} x_i x_j = \lambda^2. \tag{7.9}$$

## 7.2  The mechanics of rigid bodies: dynamics

DEFINITION 7.1  *The symmetry axes of the ellipsoid of inertia are called* the principal axes of inertia. ∎

If the ellipsoid of inertia is an ellipsoid of rotation around one of the axes, all straight lines through $O$ and orthogonal to this axis are principal axes of inertia.

A *principal reference frame* $(O, X_1, X_2, X_3)$ is the triple of the three principal axes of inertia. We denote by $\mathcal{J}_i$ the moment of inertia with respect to the principal axis $X_i$. The direction of the principal axes of inertia is determined by the eigenvectors of the matrix $(I_{ij})$ and the principal moments of inertia $\mathcal{J}_i$ are the eigenvalues of the same matrix (cf. Proposition 7.8).

PROPOSITION 7.1  *A necessary and sufficient condition for a reference frame to be the principal frame is that the inertia products with respect to each of the pairs of coordinate planes are zero.*

*Proof*
If $I_{ij} = 0$ for $i \neq j$, setting $I_{ii} = \mathcal{J}_i$, the level sets of (7.8) are determined by

$$\sum_{i=1}^{3} \mathcal{J}_i X_i^2 = \lambda^2, \qquad (7.10)$$

and the equation is put in the canonical form characteristic of the triple of the symmetry axes. Conversely, the quadratic form (7.8) written with respect to the principal frame is diagonal and the ellipsoid equation is of the form (7.10). ∎

PROPOSITION 7.2  *A necessary and sufficient condition for a straight line to be a principal axis of inertia relative to $O$ is that the products of inertia relative to the plane passing through $O$ and orthogonal to the line and to any other plane through the line are zero.*

*Proof*
Suppose that a straight line is a principal axis of inertia and consider an arbitrary frame $(O, x_1, x_2, x_3)$, with $x_3$ being the line itself. The plane of equation $x_3 = 0$ is then a symmetry plane for the ellipsoid. It follows that its equation must be invariant with respect to the variable change from $x_3$ to $-x_3$, i.e. $I_{13} = I_{23} = 0$.

Conversely, if in the same frame one has $I_{13} = I_{23} = 0$, the ellipsoid is symmetric with respect to the plane $x_3 = 0$. Hence $x_3$ is a principal axis. ∎

It is easier to determine the principal axis when there are *material symmetries*.

DEFINITION 7.2  *A rigid system has a* plane of material orthogonal symmetry *if in addition to the geometric symmetry with respect to the plane, one has the property that symmetric points have the same mass (or, for continua, the same density).* ∎

This definition can be extended to the case of material symmetry with respect to a straight line or to a point.

PROPOSITION 7.3 *Let $\pi$ be a plane of material orthogonal symmetry. Then for any $O$ in $\pi$, the straight line through $O$ and normal to $\pi$ is a principal axis of inertia relative to $O$.*

*Proof*
The ellipsoid of inertia relative to $O$ is symmetric with respect to $\pi$. ∎

*Remark 7.2*
The latter proposition can be applied to the limiting case of *plane systems*. ∎

PROPOSITION 7.4 *Suppose that a system has two distinct (non-parallel) planes of material symmetry, $\pi$ and $\pi'$. Then there are two possible cases:*
(a) *the two planes are orthogonal;*
(b) *the two planes are not orthogonal.*

*In the case (a), given $O \in r = \pi \cap \pi'$, the principal frame relative to $O$ is given by $r$ and by the two straight lines through $O$ normal to $r$ lying in the planes $\pi, \pi'$.*
*In the case (b) the ellipsoid of inertia is a surface of revolution with respect to $r$.*

*Proof*
Case (a) is trivial. In case (b) we note that the two straight lines through $O$ and normal to $\pi$ and $\pi'$ are principal axes (Proposition 7.3). It follows that the ellipsoid has a plane of symmetry containing two non-orthogonal symmetry axes. This necessarily implies the rotational symmetry with respect to $r$. ∎

Another useful property for the determination of the principal axes of inertia is the following.

PROPOSITION 7.5 *Let $(P_0, X_1^{(0)}, X_2^{(0)}, X_3^{(0)})$ be a principal frame relative to the centre of mass. The principal axes of inertia with respect to the points of the axes $X_1^{(0)}, X_2^{(0)}, X_3^{(0)}$ can be obtained from it by translation (Fig. 7.1).*

*Proof*
We recall Huygens' theorem (1.5) and use it to verify that the products of inertia in the translated frame are zero. ∎

*Problem 7.1*
Determine the principal frame for a regular material homogeneous polygon, relative to a generic point of the plane of the polygon. ∎

The following remark is in some sense the converse of Proposition 7.5.

PROPOSITION 7.6 *If a straight line is a principal axis of inertia with respect to two of its points (distinct), then it must contain the centre of mass.*

*Proof*
Let $O$ and $O'$ be the two points, and consider two frames with parallel axes $S = (O, x_1, x_2, x_3)$, $S' = (O', x'_1, x'_2, x_3)$, where $x_3$ is the line referred to in the statement. Recalling Proposition 7.2, we can write

$$\sum_{i=1}^{n} m_i x_2^{(i)} x_3^{(i)} = \sum_{i=1}^{n} m_i x_1^{(i)} x_3^{(i)} = 0,$$

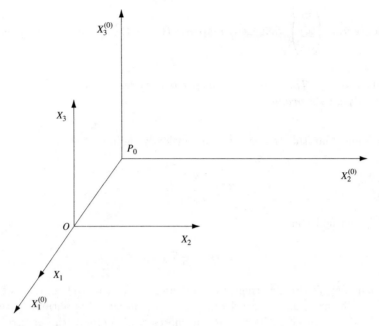

**Fig. 7.1**

and similarly

$$\sum_{i=1}^{n} m_i x_2^{(i)}(x_3^{(i)} - z) = \sum_{i=1}^{n} m_i x_1^{(i)}(x_3^{(i)} - z) = 0,$$

where $z$ is the third coordinate of $O'$ in $S$.

It follows that

$$\sum_{i=1}^{n} m_i x_2^{(i)} = \sum_{i=1}^{n} m_i x_1^{(i)} = 0,$$

and hence that $P_0$ lies on the axis $x_3$. ∎

DEFINITION 7.3 *A system such that the central ellipsoid of inertia is a surface of revolution is called* a gyroscope. *The axis of revolution of the central ellipsoid is called* the gyroscopic axis. ∎

## 7.3 Homography of inertia

Let us fix a triple $S = (O, x_1, x_2, x_3)$. Then the product of the symmetric matrix

$$\begin{pmatrix} I_{11} & I_{12} & I_{13} \\ I_{12} & I_{22} & I_{23} \\ I_{13} & I_{23} & I_{33} \end{pmatrix} \tag{7.11}$$

with the vector $\begin{pmatrix} x_1 \\ x_2 \\ x_3 \end{pmatrix}$ defines a map $\sigma : \mathbf{R}^3 \to \mathbf{R}^3$, called the *homography of inertia*.

PROPOSITION 7.7 *The map $\sigma$ is independent of the choice of the frame $S$, and depends only on its origin $O$.*

*Proof*
We now show that the vector $\sigma \mathbf{x}$ is intrinsically defined. Setting

$$f(\mathbf{x}) = \sum_{i,j=1}^{3} I_{ij} x_i x_j, \tag{7.12}$$

it is easily verified that

$$\sigma \mathbf{x} = \frac{1}{2} \nabla f. \tag{7.13}$$

If $\mathbf{x}$ is such that $f(\mathbf{x}) = \lambda^2$, then $\sigma \mathbf{x}$ is orthogonal in $\mathbf{x}$ to the ellipsoid of inertia. Since $\sigma$ is a linear map, we can deduce that in general $\sigma \mathbf{x}$ is normal to the plane tangent to the ellipsoid of inertia at the intersection point with the straight line through the origin parallel to $\mathbf{x}$ (Fig. 7.2).

We also note that for every unit vector $\mathbf{e}$:

$$\mathbf{e} \cdot \sigma \mathbf{e} = I_r, \tag{7.14}$$

where $r$ is the line passing through $O$ with unit vector $\mathbf{e}$. More generally

$$\mathbf{x} \cdot \sigma \mathbf{x} = I_r \mathbf{x}^2 \tag{7.15}$$

for any vector ($I_r$ is computed with respect to the line through $O$ and parallel to $\mathbf{x}$). Thus *the quadratic form $\mathbf{x} \cdot \sigma \mathbf{x}$ is positive definite* (we are considering the

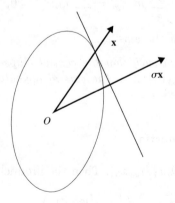

Fig. 7.2

generic case, excluding the possibility that the masses are distributed along a straight line). The relation (7.15) determines the orientation and the length of the vector $\sigma \mathbf{x}$. ∎

COROLLARY 7.1  *If the reference frame is transformed by an orthogonal matrix $A$, the components of $\sigma$ are subject to a similarity transformation*

$$\sigma'(0) = A\sigma(0)A^T. \tag{7.16}$$

∎

Because of (7.16) we can state that $\sigma$ is a covariant tensor of rank 2 (Appendix 4), called the *tensor of inertia*.

The following properties are immediate.

PROPOSITION 7.8  *The principal axes of inertia are the eigenspaces of the homography of inertia and the corresponding moments of inertia are its eigenvalues.* ∎

In other words, $\sigma \mathbf{x}$ is parallel to $\mathbf{x}$, namely

$$\sigma \mathbf{x} = \mathcal{J}\mathbf{x}, \tag{7.17}$$

*if and only if* $\mathbf{x}$ *has the direction of one of the principal axes of inertia.*

Seeking a principal triple of inertia is equivalent to the diagonalisation of (7.11), because in the principal reference frame, $\sigma(O)$ has the representation

$$\sigma(O) = \begin{pmatrix} \mathcal{J}_1 & 0 & 0 \\ 0 & \mathcal{J}_2 & 0 \\ 0 & 0 & \mathcal{J}_3 \end{pmatrix}. \tag{7.18}$$

*Example 7.1*

We solve the problem of the determination of two principal axes of inertia when the third one is known. Let $x_3$ be the known axis. Then the following three methods are equivalent.

(1) For every pair of axes $x_1$, $x_2$ in the plane $x_3 = 0$ let $\varphi$ be the angle by which it must be rotated to obtain the two axes. We know that $I_{13} = I_{23} = 0$ for any $\varphi$ (Proposition 7.2). Hence it suffices to find $\varphi$ such that $I_{12}(\varphi) = 0$.

(2) Find $\varphi$ in such a way that the matrix

$$A = \begin{pmatrix} \cos \varphi & \sin \varphi \\ -\sin \varphi & \cos \varphi \end{pmatrix}$$

diagonalises the matrix

$$\begin{pmatrix} I_{11} & I_{12} \\ I_{12} & I_{22} \end{pmatrix}.$$

(3) Find among the lines passing through $O$ in the plane $x_3 = 0$ those which are extremals for the moment of inertia

$$I(\varphi) = I_{11} \cos^2 \varphi + I_{22} \sin^2 \varphi + 2I_{12} \sin \varphi \cos \varphi. \tag{7.19}$$

Following the latter method we compute
$$I'(\varphi) = (I_{22} - I_{11})\sin 2\varphi + 2I_{12}\cos 2\varphi = 0, \qquad (7.20)$$
i.e.
$$\tan 2\varphi = \frac{2I_{12}}{I_{11} - I_{22}}, \qquad (7.21)$$
if $I_{11} \neq I_{22}$. The values of $\varphi$ which follow from (7.21) give two mutually orthogonal axes.

If $I_{11} = I_{22}$ the relation (7.20) reduces to $I_{12}\cos 2\varphi = 0$ and there are two possible cases:

(a) $I_{12} = 0 \Rightarrow$ the frame we started with is a principal frame;
(b) $I_{12} \neq 0 \Rightarrow$ the principal axes are the bisectors of the quadrants defined by the axes $(x_1, x_2)$. ∎

We conclude this section by recalling an important formula: given two unit vectors $\mathbf{e} = (\alpha_1, \alpha_2, \alpha_3)$, $\mathbf{e}' = (\alpha'_1, \alpha'_2, \alpha'_3)$ that are mutually orthogonal, we have
$$\mathbf{e}' \cdot \sigma(O)\mathbf{e} = \sum_{i,j} I_{ij}\alpha_i\alpha'_j = I_{\pi\pi'}, \qquad (7.22)$$
where $\pi$, $\pi'$ are the planes through $O$ whose normals are given by $\mathbf{e}$, $\mathbf{e}'$, respectively. The proof is left as an exercise.

## 7.4 Relevant quantities in the dynamics of rigid bodies

*(a) Angular momentum*

Using equation (6.4) in the definition of the angular momentum, we find that for a rigid system
$$\mathbf{L}(O) = m(P_0 - O) \times \mathbf{v}(O) - \sum_{i=1}^{n} m_i(P_i - O) \times [(P_i - O) \times \boldsymbol{\omega}]. \qquad (7.23)$$

Let us examine the operator $\Sigma(O)$ defined by
$$\Sigma(O)\mathbf{e} = -\sum_{i=1}^{n} m_i(P_i - O) \times [(P_i - O) \times \mathbf{e}]$$
$$= \mathbf{e}\sum_{i=1}^{n} m_i(P_i - O)^2 - \sum_{i=1}^{n} m_i(P_i - O)[(P_i - O) \cdot \mathbf{e}]. \qquad (7.24)$$

The scalar product with $\mathbf{e}$ yields
$$\mathbf{e} \cdot \Sigma(O)\mathbf{e} = I_\mathbf{e} = \mathbf{e} \cdot \sigma(O)\mathbf{e} \qquad (7.25)$$
(recall (7.14)), and hence $\Sigma(O)\mathbf{e}$ and $\sigma(O)\mathbf{e}$ have the same component on $\mathbf{e}$.

## 7.4 The mechanics of rigid bodies: dynamics

Let $\mathbf{e}'$ be any unit vector orthogonal to $\mathbf{e}$, and consider the scalar product of (7.24) with $\mathbf{e}'$; this yields by the definition of the product of inertia $I_{\mathbf{ee}'}$ and by (7.22),

$$\mathbf{e}' \cdot \Sigma(O)\,\mathbf{e} = I_{\mathbf{ee}'} = \mathbf{e}' \cdot \sigma(O)\,\mathbf{e}. \tag{7.26}$$

Equations (7.25) and (7.26) show that the operators $\Sigma(O)$ and $\sigma(O)$ coincide. We then write

$$\mathbf{L}(O) = m(P_0 - O) \times \mathbf{v}(O) + \sigma(O)\,\boldsymbol{\omega}. \tag{7.27}$$

In particular, if $O = P_0$ or if $O$ is fixed,

$$\mathbf{L}(O) = \sigma(O)\boldsymbol{\omega}. \tag{7.28}$$

**DEFINITION 7.4** *We say that quantities observed from a reference frame $\Sigma'$ with origin at the centre of mass and axes parallel to those of a fixed reference frame $\Sigma$, are* relative to the centre of mass. ∎

**PROPOSITION 7.9** *The reference frames $\Sigma$ and $\Sigma'$ measure identical values of the angular momentum relative to $P_0$.*

*Proof*
The proof is based on (7.28), which can also be written as

$$\mathbf{L}(P_0) = \sigma(P_0)\,\boldsymbol{\omega} \tag{7.29}$$

in the two systems, which measure the same value of $\boldsymbol{\omega}$ (recall equation (6.33)). ∎

### (b) Kinetic energy

It is easy to find

$$T = \frac{1}{2} m[\mathbf{v}(O)]^2 + \frac{1}{2} \sum_{i=1}^{n} m_i [(P_i - O) \times \boldsymbol{\omega}]^2 + m\mathbf{v}(O) \cdot \boldsymbol{\omega} \times (P_0 - O),$$

and hence if $\mathbf{v}(O) = 0$

$$T = \frac{1}{2} I\omega^2 \tag{7.30}$$

($I$ is the moment of inertia with respect to the axis of instantaneous rotation), and more generally, choosing $O = P_0$,

$$T = \frac{1}{2} m[\mathbf{v}(P_0)]^2 + \frac{1}{2} I_0 \omega^2, \tag{7.31}$$

which is known as the *König theorem* (the kinetic energy is the sum of the rotational energy relative to the centre of mass and of the translational energy associated with the point $(P_0, m)$).

Note that the comparison between (7.30) and (7.31), and the fact that $\mathbf{v}(P_0) = \boldsymbol{\omega} \times (P_0 - O)$ if $\mathbf{v}(O) = 0$, shows that the König theorem is equivalent to Huygens' theorem for the moments of inertia.

Finally for a precession

$$T = \frac{1}{2} \boldsymbol{\omega} \cdot \sigma(O) \boldsymbol{\omega} \tag{7.32}$$

and by exploiting (6.38), we find the expression for $T$ in the principal frame of reference as a function of the Euler angles of that frame:

$$2T = \mathcal{J}_1(\dot{\theta} \cos\varphi + \dot{\psi} \sin\theta \sin\varphi)^2 + \mathcal{J}_2(\dot{\theta} \sin\varphi - \dot{\psi} \sin\theta \cos\varphi)^2 \\ + \mathcal{J}_3(\dot{\varphi} + \dot{\psi} \cos\theta)^2. \tag{7.33}$$

## 7.5 Dynamics of free systems

The power of a system of resultant force $\mathcal{R}$ and a resultant torque $\mathbf{M}(O)$ acting on a rigid system can be computed easily by (6.4):

$$W = \mathcal{R} \cdot \mathbf{v}(O) + \mathbf{M}(O) \cdot \boldsymbol{\omega}. \tag{7.34}$$

From (7.34) we can deduce two important consequences. The first is that *a balanced system of forces* ($\mathcal{R} = 0$, $\mathbf{M} = 0$) *has zero power when acting on a rigid system*. This is the case for the system of reactions due to rigidity constraints, and hence rigid bodies belong to the category of systems with smooth fixed constraints.

The second is that the equations of motion (4.19) express the vanishing of the power of the force system $F_i - m_i a_i$, for any arbitrary choice of $(\mathbf{v}(O), \boldsymbol{\omega})$, and hence it is equivalent to the vanishing of the multipliers of $\mathbf{v}(O)$ and $\boldsymbol{\omega}$ in (7.34) (with the inclusion of the contributions of inertial forces) and leads to the *cardinal equations*, which we write in the form

$$m\mathbf{a}(P_0) = \mathcal{R}, \tag{7.35}$$

$$\dot{\mathbf{L}}(P_0) = \mathbf{M}(P_0). \tag{7.36}$$

Hence equations (7.35), (7.36) *for an unconstrained rigid body are equivalent to Lagrange's equations* (4.40) and consequently they are sufficient to study the motion and the equilibrium of the system; in the latter case they reduce to the form

$$\mathcal{R} = 0, \quad \mathbf{M} = 0. \tag{7.37}$$

From this follows a well-known property concerning the motion and equilibrium of a rigid system, namely that two systems of forces with the same resultant force and the same resultant moment are equivalent. In particular the weight force field can be replaced by its resultant force (the total weight applied at the barycentre).

A more interesting case arises when equation (7.35) is independently integrable; this is the case if $\mathcal{R}$ depends only on the coordinates of $P_0$ (not on the Euler angles). In this case it is possible to first determine the motion of the centre of mass (starting from given initial conditions) and then integrate equation (7.36), which describes the motion 'relative to the centre of mass' (Definition 7.4 and Proposition 7.9). This motion is obviously a precession. The study of precessions is therefore of particular significance. This is considered in the following sections.

## 7.6 The dynamics of constrained rigid bodies

Suppose that a rigid system is also subject to external constraints, *holonomic and frictionless*. Let $\phi^{(e)}$ be the resultant of the constraint reaction, and $\boldsymbol{\mu}^{(e)}$ be the resultant moment for the same system; then the cardinal equations take the form

$$m\mathbf{a}(P_0) = \mathcal{R}^{(e)} + \phi^{(e)}, \tag{7.38}$$

$$\dot{\mathbf{L}}(P_0) = \mathbf{M}^{(e)}(P_0) + \boldsymbol{\mu}^{(e)}(P_0) \tag{7.39}$$

(if as a result of the constraints, the system has a fixed point $O$, it is convenient to refer the latter to this point).

Since the motion can be found by means of the Lagrange equations, (7.38) and (7.39) can be used to determine $\phi^{(e)}$ and $\boldsymbol{\mu}^{(e)}$, and hence each of the constraint reactions $\phi_i$, as long as these constraints are linearly independent. On the other hand, in this case the number of scalar unknowns appearing in $\phi^{(e)}$, $\boldsymbol{\mu}^{(e)}$ in (7.38), (7.39) is equal to the number of degrees of freedom suppressed by the constraints, and hence in general equations (7.38), (7.39) can be used directly to determine simultaneously the motion and the constraint reactions. Similar considerations are valid for the equilibrium.

If the *constraints are not smooth* it is necessary to acquire additional information to balance equations (7.38), (7.39). This can be done by relating the actions due to friction with the motion of the system. In the static case it is necessary to define the maximal resistance that the constraint opposes sliding and rotation.

*Example 7.2*
A homogenous rod $AB$ of mass $m$ and length $\ell$ has the point $A$ sliding along a line $r$ in a horizontal plane (Fig. 7.3). All constraints are smooth. At time $t = 0$ the point $A$ has zero velocity, $B$ is on $r$ and the angular velocity of the rod is $\omega_0$. Study the motion of the rod and determine the constraint reaction $\phi_A$.

Fix the axis $x$ to coincide with the initial configuration of the rod and fix the origin in the initial position of $A$. Take as Lagrangian coordinates the $x$-coordinate $x$ of $A$ and the angle $\varphi$; see the figure. As initial conditions we have $x(0) = 0$, $\dot{x}(0) = 0$, $\varphi(0) = 0$, $\dot{\varphi}(0) = \omega_0$.

From the first cardinal equation we deduce that the $x$-component of the linear momentum is constant. Since the initial velocity of $P_0$ is orthogonal to the $x$-axis,

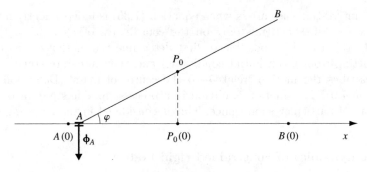

Fig. 7.3

$P_0$ moves along the line orthogonal to the $x$-axis and passing through $P_0(0)$, i.e.

$$x + \frac{1}{2}\ell \cos\varphi = \frac{1}{2}\ell. \tag{7.40}$$

Since no work is done on the system, the kinetic energy is conserved and this implies that it is easy to determine

$$\dot\varphi^2 = \frac{4\omega_0^2}{1 + 3\cos^2\varphi}. \tag{7.41}$$

The integration of this relation via elliptic functions (see Appendix 2) yields a complete description of the motion.

Since $\dot L(P_0) = \frac{1}{12} m\ell^2 \ddot\varphi$ and $\dot{\mathbf{L}}(P_0) = (A - P_0) \times \boldsymbol{\phi}_A$, we can obtain the expression for the unique component of $\phi_A$ by differentiating (7.41) and expressing $\ddot\varphi$ as a function of $\varphi$:

$$\ddot\varphi = \frac{12\omega_0^2}{(1 + 3\cos^2\varphi)^2} \sin\varphi \cos\varphi. \tag{7.42}$$

We finally find

$$\phi_A = -m\ell \frac{2\omega_0^2}{(1 + 3\cos^2\varphi)^2} \sin\varphi. \tag{7.43}$$

∎

*Example 7.3*
A material homogeneous system of linear density $\rho$ consists of a circular arc of opening angle $2\alpha$ and radius $R$. The system rolls without sliding along a horizontal rectilinear guide in a vertical plane. Write the equation of motion and find the expression for the horizontal component of the constraint reaction. What is the period of small oscillations?

Let us recall that the distance $\ell_0$ of the centre of mass $G$ of the arc from its centre $O$ is $\ell_0 = R(\sin\alpha)/\alpha$ and that the moment of inertia with respect to

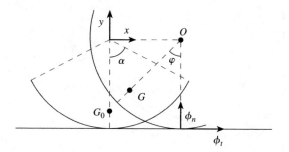

**Fig. 7.4**

the line through $O$ orthogonal to the plane of the motion is $I_O = 2\alpha R^3 \rho$. The moment with respect to the parallel through $G$ is then $I_G = I_O - 2\alpha R\rho \ell_0^2$.

Let the Lagrangian coordinate $\varphi$ be chosen as shown in Fig. 7.4 ($|\varphi| < \alpha$). Then the kinetic energy is

$$T = \frac{1}{2} I_G \dot\varphi^2 + \frac{1}{2} 2\alpha R\rho \dot G^2 = 2\alpha R^2 \rho \dot\varphi^2 (R - \ell_0 \cos\varphi),$$

given that $G = (R\varphi - \ell_0 \sin\varphi, -\ell_0 \cos\varphi)$. The conservation of energy (note that the constraint reaction has zero power) yields

$$R\left(1 - \frac{\sin\alpha}{\alpha} \cos\varphi\right)\dot\varphi^2 - g\frac{\sin\alpha}{\alpha}\cos\varphi = \text{constant}.$$

Differentiate with respect to $t$ and divide by $\dot\varphi$. This yields

$$2R\left(1 - \frac{\sin\alpha}{\alpha}\cos\varphi\right)\ddot\varphi + R\frac{\sin\alpha}{\alpha}\sin\varphi\,\dot\varphi^2 + g\frac{\sin\alpha}{\alpha}\sin\varphi = 0.$$

The component $\phi_t$ of the constraint reaction is given by

$$\phi_t = 2\alpha R\rho \ddot x_G = 2\alpha R^2 \rho \left[\ddot\varphi\left(1 + \frac{\sin\alpha}{\alpha}\cos\varphi\right) + \frac{\sin\alpha}{\alpha}\sin\varphi\,\dot\varphi^2\right].$$

If initially $\varphi(0) = 0$, $\dot\varphi(0) = \omega_0$, what must $\omega_0$ be in order for $|\varphi_{\max}| < \alpha$? Note that the motion is periodic; for small oscillations we have

$$\ddot\varphi + \frac{g}{2R}\frac{(\sin\alpha)/\alpha}{1 - (\sin\alpha)/\alpha}\varphi = 0,$$

from which the period is immediately computed.
What happens when $\alpha \to \pi$? ∎

*Example 7.4*
Consider the following model of the automatic opening of a gate. A homogeneous rod $AB$ of length $l$ and mass $m$ rotates around the point $A$ in a horizontal plane.

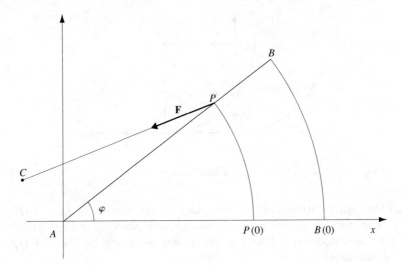

Fig. 7.5

A retractible (or extensible) arm is attached to one of its points, $P$ say. The arm is then $CP$, with $C$ different from $A$, and it is activated in such a way that its length has a constant time derivative. Compute the motion of the rod, the force applied by its extensible arm, and its power.

With reference to Fig. 7.5, suppose as an example that $\varphi(0) = 0$. The initial length of the arm is $\ell_0 = [(R-x_c)^2 + y_c^2]^{1/2}$, with $R = \overline{AP}$, and the motion of the rod can be deduced by writing that $\overline{PC}^2 = (\ell_0 - vt)^2$, where $v = \mathrm{d}\overline{PC}/\mathrm{d}t$, i.e.

$$(R\cos\varphi - x_c)^2 + (R\sin\varphi - y_c)^2 = (\ell_0 - vt)^2,$$

and hence

$$x_c \cos\varphi + y_c \sin\varphi = \frac{1}{2R}[R^2 + x_c^2 + y_c^2 - (\ell_0 - vt)^2]. \tag{7.44}$$

Obviously there is a bound on $t$, because $\ell_0 - vt$ must always be greater than the minimum possible distance between $P$ and $C$, which is $R - (x_c^2 + y_c^2)^{1/2}$.

Equation (7.44) yields $\varphi = \varphi(t)$, e.g. by using the formulae

$$\cos\varphi = \frac{1-\theta^2}{1+\theta^2}, \quad \sin\varphi = \frac{2\theta}{1+\theta^2},$$

with $\theta = \tan(\varphi/2)$.

Supposing that the constraint in $A$ is smooth (what must be changed otherwise?), the force $\mathbf{F}$ applied at $P$ can be obtained from the equation

$$\dot{\mathbf{L}}(A) = (P - A) \times \mathbf{F}, \tag{7.45}$$

by exploiting the fact that $\mathbf{F} = F(C-P)/|C-P|$, with

$$\frac{C-P}{|C-P|} = \left(\frac{x_c - R\cos\varphi}{\ell_0 - vt}, \frac{y_c - R\sin\varphi}{\ell_0 - vt}\right)$$

and knowing that $\dot{L}(A) = \frac{1}{3}m\ell^2\ddot{\varphi}$. Hence one finds

$$F(t) = \frac{1}{3}m\ell^2\ddot{\varphi}\,\frac{\ell_0 - vt}{R(y_c\cos\varphi - x_c\sin\varphi)}$$

(note that the denominator vanishes in correspondence with the extreme values of $\overline{PC}$).

For the power $W = \mathbf{F}(t)(C-P)/|C-P| \cdot \dot{P} = \dot{L}\dot{\varphi}$ we can deduce the expression

$$W = \frac{1}{3}m\ell^2\dot{\varphi}\ddot{\varphi}.$$

Try to obtain the explicit solution in the case that $x_c = 0$, $y_c = \frac{1}{4}R$.

How should one modify the problem if instead of knowing the velocity of $P$ one knows the intensity of the force $F$ (e.g. if it is known that $F$ is constant)? Or the power of $F$ as a function of time? ∎

We conclude with a brief review of the dynamics of systems consisting of more rigid components constrained among them or from the exterior. If the constraints are smooth, the problem of motion (or of the equilibrium) can be solved using the Lagrange equations. However, if one wants to determine the reactions associated with the internal constraints one must write the cardinal equations for each rigid component. A typical example is the case of a hinge between two rigid bodies, when a pair of opposite forces is applied at the hinge.

*Example 7.5*
Consider the system described in Example 6.1 (Fig. 6.3), but now suppose that the rods $AB$, $CD$ have masses equal to $m_1$ and $m_2$, respectively, that the system is in a horizontal plane and that the constraints are frictionless. Study the motion in the absence of active forces, and determine the constraint reaction at the point $D$, for generic initial conditions $\alpha(0) = \alpha_0$, $\dot{\alpha}(0) = \omega_0$.

One immediately obtains the equation of motion requiring that the kinetic energy $T = \frac{1}{6}(m_1\ell_1^2\dot{\alpha}^2 + m_2\ell_2^2\dot{\gamma}^2)$ is conserved, and recalling that $\dot{\gamma} = -(\dot{\alpha}+\dot{\beta}) = -\dot{\alpha}[1 + \xi\cos\alpha(1-\xi^2\sin^2\alpha)^{-1/2}]$, where $\xi = \ell/\ell_2$. We suppose that we are far from the critical configurations described in the original example.

We thus find

$$\frac{1}{6}\left\{m_1\ell_1^2 + m_2\ell_2^2\left[1 + \xi\frac{\cos\alpha}{(1-\xi^2\sin^2\alpha)^{1/2}}\right]^2\right\}\dot{\alpha}^2 = \text{constant},$$

and we obtain $\dot{\alpha}$ as a function of $\alpha$ and, by differentiation, we also find $\ddot{\alpha}$ as a function of $\alpha$. Let $\phi$ be the force applied on the rod $AB$ by the constraint in $D$. The second cardinal equation for the rod $AB$ with respect to the point $A$, i.e.

$$\frac{1}{3} m \ell_1^2 \ddot{\alpha} = \phi \ell_2 (\xi^2 + 1 - 2\xi \cos \gamma(\alpha))^{1/2}$$

yields the determination of $\phi$ as a function of $\alpha$. ∎

## 7.7 The Euler equations for precessions

Consider a material rigid system with a fixed point $O$. If the constraint is frictionless, the equation of motion is

$$\dot{\mathbf{L}}(O) = \mathbf{M}(O), \tag{7.46}$$

and hence is a system of three second-order differential equations for the Euler angles. By integrating equation (7.46) with prescribed initial conditions, we obtain from (7.38) the reaction applied on the constraint (or rather the equivalent resultant of the system of reactions which physically realise the constraint).

If the constraint is not smooth, it presents a *friction torque* $\boldsymbol{\mu}(O)$, which must be expressed in terms of $\boldsymbol{\omega}$. As an example,

$$\boldsymbol{\mu}(O) = -k\boldsymbol{\omega}, \tag{7.47}$$

with $k$ a positive constant. Thus the equation

$$\dot{\mathbf{L}}(O) = \mathbf{M}(O) + \boldsymbol{\mu}(O) \tag{7.48}$$

describes the motion of the system.

We now want to examine the expression for $\dot{\mathbf{L}}$ as a function of $\boldsymbol{\omega}$ and $\dot{\boldsymbol{\omega}}$. One must start from (7.28), stating that $L(O) = \sigma(O)\boldsymbol{\omega}$, but expressing $\sigma(O)$ in a *body frame*, because otherwise $\sigma(O)$ would depend on the Euler angles.

To obtain $\dot{\mathbf{L}}(O)$ recall the relation between the absolute and relative derivative (6.25):

$$\dot{\mathbf{L}}(O) = \sigma(O)\dot{\boldsymbol{\omega}} + \boldsymbol{\omega} \times \mathbf{L}(O). \tag{7.49}$$

It is convenient to choose as the body frame the principal frame of inertia relative to $O$. We thus find the *Euler equations*

$$\begin{aligned} \mathcal{J}_1 \dot{\omega}_1 &= (\mathcal{J}_2 - \mathcal{J}_3)\omega_2\omega_3 + M_1(O) + \mu_1(O), \\ \mathcal{J}_2 \dot{\omega}_2 &= (\mathcal{J}_3 - \mathcal{J}_1)\omega_3\omega_1 + M_2(O) + \mu_2(O), \\ \mathcal{J}_3 \dot{\omega}_3 &= (\mathcal{J}_1 - \mathcal{J}_2)\omega_1\omega_2 + M_3(O) + \mu_3(O). \end{aligned} \tag{7.50}$$

The initial value problem for (7.50) naturally has a unique solution, under the usual regularity assumptions for $\mathbf{M}$ and $\boldsymbol{\mu}$.

*Remark 7.3*
If **M** and **μ** depend only on **ω**, then equations (7.50) yield a *first-order* non-linear system for $\omega_1$, $\omega_2$, $\omega_3$. The phase space for equations (7.50) reduces to the space of coordinates $\omega_1$, $\omega_2$, $\omega_3$. ∎

One such case is the trivial case of *precessions by inertia*, which happens when there is zero torque with respect to the pole of the precession. This case deserves a more detailed study.

## 7.8 Precessions by inertia

Inertia precessions have particularly simple kinematic properties, which are a direct consequence of the first integrals

$$\mathbf{L}(O) = \mathbf{L}_0 \tag{7.51}$$

(vanishing moment of the forces) and

$$T = T_0 \tag{7.52}$$

(vanishing work), where $\mathbf{L}_0$ and $T_0$ are determined by the initial value of **ω**:

$$\boldsymbol{\omega}(0) = \boldsymbol{\omega}_0. \tag{7.53}$$

(Indeed $\mathbf{L}_0 = \sigma(0)\boldsymbol{\omega}_0$, $T_0 = \frac{1}{2} I \omega_0^2$.) Note that equation (7.52) is not independent of equation (7.51). Both follow from the Euler equations, which are now written as

$$\begin{aligned}
\mathcal{J}_1 \dot{\omega}_1 &= (\mathcal{J}_2 - \mathcal{J}_3)\omega_2\omega_3, \\
\mathcal{J}_2 \dot{\omega}_2 &= (\mathcal{J}_3 - \mathcal{J}_1)\omega_3\omega_1, \\
\mathcal{J}_3 \dot{\omega}_3 &= (\mathcal{J}_1 - \mathcal{J}_2)\omega_1\omega_2,
\end{aligned} \tag{7.54}$$

and simply express the vanishing of $\dot{\mathbf{L}}(O)$.[3]

The most interesting result concerning these precessions is the following, which yields a description of the motion alternative to that given by the Poinsot cones.

THEOREM 7.1 (Poinsot) *In the case of an inertia precession, the ellipsoid of inertia relative to the rod rolls without sliding on a fixed plane.*

*Proof*
At each instant, $\sigma(0)\boldsymbol{\omega} = \mathbf{L}_0$. We recall the geometric construction of $\sigma(0)\boldsymbol{\omega}$ (Section 7.3), and we can deduce that the ellipsoid of inertia, at the point where it intersects the axis of instantaneous rotation, is tangent to a plane $\pi$ orthogonal to $\mathbf{L}_0$ (Fig. 7.6).

---

[3] To obtain (7.52) from (7.54) multiply the latter by $\omega_1$, $\omega_2$, $\omega_3$, respectively, and add them term by term. This yields $\dot{T} = 0$.

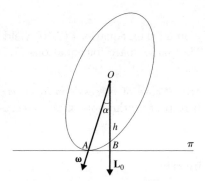

**Fig. 7.6** Poinsot motion.

To complete the proof we only need to show that the plane $\pi$ is fixed. We compute the distance $h$ from $O$:

$$h = |A - O| \cos \alpha = |A - O| \frac{\boldsymbol{\omega} \cdot \mathbf{L}_0}{\omega L_0}.$$

Recalling now (7.9), which gives the construction of the ellipsoid of inertia, we find that $|A - O| = \lambda/\sqrt{I}$, where $I$ is the moment of inertia with respect to the axis of instantaneous rotation. Since $\boldsymbol{\omega} \cdot \mathbf{L}_0 = I\omega^2 = 2T = 2T_0$, it follows that

$$h = \frac{\lambda}{L_0} \frac{\boldsymbol{\omega} \cdot \mathbf{L}_0}{(I\omega^2)^{1/2}} = \frac{\lambda}{L_0} (2T_0)^{1/2}. \tag{7.55}$$

Hence the plane $\pi$ has prescribed orientation and distance from $O$, and is therefore fixed. Since the contact point between the ellipsoid and the plane lies on the axis of instantaneous rotation, the ellipsoid does not slide. ∎

COROLLARY 7.2  *The motion is determined by the rotation of a curve moving with the ellipsoid of inertia* (polhode) *on a fixed plane curve* (herpolhode). ∎

*Remark* 7.4
The polhode is the intersection of the body Poinsot cone with the ellipsoid of inertia and the herpolhode is the intersection of the fixed Poinsot cone with the fixed plane. ∎

The equations of these curves can be obtained by remarking that a polhode is the locus of the points of the ellipsoid of inertia with the property that the plane tangent to the ellipsoid at these points has a fixed distance $h$ from $O$. In the principal frame of inertia the equations of this locus are

$$\sum_{i=1}^{3} \mathcal{J}_i x_i^2 = \lambda^2, \tag{7.56}$$

$$\sum_{i=1}^{3} \mathcal{J}_i (\mathcal{J}_i - \mathcal{J}_0) x_i^2 = 0, \tag{7.57}$$

where we define

$$\mathcal{J}_0 = (\lambda/h)^2. \qquad (7.58)$$

Equation (7.57) is the equation of the body Poinsot cone.

For a generic ellipsoid ($\mathcal{J}_1 < \mathcal{J}_2 < \mathcal{J}_3$) the maximal axis is in the $x_1$ direction, and the minimal axis is in the $x_3$ direction. Let $h_{\max}, h_{\text{med}}, h_{\min}$ be the lengths of the three semi-axes. Then the constant $h$, determined by the initial conditions, varies in the interval $[h_{\min}, h_{\max}]$ and correspondingly $\mathcal{J}_0 \in [\mathcal{J}_1, \mathcal{J}_3]$.

In the extreme cases $h = h_{\min}$ ($\mathcal{J}_0 = \mathcal{J}_3$), $h = h_{\max}$ ($\mathcal{J}_0 = \mathcal{J}_1$), equation (7.57) implies that the polhodes degenerate to the vertices of the minimal or maximal axes, respectively. The other degenerate case is $h = h_{\text{med}}(\mathcal{J}_0 = \mathcal{J}_2)$, because in this case we do not have the $x_2$ term in equation (7.57), which then represents the pair of planes

$$\mathcal{J}_1(\mathcal{J}_1 - \mathcal{J}_2)x_1 \pm \mathcal{J}_3(\mathcal{J}_3 - \mathcal{J}_2)x_3 = 0, \qquad (7.59)$$

symmetrically intersecting on the axis $x_2$. These planes produce four arcs of an ellipse on the ellipsoid of inertia, called *limiting polhodes*. These arcs are separated by the vertices on the intermediate axis, which are degenerate polhodes (Fig. 7.7).

In the generic case, the polhodes can be classified into two classes:

(a) $h_{\min} < h < h_{\text{med}}(\mathcal{J}_2 < \mathcal{J}_0 < \mathcal{J}_3)$, the body Poinsot cone has as axis the minimal axis of the ellipsoid of inertia;

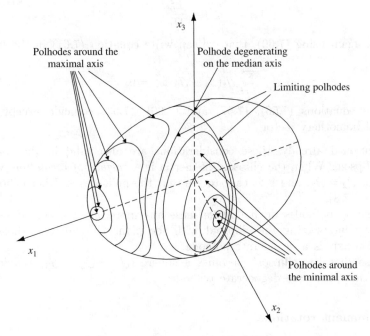

**Fig. 7.7** Classification of polhodes.

(b) $h_{med} < h < h_{max} (\mathcal{J}_1 < \mathcal{J}_0 < \mathcal{J}_2)$, the body Poinsot cone has as axis the maximal axis of the ellipsoid of inertia.

The limiting polhodes are the separatrices of the two families.

As we have already noted in Section 7.7, the phase space for Euler equations can be identified in the case of precessions by inertia with the space $(\omega_1, \omega_2, \omega_3)$. From this point of view, the study of polhodes is particularly interesting in view of the following fact.

PROPOSITION 7.10 *The polhodes represent the trajectories in the phase space of equations (7.54).*

*Proof*
We verify that (7.56), (7.57) are still valid after the substitution $\mathbf{x} \to \boldsymbol{\omega}$.

Let us write the equations requiring that $T$ and the absolute value of $\mathbf{L}(O)$ are constant:

$$\sum_{i=1}^{3} \mathcal{J}_i \omega_i^2 = 2T_0, \qquad (7.60)$$

$$\sum_{i=1}^{3} \mathcal{J}_i^2 \omega_i^2 = L_0^2. \qquad (7.61)$$

By eliminating $h$ between (7.55) and (7.58) we find

$$\mathcal{J}_0 = \frac{L_0^2}{2T_0}. \qquad (7.62)$$

It follows, again using (7.60), that we can write equation (7.61) in the form

$$\sum_{i=1}^{3} \mathcal{J}_i (\mathcal{J}_i - \mathcal{J}_0) \omega_i^2 = 0, \qquad (7.63)$$

and hence equations (7.56), (7.57) and (7.60), (7.63) coincide except for an inessential homothety factor. ∎

As mentioned already, these considerations are also valid in the case of a generic ellipsoid. When the ellipsoid of inertia is a *surface of revolution*, we have $\mathcal{J}_1 = \mathcal{J}_2$ or $\mathcal{J}_2 = \mathcal{J}_3$, and it is easy to see that the polhodes and herpolhodes are *circles* (Fig. 7.8).

The limiting polhodes do not make sense any more: every point of the circle obtained by intersecting the ellipsoid with the plane through $O$ orthogonal to the rotation axis is a degenerate polhode.

The case that the ellipsoid becomes a sphere ($\mathcal{J}_1 = \mathcal{J}_2 = \mathcal{J}_3$) is trivial: all points of the sphere are degenerate polhodes.

## 7.9 Permanent rotations

THEOREM 7.2 *If $\boldsymbol{\omega}(0) = \boldsymbol{\omega}_0$ has the direction of a principal axis of inertia, the corresponding precession by inertia is reduced to the uniform rotation $\boldsymbol{\omega} = \boldsymbol{\omega}_0$.*

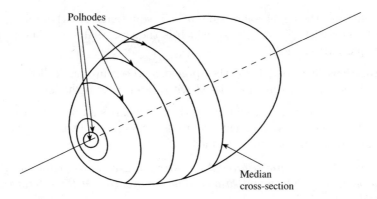

**Fig. 7.8** Polhodes for the ellipsoid of revolution.

*Conversely, if a precession by inertia is a rotation, the latter must be uniform and must be around a principal axis of inertia.*

*Proof*
The first claim is trivial, as it is immediately verified that $\boldsymbol{\omega} = \boldsymbol{\omega}_0$ is the only solution of equations (7.54).

Suppose now that the motion is a rotation and let us examine first the case when $\boldsymbol{\omega}$, which has constant direction by our hypotheses, has at least one zero component. Suppose, e.g. that $\omega_1 = 0$. Then equations (7.54) imply that $\omega_2$ and $\omega_3$ are constant, and the rotation is uniform. In addition we find that $(\mathcal{J}_2 - \mathcal{J}_3)\omega_2\omega_3 = 0$, and hence either $\mathcal{J}_2 = \mathcal{J}_3$, which implies that every diametral axis, including the axis of rotation, is a principal axis of inertia, or else one of the components $\omega_2, \omega_3$ is also zero. This implies that $\boldsymbol{\omega}$ has the direction of a principal axis of inertia.

Consider finally the case that $\omega_1\omega_2\omega_3 \neq 0$. Since our hypotheses imply that $\dot{\boldsymbol{\omega}} = 0$ or else that $\dot{\boldsymbol{\omega}} \| \boldsymbol{\omega}$, we can always write

$$\frac{\dot{\omega}_i}{\omega_i} = f(t), \quad i = 1, 2, 3, \tag{7.64}$$

and we make $f(t)$ appear on the left-hand side of equations (7.54) by rewriting them in the form

$$\mathcal{J}_i \omega_i^2 f(t) = (\mathcal{J}_j - \mathcal{J}_k) \omega_1 \omega_2 \omega_3,$$

where $\{i, j, k\}$ are the three even-order permutations. Summing term by term, we obtain

$$2T f(t) = 0,$$

which yields that $f(t) = 0$, and hence $\dot{\boldsymbol{\omega}} = 0$, and the rotation is uniform.

In conclusion, the right-hand side of all equations (7.54) vanish. This is compatible only with $\mathcal{J}_1 = \mathcal{J}_2 = \mathcal{J}_3$, which is the case that every line through $O$ is a principal axis of inertia. ∎

The rotations considered in the previous theorem are called *permanent rotations*. Such rotations are associated with degenerate polhodes.

In the case of a generic ellipsoid of inertia, there exists an important qualitative difference between the permanent rotations around the extreme axes of the ellipsoid and those around the intermediate axis.

THEOREM 7.3 *The rotations around the extreme axes of the ellipsoid of inertia are stable with respect to perturbations of $\boldsymbol{\omega}_0$; those around the intermediate axis are unstable.*

*Proof*
We use the geometrical analysis of the polhodes of the previous section, and the fact that these are the trajectories, in phase space, of equations (7.54) (Proposition 7.10). This shows that, for a fixed neighbourhood of the degenerate polhodes lying on the extreme axes, we can consider perturbations of $\boldsymbol{\omega}_0$ of such small amplitude that the corresponding polhodes remain inside the chosen neighbourhood. This is not possible for the degenerate polhodes lying on the intermediate axis, as every neighbourhood of such curves is crossed by polhodes which rotate around the maximal axis, as well as by polhodes which rotate around the minimal axis. ∎

In the case of *ellipsoids of revolution* we can easily prove that any rotation around the rotation axis of the ellipsoid is stable, while any rotation around the diametral axis is unstable.

*Remark 7.5*
Stable phenomena are not the only observable phenomena. Try to make a cylinder, of height much larger than the radius, rotate around a diametral axis; in spite of any error in initial conditions, the rotation will appear stable. This is not in contradiction with what has just been proved: if the radius of the polhode is much larger than the radius of the herpolhode the contact point must rotate many times along the latter to make any tangible progress along the polhode, in agreement with the instability of this phenomenon, although such instability can only be observed over long time intervals. ∎

## 7.10 Integration of Euler equations

We consider again the first integrals (7.60) and (7.63). Eliminating once $\omega_3$ and once $\omega_1$, we find the following equations:

$$\mathcal{J}_1(\mathcal{J}_1 - \mathcal{J}_3)\omega_1^2 = 2T_0(\mathcal{J}_0 - \mathcal{J}_3) - \mathcal{J}_2(\mathcal{J}_2 - \mathcal{J}_3)\omega_2^2, \tag{7.65}$$

$$\mathcal{J}_3(\mathcal{J}_3 - \mathcal{J}_1)\omega_3^2 = 2T_0(\mathcal{J}_0 - \mathcal{J}_1) - \mathcal{J}_2(\mathcal{J}_2 - \mathcal{J}_1)\omega_2^2. \tag{7.66}$$

In the generic case that $\mathcal{J}_1 < \mathcal{J}_2 < \mathcal{J}_3$ we deduce that
$$\omega_1^2 = A_1^2(\nu_1^2 - \omega_2^2), \quad \omega_3^2 = A_3^2(\nu_3^2 - \omega_2^2), \tag{7.67}$$
with
$$A_1^2 = \frac{\mathcal{J}_2}{\mathcal{J}_1}\frac{\mathcal{J}_3 - \mathcal{J}_2}{\mathcal{J}_3 - \mathcal{J}_1}, \quad A_3^2 = \frac{\mathcal{J}_2}{\mathcal{J}_3}\frac{\mathcal{J}_2 - \mathcal{J}_1}{\mathcal{J}_3 - \mathcal{J}_1}, \tag{7.68}$$
$$\nu_1 = \left[\frac{2T_0}{\mathcal{J}_2}\frac{\mathcal{J}_3 - \mathcal{J}_0}{\mathcal{J}_3 - \mathcal{J}_2}\right]^{1/2}, \quad \nu_3 = \left[\frac{2T_0}{\mathcal{J}_2}\frac{\mathcal{J}_0 - \mathcal{J}_1}{\mathcal{J}_2 - \mathcal{J}_1}\right]^{1/2}. \tag{7.69}$$

The dimensionless coefficients $A_1$, $A_3$ contain information only on the geometric structure of the system, while the frequencies $\nu_1$, $\nu_3$ are also determined by the initial conditions. We assume that $\mathcal{J}_0 \neq \mathcal{J}_1, \mathcal{J}_3$ (we have already considered the case that these quantities are equal).

Note also that
$$k = \frac{\nu_3^2}{\nu_1^2} = \frac{\mathcal{J}_0 - \mathcal{J}_1}{\mathcal{J}_2 - \mathcal{J}_1} \cdot \frac{\mathcal{J}_3 - \mathcal{J}_2}{\mathcal{J}_3 - \mathcal{J}_0} \tag{7.70}$$

is greater than one if $\mathcal{J}_0 > \mathcal{J}_2$ and less than one if $\mathcal{J}_0 < \mathcal{J}_2$. We exclude temporarily the case that $\mathcal{J}_0 = \mathcal{J}_2$.

Using equations (7.67) we obtain from the second of equations (7.54) a differential equation for $\omega_2$:
$$\dot{\omega}_2 = \pm A(\nu_1^2 - \omega_2^2)^{1/2}(\nu_3^2 - \omega_2^2)^{1/2}, \tag{7.71}$$
with
$$A = \left[\frac{(\mathcal{J}_2 - \mathcal{J}_1)(\mathcal{J}_3 - \mathcal{J}_2)}{\mathcal{J}_1 \mathcal{J}_3}\right]^{1/2}. \tag{7.72}$$

Note that the initial condition for $\omega_2$ must be such that $|\omega_2(0)| \leq \min(\nu_1, \nu_3)$ and the same inequality is satisfied for $|\omega_2(t)|$. In addition, the constant solution $|\omega_2| = \min(\nu_1, \nu_3)$ must be discarded; indeed if, for example, $\nu_1 < \nu_3$, it follows that $\omega_1 = 0$ (cf. (7.67)) and $\omega_2 \omega_3 \neq 0$, contradicting the first equation of (7.54).[4]

We can now compute the integral of equation (7.71) corresponding to the initial data $\omega_2(0) = 0$, and distinguish between two cases:

(a) $\mathcal{J}_0 \in (\mathcal{J}_1, \mathcal{J}_2)$, that is $k < 1$:
$$t = \pm \tau_1 F\left(\frac{\omega_2}{\nu_3}, k\right); \tag{7.73}$$

---

[4] Besides the constant solutions $\omega_2 = \pm \min(\nu_1, \nu_3)$, equation (7.71) also admits non-trivial solutions which periodically take these values.

(b) $\mathcal{J}_0 \in (\mathcal{J}_2, \mathcal{J}_3)$, that is $k > 1$:

$$t = \pm \tau_3 F\left(\frac{\omega_2}{\nu_1}, k^{-1}\right). \tag{7.74}$$

Here $F$ is the elliptic integral of the first kind (cf. Appendix 2):

$$F(z, k) = \int_0^z [(1 - \eta^2)(1 - k\eta^2)]^{-1/2} \, d\eta, \quad |z| \leq 1 \tag{7.75}$$

and

$$\tau_1 = \frac{1}{\nu_1 A}, \quad \tau_3 = \frac{1}{\nu_3 A}. \tag{7.76}$$

The sign in (7.73), (7.74) must be chosen according to the initial conditions, and must be inverted every time that $\omega_2$ reaches the extreme values (respectively, $\pm \nu_3$ and $\pm \nu_1$).

The *solution is periodic* of period $4\tau_1 \mathbf{K}(k)$ (see Appendix 2) along the polhodes in the family described by $\mathcal{J}_0 \in (\mathcal{J}_1, \mathcal{J}_2)$, and $4\tau_3 \mathbf{K}(k^{-1})$ along those in the family $\mathcal{J}_0 \in (\mathcal{J}_2, \mathcal{J}_3)$.

Finally, we examine the case that $\mathcal{J}_0 = \mathcal{J}_2$ (*motion along the limiting polhodes*). In this case the frequencies $\nu_1$ and $\nu_3$ coincide ($k = 1$):

$$\nu_1 = \nu_3 \equiv \nu = (2T_0/\mathcal{J}_2)^{1/2}. \tag{7.77}$$

Since $\lim_{k \to 1^-} \mathbf{K}(k) = \infty$ we expect that the motion is no longer periodic. Equation (7.71) can be simplified to

$$\dot{\omega}_2 = \pm A(\nu^2 - \omega_2^2), \tag{7.78}$$

where $A$ is still given by (7.72). We choose the initial data $\omega_2(0) \in (-\nu, \nu)$, since we are not interested in the extreme values, which correspond to permanent rotations. By separating variables we easily find

$$\frac{\nu + \omega_2}{\nu - \omega_2} = \frac{\nu + \omega_2(0)}{\nu - \omega_2(0)} e^{\pm 2t/\tau}, \tag{7.79}$$

with

$$\tau = \frac{1}{\nu A} = \left\{\frac{2T_0}{\mathcal{J}_1 \mathcal{J}_2 \mathcal{J}_3}(\mathcal{J}_3 - \mathcal{J}_2)(\mathcal{J}_2 - \mathcal{J}_1)\right\}^{1/2}. \tag{7.80}$$

It follows that $\omega_2(t)$ tends monotonically to $\pm \nu$, depending on the sign of $\dot{\omega}_2$, which is determined by the initial conditions and by the second of equations (7.54).

## 7.11 Gyroscopic precessions

In the previous section we integrated equations (7.54) in the generic case that $\mathcal{J}_1 < \mathcal{J}_2 < \mathcal{J}_3$. We now consider the gyroscopic precessions around a point of the gyroscopic axis. Suppose that $\mathcal{J}_1 = \mathcal{J}_2 = \mathcal{J}$ and consider initially the simple case of precessions by inertia. Setting

$$\eta = \frac{\mathcal{J}_3}{\mathcal{J}} - 1, \tag{7.81}$$

and excluding the trivial case $\eta = 0$, equations (7.54) become

$$\begin{aligned}\dot{\omega}_1 &= -\eta\omega_2\omega_3, \\ \dot{\omega}_2 &= \eta\omega_3\omega_1, \\ \dot{\omega}_3 &= 0,\end{aligned} \tag{7.82}$$

and hence the *gyroscopic component* of the rotational velocity, $\omega_3$, is constant, and the system (7.82) is *linear*; $\omega_1$ and $\omega_2$ oscillate harmonically with frequency

$$\nu = |\eta\omega_3^{(0)}|/2\pi \tag{7.83}$$

(we refer to the generic case that $\boldsymbol{\omega}(0)$ does not have the direction of the gyroscopic axis).

We also find the first integral

$$\omega_1^2 + \omega_2^2 = [\omega_1(0)]^2 + [\omega_2(0)]^2. \tag{7.84}$$

It follows that the trajectory in the phase plane $(\omega_1, \omega_2)$ is a circle centred at the origin. The vector $\boldsymbol{\omega}_e$ with components $(\omega_1, \omega_2, 0)$ is called the *equatorial component* of $\boldsymbol{\omega}$; it rotates uniformly around the gyroscopic axis with frequency $\nu$.

It is interesting to study the perturbations introduced by the presence of a moment normal to the gyroscopic axis. To illustrate the main qualitative properties of the motion, we consider a simple example for which the equations of motion are easily integrable.

Consider the case of a *moving torque*, which we suppose has the same direction as $x_2$:

$$\mathbf{M}(O) = (0, M, 0). \tag{7.85}$$

Again, $\omega_3$ is constant and we must integrate the system

$$\dot{\omega}_1 = -\eta\omega_3^{(0)}\omega_2, \quad \omega_1(0) = \omega_1^{(0)}, \tag{7.86}$$

$$\dot{\omega}_2 = \eta\omega_3^{(0)}\omega_1 + M/\mathcal{J}, \quad \omega_2(0) = \omega_2^{(0)}, \tag{7.87}$$

which is equivalent to

$$\ddot{\omega}_1 + (\eta\omega_3^{(0)})^2\omega_1 = -\eta\omega_3^{(0)} M/\mathcal{J}, \tag{7.88}$$

$$\ddot{\omega}_2 + (\eta\omega_3^{(0)})^2\omega_2 = 0, \tag{7.89}$$

with the additional conditions $\dot{\omega}_1(0) = -\eta \omega_3^{(0)} \omega_2^{(0)}$, $\dot{\omega}_2(0) = \eta \omega_3^{(0)} \omega_1^{(0)} + M/\mathcal{J}$. Instead of (7.84) we find the first integral

$$(\omega_1 - \bar{\omega}_1)^2 + \omega_2^2 = \text{constant}, \tag{7.90}$$

with

$$\bar{\omega}_1 = -M/[(\mathcal{J}_3 - \mathcal{J})\omega_3^{(0)}], \tag{7.91}$$

and we can immediately deduce the integrals for (7.88), (7.89):

$$\omega_1 - \bar{\omega}_1 = C \cos(2\pi\nu t + \alpha), \tag{7.92}$$
$$\omega_2 = C \sin(2\pi\nu t + \alpha), \tag{7.93}$$

where

$$C = \{[\omega_1^{(0)} - \bar{\omega}_1]^2 + [\omega_2^{(0)}]^2\}^{1/2}, \tag{7.94}$$

$$\tan \alpha = \frac{\omega_2^{(0)}}{\omega_1^{(0)} - \bar{\omega}_1}. \tag{7.95}$$

We summarise the properties of the perturbed motion.

(a) Amplitude and phase perturbations are measured by $\bar{\omega}_1$. If initially

$$|\omega_e|/|\omega_3^{(0)}| \ll 1 \tag{7.96}$$

and if

$$|\bar{\omega}_1|/|\omega_3^{(0)}| \ll 1 \tag{7.97}$$

then, because of (7.90), equation (7.96) is satisfied at every time. If $\mathbf{e}_3$ is the unit vector of the gyroscopic axis, its variation is described by

$$\frac{d\mathbf{e}_3}{dt} = \boldsymbol{\omega} \times \mathbf{e}_3 = \boldsymbol{\omega}_e \times \mathbf{e}_3,$$

i.e. $|d\mathbf{e}_3/dt| = |\omega_e|$, which implies that the motion of the gyroscopic axis is much slower than that around the same axis, and the effect of the torque $M$ is smaller for larger $\omega_3^{(0)}$.

(b) We note also that the vector $\boldsymbol{\omega}_e$ varies, with respect to the moving reference frame, with frequency $\nu$, proportional to $\omega_3^{(0)}$.

(c) Over a period of $\boldsymbol{\omega}_e$ the average of $\omega_2$ is zero, while the average of $\omega_1$ is $\bar{\omega}_1$. It follows that by taking the average over a period of $\boldsymbol{\omega}_e$

$$\overline{\frac{d\mathbf{e}_3}{dt}} = \boldsymbol{\omega}_e \times \mathbf{e}_3 = \frac{M}{(\mathcal{J}_3 - \mathcal{J})\omega_3^{(0)}}, \tag{7.98}$$

and hence the mean displacement of the gyroscopic axis is in the direction of the torque (*tendency to parallelism*).

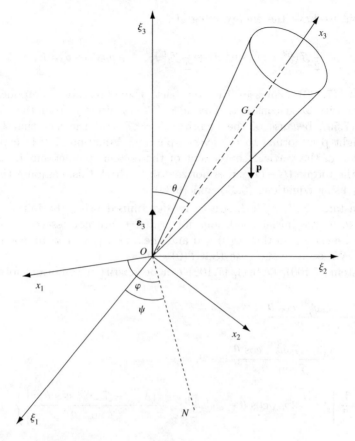

Fig. 7.9

## 7.12 Precessions of a heavy gyroscope (spinning top)

The system under consideration is depicted schematically in Fig. 7.9; the constraint at $O$ is assumed to be smooth.

Setting $\overline{OG} = \rho_0$, the gravitational potential is $U(\theta) = -p\rho_0 \cos\theta$. Recalling equation (7.33) and setting $\mathcal{J}_1 = \mathcal{J}_2 = \mathcal{J}$, we arrive at the following expression for the Lagrangian:

$$L = \frac{1}{2}\mathcal{J}(\dot{\theta}^2 + \dot{\psi}^2 \sin^2\theta) + \frac{1}{2}\mathcal{J}_3(\dot{\varphi} + \dot{\psi} \cos\theta)^2 - p\rho_0 \cos\theta. \tag{7.99}$$

This expression does not contain explicitly the two variables $\varphi$ and $\psi$. From it we can deduce the two first integrals $\partial L/\partial \dot{\varphi} = \text{constant}$, $\partial L/\partial \dot{\psi} = \text{constant}$, i.e.

$$\dot{\varphi} + \dot{\psi} \cos\theta = \omega_3^{(0)}, \tag{7.100}$$

$$\mathcal{J}\dot{\psi} \sin^2\theta + \mathcal{J}_3 \omega_3^{(0)} \cos\theta = \Lambda_3^{(0)}. \tag{7.101}$$

In addition, we have the energy integral

$$\frac{1}{2}\mathcal{J}(\dot{\theta}^2 + \dot{\psi}^2 \sin^2\theta) + \frac{1}{2}\mathcal{J}_3[\omega_3^{(0)}]^2 + p\rho_0 \cos\theta = E. \tag{7.102}$$

Equation (7.100) expresses the fact that the gyroscopic component of the angular velocity $\boldsymbol{\omega}$ is constant. This also follows directly from the third Euler equation (7.50), because of the identity $\mathcal{J}_1 = \mathcal{J}_2$ and the fact that the torque of the weight $\mathbf{p}$ is normal to the gyroscopic axis. Equation (7.101) expresses the conservation of the vertical component of the angular momentum $\mathbf{L}$, due to the fact that the torque $(G-O) \times \mathbf{p}$ is horizontal; to check this, compute the product $\sigma(0)\boldsymbol{\omega} \cdot \boldsymbol{\varepsilon}_3$ using equations (6.35) and (6.38).

The constants $\omega_3^{(0)}$, $\Lambda_3^{(0)}$, $E$ are to be determined using the initial conditions for $\dot{\varphi}$, $\dot{\psi}$, $\dot{\theta}$, $\theta$ (the initial conditions for $\varphi$ and $\psi$ are not essential, as the axes $\xi_1$, $\xi_2$ can be chosen so that $\psi(0) = 0$ and the axes $x_1$, $x_2$ can be chosen so that $\varphi(0) = 0$). We exclude the case that $\theta(0) = 0$.

The system (7.100), (7.101), (7.102) can be rewritten in normal form

$$\dot{\psi} = \frac{\Lambda_3^{(0)} - \mathcal{J}_3 \omega_3^{(0)} \cos\theta}{\mathcal{J} \sin^2\theta}, \tag{7.103}$$

$$\dot{\varphi} = \omega_3^{(0)} - \frac{\Lambda_3^{(0)} - \mathcal{J}_3 \omega_3^{(0)} \cos\theta}{\mathcal{J} \sin^2\theta} \cos\theta, \tag{7.104}$$

$$\dot{\theta} = \pm \left\{ \frac{1}{\mathcal{J}} \left[ 2E - 2p\rho_0 \cos\theta - \mathcal{J}_3[\omega_3^{(0)}]^2 - \frac{(\Lambda_3^{(0)} - \mathcal{J}_3 \omega_3^{(0)} \cos\theta)^2}{\mathcal{J} \sin^2\theta} \right] \right\}^{1/2}. \tag{7.105}$$

We analyse equation (7.105) under the assumption (consistent with $\theta(0) \neq 0$)

$$\mathcal{J}_3 \omega_3^{(0)} \neq \Lambda_3^{(0)}, \tag{7.106}$$

which implies that in equation (7.105) the expression in parentheses is always positive for the $\theta$ variable in an interval $(\theta', \theta'')$, with $0 < \theta' < \theta'' < \pi$.

If we exclude the trivial solutions $\theta = \theta'$, $\theta = \theta''$, the function $\theta$ oscillates in the interval $(\theta', \theta'')$. These oscillations can be determined by integrating (7.105), with sign inversion at the endpoints.

It is interesting to note that in equation (7.103) it may happen that $\dot{\psi}$ vanishes for some value of $\theta$. In this case the precession stops momentarily, and its direction may be inverted if $\dot{\psi}$ changes sign in a certain interval. For example, if we consider $\theta(0) = \theta_0 \in (0, \pi/2)$ and $\dot{\psi}(0) = 0$, we have $\Lambda_3^{(0)} = \mathcal{J}_3 \omega_3^{(0)} \cos\theta_0$, and equation (7.103) becomes

$$\dot{\psi} = \frac{\mathcal{J}_3 \omega_3^{(0)}}{\mathcal{J} \sin^2\theta} (\cos\theta_0 - \cos\theta). \tag{7.107}$$

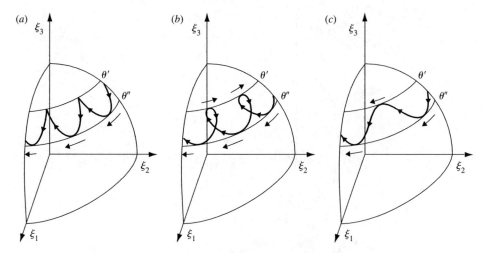

**Fig. 7.10** Motion of the trace of the gyroscopic axis on a sphere: (*a*) momentary stop, (*b*) inversion of the precession, (*c*) precession without stops.

If in addition we choose $\dot\theta(0) = 0$, equation (7.105) can be written

$$\dot\theta = \pm \left\{ \frac{2p\rho_0}{\mathcal{J}}(\cos\theta_0 - \cos\theta) - \left(\frac{\mathcal{J}_3\omega_3^{(0)}}{\mathcal{J}\sin\theta}\right)^2 (\cos\theta_0 - \cos\theta)^2 \right\}^{1/2}, \qquad (7.108)$$

which shows that necessarily $\cos\theta \leq \cos\theta_0$, or $\theta \geq \theta_0$, and hence $\theta' = \theta_0$. To fix ideas, choose

$$\mathcal{J}_3\omega_3^{(0)} > (2p\rho_0\mathcal{J}/\cos\theta_0)^{1/2}. \qquad (7.109)$$

We then have $\theta' < \theta'' < \pi/2$. It follows that this case is characterised by a momentary halt of the precession at instants when $\theta$ reaches its minimum (corresponding to an inversion of the nutation). On the other hand, when $\theta$ takes the value $\theta''$ (again the nutation is inverted), $\dot\psi$ takes its maximum value (verify that $\dot\psi$ is an increasing function of $\theta$), see Fig. 7.10a.

It is possible to choose initial conditions for which $\dot\psi$ changes sign, e.g. $0 < \Lambda_3^{(0)} < \mathcal{J}_3\omega_3^{(0)} \cos\theta'$, see Fig. 7.10b, or such that the precession is never inverted, e.g. $\Lambda_3 > \mathcal{J}_3\omega_3^{(0)} > 0$, see Fig. 7.10c.

## 7.13 Rotations

We conclude our discussion of the dynamics of rigid bodies by considering briefly the case of a rigid system with a fixed axis. The physical realisation of such a constraint can be obtained by a *spherical hinge* (triple constraint at the point $O$, see Fig. 7.11) and with a *collar* (double constraint at the point $A$).

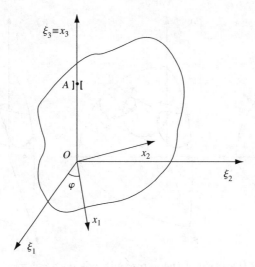

Fig. 7.11

It is convenient to fix the reference frame $(O, x_1, x_2, x_3)$ and the body frame $(O, \xi_1, \xi_2, \xi_3)$ with the axes $x_3$ and $\xi_3$ coincident with the rotation axis. We take as the Lagrangian coordinate the angle $\varphi$ between the axes $x_1$, $\xi_1$, measured counterclockwise. Hence we write

$$\boldsymbol{\omega} = \dot{\varphi}\mathbf{e}_3. \tag{7.110}$$

If the constraints are smooth, their action is expressed through a force $\boldsymbol{\phi}_O$ applied at $O$ and a force $\boldsymbol{\phi}_A$ applied at $A$ and orthogonal to the axis of rotation. Otherwise, we must additionally consider a friction couple $\boldsymbol{\mu}$, directed as the axis of rotation, to be specified (as an example, $\boldsymbol{\mu} = -k\boldsymbol{\omega}$).

The expression for the angular momentum in a rotation is given by

$$\mathbf{L}(O) = \sigma(O) \begin{pmatrix} 0 \\ 0 \\ \dot{\varphi} \end{pmatrix} = \begin{pmatrix} I_{13} \\ I_{23} \\ I_{33} \end{pmatrix} \dot{\varphi}, \tag{7.111}$$

which reduces to

$$\mathbf{L}(O) = \mathcal{J}_3 \dot{\varphi} \mathbf{e}_3 \tag{7.112}$$

if the axis of rotation is also a principal axis of inertia (Proposition 7.2).

The projection onto the axis of rotation of the second cardinal equation

$$\sigma(0)\dot{\boldsymbol{\omega}} + \boldsymbol{\omega} \times \sigma(0)\boldsymbol{\omega} = \mathbf{M}(O) + (A - O) \times \boldsymbol{\phi}_A + \boldsymbol{\mu}, \tag{7.113}$$

namely

$$I_{33}\ddot{\varphi} = M_3 + \mu, \tag{7.114}$$

can normally be integrated independently of the others. Starting from initial values $\varphi$, $\dot\varphi$, it yields the motion of the system. The other two scalar components of equation (7.113) can be written as

$$I_{13}\ddot\varphi - I_{23}\dot\varphi^2 = M_1(0) - |A-O|\phi_{A2}, \qquad (7.115)$$

$$I_{23}\ddot\varphi + I_{13}\dot\varphi^2 = M_2(0) + |A-O|\phi_{A1}, \qquad (7.116)$$

and yield the components of $\phi_A$. The reaction $\phi_O$ can be determined using the first cardinal equation.

In the simple case of uniform rotations ($\ddot\varphi = 0$), equations (7.115) and (7.116) illustrate the dynamical effect of the products of inertia, which produce an additional stress to the constraint at $A$. In this case, the constraint must balance not just the component of $\mathbf{M}$ normal to the axis of rotation, as happens when $I_{13} = I_{23} = 0$, when the axis of rotation is the principal axis of inertia. We note also that the latter case is characteristic of principal axes of inertia. Indeed, requiring that the left-hand sides of (7.115) and (7.116) vanish, one easily obtains

$$(I_{13}^2 + I_{23}^2)\dot\varphi^2 = 0, \qquad (7.117)$$

and hence $I_{13} = I_{23} = 0$.

## 7.14 Problems

**1.** Let $R$ be a rigid body with an axis $\gamma$ with the following property: $R$ takes the same geometric configuration after any rotation of $2\pi/n$ around $\gamma$, with $n > 2$ integer. Prove that:

(a) $\gamma$ contains the centre of mass;
(b) $\gamma$ is an axis of rotation of the ellipsoid of inertia with respect to any of its points.

**2.** Find the centre of mass of the following homogeneous systems: circular sector, circular segment, spherical cap, pyramid, cone and truncated cone, arc of ellipse, semi-ellipsoid.

**3.** Given a fixed reference frame, compute the tensor of inertia of the following homogeneous systems for a generic configuration: segment, general triangle, rectangle, circle, disc, regular polygon, sphere, cube.

**4.** Solve the problems in Examples 4.2, 4.3 of Chapter 4, replacing the two point particles $P_1$, $P_2$ with a rigid homogeneous rod of mass $m$.

**5.** In Example 7.2 suppose that the point $A$ is subject to a friction force given by $-\lambda\dot x$ ($\lambda > 0$). Prove that the system tends asymptotically to a configuration in which the point $A$ again takes its initial position, independently of the initial value of $\dot\varphi$.
Hint: take the projection along $x$ of the first cardinal equation, and let $x_0$ be the $x$-coordinate of $P_0$. Then $m\ddot x_0 = -\lambda\dot x$, and by integration it follows that $x$

must tend to zero (start by proving that $\dot{x}$ must tend to zero and deduce the asymptotic relation between $x_0$ and $\varphi$).
Write down the complete system of equations of motion.

**6.** Two homogeneous equal rods $AB, BC$, of length $l$ and mass $m$, are hinged at the common endpoint $B$. The system lies in a vertical plane with the point $A$ fixed and the point $C$ moving along a fixed circle passing through $A$, of radius $l$ and with centre $O$ lying on the horizontal line through $A$. All constraints are smooth. Find the configurations of stable equilibrium and the normal modes of the system.

**7.** A heavy homogeneous circle of mass $M$ and radius $R$ rotates without friction around its centre $O$. A point particle $P$ of mass $m$ is constrained to slide without friction on the circle, and it is attracted with an elastic force by a point $A$ fixed on the circle. Write down the first integrals of the cardinal equations.

**8.** Study the motion of precession of a gyroscope around its centre of mass $O$, assuming that the only torque with respect to $O$ is due to the constraint friction and that it is proportional to the angular velocity.

**9.** In a vertical plane, a homogeneous equilateral triangle $ABC$ with weight $p$ and side $l$ has the vertices $A$ and $B$ sliding without friction along a circular guide of radius $l$ (the point $C$ is located at the centre of the guide). A horizontal force of intensity $p/\sqrt{3}$ is applied at $C$. Find the equilibrium configurations and the corresponding constraint reactions. Study the stability of the (two) configurations and the small oscillations around the stable one. (Remark: the first part of the problem can be solved graphically.)

**10.** In a vertical plane, a homogeneous rod of length $l$ and mass $m$ is constrained without friction so that its endpoints lie on the parabola with equation $2ay = x^2$ ($y$ vertical, directed upwards). Study the equilibrium of the system and describe the small oscillations around the configurations of stable equilibrium.

## 7.15 Additional solved problems

*Problem 1*
In a vertical plane, two rods $AB, BC$ are hinged at the common endpoint $B$ and the point $A$ is constrained to be fixed. The respective lengths and masses are $\ell_1, \ell_2$ and $m_1, m_2$. The plane on which the rods are constrained rotates with constant angular velocity $\omega$ around the vertical line through $A$ (Fig. 7.12). Determine the configurations of relative equilibrium in the plane of the system, as well as the constraint reactions.

*Solution*
In the rotating plane, a field of centrifugal forces normal to the axis of rotation is established, whose intensity per unit mass is $\omega^2 r$, with $r$ being the distance

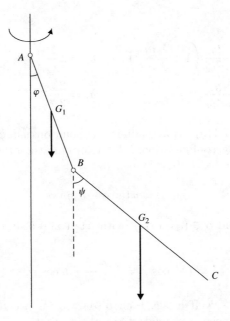

**Fig. 7.12**

from the axis of rotation. We can associate with this field the potential $\frac{1}{2}\omega^2 r^2$. It follows that the total centrifugal potential of the system is

$$\frac{1}{2}\omega^2 \frac{m_1}{\ell_1} \int_0^{\ell_1} (\xi \sin \varphi)^2 \, d\xi + \frac{1}{2}\omega^2 \frac{m_2}{\ell_2} \int_0^{\ell_2} (\ell_1 \sin \varphi + \xi \sin \psi)^2 \, d\xi.$$

In addition to this we have the gravitational potential, yielding for the total potential

$$U(\varphi, \psi) = \frac{1}{2}\ell_1 m_1 g \cos \varphi + m_2 g \ell_1 \left( \cos \varphi + \frac{1}{2}\frac{\ell_2}{\ell_1} \cos \psi \right)$$

$$+ \frac{1}{2}\omega^2 \ell_1^2 \left[ \left( \frac{1}{3} m_1 + m_2 \right) \sin^2 \varphi + \frac{\ell_2}{\ell_1} m_2 \sin \varphi \sin \psi \right.$$

$$\left. + \frac{1}{3}\left(\frac{\ell_2}{\ell_1}\right)^2 m_2 \sin^2 \psi \right].$$

Requiring to vanish the first derivatives of $U$, we find the equilibrium equations, which can be written in the form

$$\sin \psi = a_1 \tan \varphi - b_1 \sin \varphi, \qquad (7.118)$$

$$\sin \varphi = a_2 \tan \psi - b_2 \sin \psi, \qquad (7.119)$$

with

$$a_1 = \frac{2g}{\omega^2 \ell_2}\left(1 + \frac{1}{2}\frac{m_1}{m_2}\right), \quad b_1 = 2\frac{\ell_1}{\ell_2}\left(1 + \frac{1}{3}\frac{m_1}{m_2}\right),$$

$$a_2 = \frac{g}{\omega^2 \ell_1}, \quad b_2 = \frac{2}{3}\frac{\ell_2}{\ell_1}.$$

Equations (7.118), (7.119) always admit the solutions $\sin\varphi = \sin\psi = 0$. What are the corresponding configurations? To determine the other possible solutions, we study the function $\beta = \beta(\alpha)$ defined by

$$\sin\beta = a\tan\alpha - b\sin\alpha.$$

We have $\beta(0) = 0$ and $\alpha \in (-\alpha_0, \alpha_0)$, with $a\tan\alpha_0 - b\sin\alpha_0 = 1$, $\alpha_0 \in (0, \pi/2)$. Moreover

$$\beta'(\alpha)\cos\beta = \frac{a}{\cos^2\alpha} - b\cos\alpha,$$

and hence $\beta'(0) = a - b$. If $a > b$, then $\beta'(\alpha) > 0$ for $\alpha \in (0, \alpha_0)$, while if $a < b$ then $\beta'(\alpha) < 0$ in a neighbourhood of $\alpha = 0$. In addition

$$\beta''(\alpha)\cos\beta = \beta'^2\sin\beta + 2\frac{a\sin\alpha}{\cos^3\alpha} + b\sin\alpha,$$

and hence $\beta'' > 0$ for $\beta > 0$, $\alpha > 0$. We can summarise this discussion in the two graphs shown in Fig. 7.13.

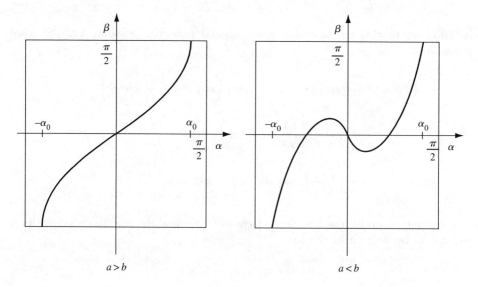

Fig. 7.13

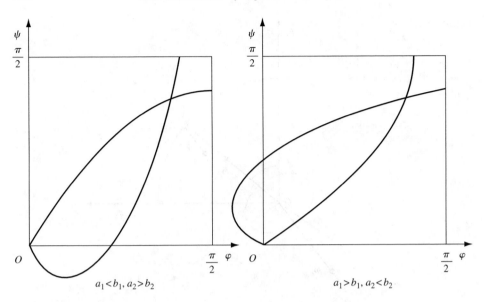

**Fig. 7.14** Graphs of the curves (7.118), (7.119).

The solution of (7.118), (7.119) yields a variety of different cases. If, for example, $a_1 < b_1$ (which happens for $\omega$ sufficiently large), but $a_2 > b_2$ (which can be achieved by diminishing $\ell_2$) there exists a unique non-zero solution (the symmetric solution corresponds to the same configuration). The same happens in the symmetric case, $a_1 > b_1$, $a_2 < b_2$ (Fig. 7.14). We leave it to the reader to complete the analysis of all remaining cases in the whole variability interval $(-\pi, \pi)$ for $\varphi$ and $\psi$.

Note that adding term by term equations (7.118), (7.119) we find the second cardinal equation written with respect to the point $A$.

To find the constraint reactions in a given configuration of relative equilibrium, we can proceed as follows:

(a) write the second cardinal equation for the rod $AB$ with respect to $A$, to obtain $\phi_B$;
(b) write the first cardinal equation for the whole system (then knowledge of $\phi_B$ is not necessary), to obtain $\phi_A$.

*Problem 2*
A homogeneous disc of mass $M$ and radius $R$ is constrained to rotate around the normal axis passing through its centre $O$. In addition, a point $A$ of the axis is constrained to rotate on a fixed circle with centre $O$ (Fig. 7.15). Determine the motion of the system starting from a generic initial condition and compute the constraint reaction at $A$.

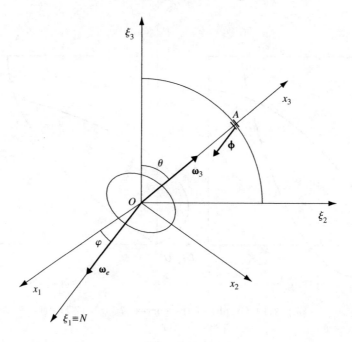

Fig. 7.15

*Solution*
The constraint imposed on the point $A$ is such that the line of nodes is fixed. Choosing the fixed and moving reference frames as in Fig. 7.15, we obtain that the vector $\boldsymbol{\omega}$ is the sum of

$$\boldsymbol{\omega} = -\dot{\varphi}\mathbf{e}_3 \tag{7.120}$$

and of

$$\boldsymbol{\omega}_e = -\dot{\theta}\mathbf{N} = -\dot{\theta}(\cos\varphi\,\mathbf{e}_1 + \sin\varphi\,\mathbf{e}_2), \tag{7.121}$$

where $\mathbf{N}$ is the unit vector of the node lines (we assume the line coincides with the $\xi_1$-axis). The signs in equations (7.120), (7.121) are due to the fact that $\theta$ and $\varphi$ increase in the negative direction of the rotations.

Recall that $\mathcal{J}_1 = \mathcal{J}_2 = \frac{1}{2}\mathcal{J}_3 = \frac{1}{4}MR^2$ and note that $(A - O) \times \boldsymbol{\phi} = \ell\mathbf{e}_3 \times \boldsymbol{\phi}(\cos\varphi\mathbf{e}_1 + \sin\varphi\mathbf{e}_2)$. Then Euler's equations can be written as

$$-\frac{1}{4}MR^2\frac{\mathrm{d}}{\mathrm{d}t}(\dot{\theta}\cos\varphi) = -\frac{1}{4}MR^2\dot{\theta}\dot{\varphi}\sin\varphi - \ell\phi\sin\varphi,$$

$$-\frac{1}{4}MR^2\frac{\mathrm{d}}{\mathrm{d}t}(\dot{\theta}\sin\varphi) = \frac{1}{4}MR^2\dot{\theta}\dot{\varphi}\cos\varphi + \ell\phi\cos\varphi,$$

the third one giving $\ddot\varphi = 0$. We can choose the reference system in such a way that $\theta(0) = \varphi(0) = 0$. In addition, $\dot\theta(0) = \omega_e^0$, $\dot\varphi(0) = \omega_3^0$. Clearly $\dot\varphi(t) = \omega_3^0$. Multiplying the first equation by $\cos\varphi$, and the second by $\sin\varphi$ and adding them, we find

$$\ddot\theta = 0 \Rightarrow \dot\theta = \omega_e^0.$$

From the latter we can easily deduce the expression for the reaction $\phi$:

$$\phi = -\frac{MR^2}{2\ell}\omega_3^0\omega_e^0 \mathbf{N}.$$

Note that the constraint contributes the torque $\mathbf{M} = \frac{1}{2}MR^2\omega_3^0\omega_e^0 \mathbf{N} \times \mathbf{e}_3$, which can be expressed as $\mathbf{M} = \mathcal{J}_3\omega_3^0\, d\mathbf{e}_3/dt$, in agreement with what was discussed regarding gyroscopic effects.

*Problem 3*
Consider the precession by inertia of a rigid body around a point $O$ with respect to which the ellipsoid of inertia is not of revolution ($\mathcal{J}_1 < \mathcal{J}_2 < \mathcal{J}_3$). Using a suitable Lyapunov function prove that the rotations around the middle axis of the ellipsoid of inertia are unstable.

*Solution*
We write the system (7.54) in the form

$$\dot{\boldsymbol{\omega}} = \mathbf{f}(\boldsymbol{\omega}),$$

with

$$f_1 = \frac{\mathcal{J}_1 - \mathcal{J}_3}{\mathcal{J}_1}\omega_2\omega_3, \quad f_2 = \frac{\mathcal{J}_3 - \mathcal{J}_1}{\mathcal{J}_2}\omega_3\omega_1, \quad f_3 = \frac{\mathcal{J}_1 - \mathcal{J}_2}{\mathcal{J}_3}\omega_1\omega_2.$$

Consider the equilibrium point $(0,\omega_2^0,0)$, $\omega_2^0 \neq 0$, and prove that the function

$$\Lambda^*(\boldsymbol{\omega}) = \mathcal{J}_1\mathcal{J}_2\mathcal{J}_3\omega_1(\omega_2 - \omega_2^0)\omega_3$$

satisfies the hypotheses of the Četaev instability theorem (Theorem 4.10). Examine first of all $\dot\Lambda^* = \nabla_\omega \Lambda^* \cdot \mathbf{f}(\boldsymbol{\omega})$:

$$\dot\Lambda^* = \mathcal{J}_2\mathcal{J}_3(\mathcal{J}_2 - \mathcal{J}_3)\omega_2\omega_3^2(\omega_2 - \omega_2^0) + \mathcal{J}_1\mathcal{J}_3(\mathcal{J}_3 - \mathcal{J}_1)\omega_1^2\omega_3^2$$
$$+ \mathcal{J}_1\mathcal{J}_2(\mathcal{J}_1 - \mathcal{J}_2)\omega_1^2\omega_2(\omega_2 - \omega_2^0).$$

The first and the last term have sign opposite to the sign of $\omega_2(\omega_2 - \omega_2^0)$, while the second term is positive if $\omega_1\omega_3 \neq 0$. To fix ideas, assume that $\omega_2^0 > 0$. Then the inequality $\omega_2(\omega_2 - \omega_2^0) < 0$ is satisfied for $0 < \omega_2 < \omega_2^0$. In this strip we have $\dot\Lambda^* > 0$. Hence if we take $\Omega_1$ as the intersection of this strip with the sets $\omega_1 > 0$, $\omega_3 > 0$, all the hypotheses of Četaev's theorem are satisfied.

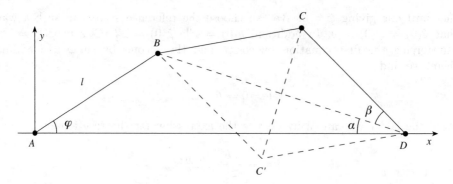

**Fig. 7.16**

*Problem 4*
In a horizontal plane a system of three equal rods $AB$, $BC$, $CD$ hinged to one another at $B$ and $C$, is constrained at the ends $A, D$ by hinges at two fixed points located at a mutual distance $a$. Let $l$ be the length of each rod. Then we assume that $0 < a < 3l$. The points $B$ and $D$ are mutually attracting by an elastic force. Study the equilibrium of the system.

*Solution*
Set $a/l = \gamma \in (0,3)$. Choose $\varphi \in (-\pi, \pi)$, the angle the rod $AB$ makes with the line $AD$ (Fig. 7.16). This angle may be subject to further limitations, as we request that $\overline{BD} \leq 2l$.

Since $\lambda = BD/l = (1 + \gamma^2 - 2\gamma \cos\varphi)^{1/2}$, it must be that $\lambda^2 \leq 4$, and hence $\gamma^2 - 2\gamma \cos\varphi - 3 \leq 0$, yielding $0 < \gamma < \cos\varphi + \sqrt{3 + \cos^2\varphi}$. The function $\cos\varphi + \sqrt{3 + \cos^2\varphi}$ takes values between 1 and 3, with maximum at $\varphi = 0$ and minima at $\varphi = \pm\pi$. It follows that for $\gamma \in (1,3)$ the above inequality defines the range $(-\varphi_0, \varphi_0)$ for $\varphi$, where $\varphi_0 \in (0, \pi)$ ($\varphi_0 \to 0$ for $\gamma \to 3$ and $\varphi_0 \to \pi$ for $\gamma \to 1$). When $\gamma \in (0,1]$ there are no prohibited values for $\varphi$. If $\varphi$ is an admissible, positive value, we find two configurations for the rods $BC$, $CD$, corresponding to the two isosceles triangles of side $l$ that can be constructed on the segment $BD$. Let $\alpha$ be the angle $B\hat{D}A$ and $\beta$ be the angle $B\hat{D}C$. Then it is easy to deduce that

$$\sin\alpha = \frac{\sin\varphi}{\lambda}, \quad \cos\beta = \frac{1}{2}\lambda,$$

and hence also that

$$\cos\alpha = \frac{(\lambda^2 - \sin^2\varphi)^{1/2}}{\lambda} = \frac{\gamma - \cos\varphi}{\lambda}, \quad \sin\beta = \frac{1}{2}(3 - \gamma^2 + 2\gamma\cos\varphi)^{1/2}.$$

In addition, there are the symmetric configurations, obtained by the substitution $\varphi \to -\varphi$. For $\gamma \in (0,1]$ at equilibrium we have the two configurations that minimise $\lambda^2$, corresponding to $\varphi = 0$, and the two configurations that maximise it ($\varphi = \pi$). Obviously, in the first case the equilibrium is stable, and in the second

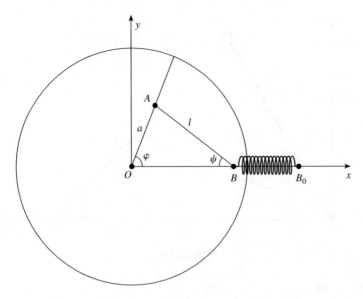

**Fig. 7.17**

case it is unstable. When $\gamma \in (1, 3)$ there still exists the solution $\varphi = 0$, with the corresponding configuration of stable equilibrium, and the solutions $\varphi = \pm\varphi_0$ for which $\lambda^2 = 4$, i.e. the segment $BD$ by the rods $BC$, $CD$ being made collinear. The corresponding equilibrium configurations are unstable.

*Problem 5*
Outline the dynamic analysis of the following plane system (Fig. 7.17): a disc (mass $M$, radius $R$) rotating around the fixed centre $O$; a rod (mass $m$, length $l$), with the end $A$ hinged at a point of the disc at a distance $a < l$ from the centre and with the end $B$ sliding along a fixed line ($x$-axis) through $O$. Suppose that the constraints are frictionless and that the only force applied is an elastic force which pulls $B$ towards the point $B_0$ on the $x$-axis at a distance $a + l$ from $O$.

*Solution*
The system has one degree of freedom. Choose as the Lagrangian coordinate the angle $\varphi$ formed by the radius $OA$ with the $x$-axis, and determine the angle $\psi = A\widehat{B}O$. By the sine theorem we find

$$\sin\psi = \frac{a}{l}\sin\varphi, \quad \cos\psi = \sqrt{1 - \frac{a^2}{l^2}\sin^2\varphi} \qquad (7.122)$$

(note that $\psi$ cannot reach $\pi/2$). The coordinates of the centre of mass $G$ of the rod are

$$G = \left(a\cos\varphi + \frac{1}{2}l\cos\psi, \frac{1}{2}l\sin\psi\right),$$

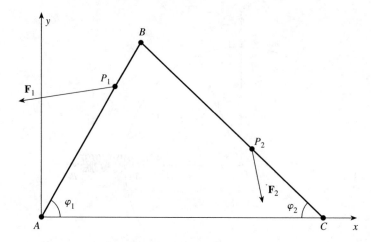

Fig. 7.18

from which we obtain

$$\dot G^2 = a^2 \sin^2 \varphi \, \dot\varphi^2 + \frac{1}{4} l^2 \dot\psi^2 - a l \sin \varphi \sin \psi \, \dot\varphi \dot\psi.$$

It follows that the kinetic energy is given by

$$T = \frac{1}{4} M R^2 \dot\varphi^2 + \frac{1}{2} m \left( a^2 \sin^2 \varphi \, \dot\varphi^2 + \frac{1}{4} l^2 \dot\phi^2 + a l \sin \varphi \sin \psi \, \dot\varphi \dot\psi + \frac{1}{12} l^2 \dot\psi^2 \right).$$

In addition to equation (7.122) we must use

$$\dot\psi = \frac{a}{l} \frac{\cos \varphi \, \dot\varphi}{\sqrt{1 - \frac{a^2}{l^2} \sin^2 \varphi}}$$

to obtain

$$T = \frac{1}{2} \dot\varphi^2 \left\{ \frac{1}{2} M R^2 + m a^2 \left( \sin^2 \varphi + \frac{1}{3} \frac{\cos^2 \varphi}{1 - \frac{a^2}{l^2} \sin^2 \varphi} + \sin^2 \varphi \frac{a}{l} \frac{\cos \varphi}{\sqrt{1 - \frac{a^2}{l^2} \sin^2 \varphi}} \right) \right\}.$$

The potential energy is

$$V = \frac{1}{2} k (B - B_0)^2 = \frac{1}{2} k [a(1 - \cos \varphi) + l(1 - \cos \psi)]^2.$$

We now have all the elements necessary to write down Lagrange's equation, which is very complicated. However it is easy to answer a question such as the following: if $\dot\varphi_0$ is the angular velocity of the disc corresponding to $\varphi = 0$, what is the velocity corresponding to $\varphi = \pi$? To answer, it is enough to use

$$T + V = E = \frac{1}{2} \dot\varphi_0^2 \left( \frac{1}{2} M R^2 + \frac{1}{3} m a^2 \right).$$

Since $V(\pi) = 2ka^2$, the new kinetic energy is

$$T = \frac{1}{2}\dot{\varphi}^2 \left(\frac{1}{2}MR^2 + \frac{1}{3}ma^2\right) = \frac{1}{2}\dot{\varphi}_0^2 \left(\frac{1}{2}MR^2 + \frac{1}{3}ma^2\right) - 2ka^2$$

(we must of course have

$$\dot{\varphi}_0^2 > \frac{4ka^2}{\frac{1}{2}MR^2 + \frac{1}{3}ma^2}$$

for $\varphi = \pi$ to be reached). The analysis of the equilibrium of the system is trivial, given that to $\varphi = 0$ and $\varphi = \pi$ there correspond the minimum and maximum of the potential energy. Since the potential energy has only variations of $\mathcal{O}(\varphi^4)$ near $\varphi = 0$, the small oscillations are not harmonic.

### Problem 6
To illustrate the use of the cardinal equations in the systems made of rigid parts mutually constrained, we examine the simplest case: two rods $AB$, $BC$, hinged at $B$ and constrained by hinges at $A, C$ (Fig. 7.18). Apply to the two rods two generic loads $\mathbf{F}_1, \mathbf{F}_2$ at the two internal points $P_1, P_2$, respectively. If $l_1, l_2$ are the lengths of the rods, determine all the constraint reactions.

### Solution
The cardinal equations

$$\mathbf{F}_1 + \mathbf{F}_2 + \boldsymbol{\phi}_A + \boldsymbol{\phi}_C = 0, \qquad (7.123)$$
$$(P_1 - A) \times \mathbf{F}_1 + (P_2 - A) \times \mathbf{F}_2 + (C - A) \times \boldsymbol{\phi}_C = 0$$

are three scalar equations in the unknowns $\phi_{Ax}, \phi_{Ay}, \phi_{Cx}, \phi_{Cy}$. Hence the system is underdetermined. This is due to the fact that the system is not a rigid system. (However, if the two rods are welded at $B$, yielding a rigid system, why are the equations still insufficient? How must the constraints be modified to make them sufficient?) To solve the problem, it is necessary to write the cardinal equations for every rigid component. In the present case, this is equivalent to writing six equations for the four unknowns mentioned plus the two components of the reaction in $B$. Let $\xi_1 = |P_1 - A|$, $\xi_2 = |P_2 - C|$. Denote by $\boldsymbol{\phi}_B$ the force that the hinge transmits to the rod $AB$ (the force transmitted to the rod $BC$ is $-\boldsymbol{\phi}_B$); we find for the equilibrium of the rod $AB$ the equations

$$F_{1x} + \phi_{Ax} + \phi_{Bx} = 0,$$
$$F_{1y} + \phi_{Ay} + \phi_{By} = 0,$$
$$\xi_1 \cos\varphi_1 F_{1y} - \xi_1 \sin\varphi_1 F_{1x} + l_1 \cos\varphi_1 \phi_{By} - l_1 \sin\varphi_1 \phi_{Bx} = 0,$$

and for $BC$ the equations

$$F_{2x} + \phi_{Cx} - \phi_{Bx} = 0,$$
$$F_{2y} + \phi_{Cy} - \phi_{By} = 0,$$
$$-\xi_2 \cos\varphi_2 F_{2y} - \xi_2 \sin\varphi_2 F_{2x} + l_2 \cos\varphi_2 \phi_{By} + l_2 \sin\varphi_2 \phi_{Bx} = 0.$$

The two torque balance equations can be solved independently to obtain $\phi_{Bx}, \phi_{By}$. Note that the determinant of the coefficients is $l_1 l_2 \sin(\varphi_1 + \varphi_2)$ and that the solvability condition is $0 < \varphi_1 + \varphi_2 < \pi$, i.e. the two rods cannot be collinear. Once $\phi_{Bx}, \phi_{By}$ are known, the remaining equations are trivial. This example includes the case that gravity is the only force acting on the system. It is also interesting to analyse the case that the rods are not carrying any load, and that the force is applied at the hinge point. This is the simplest case of a *truss* (more generally, we can have systems with more than two rods concurring at the same node). When a rod carries no load, the cardinal equations imply that the forces at the extreme points must constitute a balanced pair. Hence the rod is under pure tension or pure compression. In the elementary case that the two rods have a force **F** applied at $B$ the problem is solved immediately by decomposing **F** in the direction of the two rods.

*Problem 7*
The effect of tides caused by the Moon is to constantly increase the length of the day. Compute what is the eventual duration of the day when it coincides with the lunar month (the period of revolution of the Moon around the Earth). The radius of the Earth is $r = 6.4 \times 10^3$ km, the ratio between the masses is $m_T/m_L = 81$, the Earth–Moon distance is $R = 3.8 \times 10^5$ km and the ratio between the angular velocity $\omega_T$ of the Earth and the angular velocity $\omega_L$ of revolution of the Moon around the Earth is $\omega_T/\omega_L = 28$.

*Solution*
For simplicity suppose that the axis of rotation is orthogonal to the plane of the Moon's orbit, that the Earth is a homogeneous sphere and that the Moon is a point mass. The total angular momentum $|\mathbf{L}|$ with respect to the centre of mass of the two bodies is equal to the sum of the contributions due to the rotation of the Earth and of the Moon around the centre of mass and of the rotation of the Earth around its axis:

$$|\mathbf{L}| = m_L \left(\frac{m_T R}{m_L + m_T}\right)^2 \omega_L + m_T \left(\frac{m_L R}{m_L + m_T}\right)^2 \omega_L + \frac{2}{5} m_T r^2 \omega_T = I_L \omega_L + I_T \omega_T,$$
(7.124)

where $I_T = \frac{2}{5} m_T r^2$, $I_L = [m_L m_T / m_L + m_T] R^2$. When the day is equal to the lunar month, the angular velocity of the Earth is equal to that of the revolution of the Moon around the Earth; we denote them by $\omega$ (they both have to change to keep $|\mathbf{L}|$ constant). Under this condition, by Kepler's third law, the Moon–Earth distance is equal to $d = R(\omega_L/\omega)^{2/3}$. The angular momentum is equal to

$$|\mathbf{L}| = (I'_L + I_T)\omega, \tag{7.125}$$

where $I'_L = [m_L m_T/m_L + m_T]d^2$. Comparing (7.124) with (7.125), and setting $x = \omega/\omega_T$, since $I_L/I'_L = x^{4/3}(\omega_T/\omega_L)^{4/3}$, we find

$$x + a\left(\frac{\omega_L}{\omega_T x}\right)^{1/3} = 1 + a, \tag{7.126}$$

where $a = I_L\omega_L/I_T\omega_T$. From equation (7.126) we obtain the fourth-degree equation

$$x(1 + a - x)^3 - a^3\frac{\omega_L}{\omega_T} = 0. \tag{7.127}$$

Substituting the approximate numerical values $a \simeq 3.8$, $\omega_L/\omega_T \simeq 1/28$, we find that (7.127) has two real roots, $x_f \simeq 1/55$ and $x_p \simeq 4$. The solution $x_f$ corresponds to the future: the day has a duration of 55 present days, and the Earth–Moon distance is about $6 \times 10^5$ km. The solution $x_p$ corresponds to a past state when the day's duration was 6 present hours and the Earth–Moon distance was only $2 \times 10^4$ km. The main approximation in this computation is due to considering the rotation axes of the Moon and of the Earth as orthogonal to the plane of the orbit. In reality, the respective inclinations are 23.5° and 5°, and hence the inclination of the lunar orbit with respect to the Earth's equator varies between 18.5° and 28.5°. On the other hand, the angular momentum due to the Moon's rotation around its own axis is very small, with an approximate value of $\frac{2}{5}m_L r_L^2 \omega_L$, with $r_L = 1.7 \times 10^3$ km. Comparing this with the angular momentum of the Earth $\frac{2}{5}m_T r^2 \omega_T$ we see that the ratio is of order $10^{-3}$.

# 8 ANALYTICAL MECHANICS: HAMILTONIAN FORMALISM

## 8.1 Legendre transformations

Within the Lagrangian formalism, the phase space for the equations of motion makes use of the coordinates $(\mathbf{q}, \dot{\mathbf{q}})$. We start this chapter by studying the coordinate transformations in this space. We shall see that this study has wide and significant developments.

The first objective is the transformation of the equations of motion into a form, the so-called canonical form, whose particular structure highlights many important properties of the motion. This objective is realised by an application of a transformation due to Legendre. In this section, we study the most important properties of this transformation.

For simplicity, consider a real function $f$ of a real variable $w$, defined in an interval $(a, b)$ (not necessarily bounded), with continuous, positive second derivative,

$$f''(w) > 0. \tag{8.1}$$

Because of (8.1) the equation

$$f'(w) = p \tag{8.2}$$

uniquely defines a function $w = w(p)$, with $p$ variable in an open interval $(c, d)$ where $w'(p)$ exists and is continuous. Geometrically, we can interpret $w(p)$ as the abscissa of the point where the graph of $f(w)$ is tangent to the line of slope $p$ (Fig. 8.1).

DEFINITION 8.1   *The Legendre transform of $f(w)$ is the function*

$$g(p) = pw(p) - f[w(p)]. \tag{8.3}$$

∎

A significant property of this transform is that it is *involutive*. The meaning of this is expressed in the following.

THEOREM 8.1   *The function (8.3) in turn has a Legendre transform, which coincides with the initial function $f(w)$.*

*Proof*
To verify that $g(p)$ admits a Legendre transform it is sufficient to check that $g''(p) > 0$. Differentiating (8.3), and using (8.2), we find

$$g'(p) = w(p) + pw'(p) - f'[w(p)]w'(p) = w(p), \tag{8.4}$$

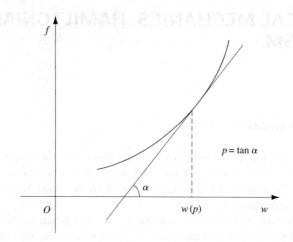

Fig. 8.1

and therefore

$$g''(p) = w'(p) = \{f''[w(p)]\}^{-1} > 0. \tag{8.5}$$

To construct the Legendre transform of $g(p)$ we must firstly define the function $p(w)$ through the equation

$$g'(p) = w. \tag{8.6}$$

Comparing this equation with (8.4), we find that $p(w)$ is the inverse function of $w(p)$. Through $p(w)$ we obtain the expression for the Legendre transform $h(w)$ of $g(p)$, simply by an application of the definition:

$$h(w) = wp(w) - g[p(w)]. \tag{8.7}$$

Finally, inserting (8.3) into (8.7) yields

$$h(w) = wp(w) - \{p(w)w - f(w)\} = f(w). \tag{8.8}$$

∎

We note that the use of Legendre transforms yields the *Young inequality*:

$$pw \leq f(w) + g(p), \tag{8.9}$$

where $f$ and $g$ are strictly convex functions which are the Legendre transform of each other.

The inequality (8.9) is proved starting from the function

$$F(w, p) = pw - f(w) \tag{8.10}$$

(where $f$ is any function which admits a Legendre transform). Indeed, note that

$$\frac{\partial F}{\partial w} = p - f'(w), \quad \frac{\partial^2 F}{\partial w^2} = -f''(w) < 0,$$

and hence the maximum of $F$ for every fixed $p$ is taken when $w = w(p)$, defined by (8.2). This maximum value thus coincides with $g(p)$, i.e.

$$F(w, p) \leq g(p),$$

which yields (8.9). This proves in particular that in equation (8.9) equality holds along the curve $w = w(p)$.

*Example* 8.1
Compute the Legendre transform of $f(w) = aw^n$, $a > 0$, $n > 1$, and prove that the Cauchy inequality

$$2pw \leq \varepsilon w^2 + \frac{1}{\varepsilon} p^2, \qquad \forall \, \varepsilon > 0 \qquad (8.11)$$

can be deduced from the Young inequality (8.9), which for this choice of $f$ has the form

$$pw \leq a \left[ w^n + (n-1) \left( \frac{p}{na} \right)^{n/(n-1)} \right] \qquad (8.12)$$

(see also Problem 6.1); it suffices to choose $n = 2$, $a = \varepsilon/2$. ∎

The previous considerations extend without difficulty to the case of a real function $f(\mathbf{w})$, with $\mathbf{w} \in \mathbf{R}^\ell$, and with continuous second partial derivatives, such that the quadratic form associated with the Hessian matrix

$$\frac{\partial^2 f}{\partial w_h \partial w_k} \qquad (8.13)$$

is positive definite. In this case it is possible to invert the system

$$\frac{\partial f}{\partial w_k} = p_k, \qquad k = 1, \ldots, \ell, \qquad (8.14)$$

and thus to define the vectorial function $\mathbf{w} = \mathbf{w}(\mathbf{p})$. It is now clear how it is possible to define the Legendre transform of $f(\mathbf{w})$.

DEFINITION 8.2 *The Legendre transform of $f(\mathbf{w})$ is*

$$g(\mathbf{p}) = \mathbf{p} \cdot \mathbf{w}(\mathbf{p}) - f[\mathbf{w}(\mathbf{p})]. \qquad (8.15)$$

∎

We can also prove that $f(\mathbf{w})$ in turn represents the Legendre transform of $g(\mathbf{p})$; it is enough to note that $\nabla g = \mathbf{w}(\mathbf{p})$, and hence that the Hessian matrix of $g(\mathbf{p})$ coincides with the Jacobian matrix of $\mathbf{w} = \mathbf{w}(\mathbf{p})$, and therefore with the inverse of the Hessian matrix of $f(\mathbf{w})$. The latter is also positive definite. We can then define the function $\mathbf{p} = \mathbf{p}(\mathbf{w})$ by inverting the system

$$\frac{\partial g}{\partial p_k} = w_k, \qquad k = 1, \ldots, \ell. \qquad (8.16)$$

We conclude that $\mathbf{p}(\mathbf{w})$ is the inverse of $\mathbf{w}(\mathbf{p})$. Finally, we proceed as for (8.7) and (8.8) to obtain the final result.

Analogously we can extend the *Young inequality*:

$$\mathbf{p} \cdot \mathbf{w} \leq f(\mathbf{w}) + g(\mathbf{p}), \tag{8.17}$$

where the equals sign holds for $\mathbf{w} = \mathbf{w}(\mathbf{p})$.

*Remark* 8.1
The Legendre transform is paired with the *invertible* variable transformation from $w_k$ to $p_k$, defined by (8.14). The inverse transform is defined by (8.16). ∎

*Remark* 8.2
According to Theorem 4.1, the Lagrangian function $L(\mathbf{q}, \dot{\mathbf{q}}, t)$ of a system admits a Legendre transform with respect to the variables $\dot{q}_k$, for every fixed $\mathbf{q}$ and $t$. ∎

*Remark* 8.3
The Legendre transform can be defined by inverting the signs in the right-hand side of (8.3); this is equivalent to considering the transform of $-f$, a common trick in thermodynamics. ∎

## 8.2 The Hamiltonian

We are ready to pass from the Lagrangian formalism to a new representation in phase space, by a Legendre transformation of the Lagrangian variables (Remark 8.2). Only the kinetic variables $\dot{q}_k$ are transformed, and replaced by the corresponding variables $p_k$, while the Lagrangian is replaced by its Legendre transform, called the *Hamilton function* or *Hamiltonian*.

Therefore the transformation is obtained by expressing explicitly the vector $\dot{\mathbf{q}} = \dot{\mathbf{q}}(\mathbf{q}, \mathbf{p}, t)$ from the system (linear in the variables $\dot{q}_k$, as clearly follows from (4.32))

$$\frac{\partial L}{\partial \dot{q}_k} = p_k, \qquad k = 1, \ldots, \ell, \tag{8.18}$$

and the Hamiltonian is then defined by

$$H(\mathbf{p}, \mathbf{q}, t) = \mathbf{p} \cdot \dot{\mathbf{q}}(\mathbf{p}, \mathbf{q}, t) - L(\mathbf{q}, \dot{\mathbf{q}}(\mathbf{p}, \mathbf{q}, t), t). \tag{8.19}$$

DEFINITION 8.3 *The variables $(p_k, q_k)$ are called* conjugated canonical variables. *The $p_k$ are called* kinetic momenta. ∎

The reason for the latter terminology is that the variables $p_k$ are the Lagrangian components of the linear momentum in $\mathbf{R}^{3n}$ (see (4.33)).

*Remark* 8.4
When there are no generalised potentials, we have $\partial L/\partial \dot{q}_k = \partial T/\partial \dot{q}_k$, and the transformation (8.18) depends only on the geometric structure of the holonomic system under consideration, and not on the system of applied forces.

Verify that if the Lagrangian $L$ is replaced by the Lagrangian $L' = cL$ or
$$L'' = L + \frac{\mathrm{d}}{\mathrm{d}t}F(\mathbf{q})$$
(Remark 4.8), we obtain the respective momenta $\mathbf{p}' = c\mathbf{p}$, $\mathbf{p}'' = \mathbf{p} + \nabla_\mathbf{q} F$, with the corresponding Hamiltonians $H' = cH$, $H'' = H$. ∎

*Example 8.2*
For an unconstrained point particle $(P, m)$, of Cartesian coordinates $q_i$, we have $T = \frac{1}{2}m \sum_{i=1}^{3} \dot{q}_i^2$, and hence $p_i = m\dot{q}_i$, $\dot{q}_i = p_i/m$. It follows that $\mathbf{p} \cdot \dot{\mathbf{q}} = p^2/m$ and if the particle is subject to a field of potential energy $V(\mathbf{q})$ we easily obtain
$$H(\mathbf{p}, \mathbf{q}) = \frac{p^2}{2m} + V(\mathbf{q}). \tag{8.20}$$
In particular, the *Hamiltonian of the harmonic oscillator* of frequency $\omega$ is
$$H(\mathbf{p}, \mathbf{q}) = \frac{p^2}{2m} + \frac{1}{2}m\omega^2 q^2. \tag{8.21}$$
∎

In the previous example the Hamiltonian coincides with the total energy. This fundamental property is valid in more general situations.

THEOREM 8.2 *In a holonomic system with fixed constraints, of Lagrangian $L(\mathbf{q}, \dot{\mathbf{q}})$, and without any generalised potential, the Hamilton function $H(\mathbf{p}, \mathbf{q})$ represents the total mechanical energy of the system.*

*Proof*
From equation (8.2) it follows immediately that $H$ does not depend explicitly on $t$. The kinetic energy $T$ is a homogeneous quadratic form in $\dot{q}_k$, and consequently
$$\mathbf{p} \cdot \dot{\mathbf{q}} = \sum_{k=1}^{\ell} \frac{\partial T}{\partial \dot{q}_k} \dot{q}_k = 2T, \tag{8.22}$$
from which it follows that
$$H(\mathbf{p}, \mathbf{q}) = 2T - L(\mathbf{q}, \mathbf{p}) = T(\mathbf{p}, \mathbf{q}) + V(\mathbf{q}). \tag{8.23}$$
∎

*Remark 8.5*
When there are generalised potentials, Theorem 8.18 is no longer valid. In this respect see Problem 8.1. ∎

*Remark 8.6*
Since $T$ (hence $L$) is a quadratic function of $\dot{q}_k$, equations (8.18) are an invertible linear system. It follows that, in the Hamiltonian formalism, $T$ becomes a quadratic function of $p_k$ (homogeneous if the constraints are fixed). ∎

## 8.3 Hamilton's equations

The main advantage in using the Legendre transformation in the phase space is that the equations of motion then take the form

$$\dot{p}_k = -\frac{\partial H}{\partial q_k}, \quad k = 1, \ldots, \ell,$$
$$\dot{q}_k = \frac{\partial H}{\partial p_k}, \quad k = 1, \ldots, \ell. \tag{8.24}$$

Equations (8.24) are called *Hamilton's canonical equations* and can be easily verified. The second group coincides with equations (8.16), and describes the transformation $\dot{\mathbf{q}} = \dot{\mathbf{q}}(\mathbf{p}, \mathbf{q}, t)$. This can also be obtained by directly differentiating the two sides of (8.19) with respect to $p_k$ and using (8.18).

Differentiating (8.19) with respect to $q_k$ we obtain

$$\frac{\partial H}{\partial q_k} = \mathbf{p} \cdot \frac{\partial \dot{\mathbf{q}}}{\partial q_k} - \frac{\partial L}{\partial q_k} - \nabla_{\dot{\mathbf{q}}} L \cdot \frac{\partial \dot{\mathbf{q}}}{\partial q_k} = -\frac{\partial L}{\partial q_k},$$

where we have once again used (8.18). Finally, recall that thanks to (8.18), Lagrange's equations (4.75) can be written in the form

$$\dot{p}_k = \frac{\partial L}{\partial q_k}, \quad k = 1, \ldots, \ell. \tag{8.25}$$

This yields immediately the first group of (8.24).

*Remark 8.7*
The Hamiltonian obviously has the same regularity as the Lagrangian. The existence and uniqueness of the solution of the initial value problem for equations (8.24) is thus guaranteed. ∎

PROPOSITION 8.1 *If $\mathbf{q} = \mathbf{q}(t), \mathbf{p} = \mathbf{p}(t)$ are solutions of the system (8.24), we have*

$$\frac{d}{dt} H(\mathbf{p}(t), \mathbf{q}(t), t) = \frac{\partial}{\partial t} H(\mathbf{p}(t), \mathbf{q}(t), t). \tag{8.26}$$

*Proof*
The proof follows directly from (8.24). ∎

*Remark 8.8*
From equations (8.26) we again find that if $\partial H/\partial t = 0$ the Hamilton function $H$ is a constant of the motion, called the *generalised energy integral*. ∎

In what follows, it is convenient to use a vector notation for equations (8.24), introducing the vector $\mathbf{x} = \begin{pmatrix} \mathbf{p} \\ \mathbf{q} \end{pmatrix}$ and the $2\ell \times 2\ell$ matrix

$$\mathbb{J} = \begin{pmatrix} 0 & -1 \\ 1 & 0 \end{pmatrix}, \tag{8.27}$$

where **1** and **0** are the identity and the null $\ell \times \ell$ matrix, respectively.

Equations (8.29) can then be written as

$$\dot{\mathbf{x}} = \mathbb{J}\nabla H. \tag{8.28}$$

*Remark 8.9*

The right-hand side of (8.28) is a vector field $\mathbf{X}(\mathbf{x}, t)$, prescribed in phase space. Equations (8.28) are the equations for the flow lines of this field.

In the autonomous case, when $\partial H/\partial t = 0$, we find that the possible trajectories of the system in the phase space are the flow lines of the field $\mathbf{X}(\mathbf{x})$, which belong to the level sets $H(\mathbf{p}, \mathbf{q}) = \text{constant}$ (and coincide with them for $l = 1$). ∎

PROPOSITION 8.2  *If $H$ has two continuous second derivatives with respect to $q_k$ and $p_k$, then*

$$\text{div } \mathbf{X} = 0, \tag{8.29}$$

where

$$\mathbf{X}(\mathbf{x}, t) = \mathbb{J}\nabla H. \tag{8.30}$$

*Proof*

It is of immediate verification. ∎

Equation (8.29) has significant consequences, which we will consider in the following section, with reference more generally to systems of the form

$$\dot{\mathbf{x}} = \mathbf{X}(\mathbf{x}, t), \qquad \text{div } \mathbf{X} = 0, \qquad \mathbf{x} \in \mathbf{R}^n. \tag{8.31}$$

*Remark 8.10*

The Hamilton system (8.24) is less generic than (8.31), since it has the peculiarity that the projections $\mathbf{X}_k$ of $\mathbf{X}$ on each of the planes $(q_k, p_k)$, have zero divergence in the geometry of the respective planes, where $\mathbf{X}_k$ is the vector of components $(\partial H/\partial p_k, -\partial H/\partial q_k)$. ∎

## 8.4 Liouville's theorem

Hamilton's equations (8.24) induce a transformation $S^t$ in the phase space into itself, depending on time. With every point $\mathbf{x}_0 \in \mathbf{R}^{2\ell}$ and for all $t > 0$ we can associate the point $\mathbf{x}(t)$ obtained by integrating the Hamilton system with the initial condition $\mathbf{x}(0) = \mathbf{x}_0$. The transformation is invertible because of the reversibility of the equations of motion.

An important property of the transformation $S^t$, which we call the *flow in phase space associated with the Hamiltonian $H$*, is the following.

THEOREM 8.3 (Liouville)  *In phase space, the Hamiltonian flow preserves volumes. This property is true in general for any system of the type (8.31).*

*Proof*
We must show that in the flow we are considering, for every $t > 0$ the image $\Omega(t)$ of any domain $\Omega \subset \mathbf{R}^{2\ell}$ with a regular boundary has the same measure as $\Omega$. Consider the flow associated with any differential system of the kind

$$\dot{\mathbf{x}} = \mathbf{X}(\mathbf{x}, t). \tag{8.32}$$

Let us ignore for the moment that div $\mathbf{X} = 0$, and let us prove that

$$\frac{\mathrm{d}}{\mathrm{d}t}|\Omega(t)| = \int_{\Omega(t)} \text{div } \mathbf{X}(\mathbf{x}, t) \, \mathrm{d}\mathbf{x}, \tag{8.33}$$

where $|\Omega|$ denotes the measure of $\Omega$. From (8.33) it clearly follows that $|\Omega(t)|$ is constant for any system of the type (8.31).

Equation (8.33) expresses the balance of volumes depicted in Fig. 8.2. The variation of the volume in time $\mathrm{d}t$ can be expressed as $\int_{\partial \Omega(t)} \mathbf{X} \cdot \mathbf{N} \, \mathrm{d}\sigma \, \mathrm{d}t$, where $\mathbf{N}$ is the outgoing normal. Therefore $\mathrm{d}/\mathrm{d}t |\Omega(t)|$ is simply the outgoing flux of the field $\mathbf{X}(\mathbf{x}, t)$ through $\partial \Omega(t)$ and equation (8.33) immediately follows. ∎

Equation (8.33) highlights the physical significance of the divergence of a velocity field $\mathbf{v}$: div $\mathbf{v}$ is *the dilation rate of the unit volume*.

We shall see that this theorem has important applications in statistical mechanics. It also gives information on the nature of *singular points* (i.e. the constant solutions) of (8.31).

COROLLARY 8.1 *A singular point of a system of the type (8.31) cannot be asymptotically stable.*

*Proof*
If $\mathbf{x}_0$ were asymptotically stable, then there would exist a sphere of centre $\mathbf{x}_0$ such that all trajectories starting inside the sphere would tend asymptotically to $\mathbf{x}_0$. The volume of the image of this sphere would therefore tend to zero as $t \to \infty$, contradicting Theorem 8.3. ∎

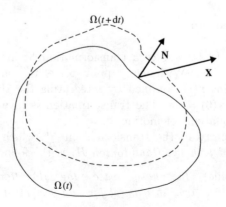

**Fig. 8.2**

*Remark* 8.11

If the points of the system are subject to *elastic collisions* with unilateral constraints (or between them), then correspondingly there exist discontinuities in the trajectories. It can be proved that Liouville's theorem is still valid in this case. It is enough to consider the elastic collisions as limits of smooth conservative interactions (see Section 2.6). ∎

*Remark* 8.12

It often happens that one needs to consider quantities such as the measure of the manifold $H = E$ in phase space, or of the norm of $\nabla_{\mathbf{x}} H$. In this case one needs to be careful about the particular metric used, as the canonical variables are not generally homogeneous quantities. The same remark applies to the components of $\nabla_{\mathbf{x}} H$. ∎

To avoid this difficulty, it is good practice to use from the start dimensionless variables. Every time we use such quantities we shall therefore consider dimensionless variables, without making an explicit note of it. As an example, for the harmonic oscillator, we can replace the Lagrangian $L = (1/2)(m\dot{q}^2 + kq^2)$ with $L' = (1/2)[(\mathrm{d}q'/\mathrm{d}t')^2 + \omega'^2 q'^2]$, where $q' = q/l$, $t' = t/t_0$ for some length $l$ and some time $t_0$, $\omega' = t_0\omega = t_0(k/m)^{1/2}$ and $L' = Lt_0^2/(ml^2)$ (recall it is always possible to multiply the Lagrangian by a constant).

Correspondingly we obtain a kinetic momentum $p' = \mathrm{d}q'/\mathrm{d}t'$ and a Hamiltonian $H' = (1/2)(p'^2 + \omega'^2 q'^2)$ which are dimensionless. It is now clear what is the (dimensionless) 'length' of an arc of a curve in phase space $(p', q')$ and what we mean by $|\nabla_{\mathbf{x}'} H'|$.

## 8.5 Poincaré recursion theorem

This celebrated theorem states that at an unknown future moment, trajectories in phase space come as close as we wish to their starting-point. We now specify the sense of this 'recurrence'.

Consider an autonomous system whose representative point in phase space is allowed to move inside a bounded region $\Omega$. This means that the point particles composing the system are confined within a bounded domain of $\mathbf{R}^3$, and the total energy is constant (hence any collisions of the particles between themselves or with unilateral constraints are elastic and the kinetic momenta are uniformly bounded).

We can now state the following theorem.

THEOREM 8.4 (Poincaré) *Consider an autonomous Hamiltonian system for which only a bounded region $\Omega$ in phase space is accessible.*

*Let $B_0$ be any sphere contained in $\Omega$ and let $B(t)$ be its image after time $t$ in the flux generated by the Hamiltonian. For any $\tau > 0$, there exists a time $t_0 > \tau$ such that $B(t_0) \cap B_0 \neq 0$.*

*Proof*

Consider the sequence of regions $B_n = B(n\tau)$, $n = 0, 1, 2, \ldots$, which, by Theorem 4.1, all have the same measure. Since the system is autonomous, $B_n$ can be obtained by applying the transformation $M$ which maps $B_0$ in $B_1 = B(\tau)$ $n$ times. (We can also define $M^0(B_0) = B_0$ and write that $M^\ell \circ M^{n-\ell} = M^n$, $\forall\, \ell = 0, 1, \ldots, n$.)

Since $B_n \subset \Omega$, $\forall\, n$, there must necessarily exist two distinct integers, which we denote by $n_0$ and $n_0 + k$ with $k > 0$, such that $B_{n_0} \cap B_{n_0+k} \neq \emptyset$. Otherwise, the measure of the set $\Gamma_N = \bigcup_{n=1,\ldots,N} B_n$ would be equal to $N$ times the measure of $B_0$, diverging for $N \to \infty$, and hence contradicting the assumption that $\Gamma_N \subset \Omega$, $\forall\, N$, and that $\Omega$ has bounded measure.

We now restrict attention to the set $B_{n_0} \cap B_{n_0+k}$. If $n_0 = 0$ the proof is finished; assume $n_0 \geq 1$. By tracing backwards the trajectories of all points for a time $\tau$, we see that they originated in $B_{n_0-1}$ and $B_{n_0+k-1}$, which must therefore intersect. Going back $n_0$ steps we find that $B_0 \cap B_k \neq 0$, which proves the theorem. ∎

COROLLARY 8.2   *All trajectories which originate in $B_0$ (except possibly a subset of $B_0$ of zero measure) must return to it infinitely many times.*

*Proof*
It is enough to note that the proof of the theorem uses only the fact that the measure of $B_0$ is positive, and not that it is a sphere. Hence for every subset of $B_0$ with positive measure, the recurrence property holds (in the same subset, and therefore in $B_0$).

Applying the theorem successively, we find that there must be infinitely many returns to $B_0$. ∎

*Remark* 8.13
In the proof of Theorem 8.4, we only used the property implying that the Hamiltonian flux preserves volumes in phase space (Theorem 8.3). This property holds for the more general flows of differential systems of the form (8.31).

We finally have the following generalisation: let $\Omega$ be an open set in the phase space of system (8.31), such that:

(1) the $n$-dimensional Lebesgue measure of $\Omega$ is finite;
(2) for any choice of the initial condition $\mathbf{x}(0) \in \Omega$, the corresponding solution $\mathbf{x}(t)$ of (8.31) belongs to $\Omega$ for every $t \in \mathbf{R}$.

It then follows that $\liminf_{t \to +\infty} \|\mathbf{x}(t) - \mathbf{x}(0)\| = 0$ for almost every initial condition $\mathbf{x}(0) \in \Omega$ (this means except for a set of initial conditions of zero Lebesgue measure). ∎

## 8.6 Problems

1. Consider the inequality (8.12) and note that, setting $n = \alpha$, $n/(n-1) = \beta$, $a = 1/\alpha$, we can deduce

$$pw \leq \frac{w^\alpha}{\alpha} + \frac{p^\beta}{\beta}, \quad \frac{1}{\alpha} + \frac{1}{\beta} = 1. \tag{8.34}$$

## 8.6 Analytical mechanics: Hamiltonian formalism

If $\xi(x)$, $\eta(x)$ are two functions defined in an interval $(a,b)$, such that the integrals

$$\left[\int_a^b |\xi(x)|^\alpha \, \mathrm{d}x\right]^{1/\alpha} \quad \text{and} \quad \left[\int_a^b |\eta(x)|^\beta \, \mathrm{d}x\right]^{1/\beta}$$

are convergent (we indicate them by $\|\xi\|_\alpha$ and $\|\eta\|_\beta$, respectively), show that from (8.34) it is possible to derive *Hölder's inequality*:

$$\int_a^b |\xi(x)\eta(x)| \, \mathrm{d}x \leq \|\xi\|_\alpha \|\eta\|_\beta. \tag{8.35}$$

*Sketch.* For fixed $x$, use (8.34) to obtain

$$\frac{|\xi|}{\|\xi\|_\alpha} \frac{|\eta|}{\|\eta\|_\beta} \leq \frac{1}{\alpha} \frac{|\xi|^\alpha}{\|\xi\|_\alpha^\alpha} + \frac{1}{\beta} \frac{|\eta|^\beta}{\|\eta\|_\beta^\beta},$$

and then one can integrate.

2. Consider the system $\dot{p} = f(p)$, $\dot{q} = g(q,p)$ and determine the structure of the function $g(q,p)$ for which the system is Hamiltonian. Repeat the problem replacing $f(p)$ by $f(q)$.

3. What are the conditions under which the system $\dot{p} = pf(q)$, $\dot{q} = g(q,p)$ is Hamiltonian?

4. Consider the motion generated by the Hamiltonian of the harmonic oscillator (8.21) with initial conditions $q(0) = q_0$, $p(0) = p_0$. Compute the functions $q(t; q_0, p_0)$ and $p(t; q_0, p_0)$, and prove directly that the area element $1/2(q \, \mathrm{d}p - p \, \mathrm{d}q)$ is invariant with respect to time, or equivalently verify that it is at every instant of time equal to $1/2(q_0 \, \mathrm{d}p_0 - p_0 \, \mathrm{d}q_0)$.

5. Formulate the theory of small oscillations around the stable equilibrium configurations in the Hamiltonian formalism.

6. Write Hamilton's equations for the system of Problem 9, Chapter 7.

7. Consider the Hamiltonian system $\dot{p} = -\alpha pq$, $\dot{q} = (\alpha/2)q^2$, with $\alpha$ a constant different from zero. Compute the solutions starting from arbitrary initial conditions. Draw the trajectories in the phase plane. Determine the nature of the point of equilibrium $p = q = 0$.

8. Find the conditions on the parameters $a, b, c, d \in \mathbf{R}$ such that the linear differential equations

$$\dot{p} = ap + bq, \quad \dot{q} = cp + dq$$

are the Hamilton equations for some function $H$, and compute $H$.
(Solution: $a = -d$, $H = c(p^2/2) - apq - b(q^2/2)$.)

9. Find the condition on $a$, $b$, $c \in \mathbf{R}$ such that the system of equations
$$\dot{p} = aq - q^2, \quad \dot{q} = bp + cq$$
is Hamiltonian, and compute the corresponding Hamilton function. Write the associated Lagrangian.

(Solution: $c = 0$, $H = (1/2)bp^2 - (1/2)aq^2 + (1/3)q^3$, $L = \dot{q}^2/(2b) + (1/2)aq^2 - (1/3)q^3$.)

10. Find the conditions on $\alpha$, $\beta$, $\delta$ (positive real constants) such that the system of equations
$$\dot{p} = -p^{\alpha+1}q^\delta, \quad \dot{q} = p^\alpha q^\beta$$
is Hamiltonian, and compute the corresponding Hamilton function. Solve the equations for $\alpha \neq -1$.

(Solution: if $\alpha = -1 \Rightarrow \beta = 0$,
$$H = \begin{cases} \log p + \dfrac{q^{\delta+1}}{\delta+1} + \text{constant}, & \text{if } \delta \neq -1, \\ \log pq + \text{constant}, & \text{if } \delta = -1. \end{cases}$$

If $\alpha \neq -1$, $\delta = \alpha$, $\beta = \alpha + 1$,
$$H = \frac{(pq)^{\alpha+1}}{\alpha+1}.$$

For $\alpha \neq -1$ we have $p(t) = p(0)e^{-\kappa t}$, $q(t) = q(0)e^{\kappa t}$, with $\kappa = [(\alpha+1)H]^{\alpha/(\alpha+1)}$.)

11. Find the conditions on the coefficients such that the system of equations
$$\dot{p}_1 = -a_1 q_1 - b_1 q_2, \qquad \dot{p}_2 = -a_2 q_1 - d_2 p_2,$$
$$\dot{q}_1 = a_3 q_1 + c_3 p_1 + d_3 p_2, \quad \dot{q}_2 = b_4 q_2 + d_4 p_2$$
is Hamiltonian and compute the corresponding Hamilton function.

(Solution: $a_3 = d_3 = 0$, $b_4 = d_2$, $a_2 = b_1$, $H = a_1 q_1^2/2 + a_2 q_1 q_2 + c_3(p_1^2/2) + d_4(p_2^2/2) + b_4 p_2 q_2$.)

12. Write down the Hamiltonian of Problem 9, Chapter 4.

13. Prove the following generalisation of formula (8.33) (*transport theorem*):
$$\frac{d}{dt}\int_{\Omega(t)} F(\mathbf{x},t)\,d\mathbf{x} = \int_{\Omega(t)} \left[\frac{\partial F}{\partial t} + \operatorname{div}(F\mathbf{X})\right] d\mathbf{x},$$
with $\Omega(t)$ being a domain with regular boundary, $F \in C^1$ and $\dot{\mathbf{x}} = \mathbf{X}(\mathbf{x},t)$.

## 8.7 Additional remarks and bibliographical notes

The Hamiltonian form of the equations of motion was introduced by W. R. Hamilton in 1835 (*Phil. Trans.*, pp. 95–144), partially anticipated by Poisson, Lagrange and Cauchy. The Legendre transformation can be generalised to functions which are not of class $\mathcal{C}^2$ (see Hörmander 1994, chapter II): let $f : \mathbf{R}^n \to \mathbf{R} \cup \{+\infty\}$ be a convex function and lower semicontinuous (i.e. $f(x) = \lim_{y \to x} \inf f(y)$ for every $x \in \mathbf{R}^n$). The *Legendre transform* $g$ of $f$ can be obtained by setting

$$g(y) = \sup_{x \in \mathbf{R}^n} (x \cdot y - f(x)).$$

It is immediate to verify that $g$ is also convex, lower semicontinuous and it can be proved that its Legendre transform is $f$. This more general formulation has numerous applications in the calculus of variations.

Poincaré's recurrence theorem can rightly be considered the first example of a theorem concerning the study of equations that preserve some measure in phase space. This is the object of ergodic theory, to which we will give an introduction in Chapter 13.

## 8.8 Additional solved problems

*Problem 1*
The Lagrangian of an electron of mass $m$ and charge $-e$ is (see (4.105) with $e \to -e$) $L = \frac{1}{2}mv^2 - (e/c)\mathbf{v} \cdot \mathbf{A}$, in the absence of an electric field. In the case of a plane motion we have $\mathbf{A} = B/2(-y, x)$. Write the Hamiltonian in polar coordinates. Study the circular orbits and their stability.

*Solution*
In polar coordinates $\mathbf{v} = \dot{r} e_r + r\dot{\varphi} e_\varphi$, and therefore $\mathbf{v} \cdot \mathbf{A} = B/2(r^2 \dot{\varphi})$. Hence the Lagrangian can be written as

$$L = \frac{1}{2}m(\dot{r}^2 + r^2 \dot{\varphi}^2) - \frac{e}{c}\frac{B}{2}r^2 \dot{\varphi}.$$

We apply the Legendre transform

$$p_r = m\dot{r}, \quad p_\varphi = mr^2\dot{\varphi} - \frac{eB}{2c}r^2.$$

Setting $\omega = eB/mc$ we have

$$\dot{r} = \frac{p_r}{m}, \quad \dot{\varphi} = \frac{p_\varphi}{mr^2} + \frac{\omega}{2},$$

and finally $H = p_r \dot{r} + p_\varphi \dot{\varphi} - L$ gives

$$H = \frac{p_r^2}{2m} + \frac{p_\varphi^2}{2mr^2} + \frac{1}{2}\omega p_\varphi + \frac{1}{8}mr^2\omega^2.$$

The coordinate $\varphi$ is cyclic, and hence $p_\varphi = $ constant. If the motion has to lie on a circular orbit, we must have $p_r = \dot{p}_r = 0$. This is equivalent to $\partial H/\partial r = 0$, or

$$-\frac{p_\varphi^2}{m}r^{-3} + \frac{1}{4}mr\omega^2 = 0.$$

The solution of this equation gives the radius of the only circular orbit corresponding to the parameters $p_\varphi, \omega$:

$$r_0 = \left(\frac{2p_\varphi}{m\omega}\right)^{1/2}.$$

Correspondingly, we have

$$\dot\varphi = \frac{\partial H}{\partial p_\varphi} = \frac{p_\varphi}{mr_0^2} + \frac{\omega}{2} = \omega,$$

and therefore $\omega$ represents the angular velocity of the circular motion. The value of $p_\varphi$ is determined by the kinetic energy. Note that the kinetic energy is

$$T = \frac{1}{2}mr_0^2\dot\varphi^2 = \frac{p_\varphi^2}{2mr_0^2} + \frac{1}{8}mr_0^2\omega^2 + \frac{1}{2}\omega p_\varphi = \omega p_\varphi$$

$((1/2)p_\varphi\dot\varphi$ is not the kinetic energy because of the presence of the generalised potential and the Hamiltonian $H = p_\varphi\dot\varphi - \frac{1}{2}mr^2\dot\varphi^2 + V$ takes the value of $T$, since we find that $p_\varphi\dot\varphi + V = 2T$). If we choose the velocity $v_0$ of the electron $(\omega r_0 = v_0)$ it follows that $T = \omega p_\varphi = \frac{1}{2}mv_0^2$, i.e. $p_\varphi = mv_0^2/2\omega$, which is consistent with the expression for $r_0 = v_0/\omega$. To study the stability of circular motion of radius $r_0$, set $r = r_0 + \rho$ and keep for $p_\varphi$ the value corresponding to $r_0$. In the Hamiltonian we must take the expansion to second order in $\rho$ of

$$\frac{1}{r^2} = \frac{1}{r_0^2}\frac{1}{(1+\rho/r_0)^2} \simeq \frac{1}{r_0^2}\left(1 - 2\frac{\rho}{r_0} + 3\left(\frac{\rho}{r_0}\right)^2\right)$$

and note that the terms linear in $\rho$ are cancelled. The remaining Hamiltonian is $(p_r = p_\rho)$

$$H = \frac{p_\rho^2}{2m} + \left(\frac{3}{2}\frac{p_\varphi^2}{mr_0^2} + \frac{1}{8}mr_0^2\omega^2\right)\frac{\rho^2}{r_0^2} + \omega p_\varphi$$

$$= \frac{p_\rho^2}{2m} + \frac{\omega p_\varphi}{r_0^2}\rho^2 + \omega p_\varphi.$$

With $p_\varphi = $ constant and $\omega p_\varphi / r_0^2 = \frac{1}{2} m \omega^2$ we obtain

$$H = \frac{p_\rho^2}{2m} + \frac{1}{2} m \omega^2 \rho^2 + \text{constant},$$

describing harmonic oscillations of the radius with frequency $\omega$.

*Problem 2*
In a horizontal plane a homogeneous rod $AB$, of length $l$ and mass $M$ is constrained to rotate around its centre $O$. A point particle $(P, m)$ can move on the rod and is attracted by the point $O$ with an elastic force of constant $k$. The constraints are frictionless.

(i) Write down Hamilton's equations.
(ii) Study the trajectories in phase space.
(iii) Study the motions with $|P - O|$ constant and the small oscillations around them.

*Solution*
(i) The kinetic energy is

$$T = \frac{1}{2} m (\dot{\xi}^2 + \xi^2 \dot{\varphi}) + \frac{1}{24} M l^2 \dot{\varphi}^2,$$

where $\xi$ is the $x$-coordinate of $P$ on $OA$ and $\varphi$ is the angle of rotation of the rod. The potential energy is

$$V = \frac{1}{2} k \xi^2.$$

The kinetic momenta are

$$p_\xi = m \dot{\xi}, \quad p_\varphi = \left( m \xi^2 + \frac{1}{12} M l^2 \right) \dot{\varphi},$$

and hence the Hamiltonian is

$$H = \frac{p_\xi^2}{2m} + \frac{1}{2} \frac{p_\varphi^2}{m \xi^2 + \frac{1}{12} M l^2} + \frac{1}{2} k \xi^2.$$

Hamilton's equations are

$$\dot{p}_\xi = -k \xi + p_\varphi^2 \frac{m \xi}{(m \xi^2 + \frac{1}{12} M l^2)^2}, \quad \dot{\xi} = \frac{p_\xi}{m},$$

$$\dot{p}_\varphi = 0 \; (\varphi \text{ is a cyclic coordinate}), \quad \dot{\varphi} = \frac{p_\varphi}{m \xi^2 + \frac{1}{12} M l^2}.$$

The first integral $p_\varphi$ = constant expresses the conservation of the total angular momentum with respect to $O$, which can clearly also be deduced from the cardinal equations (the only external force is the constraint reaction at $O$).

(ii) Excluding the trivial case $p_\varphi = 0$ (a simple harmonic motion of $P$ along the fixed rod), the trajectories in the plane $(\xi, p_\xi)$ have equation $H$ = constant. The function

$$f(\xi, p_\varphi) = \frac{1}{2} \frac{p_\varphi^2}{m\xi^2 + \frac{1}{12}Ml^2} + \frac{1}{2}k\xi^2$$

is positive and if

$$p_\varphi > \frac{M}{12}\sqrt{\frac{k}{m}}l^2,$$

(case (a)) it has a relative maximum at $\xi = 0$ ($f(0, p_\varphi) = 6p_\varphi^2/Ml^2$), it is symmetric with respect to $\xi = 0$ and it has two minima at $\xi = \pm\xi_0$, with $\xi_0^2 = 1/\sqrt{mk}\left[p_\varphi - (M/12)\sqrt{k/m}\,l^2\right]$. If, on the other hand, $p_\varphi \le (M/12)\sqrt{k/m}\,l^2$ (case (b)) there is an absolute minimum at $\xi = 0$. The phase portraits in cases (a), (b) are shown in Fig. 8.3.

Therefore there exist motions with $\xi$ a non-zero constant if and only if $p_\varphi > (M/12)\sqrt{k/ml^2}$ and the corresponding value of the radius is $\xi_0$. It is also possible to have a simple uniform rotation of the rod with $\xi = 0$, but in case (a) this is unstable, since the two separatrices between the two situations listed below pass through the origin: ($\alpha$) $E \in (E_{\min}, E^*)$, when $\xi$ oscillates around $\xi_0$ without passing through the middle point $O$; ($\beta$) $E > E^*$ (and less than some $E_{\max}$ guaranteeing $\xi < l/2$), when the point $P$ oscillates on the rod passing through the middle point $O$.

(iii) In case (b) the oscillation is around $\xi = 0$, and hence we can use the approximation

$$\frac{1}{m\xi^2 + \frac{1}{12}Ml^2} \simeq \frac{1}{\frac{1}{12}Ml^2}\left(1 - \frac{12m}{Ml^2}\xi^2\right).$$

To second order in $\xi$, we find

$$H = \frac{p_\xi^2}{2m} + \frac{1}{2}\xi^2 m\left[\frac{k}{m} - \frac{p_\varphi^2}{\left(\frac{1}{12}Ml^2\right)^2}\right],$$

describing oscillations of frequency

$$\omega = \left[\frac{k}{m} - \frac{p_\varphi^2}{\left(\frac{1}{12}Ml^2\right)^2}\right]^{1/2}.$$

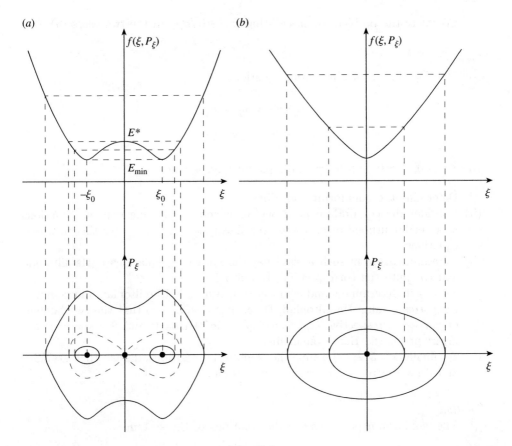

**Fig. 8.3**

In case (a) we study the perturbation $\xi = \xi_0 + \eta$ ($\eta \ll \xi_0$), by expanding $(m(\xi_0 + \eta)^2 + \frac{1}{12}Ml^2)^{-1}$ around $\xi_0$. We set $A = \frac{1}{12}Ml^2 + m\xi_0^2 = p_\varphi \sqrt{m/k}$, and then we have

$$\frac{1}{m(\xi_0 + \eta)^2 + \frac{1}{12}Ml^2} = \frac{1}{A} \frac{1}{1 + m\frac{2\xi_0\eta + \eta^2}{A}} \simeq \frac{1}{A}\left(1 - m\frac{2\xi_0\eta + \eta^2}{A} + \frac{4m^2\xi_0^2}{A^2}\eta^2\right).$$

We see that in the expression for $H$ the terms linear in $\eta$ cancel, and we are left with ($p_\xi = p_\eta$)

$$H \simeq \frac{p_\eta^2}{2m} + \frac{k}{A}4m\xi_0^2\eta^2.$$

It follows that the oscillation is harmonic, with frequency

$$\omega' = 2\xi_0 \sqrt{\frac{k}{A}}.$$

Complete the problem by integrating $\dot\varphi = \partial H/\partial p_\varphi$ in the two cases (a), (b).

*Problem 3*
Consider the system of differential equations

$$\dot p = 5p^2 q + aq^3 - bq,$$
$$\dot q = -8p^3 - cpq^2 + 6p,$$

where $a > 0$, $b > 0$, $c > 0$ are three parameters, $a < 2c$.

(i) Determine the equilibrium positions.
(ii) Consider the equilibrium positions for which $q = 0$. Linearise the equations in a neighbourhood of $q$, discuss the linear stability and solve the linearised equations.
(iii) Determine $a, b, c$ in such a way that the system of equations is Hamiltonian and compute the corresponding Hamiltonian.
(iv) Set $a = 0$; determine $b$ and $c$ so that the system is Hamiltonian and compute the corresponding Hamiltonian. Determine $\alpha$ and $\beta$ so that the two families of curves of respective equations $4p^2 + 5q^2 + \alpha = 0$ and $2p^2 + \beta = 0$ are invariant for the Hamiltonian flux.
(v) Set finally $b = 5/2$, determine the equilibrium positions, discuss their stability and draw the phase portrait of the system.

*Solution*
(i) The equilibrium positions are the solutions of the system

$$5p^2 q + aq^3 - bq = 0,$$
$$8p^3 + cpq^2 - 6p = 0,$$

which admits solutions $p = q = 0$; $p = 0$, $q = \pm\sqrt{b/a}$; $p = \pm\sqrt{3/4}$, $q = 0$. If $6a/c \le b \le 15/4$, and only then, there are four additional equilibrium points corresponding to the intersections of the two ellipses

$$5p^2 + aq^2 = b, \quad 8p^2 + cq^2 = 6.$$

(ii) The linearised equations corresponding to the equilibrium points with $q = 0$ are

$$\dot p = bq, \quad \dot q = -6p \quad \text{near } (0,0),$$

$$\dot\eta = -\left(b - \frac{15}{4}\right)q, \quad \dot q = -12\eta \quad \text{near } \left(\pm\sqrt{\frac{3}{4}}, 0\right) \text{ with } \eta = p \mp \sqrt{\frac{3}{4}}.$$

### 8.8 Analytical mechanics: Hamiltonian formalism

The first equation shows that $(0,0)$ is linearly stable, and it has the solution

$$p(t) = p(0) \cos \sqrt{6b}\, t + \sqrt{\frac{b}{6}}\, q(0) \sin \sqrt{6b}\, t,$$

$$q(t) = -\sqrt{\frac{6}{b}}\, p(0) \sin \sqrt{6b}\, t + q(0) \cos \sqrt{6b}\, t.$$

In the second case, the positions are linearly stable only if $0 < b < 15/4$. Let $\omega^2 = \left|\frac{15}{4} - b\right|$.
The solutions are

$$\left.\begin{array}{l} \eta(t) = \dfrac{q(0)\omega}{12} \sin \omega t + \eta(0) \cos \omega t \\[1mm] q(t) = q(0) \cos \omega t - \dfrac{12}{\omega} \eta(0) \sin \omega t \end{array}\right\} \quad \text{if } 0 < b < \frac{15}{4},$$

$$\left.\begin{array}{l} \eta(t) = \eta(0) \\ q(t) = q(0) - 12\eta(0)t \end{array}\right\} \quad \text{if } b = \frac{15}{4},$$

$$\left.\begin{array}{l} \eta(t) = -\dfrac{q(0)\omega}{12} \sinh \omega t + \eta(0) \cosh \omega t \\[1mm] rq(t) = q(0) \cosh \omega t - \dfrac{12}{\omega} \eta(0) \sinh \omega t \end{array}\right\} \quad \text{if } b > \frac{15}{4}.$$

(iii) For the system to be Hamiltonian, the system of first-order partial differential equations

$$-\frac{\partial H}{\partial q} = 5p^2 q + aq^3 - bq, \qquad \frac{\partial H}{\partial p} = -8p^3 - cpq^2 + 6p$$

must admit a solution. The first equation yields

$$H(p,q) = -\frac{5}{2}p^2 q^2 - \frac{a}{4}q^4 + \frac{b}{2}q^2 + f(p),$$

and substituting in the second we find that we must set $c = 5$; the Hamiltonian is

$$H(p,q) = -\frac{5}{2}p^2 q^2 - \frac{a}{4}q^4 + \frac{b}{2}q^2 - 2p^4 + 3p^2 + \text{constant}.$$

(iv) Setting $a = 0$ the previous result guarantees that the system is Hamiltonian if and only if $c = 5$, with Hamiltonian

$$H(p,q) = -\frac{5}{2}p^2 q^2 + \frac{b}{2}q^2 - 2p^4 + 3p^2 + \text{constant}$$

We then set
$$\psi(p,q,\alpha) = 4p^2 + 5q^2 + \alpha, \quad \varphi(p,q,\beta) = 2p^2 + \beta.$$
A necessary and sufficient condition for their invariance is that
$$\frac{d\psi}{dt} = \frac{\partial\psi}{\partial p}\dot{p} + \frac{\partial\psi}{\partial q}\dot{q} = -\frac{\partial\psi}{\partial p}\frac{\partial H}{\partial q} + \frac{\partial\psi}{\partial q}\frac{\partial H}{\partial p} \equiv 0; \quad \psi(p,q,\alpha) = 0,$$
and similarly for $\varphi$. We therefore find
$$\frac{d\psi}{dt} = 8p[5p^2 q - bq] + 10q[6p - 8p^3 - 5pq^2]$$
$$= 10pq\left[6 - 5q^2 - 4p^2 - \frac{4}{5}b\right].$$

Together with $\psi(p,q,\alpha) = 0$, this forces $\alpha = 4/5(b-6)$. In an analogous way we find $\beta = -\frac{2}{5}b$.
(v) Setting $b = 5/2$ the Hamiltonian can be written
$$H(p,q) = -(4p^2 + 5q^2 - 4)(2p^2 - 1)\frac{1}{4} + 1,$$
the equilibrium positions are
$$p = q = 0, \quad \text{stable},$$
$$p = \pm\sqrt{\frac{3}{4}}, \quad q = 0, \quad \text{stable},$$
$$p = \pm\frac{1}{\sqrt{2}}, \quad q = \pm\sqrt{\frac{2}{5}}, \quad \text{unstable}.$$
The phase portrait is shown in Fig. 8.4.

*Problem 4*
Consider the system of differential equations
$$\dot{x} = (a - by)x(1 - x),$$
$$\dot{y} = -(c - dx)y(1 - y),$$
with $x > 0$, $y > 0$ and $a, b, c, d$ real positive constants.
(i) Introduce new variables $p, q$ though the substitution $x = e^q/(1+e^q)$, $y = e^p/(1+e^p)$ and write the corresponding system.
(ii) Prove that the resulting system is Hamiltonian and compute the corresponding Hamiltonian.
(iii) Let $a < b$ and $c < d$. Show that the system has a unique equilibrium position; linearise the equations and solve them.

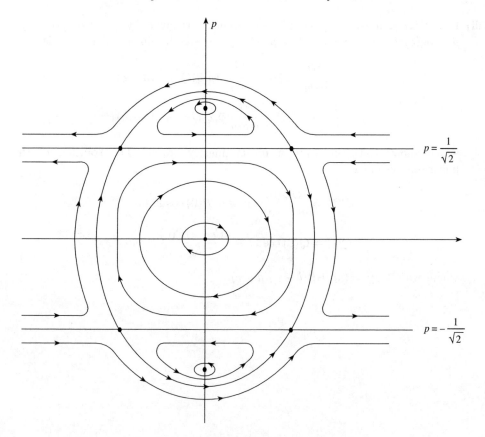

**Fig. 8.4**

*Solution*
(i) Differentiating with respect to time, we obtain

$$\dot{x} = \dot{q}\frac{e^q}{(1+e^q)^2}, \quad \dot{y} = \dot{p}\frac{e^p}{(1+e^p)^2}.$$

Replacing in the given system yields

$$\dot{p} = -(1+e^q)^{-1}[c + e^q(c-d)],$$
$$\dot{q} = (1+e^p)^{-1}[a + e^p(a-b)],$$

which is the system of Hamilton's equations associated with the Hamiltonian.
(ii) $H(p,q) = ap + cq - b\log(1+e^p) - d\log(1+e^q) + \text{constant}.$

(iii) If $a < b, c < d$ the only equilibrium solution is given by $\bar{q} = \log(c/(d-c))$, $\bar{p} = \log(a/(b-a))$. Setting $P = p - \bar{p}$, $Q = q - \bar{q}$ the linearised system is

$$\dot{P} = -\frac{\partial^2 H}{\partial p \partial q}(\bar{p}, \bar{q})P - \frac{\partial^2 H}{\partial q^2}(\bar{p}, \bar{q})Q = \frac{c}{d}(d-c)Q,$$

$$\dot{Q} = \frac{\partial^2 H}{\partial p^2}(\bar{p}, \bar{q})P + \frac{\partial^2 H}{\partial q \partial p}(\bar{p}, \bar{q})Q = -\frac{a}{b}(b-a)P,$$

which shows that the equilibrium is linearly stable. The solution of the linearised system is

$$P(t) = \frac{b\omega Q(0)}{a(b-a)} \sin \omega t + P(0) \cos \omega t,$$

$$Q(t) = Q(0) \cos \omega t - \frac{a(b-a)P(0)}{b\omega} \sin \omega t,$$

where we set $\omega^2 = (ac/bd)(b-a)(d-c)$.

# 9 ANALYTICAL MECHANICS: VARIATIONAL PRINCIPLES

## 9.1 Introduction to the variational problems of mechanics

Variational problems in mechanics are characterised by the following basic idea. For a given solution of Hamilton's equations (8.24), called the *natural motion*, we consider a family $\mathcal{F}$ of perturbed trajectories in the phase space, subject to some characterising limitations, and on it we define a *functional* $\varphi : \mathcal{F} \to \mathbf{R}$. The typical statement of a variational principle is that the functional $\varphi$ takes its minimum value in $\mathcal{F}$ corresponding to the natural motion, and conversely, that if an element of $\mathcal{F}$ has this property, then it is necessarily a solution of Hamilton's equations. The latter fact justifies the use of the term *principle*, in the sense that it is possible to assume such a variational property as an axiom of mechanics. Indeed, one can directly derive from it the correct equations of motion.

We start with a very simple example. Let $P$ be a point not subject to any force, and moving along a fixed (frictionless) line. Clearly the natural motion will be uniform. Suppose it has velocity $v_0$ and that for $t = t_0$ its coordinate on the line is equal to $x_0$. The natural motion is then represented by the function

$$x^*(t) = x_0 + v_0(t - t_0), \tag{9.1}$$

and a subsequent instant $t_1$ the function $x(t)$ reaches the value

$$x_1 = x_0 + v_0(t_1 - t_0).$$

We now fix the attention on the time interval $[t_0, t_1]$ and we define the following family $\mathcal{F}$ of perturbed motions:

$$x(t) = x^*(t) + \eta(t), \quad t_0 \leq t \leq t_1, \tag{9.2}$$

subject to the conditions (Fig. 9.1)

$$x(t_0) = x_0, \quad x(t_1) = x_1, \tag{9.3}$$

or

$$\eta(t_0) = \eta(t_1) = 0, \tag{9.4}$$

where the perturbation $\eta(t)$ is of class $\mathcal{C}^2[t_1, t_2]$.

We define the functional

$$\varphi(\eta) = \int_{t_0}^{t_1} \dot{x}^2(t)\, dt. \tag{9.5}$$

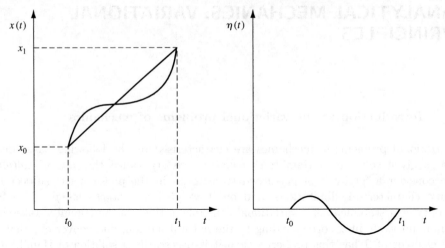

Fig. 9.1

Up to a proportionality factor, this functional represents the mean kinetic energy in the considered time interval. We compute the *variation* of the functional, i.e. the difference between its value on a generic perturbed motion and on the natural motion:

$$\delta\varphi = \varphi(\eta) - \varphi(0) = \int_{t_0}^{t_1} (2v_0 \dot{\eta}(t) + \dot{\eta}^2(t)) \, dt. \tag{9.6}$$

Due to (9.4) we find

$$\delta\varphi = \int_{t_0}^{t_1} \dot{\eta}^2(t) \, dt, \tag{9.7}$$

and we conclude that $\delta\varphi > 0$ on all the elements of $\mathcal{F}$. Hence $\varphi$ takes its minimum, relative to $\mathcal{F}$, in correspondence to the natural motion, and moreover $\delta\varphi = 0 \Leftrightarrow \eta = 0$, and hence this minimum property characterises the natural motion.

## 9.2 The Euler equations for stationary functionals

We now consider the problem from a general perspective. Let $F: \mathbf{R}^{2\ell+1} \to \mathbf{R}$ be a $\mathcal{C}^2$ function and let

$$\mathfrak{Q} = \{\mathbf{q} : \mathbf{R} \to \mathbf{R}^\ell | \mathbf{q} \in \mathcal{C}^2[t_0, t_1], \ \mathbf{q}(t_0) = \mathbf{q}_0, \ \mathbf{q}(t_1) = \mathbf{q}_1\}, \tag{9.8}$$

## 9.2  Analytical mechanics: variational principles

where $\mathbf{q}_0$, $\mathbf{q}_1$ are prescribed vectors in $\mathbf{R}^\ell$ and $[t_0, t_1]$ is a given time interval. We introduce the functional $\varphi : \mathfrak{Q} \to \mathbf{R}$:

$$\varphi(\mathbf{q}) = \int_{t_0}^{t_1} F(\mathbf{q}(t), \dot{\mathbf{q}}(t), t)\, \mathrm{d}t, \tag{9.9}$$

(note that we are using the Lagrangian formalism) and define what it means for $\varphi$ to be *stationary* on an element of $\mathbf{q}^* \in \mathfrak{Q}$. The difficulty lies in the fact that $\mathfrak{Q}$ is not a finite-dimensional space. We can simplify this concept by considering 'directions' in $\mathfrak{Q}$ along which to study the behaviour of $\varphi$, as follows.

For a given $\mathbf{q}^* \in \mathfrak{Q}$, consider the set of perturbations

$$Z = \{ \boldsymbol{\eta} : \mathbf{R} \to \mathbf{R}^\ell \,|\, \boldsymbol{\eta} \in \mathcal{C}^2[t_0, t_1],\ \boldsymbol{\eta}(t_0) = \boldsymbol{\eta}(t_1) = 0 \},$$

and for a fixed $\boldsymbol{\eta} \in Z$, consider the subset $\mathfrak{Q}_{\boldsymbol{\eta}} \subset Z$ defined by the vectors $\mathbf{q}(t)$ with components

$$q_k(t) = q_k^*(t) + \alpha_k \eta_k(t), \quad k = 1, \ldots, \ell, \tag{9.10}$$

where the vector $\boldsymbol{\alpha}$ varies in $\mathbf{R}^\ell$. The restriction of $\varphi$ to $\mathfrak{Q}_{\boldsymbol{\eta}}$ is now a function of the $\ell$ real variables $\alpha_1, \ldots, \alpha_\ell$, which we denote by $\psi(\boldsymbol{\alpha}; \boldsymbol{\eta})$.

At this point it is easy to give a precise definition.

DEFINITION 9.1  *We say that $\varphi(q)$ is stationary in $\mathfrak{Q}$ for $\mathbf{q} = \mathbf{q}^*$ if its restriction $\psi(\boldsymbol{\alpha}; \boldsymbol{\eta}) = \varphi|_{\mathfrak{Q}_{\boldsymbol{\eta}}}$ is stationary for $\boldsymbol{\alpha} = 0$, $\forall\, \boldsymbol{\eta} \in Z$.* ∎

Hence $\mathbf{q}^*$ is a stationary point for $\varphi$ if and only if

$$\nabla_{\boldsymbol{\alpha}} \psi(\boldsymbol{\alpha}, \boldsymbol{\eta})\big|_{\boldsymbol{\alpha}=0} = 0, \quad \forall\, \boldsymbol{\eta} \in Z. \tag{9.11}$$

We can now prove the following.

THEOREM 9.1  *A necessary and sufficient condition for the functional $\varphi(\mathbf{q})$ to be stationary in $\mathfrak{Q}$ for $\mathbf{q} = \mathbf{q}^*$ is that the components $q_k^*(t)$ of $\mathbf{q}^*$ are solutions of the system of differential equations*

$$\frac{\mathrm{d}}{\mathrm{d}t} \frac{\partial F}{\partial \dot{q}_k} - \frac{\partial F}{\partial q_k} = 0, \quad k = 1, \ldots, \ell \tag{9.12}$$

*(called Euler's equations).*

*Proof*
Substitute equations (9.10) into (9.9), differentiate with respect to $\alpha_k$ under the integral sign, and set $\boldsymbol{\alpha} = 0$. This yields

$$\frac{\partial}{\partial \alpha_k} \psi(0; \boldsymbol{\eta}) = \int_{t_0}^{t_1} \left( \frac{\partial F}{\partial q_k} \eta_k + \frac{\partial F}{\partial \dot{q}_k} \dot{\eta}_k \right)_{\mathbf{q} = \mathbf{q}^*} \mathrm{d}t.$$

Integrating the second term by parts, recalling that $\eta_k(t_0) = \eta_k(t_1) = 0$, we find

$$\frac{\partial}{\partial \alpha_k} \psi(0; \boldsymbol{\eta}) = \int_{t_0}^{t_1} \left( \frac{\partial F}{\partial q_k} - \frac{d}{dt} \frac{\partial F}{\partial \dot{q}_k} \right)_{\mathbf{q}=\mathbf{q}^*} \eta_k(t) \, dt. \qquad (9.13)$$

Thus the Euler equations (9.12) are a sufficient condition for the functional to be stationary.

To prove that they are also necessary, we start from the assumption that $\mathbf{q}^*$ is a stationary point, i.e. that

$$\int_{t_0}^{t_1} \left( \frac{\partial F}{\partial q_k} - \frac{d}{dt} \frac{\partial F}{\partial \dot{q}_k} \right)_{\mathbf{q}=\mathbf{q}^*} \eta_k(t) \, dt = 0, \quad \forall \boldsymbol{\eta} \in Z, \ k = 1, \ldots, \ell. \qquad (9.14)$$

We then use the following two facts:

(a) the expression in parentheses under the integral sign is a continuous function, which we henceforth denote by $\Phi_k(t)$;
(b) the functions $\eta_k(t)$ are arbitrary functions in $Z$.

If for some $\bar{t} \in (t_0, t_1)$ we have $\Phi_k(\bar{t}) \neq 0$ for at least one value of $k$, by continuity it would follow that $\Phi_k(t)$ does not change sign in an interval $(t', t'') \ni \bar{t}$. We could then choose $\eta_k(t)$ not changing its sign and with compact non-empty support in $(t', t'')$, and conclude that $\int_{t_0}^{t_1} \Phi_k(t) \eta_k(t) \, dt \neq 0$, against our assumption. It follows that $\Phi_k(t) \equiv 0$, $k = 1, \ldots, \ell$, and equations (9.12) are verified. ∎

*Remark 9.1*
In the next section we will return to the question of the formal analogy between the Euler equations (9.12) and the Lagrange equations (4.75). Here we only recall that a solvability condition for the system (9.12) is that the Hessian matrix $\|\partial^2 F/\partial \dot{q}_h \partial \dot{q}_k\|$ has non-zero determinant. In this case the function $F$ admits a Legendre transform. ∎

*Remark 9.2*
It is easy to find first integrals of equations (9.12) in the following cases:

(a) for a given $k$, $F$ does not depend on $q_k$; the integral is $\partial F/\partial \dot{q}_k = $ constant;
(b) for a given $k$, $F$ does not depend on $\dot{q}_k$; the integral is $\partial F/\partial q_k = $ constant; this is however a degenerate case, as the solvability condition just mentioned does not hold;
(c) $F$ does not depend on $t$; the conserved quantity is then

$$G = \mathbf{p} \cdot \dot{\mathbf{q}} - F,$$

## 9.2 Analytical mechanics: variational principles

with $p_k = \partial F/\partial \dot{q}_k$, i.e. the Legendre transform of $F$ with respect to $\dot{q}_k$ (Remark 9.8). We leave as an exercise the verification that $G$, evaluated along the solutions of system (9.12), has zero time derivative. ∎

*Example 9.1*
For the functional considered in Section 9.1 we have that $F(q, \dot{q}, t) = \dot{q}^2$, and hence the Euler equation is simply $\ddot{q} = 0$, coinciding with the equation of motion. ∎

*Example 9.2*
We show that the line segment between two points is the shortest path between the two points considered in the Euclidean metric.

For the case of the plane, we can reduce this to the problem of seeking the stationary points of the functional

$$\varphi(f) = \int_{x_0}^{x_1} \sqrt{1 + f'^2(x)}\, dx \tag{9.15}$$

in the class of functions $f \in C^2$ such that $f(x_0) = f(x_1) = 0$. We must write the Euler equation for $\varphi$, taking into account that $F(f(x), f'(x), x) = \sqrt{1 + f'^2(x)}$. Since $\partial F/\partial f = 0$, this can be reduced to $\partial F/\partial f' =$ constant. In addition, given that $\partial^2 F/\partial f'^2 \neq 0$, the former is equivalent to $f' =$ constant, yielding $f' = 0$ and finally $f = 0$. ∎

*Example 9.3: the brachistochrone*
Let $(P, m)$ be a point particle constrained to lie on a frictionless regular curve in a vertical plane, with endpoints $A$, $B$, and with $B$ at a lower height than $A$. We want to determine among all curves connecting the points $A$ and $B$, the one that minimises the travelling time of the particle $P$, moving under the action of its weight with initial conditions $P(0) = A$, $\mathbf{v}(0) = 0$.

Choose the coordinates in the plane of the motion as shown in Fig. 9.2, and let $x_3 = -f(x_1)$ be the equation of the curve we seek to determine. The conservation of energy implies that $v = (2gf)^{1/2}$. On the other hand, $v = \dot{s} = (1 + f'^2(x_1))^{1/2}\, dx_1/dt$. It follows that the travelling time is given by the expression

$$\theta(f) = \int_0^\alpha \left[\frac{1 + f'^2(x_1)}{2gf(x_1)}\right]^{1/2} dx_1. \tag{9.16}$$

We can then write the Euler equation for $F(f, f') = [(1 + f'^2)/f]^{1/2}$. Recall (Remark 9.2) that when $\partial F/\partial x = 0$ the Legendre transform is constant. This yields the first integral

$$f' \frac{\partial F}{\partial f'} - F = \text{constant}, \tag{9.17}$$

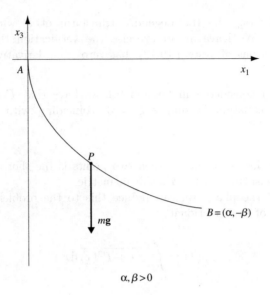

Fig. 9.2

which in our case can be written

$$f(1+f'^2) = c^2, \tag{9.18}$$

and hence $((c^2/f) - 1)^{-1/2} df = dx_1$. Using the substitution $f = c^2 \sin^2 \varphi/2$ this yields

$$x_1 = k(\varphi - \sin \varphi), \quad f = k(1 - \cos \varphi), \tag{9.19}$$

where the positive constant $k$ must be determined by imposing the condition that the point passes through $(\alpha, -\beta)$. It follows that the brachistochrone curve is an arc of a cycloid. ■

An excellent illustration of the use of the Euler equations is given by the problem of determining the 'shortest path' connecting two points on a Riemannian manifold $M$. If $(u_1, \ldots, u_l)$ are local coordinates, consider the curve $t \to \mathbf{u}(t) = (u_1(t), \ldots, u_l(t))$ on the manifold, with $0 < t < 1$.

If $g_{ij}(u_1, \ldots, u_l)$ is the Riemannian metric given on $M$, the length $\ell$ of the curve $\mathbf{u}$ is given by the functional

$$\ell(\mathbf{u}) = \int_0^1 \sqrt{\sum_{k,j=1}^l g_{kj}(u_1(t), \ldots, u_l(t)) \dot{u}_k(t) \dot{u}_j(t)} \, dt. \tag{9.20}$$

By Theorem 9.1 a curve makes the length $\ell(\mathbf{u})$ stationary if and only if it is a solution of the Euler equations (9.12), where $F = \sqrt{\sum_{k,j=1}^l g_{kj} \dot{u}_k \dot{u}_j}$, i.e. a solution of

$$\frac{d}{dt}\frac{g_{ij}\dot{u}_j}{\sqrt{g_{kj}\dot{u}_k\dot{u}_j}} = \frac{(\partial g_{kj}/\partial u_i)\dot{u}_k\dot{u}_j}{2\sqrt{g_{kj}\dot{u}_k\dot{u}_j}}, \qquad (9.21)$$

where $i = 1, \ldots, l$ and we have adopted the convention of summation over repeated indices. Take the natural parameter $s = s(t) = \int_0^t \sqrt{g_{kj}\dot{u}_k\dot{u}_j}\, dt'$ on the curve. We then find

$$\frac{d}{dt} = \sqrt{g_{kj}\dot{u}_k\dot{u}_j}\, \frac{d}{ds},$$

and substituting this into (9.21) gives

$$\frac{\partial g_{ij}}{\partial u_k}\frac{du_k}{ds}\frac{du_j}{ds} + g_{ij}\frac{d^2 u_j}{ds^2} = \frac{1}{2}\frac{\partial g_{kj}}{\partial u_i}\frac{du_k}{ds}\frac{du_j}{ds},$$

i.e.

$$g_{ij}\frac{d^2 u_j}{ds^2} + \frac{1}{2}\left(\frac{\partial g_{ij}}{\partial u_k} + \frac{\partial g_{ik}}{\partial u_j} - \frac{\partial g_{kj}}{\partial u_i}\right)\frac{du_k}{ds}\frac{du_j}{ds} = 0.$$

Multiplying both terms by $g^{ni}$ (the elements of the inverse matrix of $(g_{ij})$) and summing over $i$, we find the geodesic equation (1.68). We have proved the following.

THEOREM 9.2 *Among all paths connecting two fixed points on a Riemannian manifold, the geodesics keep the length functional (9.20) stationary.* ∎

Remark 9.3
In reality we have not proved that the geodesics make the length functional attain its *minimum*. Indeed, this is generally false. Consider as an example a pair of points not diametrically opposed on a sphere; they are connected by two arcs of a maximal circle. Both these arcs make the length functional stationary, but only one of them realises the minimal distance. If the two points are diametrically opposed there are infinitely many geodesics of equal length connecting them.

A more interesting case is the case of a flat bidimensional torus (Fig. 9.3). In this case, it is easy to verify that, given any two points, there exist infinitely many geodesics connecting them. Only one of them minimises the length. However, it can be proved (see, for example, Dubrovin et al. 1991a, chapter 5) that for any given pair of points on a Riemannian manifold, sufficiently close to each other, the shortest path connecting them is unique and it is given by a geodesic. ∎

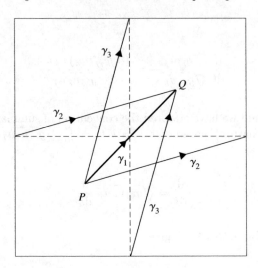

**Fig. 9.3** The curves $\gamma_1$, $\gamma_2$ and $\gamma_3$ are geodesics connecting the two points $P$ and $Q$ on the torus. The minimal length is attained by $\gamma_1$. Note that the three geodesics are not homotopic.

We now consider the problem of seeking the stationary points of a functional in the presence of a constraint. We illustrate this for the case of the functional

$$\varphi(\mathbf{q}) = \int_{t_0}^{t_1} F(\mathbf{q}(t), \dot{\mathbf{q}}(t), t) \, \mathrm{d}t, \tag{9.22}$$

with constraint

$$\int_{t_0}^{t_1} \Phi(\mathbf{q}(t), \dot{\mathbf{q}}(t), t) \, \mathrm{d}t = c, \tag{9.23}$$

where $\Phi$ is a function with the same properties as $F$, and $c$ is a constant.

The problem can be solved by writing Euler's equations for the function $G = F + \lambda \Phi$. These, together with (9.23), yield the unknown $\mathbf{q}(t)$ as well as the Lagrange multiplier $\lambda$.

*Example 9.4*
Among all plane closed curves of fixed perimeter, find the curve which encloses the maximal area (*isoperimetric* problem).

We seek the curve in the parametric form

$$x_1 = f(t), \quad x_2 = g(t), \quad 0 < t < 2\pi.$$

## 9.2  Analytical mechanics: variational principles

The constraint is

$$\int_0^{2\pi} (f'^2 + g'^2)^{1/2} \, dt = \ell, \tag{9.24}$$

and the functional to be studied is $\oint x_2 \, dx_1$, i.e.

$$\int_0^{2\pi} g(t) f'(t) \, dt. \tag{9.25}$$

Hence the function under investigation is $G = gf' + \lambda(f'^2 + g'^2)^{1/2}$, and we can write the equations

$$\frac{\partial G}{\partial f'} = g + \lambda f'(f'^2 + g'^2)^{-1/2} = c_1, \qquad \frac{d}{dt}\frac{\partial G}{\partial g'} = \frac{\partial G}{\partial g} = f'.$$

By integrating the second equation we obtain

$$g - c_1 = -\lambda f'(f'^2 + g'^2)^{-1/2}, \qquad f - c_2 = \lambda g'(f'^2 + g'^2)^{-1/2}.$$

Squaring and summing, we finally obtain the equation of a circle:

$$(g - c_1)^2 + (f - c_2)^2 = \lambda^2. \tag{9.26}$$

The constants $c_1$ and $c_2$ do not play an essential role, as their variation only produces a translation. The multiplier $\lambda$ is determined by (9.24), as we must have that $2\pi\lambda = \ell$.

To complete the solution of the problem, we must prove that by perturbing the circle, which we assume to be of radius 1, keeping the same length of the resulting curve, the enclosed area is reduced.

We write the equation of the circle of radius 1 in the form $\mathbf{x} = \mathbf{x}_0(\varphi)$ and the equations of the perturbed curves in the form

$$\mathbf{x}(\varphi) = \mathbf{x}_0(\varphi)(1 + f(\varphi)), \quad 0 < \varphi < 2\pi, \tag{9.27}$$

where $f$ is $2\pi$-periodic and such that $\|f\| \ll 1$, $\|f'\| \ll 1$, and $\|f\| = \max_{0 \leq \varphi \leq 2\pi} |f(\varphi)|$. We hence consider only perturbed curves which enclose a 'starred' domain (i.e. a domain which contains all the radii ensuing from one of its points, suitably chosen). Indeed, it is easy to realise that if a domain is not a star domain, we can modify the curve preserving its length, but enlarging the enclosed area, so we exclude such domains from our analysis. Since $\mathbf{x}_0'(\varphi)$ is the unit vector tangent to the circle, the length of the curve (9.27) is given by

$$\ell(f) = \int_0^{2\pi} [(1 + f)^2 + f'^2]^{1/2} \, d\varphi.$$

We can impose the condition $\ell(f) = 2\pi$ up to order higher than $\|f\|^2$ and $\|f'\|^2$, by writing $[(1+f)^2 + f'^2]^{1/2} \simeq 1 + f + \frac{1}{2}f'^2$ (using $\sqrt{1+x} \simeq 1 + \frac{1}{2}x - \frac{1}{8}x^2$). We find

$$\int_0^{2\pi} \left(f + \frac{1}{2}f'^2\right) d\varphi = 0. \tag{9.28}$$

Since $f$ is periodic, we can consider its Fourier expansion (Appendix 7):

$$f(\varphi) = a_0 + \sum_{n=1}^{\infty} (a_n \cos n\varphi + b_n \sin \varphi), \tag{9.29}$$

$$f'(\varphi) = \sum_{n=1}^{\infty} (-n a_n \sin n\varphi + n b_n \cos n\varphi), \tag{9.30}$$

and hence

$$\int_0^{2\pi} f \, d\varphi = 2\pi a_0, \quad \int_0^{2\pi} f'^2 \, d\varphi = \pi \sum_{n=1}^{\infty} n^2(a_n^2 + b_n^2),$$

and equation (9.28) implies the relation

$$a_0 = -\frac{1}{4} \sum_{n=1}^{\infty} n^2(a_n^2 + b_n^2). \tag{9.31}$$

We now compute the area enclosed by the perturbed curve:

$$A(f) = \frac{1}{2} \int_0^{2\pi} (1+f)^2 \, d\varphi = \pi + \int_0^{2\pi} \left(f + \frac{1}{2}f^2\right) d\varphi. \tag{9.32}$$

Again using equation (9.28) we can estimate the variation

$$A(f) - \pi = \frac{1}{2} \int_0^{2\pi} (f^2 - f'^2) \, d\varphi. \tag{9.33}$$

We now have

$$\int_0^{2\pi} f^2 \, d\varphi = \pi \left[2 a_0^2 + \sum_{n=1}^{\infty} (a_n^2 + b_n^2)\right],$$

where $a_0^2$ can be ignored. Indeed, it follows from equation (9.31), or from equation (9.28), that the average of $f$ (i.e. $a_0$) is of the same order as $\|f'\|^2$, and hence that $a_0^2$ is of the order of the error, and can be ignored. We can re-interpret equation (9.33) as

$$A(f) - \pi \simeq -\pi \sum_{n=1}^{\infty} (n^2 - 1)(a_n^2 + b_n^2). \tag{9.34}$$

We can conclude that the perturbation causes a decrease in the area, as soon as one of the Fourier coefficients with index $n > 1$ is different from zero. We must still examine the case that $f = a_0 + a_1 \cos \varphi + b_1 \sin \varphi$, when the perturbation $A(f) - \pi$ is of order greater than two. To evaluate $\ell(f)$ we must consider the expansion $\sqrt{1+x} \simeq 1 + \frac{1}{2}x - \frac{1}{8}x^2 + \frac{1}{3!}\left(\frac{3}{8}x^3\right) - \left(\frac{1}{4!}\left(\frac{5}{16}x^4\right)\right)$, which yields (keeping terms up to fourth order)

$$[(1+f)^2 + f'^2]^{1/2} \simeq 1 + f + \frac{1}{2}f'^2 - \frac{1}{2}ff'^2 + \frac{7}{24}f^4 - \frac{1}{4}f^2 f'^2.$$

To compute the integral of this expression we must take into account the fact that $a_0$ is of the same order as $a_1^2$ and $b_1^2$, and hence many terms can be discarded. Furthermore it is easy to compute

$$\frac{1}{2\pi} \int_0^{2\pi} \cos^4 \varphi \, d\varphi = \frac{3}{8}, \quad \frac{1}{2\pi} \int_0^{2\pi} \sin^2 \varphi \cos^2 \varphi \, d\varphi = \frac{1}{8}.$$

Finally, the condition $\ell(f) = 2\pi$ can be written as

$$-a_0 \left[1 - \frac{1}{4}(a_1^2 + b_1^2)\right] = \frac{1}{4}(a_1^2 + b_1^2) + \frac{5}{64}(a_1^2 + b_1^2)^2,$$

or, to the same order of approximation,

$$-a_0 = \frac{1}{4}(a_1^2 + b_1^2) + \frac{9}{64}(a_1^2 + b_1^2)^2.$$

The area relative variation is then given by

$$\frac{A(f) - \pi}{\pi} = 2a_0 + a_0^2 + \frac{1}{2}(a_1^2 + b_1^2),$$

yielding, to fourth order,

$$\frac{A(f) - \pi}{\pi} = -\frac{7}{32}(a_1^2 + b_1^2)^2 < 0. \quad \blacksquare$$

## 9.3 Hamilton's variational principle: Lagrangian form

The analogy between the Euler equations (9.12) and the Lagrange equations (4.75) is evident. The latter ones are also called the *Euler–Lagrange equations*, and we can regard them as the equations characterising when the functional

$$A(\mathbf{q}) = \int_{t_0}^{t_1} L(\mathbf{q}, \dot{\mathbf{q}}, t)\, \mathrm{d}t, \qquad (9.35)$$

called the *Hamiltonian action*, is stationary in the class $\mathcal{Q}$ of perturbed motions, defined by (9.8). These motions are called motions with *synchronous perturbations* (to stress the fact that we are not altering the time-scale).

We can summarise what we have just discussed in the following statement.

THEOREM 9.3 (Hamilton principle) *The natural motion is characterised by the property that the Hamiltonian action is stationary in the class of synchronous perturbations which preserve the configurations of the system at the initial and final time.* ∎

*Remark* 9.4
Recall that $L = T - V$. We can then state that *the natural motion makes the time average of the difference between the kinetic and potential energy stationary.* ∎

We stress the fact that the Hamilton principle is a characterisation of the motion, in the sense that it can be regarded not only as a consequence of the Lagrange equations, but it can also be assumed as the fundamental postulate of mechanics, from which the Lagrange equations can be immediately deduced.

We now examine a series of examples in which we find that the Hamiltonian action is not only stationary, but even minimised along the natural motion.

*Example* 9.5: *motion of a free point particle in the absence of forces*
It is sufficient to recall the problem solved in Section 9.1, removing the condition that the point is constrained on a line. ∎

*Example* 9.6: *motion of a point mass under gravity*
Choose the reference frame in such a way that the natural motion has equations

$$x_1^*(t) = v_{01}t, \quad x_2^*(t) = 0, \quad x_3^*(t) = v_{03}t + \frac{1}{2}gt^2 \qquad (9.36)$$

(axis $x_3$ oriented along the descending vertical, initial velocity $\mathbf{v}_0 = (v_{01}, 0, v_{03})$). The synchronous perturbations are defined by

$$x_1(t) = v_{01}t + \eta_1(t), \quad x_2(t) = \eta_2(t), \quad x_3(t) = v_{03}t + \frac{1}{2}gt^2 + \eta_3(t), \qquad (9.37)$$

with $\eta_i \in C^2[0, \theta]$, $\eta_i(0) = \eta_i(\theta) = 0$, $i = 1, 2, 3$, for a given $\theta > 0$.

The variation of the Hamiltonian action

$$A = \int_0^\theta \left(\frac{1}{2} mv^2 + mgx_3\right) dt \qquad (9.38)$$

can be easily computed:

$$\delta A = \frac{1}{2} m \int_0^\theta \sum_i \dot\eta_i^2 \, dt, \qquad (9.39)$$

and is positive for every non-zero perturbation (we can add that it is of order 2 with respect to the perturbation, in the sense that for fixed $\eta_1$, $\eta_2$, $\eta_3$ multiplied by $\alpha$, it follows that $\delta A = \mathcal{O}(\alpha^2)$). ■

*Example 9.7: the harmonic oscillator*
Choose the reference frame in such a way that we can write

$$x_1^*(t) = a \sin \omega t, \quad x_2^*(t) = x_3^*(t) = 0 \qquad (9.40)$$

and consider the variations

$$x_1(t) = a \sin \omega t + \eta_1(t), \quad x_2(t) = \eta_2(t), \quad x_3(t) = \eta_3(t), \qquad (9.41)$$

with $\eta_i$ chosen as in the previous problem.
Since $L = \frac{1}{2} mv^2 - \frac{1}{2} m\omega^2 \sum_i x_i^2$, we find

$$\delta A = \frac{1}{2} m \int_0^\theta \sum_i (\dot\eta_i^2 - \omega^2 \eta_i^2) \, dt + m \int_0^\theta (\dot x_1^* \dot\eta_1 - \omega^2 x_1^* \eta_1) \, dt. \qquad (9.42)$$

One integration by parts in the second integral yields $-\int_0^\theta (\ddot x_1^* + \omega^2 x_1^*)\eta_1 \, dt = 0$, and hence we can conclude that $\delta A$ evaluated along the natural motion is of order 2 with respect to the perturbation (implying that $A$ is stationary). Finally we note that an integral of the type $\int_0^\theta (\dot\eta^2 - \omega^2 \eta^2) \, dt$ can be estimated using

$$|\eta(t)| = \left|\int_0^t \dot\eta(\tau) \, d\tau\right| \le \sqrt{t} \left[\int_0^t \dot\eta^2(\tau) \, d\tau\right]^{1/2}.$$

We thus find

$$\int_0^\theta (\dot\eta^2 - \omega^2\eta^2)\,dt \geq \left(1 - \frac{1}{2}\omega^2\theta^2\right)\left[\int_0^\theta \dot\eta^2(t)\,dt\right]^{1/2}.$$

Hence we can conclude that $\delta A > 0$ if $\theta < \sqrt{2}/\omega$, i.e. the Hamiltonian action has a minimum when computed along the natural motion, provided that we impose a restriction on the length of the time interval over which it is computed.

As an exercise, compute $\delta A$ for $\eta_1 = \alpha \sin^2 \pi t/\theta$, $\eta_2 = \eta_3 = 0$ and note that $\delta A \gtreqless 0$ for $\theta \lesseqgtr 2\pi/3\omega$. ∎

## 9.4 Hamilton's variational principle: Hamiltonian form

As we have explicitly observed, so far we have based our analysis of variational principles on the Lagrangian formalism. This is convenient for the ease with which one can then define the synchronous perturbations in the space of Lagrangian coordinates.

Passing to the Hamiltonian formalism, we need only to express the action in the canonical variables $(\mathbf{p}, \mathbf{q})$:

$$A(\mathbf{p}, \mathbf{q}) = \int_{t_0}^{t_1} [\mathbf{p} \cdot \dot{\mathbf{q}} - H(\mathbf{p}, \mathbf{q}, t)]\,dt, \qquad (9.43)$$

where $\dot{\mathbf{q}} = \dot{\mathbf{q}}(\mathbf{p}, \mathbf{q}, t)$, but we must define the variations in the phase space. This is naturally done by perturbing $q_k^*(t)$ and in turn $p_k^*(t)$, in such a way that the formal relation $p_k = \partial L/\partial \dot{q}_k$ is preserved.

However it is more convenient to introduce independent variations for $q_k$ and $p_k$:

$$\begin{aligned} q_k(t) &= q_k^*(t) + \eta_k(t), \quad k = 1, \ldots, \ell, \\ p_k(t) &= p_k^*(t) + \zeta_k(t), \quad k = 1, \ldots, \ell, \end{aligned} \qquad (9.44)$$

with $\eta_k, \zeta_k \in C^2[t_0, t_1]$, $\eta_k(t_0) = \eta_k(t_1) = 0$, $k = 1, \ldots, \ell$, where $q_k^*(t)$ and $p_k^*(t)$ denote the solutions of the Hamilton equations. In this way we can define perturbed curves in phase space (Fig. 9.4), which in general are not admissible trajectories for the system (consider e.g. the trivial case $\ell = 1$ with $p = m\dot{q}$ and take $\zeta(t) \neq m\dot{\eta}(t)$).

The class of trajectories (9.44) is therefore larger than the class of synchronous perturbations. If we prove that the functional $A$ is stationary along the solutions of the Hamilton equations with respect to this more extended class of perturbations, it follows that it is also stationary within the more restricted class of synchronous perturbations. This is the idea in the proof of the following theorem.

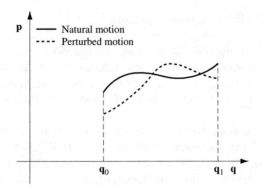

**Fig. 9.4** Sketch of the synchronous perturbations.

THEOREM 9.4  *A necessary and sufficient condition that $\delta A = 0$, to first order in the class of perturbations (9.44), is that $q_k^*$, $p_k^*$ are solutions of the Hamilton equations (8.24).*

*Proof*
We immediately find that

$$\delta A = \int_{t_0}^{t_1} (\boldsymbol{\zeta} \cdot \dot{\mathbf{q}}^* + \mathbf{p}^* \cdot \dot{\boldsymbol{\eta}} + \boldsymbol{\zeta} \cdot \dot{\boldsymbol{\eta}} - \delta H) \, dt, \tag{9.45}$$

and since we are only interested in the first-order variation we can neglect the term $\boldsymbol{\zeta} \cdot \dot{\boldsymbol{\eta}}$ and write

$$\delta H \simeq (\nabla_{\mathbf{q}} H)^* \cdot \boldsymbol{\eta} + (\nabla_{\mathbf{p}} H)^* \cdot \boldsymbol{\zeta}, \tag{9.46}$$

where $(\cdot)^*$ denotes the values taken along the natural motion.

Integrating by parts the term containing $\dot{\boldsymbol{\eta}}$, we arrive at the expression

$$\delta A = \int_{t_0}^{t_1} [\boldsymbol{\zeta} \cdot (\dot{\mathbf{q}} - \nabla_{\mathbf{p}} H)^* - \boldsymbol{\eta} \cdot (\dot{\mathbf{p}} + \nabla_{\mathbf{q}} H)^*] \, dt. \tag{9.47}$$

From this we can deduce the equivalence

$$\delta A \simeq 0 \quad \Leftrightarrow \quad (\dot{\mathbf{q}} - \nabla_{\mathbf{p}} H)^* = (\dot{\mathbf{p}} + \nabla_{\mathbf{q}} H)^* = 0, \tag{9.48}$$

if we proceed as in the final part of the proof of Theorem 9.1. ■

*Remark* 9.5
The previous theorem is still valid if we restrict to the class of perturbations (9.44), imposing the limitations $\zeta_k(t_0) = \zeta_k(t_1) = 0$ (what needs to be modified in the proof is not essential). ■

## 9.5 Principle of the stationary action

Besides Hamilton's principle, there exist several other variational principles.[1] We will discuss only one more, the most famous, which has special interest for its geometric implications. This principle is called the *principle of stationary action*, or *Maupertuis' principle*. It is valid for systems with a time-independent Hamiltonian.

It is convenient to refer to the space $(\mathbf{p}, \mathbf{q}, t)$ and to parametrise not only $\mathbf{p}$ and $\mathbf{q}$ but also $t$, thus considering the curves in $\mathbf{R}^{2\ell+1}$ given by the equations $\mathbf{p} = \mathbf{p}(u)$, $\mathbf{q} = \mathbf{q}(u)$, $t = t(u)$. To obtain a parametrisation of the natural motion, it is enough to consider a function $t = \bar{t}(u)$, $u_0 \leq u \leq u_1$, in $\mathcal{C}^2[u_0, u_1]$ with $\bar{t}'(u) \neq 0$ in $[u_0, u_1]$, and consequently define the functions $\bar{q}_k(u) = q_k^*(\bar{t}(u))$, $\bar{p}_k = p_k^*(\bar{t}(u))$.

We find the curve of equations

$$\mathbf{p} = \overline{\mathbf{p}}(u), \quad \mathbf{q} = \overline{\mathbf{q}}(u), \quad t = \bar{t}(u), \tag{9.49}$$

along which we introduce the perturbations

$$\mathbf{p} = \overline{\mathbf{p}}(u) + \boldsymbol{\zeta}(u), \quad \mathbf{q} = \overline{\mathbf{q}}(u) + \boldsymbol{\eta}(u), \quad t = \bar{t}(u) + \tau(u), \tag{9.50}$$

in such a way that the new functions $\mathbf{p}(u)$, $\mathbf{q}(u)$, $t(u)$ are also $\mathcal{C}^2$, and satisfy $\boldsymbol{\eta}(u_0) = \boldsymbol{\eta}(u_1) = 0$.

The relevant novelty is that perturbations now include a variation in the temporal scale. Therefore they are called *asynchronous perturbations* (Fig. 9.5).

We note that, in analogy with the case discussed in the previous section, only a subset of the curves (9.50) is associated with possible motions. However, every stationarity result obtained in this wider class applies to the subfamily of possible motions.

In what follows we select a particular subclass of perturbations, satisfying

$$H(\mathbf{p}(u), \mathbf{q}(u)) = H(\overline{\mathbf{p}}(u), \overline{\mathbf{q}}(u)). \tag{9.51}$$

The asynchronous perturbations subject to the condition (9.51) are called *isoenergetic*. It will soon be clear that the need to introduce asynchronous variations is due to the constraint imposed on the energy.

The functional we want to study is

$$\widehat{A} = \int_{t(u_0)}^{t(u_1)} \mathbf{p} \cdot \dot{\mathbf{q}} \, dt, \tag{9.52}$$

which is also called the *action*. The integrand must be understood in the Hamiltonian formalism. This functional is obviously linked to the time average of the kinetic energy (see (4.34)).

---

[1] See, for example, Levi-Civita and Amaldi (1927), Whittaker (1936) and Agostinelli and Pignedoli (1989).

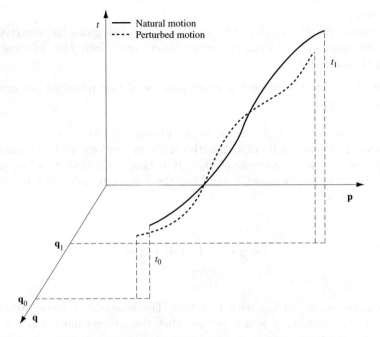

**Fig. 9.5** Sketch of the asynchronous variations.

THEOREM 9.5 (Stationary action of Maupertuis' principle) *If the Hamiltonian does not depend explicitly on time, the functional (9.52) along the natural motion is stationary with respect to the class of isoenergetic asynchronous perturbations.*

*Proof*
We make the change of variables $t = t(u)$ in (9.52) and we write

$$\widehat{A}(\zeta, \boldsymbol{\eta}, \tau) = \int_{u_0}^{u_1} (\overline{\mathbf{p}} + \boldsymbol{\zeta}(u)) \cdot \frac{\mathrm{d}}{\mathrm{d}u} (\overline{\mathbf{q}} + \boldsymbol{\eta}(u)) \, \mathrm{d}u. \tag{9.53}$$

Neglecting the higher-order term $\boldsymbol{\zeta} \cdot \dfrac{\mathrm{d}}{\mathrm{d}u} \boldsymbol{\eta}$ and integrating by parts where necessary, we arrive at the expression for the variation $\delta \widehat{A}$:

$$\delta \widehat{A} \simeq \int_{u_0}^{u_1} \left( \boldsymbol{\zeta} \boldsymbol{n} \cdot \frac{\mathrm{d}}{\mathrm{d}u} \overline{\mathbf{q}} - \boldsymbol{\eta} \cdot \frac{\mathrm{d}}{\mathrm{d}u} \overline{\mathbf{p}} \right) \mathrm{d}u. \tag{9.54}$$

Using the Hamilton equations we find immediately that, to first order,

$$\boldsymbol{\zeta} \cdot \frac{\mathrm{d}}{\mathrm{d}u} \overline{\mathbf{q}} - \boldsymbol{\eta} \cdot \frac{\mathrm{d}}{\mathrm{d}u} \overline{\mathbf{p}} \simeq \delta H \frac{\mathrm{d}t}{\mathrm{d}u}. \tag{9.55}$$

Since by hypothesis $\delta H = 0$, the proof is finished. ∎

*Remark* 9.6
The functional (9.52) contains only information on the geometric-material structure of the system. The dynamic information comes into play because of the isoenergetic constraint. ∎

Before examining the geometric consequences of this principle, we consider a few simple examples.

*Example* 9.8: *motion of a free point in the absence of forces*
Isoenergetic motions are in this case the uniform motions with the same magnitude of velocity as the natural motion. It is clear then that it is impossible to perturb the trajectory without perturbing the temporal scale. The functional $\widehat{A}$ can be written as

$$\widehat{A} = \frac{1}{2} mv \int_{t(u_0)}^{t(u_1)} v\, dt = \frac{1}{2} mvs, \tag{9.56}$$

where $s$ is the length of the path travelled. The geometric interpretation of the principle of the stationary action is then that the natural motion is the motion which makes the length of the travelled path stationary with respect to any other path with the same velocity connecting the same start and end points. ∎

*Example* 9.9 *motion of a point on an equipotential surface*
Let $P$ be constrained on the surface $V =$ constant. In this case as well the isoenergetic motions are the uniform motions with the same magnitude of velocity as the natural motion, and the conclusion is the same as in the previous case: the trajectory is a geodesic of the surface (Proposition 2.2). ∎

In fact, the minimality property of the path stressed by the previous examples holds in general, as long as the manifold of configurations is endowed with the appropriate metric. We shall develop this concept in the next section.

## 9.6 The Jacobi metric

Consider a holonomic system with fixed, smooth constraints, not subject to any force directly applied to it. For such a system the kinetic energy is constant:

$$T = \text{constant}. \tag{9.57}$$

Recall that $T = \frac{1}{2} \sum_{h,k=1}^{\ell} a_{hk} \dot{q}_h \dot{q}_k$ is a positive definite quadratic form. We interpret $(a_{hk})$ as the *metric tensor* of the manifold of configurations of the system, as we did in Theorem 4.3:

$$(\mathrm{d}s)^2 = \sum_{h,k=1}^{\ell} a_{hk}\,\mathrm{d}q_h\,\mathrm{d}q_k. \tag{9.58}$$

With this metric, the velocity of the representative point in the space is such that

$$|\dot{\mathbf{q}}|^2 = 2T, \tag{9.59}$$

and hence $|\dot{\mathbf{q}}| = $ constant.

We can apply the principle of stationary action and conclude that the natural motion is an extremal for the length of the path travelled on the Riemannian manifold $\mathcal{V}$ endowed with the metric (9.58) (this is indeed the meaning of the action). On the other hand, note that in this case Theorem 9.2 refers to the extremal for the functional $\int_{t_0}^{t_1} \sqrt{2T}\,\mathrm{d}t$, the Euler equations coinciding, for $T =$ constant, with the Lagrange equations for $L = T$ (see Problem 9.5).

Turning to the general case, when there is a conservative force field with potential energy $V(\mathbf{q})$, it is still possible to obtain an analogous result, as long as the chosen metric incorporates the function $V(\mathbf{q})$ in a suitable way, at the same time preserving the information encoded in $T$. More precisely, we write

$$(\mathrm{d}s)^2 = 2(E-V) \sum_{h,\,k=1}^{\ell} a_{hk}\,\mathrm{d}q_h\,\mathrm{d}q_k, \tag{9.60}$$

so that

$$|\dot{\mathbf{q}}| = 2T \tag{9.61}$$

and consequently the action coincides directly with the length of the arc of the trajectory travelled by the point in the space of configurations. The metric (9.60) is called the *Jacobi metric* and it is defined in the regions $\mathcal{V}_E = \{V(\mathbf{q}) \leq E\}$.

For a fixed energy $E$, the manifold $\mathcal{V}_E$ with the metric (9.60) defines a Riemannian manifold with boundary ($\partial \mathcal{V}_E = \{V(\mathbf{q}) = E\}$), and from the Maupertuis principle it follows that the natural motion travels along the geodesics of this manifold. Note that the metric (9.60) is singular on $\partial \mathcal{V}_E$.

The following examples make reference to systems with two degrees of freedom. In the space $(q_1, q_2)$ we look for the trajectories of the form $q_1 = f(q_2)$. Hence the functional we have to study is of the form

$$l(f) = \int_a^b [E - V(f, q_2)]^{1/2} \left[ a_{11} \left( \frac{\mathrm{d}f}{\mathrm{d}q_2} \right)^2 + 2a_{12} \frac{\mathrm{d}f}{\mathrm{d}q_2} + a_{22} \right]^{1/2} \mathrm{d}q_2,$$

with $a_{ij}$ functions of $f$ and $q_2$. After elimination of time, we can neglect the perturbations of this variable.

*Example 9.10*
Verify that the trajectory of a central motion with potential energy $V(r)$ is a geodesic with respect to the Jacobi metric (9.60).
We seek the extremals of the functional

$$\int_{\varphi_1}^{\varphi_2} [(\rho^2 + \rho'^2)(E - V(\rho))]^{1/2} \, d\varphi, \tag{9.62}$$

from which we obtain the Euler equation

$$\frac{\rho\rho'' - 2\rho'^2 - \rho^2}{\rho^2 + \rho'^2} + \rho \frac{V'}{2(E-V)} = 0 \tag{9.63}$$

for the trajectory $r = \rho(\varphi)$. We want to check that by integrating (9.63) we find the trajectory of the motion under consideration. Indeed, it is enough to note that, setting $u = 1/\rho$ and $\hat{V}(u) = V(\rho)$, equation (9.63) becomes

$$\frac{u'' + u}{u^2 + u'^2} = \frac{\hat{V}'(u)}{2(\hat{V} - E)}, \tag{9.64}$$

admitting the first integral

$$u'^2 + u^2 = k(E - \hat{V}), \tag{9.65}$$

with $k$ constant (it suffices to multiply the two sides of (9.64) by $2u'$). This is simply the energy integral, after identifying $k$ with $2m/L_z^2$ (see (5.27)).
Indeed, by substituting (9.65) into (9.64) we find

$$u'' + u = -\frac{m}{L_z^2} \hat{V}'(u), \tag{9.66}$$

and hence we obtain (5.26).
This proves that the solution of the variational problem, i.e. the integration of equation (9.64), is equivalent to the classical solution of the dynamical problem. ∎

*Example 9.11: motion of a point mass in a one-dimensional field*
Choose the $x_3$-axis in the direction of the field, and let $V(x_3)$ be the potential energy, with $V(0) = 0$. We study the motion in the $(x_1, x_3)$ plane, with the initial conditions $x_1(0) = x_3(0) = 0$, $\dot{x}_1(0) = v_{01}$, $\dot{x}_3(0) = v_{03}$.

Since $x_1(t) = v_{01} t$ and $\dot{x}_3 = \pm \left( -2/m \, (V(x_3) + v_{03}^2) \right)^{1/2}$, we find by separation of variables in the latter

$$ t = \pm \int_0^{x_3} \left( -\frac{2}{m} V(\zeta) + v_{03}^2 \right)^{-1/2} d\zeta. \tag{9.67} $$

The equation of the trajectory is then

$$ x_1 = \pm v_{01} \int_0^{x_3} \left( -\frac{2}{m} V(\zeta) + v_{03}^2 \right)^{-1/2} d\zeta, \tag{9.68} $$

where the sign must be changed in correspondence to the possible singularities of the integrand.

We now solve the problem using the variational technique considered in this section, by finding the extremal of the length of the arc of the trajectory with respect to the metric (9.60). Hence we find the function $x_1 = \xi(x_3)$ which is an extremal of the functional

$$ \ell(\xi) = \int_0^{\zeta_0} [-2V(x_3) + m(v_{01}^2 + v_{03}^2)]^{1/2} [m(1 + \xi'^2(x_3))]^{1/2} \, dx_3, \tag{9.69} $$

where $\xi(0) = 0$ and $\xi(\zeta_0)$ must coincide with the value taken by (9.68) for $x_3 = \zeta_0$.

Since the integrand in (9.69) does not depend on $\xi(x_3)$, the Euler equation admits the first integral

$$ [-2V(x_3) + m(v_{01}^2 + v_{03}^2)]^{1/2} \frac{\xi'}{(1 + \xi'^2)^{1/2}} = c, $$

i.e.

$$ \xi' = c[-2V(x_3) + m(v_{01}^2 + v_{03}^2) - c^2]^{-1/2}. \tag{9.70} $$

To find the desired value of $\xi(\zeta_0)$ we take $c^2 = mv_{01}^2$, and hence $c = \pm\sqrt{m}v_{01}$, and the integral of equation (9.70) then coincides with (9.68). ∎

*Example 9.12*
Consider a rod $AB$ constrained in the $(x, z)$ plane, and with the point $A$ sliding on the $x_1$-axis, without any directly applied force. The rod has length $2\ell$ and mass $m$. We seek the equation of the trajectory in the Lagrangian coordinate space. Choose the coordinates $\varphi$, $\xi = x/\ell$ as in Fig. 9.6.

We compute the kinetic energy

$$ T = \frac{1}{2} m\ell^2 \left( \dot{\xi}^2 - 2\dot{\xi}\dot{\varphi} \sin\varphi + \frac{4}{3}\dot{\varphi}^2 \right) \tag{9.71} $$

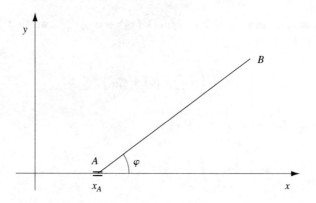

Fig. 9.6

and deduce that the correct metric to use in solving the problem is

$$ds^2 = d\xi^2 - 2 \sin \varphi \, d\xi \, d\varphi + \frac{4}{3} d\varphi^2.$$

By seeking the trajectories in the form $\xi = \xi(\varphi)$, we must find the extremal of the functional

$$\ell(\xi) = \int_{\varphi_1}^{\varphi_2} \left( \xi'^2 - 2 \sin \varphi \xi' + \frac{4}{3} \right)^{1/2} d\varphi. \tag{9.72}$$

The Euler equation admits the first integral

$$\frac{\partial}{\partial \xi'} \left( \xi'^2 - 2 \sin \varphi \xi' + \frac{4}{3} \right)^{1/2} = k,$$

from which

$$\xi' = \sin \varphi \pm (1 - k^2)^{-1/2} \left[ \frac{4}{3} - \sin^2 \varphi \right]^{1/2}, \quad \text{with } |k| < 1. \tag{9.73}$$

This equation leads to an elliptic integral.

Classically we can solve the problem by writing the conservation of the kinetic energy:

$$\dot{\xi}^2 - 2\dot{\xi}\dot{\varphi} \sin \varphi + \frac{4}{3} \dot{\varphi}^2 = c_0 \tag{9.74}$$

and of the first component of the momentum:

$$\dot{\xi} - \dot{\varphi} \sin \varphi = c_1. \tag{9.75}$$

Note that we must have $c_0 \geq c_1^2$.

Solving the system (9.74), (9.75) with respect to $\dot{\xi}$, $\dot{\varphi}$ and eliminating time, we find (9.73), with $k = c_1/\sqrt{c_0}$. ∎

## 9.7 Problems

1. Determine all plane curves of equation $y = y(x)$ passing through the origin, and through the point with coordinates $(\pi/2, 1)$ that are extremals for the functional $\int_0^{\pi/2} [(y')^2 - y^2] \, dx$.

2. Consider all plane curves $y = y(x)$ passing through two fixed points $A$ and $B$. Show that the area of the surface of rotation obtained by rotating the graph of the curve around the $x$-axis is given by $S(y, y') = 2\pi \int_{x_A}^{x_B} y\sqrt{1 + (y')^2} \, dx$. Show that the area is stationary if $y = a\cosh(x - b)/a$, and hence a catenary. The constants of integration $a$ and $b$ are determined by requiring that the curve passes through the points $A$ and $B$ (depending on the relative position of the points, the solution may or may not be unique, or may not exist. Discuss all possible cases). (*Hint*: since the integrand is independent of $x$, use Remark 9.2(c).)

3. Determine the extremals of the following functionals, for fixed values of $q(t_0)$, $q(t_1)$:

   (a) $\int_{t_0}^{t_1} (t\dot{q} + \dot{q}^2) \, dt$;

   (b) $\int_{t_0}^{t_1} (q^2 + \dot{q}^2 - 2qt) \, dt$;

   (c) $\int_{t_0}^{t_1} (\dot{q} + t^2 \dot{q}^2) \, dt$.

4. Let $h = \mathbf{R}^2 \to \mathbf{R}$ be of the form $h(x, x') = (x - x')^2/2 + u(x)$, where $u : \mathbf{R} \to \mathbf{R}$ is of class $C^\infty$. Given any finite sequence of real numbers $(x_j, \ldots, x_k)$, $j < k$, set $h(x_j, \ldots, x_k) = \sum_{i=j}^{k-1} h(x_i, x_{i+1})$. A $(k-j)$-tuple is *minimal* for $h$ if

$$h(x_j, \ldots, x_k) \leq h(x'_j, \ldots, x'_k)$$

for every $(x'_j, \ldots, x'_k)$ such that $x'_j = x_j$, $x'_k = x_k$. Prove that if $(x_j, \ldots, x_k)$ is minimal then it satisfies the following condition to be stationary:

$$x_{i+1} - 2x_i + x_{i-1} = u'(x_i) \quad \text{for all } j < i < k.$$

Determine all the stationary $n$-tuples for the case $u \equiv $ constant and $u = ax$. Which are the minimal ones?

5. Deduce from the principle of stationary action that the orbit of a point particle in a central force field of potential energy $V(r) = \frac{1}{2}kr^2$, $k > 0$, is an ellipse with centre at the origin.

6. Within special relativity theory, the Lagrangian of a point particle with mass $m$ (at rest) and in the absence of forces, is $L(\dot{q}) = -mc^2\sqrt{1 - (|\dot{\mathbf{q}}|^2/c^2)}$, where $c$ is the speed of light.

Determine the kinetic momentum $\mathbf{p}$, the Hamiltonian $H$ and show than, for any speed much smaller than the speed of light, $H \sim mc^2 + |\mathbf{p}|^2/2m$. Write the Euler equations for the relativistic action functional $S = \int L(\dot{\mathbf{q}})\,dt$ and show that, in the case $|\dot{\mathbf{q}}| \ll c$, they reduce to the equation $m\ddot{\mathbf{q}} = 0$.

## 9.8 Additional remarks and bibliographical notes

Although the first studies in the calculus of variations date back to the seventeenth century, it was only in 1736 that Euler proved Theorem 9.1, which is still today considered the fundamental result in this field. The proof we gave is due to Lagrange, who obtained it in 1756. He also introduced the principle of stationary action, without, of course, using the Hamiltonian formalism. The problem of the additional conditions that a solution of the Euler equation must satisfy, in order to effectively provide the maximum or minimum of the functional, was successfully considered by Legendre, who gave an additional necessary condition. It was only in 1837 that Jacobi succeeded in strengthening the condition of Legendre to make it a sufficient condition, when he discovered the existence of *conjugate points* at which the minimisation problem loses uniqueness. A detailed discussion of these beautiful results goes beyond the scope of this work; for an elementary and pleasant introduction, we recommend Fox (1987, chapters 1–3). We simply note that if

(1) the Euler equation is satisfied,
(2) the interval of integration $[t_1, t_2]$ is sufficiently small,
(3) the $\ell \times \ell$ matrix $\partial^2 F / \partial \dot{q}_i \partial \dot{q}_j$ is either positive definite or negative definite,

then there is a maximum or a minimum according to whether $\partial^2 F / \partial \dot{q}_i \partial \dot{q}_j$ is negative or positive definite. This is enough to show that the Hamiltonian action (9.35) is minimised along the natural motion (for sufficiently short time intervals).

## 9.9 Additional solved problems

*Problem 1*
Let $S$ be a surface given as the graph $z = f(x, y)$, with $f \in C^2(\mathbf{R}^2)$. Find the periodic function $\rho(\varphi) > 0$ such that the area of the portion of $S$ projected in the region bounded by the curve $r = \rho(\varphi)$ on the plane $(x, y)$, with prescribed length $\ell$, is an extremal.

*Solution*
The length of the curve is

$$\ell = \int_0^{2\pi} \sqrt{\rho^2(\varphi) + \rho'^2(\varphi)}\,d\varphi.$$

## 9.9  Analytical mechanics: variational principles

The area we are considering is

$$A(\rho) = \int_0^{2\pi} \left\{ \int_0^{\rho(\varphi)} \sqrt{EG - F^2}\, dr \right\} d\varphi,$$

where $E, F, G$ are obtained by the following parametrisation of $S$:

$$x = r\cos\varphi, \quad y = r\sin\varphi, \quad z = f(r\cos\varphi, r\sin\varphi).$$

One verifies that

$$EG - F^2 = r^2(1 + (\nabla f)^2),$$

and hence the functional for which we seek an extremal is

$$\int_0^{2\pi} \left\{ \int_0^{\rho(\varphi)} r(1 + (\nabla f)^2)^{1/2}\, dr - \lambda\sqrt{\rho^2 + \rho'^2} \right\} d\varphi.$$

Note that if $S$ is a plane, then $(\nabla f)^2$ is constant and $A(\rho)$ is simply the area enclosed by the curve $r = \rho(\varphi)$, and hence $\frac{1}{2}\int_0^{2\pi} \rho^2(\varphi)\, d\varphi$, divided by $1/(1+(\nabla f)^2)^{1/2}$. In this case the problem is equivalent to that of Example 9.4. More generally, we must solve the Euler equation

$$\frac{d}{d\varphi}\frac{\lambda\rho'}{\sqrt{\rho^2 + \rho'^2}} + \rho(1 + (\nabla f)^2)^{1/2}_{r=\rho(\varphi)} - \frac{\lambda\rho}{\sqrt{\rho^2 + \rho'^2}} = 0$$

requiring that the solution is periodic, with period $2\pi$, and determining $\lambda$ using the constraint on length. As an example, in the case of a surface of rotation $z = f(r)$ we find $(\nabla f)^2 = f'^2$ and the above equation becomes

$$\lambda\rho\,\frac{\rho\rho'' - 2\rho'^2 - \rho^2}{(\rho^2 + \rho'^2)^{3/2}} + \rho(1 + f'^2(\rho))^{1/2} = 0,$$

which admits the solution $\rho = R_0$, with $R_0 = \ell/2\pi$, as long as $R_0$ is inside the domain of definition of $f'$; indeed, it is enough to choose $\lambda = (1+f'^2(R_0))^{1/2}/R_0$. In the case of the sphere $f(r) = \sqrt{R^2 - r}$, to find the circular solution we need $R_0 < R$.

**Problem 2**
On the surface of rotation

$$x = \rho(z)\cos\varphi, \quad y = \rho(z)\sin\varphi, \quad z = z$$

consider the family of elicoidal curves defined by

$$\varphi = f(z), \quad f(z_1) = 0, \quad f(z_2) = 2\pi,$$

with $f \in C^2$ increasing, and the interval $(z_1, z_2)$ inside the domain of definition of $\rho(z)$. Find $f$ so that the length of the curve is stationary.

*Solution*

The length of the curve is given by the functional

$$\ell(f) = \int_{z_1}^{z_2} [(1 + \rho'^2) + \rho^2 f'^2]^{1/2} \, dz.$$

Since the integrand does not depend on $f$, we can immediately write a first integral of the Euler equation:

$$\rho^2 f' = c[(1 + \rho'^2) + \rho^2 f'^2]^{1/2}, \quad c > 0, \tag{9.76}$$

from which we find $f'$:

$$f' = \frac{c}{\rho} \left[ \frac{1 + \rho'^2}{\rho^2 - c^2} \right]^{1/2}.$$

The constant $c$ has to be determined by imposing, if possible,

$$c \int_{z_1}^{z_2} \frac{1}{\rho} \left[ \frac{1 + \rho'^2}{\rho^2 - c^2} \right]^{1/2} dz = 2\pi. \tag{9.77}$$

If the surface has a vertex in $z = z^*$ (i.e. $\rho(z^*) = 0$) and $z^*$ lies in $[z_1, z_2]$, then equation (9.76) is incompatible with $f' > 0$, because it forces $c = 0$. In this case the problem does not admit a solution. Even when this is not the case, equation (9.77) is not always solvable. Take for example the cone $\rho = z\alpha$, with opening angle $\alpha$, for which

$$f' = \frac{c}{z\alpha} \left[ \frac{1 + \alpha^2}{z^2 \alpha^2 - c^2} \right]^{1/2} = \frac{\gamma (1 + \alpha^2)^{1/2}}{z \alpha} \left( \frac{1}{z^2 \sin^2 \alpha - \gamma^2} \right)^{1/2},$$

with $\gamma = c(\sin \alpha)/\alpha$. Setting

$$z \sin \alpha = \zeta \quad \text{and} \quad \zeta = -\gamma \frac{1 + t^2}{1 - t^2},$$

the integral can easily be computed and yields

$$f(z) = \frac{2}{\alpha} (1 + \alpha^2)^{1/2} \left[ \arctan \left( \frac{z_1 \sin \alpha + \gamma}{z_1 \sin \alpha - \gamma} \right)^{1/2} - \arctan \left( \frac{z \sin \alpha + \gamma}{z \sin \alpha - \gamma} \right)^{1/2} \right]. \tag{9.78}$$

It is easy to see that the difference of the arctangents is positive for $z > z_1$, and it is always less than $\pi/4$. Hence the condition $f(z_2) = 2\pi$ cannot be satisfied if $\alpha^2 > 1/15$. If on the other hand there exists a solution, it is unique, as the right-hand side of (9.78) is an increasing function of $\gamma$ in the interval $(0, z_1 \sin \alpha)$. It is not at all intuitive that there may be cases when no solution exists. In fact a solution always exists, but when it is singular it cannot be found as a solution of the Euler equation. Indeed, in the class considered, a path that follows meridians ($f' = 0$) and parallels ($f'$ singular) may be the most economical (in terms of length).

*Problem 3*
A point particle travels along the smooth curve $z = -f(x) \leq 0$, $0 < x < a$, in the vertical plane $(x, z)$. The curve joins the two points $(x = 0, z = 0)$, $(x = a, z = 0)$. Initially the point is at $(0, 0)$ with zero velocity, and its motion is periodic. Find the curve, in the family of curves of class $C^2(a, b)$ with fixed length $\ell > a$, for which the period is an extremal.

*Solution*
Without the constraint on length, the curve would be a cycloid. The period is twice the travelling time along the curve between the points $(0,0)$ and $(a,0)$. Conservation of energy implies that $\dot{s} = \sqrt{2gf(x)}$. Since $ds = \sqrt{1 + f'^2}\, dx$ we have $dt = ((1 + f'^2)/2gf)^{1/2}\, dx$. The period is then

$$T(f) = 2 \int_0^a \left( \frac{1 + f'^2}{2gf} \right)^{1/2} dx.$$

We need to find the extremals of the functional $T(f) - \lambda \int_0^a (1 + f'^2)^{1/2}\, dx$, where $\lambda$ is the Lagrange multiplier. The corresponding Euler equation has first integral given by the Legendre transform of $F(f, f') = (1 + f'^2)^{1/2} \left( \sqrt{2/gf} - \lambda \right)$, i.e.

$$f' \frac{\partial F}{\partial f'} - F = \left( \lambda - \sqrt{\frac{2}{gf}} \right) \frac{1}{\sqrt{1 + f'^2}}.$$

Introducing the integration constant $c$ we can write

$$f'^2 = c^2 \left( \lambda - \sqrt{\frac{2}{gf}} \right)^2 - 1,$$

and separate variables:

$$\int_0^{f(x)} df \left[ c^2 \left( \lambda - \sqrt{\frac{2}{gf}} \right)^2 - 1 \right]^{-1/2} = x, \quad 0 < x < \frac{a}{2}.$$

(the branch $a/2 < x < a$ is symmetric, and equal values of $f$ correspond to opposite values of $f'$). Set $\lambda - \sqrt{2/gf} = \zeta$, i.e. $gf/2 = (\lambda - \zeta)^{-2}$ and $df = -4/g(\lambda - \zeta)^{-3} d\zeta$. This puts the indefinite integral in the form

$$-\frac{4}{g} \int (\lambda - \zeta)^{-3}(c^2\zeta^2 - 1)^{-1/2} d\zeta.$$

The transformation $(c^2\zeta^2 - 1)^{1/2} = (c\zeta - 1)t$, i.e.

$$\zeta = -\frac{1}{c}\frac{1+t^2}{1-t^2}, \quad d\zeta = -\frac{4}{c}\frac{t\, dt}{(1-t^2)^2}$$

carries the integral to a rational form

$$-\frac{8}{cg}\int \left[\lambda + \frac{1}{c}\frac{1+t^2}{1-t^2}\right]^{-3} \frac{dt}{1-t^2}.$$

For $f \downarrow 0$ we have $\zeta \to -\infty$ and $t \to 1$. The upper extremum can be deduced from $t = ((c\zeta + 1)/(c\zeta - 1))^{1/2}$ with $\zeta$ expressed through $f(x)$. We then obtain an implicit expression for $f(x)$, where the constant $c$ must be determined through the condition $f'(a/2) = 0$, i.e. $c^2 \left(\lambda - \sqrt{2/gf(a/2)}\right)^2 = 1$. As usual the multiplier $\lambda$ is found by imposing the length constraint.

*Problem 4*
Prove that if for a functional

$$\varphi(q) = \int_{t_0}^{t_1} F(q(t), \dot{q}(t), t)\, dt, \quad \text{with } q \in C^2([t_0, t_1], \mathbf{R}), \quad q(t_0) = q_0, \quad q(t_1) = q_1,$$

the Euler equation (9.12) becomes an identity, then $\varphi$ does not depend on the integration path $q$ but only on $(t_0, q_0)$ and $(t_1, q_1)$.

*Solution*
Writing explicitly equation (9.12) we find

$$\frac{\partial^2 F}{\partial \dot{q}^2}\ddot{q} + \frac{\partial^2 F}{\partial q \partial \dot{q}}\dot{q} + \frac{\partial^2 F}{\partial \dot{q}\partial t} - \frac{\partial F}{\partial q} = 0.$$

If this equation is an identity, i.e. it is satisfied by any $q$, necessarily the coefficient of $\ddot{q}$ must be identically zero, because there is no other way to eliminate $\ddot{q}$. Hence $F$ must be of the form $F = a(t,q)\dot{q} + b(t,q)$. Substituting in the equation, we find $\partial a/\partial t = \partial b/\partial q$, and hence the 1-form $F\, dt$ is exact: $F\, dt = a(t,q)\, dq + b(t,q)\, dt = df(t,q)$. From this it follows that

$$\int_{t_0}^{t_1} F(q, \dot{q}, t)\, dt = f(t_1, q_1) - f(t_0, q_0).$$

*Problem 5*
Consider the variational problem for the functional (9.9) in the function class (9.8). Find the necessary and sufficient condition for its solutions to also be solutions of the variational problem for the problem

$$\psi(\mathbf{q}) = \int_{t_0}^{t_1} G[F(\mathbf{q}, \dot{\mathbf{q}}, t)]\, \mathrm{d}t$$

with $G(F) \in C^2$, $G'' \neq 0$.

*Solution*
Setting $\mathcal{F}(\mathbf{q}, \dot{\mathbf{q}}, t) = G[F(\mathbf{q}, \dot{\mathbf{q}}, t)]$, we immediately find

$$\frac{\mathrm{d}}{\mathrm{d}t}\nabla_{\dot{\mathbf{q}}}\mathcal{F} - \nabla_{\mathbf{q}}\mathcal{F} = \nabla_{\dot{\mathbf{q}}} F \frac{\mathrm{d}^2 G}{\mathrm{d}F^2}\frac{\mathrm{d}F}{\mathrm{d}t}.$$

Hence the required condition is that the function $F(\mathbf{q}, \dot{\mathbf{q}}, t)$ is a first integral of the Euler equation for the functional (9.9).

# 10 ANALYTICAL MECHANICS: CANONICAL FORMALISM

## 10.1 Symplectic structure of the Hamiltonian phase space

Consider the real $2l \times 2l$ matrix

$$\mathcal{J} = \begin{pmatrix} 0 & -\mathbf{1} \\ \mathbf{1} & 0 \end{pmatrix} \tag{10.1}$$

(with **1** and **0** we henceforth denote the identity and the null matrix, with the obvious dimensions, e.g. $l \times l$ in (10.1)). Note that $\mathcal{J}$ is orthogonal and skew-symmetric, i.e.

$$\mathcal{J}^{-1} = \mathcal{J}^T = -\mathcal{J} \tag{10.2}$$

and that $\mathcal{J}^2 = -\mathbf{1}$. As observed in Chapter 8, setting $\mathbf{x} = (\mathbf{p}, \mathbf{q})$, the Hamilton equations can be written in the form

$$\dot{\mathbf{x}} = \mathcal{J} \nabla_{\mathbf{x}} H(\mathbf{x}, t). \tag{10.3}$$

*Example 10.1*
Let $S$ be a real symmetric constant $2l \times 2l$ matrix. A *linear Hamiltonian system with constant coefficients* is a system of $2l$ ordinary differential equations of the form (10.3), where

$$H(\mathbf{x}) = \frac{1}{2} \mathbf{x}^T S \mathbf{x}. \tag{10.4}$$

The Hamiltonian is then a quadratic form in $\mathbf{x}$ and (10.3) takes the form

$$\dot{\mathbf{x}} = \mathcal{J} S \mathbf{x}.$$

The solution of this system of differential equations with the initial condition $\mathbf{x}(0) = \mathbf{X}$ is given by

$$\mathbf{x}(t) = e^{tB} \mathbf{X}, \tag{10.5}$$

where we set

$$B = \mathcal{J} S.$$

The matrices with this structure deserve special attention. ∎

DEFINITION 10.1 A real $2l \times 2l$ matrix $B$ is called Hamiltonian (or infinitesimally symplectic) if

$$B^T \mathsf{J} + \mathsf{J} B = 0. \tag{10.6}$$

∎

THEOREM 10.1 *The following conditions are equivalent:*

*(1) the matrix $B$ is Hamiltonian;*
*(2) $B = \mathsf{J} S$, with $S$ a symmetric matrix;*
*(3) $\mathsf{J} B$ is a symmetric matrix.*

*In addition, if $B$ and $C$ are two Hamiltonian matrices, $B^T$, $\beta B$ (with $\beta \in \mathbf{R}$), $B \pm C$ and $[B, C] = BC - CB$ are Hamiltonian matrices.*

*Proof*
From the definition of a Hamiltonian matrix it follows that

$$\mathsf{J} B = -B^T \mathsf{J} = (\mathsf{J} B)^T,$$

and hence (1) and (3) are equivalent. The equivalence of (2) and (3) is immediate, as $S = -\mathsf{J} B$.

The first three statements of the second part of the theorem are obvious (for the first, note that $B^T = -S\mathsf{J} = \mathsf{J} S'$, with $S' = \mathsf{J} S \mathsf{J}$ symmetric). Setting $B = \mathsf{J} S$ and $C = \mathsf{J} R$ (with $S$ and $R$ symmetric matrices) we have

$$[B, C] = \mathsf{J}(S \mathsf{J} R - R \mathsf{J} S)$$

and

$$(S \mathsf{J} R - R \mathsf{J} S)^T = -R \mathsf{J} S + S \mathsf{J} R.$$

It follows that the matrix $[B, C]$ is Hamiltonian. ∎

*Remark* 10.1
Writing $B$ as a $2l \times 2l$ block matrix

$$B = \begin{pmatrix} a & b \\ c & d \end{pmatrix},$$

where $a$, $b$, $c$, $d$ are $l \times l$ matrices, (10.6) becomes

$$B^T \mathsf{J} + \mathsf{J} B = \begin{pmatrix} -c + c^T & -a^T - d \\ a + d^T & b - b^T \end{pmatrix},$$

and hence $B$ is Hamiltonian if and only if $b$ and $c$ are symmetric matrices and $a^T + d = 0$. If $l = 1$, $B$ is Hamiltonian if and only if it has null trace. ∎

*Remark* 10.2
From Theorem 10.1 it follows that the Hamiltonian matrices form a group (with

respect to matrix sum) called sp$(l, \mathbf{R})$. If we identify the vector space of real $2l \times 2l$ matrices with $\mathbf{R}^{4l^2}$, the Hamiltonian matrices form a linear subspace, of dimension $l(2l+1)$ (indeed, from what was previously discussed we may choose $l(l+1)/2$ elements of the matrices $b$ and $c$ and, for example, $l^2$ elements of the matrix $a$). In addition, since the *Lie product* (or *commutator*) [ , ] preserves the group of Hamiltonian matrices, sp$(l, \mathbf{R})$ has a *Lie algebra* structure (see Arnol'd 1978a). ∎

DEFINITION 10.2   A real $2l \times 2l$ matrix $A$ is called symplectic if
$$A^T \mathsf{J} A = \mathsf{J}. \tag{10.7}$$
∎

THEOREM 10.2   *Symplectic* $2l \times 2l$ *matrices form a group under matrix multiplication, denoted by* Sp$(l, \mathbf{R})$. *The transpose of a symplectic matrix is symplectic.*

*Proof*
Evidently the $2l \times 2l$ identity matrix is symplectic, and if $A$ satisfies (10.7) then it is necessarily non-singular, since from (10.7) it follows that
$$(\det(A))^2 = 1. \tag{10.8}$$

In addition, it can be easily seen that
$$A^{-1} = -\mathsf{J} A^T \mathsf{J}, \tag{10.9}$$

so that
$$(A^{-1})^T \mathsf{J} A^{-1} = (A^T)^{-1} \mathsf{J}(-\mathsf{J} A^T \mathsf{J}) = (A^T)^{-1} A^T \mathsf{J} = \mathsf{J},$$

i.e. $A^{-1}$ is symplectic. If $C$ is another symplectic matrix, we immediately have that
$$(AC)^T \mathsf{J} AC = C^T A^T \mathsf{J} AC = C^T \mathsf{J} C = \mathsf{J}.$$

In addition, $A^T = -\mathsf{J} A^{-1} \mathsf{J}$, from which it follows that
$$A \mathsf{J} A^T = A A^{-1} \mathsf{J} = \mathsf{J}.$$
∎

*Example* 10.2
The group of symplectic $2 \times 2$ matrices with real coefficients, Sp$(1, \mathbf{R})$, coincides with the group SL$(2, \mathbf{R})$ of matrices with determinant 1. Indeed, if
$$A = \begin{pmatrix} \alpha & \beta \\ \gamma & \delta \end{pmatrix},$$
the symplecticity condition becomes
$$A^T \mathsf{J} A = \begin{pmatrix} 0 & -\alpha\delta + \beta\gamma \\ -\beta\gamma + \alpha\delta & 0 \end{pmatrix} = \mathsf{J}.$$

Hence $A$ is symplectic if and only if $\det(A) = \alpha\delta - \beta\gamma = 1$. It follows that every symplectic $2 \times 2$ matrix defines a linear transformation preserving area and orientation. The orthogonal unit matrices (with determinant equal to 1) are a subgroup of $SL(2, \mathbf{R})$, and hence also of $Sp(1, \mathbf{R})$. ∎

*Remark* 10.3
Let $A$ be a symplectic $2l \times 2l$ matrix. We write it as an $l \times l$ block matrix:

$$A = \begin{pmatrix} a & b \\ c & d \end{pmatrix}. \tag{10.10}$$

The condition that the matrix is symplectic then becomes

$$A^T J A = \begin{pmatrix} -a^T c + c^T a & -a^T d + c^T b \\ -b^T c + d^T a & -b^T d + d^T b \end{pmatrix} = \begin{pmatrix} 0 & -1 \\ 1 & 0 \end{pmatrix}, \tag{10.11}$$

and hence $A$ is symplectic only if $a^T c$ and $b^T d$ are $l \times l$ symmetric matrices and $a^T d - c^T b = 1$. The symplecticity condition is therefore more restrictive in dimension $l > 1$ than in dimension $l = 1$, when it becomes simply $\det(A) = 1$. It is not difficult to prove (see Problem 1) that *symplectic matrices have determinant equal to 1 for every $l$* (we have already seen that $\det(A) = \pm 1$, see (10.8)). ∎

*Remark* 10.4
Symplectic matrices have a particularly simple inverse: from (10.9) and (10.10) it follows immediately that

$$A^{-1} = \begin{pmatrix} d^T & -b^T \\ -c^T & a^T \end{pmatrix}. \tag{10.12}$$

∎

*Remark* 10.5
If we identify the vector space of the $2l \times 2l$ matrices with $\mathbf{R}^{4l^2}$, the group $Sp(l, \mathbf{R})$ defines a regular submanifold of $\mathbf{R}^{4l^2}$ of dimension $l(2l+1)$ (this can be verified immediately in view of the conditions expressed in Remark 10.3; indeed, starting from the dimension of the ambient space, $4l^2$, we subtract $2(l(l-1))/2$, since the matrices $a^T c$ and $b^T d$ must be symmetric, and $l^2$ since $a^T d - c^T b = 1$.) ∎

PROPOSITION 10.1  *The tangent space to $Sp(l, \mathbf{R})$ at $\mathbf{1}$ is the space of Hamiltonian matrices:*

$$T_{\mathbf{1}} Sp(l, \mathbf{R}) = sp(l, \mathbf{R}). \tag{10.13}$$

## Proof

Let $A(t)$ be a curve in $\mathrm{Sp}(l, \mathbf{R})$ passing through **1** when $t = 0$, and hence such that

$$A(t)^T \mathsf{J} A(t) = \mathsf{J} \qquad (10.14)$$

for every $t$ and $A(0) = \mathbf{1}$.

By differentiating (10.14) with respect to $t$ we find

$$\dot{A}^T \mathsf{J} A + A^T \mathsf{J} \dot{A} = 0,$$

from which, setting $B = \dot{A}(0) \in T_{\mathbf{1}} \mathrm{Sp}(l, \mathbf{R})$

$$B^T \mathsf{J} + \mathsf{J} B = 0,$$

and hence $B \in \mathrm{sp}(l, \mathbf{R})$. ∎

Conversely, to every Hamiltonian matrix there corresponds a curve in $\mathrm{Sp}(l, \mathbf{R})$, as shown in the following.

**PROPOSITION 10.2** *Let $B$ be a Hamiltonian matrix. The matrix $A(t) = e^{tB}$ is symplectic for every $t \in \mathbf{R}$.*

## Proof

We must show that $A(t)$ satisfies (10.7) for every $t$, i.e.

$$(e^{tB})^T \mathsf{J} e^{tB} = \mathsf{J}.$$

It follows immediately from the definition

$$e^{tB} = \sum_{n=0}^{\infty} \frac{t^n}{n!} B^n$$

that $(e^{tB})^T = e^{tB^T}$, and $(e^{tB})^{-1} = e^{-tB}$.

Hence the condition for the matrix to be symplectic becomes

$$e^{tB^T} \mathsf{J} = \mathsf{J} e^{-tB}.$$

But

$$e^{tB^T} \mathsf{J} = \sum_{n=0}^{\infty} \frac{t^n}{n!} (B^T)^{n-1} B^T \mathsf{J} = \sum_{n=0}^{\infty} \frac{t^n}{n!} (B^T)^{n-1} (-\mathsf{J} B).$$

Iterating, we find

$$e^{tB^T} \mathsf{J} = \mathsf{J} \sum_{n=0}^{\infty} \frac{t^n}{n!} (-1)^n B^n = \mathsf{J} e^{-tB}. \qquad \blacksquare$$

**DEFINITION 10.3** *The symplectic product on a real vector space $V$ of dimension $2l$ is a skew-symmetric, non-degenerate bilinear form $\omega : V \times V \to \mathbf{R}$. The* space $V$ endowed with a symplectic product has a symplectic structure and $V$ is a symplectic space. ∎

We recall that a bilinear skew-symmetric form is *non-degenerate* if and only if $\omega(v_1, v_2) = 0$ for every $v_2 \in V$ implies $v_1 = 0$. We note also that only vector spaces of even dimension admit a symplectic structure. Indeed, all bilinear skew-symmetric forms are necessarily degenerate in a space of odd dimension.

Consider the canonical basis $\mathbf{e}_1, \ldots, \mathbf{e}_{2l}$ in $\mathbf{R}^{2l}$. The symplectic product $\omega$ has a matrix representation $W$ obtained by setting

$$W_{ij} = \omega(\mathbf{e}_i, \mathbf{e}_j).$$

Evidently the representative matrix $W$ is skew-symmetric and the non-degeneracy condition is equivalent to $\det(W) \neq 0$. Moreover, for every $\mathbf{x}, \mathbf{y} \in \mathbf{R}^{2l}$ We have

$$\omega(\mathbf{x}, \mathbf{y}) = \sum_{i,j=1}^{2l} W_{ij} x_i y_j = \mathbf{x}^T W \mathbf{y}. \tag{10.15}$$

By choosing the matrix $W = \mathfrak{J}$ we obtain the so-called *standard symplectic product* (henceforth simply referred to as symplectic product unless there is a possibility of confusion) and correspondingly the standard symplectic structure.

*Remark* 10.6

The standard symplectic product has an interesting geometric characterisation. Given two vectors $\mathbf{x}, \mathbf{y}$ we have

$$\mathbf{x}^T \mathfrak{J} \mathbf{y} = -x_1 y_{l+1} - \ldots - x_l y_{2l} + x_{l+1} y_1 + \ldots + x_{2l} y_l$$
$$= (x_{l+1} y_1 - x_1 y_{l+1}) + \ldots + (x_{2l} y_l - x_l y_{2l}),$$

corresponding to the sum of the (oriented) areas of the projection of the parallelogram with sides $\mathbf{x}, \mathbf{y}$ on the $l$ planes $(x_1, x_{l+1}), \ldots, (x_l, x_{2l})$. Hence, if $\mathbf{p}$ is the vector constructed with the first $l$ components of $\mathbf{x}$, and $\mathbf{q}$ is the one constructed with the remaining components, we have $\mathbf{x} = (\mathbf{p}, \mathbf{q})$, and analogously if $\mathbf{y} = (\mathbf{p}', \mathbf{q}')$, we have

$$\omega(\mathbf{x}, \mathbf{y}) = \mathbf{x}^T \mathfrak{J} \mathbf{y} = (q_1 p_1' - p_1 q_1') + \ldots + (q_l p_l' - p_l q_l'). \tag{10.16}$$

Note that in $\mathbf{R}^2$ the symplectic product of two vectors coincides with the unique non-zero scalar component of their vector product. ∎

DEFINITION 10.4  Suppose we are given a symplectic product in $\mathbf{R}^{2l}$. A symplectic basis *is a basis of* $\mathbf{R}^{2l}$ *with respect to which the symplectic product takes the standard form* (10.16), *and hence it has as representative matrix the matrix* $\mathfrak{J}$. ∎

Given a symplectic product $\omega$, a symplectic basis $\mathbf{e}_1, \ldots, \mathbf{e}_{2l} = \mathbf{e}_{p_1}, \ldots, \mathbf{e}_{p_l}, \mathbf{e}_{q_1}, \ldots, \mathbf{e}_{q_l}$ satisfies

$$\omega(\mathbf{e}_{q_i}, \mathbf{e}_{q_j}) = \omega(\mathbf{e}_{p_i}, \mathbf{e}_{p_j}) = 0, \tag{10.17}$$

## 10.1 Analytical mechanics: canonical formalism

for every $i, j = 1, \ldots, l$ and

$$\omega(\mathbf{e}_{q_i}, \mathbf{e}_{p_j}) = \delta_{ij}. \tag{10.18}$$

**Remark 10.7**
It follows that the choice of standard symplectic structure for $\mathbf{R}^{2l}$ coincides with the choice of the canonical basis of $\mathbf{R}^{2l}$ as symplectic basis. ∎

Using a technique similar to the Gram–Schmidt orthonormalisation for the basis in an inner product space, it is not difficult to prove the following theorem.

**THEOREM 10.3** *In any space endowed with a symplectic product it is possible to construct a symplectic basis.* ∎

As for inner product spaces, it is possible to choose as the first vector of the basis any non-zero vector.

Pursuing the analogy between an inner and a symplectic product, we can define a class of transformations that preserve the symplectic product, taking as a model the orthogonal transformations, which preserve the inner product.

**DEFINITION 10.5** *Given two symplectic spaces $V_1, \omega_1$ and $V_2, \omega_2$, a linear map $S: V_1 \to V_2$ is symplectic if $\omega_2(S(\mathbf{v}), S(\mathbf{w})) = \omega_1(\mathbf{v}, \mathbf{w})$ for every $\mathbf{v}, \mathbf{w} \in V_1$. If moreover $S$ is an isomorphism, we say that $S$ is a symplectic isomorphism.* ∎

**Remark 10.8**
From Theorem 10.3 it follows, as an obvious corollary, that all symplectic spaces of the same dimension are symplectically isomorphic. A 'canonical' isomorphism can be obtained by choosing a symplectic basis in each space, and setting a correspondence between the basis elements with the same index. In particular, all symplectic spaces of dimension $2l$ are symplectically isomorphic to $\mathbf{R}^{2l}$ with its standard structure. ∎

**THEOREM 10.4** *Let $\mathbf{R}^{2l}$ be considered with its standard structure. A linear map $S: \mathbf{R}^{2l} \to \mathbf{R}^{2l}$ is symplectic if and only if its representative matrix is symplectic.*

*Proof*
This is a simple check: given $\mathbf{x}, \mathbf{y} \in \mathbf{R}^{2l}$ we have

$$\omega(S\mathbf{x}, S\mathbf{y}) = (S\mathbf{x})^T \mathsf{J} S\mathbf{y} = \mathbf{x}^T S^T \mathsf{J} S\mathbf{y},$$

which is equal to

$$\omega(\mathbf{x}, \mathbf{y}) = \mathbf{x}^T \mathsf{J} \mathbf{y}$$

for every $\mathbf{x}, \mathbf{y}$ if and only if

$$S^T \mathsf{J} S = \mathsf{J}. \qquad \blacksquare$$

We conclude this section with the definition and characterisation of Hamiltonian vector fields (or symplectic gradient vector fields). These are useful in view of the fact that the Hamilton equations can be written in the form (10.3).

DEFINITION 10.6 *A vector field* $\mathbf{X}(\mathbf{x},t)$ *in* $\mathbf{R}^{2l}$ *is* Hamiltonian *if there exists a function* $f(\mathbf{x},t)$ *in* $\mathcal{C}^2$ *such that*

$$\mathbf{X}(\mathbf{x},t) = \mathbf{J}\nabla_{\mathbf{x}} f(\mathbf{x},t). \tag{10.19}$$

*In this case* $f$ *is called the* Hamiltonian *corresponding to the field* $\mathbf{X}$ *and the field* $\mathbf{X}$ *is called the* symplectic gradient *of* $f$. *If* $\mathbf{X}$ *is Hamiltonian, the system of differential equations*

$$\dot{\mathbf{x}} = \mathbf{X}(\mathbf{x},t) \tag{10.20}$$

*is called* Hamiltonian. ∎

The system of Example 10.1 is Hamiltonian.

*Remark* 10.9
A Hamiltonian vector field determines the corresponding Hamiltonian $f$ up to an arbitrary function $h(t)$ depending only on time $t$. This arbitrariness can be removed by requiring that the Hamiltonian associated with the field $\mathbf{X} = \mathbf{0}$ be zero. ∎

*Remark* 10.10
In $\mathbf{R}^2$ the vector $\mathbf{w} = \mathbf{J}\mathbf{v}$ can be obtained by rotating $\mathbf{v}$ by $\pi/2$ in the positive direction. It is easy to check that, in $\mathbf{R}^{2l}$, $\mathbf{J}\mathbf{v}$ is normal to $\mathbf{v}$. It follows that in a Hamiltonian field, for every fixed $t$, the Hamiltonian is constant along the lines of the field (Fig. 10.1). If the field is independent of time the Hamiltonian is constant along its integral curves, i.e. along the Hamiltonian flow (recall equation (8.26)). ∎

It is essential to characterise Hamiltonian vector fields. This is our next aim.

THEOREM 10.5 *A necessary and sufficient condition for a vector field* $\mathbf{X}(\mathbf{x},t)$ *in* $\mathbf{R}^{2l}$ *to be Hamiltonian is that the Jacobian matrix* $\nabla_{\mathbf{x}} \mathbf{X}(\mathbf{x},t)$ *is Hamiltonian for every* $(\mathbf{x},t)$.

*Proof*
The condition is necessary. Indeed, if $f$ is the Hamiltonian corresponding to $\mathbf{X}$ we have that

$$\frac{\partial X_i}{\partial x_j} = \sum_{k=1}^{l} \mathcal{J}_{ik} \frac{\partial^2 f}{\partial x_k \partial x_j},$$

and hence the matrix $\nabla_{\mathbf{x}} \mathbf{X}$ can be written as the product of the matrix $\mathcal{J}$ and the Hessian matrix of $f$, which is evidently symmetric.

The condition is also sufficient: if $\nabla_{\mathbf{x}} \mathbf{X}(\mathbf{x},t)$ is Hamiltonian for every $(\mathbf{x},t)$, setting $\mathbf{Y}(\mathbf{x},t) = \mathcal{J}\mathbf{X}(\mathbf{x},t)$, by (3) of Theorem 10.1, we have that

$$\frac{\partial Y_i}{\partial x_j} = \frac{\partial Y_j}{\partial x_i}.$$

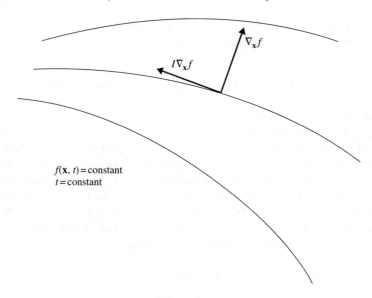

$f(\mathbf{x}, t) = $ constant
$t = $ constant

Fig. 10.1

Consequently, there exists a function $f(\mathbf{x}, t)$ such that

$$\mathbf{Y}(\mathbf{x}, t) = -\nabla_{\mathbf{x}} f(\mathbf{x}, t).$$

From this it follows that

$$\mathbf{X}(\mathbf{x}, t) = -\mathbb{J}\mathbf{Y}(\mathbf{x}, t) = \mathbb{J}\nabla_{\mathbf{x}} f(\mathbf{x}, t). \qquad \blacksquare$$

*Example* 10.3
Consider the system of differential equations

$$\dot{p} = -p^{\alpha+1} q^{\delta}, \quad \dot{q} = p^{\alpha} q^{\beta},$$

and compute for which values of the real constants $\alpha$, $\beta$ and $\delta$ this is a Hamiltonian system. Find the corresponding Hamiltonian $H(q, p)$.

Consider the second equation; if there exists a Hamiltonian $H(p, q)$ such that $\dot{q} = \partial H / \partial p$, by integrating with respect to $p$ we find:

(a) $H = q^{\beta} \log p + f(q)$ if $\alpha = -1$;
(b) $H = p^{\alpha+1} q^{\beta} / (\alpha + 1) + g(q)$ if $\alpha \neq -1$.

By substituting in the equation $\dot{p} = -\partial H / \partial q$ and comparing with the equation given for $p$, we find that, if $\alpha = -1$, necessarily $\beta = 0$ and

(a') $H = \log p + \{q^{\delta+1}/(\delta + 1) + c\}$ if $\delta \neq -1$, where $c$ is an arbitrary constant;
(a'') $H = \log p + \log q + c$ if $\delta = -1$, where $c$ is an arbitrary constant.

If on the other hand $\alpha \neq -1$ we find $H = \{(qp)^{\alpha+1}/(\alpha+1)\} + c$, where as usual $c$ is an arbitrary integration constant. ∎

## 10.2 Canonical and completely canonical transformations

A method which can sometimes be applied to integrate differential equations is to use an appropriate change of variables which makes it possible to write the equation in a form such that the solution (or some property of the solution) can be immediately obtained. The study of particular classes of coordinate transformations in the phase space for the Hamilton equations is of great importance and will be carried out in this and the next sections. In Chapters 11 and 12 we will show how, through these transformations, it is possible to solve (exactly or approximately) the Hamilton equations for a large class of systems.

Given a system of ordinary differential equations

$$\dot{\mathbf{x}} = \mathbf{v}(\mathbf{x}, t), \tag{10.21}$$

where $\mathbf{x} \in \mathbf{R}^n$ (or a differentiable manifold of dimension $n$), consider an invertible coordinate transformation (possibly depending on time $t$)

$$\mathbf{x} = \mathbf{x}(\mathbf{y}, t), \tag{10.22}$$

with inverse

$$\mathbf{y} = \mathbf{y}(\mathbf{x}, t). \tag{10.23}$$

If the function $\mathbf{y}(\mathbf{x}, t)$ has continuous first derivatives, the system (10.21) is transformed into

$$\dot{\mathbf{y}} = \mathbf{w}(\mathbf{y}, t), \tag{10.24}$$

where

$$\mathbf{w}(\mathbf{y}, t) = J\mathbf{v} + \frac{\partial \mathbf{y}}{\partial t},$$

$J$ is the Jacobian matrix of the transformation, $J_{ik} = \partial y_i/\partial x_k$, and the right-hand side is expressed in terms of the variables $(\mathbf{y}, t)$ using (10.22). Likewise we consider the system of canonical equations with Hamiltonian $H(\mathbf{x}, t)$, where $\mathbf{x} = (\mathbf{p}, \mathbf{q}) \in \mathbf{R}^{2l}$,

$$\dot{\mathbf{x}} = \mathbb{J}\nabla_{\mathbf{x}} H(\mathbf{x}, t), \tag{10.25}$$

and make the coordinate transformation

$$\mathbf{x} = \mathbf{x}(\mathbf{X}, t), \tag{10.26}$$

with $\mathbf{X} = (\mathbf{P}, \mathbf{Q}) \in \mathbf{R}^{2l}$, subject to the invertibility condition

$$\mathbf{X} = \mathbf{X}(\mathbf{x}, t), \qquad (10.27)$$

and to the condition of continuity of the first derivatives. Then the system of canonical equations (10.25) is transformed into a new system of $2l$ differential equations

$$\dot{\mathbf{X}} = \mathbf{W}(\mathbf{X}, t), \qquad (10.28)$$

where

$$\mathbf{W}(\mathbf{X}, t) = J\mathbb{J}\nabla_{\mathbf{x}} H + \frac{\partial \mathbf{X}}{\partial t}, \qquad (10.29)$$

$J$ is the Jacobian matrix of the transformation, with components $J_{ik} = \partial X_i/\partial x_k$, and the right-hand side is expressed in terms of the variables $\mathbf{X} = (\mathbf{P}, \mathbf{Q})$. In general, the system (10.28) does not have the canonical structure (10.25), as it is not necessarily true that a Hamiltonian $K(\mathbf{X}, t)$ exists such that

$$\mathbf{W} = \mathbb{J}\nabla_{\mathbf{X}} K. \qquad (10.30)$$

*Example* 10.4
We go back to Example 10.1 with $H(\mathbf{x}) = \frac{1}{2}\mathbf{x}^T S \mathbf{x}$, where $S$ is a constant symmetric matrix. Let us consider how the Hamilton equation $\dot{\mathbf{x}} = \mathbb{J}S\mathbf{x}$ is transformed when passing to the new variables $\mathbf{X} = A\mathbf{x}$, with $A$ a constant invertible matrix. We immediately find that $\dot{\mathbf{X}} = A\mathbb{J}SA^{-1}\mathbf{X}$ and in order to preserve the canonical structure we must have $A\mathbb{J}SA^{-1} = \mathbb{J}C$, with $C$ symmetric. It is important to note that this must happen *for every symmetric matrix* $S$, and hence this is a genuine restriction on the class to which $A$ must belong. We can rewrite this condition as $A^T \mathbb{J} A \mathbb{J} S = -A^T C A$. It follows that the existence of a symmetric matrix $C$ is equivalent to the symmetry condition

$$A^T \mathbb{J} A \mathbb{J} S = S \mathbb{J} A^T \mathbb{J} A, \qquad (10.31)$$

i.e. $\Lambda^T \mathbb{J} S + S \mathbb{J} \Lambda = 0$ with $\Lambda = A^T \mathbb{J} A = -\Lambda^T$, for every symmetric matrix $S$. If $A$ is symplectic then $\Lambda = \mathbb{J}$ and the condition is satisfied. The same is true if $\Lambda = a\mathbb{J}$ (with $a \neq 0$ so that $A$ is invertible). These conditions are also necessary. Indeed, using the $l \times l$ block decomposition we have $\Lambda = \begin{pmatrix} \lambda & \mu \\ -\mu^T & \nu \end{pmatrix}$ and $S = \begin{pmatrix} \alpha & \beta \\ \beta^T & \gamma \end{pmatrix}$, with the conditions $\lambda^T = -\lambda$, $\nu^T = -\nu$, $\alpha^T = \alpha$, $\gamma^T = \gamma$. The equation $\Lambda \mathbb{J} S = S \mathbb{J} \Lambda$ leads to the system

$$-\lambda \beta^T + \mu \alpha = \alpha \mu^T + \beta \lambda,$$
$$-\lambda \gamma + \mu \beta = -\alpha \nu + \beta \mu,$$
$$\mu^T \beta^T + \nu \alpha = \beta^T \mu^T + \gamma \lambda,$$
$$\mu^T \gamma + \nu \beta = -\beta^T \nu + \gamma \nu.$$

Considering the particular case $\alpha = \gamma = 0$ we find that $\mu$ must commute with every $l \times l$ matrix, and therefore $\mu = a\mathbf{1}$. Choosing $\alpha = \beta = 0$ we find $\lambda = 0$. From $\beta = \gamma = 0$ it follows that $\nu = 0$. Hence $\Lambda = a\mathbb{J}$, and in addition, from $A^T \mathbb{J} A = a\mathbb{J}$ it follows that $\mathbb{J} A \mathbb{J} = -a(A^{-1})^T$. We finally find that $C = a(A^{-1})^T S A^{-1}$ and the new Hamiltonian is $K(\mathbf{X}) = \frac{1}{2} \mathbf{X}^T C \mathbf{X}$. If $A$ is symplectic it holds that $K(\mathbf{X}) = H(\mathbf{x})$, and if $a \neq 1$ we find $K(\mathbf{X}) = aH(\mathbf{x})$. ∎

The necessity to preserve the canonical structure of the Hamilton equations, which has very many important consequences (see the following sections and Chapter 11), justifies the following definition.

DEFINITION 10.7  *A coordinate transformation* $\mathbf{X} = \mathbf{X}(\mathbf{x}, t)$ *which is differentiable and invertible (for every fixed $t$) preserves the canonical structure of Hamilton equations if for any Hamiltonian* $H(\mathbf{x}, t)$ *there exists a corresponding function* $K(\mathbf{X}, t)$, *the new Hamiltonian, such that the system of transformed equations* (10.28) *coincides with the system of Hamilton equations* (10.30) *for* $K$:

$$\dot{P}_i = -\frac{\partial K}{\partial Q_i}(\mathbf{Q}, \mathbf{P}, t), \quad i = 1, \ldots, l,$$
$$\dot{Q}_i = \frac{\partial K}{\partial P_i}(\mathbf{Q}, \mathbf{P}, t), \quad i = 1, \ldots, l. \tag{10.32}$$

∎

*Remark* 10.11
The new Hamiltonian $K(\mathbf{Q}, \mathbf{P}, t)$ is not necessarily obtained by substituting into $H(\mathbf{q}, \mathbf{p}, t)$ the transformation (10.26). This is illustrated in the following examples. ∎

*Example* 10.5
The translations of $\mathbf{R}^{2l}$ preserve the canonical structure of the Hamilton equations. The rotations $\mathbf{X} = R\mathbf{x}$, where $R$ is an orthogonal matrix $R^T = R^{-1}$, preserve the structure if and only if $R$ is a symplectic matrix (see Theorem 10.6 below). This is always true for $l = 1$, if $R$ preserves the orientation of the plane (see Example 10.2), and hence if $\det(R) = 1$. ∎

*Example* 10.6
The transformations

$$P_i = \nu_i p_i, \quad i = 1, \ldots, l,$$
$$Q_i = \mu_i q_i, \quad i = 1, \ldots, l, \tag{10.33}$$

where $\mu_1, \ldots, \mu_l$ and $\nu_1, \ldots, \nu_l$ are $2l$ real arbitrary non-zero constants satisfying the condition $\mu_i \nu_i = \lambda$ for every $i = 1, \ldots, l$, are called *scale transformations* and preserve the canonical structure of the Hamilton equations. Indeed, it can be verified that the new Hamiltonian $K$ is related to the old one $H$ through

$$K(\mathbf{P}, \mathbf{Q}, t) = \lambda H(\nu_1^{-1} P_1, \ldots, \nu_l^{-1} P_l, \mu_1^{-1} Q_1, \ldots, \mu_l^{-1} Q_l, t).$$

Note that $K$ is the transform of $H$ only in the case that $\mu_i \nu_i = 1$, $i = 1, \ldots, l$, and hence if $\lambda = 1$ (in this case the Jacobian matrix of the transformation is symplectic). When $\lambda \neq 1$ we say that the scale transformation is *not natural*. Note that the Jacobian determinant of (10.33) is $\lambda^l$, and hence the transformation (10.33) preserves the measure if and only if $\lambda = 1$. The scale transformations are commonly used to change to dimensionless coordinates. ∎

*Example 10.7*
Let $a(t)$ be a differentiable non-zero function. The transformation
$$\mathbf{Q} = a(t)\mathbf{q}, \quad \mathbf{P} = \frac{1}{a(t)}\mathbf{p}$$
preserves the canonical structure of the Hamilton equations. Indeed, the Hamilton equations become
$$\dot{\mathbf{P}} = -\frac{1}{a(t)}\nabla_\mathbf{q} H - \frac{\dot{a}(t)}{a^2(t)}\mathbf{P},$$
$$\dot{\mathbf{Q}} = a(t)\nabla_\mathbf{p} H + \dot{a}(t)\mathbf{q},$$
corresponding to the Hamilton equations for the function
$$K(\mathbf{P}, \mathbf{Q}, t) = H\left(a(t)\mathbf{P}, \frac{\mathbf{Q}}{a(t)}, t\right) + \frac{\dot{a}(t)}{a(t)}\mathbf{P} \cdot \mathbf{Q}. \quad \blacksquare$$

*Example 10.8*
The transformation exchanging (up to sign) the coordinates $q_i$ with the corresponding kinetic moments $p_i$ preserves the canonical structure of the Hamilton equations
$$\mathbf{P} = -\mathbf{q}, \quad \mathbf{Q} = \mathbf{p}. \tag{10.34}$$
The new Hamiltonian is related to the old Hamiltonian through
$$K(\mathbf{P}, \mathbf{Q}, t) = H(\mathbf{Q}, -\mathbf{P}, t).$$
This transformation shows how, within the Hamiltonian formalism, there is no essential difference between the role of the coordinates $\mathbf{q}$ and of the conjugate momenta $\mathbf{p}$. ∎

*Example 10.9*
The *point transformations* preserve the canonical structure of the Hamilton equations. Indeed, let
$$\mathbf{Q} = \mathbf{Q}(\mathbf{q}) \tag{10.35}$$
be an invertible Lagrangian coordinate transformation. The generalised velocities are transformed linearly:
$$\dot{Q}_i = \frac{\partial Q_i}{\partial q_j}(\mathbf{q})\dot{q}_j = J_{ij}(\mathbf{q})\dot{q}_j,$$

where $i = 1, \ldots, l$ and we have adopted the convention of summation over repeated indices. Here $J(\mathbf{q}) = (J_{ij}(\mathbf{q}))$ is the Jacobian matrix of the transformation (10.35). If $L(\mathbf{q}, \dot{\mathbf{q}}, t)$ is the Lagrangian of the system, we denote by

$$\hat{L}(\mathbf{Q}, \dot{\mathbf{Q}}, t) = L(\mathbf{q}(\mathbf{Q}), J^{-1}(\mathbf{q}(\mathbf{Q}))\dot{\mathbf{Q}}, t)$$

the Lagrangian expressed through the new coordinates, and by $\mathbf{P}$ the corresponding kinetic momentum, whose components are given by

$$P_i = \frac{\partial \hat{L}}{\partial \dot{Q}_i} = J_{ji}^{-1} \frac{\partial L}{\partial \dot{q}_j} = J_{ji}^{-1} p_j,$$

for $i = 1, \ldots, l$. The transformation (10.35) induces a transformation of the conjugate kinetic momenta:

$$\mathbf{P} = (J^T)^{-1}\mathbf{p}, \tag{10.36}$$

and Hamilton's equations associated with the Hamiltonian $H(\mathbf{p}, \mathbf{q}, t)$ become

$$\begin{aligned} \dot{P}_i &= -J_{ji}^{-1}\frac{\partial H}{\partial q_j} + p_j \frac{\partial J_{ji}^{-1}}{\partial Q_k} J_{kn} \frac{\partial H}{\partial p_n}, \\ \dot{Q}_i &= J_{ij}\frac{\partial H}{\partial p_j}, \end{aligned} \tag{10.37}$$

where $i = 1, \ldots, l$.

Point transformations necessarily preserve the canonical structure. For the Hamiltonian systems originating from a Lagrangian, the proof is easy. Indeed, starting from the new Lagrangian $\hat{L}(\mathbf{Q}, \dot{\mathbf{Q}}, t)$ we can construct the Legendre transform $\hat{H}(\mathbf{P}, \mathbf{Q}, t)$ to take the role of the Hamiltonian in the equations thus obtained. It is easy to check that $\hat{H}$ is the transform of $H$:

$$\hat{H}(\mathbf{P}, \mathbf{Q}, t) = H(J^T(\mathbf{q}(\mathbf{Q}))\mathbf{P}, \mathbf{q}(\mathbf{Q}), t).$$

Indeed, to obtain the Legendre transform (8.19) of $\hat{L}(\mathbf{Q}, \dot{\mathbf{Q}}, t)$ we must compute

$$\hat{H}(\mathbf{P}, \mathbf{Q}, t) = \mathbf{P}^T \dot{\mathbf{Q}} - \hat{L}(\mathbf{Q}, \dot{\mathbf{Q}}, t),$$

and reintroducing the variables $(\mathbf{p}, \mathbf{q})$ we note that $\hat{L}$ goes to $L$, while $\mathbf{P}^T\dot{\mathbf{Q}} = \mathbf{p}^T J^{-1} J \dot{\mathbf{q}} = \mathbf{p}^T \dot{\mathbf{q}}$. It follows that $\hat{H}(\mathbf{P}, \mathbf{Q}, t) = H(\mathbf{p}, \mathbf{q}, t)$. We leave it to the reader to verify that (10.37) are the Hamilton equations associated with $\hat{H}$. ∎

DEFINITION 10.8 *A differentiable and invertible coordinate transformation* $\mathbf{X} = \mathbf{X}(\mathbf{x}, t)$ *(for every fixed $t$) is called* canonical *if the Jacobian matrix*

$$J(\mathbf{x}, t) = \nabla_{\mathbf{x}} \mathbf{X}(\mathbf{x}, t)$$

*is symplectic for every choice of* $(\mathbf{x}, t)$ *in the domain of definition of the transformation. A time-independent canonical transformation* $\mathbf{X} = \mathbf{X}(\mathbf{x})$ *is called* completely canonical. ∎

We systematically assume in what follows that the matrix $J$ is sufficiently regular (at least $C^1$). All arguments are local (i.e. are valid in an open connected subset of $\mathbf{R}^{2l}$).

*Example* 10.10
It can immediately be verified that the transformation considered in Example 10.7 is canonical, and those considered in Examples 10.22, 10.25 and 10.26 are completely canonical. The scale transformations (Example 10.5) are not canonical, except when $\lambda = 1$. ∎

*Remark* 10.12
Recall that symplectic matrices form a group under matrix multiplication. Then we immediately deduce that *the canonical transformations form a group*. The completely canonical transformations form a subgroup, usually denoted by $\text{SDiff}(\mathbf{R}^{2l})$. We also note that $\det J = 1$, and hence *canonical transformations preserve the Lebesgue measure* in phase space. ∎

THEOREM 10.6 *The canonical transformations preserve the canonical structure of the Hamilton equations.* ∎

Before proving Theorem 10.6 it is convenient to digress and introduce a short lemma frequently used in the remainder of this chapter. We define first of all a class of $2l \times 2l$ matrices that generalises the class of symplectic matrices, by replacing the equation $J^T \mathbb{J} J = \mathbb{J}$ by

$$J^T \mathbb{J} J = a\mathbb{J}, \tag{10.38}$$

where $a$ is a *constant* different from zero. It is immediately verified that these matrices have as inverse $J^{-1} = -(1/a)\mathbb{J} J^T \mathbb{J}$. This inverse belongs to the analogous class with $a^{-1}$ instead of $a$. Therefore $J^T = -a\mathbb{J} J^{-1} \mathbb{J}$ and we can verify that $J^T$ belongs to the same class of $J$, i.e. $JJJ^T = a\mathbb{J}$. Obviously the class (10.38) includes as a special case (for $a = 1$) the symplectic matrices. An important property of the time-dependent matrices that satisfy the property (10.38) (with $a$ constant) is the following.

LEMMA 10.1 *If $J(\mathbf{X}, t)$ is a matrix in the class (10.38) then the matrix $B = (\partial J/\partial t) J^{-1}$ is Hamiltonian.*

*Proof*
Recalling Theorem 10.1, it is sufficient to prove that the matrix

$$A = \mathbb{J} \frac{\partial J}{\partial t} J^{-1} \tag{10.39}$$

is symmetric. Differentiating with respect to $t$ the two sides of (10.38) we obtain

$$\frac{\partial J^T}{\partial t} \mathbb{J} J + J^T \mathbb{J} \frac{\partial J}{\partial t} = 0. \tag{10.40}$$

Multiplying this on the left by $(J^{-1})^T$ and on the right by $J^{-1}$ then yields

$$A^T = -(J^{-1})^T \frac{\partial J^T}{\partial t} \mathbb{J} = \mathbb{J} \frac{\partial J}{\partial t} J^{-1} = A. \qquad \blacksquare$$

We now turn to Theorem 10.6.

*Proof of Theorem 10.6*
Let $\mathbf{X} = \mathbf{X}(\mathbf{x}, t)$ be a canonical transformation.

By differentiating $\mathbf{X}$ with respect to $t$ and using $\dot{\mathbf{x}} = \mathbb{J} \nabla_{\mathbf{x}} H(\mathbf{x}, t)$ we find

$$\dot{\mathbf{X}} = \frac{\partial \mathbf{X}}{\partial t} + J \mathbb{J} \nabla_{\mathbf{x}} H. \tag{10.41}$$

Setting

$$\hat{H}(\mathbf{X}, t) = H(\mathbf{x}(\mathbf{X}, t), t), \tag{10.42}$$

we have that

$$\nabla_{\mathbf{x}} H = J^T \nabla_{\mathbf{X}} \hat{H}, \tag{10.43}$$

from which it follows that equation (10.41) can be written as

$$\dot{\mathbf{X}} = \frac{\partial \mathbf{X}}{\partial t} + J \mathbb{J} J^T \nabla_{\mathbf{X}} \hat{H}. \tag{10.44}$$

But $J$ is by hypothesis symplectic, and therefore we arrive at the equation

$$\dot{\mathbf{X}} = \frac{\partial \mathbf{X}}{\partial t} + \mathbb{J} \nabla_{\mathbf{X}} \hat{H}, \tag{10.45}$$

which stresses the fact that the field $\mathbb{J} \nabla_{\mathbf{X}} \hat{H}$ is Hamiltonian.

To complete the proof we must show that $\partial \mathbf{X}/\partial t$ is also a Hamiltonian vector field. By Theorem 10.5, a necessary and sufficient condition is that $B = \nabla_{\mathbf{X}}((\partial \mathbf{X}(\mathbf{x}(\mathbf{X}, t), t))/\partial t)$ is Hamiltonian.

We see immediately that

$$B_{ij} = \frac{\partial}{\partial X_j} \frac{\partial X_i}{\partial t} = \sum_{n=1}^{2l} \frac{\partial^2 X_i}{\partial t \partial x_n} \frac{\partial x_n}{\partial X_j},$$

and hence

$$B = \frac{\partial J}{\partial t} J^{-1}. \tag{10.46}$$

Now Lemma 10.1 ends the proof. $\blacksquare$

*Remark* 10.13
The new Hamiltonian $K$ corresponding to the old Hamiltonian $H$ is given by

$$K = \hat{H} + K_0, \qquad (10.47)$$

where $\hat{H}$ is the old Hamiltonian expressed through the new variables (see (10.42)) and $K_0$ is the Hamiltonian of the Hamiltonian vector field $\partial \mathbf{X}/\partial t$, and hence satisfying

$$\frac{\partial \mathbf{X}}{\partial t} = \mathbb{J} \nabla_{\mathbf{X}} K_0. \qquad (10.48)$$

It follows that $K_0$ depends only on the transformation $\mathbf{X}(\mathbf{x},t)$ and it is uniquely determined by it, up to an arbitrary function $h(t)$ which we always assume to be identically zero (see Remark 10.9). Here $K_0$ can be identified with the Hamiltonian corresponding to $H \equiv 0$. If the transformation is completely canonical we have that $K_0 \equiv 0$, and the new Hamiltonian is simply obtained by expressing the old Hamiltonian in terms of the new coordinates (consistent with the interpretation of the Hamiltonian as the total mechanical energy of the system). ∎

We then have the following.

COROLLARY 10.1  *For a completely canonical transformation the new Hamiltonian is simply the transformation of the original Hamiltonian. A time-dependent canonical transformation* $\mathbf{X} = \mathbf{X}(\mathbf{x}, t)$ *is necessarily a Hamiltonian flow, governed by the equation* $\partial \mathbf{X}/\partial t = \mathbb{J} \nabla_{\mathbf{X}} K_0(\mathbf{X}, t)$. ∎

We shall see that to every Hamiltonian flow $\mathbf{X} = S^t \mathbf{x}$ we can associate a canonical transformation. Hence we can identify the class of time-dependent canonical transformations with the class of Hamiltonian flows.

*Example* 10.11
Consider the *time-dependent* transformation

$$p = P - at, \quad q = Q + Pt - \frac{1}{2}at^2, \qquad (10.49)$$

where $a$ is a fixed constant. We can immediately check that the transformation is canonical, with inverse given by

$$P = p + at, \quad Q = q - pt - \frac{1}{2}at^2.$$

The Hamiltonian $K_0$ is the solution of (see (10.48))

$$\frac{\partial P}{\partial t} = a = -\frac{\partial K_0}{\partial Q}, \quad \frac{\partial Q}{\partial t} = -p - at = -P = \frac{\partial K_0}{\partial P},$$

from which it follows that

$$K_0(P, Q) = -\frac{P^2}{2} - aQ, \qquad (10.50)$$

and the new Hamiltonian $K(P,Q,t)$ corresponding to $H(p,q,t)$ is:

$$K(P,Q,t) = H\left(P - at, Q + Pt - \frac{1}{2}at^2, t\right) + K_0(P,Q) = \hat{H}(P,Q,t) - \frac{P^2}{2} - aQ.$$

■

The next theorem includes Theorem 10.6, and characterises the whole class of transformations which preserve the canonical structure of the Hamilton equations. Moreover, it characterises how these transformations act on the Hamiltonian.

THEOREM 10.7  *A necessary and sufficient condition for a differentiable and invertible (for every fixed t) coordinate transformation* $\mathbf{X} = \mathbf{X}(\mathbf{x},t)$ *to preserve the canonical structure of the Hamilton equations is that its Jacobian matrix belongs to the class (10.38), i.e.*

$$J\mathfrak{J}J^T = J^T \mathfrak{J} J = a\mathfrak{J} \qquad (10.51)$$

*for some constant a different from zero. The transformation acts on the Hamiltonian as follows:*

$$K(\mathbf{X},t) = a\hat{H}(\mathbf{X},t) + K_0(\mathbf{X},t), \qquad (10.52)$$

*where* $\hat{H}(\mathbf{X},t) = H(\mathbf{x}(\mathbf{X},t),t)$ *is the transform of the original Hamiltonian and* $K_0$ *(corresponding to $H = 0$) is the Hamiltonian of the vector field $\partial \mathbf{X}/\partial t$. The transformation is canonical if and only if $a = 1$.* ■

COROLLARY 10.2  *The canonical transformations are the only ones leading to a new Hamiltonian of the form* $K = \hat{H} + K_0$, *and the completely canonical ones are the only ones for which* $K = \hat{H}$. ■

In addition, note that when $a \neq 1$ the transformation can be made into a canonical transformation by composing it with an appropriate scale change.

The proof of Theorem 10.7 makes use of a lemma. We present the proof of this lemma as given in Benettin *et al.* (1991).

LEMMA 10.2  *Let* $A(\mathbf{x},t)$ *be a regular function of* $(\mathbf{x},t) \in \mathbf{R}^{2l+1}$ *with values in the space of real non-singular $2l \times 2l$ matrices. If for any regular function $H(\mathbf{x},t)$, the vector field* $A\nabla_\mathbf{x} H$ *is irrotational, then there exists a function* $a : \mathbf{R} \to \mathbf{R}$ *such that* $A = a(t)\mathbf{1}$.

*Proof*
If $A\nabla_\mathbf{x} H$ is irrotational, for every $i, j = 1, \ldots, 2l$, we have that

$$\frac{\partial}{\partial x_i}(A\nabla_\mathbf{x} H)_j = \frac{\partial}{\partial x_j}(A\nabla_\mathbf{x} H)_i. \qquad (10.53)$$

Let $H = x_i$. Then

$$\frac{\partial}{\partial x_i} A_{ji} = \frac{\partial}{\partial x_j} A_{ii} \qquad (10.54)$$

(note that we are *not* using the convention of summation over repeated indices!), while if we let $H = x_i^2$ then

$$\frac{\partial}{\partial x_i}(A_{ji}x_i) = \frac{\partial}{\partial x_j}(A_{ii}x_i). \tag{10.55}$$

It follows using (10.54) that

$$A_{ji} = A_{ii}\delta_{ij},$$

i.e. the matrix $A$ is diagonal. From (10.54) it also follows that

$$\frac{\partial A_{ii}}{\partial x_j} = 0, \quad \text{if } j \neq i,$$

and therefore $A$ has the form

$$A_{ij}(\mathbf{x}, t) = a_i(x_i, t)\delta_{ij},$$

for suitable functions $a_i$. Using (10.53) we find that

$$a_j \frac{\partial^2 H}{\partial x_i \partial x_j} = a_i \frac{\partial^2 H}{\partial x_i \partial x_j}, \quad \text{for } j \neq i,$$

from which it follows that $a_j = a_i = a(t)$. ∎

*Proof of Theorem 10.7*
Suppose that the transformation preserves the canonical structure, so that

$$\dot{\mathbf{X}} = \mathsf{J}\nabla_\mathbf{X} K(\mathbf{X}, t). \tag{10.56}$$

Comparing (10.56) with the general form (10.44) of the transformed equation

$$\dot{\mathbf{X}} = \frac{\partial \mathbf{X}}{\partial t} + \mathsf{J}\mathsf{J}\mathsf{J}^T \nabla_\mathbf{X} \hat{H} \tag{10.57}$$

we deduce

$$\frac{\partial \mathbf{X}}{\partial t} = \mathsf{J}\nabla_\mathbf{X} K - \mathsf{J}\mathsf{J}\mathsf{J}^T \nabla_\mathbf{X} \hat{H}. \tag{10.58}$$

We also know (by hypothesis) that to $H = 0$ there corresponds a Hamiltonian $K_0$, for which (10.58) becomes

$$\frac{\partial \mathbf{X}}{\partial t} = \mathsf{J}\nabla_\mathbf{X} K_0. \tag{10.59}$$

By substituting (10.59) into (10.58) and multiplying by $\mathsf{J}$ we find

$$\nabla_\mathbf{X}(K - K_0) = -\mathsf{J}\mathsf{J}\mathsf{J}\mathsf{J}^T \nabla_\mathbf{X} \hat{H}. \tag{10.60}$$

Hence the matrix $-\mathbb{J}J\mathbb{J}J^T$ satisfies the assumptions of Lemma 10.2 (because $\hat{H}$ is arbitrary). It follows that there exists a function $a(t)$ such that

$$-\mathbb{J}J\mathbb{J}J^T = a(t)\mathbf{1}. \qquad (10.61)$$

Equation (10.61) shows clearly that $J$ satisfies equation (10.51), with $a$ possibly depending on time. To prove that $a$ is constant we note that, since $\partial \mathbf{X}/\partial t$ is a Hamiltonian vector field (see (10.59)), its Jacobian matrix

$$B = \nabla_{\mathbf{X}}\frac{\partial \mathbf{X}}{\partial t} = \frac{\partial J}{\partial t}J^{-1}$$

is Hamiltonian (see Theorem 10.5 and equation (10.46)). Therefore we can write (Definition 10.1)

$$\left(\frac{\partial J}{\partial t}J^{-1}\right)^T \mathbb{J} + \mathbb{J}\frac{\partial J}{\partial t}J^{-1} = 0. \qquad (10.62)$$

This is equivalent to the statement that $(\partial/\partial t)(J^T\mathbb{J}J) = 0$, yielding $a = $ constant. Now from (10.57) and (10.59), we can deduce the expression (10.52) for the new Hamiltonian $K$.

Conversely, suppose that the matrix $J$ satisfies the condition (10.51). Then (Lemma 10.1) $(\partial J/\partial t)J^{-1} = \nabla_{\mathbf{X}}\partial \mathbf{X}/\partial t$ is a Hamiltonian matrix. Therefore, the field $\partial \mathbf{X}/\partial t$ is Hamiltonian, and we can conclude that equation (10.57) takes the form

$$\dot{\mathbf{X}} = \mathbb{J}\nabla_{\mathbf{X}}(K_0 + a\hat{H}).$$

It follows that the transformation preserves the canonical structure, and the new Hamiltonian $K$ is given by (10.52). ∎

For the case $l = 1$, Theorem 10.7 has the following simple interpretation.

COROLLARY 10.3  *For $l = 1$ the condition of Theorem 10.7 reduces to*

$$\det J = \text{constant} \neq 0. \qquad (10.63)$$

*Proof*
It is enough to note that for $l = 1$ we have $J^T\mathbb{J}J = \mathbb{J}\det J$. ∎

*Example* 10.12
The transformation

$$p = \alpha\sqrt{P}\cos\gamma Q, \quad q = \beta\sqrt{P}\sin\gamma Q, \quad \alpha\beta\gamma \neq 0,$$

with $\alpha, \beta, \gamma$ constants, satisfies condition (10.63), since $\det J = \frac{1}{2}\alpha\beta\gamma$. It is (completely) canonical if and only if $\frac{1}{2}\alpha\beta\gamma = 1$. ∎

## 10.2  Analytical mechanics: canonical formalism

It is useful to close this section with a remark on the transformations which are inverses of those preserving the canonical structure. These inverse transformations clearly have the same property. If $\mathbf{X} = \mathbf{X}(\mathbf{x}, t)$ is a transformation in the class (10.51), its inverse $\mathbf{x} = \mathbf{x}(\mathbf{X}, t)$ has Jacobian matrix $J^{-1} = -(1/a)\mathfrak{J}J^T\mathfrak{J}$, such that $(J^{-1})^T \mathfrak{J} J^{-1} = (1/a)\mathfrak{J}$ (as we have already remarked). The inverse transformation reverts the Hamiltonian (10.52) to the original Hamiltonian $H$. For the case of the inverse transformation, the same relation (10.52) is then applied as follows:

$$H(\mathbf{x}, t) = K'_0(\mathbf{x}, t) + \frac{1}{a}\left[\hat{K}_0(\mathbf{x}, t) + aH(\mathbf{x}, t)\right], \qquad (10.64)$$

where $\hat{K}_0(\mathbf{x}, t)$ denotes the transform of $K_0(\mathbf{X}, t)$, and $K'_0(\mathbf{x}, t)$ is the Hamiltonian of the inverse flow $\partial \mathbf{x}/\partial t$. Equation (10.64) shows that $K'_0$ and $\hat{K}_0$ are related by

$$K'_0(\mathbf{x}, t) = -\frac{1}{a}\hat{K}_0(\mathbf{x}, t). \qquad (10.65)$$

Hence in the special case of the canonical transformations ($a = 1$) we have

$$K'_0(\mathbf{x}, t) = -\hat{K}_0(\mathbf{x}, t). \qquad (10.66)$$

This fact can easily be interpreted as follows. To produce a motion that is retrograde with respect to the flow $\partial \mathbf{X}/\partial t = \mathfrak{J}\nabla_{\mathbf{X}} K_0(\mathbf{X}, t)$ there are two possibilities:

(a) reverse the orientation of time ($t \to -t$), keeping the Hamiltonian fixed;
(b) keep the time orientation, but change $K_0$ into $-K_0$.

The condition (10.66) expresses the second possibility.

*Example* 10.13
The transformation

$$P = \alpha p \cos\omega t + \beta q \sin\omega t, \quad Q = -\frac{a}{\beta}p\sin\omega t + \frac{a}{\alpha}q\cos\omega t, \qquad (10.67)$$

with $\alpha, \beta, \omega, a$ non-zero constants, preserves the canonical structure of the Hamilton equations (check that $\det J = a$). It is canonical if and only if $a = 1$. In this case, it is the composition of a rotation with a 'natural' change of scale. The inverse of (10.67) is given by

$$p = \frac{1}{\alpha}P\cos\omega t - \frac{\beta}{a}Q\sin\omega t, \quad q = \frac{\alpha}{a}Q\cos\omega t + \frac{1}{\beta}P\sin\omega t. \qquad (10.68)$$

By differentiating (10.67) with respect to time, and inserting (10.68) we find the equations for the Hamiltonian flow $\mathbf{X} = \mathbf{X}(\mathbf{x}, t)$:

$$\frac{\partial P}{\partial t} = \frac{\alpha\beta\omega}{a}Q, \quad \frac{\partial Q}{\partial t} = -\frac{a\omega}{\alpha\beta}P, \qquad (10.69)$$

with which we associate the Hamiltonian

$$K_0 = -\frac{1}{2}\frac{\alpha\beta\omega}{a}Q^2 - \frac{1}{2}\frac{a\omega}{\alpha\beta}P^2. \qquad (10.70)$$

Performing the corresponding manipulations for the inverse transformation (10.68) we find the equations for the retrograde flow:

$$\frac{\partial p}{\partial t} = -\frac{\beta\omega}{\alpha}q, \quad \frac{\partial q}{\partial t} = \frac{\alpha\omega}{\beta}p, \qquad (10.71)$$

which is derived from the Hamiltonian

$$K_0' = \frac{1}{2}\frac{\beta\omega}{\alpha}q^2 + \frac{1}{2}\frac{\alpha\omega}{\beta}p^2. \qquad (10.72)$$

Expressing $K_0$ in the variables $(p,q)$ we obtain

$$\hat{K}_0 = -aK_0', \qquad (10.73)$$

which is in agreement with equation (10.65). ∎

## 10.3 The Poincaré–Cartan integral invariant. The Lie condition

In this section we want to focus on the geometric interpretation of canonical transformations. In the process of doing this, we derive a necessary and sufficient condition for a transformation to be canonical. This condition is very useful in practice, as we shall see in the next section.

Let us start by recalling a few definitions and results concerning differential forms.

DEFINITION 10.9   *A differential form $\omega$ in $\mathbf{R}^{2l+1}$*

$$\omega = \sum_{i=1}^{2l+1} \omega_i(\mathbf{x})\, \mathrm{d}x_i, \qquad (10.74)$$

*is* non-singular *if the $(2l+1) \times (2l+1)$ skew-symmetric matrix $A(\mathbf{x})$, defined by*

$$A_{ij} = \frac{\partial \omega_i}{\partial x_j} - \frac{\partial \omega_j}{\partial x_i}, \qquad (10.75)$$

*has maximal rank $2l$. The kernel of $A(\mathbf{x})$, characterised by $\{\mathbf{v} \in \mathbf{R}^{2l+1} | A(\mathbf{x})\mathbf{v} = \mathbf{0}\}$, as $\mathbf{x}$ varies determines a field of directions in $\mathbf{R}^{2l+1}$ called* characteristic directions. *The integral curves of the field of characteristic directions are called* characteristics *of $\omega$.* ∎

*Remark* 10.14
For $l = 1$, setting $\boldsymbol{\omega} = (\omega_1, \omega_2, \omega_3)$ the matrix $A(\mathbf{x})$ is simply

$$A(\mathbf{x}) = \begin{pmatrix} 0 & -(\omega)_3 & (\omega)_2 \\ (\omega)_3 & 0 & -(\omega)_1 \\ -(\omega)_2 & (\omega)_1 & 0 \end{pmatrix}$$

and $A(\mathbf{x})\mathbf{v} = \boldsymbol{\omega}(\mathbf{x}) \times \mathbf{v}$. Therefore the characteristics of the form $\omega$ can be indentified with those of the field $\boldsymbol{\omega}$. ■

*Example* 10.14
The form $\omega = x_2 \, \mathrm{d}x_1 + x_3 \, \mathrm{d}x_2 + x_1 \, \mathrm{d}x_3$ in $\mathbf{R}^3$ is non-singular. The associated characteristic direction is constant and is determined by the line $x_1 = x_2 = x_3$. ■

*Example* 10.15
The form $\omega = x_1 \, \mathrm{d}x_2 + \frac{1}{2}(x_1^2 + x_2^2) \, \mathrm{d}x_3$ is non-singular. The associated field of characteristic directions is $(x_2, -x_1, 1)$. ■

*Remark* 10.15
The reader familiar with the notion of a differential 2-form (see Appendix 4) will recognise in the definition of the matrix $A$ the representative matrix of the 2-form

$$-\mathrm{d}\omega = \sum_{i,j=1}^{2l+1} \frac{\partial \omega_i}{\partial x_j} \mathrm{d}x_i \wedge \mathrm{d}x_j.$$

■

The following result can be easily deduced from Definition 10.9.

PROPOSITION 10.3  *Two non-singular forms differing by an exact form have the same characteristics.* ■

Consider any regular closed curve $\gamma$. The characteristics of $\omega$ passing through the points of $\gamma$ define a surface in $\mathbf{R}^{2l+1}$ (i.e. a regular submanifold of dimension 2) called the *tube of characteristics*. The significance of non-singular differential forms, and of the associated tubes of characteristics, is due to the following property.

THEOREM 10.8 (Stokes' lemma)  *Let $\omega$ be a non-singular differential form, and let $\gamma_1$ and $\gamma_2$ be any two homotopic closed curves belonging to the same tube of characteristics. Then*

$$\oint_{\gamma_1} \omega = \oint_{\gamma_2} \omega. \tag{10.76}$$

■

Equation (10.76) expresses the invariance of the *circulation* of the field $\mathbf{X}(\mathbf{x})$, whose components are the $\omega_i$, along the closed lines traced on a tube of characteristics.

The previous theorem is a consequence of Stokes' lemma, discussed in Appendix 4. Note that this is natural generalisation of the Stokes formula, well known from basic calculus (see Giusti 1989).

We now consider a system with Hamiltonian $H(\mathbf{p},\mathbf{q},t)$ and its 'extended' phase space, where together with the canonical coordinates we consider the time $t : (\mathbf{p},\mathbf{q},t) \in \mathbf{R}^{2l+1}$.

**THEOREM 10.9** *The differential form*

$$\omega = \sum_{i=1}^{l} p_i \, dq_i - H(\mathbf{p},\mathbf{q},t) \, dt \tag{10.77}$$

*in $\mathbf{R}^{2l+1}$ is non-singular and it is called the Poincaré–Cartan form. Its characteristics are the integral curves of the system of Hamilton's equations associated with the Hamiltonian $H$.*

*Proof*
The matrix associated with the form $\omega$ is

$$A(\mathbf{p},\mathbf{q},t) = \begin{pmatrix} 0 & -1 & \nabla_{\mathbf{p}} H \\ 1 & 0 & \nabla_{\mathbf{q}} H \\ -(\nabla_{\mathbf{p}} H)^T & -(\nabla_{\mathbf{q}} H)^T & 0 \end{pmatrix}.$$

Evidently the rank of the matrix $A$ is equal to $2l$ for every $(\mathbf{p},\mathbf{q},t)$ (note that one of its $2l \times 2l$ submatrices coincides with the matrix $\mathfrak{I}$). It follows that the form $\omega$ is non-singular. Moreover, the vector

$$\mathbf{v}(\mathbf{p},\mathbf{q},t) = (-\nabla_{\mathbf{q}} H, \nabla_{\mathbf{p}} H, 1)$$

is in the kernel of $A$ for every $(\mathbf{p},\mathbf{q},t)$, and therefore it determines the characteristics of $\omega$. The integral curves of $\mathbf{v}$ are the solutions of

$$\dot{\mathbf{p}} = -\nabla_{\mathbf{q}} H,$$
$$\dot{\mathbf{q}} = \nabla_{\mathbf{p}} H,$$
$$\dot{t} = 1,$$

and hence they are precisely the integral curves of Hamilton's system of equations for $H$, expressed in the extended phase space $\mathbf{R}^{2l+1}$. ∎

The application of Stokes' lemma to the Poincaré–Cartan form (10.77) has a very important consequence.

**THEOREM 10.10** (Integral invariant of Poincaré–Cartan) *Let $\gamma_1$ and $\gamma_2$ be any two homotopic closed curves in $\mathbf{R}^{2l+1}$ belonging to the same tube of characteristics relative to the form (10.77). Then*

$$\oint_{\gamma_1} \left( \sum_{i=1}^{l} p_i \, dq_i - H(\mathbf{p},\mathbf{q},t) \, dt \right) = \oint_{\gamma_2} \left( \sum_{i=1}^{l} p_i \, dq_i - H(\mathbf{p},\mathbf{q},t) \, dt \right). \tag{10.78}$$

∎

## 10.3     Analytical mechanics: canonical formalism

*Remark* 10.16

Denote by $\gamma_0$ a closed curve belonging to the same tube of characteristics as $\gamma$, lying in the plane $t = t_0$, for fixed $t_0$. Then the result of Theorem 10.10 yields as a consequence the fact that

$$\oint_\gamma \left( \sum_{i=1}^l p_i \, dq_i - H(\mathbf{p},\mathbf{q},t) \, dt \right) = \oint_{\gamma_0} \sum_{i=1}^l p_i \, dq_i. \tag{10.79}$$

We shall see how the integral (10.79) completely characterises the canonical transformations, highlighting the relation with the geometry of the Hamiltonian flow (i.e. of the tubes of characteristics of the Poincaré–Cartan form). Indeed, starting from a system of Hamilton's equations for a Hamiltonian $H$ and going to a new system of Hamilton's equations for a new Hamiltonian $K$, the canonical transformations map the tubes of characteristics of the Poincaré–Cartan form (10.77) associated with $H$ onto the tubes of characteristics of the corresponding form associated with $K$. ■

We can state the following corollary to Theorem 10.12.

COROLLARY 10.4   *A canonical transformation maps the tubes of characteristics of the Poincaré–Cartan form (10.80) into the tubes of characteristics of the corresponding form*

$$\Omega = \sum_{i=1}^l P_i dQ_i - K(P,Q,t) \, dt. \tag{10.80}$$

■

*Example* 10.16

Consider the transformation of Example 10.12, which we rewrite as

$$p = \alpha \sqrt{P} \cos \gamma Q, \quad q = \beta \sqrt{P} \sin \gamma Q.$$

For $\alpha\beta\gamma = 2$ this transformation is completely canonical. We compare the Poincaré–Cartan forms written in the two coordinate systems:

$$\omega = p \, dq - H(p,q,t) \, dt, \quad \Omega = P \, dQ - \hat{H}(P,Q,t) \, dt.$$

The difference is

$$\omega - \Omega = p \, dq - P \, dQ.$$

Expressing it in the variables $P, Q$ we obtain

$$\omega - \Omega = d\left( \frac{1}{2\gamma} P \sin 2\gamma Q \right).$$

Since $\omega$ and $\Omega$ differ by an exact differential, they have the same tubes of characteristics. ■

We now want to show that the result discussed in the previous example ($\omega - \Omega = \mathrm{d}f$) is entirely general and constitutes a necessary and sufficient condition for a transformation to be canonical. We start by analysing the difference $\omega - \Omega$ when we 'fix time' (freezing the variable $t$).

Consider a differentiable, invertible transformation $\mathbf{X} = \mathbf{X}(\mathbf{x}, t)$ from the coordinates $\mathbf{x} = (\mathbf{p}, \mathbf{q})$ to $\mathbf{X} = (\mathbf{P}, \mathbf{Q})$:

$$p_i = p_i(\mathbf{P}, \mathbf{Q}, t), \quad q_i = q_i(\mathbf{P}, \mathbf{Q}, t), \tag{10.81}$$

where $i = 1, \ldots, l$. Consider the differential form

$$\tilde{\omega} = \sum_{i=1}^{l} p_i(\mathbf{P}, \mathbf{Q}, t) \, \tilde{\mathrm{d}} q_i(\mathbf{P}, \mathbf{Q}, t), \tag{10.82}$$

where, given any regular function $f(\mathbf{P}, \mathbf{Q}, t)$, we set

$$\tilde{\mathrm{d}} f = \mathrm{d} f - \frac{\partial f}{\partial t} \, \mathrm{d}t = \sum_{i=1}^{l} \left( \frac{\partial f}{\partial P_i} \, \mathrm{d}P_i + \frac{\partial f}{\partial Q_i} \, \mathrm{d}Q_i \right). \tag{10.83}$$

Here $\tilde{\mathrm{d}}$ is the so-called 'virtual differential' or 'time frozen differential' (see Levi-Civita and Amaldi 1927).

THEOREM 10.11 (Lie condition) *The transformation (10.81) is canonical if and only if the difference between the differential forms $\tilde{\omega}$ and $\tilde{\Omega}$ is exact, and hence if there exists a regular function $f(\mathbf{P}, \mathbf{Q}, t)$ such that*

$$\tilde{\omega} - \tilde{\Omega} = \sum_{i=1}^{l} (p_i \, \tilde{\mathrm{d}} q_i - P_i \, \tilde{\mathrm{d}} Q_i) = \tilde{\mathrm{d}} f. \tag{10.84}$$

*Proof*
Consider the difference

$$\tilde{\vartheta} = \tilde{\omega} - \tilde{\Omega}$$

and write it as

$$2\tilde{\vartheta} = \sum_{i=1}^{l} \left( p_i \, \tilde{\mathrm{d}} q_i - q_i \, \tilde{\mathrm{d}} p_i \right) - \sum_{i=1}^{l} \left( P_i \, \tilde{\mathrm{d}} Q_i - Q_i \, \tilde{\mathrm{d}} P_i \right) + \tilde{\mathrm{d}} \sum_{i=1}^{l} (p_i q_i - P_i Q_i)$$

$$= \tilde{\eta} + \tilde{\mathrm{d}} \sum_{i=1}^{l} (p_i q_i - P_i Q_i).$$

The form $\tilde{\eta}$ can be rewritten as

$$\tilde{\eta} = \mathbf{X}^T \mathbb{J} \, \tilde{\mathrm{d}} \mathbf{X} - \mathbf{x}^T \mathbb{J} \, \tilde{\mathrm{d}} \mathbf{x}.$$

Recalling that $\tilde{\mathrm{d}} \mathbf{X} = J \, \tilde{\mathrm{d}} \mathbf{x}$, we see that

$$\tilde{\eta} = \left( \mathbf{X}^T \mathbb{J} J - \mathbf{x}^T \mathbb{J} \right) \tilde{\mathrm{d}} \mathbf{x} = \mathbf{g}^T \, \tilde{\mathrm{d}} \mathbf{x},$$

with $\mathbf{g} = -J^T\mathfrak{I}\mathbf{X} + \mathfrak{I}\mathbf{x}$. Therefore, the form $\tilde{\eta}$ is exact if and only if $\partial g_i/\partial x_j = \partial g_j/\partial x_i$. We now compute (using the convention of summation over repeated indices)

$$\frac{\partial g_i}{\partial x_j} = \mathfrak{I}_{ij} - \frac{\partial J_{ki}}{\partial x_j}\mathfrak{I}_{kh}X_h - J_{ki}\mathfrak{I}_{kh}J_{hj},$$

$$\frac{\partial g_j}{\partial x_i} = \mathfrak{I}_{ji} - \frac{\partial J_{kj}}{\partial x_i}\mathfrak{I}_{kh}X_h - J_{kj}\mathfrak{I}_{kh}J_{hi},$$

and note that

$$\frac{\partial J_{ki}}{\partial x_j} = \frac{\partial^2 X_k}{\partial x_i \partial x_j} = \frac{\partial J_{kj}}{\partial x_i},$$

and hence

$$\frac{\partial g_i}{\partial x_j} - \frac{\partial g_j}{\partial x_i} = (\mathfrak{I} - J^T\mathfrak{I}J)_{ij} - (\mathfrak{I} - J^T\mathfrak{I}J)_{ji} = 2(\mathfrak{I} - J^T\mathfrak{I}J)_{ij},$$

where $\mathfrak{I}-J^T\mathfrak{I}J$ is skew-symmetric. We can conclude that the form $\tilde{\eta}$, and therefore $\tilde{\omega} - \tilde{\Omega}$, is exact if and only if $J$ is symplectic, or equivalently if and only if the transformation is canonical. ∎

*Remark* 10.17
If the transformation is completely canonical, it is immediate to check that in the expression (10.84) $\tilde{\mathrm{d}} = \mathrm{d}$, and $f$ can be chosen to be independent of $t$. ∎

*Example* 10.17
Using the Lie condition it is easy to prove that point transformations (Example 10.9) are canonical. It follows from (10.35), (10.36) that

$$\sum_{i=1}^{l}(p_i\ \tilde{\mathrm{d}}q_i - P_i\ \tilde{\mathrm{d}}Q_i) = \sum_{i=1}^{l}p_i\ \tilde{\mathrm{d}}q_i - \sum_{i,j,k=1}^{l}J_{ji}^{-1}p_jJ_{ik}\ \tilde{\mathrm{d}}q_k$$

$$= \sum_{i=1}^{l}p_i\ \tilde{\mathrm{d}}q_i - \sum_{j,k=1}^{l}p_j\delta_{jk}\ \tilde{\mathrm{d}}q_k = 0.$$  ∎

*Example* 10.18
Using the Lie condition let us check that the transformation (see Gallavotti 1986)

$$q_1 = \frac{P_1P_2 - Q_1Q_2}{P_1^2 + Q_2^2}, \qquad q_2 = \frac{P_2Q_2 + P_1Q_1}{P_1^2 + Q_2^2},$$

$$p_1 = -P_1Q_2, \qquad p_2 = \frac{P_1^2 - Q_2^2}{2}$$

is completely canonical. Setting

$$\mathcal{P} = p_1 + ip_2, \qquad \mathcal{Q} = q_1 + iq_2,$$

where $i = \sqrt{-1}$, note that

$$\mathcal{P} = \frac{i}{2}(P_1 + iQ_2)^2, \qquad \mathcal{Q} = \frac{P_2 + iQ_1}{P_1 - iQ_2},$$

from which it follows that

$$p_1\,dq_1 + p_2\,dq_2 = \operatorname{Re}(\mathcal{P}\,d\overline{\mathcal{Q}}) = P_1\,dQ_1 + P_2\,dQ_2 - \frac{1}{2}d(P_1Q_1 + P_2Q_2);$$

hence the Lie condition is satisfied with $f = -\frac{1}{2}(P_1Q_1 + P_2Q_2)$. ∎

*Remark* 10.18

We can see that the Lie condition (10.84) is equivalent to the statement that there exists a regular function $f(\mathbf{P}, \mathbf{Q}, t)$, defined up to an arbitrary function of time, such that, for every $i = 1, \ldots, l$,

$$\begin{aligned}
\frac{\partial f}{\partial P_i}(\mathbf{P}, \mathbf{Q}, t) &= \sum_{j=1}^{l} p_j(\mathbf{P}, \mathbf{Q}, t) \frac{\partial q_j}{\partial P_i}(\mathbf{P}, \mathbf{Q}, t), \\
\frac{\partial f}{\partial Q_i}(\mathbf{P}, \mathbf{Q}, t) &= \sum_{j=1}^{l} p_j(\mathbf{P}, \mathbf{Q}, t) \frac{\partial q_j}{\partial Q_i}(\mathbf{P}, \mathbf{Q}, t) - P_i.
\end{aligned} \qquad (10.85)$$

∎

The Lie condition has as a corollary an interesting result that characterises the canonical transformations through the Poincaré–Cartan integral invariant.

COROLLARY 10.5 *The transformation (10.81) is canonical if and only if, for every closed curve $\gamma_0$ in $\mathbf{R}^{2l+1}$ made of simultaneous states $(\mathbf{p}, \mathbf{q}, t_0)$, if $\Gamma_0$ is its image under the given transformation (in turn made of simultaneous states $(\mathbf{P}, \mathbf{Q}, t_0)$), then*

$$\oint_{\gamma_0} \sum_{i=1}^{l} p_i\,dq_i = \oint_{\Gamma_0} \sum_{i=1}^{l} P_i\,dQ_i. \qquad (10.86)$$

*Proof*

From the definition of a fixed time differential, it follows that

$$\oint_{\gamma_0} \sum_{i=1}^{l} p_i\,dq_i = \oint_{\Gamma_0} \tilde{\omega}, \qquad \oint_{\Gamma_0} \sum_{i=1}^{l} P_i\,dQ_i = \oint_{\Gamma_0} \tilde{\Omega},$$

where $\tilde{\omega}$ and $\tilde{\Omega}$ are computed fixing $t = t_0$. Note that on $\Gamma_0$ we assume that $\tilde{\omega}$ is expressed in the new variables. Therefore the condition is necessary. Indeed, if the transformation is canonical, by the Lie condition the difference $\tilde{\omega} - \tilde{\Omega}$ is an exact form, whose integral along any closed path vanishes.

Evidently the condition is also sufficient. Indeed, if

$$\oint_{\Gamma_0} (\tilde{\omega} - \tilde{\Omega}) = 0$$

along any closed path $\Gamma_0$ then the form $\tilde{\omega} - \tilde{\Omega}$ is exact (see Giusti 1989, Corollary 8.2.1). ∎

For $l = 1$ equation (10.86) is simply the area conservation property, which we already know (in the form $\det J = 1$) to be the characteristic condition for a transformation to be canonical.

We can now prove the important result, stated previously: the conservation of the Poincaré–Cartan integral invariant is exclusively a property of canonical transformations.

THEOREM 10.12  *If the transformation (10.81) is canonical, denote by*

$$\Omega = \sum_{i=1}^{l} P_i \, dQ_i - K(\mathbf{P}, \mathbf{Q}, t) \, dt \tag{10.87}$$

*the new Poincaré–Cartan form. Then there exists a regular function* $\mathcal{F}(\mathbf{P}, \mathbf{Q}, t)$ *such that*

$$\sum_{i=1}^{l} (p_i \, dq_i - P_i \, dQ_i) + (K - H) \, dt = \omega - \Omega = d\mathcal{F}. \tag{10.88}$$

*Hence the difference between the two Poincaré–Cartan forms is exact. Conversely, if (10.81) is a coordinate transformation such that there exist two functions* $K(\mathbf{P}, \mathbf{Q}, t)$ *and* $\mathcal{F}(\mathbf{P}, \mathbf{Q}, t)$ *which, for* $\Omega$ *defined as in (10.87), satisfy (10.88), then the transformation is canonical and $K$ is the new Hamiltonian.*

*Proof*
We prove that if the transformation is canonical, then condition (10.88) is satisfied. Consider any regular closed curve $\gamma$ in $\mathbf{R}^{2l+1}$, and let $\Gamma$ be its image under the canonical transformation (10.81).

Since the transformation is canonical the tube of characteristics of $\omega$ through $\gamma$ is mapped to the tube of characteristics of $\Omega$ through $\Gamma$ (Corollary 10.74). Therefore it is possible to apply Stokes' lemma to write

$$\oint_{\Gamma} (\omega - \Omega) = \oint_{\Gamma_0} (\omega - \Omega) = \oint_{\gamma_0} \sum_{i=1}^{l} p_i \, dq_i - \oint_{\Gamma_0} \sum_{i=1}^{l} P_i \, dQ_i = 0,$$

where $\gamma_0, \Gamma_0$ are the intersections of the respective tubes of characteristics with $t = t_0$ (Fig. 10.2). It follows that the integral of $\omega - \Omega$ along any closed path in $\mathbf{R}^{2l+1}$ is zero, and therefore the form is exact.

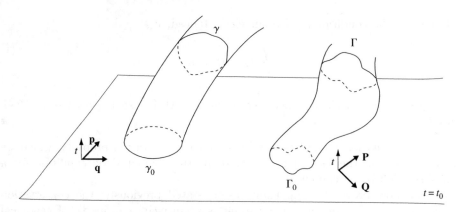

**Fig. 10.2**

We now prove the second part of the theorem. Since the difference $\omega - \Omega$ is exact we have

$$\oint_{\gamma_0} \sum_{i=1}^{l} p_i \, \mathrm{d}q_i - \oint_{\Gamma_0} \sum_{i=1}^{l} P_i \, \mathrm{d}Q_i = \oint_{\Gamma_0} (\omega - \Omega) = 0,$$

and the transformation is canonical. Therefore the characteristic directions of the form $\omega$ coincide, after the transformation, with those of the form $\Omega' = \sum_{i=1}^{l} P_i \, \mathrm{d}Q_i - K' \, \mathrm{d}t$, where $K'$ is the new Hamiltonian. On the other hand, the characteristic directions of $\omega$ coincide with those of $\Omega + \mathrm{d}\mathcal{F}$, and hence of $\Omega$. In addition $\Omega' - \Omega = (K' - K) \, \mathrm{d}t$ and the coincidence of characteristics implies that $K' - K$ may depend only on $t$. Hence, following our convention, $K' = K$. ∎

*Example* 10.19
We consider again Example 10.11 in the light of the results of this section. By equation (10.49), the Lie condition (10.84) can be written as

$$p \, \tilde{\mathrm{d}}q - P \, \tilde{\mathrm{d}}Q = (P - at)(\mathrm{d}Q + t \, \mathrm{d}P) - P \, \mathrm{d}Q = \tilde{\mathrm{d}}f(P, Q, t),$$

from which it follows that

$$f(P, Q, t) = t\frac{P^2}{2} - at^2 P - atQ + f_1(t),$$

where $f_1$ is an arbitrary function of time.

The condition (10.88) for the transformation (10.49), taking into account (10.50), can be written as

$$(P - at)(\mathrm{d}Q + P \, \mathrm{d}t + t \, \mathrm{d}P - at \, \mathrm{d}t) - P \, \mathrm{d}Q + \left(-\frac{P^2}{2} - aQ\right) \mathrm{d}t = \mathrm{d}\mathcal{F}(P, Q, t),$$

## 10.3  Analytical mechanics: canonical formalism

and after some simple manipulations we find

$$\mathcal{F}(P,Q,t) = \frac{1}{2}tP^2 - at^2P - atQ + \frac{1}{3}a^2t^3.$$  ∎

We conclude this section by proving that the Hamiltonian flow defines a canonical transformation.

Let $H(\mathbf{p},\mathbf{q},t)$ be a Hamiltonian function, and consider the associated Hamiltonian flow $\mathbf{x} = S^t\mathbf{X}$:

$$p_i = p_i(\mathbf{P},\mathbf{Q},t), \quad q_i = q_i(\mathbf{P},\mathbf{Q},t), \tag{10.89}$$

where $i = 1,\ldots,l$. Equations (10.89) are therefore the solutions of the system of equations

$$\frac{\partial p_i}{\partial t} = -\frac{\partial H}{\partial q_i}, \quad \frac{\partial q_i}{\partial t} = \frac{\partial H}{\partial p_i}, \tag{10.90}$$

with initial conditions $p_i(0) = P_i$, $q_i(0) = Q_i$, $i = 1,\ldots,l$. By the theorem of existence, uniqueness and continuous dependence on the initial data for ordinary differential equations (see Appendix 1) equation (10.89) defines a coordinate transformation which is regular and invertible.

**THEOREM 10.13**  *The Hamiltonian flow (10.89) is a time-dependent canonical transformation, that at every time instant $t$ maps $\mathbf{X}$ to $S^t\mathbf{X}$. In addition, the new Hamiltonian associated with $H$ in the variables $\mathbf{X}$ is $K \equiv 0$.*

*Proof*
We verify that the Lie condition (10.84) is satisfied, with

$$f(\mathbf{P},\mathbf{Q},t) = \int_0^t \left[\sum_{j=1}^l p_j(\mathbf{P},\mathbf{Q},\tau)\frac{\partial q_j}{\partial t}(\mathbf{P},\mathbf{Q},\tau) - H(\mathbf{p}(\mathbf{P},\mathbf{Q},\tau),\mathbf{q}(\mathbf{P},\mathbf{Q},\tau),\tau)\right] d\tau. \tag{10.91}$$

By Remark 10.18, it is enough to show that for every $i = 1,\ldots,l$ we have

$$\frac{\partial f}{\partial P_i}(\mathbf{P},\mathbf{Q},t) = \sum_{j=1}^l p_j(\mathbf{P},\mathbf{Q},t)\frac{\partial q_j}{\partial P_i}(\mathbf{P},\mathbf{Q},t),$$

$$\frac{\partial f}{\partial Q_i}(\mathbf{P},\mathbf{Q},t) = \sum_{j=1}^l p_j(\mathbf{P},\mathbf{Q},t)\frac{\partial q_j}{\partial Q_i}(\mathbf{P},\mathbf{Q},t) - P_i.$$

We prove the second relation. The first one can be shown in an analogous manner. We have

$$\frac{\partial f}{\partial Q_i} = \int_0^t \sum_{j=1}^l \left[\frac{\partial p_j}{\partial Q_i}\frac{\partial q_j}{\partial t} + p_j\frac{\partial^2 q_j}{\partial t \partial Q_i} - \frac{\partial H}{\partial p_j}\frac{\partial p_j}{\partial Q_i} - \frac{\partial H}{\partial q_j}\frac{\partial q_j}{\partial Q_i}\right] d\tau,$$

but since (10.89) is the transformation generated by the Hamiltonian flow, it follows from equations (10.90) that

$$\begin{aligned}
\frac{\partial f}{\partial Q_i} &= \int_0^t \sum_{j=1}^l \left[ \frac{\partial p_j}{\partial Q_i} \frac{\partial q_j}{\partial t} + p_j \frac{\partial^2 q_j}{\partial t \partial Q_i} - \frac{\partial q_j}{\partial t} \frac{\partial p_j}{\partial Q_i} + \frac{\partial p_j}{\partial t} \frac{\partial q_j}{\partial Q_i} \right] d\tau \\
&= \int_0^t \frac{\partial}{\partial t} \sum_{j=1}^l p_j \frac{\partial q_j}{\partial Q_i} \, d\tau \\
&= \sum_{j=1}^l p_j(\mathbf{P},\mathbf{Q},t) \frac{\partial q_j}{\partial Q_i}(\mathbf{P},\mathbf{Q},t) - \sum_{j=1}^l p_j(\mathbf{P},\mathbf{Q},0) \frac{\partial q_j}{\partial Q_i}(\mathbf{P},\mathbf{Q},0) \\
&= \sum_{j=1}^l p_j(\mathbf{P},\mathbf{Q},t) \frac{\partial q_j}{\partial Q_i}(\mathbf{P},\mathbf{Q},t) - \sum_{j=1}^l P_j \delta_{ji} \\
&= \sum_{j=1}^l p_j(\mathbf{P},\mathbf{Q},t) \frac{\partial q_j}{\partial Q_i}(\mathbf{P},\mathbf{Q},t) - P_i.
\end{aligned}$$

By what we have just computed,

$$\tilde{d}f = \sum_{i=1}^l \left( p_i \, \tilde{d}q_i - P_i \, dQ_i \right),$$

while from (10.83) it obviously follows that

$$df = \tilde{d}f + \frac{\partial f}{\partial t} dt = \sum_{i=1}^l \left( p_i \, \tilde{d}q_i - P_i \, \tilde{d}Q_i \right) + \left[ \sum_{j=1}^l p_j \frac{\partial q_j}{\partial t} - H \right] dt$$

$$= \sum_{i=1}^l (p_i \, dq_i - P_i \, dQ_i) - H \, dt.$$

Taking into account Theorem 10.6, it follows from this that the new Hamiltonian associated with $H$ is exactly $K \equiv 0$. ∎

*Remark* 10.19

From the expression (10.91) for $f$, since $\dot{p}_i = \partial p_i / \partial t$ and $\dot{q}_i = \partial q_i / \partial t$, we see that $f(\mathbf{P},\mathbf{Q},t)$ is the Hamiltonian action $A(\mathbf{P},\mathbf{Q},t)$ (see (9.43)) computed by an integration along the Hamiltonian flow (10.89), i.e. the natural motion. ∎

Recalling the result of Corollary 10.1, we can now state that *the canonical transformations depending on time are all and exclusively the Hamiltonian flows*. If we apply the canonical transformation $\mathbf{x} = \mathbf{x}(\mathbf{x}^*, t)$ generated by the Hamiltonian $H(\mathbf{x}, t)$, to a system with Hamiltonian $H^*(\mathbf{x}^*, t)$, we obtain the new Hamiltonian $K^*(\mathbf{x}, t) = \hat{H}^*(\mathbf{x}, t) + H(\mathbf{x}, t)$ (here $H$ plays the role of the function indicated by $K_0$ in the previous section). Consider now the Hamiltonian flow $\mathbf{x} = S^t \mathbf{X}$, with Hamiltonian $H(\mathbf{x}, t)$. The inverse transformation, mapping $S^t \mathbf{X}$ in $\mathbf{X}$ for every $t$, corresponds to the retrograde motion (with Hamiltonian $-H$) and it is naturally canonical. For the canonical transformation $\mathbf{x} = S^t \mathbf{X}$ the variables $\mathbf{X}$ play the role of *constant canonical coordinates* ($\dot{\mathbf{X}} = 0$). In agreement with this fact, we note that the composition of the two flows yields the Hamiltonian $K(\mathbf{X}, t) = 0$ and therefore precisely constant canonical coordinates. As an example, note that the transformation (10.49) is the flow with Hamiltonian $H = p^2/2 + aq$. This is independent of time, and hence it is a constant of the motion, implying that $p^2/2 + aq = P^2/2 + aQ$. This is the equation for the trajectories, travelled 'forwards' $(P, Q) \to (p, q)$ through the flow with Hamiltonian $H(p, q)$, and 'backwards' $(p, q) \to (P, Q)$ with Hamiltonian (10.50), i.e. $-H(P, Q)$. The superposition of the two yields $(P, Q) \to (P, Q)$ for every $t$, and hence $\dot{P} = \dot{Q} = 0$ (corresponding to the null Hamiltonian).

*Remark 10.20*
The apparent lack of symmetry between the condition

$$\sum_{i=1}^{l}(p_i \, \mathrm{d}q_i - P_i \, \mathrm{d}Q_i) = \mathrm{d}\mathcal{F},$$

where $\mathcal{F}$ is independent of $t$, for a transformation to be *completely* canonical, and the relation

$$\sum_{i=1}^{l}(p_i \, \mathrm{d}q_i - P_i \, \mathrm{d}Q_i) + (K - H) \, \mathrm{d}t = \mathrm{d}\mathcal{F},$$

where $\mathcal{F}$ depends also on $t$, for a time-dependent transformation to be canonical, can be eliminated by using a significant extension of the Hamiltonian formalism.

Indeed, given a non-autonomous Hamiltonian system $H(\mathbf{p}, \mathbf{q}, t)$, we consider, in addition to the canonical equations (10.90), the equations (see (8.26))

$$-\dot{H} = -\frac{\mathrm{d}H}{\mathrm{d}t} = -\frac{\partial H}{\partial t}, \quad \dot{t} = 1. \qquad (10.92)$$

The system of equations (10.90), (10.92) corresponds to the canonical equations for the Hamiltonian $\mathcal{H} : \mathbf{R}^{2l+2} \to \mathbf{R}$,

$$\mathcal{H}(\mathbf{p}, \pi, \mathbf{q}, \tau) = H(\mathbf{p}, \mathbf{q}, \tau) + \pi, \qquad (10.93)$$

where

$$\pi = -H, \quad \tau = t, \qquad (10.94)$$

and hence the Hamiltonian and time are considered as a new pair of canonically conjugate variables. This is possible since $\nabla_\mathbf{p} \mathcal{H} = \nabla_\mathbf{p} H$, $\nabla_\mathbf{q} \mathcal{H} = \nabla_\mathbf{q} H$ and

$$\dot{\pi} = -\frac{\partial \mathcal{H}}{\partial \tau} = -\frac{\partial H}{\partial t}, \quad \dot{\tau} = \frac{\partial \mathcal{H}}{\partial \pi} = 1.$$

By (10.94) we also have that $\mathcal{H} = 0$, and the Poincaré–Cartan form (10.77) becomes

$$\sum_{i=1}^{l} p_i \, \mathrm{d}q_i - H \, \mathrm{d}t = \sum_{i=1}^{l} p_i \, \mathrm{d}q_i + \pi \, \mathrm{d}\tau = \sum_{i=1}^{l+1} p_i \, \mathrm{d}q_i, \qquad (10.95)$$

where we set $p_{l+1} = \pi$, $q_{l+1} = \tau$.

The canonical transformations (10.81) are therefore always completely canonical in $\mathbf{R}^{2l+2}$, and they associate with the variables $(\mathbf{p}, \pi, \mathbf{q}, \tau)$ new variables $(\mathbf{P}, \Pi, \mathbf{Q}, T)$, with the constraint $T = \tau$. The Hamiltonian $\mathcal{H}$ is always zero.

Conversely, transformations such as

$$\tau = a(T), \quad \pi = \frac{1}{a'(T)} \Pi \qquad (10.96)$$

can be included in the canonical formalism, since

$$\pi \, \mathrm{d}\tau = \frac{1}{a'(T)} \Pi \, a'(T) \, \mathrm{d}T = \Pi \, \mathrm{d}T.$$

The effect of equation (10.96) is a re-parametrisation of time, and by using the fact that it is canonical one can show that the canonical structure of Hamilton's equations is preserved, by appropriately rescaling the Hamiltonian $H = -\pi$. ∎

## 10.4 Generating functions

In the previous sections we completely described the class of canonical transformations. We now study a procedure to generate *all* canonical transformations.

As we saw in the previous section, the Lie condition (10.84), or its equivalent formulation (10.88), is a necessary and sufficient condition for a coordinate transformation to be canonical. In the form (10.88), it allows the introduction of an efficient way to construct other canonical transformations.

Assume that

$$\mathbf{p} = \mathbf{p}(\mathbf{P}, \mathbf{Q}, t), \quad \mathbf{q} = \mathbf{q}(\mathbf{P}, \mathbf{Q}, t) \qquad (10.97)$$

defines a canonical transformation in an open domain of $\mathbf{R}^{2l}$, with inverse

$$\mathbf{P} = \mathbf{P}(\mathbf{p}, \mathbf{q}, t), \quad \mathbf{Q} = \mathbf{Q}(\mathbf{p}, \mathbf{q}, t). \qquad (10.98)$$

A canonical transformation of the type (10.97) satisfying

$$\det\left(\frac{\partial q_i}{\partial P_j}\right) \neq 0 \qquad (10.99)$$

is called *free*. Applying the implicit function theorem to the second of equations (10.97), the condition (10.99) ensures that the variables $\mathbf{P}$ can be naturally expressed as functions of the variables $\mathbf{q}$, $\mathbf{Q}$, as well as of time. Therefore, if

$$\mathbf{P} = \hat{\mathbf{P}}(\mathbf{q}, \mathbf{Q}, t), \qquad (10.100)$$

## 10.4  Analytical mechanics: canonical formalism

by substituting this relation into the first of equations (10.97) we find

$$\mathbf{p} = \hat{\mathbf{p}}(\mathbf{q}, \mathbf{Q}, t). \tag{10.101}$$

The condition (10.88)

$$\sum_{i=1}^{l} p_i \, dq_i - H \, dt - \left( \sum_{i=1}^{l} P_i \, dQ_i - K \, dt \right) = d\mathcal{F}$$

can therefore be written

$$\sum_{i=1}^{l} \hat{p}_i(\mathbf{q}, \mathbf{Q}, t) \, dq_i - H(\mathbf{q}, \hat{\mathbf{p}}(\mathbf{q}, \mathbf{Q}, t), t) \, dt$$

$$- \left( \sum_{i=1}^{l} \hat{P}_i(\mathbf{q}, \mathbf{Q}, t) \, dQ_i - K(\hat{\mathbf{P}}(\mathbf{q}, \mathbf{Q}, t), \mathbf{Q}, t) \, dt \right) = dF(\mathbf{q}, \mathbf{Q}, t), \tag{10.102}$$

where the variables $(\mathbf{q}, \mathbf{Q})$ are considered to be independent and $F(\mathbf{q}, \mathbf{Q}, t)$ is obtained from $\mathcal{F}(\mathbf{P}, \mathbf{Q}, t)$ through equation (10.100). From (10.102) it follows that

$$p_i = \frac{\partial F}{\partial q_i}, \tag{10.103}$$

$$P_i = -\frac{\partial F}{\partial Q_i}, \tag{10.104}$$

$$K = H + \frac{\partial F}{\partial t}, \tag{10.105}$$

where $i = 1, \ldots, l$.

Equation (10.104) shows that the matrix $-(\partial q_i/\partial P_j)$ is the inverse matrix of $(\partial^2 F/(\partial q_i \partial Q_j))$. Therefore the condition (10.99) is clearly equivalent to requiring that

$$\det \left( \frac{\partial^2 F}{\partial q_i \partial Q_j} \right) \neq 0. \tag{10.106}$$

We now follow the converse path, starting from the choice of a function of the type (10.106).

DEFINITION 10.10  *A function $F(\mathbf{q}, \mathbf{Q}, t)$ satisfying condition (10.106) is called a generating function (of the first kind, and it is often denoted by $F = F_1$) of the canonical transformation defined implicitly by equations (10.103)–(10.105).* ∎

*Remark* 10.21
Given the generating function $F$, equations (10.103)–(10.105) define the canonical transformation implicitly. However the condition (10.106) ensures that the variables $\mathbf{Q}$ can be expressed as functions of $(\mathbf{q}, \mathbf{p})$ and of time $t$, by inverting equation (10.103). The expression of $\mathbf{P}$ as a function of $(\mathbf{q}, \mathbf{p})$ and of the time $t$ can be obtained by substituting the relation $Q_i = Q_i(\mathbf{q}, \mathbf{p}, t)$ into equation (10.104). The invertibility of the transformation thus obtained is again guaranteed

by the implicit function theorem. Indeed, equation (10.106) also ensures that it is possible to express $\mathbf{q} = \mathbf{q}(\mathbf{Q},\mathbf{P},t)$ by inverting (10.104). Substituting these into equation (10.103) we finally find $\mathbf{p} = \mathbf{p}(\mathbf{Q},\mathbf{P},t)$. ∎

*Example* 10.20
The function $F(q,Q) = m\omega/2q^2 \cot Q$ generates a canonical transformation

$$p = \sqrt{2P\omega m} \cos Q, \qquad q = \sqrt{\frac{2P}{\omega m}} \sin Q,$$

which transforms the Hamiltonian of the harmonic oscillator

$$H(p,q) = \frac{p^2}{2m} + \frac{m\omega^2 q^2}{2}$$

into

$$K(P,Q) = \omega P.$$

∎

*Example* 10.21
The identity transformation $p = P$, $q = Q$ is not free. Hence it does not admit a generating function of the first kind. ∎

After setting $\mathbf{x} = (\mathbf{p},\mathbf{q})$ and $\mathbf{X} = (\mathbf{P},\mathbf{Q})$, we see that a generating function can also depend on $x_{m_1},\ldots,x_{m_l}, X_{n_1},\ldots,X_{n_l}$ for an arbitrary choice of the indices $m_i$ and $n_i$ (all different). We quickly analyse all possible cases.

DEFINITION 10.11  A function $F(\mathbf{q},\mathbf{P},t)$ satisfying the condition

$$\det\left(\frac{\partial^2 F}{\partial q_i \partial P_j}\right) \neq 0 \qquad (10.107)$$

is called a generating function of the second kind (and it is often denoted by $F = F_2$) of the canonical transformation implicitly defined by

$$p_i = \frac{\partial F}{\partial q_i}, \quad i = 1,\ldots,l, \qquad (10.108)$$

$$Q_i = \frac{\partial F}{\partial P_i}, \quad i = 1,\ldots,l. \qquad (10.109)$$

∎

*Example* 10.22
Point transformations (see Example 10.9)

$$\mathbf{Q} = \mathbf{Q}(\mathbf{q},t)$$

are generated by

$$F_2(\mathbf{q},\mathbf{P},t) = \sum_{i=1}^{l} P_i Q_i(\mathbf{q},t).$$

Setting $\mathbf{Q} = \mathbf{q}$ we find that $F_2 = \sum_{i=1}^{l} P_i q_i$ is the generating function of the identity transformation. ∎

DEFINITION 10.12 *A function $F(\mathbf{p},\mathbf{Q},t)$ which satisfies the condition*

$$\det\left(\frac{\partial^2 F}{\partial p_i \partial Q_j}\right) \neq 0 \qquad (10.110)$$

*is called a generating function of the third kind (and it is often denoted by $F = F_3$) of the canonical transformation implicitly defined by*

$$q_i = -\frac{\partial F}{\partial p_i}, \quad i = 1,\ldots,l, \qquad (10.111)$$

$$P_i = -\frac{\partial F}{\partial Q_i}, \quad i = 1,\ldots,l. \qquad (10.112)$$

∎

*Example* 10.23
It is immediate to check that the function $F(p,Q) = -p(e^Q - 1)$ generates the canonical transformation

$$P = p(1+q), \qquad Q = \log(1+q). \qquad \blacksquare$$

DEFINITION 10.13 *A function $F(\mathbf{p},\mathbf{P},t)$ which satisfies the condition*

$$\det\left(\frac{\partial^2 F}{\partial p_i \partial P_j}\right) \neq 0 \qquad (10.113)$$

*is called a generating function of the fourth kind (and it is often denoted by $F = F_4$) of the canonical transformation implicitly defined by*

$$q_i = -\frac{\partial F}{\partial p_i}, \quad i = 1,\ldots,l, \qquad (10.114)$$

$$Q_i = \frac{\partial F}{\partial P_i}, \quad i = 1,\ldots,l. \qquad (10.115)$$

∎

*Example* 10.24
The canonical transformation of Example 10.8, exchanging the coordinates and the kinetic momenta, admits as generating function $F(\mathbf{p},\mathbf{P}) = \sum_{i=1}^{l} p_i P_i$. ∎

**Theorem 10.14** *The generating functions of the four kinds $F_1$, $F_2$, $F_3$ and $F_4$ satisfy, respectively,*

$$\sum_{i=1}^{l}(p_i\,dq_i - P_i\,dQ_i) + (K-H)\,dt = dF_1(\mathbf{q},\mathbf{Q},t), \tag{10.116}$$

$$\sum_{i=1}^{l}(p_i\,dq_i + Q_i\,dP_i) + (K-H)\,dt = dF_2(\mathbf{q},\mathbf{P},t), \tag{10.117}$$

$$\sum_{i=1}^{l}(-q_i\,dp_i - P_i\,dQ_i) + (K-H)\,dt = dF_3(\mathbf{p},\mathbf{Q},t), \tag{10.118}$$

$$\sum_{i=1}^{l}(-q_i\,dp_i + Q_i\,dP_i) + (K-H)\,dt = dF_4(\mathbf{p},\mathbf{P},t). \tag{10.119}$$

*If a canonical transformation admits more than one generating function of the previous kinds, then these are related by a Legendre transformation:*

$$F_2 = F_1 + \sum_{i=1}^{l} P_i Q_i,$$

$$F_3 = F_1 - \sum_{i=1}^{l} p_i q_i, \tag{10.120}$$

$$F_4 = F_1 - \sum_{i=1}^{l} p_i q_i + \sum_{i=1}^{l} P_i Q_i = F_2 - \sum_{i=1}^{l} p_i q_i = F_3 + \sum_{i=1}^{l} P_i Q_i.$$

*Proof*

The first part of the theorem is a consequence of Definitions 10.10–10.13. The proof of the second part is immediate, and can be obtained by adding or subtracting $\sum_{i=1}^{l} P_i Q_i$ and $\sum_{i=1}^{l} p_i q_i$ from (10.116). ∎

*Remark 10.22*

At this point it should be clear how, in principle, there exist $2\binom{2l}{l}$ different kinds of generating functions, each corresponding to a different arbitrary choice of $l$ variables among $\mathbf{q}$, $\mathbf{p}$ and of $l$ variables among $\mathbf{Q}$, $\mathbf{P}$. However, it is always possible to reduce it to one of the four previous kinds, by taking into account that the exchanges of Lagrangian coordinates and kinetic momenta are canonical transformations (see Example 10.8). ∎

The transformations associated with generating functions exhaust all canonical transformations.

**Theorem 10.15** *It is possible to associate with every canonical transformation a generating function, and the transformation is completely canonical if and only if its generating function is time-independent. The generating function is of one of the four kinds listed above, up to possible exchanges of Lagrangian coordinates with kinetic moments.*

## 10.4 Analytical mechanics: canonical formalism

*Proof*
Consider a canonical transformation, and let $\mathcal{F}$ the function associated with it by Theorem 10.12. If it is possible to express the variables $\mathbf{p}$, $\mathbf{P}$ as functions of $\mathbf{q}$, $\mathbf{Q}$, and hence if (10.99) holds, then, as we saw at the beginnning of this section, it is enough to set

$$F_1(\mathbf{q}, \mathbf{Q}, t) = \mathcal{F}(\hat{\mathbf{P}}(\mathbf{q}, \mathbf{Q}, t), \mathbf{Q}, t)$$

and the conditions of Definition 10.10 are satisfied.

If, on the other hand, we have

$$\det\left(\frac{\partial q_i}{\partial Q_j}\right) \neq 0, \tag{10.121}$$

we can deduce $\mathbf{Q} = \hat{\mathbf{Q}}(\mathbf{q}, \mathbf{P}, t)$ from the second of equations (10.97) and, by substitution into the first of equations (10.97), we find that the variables $\mathbf{p}$ can also be expressed through $\mathbf{q}$, $\mathbf{P}$. Hence we set

$$F_2(\mathbf{q}, \mathbf{P}, t) = \mathcal{F}(\mathbf{P}, \hat{\mathbf{Q}}(\mathbf{q}, \mathbf{P}, t), t) + \sum_{i=1}^{l} P_i \hat{Q}_i(\mathbf{q}, \mathbf{P}, t).$$

The condition (10.107) is automatically satisfied, since $(\partial^2 F / \partial q_i \partial P_j)$ is the inverse matrix of $(\partial q_i / \partial Q_j)$.

Analogously, if

$$\det\left(\frac{\partial p_i}{\partial P_j}\right) \neq 0, \tag{10.122}$$

the variables $\mathbf{q}$, $\mathbf{P}$ can be expressed through $\mathbf{p}$, $\mathbf{Q}$, and we set

$$F_3(\mathbf{p}, \mathbf{Q}, t) = \mathcal{F}(\hat{\mathbf{P}}(\mathbf{p}, \mathbf{Q}, t), \mathbf{Q}, t) - \sum_{i=1}^{l} p_i \hat{q}_i(\mathbf{p}, \mathbf{Q}, t).$$

Then the conditions of Definition 10.12 are satisfied.

Finally, if

$$\det\left(\frac{\partial p_i}{\partial Q_j}\right) \neq 0, \tag{10.123}$$

by expressing $\mathbf{q}$, $\mathbf{Q}$ as functions of $\mathbf{p}$, $\mathbf{P}$, we find that the generating function is given by

$$F_4(\mathbf{p}, \mathbf{P}, t) = \mathcal{F}(\mathbf{P}, \hat{\mathbf{Q}}(\mathbf{p}, \mathbf{P}, t), t) - \sum_{i=1}^{l} p_i \hat{q}_i(\mathbf{p}, \mathbf{P}, t) + \sum_{i=1}^{l} P_i \hat{Q}_i(\mathbf{p}, \mathbf{P}, t).$$

It is always possible to choose $l$ variables among $\mathbf{p}$, $\mathbf{q}$ and $l$ variables among $\mathbf{P}$, $\mathbf{Q}$ as independent variables. As a matter of fact, the condition that the Jacobian matrix of the transformation is symplectic, and therefore non-singular, guarantees the existence an $l \times l$ submatrix with a non-vanishing determinant. If the selected independent variables are not in any of the four groups already considered, we can proceed in a similar way, and obtain a generating function of a different kind. On the other hand, it is always possible to reduce to one of the previous cases by a suitable exchange of variables. ∎

*Remark 10.23*
An alternative proof of the previous theorem, that is maybe more direct and certainly more practical in terms of applications, can be obtained simply by remarking how conditions (10.99), (10.121)–(10.123) ensure that the Lie condition can be rewritten in the form (10.116)–(10.119), respectively. The functions $F_1, \ldots, F_4$ can be determined by integration along an arbitrary path in the domain of definition and the invertibility of the transformation. ∎

*Example 10.25*
Consider the canonical transformation

$$p = 2e^t \sqrt{PQ} \log P, \quad q = e^{-t} \sqrt{PQ},$$

defined in $D = \{(P, Q) \in \mathbf{R}^2 | P > 0, Q \geq 0\} \subset \mathbf{R}^2$. Evidently it is possible to choose $(q, P)$ as independent variables and write

$$p = 2e^{2t} q \log P, \quad Q = \frac{e^{2t} q^2}{P}.$$

The generating function $F_2(q, P, t)$ can be found, for example, by integrating the differential form

$$\hat{p}(q, P, t) \, dq + \hat{Q}(q, P, t) \, dP$$

along the path $\gamma = \{(x, 1) | 0 \leq x \leq q\} \cup \{(q, y) | 1 \leq y \leq P\}$ in the plane $(q, P)$. Since along the first horizontal part of the path $\gamma$ one has $p(x, 1, t) \equiv 0$ (this simplification motivates the choice of the integration path $\gamma$), we have

$$F_2(q, P, t) = e^{2t} q^2 \int_1^P \frac{dy}{y} + \tilde{F}_2(t) = e^{2t} q^2 \log P + \tilde{F}_2(t),$$

where $\tilde{F}_2$ is an arbitrary function of time. ∎

*Remark 10.24*
Every generating function $F$ is defined up to an arbitrary additive term, a function only of time. This term does not change the transformation generated by $F$, but it modifies the Hamiltonian (because of (10.105)) and it arises from the corresponding indetermination of the difference between the Poincaré–Cartan forms associated with the transformation (see Remark 10.18). Similarly

to what has already been seen, this undesired indetermination can be overcome by requiring that the function $F$ does not contain terms that are only functions of $t$. ∎

We conclude this section by proving a uniqueness result for the generating function (once the arbitrariness discussed in the previous remark is resolved).

PROPOSITION 10.4  *All the generating functions of a given canonical transformation, depending on the same group of independent variables, differ only by a constant.*

*Proof*
Consider as an example the case of two generating functions $F(\mathbf{q}, \mathbf{Q}, t)$ and $G(\mathbf{q}, \mathbf{Q}, t)$. The difference $F - G$ satisfies the conditions

$$\frac{\partial}{\partial q_i}(F - G) = 0, \qquad \frac{\partial}{\partial Q_i}(F - G) = 0,$$

for every $i = 1, \ldots, l$. Hence, since by Remark 10.24 we have neglected additive terms depending only on time, $F - G$ is necessarily constant. ∎

## 10.5 Poisson brackets

Consider two funtions $f(\mathbf{x}, t)$ and $g(\mathbf{x}, t)$ defined in $\mathbf{R}^{2l} \times \mathbf{R}$ with sufficient regularity, and recall the definition (10.16) of a standard symplectic product.

DEFINITION 10.14  The *Poisson bracket of the two functions*, denoted by $\{f, g\}$, *is the function defined by the symplectic product of the gradients of the two functions:*

$$\{f, g\} = (\nabla_{\mathbf{x}} f)^T \mathbf{J} \nabla_{\mathbf{x}} g. \tag{10.124}$$

∎

*Remark* 10.25
If $\mathbf{x} = (\mathbf{p}, \mathbf{q})$, the Poisson bracket of two functions $f$ and $g$ is given by

$$\{f, g\} = \sum_{i=1}^{l} \left( \frac{\partial f}{\partial q_i} \frac{\partial g}{\partial p_i} - \frac{\partial g}{\partial q_i} \frac{\partial f}{\partial p_i} \right). \tag{10.125}$$

∎

*Remark* 10.26
Using the Poisson brackets, Hamilton's equations in the variables $(\mathbf{p}, \mathbf{q})$ can be written in a perfectly symmetric form as

$$\dot{p}_i = \{p_i, H\}, \qquad \dot{q}_i = \{q_i, H\}, \quad i = 1, \ldots, l. \tag{10.126}$$

∎

*Remark* 10.27
From equation (10.125) we derive the *fundamental Poisson brackets*

$$\{p_i, p_j\} = \{q_i, q_j\} = 0, \quad \{q_i, p_j\} = -\{p_i, q_j\} = \delta_{ij}. \tag{10.127}$$

∎

*Example* 10.26
If we consider the phase space $\mathbf{R}^6$ of a free point particle, if $L_1, L_2$ and $L_3$ are the three components of its angular momentum, and $p_1$, $p_2$, $p_3$ are the kinetic momenta, conjugate with the Cartesian cordinates of the point, we have:

$$\{p_1, L_3\} = -p_2, \quad \{p_2, L_3\} = p_1, \quad \{p_3, L_3\} = 0,$$

and similarly for $L_1$ and $L_2$. Using the Ricci tensor $\epsilon_{ijk}$, the previous relations take the more concise form

$$\{p_i, L_j\} = \epsilon_{ijk} p_k$$

($\epsilon_{ijk} = 0$ if the indices are not all different, otherwise $\epsilon_{ijk} = (-1)^n$, where $n$ is the number of permutations of pairs of elements to be performed on the sequence $\{1, 2, 3\}$ to obtain $\{i, j, k\}$). It can be verified in an analogous way that

$$\{L_i, L_j\} = \epsilon_{ijk} L_k,$$

and that

$$\{L_i, L^2\} = 0,$$

where

$$L^2 = L_1^2 + L_2^2 + L_3^2.$$
■

The Poisson brackets are an important tool, within the Hamiltonian formalism, for the analysis of the first integrals of the motion (also, as we shall see, to characterise the canonical transformations). Indeed, let $H : \mathbf{R}^{2l} \times \mathbf{R} \to \mathbf{R}$, $H = H(\mathbf{x}, t)$ be a Hamiltonian function and consider the corresponding canonical equations

$$\dot{\mathbf{x}} = \mathsf{J} \nabla_{\mathbf{x}} H, \tag{10.128}$$

with initial conditions $\mathbf{x}(0) = \mathbf{x}_0$. Suppose that the solution of Hamilton's equations can be continued for all times $t \in \mathbf{R}$, for any initial condition. In this case, the Hamiltonian flow $\mathbf{x}(t) = \mathbf{S}^t(\mathbf{x}_0)$ defines an *evolution operator* $U^t$ acting on the observables of the system, i.e. on every function $f : \mathbf{R}^{2l} \times \mathbf{R} \to \mathbf{R}$, $f = f(\mathbf{x}, t)$:

$$(U^t f)(\mathbf{x}_0, 0) = f(\mathbf{S}^t \mathbf{x}_0, t) = f(\mathbf{x}(t), t). \tag{10.129}$$

DEFINITION 10.15 *A function $f(\mathbf{x}, t)$ is a* first integral *for the Hamiltonian flow $\mathbf{S}^t$ if and only if for every choice of $\mathbf{x}_0 \in \mathbf{R}^{2l}$ and $t \in \mathbf{R}$, it holds that*

$$f(\mathbf{S}^t \mathbf{x}_0, t) = f(\mathbf{x}_0, 0). \tag{10.130}$$
■

The total derivative of $f$ with respect to time $t$, computed along the Hamiltonian flow $\mathbf{S}^t$, is given by

$$\frac{df}{dt} = \frac{\partial f}{\partial t} + (\nabla_{\mathbf{x}} f)^T \dot{\mathbf{x}} = \frac{\partial f}{\partial t} + (\nabla_{\mathbf{x}} f)^T \mathfrak{I} \nabla_{\mathbf{x}} H.$$

Then using equation (10.124) we have

$$\frac{df}{dt} = \frac{\partial f}{\partial t} + \{f, H\}, \qquad (10.131)$$

which yields the following.

THEOREM 10.16 *A function $f(\mathbf{x})$, independent of time, is a first integral for the Hamiltonian flow $\mathbf{S}^t$ if and only if its Poisson bracket with the Hamiltonian vanishes.* ∎

This characterisation of first integrals is one of the most important properties of the Poisson brackets. However, since Definition 10.14 is made with reference to a specific coordinate system, while a first integral depends only on the Hamiltonian flow and is evidently invariant under canonical transformations, we must consider the question of the invariance of the Poisson brackets under canonical transformations.

THEOREM 10.17 *The following statements are equivalent.*

*(1) The transformation*

$$\mathbf{x} = \mathbf{x}(\mathbf{X}, t), \qquad (10.132)$$

*is canonical.*
*(2) For every pair of functions $f(\mathbf{x}, t)$ and $g(\mathbf{x}, t)$, if $F(\mathbf{X}, t) = f(\mathbf{x}(\mathbf{X}, t), t)$ and $G(\mathbf{X}, t) = g(\mathbf{x}(\mathbf{X}, t), t)$ are the corresponding transforms, then*

$$\{f, g\}_{\mathbf{x}} = \{F, G\}_{\mathbf{X}} \qquad (10.133)$$

*at every instant $t$. Here $\{f, g\}_{\mathbf{x}}$ indicates the Poisson bracket computed with respect to the original canonical variables $\mathbf{x} = (\mathbf{p}, \mathbf{q})$, and $\{F, G\}_{\mathbf{X}}$ indicates that computed with respect to the new variables $\mathbf{X} = (\mathbf{P}, \mathbf{Q})$.*
*(3) For every $i, j = 1, \ldots, l$ and at every instant $t$ it holds that*

$$\begin{aligned} \{P_i, P_j\}_{\mathbf{x}} &= \{Q_i, Q_j\}_{\mathbf{x}} = 0, \\ \{Q_i, P_j\}_{\mathbf{x}} &= \delta_{ij}, \end{aligned} \qquad (10.134)$$

*i.e. the transformation (10.132) preserves the fundamental Poisson brackets.*

Proof
We start by checking that (1) ⇒ (2). We know that a transformation is canonical if and only if its Jacobian matrix

$$J = \nabla_{\mathbf{x}} \mathbf{X}$$

is at every instant a symplectic matrix. Using equation (10.124) and recalling the transformation rule for the gradient, we find $\nabla_{\mathbf{x}} f = J^T \nabla_{\mathbf{X}} F$,

$$\{f,g\}_{\mathbf{x}} = (\nabla_{\mathbf{x}} f)^T \mathbb{J} \nabla_{\mathbf{x}} g = (J^T \nabla_{\mathbf{X}} F)^T \mathbb{J} J^T \nabla_{\mathbf{X}} G = (\nabla_{\mathbf{X}} F)^T J \mathbb{J} J^T \nabla_{\mathbf{X}} G = \{F,G\}_{\mathbf{X}}.$$

That (2) $\Rightarrow$ (3) is obvious ((3) is a special case of (2)). To conclude, we prove then that (3) $\Rightarrow$ (1). For this it is enough to note that equations (10.134) imply that the Jacobian matrix $J$ is symplectic. Indeed, it is immediate to verify that, for any transformation, the matrix $J\mathbb{J}J^T$ has an $l \times l$ block representation

$$J\mathbb{J}J^T = \begin{pmatrix} A & B \\ C & D \end{pmatrix},$$

where $A, B, C, D$ have as entries

$$A_{ij} = \{P_i, P_j\}, \quad B_{ij} = \{P_i, Q_j\}, \quad C_{ij} = \{Q_i, P_j\}, \quad D_{ij} = \{Q_i, Q_j\}.$$

Note that if $l = 1$, then $\{Q, P\} = \det J$ and equations (10.134) reduce to $\det J = 1$. ∎

The formal properties of the Poisson brackets will be summarised at the end of the next section.

## 10.6 Lie derivatives and commutators

DEFINITION 10.16 *A Lie derivative associates with the vector field* $\mathbf{v}$ *the differentiation operator*

$$L_{\mathbf{v}} = \sum_{i=1}^{N} v_i \frac{\partial}{\partial x_i}. \tag{10.135}$$

∎

Evidently the Lie derivative is a linear operator and it satisfies the Leibniz formula: if $f$ and $g$ are two functions on $\mathbf{R}^N$ with values in $\mathbf{R}$ then

$$L_{\mathbf{v}}(fg) = f L_{\mathbf{v}} g + g L_{\mathbf{v}} f. \tag{10.136}$$

Consider the differential equation

$$\dot{\mathbf{x}} = \mathbf{v}(\mathbf{x}), \tag{10.137}$$

associated with the field $\mathbf{v}$, and denote by $\mathbf{g}^t(\mathbf{x}_0)$ the solution passing through $\mathbf{x}_0$ at time $t = 0$, i.e. the flow associated with $\mathbf{v}$. The main property of the Lie derivative is given by the following proposition. This proposition also justifies the name 'derivative along the vector field $\mathbf{v}$' that is sometimes used for $L_{\mathbf{v}}$.

PROPOSITION 10.5 *The Lie derivative of a function* $f : \mathbf{R}^N \to \mathbf{R}$ *is given by*

$$(L_{\mathbf{v}} f)(\mathbf{x}) = \frac{d}{dt} f \circ \mathbf{g}^t(\mathbf{x})|_{t=0}. \tag{10.138}$$

## 10.6 Analytical mechanics: canonical formalism

*Proof*
This fact is of immediate verification: since $\mathbf{g}^t(\mathbf{x})$ is the solution of (10.137) passing through $\mathbf{x}$ for $t = 0$,

$$\left.\frac{d}{dt}\right|_{t=0} f \circ \mathbf{g}^t(\mathbf{x}) = \sum_{i=1}^{N} \frac{\partial f}{\partial x_i} \dot{g}_i^0(\mathbf{x}) = \sum_{i=1}^{N} \frac{\partial f}{\partial x_i} v_i(\mathbf{x}) = \mathbf{v} \cdot \nabla f. \qquad (10.139)$$

■

From the previous proposition it follows that a function $f(\mathbf{x})$ is a first integral of the motion for the flow $\mathbf{g}^t$ associated with the equation (10.137) if and only if its Lie derivative is zero.

If $\mathbf{v} = \mathfrak{I}\nabla_\mathbf{x} H$ is a Hamiltonian field, then, as we saw, $L_\mathbf{v} f = \{f, H\}$. Suppose now that two vector fields $\mathbf{v}_1$ and $\mathbf{v}_2$ are given, and denote by $\mathbf{g}_1^t$ and $\mathbf{g}_2^s$ the respective flows. In general, the flows of two vector fields do not commute, and hence

$$\mathbf{g}_1^t \mathbf{g}_2^s(\mathbf{x}) \neq \mathbf{g}_2^s \mathbf{g}_1^t(\mathbf{x}).$$

*Example* 10.27
Consider the flows

$$\mathbf{g}_1^t(\mathbf{x}) = (x_1 \cos t - x_2 \sin t, x_1 \sin t + x_2 \cos t),$$
$$\mathbf{g}_2^t(\mathbf{x}) = (x_1 + t, x_2),$$

associated with the two vector fields in $\mathbf{R}^2$ given by

$$\mathbf{v}_1(\mathbf{x}) = (-x_2, x_1),$$
$$\mathbf{v}_2(\mathbf{x}) = (1, 0).$$

One can immediately verify in this case that they do not commute (Fig. 10.3). In addition, the function $f_1(x_1, x_2) = \frac{1}{2}(x_1^2 + x_2^2)$, such that $\mathfrak{I}\nabla f_1 = \mathbf{v}_1$, is a first integral of the motion for $\mathbf{g}_1^t$, and its Lie derivative is $L_{\mathbf{v}_1} f_1 = 0$, while it is not constant along $\mathbf{g}_2^t$ and $L_{\mathbf{v}_2} f_1 \neq 0$. By symmetry, for $f_2(x_1, x_2) = -x_2$, such that $\mathfrak{I}\nabla f_2 = \mathbf{v}_2$, we have $L_{\mathbf{v}_2} f_2 = 0$ and $L_{\mathbf{v}_1} f_2 \neq 0$.

■

Using the Lie derivative it is possible to measure the *degree of non-commutativity* of two flows. To this end, we consider any regular function $f$, defined on $\mathbf{R}^N$ and we compare the values it assumes at the points $\mathbf{g}_1^t \mathbf{g}_2^s(\mathbf{x})$ and $\mathbf{g}_2^s \mathbf{g}_1^t(\mathbf{x})$. The lack of commutativity is measured by the difference

$$(\Delta f)(t, s, \mathbf{x}) = f(\mathbf{g}_2^s \mathbf{g}_1^t(\mathbf{x})) - f(\mathbf{g}_1^t \mathbf{g}_2^s(\mathbf{x})). \qquad (10.140)$$

Clearly $(\Delta f)(0, 0, \mathbf{x}) \equiv 0$ and it is easy to check that the first non-zero term (with starting-point $s = t = 0$) in the Taylor series expansion of $\Delta f$ with respect to $s$ and $t$ is given by

$$\frac{\partial^2 (\Delta f)}{\partial t \partial s}(0, 0, \mathbf{x}) st,$$

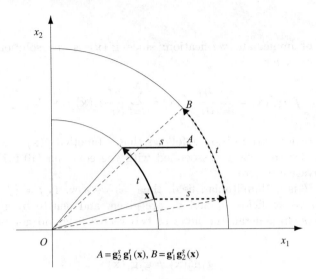

Fig. 10.3

while the other terms of degree 2 are zero. We now seek an explicit expression for it.

DEFINITION 10.17   The commutator *of two vector fields* $\mathbf{v}_1$, $\mathbf{v}_2$ *is the vector field* $\mathbf{w}$, *denoted by* $\mathbf{w} = [\mathbf{v}_1, \mathbf{v}_2]$, *with components*

$$w_i = \sum_{j=1}^{N} \left( (\mathbf{v}_1)_j \frac{\partial (\mathbf{v}_2)_i}{\partial x_j} - (\mathbf{v}_2)_j \frac{\partial (\mathbf{v}_1)_i}{\partial x_j} \right) = L_{\mathbf{v}_1}(\mathbf{v}_2)_i - L_{\mathbf{v}_2}(\mathbf{v}_1)_i. \qquad (10.141)$$

∎

For the fields of Example 10.26 the commutator is $[\mathbf{v}_1, \mathbf{v}_2] = (0, -1)$.

Remark 10.28
For any function $f$ we find:

$$\begin{aligned}
L_{\mathbf{v}_1} L_{\mathbf{v}_2} f - L_{\mathbf{v}_2} L_{\mathbf{v}_1} f &= (\mathbf{v}_1^T \nabla_{\mathbf{x}}) \mathbf{v}_2^T \nabla_{\mathbf{x}} f - (\mathbf{v}_2^T \nabla_{\mathbf{x}}) \mathbf{v}_1^T \nabla_{\mathbf{x}} f \\
&= \sum_{i,j=1}^{N} (\mathbf{v}_1)_j \frac{\partial}{\partial x_j} \left[ (\mathbf{v}_2)_i \frac{\partial f}{\partial x_i} \right] - (\mathbf{v}_2)_j \frac{\partial}{\partial x_j} \left[ (\mathbf{v}_1)_i \frac{\partial f}{\partial x_i} \right] \\
&= \sum_{i,j=1}^{N} \left[ (\mathbf{v}_1)_j (\mathbf{v}_2)_i \frac{\partial^2 f}{\partial x_i \partial x_j} + (\mathbf{v}_1)_j \frac{\partial (\mathbf{v}_2)_i}{\partial x_j} \frac{\partial f}{\partial x_i} \right. \\
&\quad \left. - (\mathbf{v}_2)_j \frac{\partial (\mathbf{v}_1)_i}{\partial x_j} \frac{\partial f}{\partial x_i} - (\mathbf{v}_2)_j (\mathbf{v}_1)_i \frac{\partial^2 f}{\partial x_i \partial x_j} \right] \\
&= \mathbf{w} \cdot \nabla f,
\end{aligned} \qquad (10.142)$$

since the terms containing the second derivatives cancel each other. Hence we obtain the important result

$$[L_{\mathbf{v}_1}, L_{\mathbf{v}_2}] = L_{\mathbf{w}} = L_{[\mathbf{v}_1,\mathbf{v}_2]}, \tag{10.143}$$

so that the commutator of the Lie derivatives $L\mathbf{v}_1, L\mathbf{v}_2$ is the Lie derivative $L_{\mathbf{w}}$ associated with the commutator $[\mathbf{v}_1, \mathbf{v}_2]$. ∎

PROPOSITION 10.6  Let $(\Delta f)(t, s, \mathbf{x})$ be defined as in (10.140). Then

$$\frac{\partial^2 (\Delta f)}{\partial t \partial s}(0, 0, \mathbf{x}) = (L_{[\mathbf{v}_1, \mathbf{v}_2]} f)(\mathbf{x}). \tag{10.144}$$

*Proof*
From equation (10.138) it follows that

$$(L_{\mathbf{v}_1} f)(\mathbf{g}_2^s(\mathbf{x})) = \left.\frac{\partial}{\partial t}\right|_{t=0} f(\mathbf{g}_1^t \mathbf{g}_2^s(\mathbf{x})).$$

Therefore, applying (10.138) to the function $(L_{\mathbf{v}_1} f)$ we find

$$(L_{\mathbf{v}_2}(L_{\mathbf{v}_1} f))(\mathbf{x}) = \left.\frac{\partial^2}{\partial t \partial s}\right|_{t=s=0} f(\mathbf{g}_1^t \mathbf{g}_2^s(\mathbf{x})).$$

Interchanging the order of $\mathbf{g}_1^t$ and $\mathbf{g}_2^s$ and subtracting we reach the conclusion. ∎

For the fields of Example 10.26, we have $L_{[\mathbf{v}_1,\mathbf{v}_2]} f = -\partial f/\partial x_2$.

To define the degree of non-commutativity $\delta$ we can now apply the operator $L_{\mathbf{w}}$ to the functions $x_i$ and set $\delta = \sup_{i=1,\ldots,2l} |L_{\mathbf{w}} x_i|$.

REMARK 10.29
Two flows commute if and only if their commutator is zero (see Arnol'd 1978a, section 39e). ∎

We now seek a characterisation of a pair of *Hamiltonian* flows which commute. This analysis has important consequences, which will be considered in the next chapter. To this end, we define

$$D_f = \{f, \cdot\}, \tag{10.145}$$

the first-order differential operator which to each function $g : \mathbf{R}^{2l} \to \mathbf{R}$, $g = g(\mathbf{x})$ associates its Poisson bracket with $f : \mathbf{R}^{2l} \to \mathbf{R}$:

$$D_f g = \{f, g\}.$$

If $\mathbf{v}_f$ denotes a Hamiltonian vector field associated with $f$, i.e.

$$\mathbf{v}_f = \mathbb{J} \nabla_{\mathbf{x}} f,$$

we have
$$L_{\mathbf{v}_f} = (\mathfrak{I}\nabla_{\mathbf{x}}f)^T\nabla_{\mathbf{x}} = -(\nabla_{\mathbf{x}}f)^T\mathfrak{I}\nabla_{\mathbf{x}} = -D_f. \qquad (10.146)$$

We can now prove our main result.

THEOREM 10.18 *Consider two Hamiltonian fields* $\mathbf{v}_i = \mathfrak{I}\nabla_{\mathbf{x}}f_i$, $i = 1, 2$. *The respective Hamiltonian flows* $g_i^t$, $i = 1, 2$, *commute if and only if* $\{f_1, f_2\} =$ *constant*.

*Proof*
Compute the commutator $\mathbf{w}$ of the two Hamiltonian fields. Following the definition (10.141) and applying equation (10.145), we obtain

$$w_i = (\nabla_{\mathbf{x}}\{f_1, f_2\})_{i+l}, \quad 1 \le i \le l, \qquad (10.147)$$
$$w_i = -(\nabla_{\mathbf{x}}\{f_1, f_2\})_{i-l}, \quad l+1 \le i \le 2l, \qquad (10.148)$$

and hence

$$\mathbf{w} = -\mathfrak{I}\nabla_{\mathbf{x}}\{f_1, f_2\}. \qquad (10.149)$$

From (10.149) and Remark 10.29 the conclusion follows immediately. ∎

DEFINITION 10.18 *Two regular functions* $f_1, f_2 : \mathbf{R}^{2l} \to \mathbf{R}$ *such that*

$$\{f_1, f_2\} = 0 \qquad (10.150)$$

*are said to be* in involution. ∎

*Remark* 10.30
From Theorem 10.18 it follows in particular that pairs of Hamiltonians in involution generate commuting flows. If $\{f_1, f_2\} = 0$ then not only $\mathbf{w} = 0$ but also $L_{\mathbf{v}_1}f_2 = L_{\mathbf{v}_2}f_1 = 0$. ∎

Note that in Example 10.26 the two non-commuting flows have Hamiltonians that satisfy $\{f_1, f_2\} = x_1$, which is non-constant. For two uniform rectilinear motions in orthogonal directions in $\mathbf{R}^2$, generated by $f_1 = x_1, f_2 = x_2$, we have $\{f_1, f_2\} = 0$ and hence commutativity.

To conclude, we summarise the main properties of the Poisson brackets.

THEOREM 10.19 *The Poisson brackets satisfy the following properties*:
(1) *they define a bilinear skew-symmetric form on functions, i.e.* $(f, g) \to \{f, g\}$ *is linear in both arguments and* $\{f, g\} = -\{g, f\}$;
(2) *Leibniz's rule*:

$$\{f_1 f_2, g\} = f_1\{f_2, g\} + f_2\{f_1, g\}; \qquad (10.151)$$

(3) *the Jacobi identity*:

$$\{f, \{g, h\}\} + \{g, \{h, f\}\} + \{h, \{f, g\}\} = 0; \qquad (10.152)$$

(4) if $D_f$ and $D_g$ are operators associated with $f$ and $g$ by equation (10.145) then

$$[D_f, D_g] = D_{\{f,g\}}; \tag{10.153}$$

(5) *non-degeneration*: if a point $\mathbf{x}_0 \in \mathbf{R}^{2l}$ is not a critical point of $f$, there exists a function $g$ such that $\{f, g\}(\mathbf{x}_0) \neq 0$.

*Proof*
Properties (1) and (2) have a trivial verification, left to the reader. The Jacobi identity can be shown without any lengthy calculation, by noting that if we take the expansion of (10.152) we find a sum of terms each containing a second derivative of one of the functions $f$, $g$ and $h$ multiplied by the first derivatives of the other two. If we prove that in the expansion of (10.152) no second derivative of any of the three functions appear, than we prove that all these terms cancel out, and their sum is then equal to zero. On the other hand, if we consider for example the terms containing the second derivatives of $h$, we note that they necessarily come from the first two terms of (10.152). However,

$$\begin{aligned}\{f, \{g, h\}\} + \{g, \{h, f\}\} &= \{f, \{g, h\}\} - \{g, \{f, h\}\} = D_f D_g h - D_g D_f h \\ &= [D_f, D_g]h = [L_{\mathbf{v}_f}, L_{\mathbf{v}_g}].\end{aligned} \tag{10.154}$$

Remark 10.28 ensures that the commutator of two Lie derivatives is again a Lie derivative, and hence it does not contain second derivatives. The Jacobi identity is then proved.

The property (4) is an immediate consequence of the Jacobi identity. Indeed, from (10.152) and (10.154) it follows that for any $h$ we have

$$D_{\{f,g\}}h = \{\{f, g\}, h\} = \{f, \{g, h\}\} + \{g, \{h, f\}\} = [D_f, D_g]h.$$

Finally, the non-degeneration property is an obvious consequence of the non-degeneration of the symplectic product. ∎

The Jacobi identity yields another interesting result.

COROLLARY 10.6 *If $f$ and $g$ are two first integrals then the same holds for $\{f, g\}$.* ∎

*Proof*
If in equation (10.152) we set $h = H$, it follows from $\{f, H\} = \{g, H\} = 0$ that $\{\{f, g\}, H\} = 0$. ∎

*Example* 10.28
Recall the formula $\{L_i, L_j\} = \epsilon_{ijk} L_k$ of Example 10.25. Then Corollary 10.6 guarantees that if two components of the angular momentum of a point are constant, then the third component must also be constant. ∎

## 10.7 Symplectic rectification

The canonical transformations are a powerful tool, allowing the construction of new canonical variables with the aim of writing the Hamiltonian in some desired form. A significant example of such an application is given by the following theorem.

THEOREM 10.20 (Symplectic rectification) *Let $f(\mathbf{x})$ be a function $C^1$ in $\mathbf{R}^{2l}$ and $\mathbf{x}_0$ a point which is not critical for $f$, so that $\nabla_{\mathbf{x}} f(\mathbf{x}_0) \neq 0$. Then there exists a completely canonical transformation $\mathbf{X} = \mathbf{X}(\mathbf{x})$, defined in a neighbourhood of $\mathbf{x}_0$, such that $\hat{f}(\mathbf{X}) = f(\mathbf{x}(\mathbf{X})) = X_i$ for some $i$.* ∎

To understand the meaning of this theorem, and the technique of its proof, we start by analysing a simple non-trivial example.

*Example* 10.29 *rectification of the harmonic oscillator*
Consider the Hamiltonian $H = \frac{1}{2}p^2 + \omega^2/2 q^2$ and the flow that it generates, given by

$$\dot{p} = -\omega^2 q, \quad \dot{q} = p.$$

Endowed with the initial conditions $p(0) = \eta$, $q(0) = \xi$ this gives

$$p = \eta \cos \omega t - \omega \xi \sin \omega t, \quad q = \frac{\eta}{\omega} \sin \omega t + \xi \cos \omega t. \tag{10.155}$$

The retrograde flow (with Hamiltonian $-\frac{1}{2}\eta^2 - \omega^2/2\xi^2$) with initial conditions $\eta(0) = p$, $\xi(0) = q$ is

$$\eta = p \cos \omega t + \omega q \sin \omega t, \quad \xi = -\frac{p}{\omega} \sin \omega t + q \cos \omega t. \tag{10.156}$$

Our goal is to find a completely canonical transformation (in an open set excluding the origin), such that the new coordinate $P$ is given by

$$P = \frac{1}{2}p^2 + \frac{\omega^2}{2}q^2. \tag{10.157}$$

Since the transformation is completely canonical, $P$ is also the new Hamiltonian, so that the Hamilton equations yield the solution

$$P = \text{constant}, \quad Q = t - t_0. \tag{10.158}$$

After imposing equation (10.157), the problem is reduced to making a correct choice for $Q = Q(p, q)$. Its value along the motion must coincide with time (up to translations). Hence in the plane $(\xi, \eta)$ we take a regular curve $\varphi(\xi, \eta) = 0$, with $\{\varphi, -\frac{1}{2}\eta^2 - \frac{1}{2}\omega^2 \xi^2\} \neq 0$ (so that the curve cannot be a trajectory of (10.156)). Fix a point $(q, p)$ such that the trajectory (10.156) intersects the curve and denote by $\bar{t}(p,q)$ the time of impact. If we denote the functions (10.156) by $\xi(p,q,t)$, $\eta(p,q,t)$, then the function $\bar{t}(p,q)$ is defined implicitly by

$$\varphi[\xi(p,q,\bar{t}),\eta(q,p,\bar{t})] = 0. \tag{10.159}$$

We now complete the transformation (10.157), that is independent of time, with

$$Q = \bar{t}(p,q) \tag{10.160}$$

and check that the variables $P,Q$ are canonical. Let us compute the Poisson bracket

$$\left\{\bar{t}(p,q), \frac{1}{2}(p^2+\omega^2 q^2)\right\} = \frac{\partial \bar{t}}{\partial q}p - \frac{\partial \bar{t}}{\partial p}\omega^2 q. \tag{10.161}$$

Equation (10.159) yields for the derivatives of $\bar{t}$:

$$\frac{\partial \bar{t}}{\partial p} = -\left(\frac{\partial \varphi}{\partial \xi}\frac{\partial \xi}{\partial p} + \frac{\partial \varphi}{\partial \eta}\frac{\partial \eta}{\partial p}\right)\frac{1}{D}, \quad \frac{\partial \bar{t}}{\partial q} = -\left(\frac{\partial \varphi}{\partial \xi}\frac{\partial \xi}{\partial q} + \frac{\partial \varphi}{\partial \eta}\frac{\partial \eta}{\partial q}\right)\frac{1}{D},$$

where

$$D = \frac{\partial \varphi}{\partial \xi}\frac{\partial \xi}{\partial t} + \frac{\partial \varphi}{\partial \eta}\frac{\partial \eta}{\partial t} = \{\varphi(\xi,\eta), -H(\eta,\xi)\} \neq 0.$$

Now it is easy to note that

$$\frac{\partial \bar{t}}{\partial q}p - \frac{\partial \bar{t}}{\partial p}\omega^2 q = -\frac{1}{D}\left\{\frac{\partial \varphi}{\partial \xi}\left(p\frac{\partial \xi}{\partial q} - \omega^2 q\frac{\partial \xi}{\partial p}\right) + \frac{\partial \varphi}{\partial \eta}\left(p\frac{\partial \eta}{\partial q} - \omega^2 q\frac{\partial \eta}{\partial p}\right)\right\}$$

$$= -\frac{1}{D}\left[\frac{\partial \varphi}{\partial \xi}\{\xi(p,q,t), H(p,q)\} + \frac{\partial \varphi}{\partial \eta}\{\eta(p,q,t), H(p,q)\}\right].$$

Since the transformation $(p,q) \rightleftarrows (\eta,\xi)$ is canonical and the Poisson brackets are preserved,

$$-\{\xi(p,q,t), H(p,q)\} = \{\xi, -H(\eta,\xi)\} = \frac{\partial \xi}{\partial t},$$

$$-\{\eta(p,q,t), H(p,q)\} = \{\eta, -H(\eta,\xi)\} = \frac{\partial \eta}{\partial t}.$$

We finally find

$$\left\{\bar{t}(p,q), \frac{1}{2}(p^2+\omega^2 q^2)\right\} = 1, \tag{10.162}$$

which shows that the transformation is completely canonical. The time $\bar{t}(p,q)$ is also equal to the time, on the direct flow, necessary to reach the point $(p,q)$ starting from the curve $\varphi(\xi,\eta) = 0$. Note that the curve $\varphi(\xi,\eta) = 0$ in this procedure is arbitrary. Therefore there exist infinitely many transformations of the kind sought. For example, for $\varphi(\xi,\eta)=\xi$ we find $Q=(1/w)\mathrm{arccot}(p/wq)$, for $\varphi(\xi,\eta)=\eta$ we find $Q=-(1/w)\mathrm{arccot}(wq/p)$, for $\varphi(\xi,\eta)=w\xi+\eta$ we find $Q=(1/w)\mathrm{arccot}[(p-wq)/(p+wq)]$, and finally for $\varphi(\xi,\eta)=w\xi-\eta$ we have $Q=-(1/w)\mathrm{arccot}[(p+wq)/(p-wq)]$. Each of these formulae, together with (10.157), gives a symplectic of the harmonic oscillator. Verify that in all cases, knowledge of integrals (10.158) leads, through the transformation, to the known integral of the harmonic motion. ■

*Proof of Theorem 10.14*
By hypothesis, in a neighbourhood of $\mathbf{x}_0$ at least one of the first derivatives of $f$ is different from zero. We can assume without loss of generality that $\partial f/\partial p_1 \neq 0$. As in the example, we consider the direct flow $\dot{\mathbf{p}}=-\nabla_\mathbf{q} f$, $\dot{\mathbf{q}}=\nabla_\mathbf{p} f$ with initial conditions $\mathbf{p}=\boldsymbol{\eta}$, $\mathbf{q}=\boldsymbol{\xi}$ and the inverse flow with Hamiltonian $-f(\boldsymbol{\eta},\boldsymbol{\xi})$ and initial conditions $\boldsymbol{\eta}=\mathbf{p}$, $\boldsymbol{\xi}=\mathbf{q}$. The condition $\partial f/\partial p_1\neq 0$ can be interpreted as $\partial\xi_1/\partial t=-\partial f/\partial\eta_1\neq 0$ in the retrograde flow. Hence it is possible to express the function $\bar{t}(\mathbf{p},\mathbf{q})$ explicitly from the equation

$$\xi_1(\mathbf{p},\mathbf{q},t)=0. \tag{10.163}$$

As in the example, the function $\bar{t}(\mathbf{p},\mathbf{q})$ can also be defined in alternative ways, but for simplicity we consider only (10.163), keeping in mind that for every different choice of $\bar{t}(\mathbf{p},\mathbf{q})$ we obtain a different transformation. From (10.163) we deduce

$$\frac{\partial \bar{t}}{\partial p_i}=-\frac{\partial \xi_1}{\partial p_i}\bigg/\frac{\partial \xi_1}{\partial t}, \quad \frac{\partial \bar{t}}{\partial q_i}=-\frac{\partial \xi_1}{\partial q_i}\bigg/\frac{\partial \xi_1}{\partial t}, \quad \text{with} \quad \frac{\partial \xi_1}{\partial t}=-\frac{\partial f(\boldsymbol{\eta},\boldsymbol{\xi})}{\partial \eta_1},$$

and hence

$$\{\bar{t},f\}_{(\mathbf{p},\mathbf{q})}=\frac{1}{\partial f/\partial\eta_1}\{\xi_1,f\}_{(\boldsymbol{\xi},\boldsymbol{\eta})}=1$$

(we used the invariance of $\{\xi_1,f\}$ passing from $(\mathbf{p},\mathbf{q})$ to $(\boldsymbol{\eta},\boldsymbol{\xi})$ and in addition $\{\xi_1,-f\}=\partial\xi_1/\partial t$). Hence also on the direct flow the function $\bar{t}$ takes the values of time $t$. Therefore, if $l=1$, the transformation

$$P=f(p,q), \quad Q=\bar{t}(p,q)$$

is the one we were seeking, and the theorem is proved. If $l>1$, we set

$$P_1=f(\mathbf{p},\mathbf{q}), \quad Q_1=\bar{t}(\mathbf{p},\mathbf{q}) \tag{10.164}$$

and complete the transformation (independent of time) by defining

$$P_i=X_i=\eta_i(\mathbf{p},\mathbf{q},\bar{t}(\mathbf{p},\mathbf{q})), \quad Q_i=X_{i+l}=\xi_i(\mathbf{p},\mathbf{q},\bar{t}(\mathbf{p},\mathbf{q})), \quad i=2,\ldots,l. \tag{10.165}$$

To verify that the transformation is canonical we must compute all the fundamental Poisson brackets. We start with $\{P_i, P_j\}$, with $i, j \neq 1$:

$$\{P_i, P_j\} = \sum_{k=1}^{l} \left( \frac{\partial \eta_i}{\partial q_k} + \frac{\partial \eta_i}{\partial t} \frac{\partial \bar{t}}{\partial q_k} \right) \left( \frac{\partial \eta_j}{\partial p_k} + \frac{\partial \eta_j}{\partial t} \frac{\partial \bar{t}}{\partial p_k} \right)$$

$$- \sum_{k=1}^{l} \left( \frac{\partial \eta_j}{\partial q_k} + \frac{\partial \eta_j}{\partial t} \frac{\partial \bar{t}}{\partial q_k} \right) \left( \frac{\partial \eta_i}{\partial p_k} + \frac{\partial \eta_i}{\partial t} \frac{\partial \bar{t}}{\partial p_k} \right)$$

$$= \{\eta_i, \eta_j\} - \frac{1}{\partial \xi_1 / \partial t} \left[ \frac{\partial \eta_i}{\partial t} \{\xi_1, \eta_j\} - \frac{\partial \eta_j}{\partial t} \{\xi_1, \eta_i\} \right].$$

Since $(\eta, \xi)$ are canonical variables, then $\{\eta_i, \eta_j\} = 0$ for every $i, j$ and $\{\xi_1, \eta_i\} = 0$ for $i > 1$. Therefore $\{P_i, P_j\} = 0$ for $i, j \neq 1$. The expression for $\{Q_i, Q_j\}$ for $i, j \neq 1$ is analogous, with $\xi_i$ and $\xi_j$ in place of $\eta_i, \eta_j$; the conclusion is the same. The evaluation of $\{P_i, f\}$ and $\{Q_i, f\}$ with $i > 1$ is made easy by the fact that these are the derivatives of the functions $\eta_i(\mathbf{p}(t), \mathbf{q}(t), t)$, $\xi_i(\mathbf{p}(t), \mathbf{q}(t), t)$ in the direct Hamiltonian flow, along which the functions $\eta_i, \xi_i$ are all constant. The Poisson brackets are again zero. It is not difficult to check that

$$\{Q_1, P_i\} = \{\bar{t}, P_i\} = -\frac{1}{\partial \xi_1 / \partial t} \{\xi_1, \eta_i\} = 0, \quad \text{for } i > 1,$$

and similarly

$$\{Q_1, Q_i\} = \{\bar{t}, Q_i\} = -\frac{1}{\partial \xi_1 / \partial t} \{\xi_1, \xi_i\} = 0.$$

We must finally check that $\{Q_i, P_j\} = \delta_{ij}$ for $i, j \neq 1$. Proceeding as usual, we find

$$\{Q_i, P_j\} = \{\xi_i, \eta_j\} - \frac{1}{\partial \xi_1 / \partial t} \left[ \frac{\partial \xi_i}{\partial t} \{\xi_1, \eta_j\} - \frac{\partial \eta_j}{\partial t} \{\xi_1, \xi_i\} \right] = \delta_{ij}, \quad i, j \neq 1.$$

This concludes the proof of the theorem. ∎

**Remark 10.31**
Symplectic rectification leads to a pair of conjugate variables taking respectively the values of the Hamiltonian (constant) and of time, while (for $l > 1$) the remaining coordinates are all first integrals of the motion. ∎

**Example 10.30**
We seek a symplectic rectification of the system with Hamiltonian

$$f(\mathbf{p}, \mathbf{q}) = p_1 q_2 - p_2 q_1.$$

Considering directly the equations of the retrograde motion, with Hamiltonian $-f(\boldsymbol{\eta}, \boldsymbol{\xi}) = -\eta_1 \xi_2 + \eta_2 \xi_1$, and initial values $\eta_i(0) = p_i$, $\xi_i(0) = q_i$, $i = 1, 2$, we find

$$\xi_1(\mathbf{p}, \mathbf{q}, t) = q_1 \cos t - q_2 \sin t, \quad \xi_2(\mathbf{p}, \mathbf{q}, t) = q_2 \cos t + q_1 \sin t,$$
$$\eta_1(\mathbf{p}, \mathbf{q}, t) = p_1 \cos t - p_2 \sin t, \quad \eta_2(\mathbf{p}, \mathbf{q}, t) = p_2 \cos t + p_1 \sin t.$$

Assuming, for example, that $q_2 \neq 0$, we find that the transformation we seek (setting $\xi_1 = 0$) is

$$P_1 = p_1 q_2 - p_2 q_1, \quad Q_1 = \operatorname{arccot} \frac{q_1}{q_2},$$

$$P_2 = p_2 \cos\left(\operatorname{arccot} \frac{q_1}{q_2}\right) + p_1 \sin\left(\operatorname{arccot} \frac{q_1}{q_2}\right),$$

$$Q_2 = q_2 \cos\left(\operatorname{arccot} \frac{q_1}{q_2}\right) + q_1 \sin\left(\operatorname{arccot} \frac{q_1}{q_2}\right).$$

Therefore, possible expressions for $P_2, Q_2$ are

$$P_2 = \frac{p_1 q_1 + p_2 q_2}{\sqrt{q_1^2 + q_2^2}}, \quad Q_2 = \sqrt{q_1^2 + q_2^2}.$$

The new coordinates characterise the first integrals $P_1$ (constant Hamiltonian), $P_2, Q_2$. Since the motion generated in the space $(\mathbf{p}, \mathbf{q})$ is a uniform rotation in the plane $q_1, q_2$ together with a uniform rotation in the plane $p_1, p_2$, it is clear that $q_1^2 + q_2^2 = R^2 = \text{constant}$ (and also $p_1^2 + p_2^2$), while $P_2 = \text{constant}$ is equivalent to $\mathbf{p} \cdot \mathbf{q} = \text{constant}$. ∎

## 10.8 Infinitesimal and near-to-identity canonical transformations. Lie series

The canonical transformations that are 'near' (in a sense to be made precise) to the identity transformation have great importance. Indeed, as we shall see in Chapter 12 when we treat the canonical theory of perturbations, using these nearly identical transformations one can study the dynamics of many interesting mechanical systems. For most applications, we only use completely canonical transformations 'near to identity'. Hence in this and the following sections we study only time-independent transformations. Due to Remark 10.20 this is not a real restriction.

DEFINITION 10.19 *Let* $\mathbf{f}$ *and* $\mathbf{g}$ *be two functions of class* $\mathcal{C}^\infty$ *defined on an open set* $A \subset \mathbf{R}^{2l}$, *with values in* $\mathbf{R}^l$. *Consider* $\varepsilon \in \mathbf{R}$, $|\varepsilon| \ll 1$. *An* infinitesimal coordinate transformation *can be expressed as*

$$\mathbf{p} = \mathbf{P} + \varepsilon \mathbf{f}(\mathbf{P}, \mathbf{Q}),$$
$$\mathbf{q} = \mathbf{Q} + \varepsilon \mathbf{g}(\mathbf{P}, \mathbf{Q}). \tag{10.166}$$

∎

THEOREM 10.21 *If* $\varepsilon$ *is sufficiently small, then the transformation defined by* (10.166) *is invertible, i.e. for every open bounded subset* $C$ *of* $A$, *with* $\overline{C} \subset A$, *there exists* $\varepsilon_0 > 0$ *such that for every* $\varepsilon \in \mathbf{R}$, $|\varepsilon| < \varepsilon_0$, *the transformation* (10.166) restricted to $C$ is invertible. The inverse transformation is given, to first order in $\varepsilon$, by

$$\mathbf{P} = \mathbf{p} - \varepsilon \mathbf{f}(\mathbf{p}, \mathbf{q}) + \mathcal{O}(\varepsilon^2),$$
$$\mathbf{Q} = \mathbf{q} - \varepsilon \mathbf{g}(\mathbf{p}, \mathbf{q}) + \mathcal{O}(\varepsilon^2). \tag{10.167}$$

*Proof*
The Jacobian matrix of the transformation (10.166) is

$$\frac{\partial(\mathbf{p}, \mathbf{q})}{\partial(\mathbf{P}, \mathbf{Q})} = \mathbf{1} + \varepsilon \begin{pmatrix} \nabla_\mathbf{P} \mathbf{f} & \nabla_\mathbf{Q} \mathbf{f} \\ \nabla_\mathbf{P} \mathbf{g} & \nabla_\mathbf{Q} \mathbf{g} \end{pmatrix},$$

where $\mathbf{1}$ indicates the $2l \times 2l$ identity matrix. Since $\mathbf{f}$ and $\mathbf{g}$ are in $\mathcal{C}^\infty$, their first derivatives are uniformly bounded on each compact subset of $A$. Therefore there exists a constant $M > 0$ such that

$$\left| \det \frac{\partial(\mathbf{p}, \mathbf{q})}{\partial(\mathbf{P}, \mathbf{Q})} \right| > 1 - \varepsilon M$$

on $C$. It follows that if $|\varepsilon| < \varepsilon_0 = 1/M$, the Jacobian is non-singular and the transformation is invertible. Since in addition

$$\mathbf{f}(\mathbf{P}, \mathbf{Q}) = \mathbf{f}(\mathbf{p}, \mathbf{q}) + \mathcal{O}(\varepsilon), \quad \mathbf{g}(\mathbf{P}, \mathbf{Q}) = \mathbf{g}(\mathbf{p}, \mathbf{q}) + \mathcal{O}(\varepsilon),$$

from (10.166) we can immediately deduce (10.167). ■

DEFINITION 10.20 *An infinitesimal transformation (10.166) defines a canonical infinitesimal transformation if*

$$\{p_i, p_j\} = \{q_i, q_j\} = \mathcal{O}(\varepsilon^2),$$
$$\{q_i, p_j\} = \delta_{ij} + \mathcal{O}(\varepsilon^2), \tag{10.168}$$

*where $i, j = 1, \ldots, l$, and the Poisson brackets are computed with respect to the variables $(\mathbf{P}, \mathbf{Q})$.* ■

The infinitesimal canonical transformations are the transformations which preserve the fundamental Poisson brackets, up to terms of order $\mathcal{O}(\varepsilon^2)$.

THEOREM 10.22 *The infinitesimal transformation (10.166) is canonical if and only if there exists a function $K : A \to \mathbf{R}$ of class $\mathcal{C}^\infty$ such that*

$$f_i(\mathbf{P}, \mathbf{Q}) = -\frac{\partial K}{\partial Q_i},$$
$$g_i(\mathbf{P}, \mathbf{Q}) = \frac{\partial K}{\partial P_i}, \tag{10.169}$$

*where $i = 1, \ldots, l$. We say that $K$ is the Hamiltonian associated with the infinitesimal canonical transformation (10.166).*

*Proof*
The Jacobian matrix of the system (10.166) is $\tilde{J} = \mathbf{1} + \varepsilon J$, where
$$J = \begin{pmatrix} \nabla_{\mathbf{P}}\mathbf{f} & \nabla_{\mathbf{Q}}\mathbf{f} \\ \nabla_{\mathbf{P}}\mathbf{g} & \nabla_{\mathbf{Q}}\mathbf{g} \end{pmatrix}$$
is the matrix $\nabla_{\mathbf{X}}\phi(\mathbf{X})$, where $\phi$ is the vector field $(\mathbf{f}(\mathbf{X}), \mathbf{g}(\mathbf{X}))$. The condition that the transformation is canonical can be written as
$$(\mathbf{1} + \varepsilon J^T)\mathfrak{I}(\mathbf{1} + \varepsilon J) = \mathfrak{I} + \mathcal{O}(\varepsilon^2),$$
which is equivalent to
$$J^T \mathfrak{I} + \mathfrak{I} J = 0,$$
and hence to the fact that $J$ is Hamiltonian (Definition 10.1). It follows from Theorem 10.5 that the transformation is canonical if and only if the field $(\mathbf{f}, \mathbf{g})$ is Hamiltonian, i.e. if it is generated by a Hamiltonian $K(\mathbf{P}, \mathbf{Q})$. ∎

Hamiltonian matrices are sometimes also called *infinitesimally symplectic matrices*. This is due to the property just seen, that if $J$ is Hamiltonian then $\mathbf{1} + \varepsilon J$ is symplectic to first order in $\varepsilon$. Due to (10.169) we note that, by interpreting $\varepsilon$ as an infinitesimal 'time', the transformation (10.166) is a canonical infinitesimal transformation if and only if (up to terms of order $\mathcal{O}(\varepsilon^2)$) it has the structure of a Hamiltonian flow with respect to the parameter $\varepsilon$.

*Example 10.31*
The infinitesimal transformation
$$q_i = Q_i + \varepsilon Q_i, \quad p_i = P_i - \varepsilon P_i, \quad i = 1, \ldots, l \quad (10.170)$$
is canonical. Indeed $\{q_i, q_j\} = \{p_i, p_j\} = 0$ and
$$\{q_i, p_j\} = \{Q_i + \varepsilon Q_i, P_j - \varepsilon P_j\} = \{Q_i, P_j\} - \varepsilon^2 \{Q_i, P_j\} = \delta_{ij}(1 - \varepsilon^2).$$
The inverse transformation is
$$Q_i = \frac{q_i}{1+\varepsilon} = q_i - \varepsilon q_i + \mathcal{O}(\varepsilon^2), \quad P_i = \frac{p_i}{1-\varepsilon} = p_i + \varepsilon p_i + \mathcal{O}(\varepsilon^2),$$
and the function $K$ is $K = \sum_{i=1}^{l} P_i Q_i$. ∎

*Example 10.32*
The infinitesimal transformation
$$p = P + 2\varepsilon Q(1 + \cos P), \quad q = Q + \varepsilon Q^2 \sin P \quad (10.171)$$
is canonical. Indeed,
$$\{q, p\} = (1 + 2\varepsilon Q \sin P)(1 - 2\varepsilon Q \sin P) - \varepsilon(Q^2 \cos P)2\varepsilon(1 + \cos P)$$
$$= 1 - 2\varepsilon^2 Q^2(1 + \cos P + \sin^2 P),$$

and since $l = 1$, the Poisson bracket $\{q, p\}$ is equal to the Jacobian determinant of the transformation. Therefore, if $(P, Q) \in C$, where $C$ is the rectangle $(-\pi, \pi) \times (-1, 1)$, the condition $\varepsilon < 1/\sqrt{6}$ is sufficient to ensure the invertibility of the transformation. Evidently the associated Hamiltonian is

$$K = -Q^2(1 + \cos P). \tag{10.172}$$

■

While the infinitesimal canonical transformations are canonical in the approximation $\mathcal{O}(\varepsilon^2)$, the near-to-identity canonical transformations which we are about to define depend on a small parameter $\varepsilon$, but are exactly canonical as $\varepsilon$ varies.

DEFINITION 10.21 *A one-parameter family of completely canonical transformations* $\mathbf{x} = \mathbf{x}(\mathbf{X}, \varepsilon)$ *from the variables* $\mathbf{x} = (\mathbf{p}, \mathbf{q})$ *to* $\mathbf{X} = (\mathbf{P}, \mathbf{Q})$ *is near to identity if it has the form*

$$\begin{aligned}\mathbf{p} &= \mathbf{P} + \varepsilon \mathbf{f}(\mathbf{P}, \mathbf{Q}, \varepsilon), \\ \mathbf{q} &= \mathbf{Q} + \varepsilon \mathbf{g}(\mathbf{P}, \mathbf{Q}, \varepsilon),\end{aligned} \tag{10.173}$$

*where* $\varepsilon$ *is a parameter that varies in an open interval* $I = (-\varepsilon_0, \varepsilon_0)$, *with* $0 < \varepsilon_0 \ll 1$, *the functions* $\mathbf{f}$, $\mathbf{g}$, $A \times I \to \mathbf{R}^l$ *are of class* $\mathcal{C}^\infty$ *in all their arguments, and $A$ is an open set in* $\mathbf{R}^{2l}$. ■

THEOREM 10.23 *Let $C$ be a compact subset of* $\mathbf{R}^{2l}$. *Every near-to-identity canonical transformation defined in an open neighbourhood of $C$ admits a generating function $F(\mathbf{q}, \mathbf{P}, \varepsilon)$ of the form*

$$F(\mathbf{q}, \mathbf{P}, \varepsilon) = \sum_{i=1}^{l} q_i P_i + \varepsilon \mathcal{F}(\mathbf{q}, \mathbf{P}, \varepsilon), \tag{10.174}$$

*and vice versa. Here $\mathcal{F}$ is a function of class* $\mathcal{C}^\infty$ *in all its arguments, for every* $\varepsilon \in (-\varepsilon_0, \varepsilon_0)$, *where $\varepsilon_0$ is a sufficiently small positive constant.*

*Proof*
From the second of equations (10.173) it follows that

$$\frac{\partial q_i}{\partial Q_j} = \delta_{ij} + \varepsilon \frac{\partial g_i}{\partial Q_j},$$

and therefore as $(\mathbf{P}, \mathbf{Q}) \in C$ varies, the condition (10.121), i.e. $\det(\nabla_{\mathbf{Q}} \mathbf{q}) \neq 0$, is certainly satisfied, if $\varepsilon_0$ is sufficiently small. Hence there exists a regular function $\mathcal{Q}(\mathbf{q}, \mathbf{P}, \varepsilon)$ such that

$$\mathbf{Q} = \mathbf{q} + \varepsilon \mathcal{Q}(\mathbf{q}, \mathbf{P}, \varepsilon).$$

Substituting it into the first of equations (10.173) we find

$$\mathbf{p} = \mathbf{P} + \varepsilon \mathcal{P}(\mathbf{q}, \mathbf{P}, \varepsilon),$$

where $\mathcal{P}(\mathbf{q},\mathbf{P},\varepsilon) = \mathbf{f}(\mathbf{P}, \mathbf{q}+\varepsilon\mathcal{Q}(\mathbf{q},\mathbf{P},\varepsilon),\varepsilon)$. We now recall that if the transformation (10.173) is canonical, the form $\sum_{i=1}^{l}(p_i\,dq_i + Q_i\,dP_i)$ can be integrated to find the generating function (10.175) (see (10.117)).

Conversely, since $(\mathbf{q},\mathbf{P})$ varies in a compact subset of $\mathbf{R}^{2l}$, if $\varepsilon_0$ is sufficiently small then

$$\det\left(\frac{\partial^2 F}{\partial q_i \partial P_j}\right) = \det\left(\delta_{ij} + \varepsilon\frac{\partial^2 \mathcal{F}}{\partial q_i \partial P_j}\right) > 0,$$

and the equations

$$\mathbf{p} = \nabla_{\mathbf{q}}F = \mathbf{P} + \varepsilon\nabla_{\mathbf{q}}\mathcal{F}, \quad \mathbf{Q} = \nabla_{\mathbf{P}}F = \mathbf{q} + \varepsilon\nabla_{\mathbf{P}}\mathcal{F} \tag{10.175}$$

generate a near-to-identity canonical transformation. ∎

**Example 10.33**
Consider the function

$$F(q,P,\varepsilon) = qP + \varepsilon q^2(1+\cos P). \tag{10.176}$$

Since $\partial^2 F/\partial q \partial P = 1 - 2\varepsilon q \sin P$, as $(q,P)$ varies in a compact subset, if $|\varepsilon| < \varepsilon_0$ is sufficiently small, the function $F$ generates a near-to-identity canonical transformation. For example if $(q,P) \in [-a,a] \times [-\pi,\pi]$ we set $\varepsilon_0 < 1/2a$. The transformation generated by $F$ is defined implicitly by

$$Q = q - \varepsilon q^2 \sin P, \quad p = P + 2\varepsilon q(1+\cos P).$$

Solving the first equation for $q$ we find

$$q = \frac{1}{2\varepsilon \sin P}\left(1 - \sqrt{1 - 4\varepsilon Q \sin P}\right),$$

where the choice of negative sign for the determination of the square root is fixed by the requirement that $q \to Q$ when $\varepsilon \to 0$. Taking the Taylor series of the square root:

$$-\sqrt{1-x} = -1 + \frac{x}{2} + \sum_{n=2}^{\infty}\frac{(2n-3)!!}{2^n n!}x^n,$$

and taking into account that $x = 4\varepsilon Q \sin P$ we find

$$q = Q + \varepsilon Q^2 \sin P + \sum_{j=2}^{\infty}\frac{(2j-1)!!}{(j+1)!}2^j \varepsilon^j(\sin P)^j Q^{j+1}. \tag{10.177}$$

The same result is obtained by an application of Lagrange's formula (see Theorem 5.5):

$$q = Q + \varepsilon Q^2 \sin P + \sum_{j=2}^{\infty}\frac{\varepsilon^j}{j!}(\sin P)^j \frac{d^{j-1}}{dQ^{j-1}}Q^{2j},$$

considering that

$$\frac{d^{j-1}}{dQ^{j-1}}Q^{2j} = \frac{2^j(2j-1)!!}{j+1}Q^{j+1},$$

as is easily verified by induction. Substituting the expression for $q$ into the expression for $p$ we arrive at

$$p = P + 2\varepsilon(1+\cos P)\left(Q + \varepsilon Q^2 \sin P + \sum_{j=2}^{\infty}\frac{(2j-1)!!}{(j+1)!}2^j\varepsilon^j(\sin P)^j Q^{j+1}\right).$$

(10.178)

∎

The comparison between Examples 10.32 and 10.33 sheds light on the difference between infinitesimal canonical transformations and near-identical canonical transformations. Clearly, the transformation (10.171) coincides with equations (10.177), (10.178) up to terms of order $\mathcal{O}(\varepsilon^2)$. Since the Hamiltonian associated with (10.171) is $K = -Q^2(1+\cos P)$, comparing this with (10.176) suggests that by setting $\mathcal{F}(\mathbf{q},\mathbf{P}) = -K(\mathbf{q},\mathbf{P})$ in Theorem 10.17 we obtain the generating function of a near-to-identity canonical transformation starting from an infinitesimal canonical transformation. This is precisely the conclusion of the following theorem.

THEOREM 10.24 *To every infinitesimal canonical transformation (10.166) with associated Hamiltonian $K$ (see (10.169)) there corresponds a near-to-identity canonical transformation. The latter coincides with (10.166) up to terms of order $\mathcal{O}(\varepsilon^2)$. The transformation can be obtained starting from the generating function (10.174) by setting $\mathcal{F} = -K(\mathbf{q},\mathbf{P})$. Conversely, to every near-to-identity canonical transformation (10.173) there corresponds an infinitesimal canonical transformation (10.166) obtained by neglecting terms of order $\mathcal{O}(\varepsilon^2)$ in (10.173). The associated Hamiltonian is given by $K = -\mathcal{F}(\mathbf{Q},\mathbf{P},0)$.*

*Proof*
To prove the first statement it is enough to note that from equation (10.176), setting $\mathcal{F} = -K(\mathbf{q},\mathbf{P})$, it follows that

$$\mathbf{p} = \mathbf{P} - \varepsilon\nabla_{\mathbf{q}}K, \quad \mathbf{Q} = \mathbf{q} - \varepsilon\nabla_{\mathbf{P}}K,$$

and hence

$$\mathbf{p} = \mathbf{P} - \varepsilon\nabla_{\mathbf{Q}}K + \mathcal{O}(\varepsilon^2), \quad \mathbf{q} = \mathbf{Q} + \varepsilon\nabla_{\mathbf{P}}K.$$

The second part of the theorem has an analogous proof. ∎

We saw in Theorem 10.13 that the Hamiltonian flow is canonical. Considering time as a parameter and setting $\varepsilon = t$ in an interval $(-\varepsilon_0, \varepsilon_0)$, this flow gives an example of a near-to-identity canonical transformation, while neglecting terms of order $\mathcal{O}(\varepsilon^2)$ it provides an example of an infinitesimal canonical transformation. Indeed, consider the canonical equations (10.90) for a Hamiltonian $H(\mathbf{p}, \mathbf{q})$, where $(\mathbf{P}, \mathbf{Q})$ denote the initial conditions (at time $t = \varepsilon = 0$) and $(\mathbf{p}, \mathbf{q})$ denote the solutions of (10.90) at time $t = \varepsilon$. An integration of equations (10.90) that is accurate to first order in $\varepsilon$ yields

$$\begin{aligned} \mathbf{p} &= \mathbf{P} - \varepsilon \nabla_\mathbf{Q} H(\mathbf{P}, \mathbf{Q}) + \mathcal{O}(\varepsilon^2), \\ \mathbf{q} &= \mathbf{Q} + \varepsilon \nabla_\mathbf{P} H(\mathbf{P}, \mathbf{Q}) + \mathcal{O}(\varepsilon^2), \end{aligned} \tag{10.179}$$

and hence equations (10.169) are satisfied with $K = H$.

We now show how it is always possible, at least in principle, to formally construct a near-to-identity transformation associated with a given Hamiltonian $H$.

Let

$$(\mathbf{p}, \mathbf{q}) = \mathbf{S}^t(\mathbf{P}, \mathbf{Q}) = (\mathbf{p}(\mathbf{P}, \mathbf{Q}, t), \mathbf{q}(\mathbf{P}, \mathbf{Q}, t)) \tag{10.180}$$

be the Hamiltonian flow associated with $H$. As we saw in Section 10.5, $\mathbf{S}^t$ defines an evolution operator $U^t$ acting on the observables of the system. If the Hamiltonian $H(\mathbf{p}, \mathbf{q})$ is independent of time, and we consider the action of $U^t$ on the functions $f(\mathbf{P}, \mathbf{Q}) \in \mathcal{C}^\infty(\mathbf{R}^{2l})$, then we have

$$\frac{d}{dt}(U^t f)(\mathbf{P}, \mathbf{Q}) = (\{f, H\} \circ \mathbf{S}^t)(\mathbf{P}, \mathbf{Q}) = \{f, H\}(\mathbf{p}, \mathbf{q}) = (-D_H f)(\mathbf{p}, \mathbf{q}), \tag{10.181}$$

where $D_H = \{H, \cdot\}$ (see (10.145)) is called an *infinitesimal generator* of $U^t$.

THEOREM 10.25  *For every $t \in \mathbf{R}$ we formally have that*

$$U^t = e^{-tD_H}, \tag{10.182}$$

*i.e.*

$$(U^t f)(\mathbf{P}, \mathbf{Q}) = \sum_{j=0}^\infty \frac{(-t)^j}{j!}(D_H^j f)(\mathbf{P}, \mathbf{Q}) \tag{10.183}$$

*as long as the series converges.* ∎

*Remark* 10.32
Here $D_H^j$ denotes the operator $D_H$ applied $j$ times if $j \geq 1$, and the identity operator $D_H^0 f = f$ if $j = 0$. The series expansion (10.183) for the evolution operator $U^t$ is called the *Lie series*. ∎

*Proof of Theorem 10.25*
By the theorem of the existence and uniqueness for ordinary differential equations, the Hamiltonian flow is uniquely determined and it is a one-parameter group

of diffeomorphisms. Therefore it is sufficient to check that the series (10.183) solves equation (10.181). Formally, this has an immediate verification: indeed, by differentiating the series (10.183) term by term we find

$$\frac{d}{dt}\sum_{j=0}^{\infty}\frac{(-t)^j}{j!}(D_H^j f)(\mathbf{P},\mathbf{Q})$$

$$= -\sum_{j=1}^{\infty}\frac{(-t)^{j-1}}{(j-1)!}(D_H^j f)(\mathbf{P},\mathbf{Q})$$

$$= -D_H\left(\sum_{j=1}^{\infty}\frac{(-t)^{j-1}}{(j-1)!}(D_H^{j-1} f)(\mathbf{P},\mathbf{Q})\right)$$

$$= -D_H\left(\sum_{n=0}^{\infty}\frac{(-t)^n}{n!}(D_H^n f)(\mathbf{P},\mathbf{Q})\right)$$

$$= -D_H(e^{-tD_H}f)(\mathbf{P},\mathbf{Q}) = \{f,H\}(\mathbf{p},\mathbf{q}),$$

where we set $n = j - 1$. ∎

*Example 10.34*
Let $l = 1$, $H(q,p) = qp$. The Hamiltonian flow is clearly given by

$$p = e^{-t}P, \quad q = e^t Q.$$

Consider the functions $f_1(p,q) = q$ and $f_2(p,q) = p$ and apply equations (10.183), to obtain

$$p = U^t f_2 = \sum_{j=0}^{\infty}\frac{(-t)^j}{j!}(D_H^j p)|_{(p,q)=(P,Q)},$$

$$q = U^t f_1 = \sum_{j=0}^{\infty}\frac{(-t)^j}{j!}(D_H^j q)|_{(p,q)=(P,Q)}.$$

On the other hand, $-D_H q = \{q,H\} = q$, $D_H^2 q = -D_H(-D_H q) = q$, and hence $(-D_H)^j q = q$ for every $j \geq 1$. In addition $(-D_H)^j p = (-1)^j p$, and substituting into the series we find

$$p = P\sum_{j=0}^{\infty}\frac{(-t)^j}{j!} = e^{-t}P, \quad q = Q\sum_{j=0}^{\infty}\frac{t^j}{j!} = e^t Q.$$

∎

*Example 10.35*
Let $l = 1$, $H(p,q) = (q^2 + p^2)/2$. The associated Hamiltonian flow is

$$p = -Q\sin t + P\cos t, \quad q = Q\cos t + P\sin t.$$

Applying (10.183) to $q$ and $p$, and observing that $-D_H q = p$, $D_H^2 q = -D_H p = -q$, $-D_H^3 q = -D_H^2 p = -p$ and $D_H^4 q = -D_H^3 p = q$, we find

$$p = \sum_{j=0}^{\infty} \frac{t^j}{j!}(-D_H)^j p = P \sum_{j=0}^{\infty} \frac{(-1)^j t^{2j}}{(2j)!} - Q \sum_{j=0}^{\infty} \frac{(-1)^j t^{2j+1}}{(2j+1)!} = P \cos t - Q \sin t,$$

$$q = \sum_{j=0}^{\infty} \frac{t^j}{j!}(-D_H)^j q = Q \sum_{j=0}^{\infty} \frac{(-1)^j t^{2j}}{(2j)!} + P \sum_{j=0}^{\infty} \frac{(-1)^j t^{2j+1}}{(2j+1)!} = Q \cos t + P \sin t.$$

For example, if we consider a function $f(p,q) = qp$, on the one hand, we have

$$(U^t f)(P, Q) = (Q \cos t + P \sin t)(-Q \sin t + P \cos t)$$
$$= (P^2 - Q^2) \sin t \cos t + PQ(\cos^2 t - \sin^2 t),$$

while, on the other, from $-D_H f = \{qp, H\} = p^2 - q^2$, $D_H^2 f = -4pq$ and from equation (10.183) it follows that

$$(U^t f)(P, Q) = \sum_{j=0}^{\infty} \frac{t^j}{j!}(-D_H)^j f = PQ + t(P^2 - Q^2)$$

$$+ \frac{t^2}{2}(-4PQ) + \frac{t^3}{3!}(-4)(P^2 - Q^2) + \cdots.$$

This coincides with the series expansion of $(P^2 - Q^2) \sin t \cos t + PQ (\cos^2 t - \sin^2 t)$. ∎

**Example 10.36**
Let $l = 1$, $H = p^2/2$. Then

$$p = P,$$
$$q = Pt + Q.$$

If $f(p,q) = p^n q^m$, with $n$ and $m$ non-negative integers, then $(U^t f)(P, Q) = (Pt + Q)^m P^n$. On the other hand, $(-D_H)^j f = m(m-1)\ldots(m-j+1)p^{n+j}q^{m-j}$ for $j = 1, \ldots, m$, and $(-D_H)^j f = 0$ for all $j > m$. Applying equations (10.183) we find

$$(U^t f)(P, Q) = \sum_{j=0}^{m} \frac{t^j}{j!} m(m-1)\ldots(m-j+1) P^{n+j} Q^{m-j}$$

$$= P^n \sum_{j=0}^{m} \binom{m}{j} t^j P^j Q^{m-j} = P^n (Pt + Q)^m$$

(in the last equality we used Newton's binomial formula). ∎

**Example 10.37**
Let $H = (q^2 p^2)/2$. Since $pq = \sqrt{2E}$ is constant, the canonical equations can immediately be integrated:

$$p = Pe^{-PQt}, \quad q = Qe^{PQt}.$$

On the other hand, $(-D_H)^j p = (-1)^j q^j p^{j+1}$, $(-D_H)^j q = p^j q^{j+1}$, and hence by equations (10.183) we have

$$p = \sum_{j=0}^{\infty} \frac{(-t)^j}{j!} Q^j P^{j+1} = P e^{-PQt},$$

$$q = \sum_{j=0}^{\infty} \frac{t^j}{j!} Q^{j+1} P^j = Q e^{PQt}.$$

∎

## 10.9 Symmetries and first integrals

In this section we briefly consider the relations between the invariance properties of the Hamiltonian for groups of canonical transformations and the first integrals. For a more detailed study of this important topic in analytical mechanics, see Arnol'd (1979a) (Appendix 5).

Let $H : \mathbf{R}^{2l} \to \mathbf{R}$ be a regular Hamiltonian.

DEFINITION 10.22 *A completely canonical transformation* $\mathbf{x} = \tilde{\mathbf{x}}(\mathbf{X})$ *of* $\mathbf{R}^{2l}$ *is a symmetry of* $H$ *if the Hamiltonian is invariant for the transformation, and hence if*

$$H(\tilde{\mathbf{x}}(\mathbf{X})) = H(\mathbf{X}). \tag{10.184}$$

∎

*Example* 10.38
If $H$ has one cyclic coordinate $x_i$ (note that $x_i$ can be either a coordinate $q$ or a kinetic momentum $p$), $H$ is invariant for the translations $x_i \to x_i + \alpha$. ∎

*Example* 10.39
The rotations around the origin in $\mathbf{R}^2$:

$$p = P \cos \alpha + Q \sin \alpha, \quad q = -P \sin \alpha + Q \cos \alpha,$$

are a symmetry of $H = (p^2 + q^2)/2$. ∎

Another class of interesting examples is given by the following proposition.

PROPOSITION 10.7 *If* $H(\mathbf{p}, \mathbf{q})$ *is the Legendre transform of the Lagrangian* $L(\mathbf{q}, \dot{\mathbf{q}})$, *and the point transformation* $\mathbf{q} = \tilde{\mathbf{q}}(\mathbf{Q})$ *is admissible for* $L$ *(see Definition 9.1), the associated completely canonical transformation*

$$\begin{aligned} \mathbf{q} &= \tilde{\mathbf{q}}(\mathbf{Q}), \\ \mathbf{p} &= (J^T(\mathbf{Q}))^{-1} \mathbf{P}, \end{aligned} \tag{10.185}$$

*where* $J = (J_{ij}) = (\partial \tilde{q}_i / \partial Q_j)$, *is a symmetry of the Hamiltonian* $H$.

*Proof*
In the new variables the Hamiltonian $\hat{H}(\mathbf{P},\mathbf{Q})$ is obtained as the transform of $H(\mathbf{p},\mathbf{q})$ and it is also the Legendre transform of the Lagrangian $\tilde{L}(\mathbf{Q},\dot{\mathbf{Q}}) = L(\tilde{\mathbf{q}}(\mathbf{Q}), J(\mathbf{Q})\dot{\mathbf{Q}})$, i.e.

$$\hat{H}(\mathbf{P},\mathbf{Q}) = H((J^T)^{-1}\mathbf{P},\tilde{\mathbf{q}}(\mathbf{Q})) = \mathbf{P}\cdot\dot{\mathbf{Q}} - \tilde{L}(\mathbf{Q},\dot{\mathbf{Q}}).$$

We now satisfy the hypothesis $\tilde{L}(\mathbf{Q},\dot{\mathbf{Q}}) = L(\mathbf{Q},\dot{\mathbf{Q}})$; hence to compare the new with the old Hamiltonian we must compare $\mathbf{P}\cdot\dot{\mathbf{Q}}$ with $\mathbf{p}\cdot\dot{\mathbf{q}}$. We already know that they take the same values (see Section 10.2), but we want to see that if $\mathbf{p}\cdot\dot{\mathbf{q}} = F(\mathbf{p},\mathbf{q})$ then $\mathbf{P}\cdot\dot{\mathbf{Q}} = F(\mathbf{P},\mathbf{Q})$. Obviously it is enough to show that if $\dot{\mathbf{q}} = \mathbf{f}(\mathbf{p},\mathbf{q})$ then $\dot{\mathbf{Q}} = \mathbf{f}(\mathbf{P},\mathbf{Q})$. This holds because $\dot{\mathbf{q}} = \mathbf{f}(\mathbf{p},\mathbf{q})$ can be obtained by inverting the system $\mathbf{p} = \nabla_{\dot{\mathbf{q}}}L(\mathbf{q},\dot{\mathbf{q}})$. Because of the admissibility of the tramsformation, this sytem is formally identical to $\mathbf{P} = \nabla_{\dot{\mathbf{Q}}}L(\mathbf{Q},\dot{\mathbf{Q}})$. In conclusion $\hat{H}(\mathbf{P},\mathbf{Q}) = H(\mathbf{P},\mathbf{Q})$. ■

DEFINITION 10.23 *A one-parameter family $s \in \mathbf{R}$ of completely canonical transformations $\mathbf{x} = \tilde{\mathbf{x}}(\mathbf{X},s)$ of $\mathbf{R}^{2l}$ is called a* one-parameter group *(of completely canonical transformations) if it possesses the following properties:*

*(1)* $\tilde{\mathbf{x}}(\mathbf{X},0) = \mathbf{X}$ *for all $\mathbf{X} \in \mathbf{R}^{2l}$;*
*(2)* $\tilde{\mathbf{x}}(\tilde{\mathbf{x}}(\mathbf{X},s_1),s_2) = \tilde{\mathbf{x}}(\mathbf{X},s_1+s_2)$ *for every $s_1, s_2 \in \mathbf{R}$ and for every $\mathbf{X} \in \mathbf{R}^{2l}$.*

*If for every $s \in \mathbf{R}$ the transformation $\tilde{\mathbf{x}}(\mathbf{X},s)$ is a symmetry of $H$, the group is a* one-parameter group of symmetries *of $H$.* ■

*Remark* 10.33
For the groups of point transformations see Problem 9 of Section 10.14. ■

We now examine how it is possible to interpret any one-parameter group of completely canonical transformations as a Hamiltonian flow.

DEFINITION 10.24 *Let $\tilde{\mathbf{x}}(\mathbf{X},s)$ be a one-parameter group of completely canonical transformations of $\mathbf{R}^{2l}$. The vector field*

$$\mathbf{v}(\mathbf{x}) = \frac{\partial \tilde{\mathbf{x}}}{\partial s}(\mathbf{x},0) \tag{10.186}$$

*is called an* infinitesimal generator *of the group of transformations.* ■

The following theorem clarifies the role of the infinitesimal generator.

THEOREM 10.26 *The infinitesimal generator $\mathbf{v}(\mathbf{x})$ of a one-parameter group $\tilde{\mathbf{x}}(\mathbf{X},s)$ of completely canonical transformations is a Hamiltonian field. In addition the group of transformations coincides with the corresponding Hamiltonian flow, and hence it is a solution of the system*

$$\dot{\mathbf{x}}(t) = \mathbf{v}(\mathbf{x}(t)), \quad \mathbf{x}(0) = \mathbf{X}. \tag{10.187}$$

*Proof*
We first check that $\tilde{\mathbf{x}}(\mathbf{X},t)$ is a solution of equations (10.187). Because of the group properties we have, setting $\mathbf{x}(t) = \tilde{\mathbf{x}}(\mathbf{X},t)$, that

$$\dot{\mathbf{x}}(t) = \lim_{\Delta t \to 0} \frac{\tilde{\mathbf{x}}(\mathbf{X}, t+\Delta t) - \tilde{\mathbf{x}}(\mathbf{X},t)}{\Delta t} = \lim_{\Delta t \to 0} \frac{\tilde{\mathbf{x}}(\mathbf{x}(t), \Delta t) - \tilde{\mathbf{x}}(\mathbf{x}(t), 0)}{\Delta t} = \mathbf{v}(\mathbf{x}(t)).$$

Since by the hypothesis, the Jacobian matrix $J = \nabla_{\mathbf{X}}\tilde{\mathbf{x}}(\mathbf{X},t)$ is symplectic for every $t$, we deduce by Lemma 10.1 that the matrix $B = (\partial J/\partial t)J^{-1}$ is Hamiltonian. Now note that we can write

$$\frac{\partial J}{\partial t} = \frac{\partial}{\partial t}\nabla_{\mathbf{X}}\tilde{\mathbf{x}}(\mathbf{X},t) = \nabla_{\mathbf{X}}\mathbf{v}(\tilde{\mathbf{x}}(\mathbf{X},t)) = (\nabla_{\mathbf{x}}\mathbf{v})J.$$

Using the fact that $\tilde{\mathbf{x}}(\mathbf{X},t)$ solves equations (10.187), it follows that $\nabla_{\mathbf{x}}\mathbf{v}(\mathbf{x}) = (\partial J/\partial t)J^{-1}$ and hence the field $\mathbf{v}(\mathbf{x})$ is Hamiltonian (Theorem 10.5). ∎

We can now prove the following extension of Noether's theorem. Recall how in the Lagrangian formulation (Theorem 4.4) the validity of this theorem was limited to symmetry groups associated to point transformations.

THEOREM 10.27 (Noether, Hamiltonian formulation) *If a system with Hamiltonian $H(\mathbf{x})$ has a one-parameter group of symmetries $\tilde{\mathbf{x}}(\mathbf{X},t)$, the Hamiltonian $K(\mathbf{x})$ of which the group is the flow is a first integral for the flow associated with $H$.*

*Proof*
The invariance of $H$ can be interpreted as its being constant along the flow generated by $K$. Therefore $L_{\mathbf{v}}H = \{H, K\} = 0$.
Conversely this implies that $K$ is a first integral for the flow generated by $H$. ∎

In summary, if $f(\mathbf{x}), g(\mathbf{x})$ are in involution, recalling Remark 10.30 we see that:

(i) the Hamiltonian flow generated by $f(\mathbf{x})$ has $g(\mathbf{x})$ as first integral and vice versa ;
(ii) the two flows associated with $f$ and $g$ commute;
(iii) the flow generated by $f(\mathbf{x})$ represents a symmetry for the Hamiltonian $g(\mathbf{x})$ and vice versa.

## 10.10 Integral invariants

In this section, which can be omitted at a first reading, we want to characterise the canonical transformations using the language of differential forms (see Appendix 4). For simplicity, we limit the exposition to the case of differential forms in $\mathbf{R}^{2l}$, while in the next section we introduce the notion of a *symplectic manifold* which allows us to extend the Hamiltonian formalism to a wider context.

**Theorem 10.28** *A transformation* $(\mathbf{p},\mathbf{q}) = (\mathbf{p}(\mathbf{P},\mathbf{Q}),\mathbf{q}(\mathbf{P},\mathbf{Q}))$ *is completely canonical if and only if*

$$\sum_{i=1}^{l} dp_i \wedge dq_i = \sum_{i=1}^{l} dP_i \wedge dQ_i. \tag{10.188}$$

∎

*Remark* 10.34
A transformation satisfying (10.188) is also called a *symplectic diffeomorphism* as it preserves the *symplectic* 2-form[1]

$$\omega = \sum_{i=1}^{l} dp_i \wedge dq_i. \tag{10.189}$$

∎

*Proof of Theorem 10.28*
The proof follows from an immediate application of the Lie condition (10.190). From

$$\sum_{i=1}^{l} p_i \, dq_i - \sum_{i=1}^{l} P_i \, dQ_i = df \tag{10.190}$$

(note that, since the transformation is independent of time, $\tilde{d} = d$), if we perform an external differentiation of both sides and take into account $d^2 f = 0$ we find (10.188). Conversely, since (10.188) is equivalent to

$$d\left( \sum_{i=1}^{l} p_i \, dq_i - \sum_{i=1}^{l} P_i \, dQ_i \right) = 0, \tag{10.191}$$

we immediately deduce (10.190) because of Poincaré's lemma (Theorem 2.2, Appendix 4): every closed form in $\mathbf{R}^{2l}$ is exact. ∎

From Theorem 10.188 we easily deduce some interesting corollaries.

**Corollary 10.7** *A canonical transformation preserves the differential $2k$-forms:*

$$\omega^{2k} = \sum_{1 \leq i_1 < i_2 < \ldots < i_k \leq l} dp_{i_1} \wedge \ldots \wedge dp_{i_k} \wedge dq_{i_1} \wedge \ldots \wedge dq_{i_k}, \tag{10.192}$$

*where* $k = 1, \ldots, l$.

*Proof*
If the transformation is canonical, it preserves the 2-form $\omega$, and hence it also preserves the external product of $\omega$ with itself $k$ times (see (A4.20)):

$$\Omega^k = \omega \wedge \ldots \wedge \omega = \sum_{i_1,\ldots,i_k} dp_{i_1} \wedge dq_{i_1} \wedge \ldots \wedge dp_{i_k} \wedge dq_{i_k}, \tag{10.193}$$

---

[1] Be careful: in spite of the same notation, this is not to be confused with the Poincaré–Cartan form.

which is proportional to $\omega^{2k}$:

$$\Omega^k = (-1)^{k-1} k! \omega^{2k}. \tag{10.194}$$

∎

Evidently Corollary 10.21 for $k = l$ can be stated as follows.

COROLLARY 10.8  *A canonical transformation preserves the volume form*

$$\omega^{2l} = dp_1 \wedge \ldots \wedge dp_l \wedge dq_1 \wedge \ldots \wedge dq_l. \tag{10.195}$$

∎

*Remark* 10.35
The forms $\omega^{2k}$ have a significant geometrical interpretation. If $k = 1$, the integral of the form $\omega$ on a submanifold $S$ of $\mathbf{R}^{2l}$ is equal to the sum of the areas (the sign keeps track of the orientation) of the projections of $S$ onto the planes $(p_i, q_i)$. Analogously, the integral of $\omega^{2k}$ is equal to the sum of the measures (with sign) of the projections of $S$ onto all the hyperplanes $(p_{i_1}, \ldots, p_{i_k}, q_{i_1}, \ldots, q_{i_k})$, with $1 \leq i_1 < \ldots < i_k \leq l$. It follows that a completely canonical transformation preserves the sum of the measures of the projections onto all coordinate planes and hyperplanes $(p_{i_1}, \ldots, p_{i_k}, q_{i_1}, \ldots, q_{i_k})$.

## 10.11 Symplectic manifolds and Hamiltonian dynamical systems

DEFINITION 10.25  *A differentiable manifold $M$ of dimension $2l$ is a symplectic manifold if there exists a closed non-degenerate differential 2-form[2] $\omega$, i.e. such that $d\omega = 0$ and, for every non-zero vector $v \in T_m M$ tangent to $M$ at a point $m$, there exists a vector $w \in T_m M$ such that $\omega(v, w) \neq 0$.*  ∎

*Example* 10.40
As seen in Section 10.1, every real vector space of even dimension can be endowed with a symplectic structure which makes it into a symplectic manifold. An example is $\mathbf{R}^{2l}$ with the standard structure $\omega = \sum_{i=1}^{l} dp_i \wedge dq_i$.  ∎

*Example* 10.41
The tori of even dimension $\mathbf{T}^{2l}$ with the 2-form $\omega = \sum_{i=1}^{l} d\phi_i \wedge d\phi_{i+l}$ are symplectic manifolds. This is an interesting case. Indeed, because of the particular topology (the tori are not simply connected) it is easy to construct examples of vector fields which are locally Hamiltonian, but not globally a Hamiltonian. For example, the infinitesimal generator of the one-parameter group of translations $\phi_i \to \phi_i + \alpha_i t$, where $\alpha \in \mathbf{R}^{2l}$ is fixed, is evidently $\mathbf{v} = \alpha$ and it is a locally, but *not* globally, a Hamiltonian vector field, since every function $H : \mathbf{T}^{2l} \to \mathbf{R}$ necessarily has at least two points where the gradient, and hence the field $\mathbb{J} \nabla_\phi H$, vanishes.  ∎

---

[2] Be careful: not to be confused, in spite of the same notation, with the Poincaré–Cartan form, which is a 1-form on $M \times \mathbf{R}$.

*Example* 10.42

A natural Lagrangian system is a mechanical system subject to ideal constraints, frictionless, fixed and holonomous, and subject to conservative forces. Its space of configurations is a differentiable manifold $S$ of dimension $l$ (the number of degrees of freedom of the system) and the Lagrangian $L$ is a real function defined on the tangent bundle of $S$, $L: TS \to \mathbf{R}$. The kinetic energy $T(\mathbf{q}, \dot{\mathbf{q}}) = \frac{1}{2}\sum_{i,j=1}^{l} a_{ij}\dot{q}_i\dot{q}_j$ defines a Riemannian metric on $S$:$(\mathrm{d}s)^2 = \sum_{i,j=1}^{l} a_{ij}\mathrm{d}q_i\mathrm{d}q_j$, and the Lagrangian can be written then as $L = \frac{1}{2}|\mathrm{d}s/\mathrm{d}t|^2 - V(\mathbf{q})$, where $V$ is the potential energy of the conservative forces. The Hamiltonian $H$ is a real function defined on the cotangent bundle $M = T^*S$ of the space of configurations $H: M \to \mathbf{R}$, through a Legendre transformation applied to the Lagrangian $L: H = \mathbf{p} \cdot \dot{\mathbf{q}} - L$. If $A$ denotes the matrix $a_{ij}$ of the kinetic energy, $H(\mathbf{p}, \mathbf{q}) = \frac{1}{2}\mathbf{p} \cdot A^{-1}\mathbf{p} + V(\mathbf{q})$. The cotangent bundle $M = T^*S$ has a natural symplectic manifold structure with the symplectic 2-from $\omega = \mathrm{d}\left(\sum_{i=1}^{l} p_i \mathrm{d}q_i\right)$. ∎

*Remark* 10.36

The 2-form $\omega$ induces an isomorphism between the tangent space $T_sS$ and the cotangent space $T_s^*S$ to $S$ at any point $s \in S$. Indeed, it is enough to associate with every vector $v \in T_sS$ the covector $\omega(v, \cdot)$. If the chosen system of local coordinates $(\mathbf{p}, \mathbf{q})$ of $T_s^*S$ is such that $\omega = \sum_{i=1}^{l} \mathrm{d}p_i \wedge \mathrm{d}q_i$, the representative matrix of the isomorphism is the matrix $\mathsf{J}$ given by (10.1). ∎

It is possible to construct an atlas of $M$ in which the symplectic 2-form $\omega$ has a particularly simple structure (just as in a vector space endowed with a symplectic product, there exists a symplectic basis with respect to which the product takes the standard form (10.16)).

THEOREM 10.29 (Darboux)  *Let $M$ be a symplectic manifold. There always exists an atlas of $M$, called the symplectic atlas, with respect to which the 2-form $\omega$ is written as $\omega = \sum_{i=1}^{l} \mathrm{d}p_i \wedge \mathrm{d}q_i$.* ∎

For the proof see Abraham and Marsden (1978) or Arnol'd (1978a).

In analogy with Definition 10.8 of a completely canonical transformation in $\mathbf{R}^{2l}$, we have the following.

DEFINITION 10.26  *A local coordinate transformation of $M$ is called* (completely) *canonical if its Jacobian matrix is at every point a symplectic matrix.* ∎

*Remark* 10.37

The transformations from one chart to another chart of the symplectic atlas, whose existence is guaranteed by the theorem of Darboux, are automatically canonical transformations. ∎

DEFINITION 10.27  *Let $g: M \to M$ be a diffeomorphism of a symplectic manifold. Then $g$ is a symplectic diffeomorphism if $g^*\omega = \omega$.* ∎

*Remark 10.38*
The symplectic diffeomorphisms of a symplectic manifold $M$ constitute a subgroup $\text{SDiff}(M)$ of the group $\text{Diff}(M)$ of diffeomorphisms of $M$. ∎

DEFINITION 10.28  *A Hamiltonian dynamical system is the datum of a symplectic manifold $M$ endowed with a 2-form $\omega$ and of a function $H : M \to \mathbf{R}$, the Hamiltonian, inducing the Hamilton equations*

$$\dot{\mathbf{x}} = \mathfrak{I}\, dH(\mathbf{x}), \tag{10.196}$$

*where $\mathbf{x} = (\mathbf{p}, \mathbf{q}) \in M$, and $\mathfrak{I}$ denotes the isomorphism between the cotangent bundle $T^*_{\mathbf{x}}M$ and the space tangent to $T_{\mathbf{x}}M$.* ∎

*Remark 10.39*
Theorem 3.6 regarding a canonical Hamiltonian flow, together with the previous definitions, guarantees that the one-parameter group of symplectic diffeomorphisms $S^t$, solutions of (10.196), is a group of symplectic diffeomorphisms of $M$, and hence that for every $t \in \mathbf{R}$ we have

$$(S^t)^*\omega = \omega. \tag{10.197}$$

In constrast with the case when $M = \mathbf{R}^{2l}$, in general it is not true that every one-parameter group of diffeomorphisms is the Hamiltonian flow of a Hamiltonian $H : M \to \mathbf{R}$ (see Theorem 9.1). Example 11.2 provides a significant counterexample to the extension of Theorem 9.1 to any symplectic manifold. ∎

## 10.12 Problems

1. Find the conditions ensuring that the linear transformation of $\mathbf{R}^4$ given by

$$\begin{pmatrix} P_1 \\ P_2 \\ Q_1 \\ Q_2 \end{pmatrix} = \begin{pmatrix} a_{11} & a_{12} & a_{13} & 0 \\ a_{21} & a_{22} & 0 & a_{24} \\ 0 & a_{32} & a_{33} & 0 \\ a_{41} & 0 & a_{43} & 0 \end{pmatrix} \begin{pmatrix} p_1 \\ p_2 \\ q_1 \\ q_2 \end{pmatrix}$$

(a) preserves orientation and volume;
(b) is symplectic.

2. Let $A$ be a symplectic matrix, $A \in \text{Sp}(l, \mathbf{R})$. Prove that the characteristic polynomial of $A$:

$$\mathcal{P}_A(\lambda) = \det(A - \lambda \mathbf{1})$$

is reciprocal, and hence that it satisfies the condition

$$\mathcal{P}_A(\lambda) = \lambda^{2l} \mathcal{P}_A(\lambda^{-1}).$$

Deduce that if $\lambda$ is an eigenvalue of $A$, $\lambda^{-1}$ is also an eigenvalue of $A$ (see Arnol'd 1978b).

Analogously prove that the characteristic polynomial of a Hamiltonian matrix is even.

**3.** Let $B$ be a Hamiltonian matrix and let $\lambda, \mu$ be two of its eigenvalues such that $\lambda + \mu \neq 0$. Prove that the corresponding eigenspaces are $J$-orthogonal (i.e. if $Bv = \lambda v$ and $Bw = \mu w$ then $v^T J w = 0$).

**4.** Assume that the Hamiltonian matrix $B$ has $2n$ distinct eigenvalues $\lambda_1, \ldots, \lambda_n, -\lambda_1, \ldots, -\lambda_n$ (see Problem 2). Prove that there exists a symplectic matrix $S$ (possibly complex) such that $S^{-1}BS = \text{diag}(\lambda_1, \ldots, \lambda_n, -\lambda_1, \ldots, -\lambda_n)$.

**5.** Prove that a real $2l \times 2l$ matrix $P$ is symmetric, positive definite and symplectic if and only if $P = \exp(B)$, where the matrix $B$ is $\begin{pmatrix} a & b \\ b & -a \end{pmatrix}$, $a = a^T$ and $b = b^T$.

**6.** Prove Theorem 10.3.

**7.** Find the completely canonical linear transformation which maps the Hamiltonian $H = \frac{1}{2}(P_1Q_1 + P_2Q_2)^2$ into $H = \frac{1}{8}(p_1^2 - q_1^2 + p_2^2 - q_2^2)^2$.

**8.** Prove that the transformation

$$q_1 = \frac{Q_1^2 - Q_2^2}{2},$$

$$q_2 = Q_1 Q_2,$$

$$p_1 = \frac{P_1 Q_1 - P_2 Q_2}{Q_1^2 + Q_2^2},$$

$$p_2 = \frac{P_2 Q_1 + P_1 Q_2}{Q_1^2 + Q_2^2}$$

is completely canonical, and check that it transforms the Hamiltonian $H = 1/2m(p_1^2 + p_2^2) - k/\sqrt{q_1^2 + q_2^2}$ to $K = 1/2m(Q_1^2 + Q_2^2)(P_1^2 + P_2^2 - 4mk)$.

**9.** Let $B$ and $C$ be two Hamiltonian matrices. Prove that if $[B, C] = 0$, for every $s, t \in \mathbf{R}$, the symplectic matrices $e^{tB}$ and $e^{sC}$ commute:

$$e^{tB} e^{sC} = e^{tB+sC} = e^{sC} e^{tB}.$$

**10.** Let $a \in \mathbf{R}$ be fixed. Prove that the following transformation of $\mathbf{R}^2$:

$$x_1' = a - x_2 - x_1^2, \quad x_2' = x_1',$$

is invertible and preserves the standard symplectic structure of $\mathbf{R}^2$. Compute the inverse transformation.

**11.** Consider the transformation $\mathbf{P} = \mathbf{P}(\mathbf{p})$ with non-singular Jacobian matrix $J$. How can it be completed to obtain a canonical transformation? (Answer: $\mathbf{Q} = (J^T)^{-1}\mathbf{q}$.)

**12.** Prove that the transformation
$$P = \frac{p}{\sqrt{1+q^2p^2}}, \quad Q = q\sqrt{1+q^2p^2}$$
is completely canonical and find a generating function $F(q,P)$. (Answer: $F(q,P) = \arcsin Pq$.) Compute all other admissible generating functions.

**13.** Let $P = p^\alpha$, where $\alpha \neq 0$ is a real parameter. Determine $Q$ as a function of $(q,p)$ in such a way that the transformation $(q,p) \to (Q,P)$ thus obtained is completely canonical and find a generating function for it. (Answer: $Q = p^{1-\alpha}q + g(p)$, where $g$ is an arbitrary regular function; $F(q,P) = qP^{1/\alpha} + \int_0^P \hat{g}(P')\,dP'$, where $\hat{g}(P) = g(P^{1/\alpha})$.)

**14.** Determine the real parameters $k$, $l$, $m$, $n$ such that the transformation
$$P = p^k q^l, \quad Q = p^m q^n$$
is completely canonical and find all generating functions. (Answer: $k = 1-m$, $l = -m$, $n = 1+m$; $F_1(q,Q) = -m\,(Q/q)^{1/m}$ with the condition $m \neq 0$, $F_2(q,P) = (1-m)(qP)^{1/(1-m)}$ if $m \neq 1$, $F_3(p,Q) = -(1+m)(pQ)^{1/(1+m)}$ with the condition $m \neq -1$ and $F_4(p,P) = -m\,(p/P)^{1/m}$ if $m \neq 0$.)

**15.** Prove that the transformation
$$P = q\cot p, \quad Q = \log\left(\frac{\sin p}{q}\right)$$
is completely canonical. Determine the generating functions $F_1(q,Q)$ and $F_2(q,P)$. (Answer: $F_1(q,Q) = q\arcsin p(qe^Q) + (e^{-2Q} - q^2)^{1/2}$, $F_2(q,P) = q\arctan(q/P) + P[1 - \frac{1}{2}\log(q^2 + P^2)]$.)

**16.** Determine which among the following transformations is canonical ($k$ is a real parameter):

| | | |
|---|---|---|
| $Q = \dfrac{q^2}{2}$, | $P = \dfrac{p}{q}$, | (10.198) |
| $Q = \tan q$, | $P = (p-k)\cos^2 q$, | (10.199) |
| $Q = \sin q$, | $P = \dfrac{p-k}{\cos q}$, | (10.200) |
| $Q = \sqrt{2q}e^t \cos p$, | $P = \sqrt{2q}e^{-t}\sin p$, | (10.201) |

and find the generating functions corresponding to each transformation.

**17.** Determine the real parameters $\alpha$, $\beta$, $\gamma$, $\delta$ such that the transformation
$$P_1 = \alpha q_1 + \beta p_1, \quad P_2 = \gamma q_2 + \delta p_2, \quad Q_1 = p_1, \quad Q_2 = p_2$$
is completely canonical. Find a generating function.

**18.** Consider the transformation

$$P = -q - \sqrt{p+q^2}, \quad Q = -q^2 - aq\sqrt{q^2+p}.$$

Find its domain of definition and determine for which values of the real parameter $a$ the transformation is completely canonical. Compute the generating function $F(q,P)$. (Answer: $a=2$, $F(q,P) = qP(q+P)$.)

**19.** Consider the transformation

$$P = -p^\alpha q^\beta, \quad Q = \gamma \log p.$$

Determine for which values of the parameters $\alpha$, $\beta$, $\gamma$ the transformation is completely canonical and compute the generating function $F(q,Q)$. (Answer: $\alpha = \beta = \gamma = 1$, $F(q,Q) = qe^Q$.)

**20.** Prove that the transformation

$$Q = \log(1 + \sqrt{q}\cos p), \quad P = 2(1 + \sqrt{q}\cos p)\sqrt{q}\sin p$$

is completely canonical and find the generating functions.

**21.** Consider the transformation

$$p = a(e^{\alpha P(1+\beta Q)} - 1), \quad q = b\log(1 + \beta Q)e^{-\alpha P(1+\beta Q)},$$

where $a$, $b$, $\alpha$, $\beta$ are real parameters.

(a) Determine the domain of definition of the transformation, and compute the inverse transformation and its domain.

(b) Determine the conditions on the parameters $a$, $b$, $\alpha$, $\beta$ that ensure that the transformation is completely canonical and compute the generating function $F(p,Q)$.

**22.** Prove that the completely canonical transformations of $\mathbf{R}^2$ admitting a generating function of the form

$$F(q,P) = \frac{1}{2}(aq^2 + 2bqP + cP^2)$$

are not a subgroup of $\mathrm{Sp}(1,\mathbf{R})$.

**23.** Given the transformation

$$p = \tan(\alpha P)e^{\delta t}, \quad q = \frac{\beta Q}{1 + (\tan(\gamma P))^2}e^{\eta t},$$

where $\alpha$, $\beta$, $\gamma$, $\delta$, $\eta$ are real parameters:

(a) determine the domain of the transformation, compute the inverse transformation and its domain;

(b) determine the conditions on the parameters $\alpha$, $\beta$, $\gamma$, $\delta$, $\eta$ ensuring that the transformation is canonical and compute the generating function $F(q,P,t)$.

24. Consider the transformation
$$q = e^{-t}(PQ)^\alpha, \quad p = 2e^t(PQ)^\gamma \log P^\beta,$$
where $\alpha$, $\beta$ and $\gamma$ are real positive constants.
   (a) Determine for which values of $\alpha$, $\beta$ and $\gamma$ the transformation is canonical.
   (b) If $\alpha = \frac{1}{2}$, compute the generating function $F(q, P)$.
   (c) For $\alpha = \frac{1}{2}$, how is the Hamiltonian $H(q,p) = -qp$ transformed?

25. Prove that the transformation
$$q_1 = \sqrt{\frac{2Q_1}{\lambda_1}} \cos P_1 + \sqrt{\frac{2Q_2}{\lambda_2}} \cos P_2,$$
$$q_2 = -\sqrt{\frac{2Q_1}{\lambda_1}} \cos P_1 + \sqrt{\frac{2Q_2}{\lambda_2}} \cos P_2,$$
$$p_1 = \frac{1}{2}\sqrt{2Q_1\lambda_1}\sin P_1 + \frac{1}{2}\sqrt{2Q_2\lambda_2}\sin P_2,$$
$$p_2 = -\frac{1}{2}\sqrt{2Q_1\lambda_1}\sin P_1 + \frac{1}{2}\sqrt{2Q_2\lambda_2}\sin P_2$$
is completely canonical and that it transforms the Hamiltonian
$$H = p_1^2 + p_2^2 + \frac{1}{8}\lambda_1^2(q_1 + q_2)^2 + \frac{1}{8}\lambda_2^2(q_1 + q_2)^2$$
to $K = \lambda_1 Q_1 + \lambda_2 Q_2$. Use this transformation to find the solution of Hamilton's equations in the variables $(q_1, q_2, p_1, p_2)$.

26. Prove that the transformation
$$q_1 = \frac{1}{\sqrt{m\omega}}\left(\sqrt{2P_1}\sin Q_1 + P_2\right), \quad q_2 = \frac{1}{\sqrt{m\omega}}\left(\sqrt{2P_1}\cos Q_1 + Q_2\right),$$
$$p_1 = \frac{\sqrt{m\omega}}{2}\left(\sqrt{2P_1}\cos Q_1 - Q_2\right), \quad p_2 = \frac{\sqrt{m\omega}}{2}\left(-\sqrt{2P_1}\sin Q_1 + P_2\right)$$
is completely canonical.

27. Using the method of Lie series, compute the flow associated with the Hamiltonian $H(p_1, p_2, q_1, q_2) = p_1 q_2 + p_2^n$, where $n \in \mathbf{N}$.

28. Given two Hamiltonians $H$ and $K$, prove that
$$e^{-tD_H}e^{-sD_K} = e^{-sD_K}e^{-tD_H} + stD_{\{H,K\}} + \mathcal{O}(3),$$
where $\mathcal{O}(3)$ denotes terms of order $s^3$, $s^2 t$, $st^2$, $t^3$ or higher.

29. Prove that the transformation
$$x' = [x + y + f(x)] \mathrm{mod}(2\pi), \quad y' = y + f(x),$$
where $(x, y) \in \mathbf{S}^1 \times \mathbf{R}$, and $f : \mathbf{S}^1 \to \mathbf{R}$ is a regular function, preserves the symplectic 2-form $dy \wedge dx$ on the cylinder $T^*\mathbf{S}^1 = \mathbf{S}^1 \times \mathbf{R}$.

## 10.13 Additional remarks and bibliographical notes

In this chapter we started the study of the canonical formalism of analytical mechanics. This formalism will prove to be a powerful tool for solving the equations of motion, as we shall see in the next two chapters.

Our exposition adopts the viewpoint and general influence of the beautiful book of Arnol'd (1978a), which is to be considered the fundamental reference for any further study. We differ from Arnol'd in the initial definition of the transformations that preserve the canonical structure of Hamilton's equations, as we prefer to stress the importance of the latter rather than the geometric aspect. Indeed, we believe that to fully appreciate the geometric picture, one needs a good knowledge of the techniques of modern differential geometry, going beyond the scope of the present exposition. Another useful text is the book by Abraham and Marsden (1978), which is encyclopaedic in character.

However, Theorems 10.6 and 10.7 indicate that there is a substantial equivalence between the two methods, as they identify the canonical transformations as the *natural* transformations that leave the canonical structure of Hamilton's equations invariant.

We recommend as supplementary reading, at approximately the same level as the present book, the texts by Cercignani (1976a, 1976b) and Benettin *et al.* (1991).

The book by Levi-Civita and Amaldi (1927), although quite old, is still very useful for depth and clarity, as well as for the reasonable mathematical level required of the reader. The section on Pfaffian systems (non-singular differential forms) and their use in the canonical formalism is especially recommended.

The reading of the text of Gallavotti (1980) is more difficult, but certainly useful, also because of the many interesting problems which stimulate the reader to critically study the material.

Another reference text is Meyer and Hall (1992), which adopts from the beginning a 'dynamical systems' point of view, and considers only transformations that are independent of time.

It has not been possible to introduce the study of symplectic geometry and topology, both active research fields with rich interesting results. The most serious consequence is the extreme conciseness of our section on Hamiltonian systems with symmetries, and the lack of a discussion on the so-called reduction of phase space, and hence of the practical use of first integrals in reducing the order of the equations of motion. The book by Arnol'd *et al.* (1983) is full of examples and applications, although it may be hard to follow as a first reading. The symmetry argument and the so-called 'momentum map', which yields a formulation of Noether's theorem in the more general context of symplectic manifolds, are also discussed in depth in Abraham and Marsden (1978).

## 10.14 Additional solved problems

*Problem 1*
Prove that the determinant of a symplectic matrix is equal to $+1$.

*Solution*
This result can be obtained in various ways. We give here a proof which only uses the definition of a symplectic matrix and an elementary knowledge of linear algebra. Every real $m \times m$ invertible matrix $A$ can be uniquely written as the product of a symmetric positive definite matrix and of an orthogonal matrix (*polar decomposition*):

$$A = P_1 O_1 = O_2 P_2, \quad \text{with } P_i = P_i^T > 0, \ O_i^T = O_i^{-1}, \ i = 1, 2. \quad (10.202)$$

Indeed, note that the matrices $AA^T$ and $A^T A$ are both symmetric and positive definite. Then, using the results of Chapter 4, Section 4.10, we set $P_1 = \sqrt{AA^T}$, $P_2 = \sqrt{A^T A}$, $O_1 = P_1^{-1} A$, $O_2 = AP_2^{-1}$. It is immediately verified that $O_1$ and $O_2$ are orthogonal: for example $O_1 O_1^T = P_1^{-1} AA^T (P_1^{-1})^T = (\sqrt{AA^T})^{-1} AA^T (\sqrt{AA^T})^{-1} = 1$. We leave it for the reader to verify that the polar decomposition is unique. If the matrix $A$ is symplectic, then the matrices $P_1, P_2, O_1, O_2$ are also symplectic. Indeed, from $A^{-1} = -\mathfrak{I} A^T \mathfrak{I}$ it follows that

$$O_1^{-1} P_1^{-1} = -\mathfrak{I} O_1^T P_1^T \mathfrak{I} = (\mathfrak{I}^{-1} O_1^T \mathfrak{I})(\mathfrak{I}^T P_1^T \mathfrak{I}). \quad (10.203)$$

On the other hand, $\mathfrak{I}^{-1} O_1^T \mathfrak{I}$ is an orthogonal matrix: $\mathfrak{I}^{-1} O_1^T \mathfrak{I} (\mathfrak{I}^{-1} O_1^T \mathfrak{I})^T = -\mathfrak{I} O_1^T (O_1^T)^{-1} \mathfrak{I} = 1$. The matrix $\mathfrak{I}^T P_1^T \mathfrak{I}$ is symmetric and positive definite: $(\mathfrak{I}^T P_1^T \mathfrak{I})^T = \mathfrak{I}^T P_1 \mathfrak{I} = \mathfrak{I}^T P_1^T \mathfrak{I}$ for any vector $v$, and because $P_1$ is positive definite, we have

$$\mathfrak{I}^T P_1^T \mathfrak{I} v \cdot v = P_1^T \mathfrak{I} v \cdot (\mathfrak{I} v) \geq a^2 \mathfrak{I} v \cdot \mathfrak{I} v = a^2 v \cdot v,$$

for some constant $a \neq 0$.

The uniqueness of the polar decomposition applied to (10.203) implies that $O_1, P_1$ are symplectic. Since we already know that the determinant of a real symplectic matrix is $\pm 1$, the fact that the polar decomposition is symplectic shows that to deduce that the determinant is $+1$ it is not restrictive to assume that the given matrix $A$ is symplectic and orthogonal. Since $A^{-1} = -\mathfrak{I} A^T \mathfrak{I}$, denoting by $a, b, c, d$ the $l \times l$ blocks that constitute $A = \begin{pmatrix} a & b \\ c & d \end{pmatrix}$, and requiring that $A^{-1} = A^T$ we find that $A$ must be of the form $A = \begin{pmatrix} a & b \\ -b & a \end{pmatrix}$, with $a^T b$ a symmetric matrix and

$$a^T a + b^T b = 1 \quad (10.204)$$

(see Remarks 10.3 and 10.4). Consider a complex $2l \times 2l$ matrix $Q$, whose block structure is $Q = \frac{1}{\sqrt{2}} \begin{pmatrix} 1 & i1 \\ 1 & -i1 \end{pmatrix}$. Then $Q$ is unitary, i.e. $Q^{-1} = Q^*$, where $Q^*$

denotes the conjugate transpose matrix of $Q$. On the other hand, $QAQ^{-1} = \begin{pmatrix} a - bi & 0 \\ 0 & a + bi \end{pmatrix}$ and since the determinant is invariant under conjugation, we find

$$\det A = \det(QAQ^{-1}) = \det(a - bi)\det(a + bi) = |\det(a - bi)|^2 > 0,$$

contradicting $\det A = -1$. We note that the proof also shows that the group $\mathrm{Sp}(l, \mathbf{R}) \cap O(2l, \mathbf{R})$ is isomorphic to the group $\mathbf{U}(l, \mathbf{C})$ of unitary matrices: from (10.204) it follows that the matrix $a - ib$ is unitary and, conversely, given a unitary matrix $U$ we can associate with it a symplectic orthogonal matrix $A$, whose blocks $a$ and $b$ are the real and imaginary parts, respectively. We leave to the reader the verification that this is indeed a group isomorphism.

*Problem 2*
Let $(V, \omega)$ be a symplectic vector space. Show that the map $: V \to V^*$, $v \mapsto v^\flat = \omega(v, \cdot)$, is an isomorphism whose inverse we denote by $\# : V^* \to V$. Let $U$ be a linear subspace of $V$; its *orthogonal symplectic complement* is

$$U^{\perp, \omega} = \{v \in V \mid \omega(v, U) = 0\}.$$

We say that a linear subspace $U$ of $V$ is a *Lagrangian subspace* if $\dim U = \dim V/2$ and $\omega|_U \equiv 0$, and a *symplectic subspace* if $\omega|_U \equiv 0$ is non-degenerate.

Clearly, if $U$ is a symplectic subspace then its dimension is even and $(U, \omega|_U)$ is a linear symplectic subspace. Two linear subspaces $U_1$ and $U_2$ of $V$ provide a *symplectic decomposition* or a *Lagrangian decomposition* of $V$ if $V = U_1 \oplus U_2$ and $U_1, U_2$ are a symplectic or Lagrangian subspace, respectively. Prove the following.

(i) If $U$ is a symplectic subspace, then $V = U \oplus U^{\perp, \omega}$ is a symplectic decomposition. Conversely, if $V = U_1 \oplus U_2$ and $\omega(U_1, U_2) = 0$ then the decomposition is symplectic.
(ii) For every Lagrangian subspace $U_1$ of $V$ there exists at least one Lagrangian decomposition of $V = U_1 \oplus U_2$.
(iii) Let $V = U_1 \oplus U_2$ be a Lagrangian decomposition. For every basis $(e_1, \ldots, e_n)$ of $U_1$ there exists a basis $(e_{n+1}, \ldots, e_{2n})$ of $U_2$ such that $(e_1, \ldots, e_n, e_{n+1}, \ldots, e_{2n})$ is a symplectic basis of $V$.

*Solution*
The map is an isomorphism because $\omega$ is non-degenerate. To prove (i): if $x \in U \cap U^{\perp, \omega}$ then $\omega(x, y) = 0$ for every $u \in U$. Since $U$ is a symplectic subspace, the form $\omega|_U$ is non-degenerate. Therefore $x = 0$. Hence $U \cap U^{\perp, \omega} = \{0\}$, from which it follows that $V = U \oplus U^{\perp, \omega}$. Conversely, if $V = U_1 \oplus U_2$ and $\omega(U_1, U_2) = 0$ it is possible to prove by contradiction that $\omega|_{U_1}$ is necessarily non-degenerate. Hence $U_1$ and $U_2$ are symplectic subspaces and the decomposition is symplectic. The proof of (ii) is immediate: for example $\mathfrak{J}U_1$ is also a Lagrangian subspace and $V = U_1 \oplus \mathfrak{J}U_1$. Finally, to prove (iii): let $f_1, \ldots, f_n$ be the $n$ elements of $U_2^*$ defined

by $f_i(y) = \omega(e_i, y)$ for every $y \in U_2$. It is easy to check that $(f_1, \ldots, f_n)$ is a basis of $U_2^*$: we denote by $(e_{n+1}, \ldots, e_{2n})$ the dual basis in $U_2$. If $1 \le i \le n < j \le 2n$ we have by construction that $\omega(e_i, e_j) = f_i(e_j) = \delta_{i,j-n}$, while $\omega(e_i, e_j) = 0$ if $1 \le i, j \le n$ or $n+1 \le i, j \le 2n$, and therefore the basis $(e_1, \ldots, e_{2n})$ is symplectic.

### Problem 3
Consider a system of canonical coordinates $(\mathbf{p}, \mathbf{q}) \in \mathbf{R}^{2l}$ and the transformation
$$\mathbf{P} = \mathbf{f}(\mathbf{p}, \mathbf{q}), \quad \mathbf{Q} = \mathbf{q}.$$
Determine the structure $\mathbf{f}$ must have for the transformation to be canonical and find a generating function of the transformation.

### Solution
If the transformation is canonical then
$$\{f_i, f_j\} = 0, \quad \{q_i, f_j\} = \delta_{ij}, \quad \forall\ i, j.$$
Since $\{q_i, f_j\} = \partial f_j / \partial p_i$, we find that $f_j = p_j + g_j(\mathbf{q})$, and the conditions $\{f_i, f_j\} = 0$ yield $\partial g_j / \partial q_i = \partial g_i / \partial q_j$, then $\mathbf{g}(\mathbf{q}) = \nabla_\mathbf{q} U(\mathbf{q})$. In conclusion, we must have
$$\mathbf{f}(\mathbf{p}, \mathbf{q}) = \mathbf{p} + \nabla_\mathbf{q} U(\mathbf{q}).$$
A generating function is $F(\mathbf{P}, \mathbf{q}) = \mathbf{P} \cdot \mathbf{q} + U(\mathbf{q})$.

### Problem 4
Consider a group of orthogonal matrices $A(s)$ commuting with the matrix $\mathbf{J}$, with $A(0)$ being the identity matrix.

(i) Prove that the matrices $A(s)$ are symplectic.
(ii) Find the infinitesimal generator of the group of canonical transformations $\mathbf{x} = A(s)\mathbf{X}$.
(iii) Find the Hamiltonian of the group of transformations.

### Solution
(i) $A^T \mathbf{J} A = A^T A \mathbf{J} = \mathbf{J}$.
(ii) $\partial/\partial s A(s)\mathbf{X}|_{s=0} = A'(0)\mathbf{x} = \mathbf{v}(\mathbf{x})$, the infinitesimal generator.
(iii) The matrix $A'(0)$ is Hamiltonian. Setting $A'(0) = \mathbf{J}S$, with $S$ symmetric, the Hamiltonian generating the group of transformations is such that $\mathbf{J}S\mathbf{x} = \mathbf{J}\nabla_\mathbf{x} H(\mathbf{x})$. Hence $H = \frac{1}{2}\mathbf{x}^T S \mathbf{x}$.

### Problem 5
Determine the functions $f, g, h$ in such a way that the transformation
$$Q = g(t) f(p - 2q), \quad P = h(t)(2q^2 - qp)$$
is canonical. Write down the generating function $F(q, Q, t)$. Use it to solve Hamilton's equations associated with $H(p, q) = G(2q^2 - pq)$, where $G$ is a prescribed function (all functions are assumed sufficiently regular).

*Solution*
The condition $\{Q, P\} = 1$ can be written
$$(p - 2q)ghf' = 1.$$

This implies that the product $gh$ must be constant. Setting $gh = 1/c$ and $\xi = p - 2q$, we arrive at the equation $\xi f'(\xi) = c$, and hence $f(\xi) = c\log(|\xi|/\xi_0)$ with $\xi_0 > 0$ constant. Therefore the transformation is of the form
$$Q = c\, g(t) \log \frac{|p - 2q|}{\xi_0}, \quad P = \frac{-q}{c\, g(t)}(p - 2q). \tag{10.205}$$

The function $g(t)$ is arbitrary. To find the generating function $F(q, Q, t)$ we set
$$p = \frac{\partial F}{\partial q}, \quad P = -\frac{\partial F}{\partial Q}.$$

Since $p = 2q + \xi_0 e^{Q/(c\, g(t))}$ (assuming $p - 2q > 0$), integrating with respect to $q$ we find $F(q, Q, t) = q^2 + q\xi_0 e^{Q/(c\, g(t))} + \varphi(Q)$, and differentiating with respect to $Q$ we arrive, after requiring that the result is equal to $-P$, at the conclusion that $\varphi'(Q) = 0$, or $\varphi = 0$. The generating function is therefore $F(q, Q, t) = q^2 + q\xi_0 e^{Q/(c\, g(t))}$. We consider now the Hamiltonian $H = G(2q^2 - pq)$. Applying to it the transformation (10.205) with $c\, g(t) = -1$ (completely canonical transformation) we find the new Hamiltonian $K = G(P)$, and hence the solutions of Hamilton's equations are
$$P = P_0, \quad Q = G'(P_0)t + Q_0, \tag{10.206}$$

with constant $P_0, Q_0$. Now it is sufficient to invert (10.205), written as $(p - 2q > 0)$:
$$Q = -\log\frac{p - 2q}{\xi_0}, \quad P = q(p - 2q),$$

and hence
$$q = \frac{1}{\xi_0}Pe^Q, \quad p = \frac{2}{\xi_0}Pe^Q + \xi_0 e^{-Q}.$$

From equations (10.205) we arrive at
$$q = \frac{P_0}{\xi_0}e^{G'(P_0)t + Q_0}, \quad p = \frac{P_0}{2\xi_0}e^{G'(P_0)t + Q_0} + \xi_0 e^{-(G'(P_0)t + Q_0)}.$$

We can determine the constants $P_0, Q_0$ so that the initial conditions for $p, q$ (compatible with $p - 2q > 0$, otherwise substitute $\xi_0$ with $-\xi_0$) are satisfied.

## Problem 6
Find a symmetry for the Hamiltonian

$$H(p_1, p_2, q_1, q_2) = \frac{p_1^2 + q_1^2 p_2 q_2}{2}$$

and the corresponding first integral of the motion. Use the result to integrate Hamilton's equations.

### Solution
We seek a one-parameter group of completely canonical transformations which leaves the coordinates $p_1, q_1$ and the product $p_2 q_2$ invariant. We try the transformation

$$p_1 = P_1, \quad q_1 = Q_1, \quad p_2 = f(s) P_2, \quad q_2 = \frac{1}{f(s)} Q_2,$$

which is canonical for every $f(s)$, requiring that $f(0) = 1$ and $f(s_1) f(s_2) = f(s_1 + s_2)$. This forces the choice $f(s) = e^{\alpha s}$, with $\alpha$ constant. The infinitesimal generator of the group is

$$\mathbf{v}(\mathbf{x}) = \left. \frac{\partial \mathbf{x}(\mathbf{X}, s)}{\partial s} \right|_{s=0},$$

and hence $\mathbf{v}(p_1, p_2, q_1, q_2) = (0, \alpha p_2, 0, -\alpha q_2)$. The corresponding Hamiltonian $K(p_1, p_2, q_1, q_2)$ must be such that $-\partial K/\partial q_1 = 0$, $-\partial K/\partial q_2 = \alpha p_2$, $\partial K/\partial p_1 = 0$, $\partial K/\partial p_2 = -\alpha q_2$, yielding $K = -\alpha p_2 q_2$. Hence this is a constant of the flow generated by $H$. It is easy to check that $\{H, K\} = 0$. Since $p_2 q_2 = c$ we can integrate Hamilton's equations for $p_1, q_1$ and then for $p_2, q_2$.

## Problem 7
In $\mathbf{R}^2$ consider the flow

$$\dot{\mathbf{x}} = \nabla \xi(\mathbf{x}), \tag{10.207}$$

with $\xi(\mathbf{x})$ a regular function and $\nabla \xi \neq 0$. In which cases is this flow Hamiltonian?

### Solution
We must have

$$\nabla \cdot \nabla \xi = \nabla^2 \xi = 0. \tag{10.208}$$

The operator $\nabla^2 = \partial^2/\partial x_1^2 + \partial^2/\partial x_2^2$ is called the Laplacian, and equation (10.208) is *Laplace's equation*. Its solutions are called *harmonic functions*. There is a vast literature on them (see for example Ladyzenskaya and Ural'ceva (1968), Gilbar and Trudinger (1977)).

If the system (10.207) is Hamiltonian, then it can be written in the form

$$\dot{\mathbf{x}} = -\mathbb{J} \nabla \eta(\mathbf{x}), \tag{10.209}$$

with $\mathcal{J} = \begin{pmatrix} 0 & -1 \\ 1 & 0 \end{pmatrix}$, where the Hamiltonian $-\eta$ is determined by $\nabla\xi = -\mathcal{J}\nabla\eta$, i.e.

$$\frac{\partial \xi}{\partial x_1} = +\frac{\partial \eta}{\partial x_2}, \quad \frac{\partial \xi}{\partial x_2} = -\frac{\partial \eta}{\partial x_1}, \tag{10.210}$$

which are the celebrated *Cauchy–Riemann equations*.

The trajectories orthogonal to $\xi = $ constant are identified with $\eta = $ constant (Problem 1.15). Symmetrically, $-\xi$ plays the role of Hamiltonian ($\nabla\eta = -\mathcal{J}\nabla\xi$) for the flow orthogonal to $\eta = $ constant. Clearly the function $\eta$ is harmonic. It is called the *conjugate harmonic* of $\xi$.

Equations (10.210) are of central importance in the theory of complex holomorphic functions.

Indeed, it can be shown that if $\xi, \eta$ are $C^1$ functions satisfying the Cauchy–Riemann equations, then the function $f : \mathbf{C} \to \mathbf{C}$

$$f(z) = \xi(x_1, x_2) + i\eta(x_1, x_2) \tag{10.211}$$

of the complex variable $z = x_1 + ix_2$ is holomorphic (i.e. the derivative $f'(z)$ exists).

Holomorphic functions have very important properties (for example they admit a power series expansion, are $C^\infty$, and so on, see Lang (1975)).

The converse is also true: if $f(z)$ is holomorphic then its real and imaginary parts are conjugate harmonic functions.

A simple example is given by $\xi = \log r$, $r = (x_1^2 + x_2^2)^{1/2}$, whose harmonic conjugate is $\eta = \arctan(x_2/x_1)$, as is easily verified. The curves $\xi = $ constant are circles centred at the origin, $\eta = $ constant are the radii. Because of the Cauchy–Riemann conditions, for any holomorphic $f(z)$, the curves Re $f = $ constant intersect orthogonally the curves Im $f = $ constant. This fact can be exploited to determine the plane fields satisfying special conditions. For example, if seeking a field of the form $\mathbf{E} = -\nabla\phi$ with the property div $\mathbf{E} = 0$ (i.e. $\nabla^2\phi = 0$), we can view the field lines as orthogonal trajectories of the equipotential lines $\phi = $ constant, and hence as the level sets of the conjugate harmonic $\psi$. This is the case of a plane electrostatic field in a region without charges. If we require that the circle $r = 1$ be equipotential ($\phi = 0$) and that at infinity the field be $E_0 \mathbf{e}_2$, then it is easy to verify that $\phi, \psi$ are the real and imaginary parts of the function $-iE_0 f_J(z)$, where $f_J(z)$ is the *Jukowski function*

$$f_J(z) = z + \frac{1}{z}. \tag{10.212}$$

*Problem 8*
Consider the harmonic conjugate Hamiltonians $\xi(p,q), \eta(p,q)$, generating flows with mutually orthogonal trajectories (see Problem 7). Do the respective flows commute?

## 10.14

*Solution*
The answer is in general negative. Indeed, using the Cauchy–Riemann equations we find that $\{\xi, \eta\} = |\nabla \xi|^2 = |\nabla \eta|^2$ is not constant.

The case $\nabla \xi = \mathbf{a}$(constant), corresponding to $f(z) = a_1 z - i a_2 z$, is an exception. The reader can complete the discussion by considering the case $|\nabla \xi|^2 = $ constant.

*Problem 9*
Let $\mathbf{q} = \tilde{\mathbf{q}}(\mathbf{Q}, s)$ be a group of point transformations. Consider the corresponding group of canonical transformations and find its infinitesimal generator and the corresponding Hamiltonian.

*Solution*
The group under study is $\mathbf{p} = [J^T(\mathbf{Q}, s)]^{-1} \mathbf{P}$, $\mathbf{q} = \tilde{\mathbf{q}}(\mathbf{Q}, s)$, with $J = \nabla_\mathbf{Q} \tilde{\mathbf{q}}$. The infinitesimal generator is the field

$$\mathbf{v}(\mathbf{p}, \mathbf{q}) = \left( \frac{\partial}{\partial s}(J^T)^{-1}\bigg|_{s=0} \mathbf{p}, \frac{\partial \tilde{\mathbf{q}}}{\partial s}\bigg|_{s=0} \right).$$

The corresponding Hamiltonian is $K = \mathbf{p} \cdot \partial \tilde{\mathbf{q}} / \partial s\big|_{s=0}$. It is sufficient to note that $\nabla_\mathbf{q} K = \partial J^T / \partial s\big|_{s=0} \mathbf{p}$ and that $\partial J^T / \partial s\big|_{s=0} = -\partial/\partial s (J^T)^{-1}\big|_{s=0}$, since $(J^T)^{-1} J^T = 1$ and $J\big|_{s=0} = 1$. Note that if the group is a symmetry for some Hamiltonian $H(\mathbf{p}, \mathbf{q})$ then $K = $ constant along the corresponding flow, in agreement with (4.123).

# 11 ANALYTIC MECHANICS: HAMILTON–JACOBI THEORY AND INTEGRABILITY

## 11.1 The Hamilton–Jacobi equation

We have discussed (see Theorem 10.13) how the Hamiltonian flow corresponding to a Hamiltonian $H$ is a canonical transformation which associates with $H$ a new Hamiltonian $K$ that is identically zero. We now consider essentially the question of finding the corresponding generating function.

The problem of the integration of the equations of motion in a Hamiltonian system described by the Hamiltonian $H(\mathbf{p}, \mathbf{q}, t)$ can be reduced to the following: find a canonical transformation from the variables $(\mathbf{p}, \mathbf{q})$ to new variables $(\mathbf{P}, \mathbf{Q})$, generated by a function $F(\mathbf{q}, \mathbf{P}, t)$ in such a way that the new Hamiltonian $K(\mathbf{P}, \mathbf{Q}, t)$ is identically zero:

$$K(\mathbf{P}, \mathbf{Q}, t) = 0. \tag{11.1}$$

Indeed, in this case the canonical equations can immediately be integrated: for every $t \in \mathbf{R}$ we have

$$P_j(t) = \eta_j, \quad Q_j(t) = \xi_j, \quad j = 1, \ldots, l, \tag{11.2}$$

where $(\boldsymbol{\eta}, \boldsymbol{\xi})$ are constant vectors that can be determined starting from the initial conditions. From equations (11.2) we can then reconstruct the integrals of the canonical equations in terms of the original variables through the inverse transformation:

$$\mathbf{p} = \mathbf{p}(\boldsymbol{\eta}, \boldsymbol{\xi}, t), \quad \mathbf{q} = \mathbf{q}(\boldsymbol{\eta}, \boldsymbol{\xi}, t). \tag{11.3}$$

Note that the Hamiltonian flow associated with $H$ is not the only canonical transformation leading to (11.1): for example, by composing the Hamiltonian flow with any completely canonical transformation the new Hamiltonian is still zero.

Suppose that $\nabla_{\mathbf{x}} H \neq \mathbf{0}$, and hence that we are not near a singular point. Since the transformation which interchanges pairs of the variables $(\mathbf{p}, \mathbf{q})$ is canonical there is no loss of generality in assuming that $\nabla_{\mathbf{p}} H \neq \mathbf{0}$ (the latter condition is automatically satisfied by the Hamiltonians of systems with fixed holonomic constraints far from the subspace $\mathbf{p} = \mathbf{0}$).

Recalling equations (10.105), (10.107)–(10.109) of Chapter 10, we know that to realise such a transformation we need to find a generating function

$$S = S(\mathbf{q}, \boldsymbol{\eta}, t), \tag{11.4}$$

solving identically the equation

$$H(\nabla_{\mathbf{q}} S, \mathbf{q}, t) + \frac{\partial S}{\partial t} = 0, \tag{11.5}$$

as $\boldsymbol{\eta}$ varies in an appropriate open subset of $\mathbf{R}^l$, and satisfying the condition

$$\det\left(\frac{\partial^2 S}{\partial q_i \partial \eta_j}\right) \neq 0. \tag{11.6}$$

Equation (11.5) is known as the *Hamilton–Jacobi equation*. It is a non-linear partial differential equation of the first order. The independent variables are $q_1, \ldots, q_l, t$. We do not need to find its *general integral* (i.e. a solution depending on an arbitrary function); we are interested instead in ensuring that the equation admits a *complete integral*, i.e. a solution depending on as many constants as the number of independent variables, that is $l+1$. A solution of the type $S(\mathbf{q}, \boldsymbol{\eta}, t) + \eta_0$ (with $S$ satisfying the invertibility condition (11.6)) is a complete integral of the Hamilton–Jacobi equation. One of the arbitrary constants is always additive, because $S$ appears in (11.5) only through its derivatives, and hence if $S$ is a solution of (11.5) then $S + \eta_0$ is also a solution.

THEOREM 11.1 (Jacobi)  *Given the Hamiltonian $H(\mathbf{q}, \mathbf{p}, t)$, let $S(\mathbf{q}, \boldsymbol{\eta}, t)$ be a complete integral of the Hamilton–Jacobi equation (11.5), depending on $l$ arbitrary constants $\eta_1, \ldots, \eta_l$ and satisfying the condition (11.6). Then the solutions of the system of Hamilton's equations for $H$ can be deduced from the system*

$$p_j = \frac{\partial S}{\partial q_j}, \quad \xi_j = \frac{\partial S}{\partial \eta_j}, \quad j = 1, \ldots, l, \tag{11.7}$$

*where $\xi_1, \ldots, \xi_l$ are constants.*

*Proof*
The function $S$ meets the requirements of Definition 10.11, and hence the system of new coordinates $(\boldsymbol{\eta}, \boldsymbol{\xi})$ is canonical. Equation (11.5) implies that the new Hamiltonian is identically zero, and hence that Hamilton's equations are

$$\dot{\eta}_j = 0, \quad \dot{\xi}_j = 0, \quad j = 1, \ldots, l.$$

Inverting the relations (11.7) (this is possible because of (11.6) and of the implicit function theorem) we deduce equations (11.3) for $(\mathbf{p}, \mathbf{q})$. ∎

The function $S$ is known as *Hamilton's principal function*.

*Remark 11.1*
Every time that the Hamiltonian flow is known, it is possible to compute Hamilton's principal function: since $K = 0$ it is enough to compute the generating function $F_2(\mathbf{q}, \mathbf{P}, t)$ using (10.117), in which we substitute $\mathbf{p} = \hat{\mathbf{p}}(\mathbf{q}, \mathbf{P}, t)$ and $\mathbf{Q} = \hat{\mathbf{Q}}(\mathbf{q}, \mathbf{P}, t)$ deduced from equation (10.89) (which we suppose to be explicitly known). This procedure is possible away from the singular points of

## 11.1  Analytic mechanics: Hamilton–Jacobi theory and integrability

$H$ and for sufficiently small times $t$. Indeed, for $t = 0$ the Hamiltonian flow is reduced to the identical transformation, admitting $F_2 = \mathbf{q} \cdot \mathbf{P}$ as generating function.

It is interesting to remark that the function $S$ has a physical meaning. Computing the derivative along the motion, we find

$$\frac{dS}{dt} = \sum_{j=1}^{l} \frac{\partial S}{\partial q_j} \dot{q}_j + \frac{\partial S}{\partial t} = \sum_{j=1}^{l} p_j \dot{q}_j - H = L.$$

It follows that $S|_{t_0}^{t_1} = \int_{t_0}^{t_1} L\, dt$ is the Hamiltonian action, and hence the values taken on by $S$ in correspondence with the natural motion are those of the Hamiltonian action. ∎

*Remark* 11.2
Theorem 11.1 shows how the knowledge of a complete integral of the Hamilton–Jacobi equation ensures the integrability of Hamilton's equations 'by quadratures': the solution can be obtained by a finite number of algebraic operations, functional inversions and the computation of integrals of known functions.

On the other hand, the Hamilton–Jacobi equation does not always admit a complete integral: for example, this is the case in a neighbourhood of an equilibrium point.

The study of non-linear first-order partial differential equations (such as equation (11.5)) is rather difficult and cannot be considered here. There exists a very elegant and well-developed classical theory (see Courant and Hilbert 1953 and Arnol'd 1978b, Chapter 6), which highlights even more clearly the link between the existence of a solution of the Hamilton–Jacobi equation and of a solution of Hamilton's system. ∎

If the Hamiltonian $H$ does not depend explicitly on time, we can seek a solution $S$ of (11.5) in the form

$$S = -E(\boldsymbol{\alpha})t + W(\mathbf{q}, \boldsymbol{\alpha}), \tag{11.8}$$

where $\boldsymbol{\alpha} = (\alpha_1, \ldots, \alpha_l)$ denotes the vector of $l$ arbitrary constants on which the solution depends (we neglect the additive constant), and $E(\boldsymbol{\alpha})$ is a function of class at least $\mathcal{C}^2$ such that $\nabla_\alpha E \neq 0$ (note that $\nabla_\mathbf{p} H^T(\partial^2 W/\partial \mathbf{q} \partial \boldsymbol{\alpha}) = \nabla_\alpha E$). Equation (11.5) is then reduced to

$$H(\nabla_\mathbf{q} W, \mathbf{q}) = E(\boldsymbol{\alpha}). \tag{11.9}$$

Hence $E$ is identified with the total energy. Equation (11.9) is also called the Hamilton–Jacobi equation.

The function $W$ is called *Hamilton's characteristic function*. Note also that

$$\left(\frac{\partial^2 S}{\partial q_i \partial \alpha_j}\right) = \left(\frac{\partial^2 W}{\partial q_i \partial \alpha_j}\right),$$

and thus $W$ is the generating function of a completely canonical transformation in the new variables $(\boldsymbol{\alpha}, \boldsymbol{\beta})$. With respect to these variables, the new Hamiltonian, as seen in (11.9), is $E(\boldsymbol{\alpha})$. Since the new generalised coordinates $\beta_1,\ldots,\beta_l$ are cyclic, we have

$$\dot{\alpha}_j = 0, \tag{11.10}$$

and the new kinetic momenta $\alpha_1,\ldots,\alpha_l$ are first integrals of the motion. In addition, Hamilton's equations for $\beta_j$, namely

$$\dot{\beta}_j = \frac{\partial E}{\partial \alpha_j} = \gamma_j(\boldsymbol{\alpha}), \quad j = 1,\ldots,l, \tag{11.11}$$

are immediately integrable:

$$\beta_j(t) = \gamma_j(\boldsymbol{\alpha})t + \beta_j(0), \quad j = 1,\ldots,l. \tag{11.12}$$

It can be checked that the transformation $\beta_j - \gamma_j t = \xi_j, \alpha_j = \eta_j$ is canonical, highlighting the relation between the variables $(\boldsymbol{\eta}, \boldsymbol{\xi})$ used previously and $(\boldsymbol{\alpha}, \boldsymbol{\beta})$.

We have proved the following theorem, analogous to Theorem 11.1.

**THEOREM 11.2** *Given the Hamiltonian $H(\mathbf{p}, \mathbf{q})$, let $W(\mathbf{q}, \boldsymbol{\alpha})$ be a complete integral of the Hamilton–Jacobi equation (11.9), depending on $l$ arbitrary constants $\boldsymbol{\alpha} = (\alpha_1, \ldots, \alpha_l)$ and satisfying the condition*

$$\det\left(\frac{\partial^2 W}{\partial q_i \partial \alpha_j}\right) \neq 0. \tag{11.13}$$

*Then $W$ is the generating function of a completely canonical transformation. The new Hamiltonian $E(\boldsymbol{\alpha})$ has $l$ cyclic coordinates, which are linear functions of time, given by (11.12), while the new kinetic momenta $\alpha_1,\ldots,\alpha_l$ are first integrals of the motion.* ∎

*Remark* 11.3

The condition (11.13) guarantees the invertibility of the transformation generated by $W$, and hence the solution of Hamilton's equations associated with $H$ have the form

$$p_j(t) = p_j(\alpha_1,\ldots,\alpha_l, \gamma_1 t + \beta_1(0),\ldots,\gamma_l t + \beta_l(0)), \quad j = 1,\ldots,l,$$
$$q_j(t) = q_j(\alpha_1,\ldots,\alpha_l, \gamma_1 t + \beta_1(0),\ldots,\gamma_l t + \beta_l(0)), \quad j = 1,\ldots,l,$$

and can be obtained from the relations

$$\beta_j = \frac{\partial W}{\partial \alpha_j}, \quad p_j = \frac{\partial W}{\partial q_j}, \quad j = 1,\ldots,l.$$

The initial values of the variables $(\mathbf{p}, \mathbf{q})$ are in one-to-one correspondence with the constants $(\boldsymbol{\alpha}, \boldsymbol{\beta}(0))$. ∎

## 11.1 Analytic mechanics: Hamilton–Jacobi theory and integrability

*Remark 11.4*
If the Hamiltonian $H$ is independent of time and has $n < l$ cyclic coordinates $(q_1, \ldots, q_n)$, equation (11.9) becomes

$$H\left(\frac{\partial W}{\partial q_1}, \ldots, \frac{\partial W}{\partial q_l}, q_{n+1}, \ldots, q_l\right) = E(\alpha_1, \ldots, \alpha_l). \tag{11.14}$$

From this we can deduce that $W$ is linear in the $n$ cyclic variables:

$$W = \sum_{i=1}^{n} \alpha_i q_i + W_0(q_{n+1}, \ldots, q_l, \alpha_1, \ldots, \alpha_l),$$

and (11.14) reduces to one equation in $l - n$ variables. The constants $\alpha_1, \ldots, \alpha_n$ coincide with the momenta $p_1, \ldots, p_n$ conjugate to the cyclic coordinates. ∎

*Remark 11.5*
A specific version of the method just described (known as Poincaré's method) consists of assuming that, for example, $E(\alpha_1, \ldots, \alpha_l) = \alpha_1$ (Jacobi's method). It then follows from equations (11.11) that the coordinates $\beta_j$, conjugate to $\alpha_j$, are constant for every $j = 2, \ldots, l$, while the coordinate conjugate to $\alpha_1$, i.e. to the energy, is $\beta_1 = t - t_0$ with $t_0$ constant. The equations

$$\beta_j = \frac{\partial W}{\partial \alpha_j}(q_1, \ldots, q_l, E, \alpha_2, \ldots, \alpha_l), \quad j = 2, \ldots, l$$

represent the trajectory of the system in the configuration space. ∎

*Remark 11.6*
The transformation described in the previous remark is just a symplectic rectification. We knew that this was possible (Theorem 10.20), although the explicit computation assumed that the Hamiltonian flow be known.

From the corresponding system of coordinates $(\boldsymbol{\alpha}, \boldsymbol{\beta})$, with respect to which the Hamiltonian is $K = \alpha_1$, we can transform to another system in which the Hamiltonian has the generic form $K = K(\boldsymbol{\alpha}')$, using a completely canonical transformation (see Problem 11 of Section 10.12):

$$\boldsymbol{\alpha}' = \boldsymbol{\alpha}'(\boldsymbol{\alpha}), \quad \boldsymbol{\beta}' = (J^{-1}(\boldsymbol{\alpha}))^T \boldsymbol{\beta},$$

where $J = \nabla_{\boldsymbol{\alpha}} \boldsymbol{\alpha}'$. Note that the new variables $\beta'_i$ are linear functions of time (which becomes identified with $\beta_1$). ∎

*Example 11.1: a free point particle*
Starting from the Hamiltonian

$$H = \frac{1}{2m}(p_x^2 + p_y^2 + p_z^2),$$

we obtain the equation

$$\frac{1}{2m}\left[\left(\frac{\partial S}{\partial x}\right)^2 + \left(\frac{\partial S}{\partial y}\right)^2 + \left(\frac{\partial S}{\partial z}\right)^2\right] + \frac{\partial S}{\partial t} = 0.$$

It is natural to proceed by separation of variables, and look for a solution in the form
$$S(x, y, z, t) = X(x) + Y(y) + Z(z) + T(t).$$

The equation becomes
$$\frac{1}{2m}\left[\left(\frac{dX}{dx}\right)^2(x) + \left(\frac{dY}{dy}\right)^2(y) + \left(\frac{dZ}{dz}\right)^2(z)\right] + \frac{dT}{dt}(t) = 0,$$

and hence
$$\frac{dX}{dx} = \eta_1, \quad \frac{dY}{dy} = \eta_2, \quad \frac{dZ}{dz} = \eta_3, \quad \frac{dT}{dt} = -\frac{\eta_1^2 + \eta_2^2 + \eta_3^2}{2m},$$

where $\eta_1, \eta_2, \eta_3$ are arbitrary integration constants. By integration, we obtain the solution
$$S(x, y, z, \eta_1, \eta_2, \eta_3, t) = \eta_1 x + \eta_2 y + \eta_3 z - \frac{\eta_1^2 + \eta_2^2 + \eta_3^2}{2m} t,$$

which clearly satisfies condition (11.6) and generates the transformation (11.7):
$$p_x = \eta_1, \quad p_y = \eta_2, \quad p_z = \eta_3,$$
$$\xi_x = x - \frac{\eta_1}{m} t, \quad \xi_y = y - \frac{\eta_2}{m} t, \quad \xi_z = z - \frac{\eta_3}{m} t.$$
∎

*Example* 11.2: *the harmonic oscillator*
The Hamiltonian of the harmonic oscillator is
$$H(p, q) = \frac{1}{2m}(p^2 + m^2\omega^2 q^2),$$

from which it follows that the Hamilton–Jacobi equation (11.5) takes the form
$$\frac{1}{2m}\left[\left(\frac{\partial S}{\partial q}\right)^2 + m^2\omega^2 q^2\right] + \frac{\partial S}{\partial t} = 0.$$

We set
$$S = S(q, E, t) = W(q, E) - Et.$$

The Hamilton–Jacobi equation (11.9) then becomes
$$\frac{1}{2m}\left[\left(\frac{\partial W}{\partial q}\right)^2 + m^2\omega^2 q^2\right] = E,$$

## 11.1 Analytic mechanics: Hamilton–Jacobi theory and integrability

and hence

$$W(q, E) = \sqrt{2mE} \int_{q_0}^{q} \sqrt{1 - \frac{m\omega^2 x^2}{2E}}\, dx.$$

It is possible to choose $q_0 = 0$. Then we find

$$W(q, E) = \frac{1}{2}\sqrt{2mE}\left[q\sqrt{1 - \frac{m\omega^2 q^2}{2E}} + \sqrt{\frac{2E}{m\omega^2}}\arcsin\left(\sqrt{\frac{m\omega^2}{2E}}q\right)\right].$$

It follows that

$$\beta = \frac{\partial W}{\partial E} = \frac{1}{2}\sqrt{\frac{2m}{E}}\int_0^q \frac{dx}{\sqrt{1 - m\omega^2 x^2/2E}} = \frac{1}{\omega}\arcsin\left(\sqrt{\frac{m\omega^2}{2E}}q\right),$$

and by inverting the relation between $\beta$ and $q$ we find

$$p = \frac{\partial W}{\partial q} = \sqrt{2mE}\sqrt{1 - \frac{m\omega^2 q^2}{2E}} = \sqrt{2mE}\cos(\omega\beta),$$

$$q = \sqrt{\frac{2E}{m\omega^2}}\sin(\omega\beta),$$
(11.15)

illustrating how the Hamilton–Jacobi method yields the solution of the equations of motion. Indeed, since $\alpha = E$, from (11.11) it follows that $\beta = t + \beta(0)$ and by imposing the initial conditions we find

$$2mE = p(0)^2 + m^2\omega^2 q(0)^2,$$

$$\tan(\omega\beta(0)) = m\omega\frac{q(0)}{p(0)}.$$

We thus obtain the well-known solution $(p(t), q(t))$. Substituting $q(t)$ into $W$, and after some manipulations we find that along the motion the function $S$ takes the value

$$S = 2E\int_0^t \left[\cos^2(\omega x + \omega\beta(0)) - \frac{1}{2}\right] dx.$$

This coincides with the integral of the Lagrangian

$$L = \frac{1}{2}m\dot{q}^2 - \frac{1}{2}m\omega^2 q^2 = 2E\left[\cos^2\omega(t + \beta(0)) - \frac{1}{2}\right],$$

computed along the natural motion.

On the other hand, the problem of the motion can be solved starting from the function $S(q, E, t)$:

$$S(q, E, t) = W(q, E) - Et.$$

Indeed, the equations

$$p = \frac{\partial S}{\partial q} = \frac{\partial W}{\partial q}, \quad \xi = \frac{\partial S}{\partial E} = \frac{\partial W}{\partial E} - t$$

are equivalent to equations (11.15). In particular, the second one gives the equation of motion in the form

$$q = \sqrt{\frac{2E}{m\omega^2}} \sin[\omega(t + \xi)].$$

∎

*Example* 11.3: *conservative autonomous systems in one dimension*
Consider a point particle of mass $m$ in motion along a line, and subject to a conservative force field with potential energy $V(x)$. The Hamiltonian of the system is

$$H = \frac{p^2}{2m} + V(x),$$

and the associated Hamilton–Jacobi equation is

$$\frac{1}{2m} \left( \frac{\partial W}{\partial x} \right)^2 + V(x) = E.$$

This can immediately be integrated:

$$W(x, E) = \sqrt{2m} \int_{x_0}^{x} \sqrt{E - V(\xi)} \, d\xi.$$

The canonical transformation generated by it is

$$p = \frac{\partial W}{\partial x} = \sqrt{2m[E - V(x)]},$$

$$\beta = \frac{\partial W}{\partial E} = \sqrt{\frac{m}{2}} \int_{x_0}^{x} \frac{d\xi}{\sqrt{E - V(\xi)}}.$$

Recall that $\beta = t - t_0$. Hence we have again derived equation (3.4). ∎

## 11.2 Separation of variables for the Hamilton–Jacobi equation

The technique of *separation of variables* is a technique that often yields an explicit complete integral of the Hamilton–Jacobi equation. The method is very well described in the book by Landau and Lifschitz (1976, Section 48). We shall closely follow their description.

Consider the particularly simple case that the Hamiltonian $H$ of the system is independent of time and is given by the sum of $l$ functions, each depending only on a pair of variables $(p_j, q_j)$:

$$H = h_1(p_1, q_1) + \cdots + h_l(p_l, q_l). \tag{11.16}$$

The Hamilton–Jacobi equation (11.9) clearly admits a solution

$$W = \sum_{j=1}^{l} W_j(q_j, \alpha_j), \tag{11.17}$$

where each function $W_j$ is determined by solving the equation

$$h_j\left(\frac{\partial W}{\partial q_j}, q_j\right) = e_j(\alpha_j), \tag{11.18}$$

with $e_j$ an arbitrary (regular) function. From this it follows that

$$E(\alpha_1, \ldots, \alpha_l) = \sum_{j=1}^{l} e_j(\alpha_j). \tag{11.19}$$

An example of a system satisfying (11.16) is a free point particle (see Example 11.1); in a similar way one can consider the harmonic oscillator in space, with Hamiltonian

$$H = \frac{p_1^2 + p_2^2 + p_3^2}{2m} + \frac{m}{2}(\omega_1^2 q_1^2 + \omega_2^2 q_2^2 + \omega_3^2 q_3^2),$$

or any sum of uncoupled one-dimensional systems.

An immediate generalisation of (11.16) is given by Hamiltonians of the kind

$$H = \mathcal{H}(h_1(p_1, q_1), \ldots, h_l(p_l, q_l)). \tag{11.20}$$

The characteristic function $W$ has the form (11.17) and can be computed by solving the system of equations (11.18), but the energy $E$ is now given by

$$E(\alpha_1, \ldots, \alpha_l) = \mathcal{H}(e_1(\alpha_1), \ldots, e_l(\alpha_l)). \tag{11.21}$$

These simple observations lead us to consider a more general case, very significant for interesting physical applications.

Suppose that one coordinate, e.g. $q_1$, and its corresponding derivative $\partial S/\partial q_1$ enter the Hamilton–Jacobi equation (11.5) only as a combination of the form

$h_1(\partial S/\partial q_1, q_1)$, not depending on other coordinates or on time, or on the other derivatives. This happens if the Hamiltonian is of the form

$$H = \mathcal{H}(h_1(p_1, q_1), p_2, \ldots, p_l, q_2, \ldots, q_l, t), \tag{11.22}$$

so that the Hamilton–Jacobi equation is written as

$$\mathcal{H}\left(h_1\left(\frac{\partial S}{\partial q_1}, q_1\right), \frac{\partial S}{\partial q_2}, \ldots, \frac{\partial S}{\partial q_l}, q_2, \ldots, q_l, t\right) + \frac{\partial S}{\partial t} = 0. \tag{11.23}$$

In this case, we seek a solution of the form

$$S = S_1(q_1, \alpha_1) + S'(q_2, \ldots, q_l, \alpha_1, \alpha_2, \ldots, \alpha_l, t), \tag{11.24}$$

and (11.23) is transformed into the system

$$h_1\left(\frac{\partial S}{\partial q_1}, q_1\right) = e_1(\alpha_1),$$

$$\mathcal{H}\left(e_1(\alpha_1), \frac{\partial S'}{\partial q_2}, \ldots, \frac{\partial S'}{\partial q_l}, q_2, \ldots, q_l, t\right) + \frac{\partial S'}{\partial t} = 0. \tag{11.25}$$

The first of equations (11.25) is a first-order ordinary differential equation from which we can compute $S_1$ via quadratures. The second is still a Hamilton–Jacobi equation, but in $l$ rather than $l+1$ variables.

If this procedure can be iterated $l+1$ times, successively separating the coordinates and time, the computation of the complete integral of the Hamilton–Jacobi equation is reduced to $l+1$ quadratures, and the Hamiltonian system under consideration is said to be *separable*. For this to be possible, the Hamiltonian we started with must be independent of the time $t$ and $S$ must be of the form

$$S = W_1(q_1, \alpha_1) + W_2(q_2, \alpha_1, \alpha_2) + \cdots + W_l(q_l, \alpha_1, \ldots, \alpha_l) - E(\alpha_1, \ldots, \alpha_l)t. \tag{11.26}$$

To this category belong the Hamiltonian systems such that

$$H = h_l(h_{l-1}(\ldots(h_2(h_1(p_1, q_1), p_2, q_2) \ldots), p_{l-1}, q_{l-1}), p_l, q_l). \tag{11.27}$$

For these systems, the Hamilton–Jacobi equation becomes

$$h_l\left(h_{l-1}\left(\ldots\left(h_2\left(h_1\left(\frac{\partial W}{\partial q_1}, q_1\right), \frac{\partial W}{\partial q_2}, q_2\right)\right)\ldots\right), \frac{\partial W}{\partial q_{l-1}}, q_{l-1}\right), \frac{\partial W}{\partial q_l}, q_l\right)$$
$$= E(\alpha_1, \ldots, \alpha_l). \tag{11.28}$$

For separation of variables to be possible, it is often necessary to choose appropriately the Lagrangian coordinate system to be used.

## 11.2 Analytic mechanics: Hamilton–Jacobi theory and integrability

*Example 11.4: systems that are separable with respect to spherical coordinates*

Consider a point particle of mass $m$ moving in Euclidean three-dimensional space, under the action of external conservative forces with potential energy $V$. Its Hamiltonian is

$$H = \frac{1}{2m}(p_x^2 + p_y^2 + p_z^2) + V. \tag{11.29}$$

Introducing spherical coordinates:

$$x = r\sin\vartheta\cos\varphi, \quad y = r\sin\vartheta\sin\varphi, \quad z = r\cos\vartheta,$$

where $r > 0$, $0 \leq \varphi \leq 2\pi$ and $0 < \vartheta < \pi$, the Hamiltonian (11.29) can be written as

$$H = \frac{1}{2m}\left(p_r^2 + \frac{p_\vartheta^2}{r^2} + \frac{p_\varphi^2}{r^2 \sin^2\vartheta}\right) + V(r,\vartheta,\varphi).$$

Suppose now that the potential $V$ expressed with respect to spherical coordinates has the following form:

$$V(r,\vartheta,\varphi) = a(r) + \frac{b(\vartheta)}{r^2} + \frac{c(\varphi)}{r^2 \sin^2\vartheta}. \tag{11.30}$$

The Hamilton–Jacobi equation for this system

$$\frac{1}{2m}\left(\left(\frac{\partial S}{\partial r}\right)^2 + \frac{1}{r^2}\left(\frac{\partial S}{\partial \vartheta}\right)^2 + \frac{1}{r^2\sin^2\vartheta}\left(\frac{\partial S}{\partial \varphi}\right)^2\right) + V(r,\vartheta,\varphi) + \frac{\partial S}{\partial t} = 0 \tag{11.31}$$

can be separated by choosing

$$S(r,\vartheta,\varphi,\alpha_r,\alpha_\vartheta,\alpha_\varphi,t)$$
$$= W_1(\varphi,\alpha_\varphi) + W_2(\vartheta,\alpha_\vartheta,\alpha_\varphi) + W_3(r,\alpha_r,\alpha_\vartheta,\alpha_\varphi) - E(\alpha_\varphi,\alpha_\vartheta,\alpha_r)t. \tag{11.32}$$

Indeed, by substituting (11.32) into the Hamilton–Jacobi equation, we find

$$\frac{1}{2m}\left(\frac{\partial W_3}{\partial r}\right)^2 + a(r) + \frac{1}{2mr^2}\left\{\left(\frac{\partial W_2}{\partial \vartheta}\right)^2 + 2mb(\vartheta)\right.$$
$$\left. + \frac{1}{\sin^2\vartheta}\left[\left(\frac{\partial W_1}{\partial \varphi}\right)^2 + 2mc(\varphi)\right]\right\} = E,$$

and the separation of the equation can be obtained by solving the system

$$\left(\frac{\partial W_1}{\partial \varphi}\right)^2 + 2mc(\varphi) = e_1(\alpha_\varphi),$$

$$\left(\frac{\partial W_2}{\partial \vartheta}\right)^2 + 2mb(\vartheta) + \frac{e_1(\alpha_\varphi)}{\sin^2 \vartheta} = e_2(\alpha_\varphi, \alpha_\vartheta), \qquad (11.33)$$

$$\frac{1}{2m}\left(\frac{\partial W_3}{\partial r}\right)^2 + a(r) + \frac{e_2(\alpha_\varphi, \alpha_\vartheta)}{2mr^2} = E(\alpha_\varphi, \alpha_\vartheta, \alpha_r).$$

The solutions of the system (11.33) are clearly given by

$$W_1 = \int \sqrt{e_1(\alpha_\varphi) - 2mc(\varphi)} \, d\varphi,$$

$$W_2 = \int \sqrt{e_2(\alpha_\varphi, \alpha_\vartheta) - 2mb(\vartheta) - \frac{e_1(\alpha_\varphi)}{\sin^2 \vartheta}} \, d\vartheta, \qquad (11.34)$$

$$W_3 = \int \sqrt{2m\left[E(\alpha_\varphi, \alpha_\vartheta, \alpha_r) - a(r) - \frac{e_2(\alpha_\varphi, \alpha_\vartheta)}{2mr^2}\right]} \, dr.$$

An important example of a system that satisfies the condition (11.30) is the motion of a point particle subject to a central potential $V(r)$. In this case the variable $\varphi$ is cyclic, $W_1 = p_\varphi \varphi$ (see Remark 11.4) and $p_\varphi$ is the $z$-component of the angular momentum of the particle, which plays the role of the constant $\sqrt{e_1}$. In addition, since $e_2 = p_\phi^2/(\sin^2 \vartheta) + p_\vartheta^2$, $e_2$ is identified with the square of the norm of the angular momentum vector. ∎

*Example 11.5: systems that are separable with respect to parabolic coordinates*
The so-called *parabolic coordinates* are given by

$$x = \frac{u^2 - v^2}{2}, \quad y = uv \cos \varphi, \quad z = uv \sin \varphi,$$

where $(u, v) \in \mathbf{R}^2$, $0 \leq \varphi \leq 2\pi$. The surfaces obtained by fixing a constant value for $u$ or for $v$ correspond to circular paraboloids whose axis coincides with the $x$-axis (Fig. 11.1):

$$x = \frac{u^2}{2} - \frac{y^2 + z^2}{2u^2}, \quad x = -\frac{v^2}{2} + \frac{y^2 + z^2}{2v^2}.$$

With respect to this system of coordinates, the Hamiltonian (11.29) can be written as

$$H = \frac{1}{2m}\frac{p_u^2 + p_v^2}{u^2 + v^2} + \frac{1}{2m}\frac{p_\varphi^2}{u^2 v^2} + V(u, v, \varphi).$$

## 11.2 Analytic mechanics: Hamilton–Jacobi theory and integrability

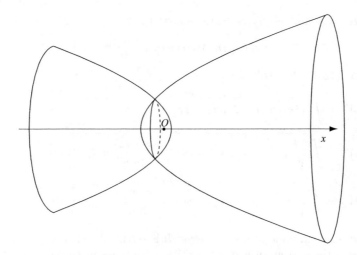

**Fig. 11.1**

Suppose that the potential energy $V$, expressed in parabolic coordinates, has the form

$$V(u, v, \varphi) = \frac{a(u) + b(v)}{u^2 + v^2} + \frac{c(\varphi)}{u^2 v^2}. \tag{11.35}$$

By choosing

$$\begin{aligned} S(u, v, \varphi, \alpha_u, \alpha_v, p_\varphi, t) \\ = W_1(\varphi, \alpha_\varphi) + W_2(u, \alpha_u, \alpha_\varphi) + W_3(v, \alpha_v, \alpha_\varphi) - E(\alpha_\varphi, \alpha_u, \alpha_v)t, \end{aligned} \tag{11.36}$$

the Hamilton–Jacobi equation for the system is

$$\frac{1}{2m(u^2+v^2)} \left[ \left(\frac{\partial W_2}{\partial u}\right)^2 + \left(\frac{\partial W_3}{\partial v}\right)^2 \right] + \frac{(\partial W_1/\partial \varphi)^2 + 2mc(\varphi)}{2mu^2v^2} + \frac{a(u)+b(v)}{u^2+v^2} = E, \tag{11.37}$$

where $E = E(\alpha_u, \alpha_v, \alpha_\varphi)$, and it can immediately be separated by multiplying both sides by $u^2 + v^2$; thus we find the system

$$\left(\frac{\partial W_1}{\partial \varphi}\right)^2 + 2mc(\varphi) = e_1(\alpha_\varphi),$$

$$\frac{1}{2m}\left(\frac{\partial W_2}{\partial u}\right)^2 + a(u) + \frac{e_1(\alpha_\varphi)}{2mu^2} - Eu^2 = e_2(\alpha_\varphi, \alpha_u), \tag{11.38}$$

$$\frac{1}{2m}\left(\frac{\partial W_3}{\partial v}\right)^2 + b(v) + \frac{e_1(\alpha_\varphi)}{2mv^2} - Ev^2 = e_3(\alpha_\varphi, \alpha_v),$$

where $e_2(\alpha_\varphi, \alpha_u)$ and $e_3(\alpha_\varphi, \alpha_v)$ are related by

$$e_2(\alpha_\varphi, \alpha_u) + e_3(\alpha_\varphi, \alpha_v) = 0.$$

The system (11.38) has solutions

$$W_1 = \int \sqrt{e_1(\alpha_\varphi) - 2mc(\varphi)} \, d\varphi,$$

$$W_2 = \int \sqrt{2m\left[e_2(\alpha_\varphi, \alpha_u) - a(u) - \frac{e_1(\alpha_\varphi)}{2mu^2} + Eu^2\right]} \, du, \qquad (11.39)$$

$$W_3 = \int \sqrt{2m\left[e_3(\alpha_\varphi, \alpha_u) - b(v) - \frac{e_1(\alpha_\varphi)}{2mv^2} + Ev^2\right]} \, dv.$$

An interesting example of a system which is separable with respect to parabolic coordinates is the system of a point particle with mass $m$ subject to a Newtonian potential and to a uniform, constant force field of intensity $F$ directed along the $x$-axis. In this case the potential energy in Cartesian coordinates has the following expression:

$$V(x, y, z) = -\frac{k}{\sqrt{x^2 + y^2 + z^2}} + Fx. \qquad (11.40)$$

This problem originates in the study of celestial mechanics. Indeed, the potential (11.40) describes the motion of a spaceship around a planet, under the propulsion of an engine providing a (small) acceleration that is constant in direction and intensity, or the effect of solar radiation pressure upon the trajectory of an artificial satellite. For some satellites the radiation pressure is the principal perturbation to the Keplerian motion. If one considers time intervals sufficiently small relative to the period of revolution of the Earth around the Sun, to a first approximation we can neglect the motion of the Earth, and hence we can assume that the radiation pressure produces an acceleration which is of constant intensity and direction.

In parabolic coordinates the potential energy (11.40) becomes

$$V(u, v) = -\frac{2k}{u^2 + v^2} + \frac{F}{2}(u^2 - v^2) = \frac{-2k + (F/2)(u^4 - v^4)}{u^2 + v^2},$$

from which it follows that

$$a(u) = -k + \frac{F}{2}u^4, \quad b(v) = -k - \frac{F}{2}v^4.$$

∎

*Example 11.6: systems that are separable with respect to elliptic coordinates*
The so-called *elliptic coordinates* are given by

$$x = d\cosh\xi \cos\eta, \quad y = d\sinh\xi \sin\eta \cos\varphi, \quad z = d\sinh\xi \sin\eta \sin\varphi,$$

where $d > 0$ is a fixed positive constant, $\xi \in \mathbf{R}^+$, $0 \leq \eta \leq \pi$ and $0 \leq \varphi \leq 2\pi$. Note that the surface $\xi$ = constant corresponds to an ellipsoid of revolution around the $x$-axis:

$$\frac{x^2}{d^2 \cosh^2 \xi} + \frac{y^2 + z^2}{d^2 \sinh^2 \xi} = 1,$$

and the surface $\eta$ = constant corresponds to a two-sheeted hyperboloid of revolution around the $x$-axis (Fig. 11.2):

$$\frac{x^2}{d^2 \cos^2 \eta} - \frac{y^2 + z^2}{d^2 \sin^2 \eta} = 1.$$

The Hamiltonian (11.29) in elliptic coordinates can be written as

$$H = \frac{1}{2md^2(\cosh^2 \xi - \cos^2 \eta)} \left[ p_\xi^2 + p_\eta^2 + \left( \frac{1}{\sinh^2 \xi} + \frac{1}{\sin^2 \eta} \right) p_\varphi^2 \right] + V(\xi, \eta, \varphi).$$

Suppose that the potential $V$ expressed in elliptic coordinates has the following form:

$$V(\xi, \eta, \varphi) = \frac{a(\xi) + b(\eta) + \left( (1/\sinh^2 \xi) + (1/\sin^2 \eta) \right) c(\varphi)}{d^2(\cosh^2 \xi - \cos^2 \eta)}. \tag{11.41}$$

By choosing

$$S(\xi, \eta, \varphi, \alpha_\xi, \alpha_\eta, \alpha_\varphi, t)$$
$$= W_1(\varphi, \alpha_\varphi) + W_2(\xi, \alpha_\varphi, \alpha_\xi) + W_3(\eta, \alpha_\varphi, \alpha_\xi, \alpha_\eta) - E(\alpha_\varphi, \alpha_\eta, \alpha_\xi)t \tag{11.42}$$

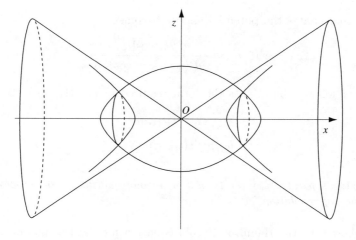

Fig. 11.2

the equation for the system under consideration becomes

$$\left(\frac{\partial W_2}{\partial \xi}\right)^2 + \left(\frac{\partial W_3}{\partial \eta}\right)^2 + \left(\frac{1}{\sinh^2 \xi} + \frac{1}{\sin^2 \eta}\right)\left[\left(\frac{\partial W_1}{\partial \varphi}\right)^2 + 2mc(\varphi)\right] + 2m(a(\xi) + b(\eta))$$
$$= 2md^2(\cosh^2 \xi - \cos^2 \eta)E, \qquad (11.43)$$

where $E = E(\alpha_\xi, \alpha_\eta, \alpha_\varphi)$. This can be separated:

$$\left(\frac{\partial W_1}{\partial \varphi}\right)^2 + 2mc(\varphi) = e_1(\alpha_\varphi),$$

$$\frac{1}{2m}\left[\left(\frac{\partial W_2}{\partial \xi}\right)^2 + \frac{e_1(\alpha_\varphi)}{\sinh^2 \xi}\right] + a(\xi) - Ed^2 \cosh^2 \xi = e_2(\alpha_\varphi, \alpha_\xi), \qquad (11.44)$$

$$\frac{1}{2m}\left[\left(\frac{\partial W_3}{\partial \eta}\right)^2 + \frac{e_1(\alpha_\varphi)}{\sin^2 \eta}\right] + b(\eta) + Ed^2 \cos^2 \eta = -e_2(\alpha_\varphi, \alpha_\xi).$$

An example of a potential for which the Hamilton–Jacobi equation is separable, with respect to elliptic coordinates, is given by the so-called problem of two centres of force. Consider a point particle subject to the gravitational attraction of two centres of force placed at $(d, 0, 0)$ and $(-d, 0, 0)$. In Cartesian coordinates, the potential energy is given by

$$V(x, y, z) = -k\left(\frac{1}{[(x-d)^2 + y^2 + z^2]^{1/2}} + \frac{1}{[(x+d)^2 + y^2 + z^2]^{1/2}}\right). \qquad (11.45)$$

Since

$$(x \pm d)^2 + y^2 + z^2 = d^2 \sinh^2 \xi \sin^2 \eta + d^2(\cosh \xi \cos \eta \pm 1)^2 = d^2(\cosh \xi \pm \cos \eta)^2,$$

in elliptic coordinates the potential energy becomes

$$V(\xi, \eta) = -\frac{2kd \cosh \xi}{d^2(\cosh^2 \xi - \cos^2 \eta)}.$$

From this it follows that $V$ has the form required in (11.41), with

$$a(\xi) = -2kd \cosh \xi,$$
$$b(\eta) = c(\varphi) = 0.$$

∎

**Example 11.7:** *separability of the Hamilton–Jacobi equation for the geodesic motion on a surface of revolution*

We now show that the Hamilton–Jacobi equation for the free motion of a point particle of mass $m$ on a surface of revolution is separable.

If
$$\mathbf{x} = (u\cos v, u\sin v, \psi(u))$$

is a parametric expression for the surface, with $0 \le v \le 2\pi$ and $u \in \mathbf{R}$, the momenta conjugate to the Lagrangian variables $u$ and $v$ are
$$p_u = m[1 + (\psi'(u))^2]\dot{u}, \quad p_v = mu^2\dot{v},$$

and the Hamiltonian of the problem is
$$H(p_u, p_v, u, v) = \frac{1}{2m}\left(\frac{p_u^2}{1 + (\psi'(u))^2} + \frac{p_v^2}{u^2}\right).$$

Note that the angular coordinate $v$ is cyclic. Hence by choosing
$$S(u, v, \alpha_u, p_v, t) = vp_v + W(u) - Et,$$

the Hamilton–Jacobi equation for the system is reduced to
$$\frac{1}{2m}\left[\frac{1}{1 + (\psi'(u))^2}\left(\frac{\partial W}{\partial u}\right)^2 + \frac{p_v^2}{u^2}\right] = E,$$

where $E = E(\alpha_u, p_v)$. Thus we find
$$W = \pm \int \sqrt{\left(2mE - \frac{p_v^2}{u^2}\right)(1 + (\psi'(u))^2)}\, du.$$
∎

*Example* 11.8: *separability of the Hamilton–Jacobi equation for the geodesic motion on an ellipsoid*

Consider a point particle of mass $m$ moving, in the absence of external forces, on the ellipsoid
$$\frac{x^2}{a^2} + \frac{y^2}{b^2} + \frac{z^2}{c^2} = 1,$$

with the condition $0 < a \le b < c$. Setting $\varepsilon = (b-a)/(c-a)$, we consider the parametrisation
$$x = \sqrt{a}\cos\vartheta\sqrt{\varepsilon + (1-\varepsilon)\cos^2\varphi},$$
$$y = \sqrt{b}\sin\vartheta\cos\varphi,$$
$$z = \sqrt{c}\sin\varphi\sqrt{1 - \varepsilon\cos^2\vartheta}.$$

Note that as $0 < \vartheta \le 2\pi$, $0 < \varphi \le 2\pi$, the ellipsoid is covered twice. Setting
$$u = a + (b-a)\cos^2 \vartheta \in [a,b],$$
$$v = b + (c-b)\cos^2 \varphi \in [b,c],$$
we find Jacobi's original parametrisation
$$x = \pm\sqrt{a}\sqrt{\frac{(u-a)(v-a)}{(c-a)(b-a)}},$$
$$y = \pm\sqrt{b}\sqrt{\frac{(b-u)(v-b)}{(c-b)(b-a)}},$$
$$z = \pm\sqrt{c}\sqrt{\frac{(c-u)(c-v)}{(c-a)(c-b)}}.$$

The Lagrangian of the system is
$$L(\vartheta,\varphi,\dot\vartheta,\dot\varphi) = \frac{1}{2}[\dot\vartheta^2 A(\vartheta) + \dot\varphi^2 B(\varphi)][C(\vartheta) + D(\varphi)],$$
where
$$A(\vartheta) = \frac{(c-a)+(b-a)\cos^2 \vartheta}{a+(b-a)\cos^2 \vartheta},$$
$$B(\varphi) = \frac{(b-a)+(c-b)\cos^2 \varphi}{b\sin^2 \varphi + c\cos^2 \varphi},$$
$$C(\vartheta) = (b-a)\sin^2 \vartheta,$$
$$D(\varphi) = (c-b)\cos^2 \varphi.$$

The Hamiltonian of the system is thus given by
$$H = \frac{1}{2}\left(\frac{p_\vartheta^2}{A(\vartheta)} + \frac{p_\varphi^2}{B(\varphi)}\right)\frac{1}{C(\vartheta) + D(\varphi)}.$$

Setting
$$S(\vartheta,\varphi,\alpha_\vartheta,\alpha_\varphi,t) = W_1(\vartheta) + W_2(\varphi) - Et,$$
the Hamilton–Jacobi equation
$$\frac{1}{2(C(\vartheta)+D(\varphi))}\left[\frac{1}{A(\vartheta)}\left(\frac{\partial S}{\partial \vartheta}\right)^2 + \frac{1}{B(\varphi)}\left(\frac{\partial S}{\partial \varphi}\right)^2\right] + \frac{\partial S}{\partial t} = 0$$

yields the system

$$\frac{1}{2A(\vartheta)}\left(\frac{\partial W_1}{\partial \vartheta}\right)^2 - EC(\vartheta) = \alpha,$$

$$\frac{1}{2B(\varphi)}\left(\frac{\partial W_2}{\partial \varphi}\right)^2 - ED(\varphi) = -\alpha.$$

By integration we obtain a complete integral of the Hamilton–Jacobi equation. ∎

## 11.3 Integrable systems with one degree of freedom: action-angle variables

Consider an autonomous Hamiltonian system with one degree of freedom:

$$H = H(p, q). \tag{11.46}$$

The trajectories of the system in the phase plane $(q, p) \in \mathbf{R}^2$ are the curves $\gamma$ defined implicitly by the equation $H(q, p) = E$. Since they depend on the fixed value of the energy $E$, we denote them by $\gamma = \gamma_E$.

Suppose that, as $E$ varies (in an open interval $I \subset \mathbf{R}$) the curves $\gamma_E$ are simple, connected, closed and non-singular, and hence that the gradient of the Hamiltonian never vanishes:

$$\left(\frac{\partial H}{\partial p}, \frac{\partial H}{\partial q}\right)\bigg|_{\gamma_E} \neq (0, 0).$$

In this case we call the motion *libration*, or *oscillatory* motion (Fig. 11.3).

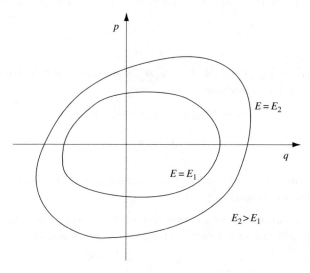

Fig. 11.3

We saw in Chapter 3 that this motion is periodic of period $T$. The period is in general a function of the energy: $T = T(E)$ (it can also be constant, in which case the motion is called *isochronous*; an example is given by the harmonic oscillator). The length of the curve and the area it encloses are also functions of the energy.

The librations typically arise in a neighbourhood of a point of stable equilibrium, corresponding to a local minimum of the Hamiltonian $H$. The non-singularity condition of the phase curves $\gamma_E$ excludes the possibility of separatrices.

With these hypotheses, every phase curve $\gamma_E$ is diffeomorphic to a circle enclosing the same area. Indeed, since $\gamma_E$ is rectifiable, it can also be parametrised (in dimensionless variables) by $p = p_E(s)$, $q = q_E(s)$. If we denote by $\lambda_E$ the length of $\gamma_E$, we can also introduce the angular coordinate $\psi = 2\pi(s/\lambda_E)$ and consider the circle $p = R_E \cos\psi$, $q = R_E \sin\psi$ that is diffeomorphic to $\gamma_E$, choosing $R_E$ so that the areas enclosed are equal. Thus we have an invertible transformation from $(p, q)$ to $(R_E, \psi)$: to $R_E$ there corresponds a curve $\gamma_E$ and to $\psi$ a point on it. Note however that in general the variables $(R_E, \psi)$ just defined, or more generally variables $(f(R_E), \psi)$ with $f' \neq 0$, are not canonical (see Example 11.10).

A natural question is whether there exists a transformation leading to a new pair of canonical variables $(J, \chi) \in \mathbf{R} \times \mathbf{S}^1$ satisfying the following conditions: the variable $\chi$ is an angle, and hence its value increases by $2\pi$ when the curve $\gamma_E$ is traced once, while the variable $J$ depends only on the energy, and characterises the phase curve under consideration (hence the Hamiltonian (11.46) expressed in the new variables is only a function of $J$).

These preliminary observations justify the following definition.

DEFINITION 11.1 *If there exists a completely canonical transformation*

$$p = p(J, \chi), \qquad (11.47)$$

$$q = q(J, \chi) \qquad (11.48)$$

*(where the dependence of $p$ and $q$ on $\chi$ is $2\pi$-periodic) to new variables $(J, \chi) \in \mathbf{R} \times \mathbf{S}^1$ satisfying the conditions*

$$E = H(p(J, \chi), q(J, \chi)) = K(J), \qquad (11.49)$$

$$\oint_{\gamma_E} \mathrm{d}\chi = 2\pi, \qquad (11.50)$$

*the system (11.46) is called* completely canonically integrable, *and the variables $(J, \chi)$ are called* action-angle *variables.* ∎

If a system is completely canonically integrable, then from equation (11.49) it follows that Hamilton's equations in the new variables are

$$\dot{J} = -\frac{\partial K}{\partial \chi} = 0, \quad \dot{\chi} = \frac{\partial K}{\partial J}. \qquad (11.51)$$

Setting

$$\omega = \omega(J) = \frac{dK}{dJ}, \qquad (11.52)$$

this yields

$$J(t) = J(0), \quad \chi(t) = \chi(0) + \omega(J(0))t, \qquad (11.53)$$

for every $t \in \mathbf{R}$. The action variable is therefore a constant of the motion, and substituting (11.53) into (11.47) and (11.48) and recalling that $p$ and $q$ are $2\pi$-periodic in $\chi$, we again find that the motion is periodic, with period

$$T = \frac{2\pi}{\omega(J)}. \qquad (11.54)$$

*Example 11.9*
The harmonic oscillator (Example 11.2) is completely canonically integrable. The transformation to action-angle variables (we shall derive it in Example 11.10) is given by

$$p = \sqrt{2m\omega J}\cos\chi, \quad q = \sqrt{\frac{2J}{m\omega}}\sin\chi. \qquad (11.55)$$

Indeed, one immediately verifies that the condition (11.50) is satisfied and that the new Hamiltonian obtained by substituting (11.55) into $H(p,q)$ is given by

$$K(J) = \omega J. \qquad (11.56)$$

■

We shall soon see that if the Hamiltonian (11.46) supports oscillatory motions, then the system is completely canonically integrable. There exists, however, another class of systems with one degree of freedom that admits action-angle variables. Assume that the Hamiltonian (11.46) has a periodic dependence on the variable $q$, so that there exists a $\lambda > 0$ such that $H(p, q+\lambda) = H(p, q)$ for every $(p, q)$. Assume also that as the energy $E$ varies, the curves $\gamma_E$ are simple and non-singular. If these curves are also closed then the motion is a libration. If they are the graph of a regular function,

$$p = \hat{p}(q, E),$$

the motion is called a *rotation* (Fig. 11.4). We assume that $\partial \hat{p}/\partial E \neq 0$. Evidently, because of the periodicity hypothesis for the Hamiltonian $H$, the function $\hat{p}$ is also periodic with respect to $q$, with period $\lambda$ (independent of $E$).

For example, in the case of the pendulum there appear both oscillations (for values of the energy less than the value on the separatrix) and rotations (for larger values). Rotations can also appear in many systems for which the Lagrangian coordinate $q$ is in fact an angle.

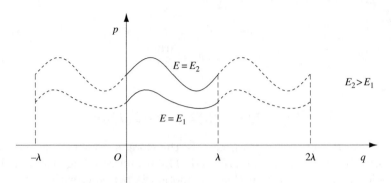

Fig. 11.4

For systems involving rotations it is also possible to seek action-angle variables, satisfying the conditions (11.49) and (11.50). The dependence of $p$ on $\chi$ is then $2\pi$-periodic, while

$$q(J, \chi + 2\pi) = q(J, \chi) + \lambda.$$

This apparent difference can be easily eliminated. It is enough to recall that the assumption of periodicity in $q$ of the Hamiltonian $H$ allows one to identify all the points in the phase space $\mathbf{R}^2$ for which the coordinate $q$ differs by an integer multiple of $\lambda$. The natural phase space for these systems is therefore the cylinder $(p, q) \in \mathbf{R} \times \mathbf{S}^1$, since $\mathbf{S}^1 = \mathbf{R}/(\lambda\mathbf{Z})$.

We now construct the canonical transformation to action-angle variables for systems with rotations or librations. Hence we seek a generating function $F(q, J)$ satisfying

$$p = \frac{\partial F}{\partial q}, \quad \chi = \frac{\partial F}{\partial J}, \tag{11.57}$$

as well as the invertibility condition

$$\frac{\partial^2 F}{\partial q \partial J} \neq 0. \tag{11.58}$$

In the case of rotations or oscillations it is possible to express the canonical variable $p$ locally as a function $\hat{p}(q, E)$. Since the action variable $J$ must satisfy the condition (11.49), we assume—as is true outside the separatrices—that $\mathrm{d}K/\mathrm{d}J \neq 0$, so that the invertibility of the relation between energy and action is guaranteed. We temporarily leave the function $E = K(J)$ undetermined. Then the generating function we are seeking is given by

$$F(q, J) = \int_{q_0}^{q} \hat{p}(q', K(J)) \, \mathrm{d}q', \tag{11.59}$$

corresponding to the integration of the differential form $p \, dq$ along $\gamma_E$. Indeed $p = \partial F/\partial q$ by construction, and hence

$$\frac{\partial^2 F}{\partial q \partial J} = \frac{\partial \hat{p}}{\partial E} \frac{dK}{dJ} \neq 0.$$

In addition, setting

$$\Delta F(J) = \oint_{\gamma_E} p(q,J) \, dq, \qquad (11.60)$$

where $E = K(J)$ and $p(q,J) = \hat{p}(q, K(J))$, from (11.57) and (11.59) it follows that

$$\oint_{\gamma_E} d\chi = \frac{d}{dJ} \Delta F(J).$$

The quantity $\Delta F(J)$ represents the increment of the generating function $F(q,J)$ when going along a phase curve $\gamma_J = \gamma_{E=K(J)}$ for a whole period.

*Remark* 11.7
It is not surprising that the generating function $F$ is multivalued, and defined up to an integer multiple of (11.60). This is due to the fact that the differential form $p \, dq$ is not exact. ∎

*Remark* 11.8
The geometric interpretation of (11.60) is immediate. For librations, $\Delta F(J)$ is equal to the area $\mathcal{A}(E)$ enclosed by the phase curve $\gamma_E$ (where $E = K(J)$). For rotations, $\oint_{\gamma_E} p(q,J) \, dq = \int_{q_0}^{q_0+\lambda} p(q,J) \, dq$ is the area under the graph of $\gamma_E$. ∎

Even if $K(J)$ in the definition of $F(q,J)$ is undetermined, we can still perform the symbolic calculation of $p = \partial F/\partial q$, but to ensure that condition (11.50) is verified, we need to impose

$$\frac{d}{dJ} \Delta F(J) = 2\pi.$$

This fact, and Remark 3.2, justify the following.

DEFINITION 11.2    *An action variable is the quantity*

$$J = \frac{1}{2\pi} \oint_{\gamma_E} p \, dq = \frac{\mathcal{A}(E)}{2\pi}. \qquad (11.61)$$

∎

It can be easily checked that

$$\frac{d\mathcal{A}}{dE} \neq 0. \qquad (11.62)$$

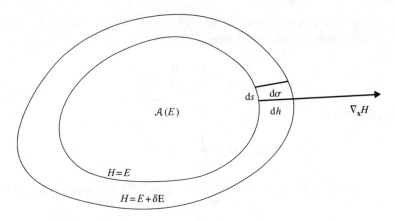

**Fig. 11.5**

Indeed (see Fig. 11.5 and recall Remark 8.12), we have $d\sigma = ds\, dh$, with $dh|\nabla_\mathbf{x} H| = \delta E$, so that

$$\mathcal{A}(E + \delta E) = \mathcal{A}(E) + \delta E \oint_{\gamma_E} \frac{ds}{|\nabla_\mathbf{x} H|} + \mathcal{O}((\delta E)^2),$$

where $\mathbf{x} = (p, q)$ and by our hypotheses $|\nabla_\mathbf{x} H| \neq 0$ on $\gamma_E$. Hence

$$\frac{d\mathcal{A}}{dE} = \lim_{\delta E \to 0} \frac{\mathcal{A}(E + \delta E) - \mathcal{A}(E)}{\delta E} = \oint_{\gamma_E} \frac{ds}{|\nabla_\mathbf{x} H|} \neq 0.$$

From equation (11.62) we have that $dJ/dE \neq 0$, and therefore the existence of the inverse function $E = K(J) = \mathcal{A}^{-1}(2\pi J)$ follows. Substituting it into (11.59) we obtain the generating function of the canonical transformation to the action-angle variables. The latter is $F(q, J) = W(q, K(J))$ (Example 11.3). Thus we have proved the following.

**THEOREM 11.3** *Every Hamiltonian system* (11.46) *with one degree of freedom and with motions of librations or rotations is completely canonically integrable.* ∎

As a consequence of (11.52), (11.54) and (11.61), the period of the motion has the following simple expression:

$$T = \frac{d\mathcal{A}}{dE} = 2\pi \frac{dJ}{dE}. \tag{11.63}$$

*Example 11.10*
Consider the harmonic oscillator with Hamiltonian

$$H = \frac{p^2}{2m} + \frac{1}{2} m\omega^2 q^2.$$

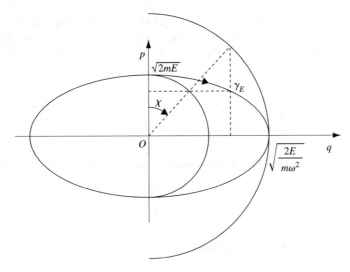

Fig. 11.6

In the phase plane, the cycles $\gamma_E$ of the equation

$$\frac{p^2}{2m} + \frac{1}{2}m\omega^2 q^2 = E$$

enclose the area $(2\pi/\omega)E$ (Fig. 11.6), and hence it follows from (11.61) that the action variable is $J = E/\omega$, i.e. $K(J) = \omega J$; we have rederived equation (11.56). The generating function is (see Example 11.2)

$$F(q, J) = W(q, \omega J) = J \arcsin\left(\sqrt{\frac{\omega m}{2J}} q\right) + \sqrt{\frac{m\omega J}{2}} q \sqrt{1 - \frac{m\omega}{2J} q^2},$$

and hence

$$\chi = \frac{\partial F}{\partial J} = \arcsin\left(\sqrt{\frac{m\omega}{2J}} q\right),$$

from which we obtain the relations

$$q = \sqrt{\frac{2J}{m\omega}} \sin \chi, \quad p = \sqrt{2m\omega J}\sqrt{1 - \frac{m\omega}{2J} q^2} = \sqrt{2m\omega J} \cos \chi,$$

which coincide with (11.55). Figure 11.6 shows the geometric meaning of the variable $\chi$. ∎

The example of the harmonic oscillator illustrates well the statement made at the beginning of this section: the transformation from the variables $(p, q)$ to variables of the kind $(f(R_E), \psi)$ is not in general canonical.

Let us compute $R_E$ and $\psi$ as functions of $(J,\chi)$ in dimensionless variables (therefore setting $m = 1$). We find that $\pi R_E^2 = 2\pi J$, and hence $R_E = \sqrt{2J}$. Having chosen the point $(\sqrt{2\omega J}, 0)$ on $\gamma_E$ to be the origin of the arcs, for $\chi = 0$, we have

$$s(J,\chi) = \int_0^\chi \left[\left(\frac{\partial p}{\partial \chi}\right)^2 + \left(\frac{\partial q}{\partial \chi}\right)^2\right]^{1/2} d\chi' = \int_0^\chi \sqrt{2\omega J \sin^2 \chi' + \frac{2J}{\omega} \cos^2 \chi'} \, d\chi'.$$

In particular, the length $\lambda_E$ of $\gamma_E$ is only a function of $J$. Finally, we find

$$\psi(J,\chi) = 2\pi \frac{s(J,\chi)}{\lambda_E(J)}.$$

We now compute the Poisson bracket:

$$\{\psi, f(R_E)\}_{(J,\chi)} = \frac{\partial \psi}{\partial \chi} \frac{\mathrm{d} R_E}{\mathrm{d} J} f'(R_E) = \frac{2\pi}{\lambda_E(J)} \frac{f'(R_E)}{\sqrt{2J}} \left(2\omega J \sin^2 \chi + \frac{2J}{\omega} \cos^2 \chi\right)^{1/2}.$$

It follows that if $\omega \neq 1$ (hence if $\gamma_E$ is not a circle) then $\{\psi, f(R_E)\} \neq 1$, independent of the choice of $f(R_E)$, and the variables $(f(R_E), \psi)$ are not canonical. If instead $\omega = 1$ we have that $\{\psi, f(R_E)\} = f'(R_E)/R_E$. Therefore, choosing $f(R_E) = \frac{1}{2} R_E^2 = J$ we naturally obtain the same canonical variables $(J, \chi)$.

*Example* 11.11
The Hamiltonian of a simple pendulum is (see Section 3.3)

$$H(p, \vartheta) = \frac{p^2}{2ml^2} - mgl \cos \vartheta.$$

Setting $e = E/mgl$, if $|e| < 1$ the motion is oscillatory and the action is equal to

$$J = \frac{2}{\pi} \sqrt{2m^2 gl^3} \int_0^{\vartheta_m} \sqrt{e + \cos \vartheta} \, d\vartheta,$$

where $\vartheta_m = \arccos(-e)$. Setting $k^2 = (e+1)/2$ and $\sin \vartheta/2 = k \sin \psi$, we find

$$J = ml\sqrt{gl}\frac{8}{\pi} \int_0^{\pi/2} \frac{k^2 \cos^2 \psi}{\sqrt{1 - k^2 \sin^2 \psi}} \, d\psi = ml\sqrt{gl}\frac{8}{\pi}[(k^2 - 1)\mathbf{K}(k) + \mathbf{E}(k)],$$

(11.64)

where $\mathbf{K}(k)$ and $\mathbf{E}(k)$ are the complete elliptic integrals of the first and second kind, respectively.

If $e > 1$ the motions are rotations, and the action is equal to

$$J = \frac{1}{\pi}\sqrt{2m^2gl^3}\int_0^\pi \sqrt{e + \cos\vartheta}\, d\vartheta.$$

Setting $k^2 = 2/(e+1)$ and $\psi = \vartheta/2$, we find

$$J = \frac{2}{\pi}\sqrt{2m^2gl^3}\int_0^{\pi/2}\sqrt{e + 1 - 2\sin^2\psi}\, d\psi = \frac{4}{\pi}\sqrt{\frac{m^2gl^3}{k^2}}\mathbf{E}(k). \tag{11.65}$$

The function $K(J)$ can be found by inverting the function $J(E)$; $J$ depends on $E$ and $e$. Writing the formula for the period

$$T = \frac{2\pi}{\omega} = 2\pi\frac{dJ}{dE}$$

and computing

$$\frac{dJ}{dE} = \frac{dJ}{de}\frac{1}{mgl},$$

we easily find the formulae (3.18) and (3.21) of Chapter 3.

We take into account in the calculations the relations

$$\frac{d\mathbf{E}(k)}{dk} = \frac{\mathbf{E}(k) - \mathbf{K}(k)}{k}, \quad \frac{d\mathbf{K}(k)}{dk} = \frac{1}{k}\left[\frac{\mathbf{E}(k)}{1 - k^2} - \mathbf{K}(k)\right]$$

(see Whittaker and Watson 1927, p. 521). ∎

Introducing action-angle variables for systems with more degrees of freedom requires some preliminary ideas. These are discussed in the following sections.

## 11.4 Integrability by quadratures. Liouville's theorem

Integrating a system of $2l$ ordinary differential equations of first order requires more than just knowledge of the $l$ first integrals. However, if the system of equations is canonical, the fact that the flow preserves the symplectic structure of the phase space has among its consequences that it is enough to know $l$ independent integrals in order to solve the Hamilton–Jacobi equations, thus leading to integration of the equations of motion. It is necessary, however, for the $l$ first integrals to be in involution (Definition 10.18). This concept is clarified in the following.

THEOREM 11.4 (Liouville) *Consider an autonomous Hamiltonian system with Hamiltonian $H(\mathbf{p},\mathbf{q})$ having $l$ degrees of freedom. Assume that the system admits*

$l$ first integrals $f_1(\mathbf{p},\mathbf{q}),\ldots,f_l(\mathbf{p},\mathbf{q})$ which are independent (hence such that for every $(\mathbf{p},\mathbf{q})$ the gradients $\nabla_{(\mathbf{p},\mathbf{q})} f_i$ are $l$ linearly independent vectors) and that they are in involution. Consider the level set

$$M_{\mathbf{a}} = \{\mathbf{x} = (\mathbf{p},\mathbf{q}) \in \mathbf{R}^{2l} | f_i(\mathbf{p},\mathbf{q}) = a_i,\ i = 1,\ldots,l\}, \tag{11.66}$$

where $\mathbf{a} \in \mathbf{R}^l$ is fixed. If $M_{\mathbf{a}}$ is not empty:

(1) $M_{\mathbf{a}}$ is a regular submanifold of dimension $l$, invariant with respect to the Hamiltonian flow $S^t$ and the phase flows $g_1^t,\ldots,g_l^t$ associated with $f_1,\ldots,f_l$;
(2) the flows $g_1^t,\ldots,g_l^t$ commute with each other.

In addition, if

$$\det\left(\frac{\partial f_i}{\partial p_j}\right) \neq 0, \tag{11.67}$$

then locally there exists a function $S = S(\mathbf{f},\mathbf{q},t)$ such that

$$\left(\sum_{i=1}^{l} p_i\, \mathrm{d}q_i - H(\mathbf{p},\mathbf{q})\, \mathrm{d}t\right)\bigg|_{(\mathbf{p},\mathbf{q}) \in M_{\mathbf{a}}} = \mathrm{d}S(\mathbf{a},\mathbf{q},t). \tag{11.68}$$

The function is a complete integral of the Hamilton–Jacobi equation (11.5) corresponding to $H$. The system is therefore integrable by quadratures. ∎

Before giving the proof, we list some remarks.

*Remark* 11.9
The system is autonomous, and hence we can include the Hamiltonian $H$ among the $l$ integrals of the motion considered in Theorem 11.4. In all cases, both $H$ and all of the first integrals $f_i$ are constant not only along their own flow $g_i^t$, but also along the flow generated by other integrals. This is due to the mutual involution condition.

In addition $H$ is always constant on each connected component of the manifold $M_{\mathbf{a}}$. Indeed, even if $f_i \neq H$ for every $i = 1,\ldots,l$ it is always possible to connect any pair of points belonging to the same connected component of $M_{\mathbf{a}}$ through successive applications (in any order) of the flows $g_i^t$. This intuitive concept will be rigorously proven in Lemma 11.2 below.

Since $H|_{M_{\mathbf{a}}}$ is constant, equation (11.68) takes locally the form

$$\sum_{i=1}^{l} p_i\, \mathrm{d}q_i|_{M_{\mathbf{a}}} = \mathrm{d}W(\mathbf{a},\mathbf{q}), \tag{11.69}$$

where

$$W(\mathbf{a},\mathbf{q}) = S(\mathbf{a},\mathbf{q},t) + E(\mathbf{a})t, \tag{11.70}$$

and $E(\mathbf{a})$ is the value taken by $H$ on $M_{\mathbf{a}}$.

## 11.4  Analytic mechanics: Hamilton–Jacobi theory and integrability

An $l$-dimensional submanifold of the phase space satisfying condition (11.69) is called *Lagrangian*. The significance of this property will be made clear when we construct the action-angle variables for systems with several degrees of freedom (see Section 11.6). ∎

*Remark* 11.10
The condition (11.67) is not restrictive, because the condition that the first integrals $f_1, \ldots, f_l$ are independent ensures that there exist $l$ canonical coordinates $x_{i_1}, \ldots, x_{i_l}$ such that

$$\det\left(\frac{\partial(f_1, \ldots, f_l)}{\partial(x_{i_1}, \ldots, x_{i_l})}\right) \neq 0. \tag{11.71}$$

We saw (see Example 10.8) that the exchange of canonical coordinates is a completely canonical transformation. Hence if $(i_1, \ldots, i_l) \neq (1, \ldots, l)$, exchanging some of the coordinates $q_{i_k}$ with the corresponding kinetic momenta $-p_{i_k}$, we can always write (11.71) in the form (11.67). ∎

*Remark* 11.11
In general the condition (11.67) cannot be globally satisfied on all of $M_\mathbf{a}$: consider for example what happens in the case of the harmonic oscillator. ∎

*Proof of Theorem 11.4*
The properties (1) and (2) are an immediate consequence of the linear independence of the integrals, and of the fact that they are in involution (see Theorem 10.18).

The condition (11.67) and the implicit function theorem guarantee the local existence of $l$ regular functions $\hat{p}_1(\mathbf{f}, \mathbf{q}), \ldots, \hat{p}_l(\mathbf{f}, \mathbf{q})$ such that

$$p_i = \hat{p}_i(\mathbf{f}, \mathbf{q}) \quad \text{and} \quad f_i(\hat{\mathbf{p}}(\mathbf{f}, \mathbf{q}), \mathbf{q}) = a_i, \tag{11.72}$$

for all $i = 1, \ldots, l$. By Remark 11.9, equation (11.68) is equivalent to

$$\sum_{i=1}^{l} \hat{p}_i(\mathbf{f}, \mathbf{q}) dq_i = dW(\mathbf{f}, \mathbf{q}), \tag{11.73}$$

with $\mathbf{f} = \mathbf{a}$ and d acting on $\mathbf{q}$ only. The existence of a function $W$ satisfying (11.73) is guaranteed if for every $j, k = 1, \ldots, l$ we have

$$\frac{\partial \hat{p}_k}{\partial q_j} = \frac{\partial \hat{p}_j}{\partial q_k}, \tag{11.74}$$

and hence if the matrix $B = (\partial \hat{p}_j / \partial q_k)$ is symmetric. On the other hand, by differentiating with respect to $q_k$ the second of equations (11.72), we find

$$\sum_{j=1}^{l} \frac{\partial f_i}{\partial p_j} \frac{\partial \hat{p}_j}{\partial q_k} + \frac{\partial f_i}{\partial q_k} = 0. \tag{11.75}$$

It follows that, setting $A = (\partial f_i / \partial p_j)$ and $C = (\partial f_i / \partial q_k)$ we have

$$B = -A^{-1}C, \qquad (11.76)$$

and equation (11.74) becomes

$$-A^{-1}C = -C^T(A^T)^{-1},$$

and hence $CA^T - AC^T = 0$. In componentwise form, this reads

$$\sum_{k=1}^{l}\left(\frac{\partial f_i}{\partial q_k}\frac{\partial f_j}{\partial p_k} - \frac{\partial f_i}{\partial p_k}\frac{\partial f_j}{\partial q_k}\right) = \{f_i, f_j\} = 0, \qquad (11.77)$$

confirming the validity of (11.73).

The function $W$ is therefore defined on $M_\mathbf{a}$ as

$$W(\mathbf{f}, \mathbf{q}) = \int_{\mathbf{q}_0}^{\mathbf{q}} \sum_{i=1}^{l} \hat{p}_i(\mathbf{f}, \boldsymbol{\xi}) \, d\xi_i, \qquad (11.78)$$

computed along an arbitrary path belonging to $M_\mathbf{a}$ joining $\mathbf{q}_0$ and $\mathbf{q}$. The extension of $W$ to non-constant values of $\mathbf{f}$ is possible because of the arbitrariness of $\mathbf{a}$. Consider now

$$\hat{H}(\mathbf{f}, \mathbf{q}) = H(\hat{\mathbf{p}}(\mathbf{f}, \mathbf{q}), \mathbf{q}). \qquad (11.79)$$

For fixed $\mathbf{f} = \mathbf{a}$, from (11.74) it follows that

$$\frac{\partial \hat{H}}{\partial q_i} = \frac{\partial H}{\partial q_i} + \sum_{j=1}^{l} \frac{\partial H}{\partial p_j}\frac{\partial \hat{p}_j}{\partial q_i} = \frac{\partial H}{\partial q_i} + \sum_{j=1}^{l} \frac{\partial \hat{p}_i}{\partial q_j}\frac{\partial H}{\partial p_j} = -\dot{p}_i + \sum_{i=1}^{l} \frac{\partial \hat{p}_i}{\partial q_j}\dot{q}_j = 0,$$

by the first of equations (11.72). Thus $\hat{H}$ is independent of $\mathbf{q}$, and the Hamiltonian can be expressed through the integrals $f_1, \ldots, f_l$:

$$H(\hat{\mathbf{p}}(\mathbf{f}, \mathbf{q}), \mathbf{q}) = \hat{H}(\mathbf{f}). \qquad (11.80)$$

It follows that setting

$$S(\mathbf{f}, \mathbf{q}, t) = W(\mathbf{f}, \mathbf{q}) - \hat{H}(\mathbf{f})t, \qquad (11.81)$$

it can immediately be verified that (11.68) is satisfied and $S$ is a solution of the Hamilton–Jacobi equation.

Indeed, by hypothesis

$$\det\left(\frac{\partial^2 W}{\partial q_i \partial f_j}\right) = \det\left(\partial p_i \partial f_j\right) \neq 0.$$

## 11.4 Analytic mechanics: Hamilton–Jacobi theory and integrability

From equation (11.73) it follows that

$$p_i = \frac{\partial W}{\partial q_i} = \frac{\partial S}{\partial q_i},$$

while equation (11.81) implies that $H + \partial S/\partial t = 0$. In addition, $S$ is a complete integral, because it depends on the $l$ arbitrary constants $a_1, \ldots, a_l$ (the fixed values of $f_1, \ldots, f_l$) and $H$ is independent of $t$. ∎

*Example* 11.12
Consider a system of $l$ non-interacting harmonic oscillators, with Hamiltonian

$$H(\mathbf{p}, \mathbf{q}) = \sum_{i=1}^{l} \left( \frac{p_i^2}{2m_i} + \frac{1}{2} m_i \omega_i^2 q_i^2 \right). \tag{11.82}$$

Evidently

$$H(\mathbf{p}, \mathbf{q}) = \sum_{i=1}^{l} f_i(p_i, q_i), \tag{11.83}$$

where

$$f_i(p_i, q_i) = \frac{p_i^2}{2m_i} + \frac{1}{2} m_i \omega_i^2 q_i^2 \tag{11.84}$$

is the energy of the $i$th oscillator. The functions $f_i$ are integrals of the motion, independent and in involution. The level manifold $M_\mathbf{a}$ is compact, connected, and diffeomorphic to an $l$-dimensional torus. The condition (11.67) is satisfied (as long as $p_i \neq 0$ for every $i = 1, \ldots, l$). Note that this property is not globally satisfied on $M_\mathbf{a}$, see Remark 11.11; however the condition (11.71) is certainly globally satisfied. The function $S$ is then given by

$$S(\mathbf{f}, \mathbf{q}, t) = \sum_{i=1}^{l} \int_{(q_i)_0}^{q_i} \pm \sqrt{2m_i f_i - m_i^2 \omega_i^2 \xi_i^2} \, d\xi_i - t \sum_{i=1}^{l} f_i. \tag{11.85}$$

Note that since the condition (11.67) is not globally satisfied on $M_\mathbf{a}$, $S$ is not a single-valued function. ∎

*Remark* 11.12
Liouville's theorem ensures that the integrals $f_1, \ldots, f_l$ can play the role of new canonical coordinates, together with the variables

$$\beta_i = \frac{\partial W}{\partial f_i}, \quad i = 1, \ldots, l. \tag{11.86}$$

The function $W(\mathbf{f}, \mathbf{q})$ is thus the generating function of a completely canonical transformation of the variables $(\mathbf{p}, \mathbf{q})$ into $(\mathbf{f}, \boldsymbol{\beta})$. Therefore it satisfies

$$\sum_{i=1}^{l} (p_i \mathrm{d} q_i + \beta_i \mathrm{d} f_i) = \mathrm{d} W(\mathbf{f}, \mathbf{q}). \tag{11.87}$$

Note that on $M_\mathbf{a}$ we have $d\mathbf{f} = 0$ and equation (11.87) reduces to (11.69). The new Hamiltonian takes the form

$$\hat{H} = \hat{H}(\mathbf{f}) = H(\hat{\mathbf{p}}(\mathbf{f},\mathbf{q}), \mathbf{q}), \qquad (11.88)$$

and hence Hamilton's equations become

$$\dot{\mathbf{f}} = \mathbf{0}, \quad \dot{\boldsymbol{\beta}} = \nabla_\mathbf{f} \hat{H}(\mathbf{f}). \qquad (11.89)$$

From this it follows that $\mathbf{f}$ is constant (as was known) and

$$\boldsymbol{\beta}(t) = \boldsymbol{\beta}(0) + \nabla_\mathbf{f} \hat{H}(\mathbf{f}(0)) t. \qquad (11.90)$$

∎

*Remark* 11.13

We saw that every time we can solve the Hamilton–Jacobi equation (11.9) and compute the Hamilton characteristic function (as is the case, for example, when we can apply the method of separation of variables) then we determine $l$ first integrals, independent and in involution. These are precisely the new canonical coordinates $\alpha_1, \ldots, \alpha_l$.

The theorem of Liouville gives the converse: the knowledge of $l$ integrals, independent and in involution, yields Hamilton's characteristic function.

Note finally that in the separable cases equation (11.73) is simplified in a similar way to the Hamilton–Jacobi equation, as each function $\hat{p}_i$ depends on $\mathbf{f}$ and only on the corresponding $q_i$.

∎

*Example* 11.13

Consider a point particle of mass $m$ in free motion on an $(l-1)$ dimensional ellipsoid embedded in $\mathbf{R}^l$, described by the implicit equation

$$\sum_{i=1}^{l} \frac{x_i^2}{a_i} = 1, \qquad (11.91)$$

where $0 < a_1 < a_2 < \ldots < a_l$, and $\sqrt{a_i}$ is the length of the $i$th semi-axis of the ellipsoid. We introduce a convenient parametrisation of the ellipsoid (due to Jacobi, see Arnol'd et al. (1983), p. 126–9) via the equation

$$f(\mathbf{x}, \lambda) = \sum_{i=1}^{l} \frac{x_i^2}{a_i - \lambda} = 1. \qquad (11.92)$$

This associates to any generic point $\mathbf{x} = (x_1, \ldots, x_l) \in \mathbf{R}^l$, $l$ real numbers $\lambda_1 \leq \ldots \leq \lambda_l$ (the $l$ roots of equation (11.92)) which evidently alternate with the $a_i$: $\lambda_1 < a_1 \leq \lambda_2 < a_2 \leq \ldots \leq \lambda_l < a_l$. To show this, it is enough to note that for every fixed non-zero point $\mathbf{x}$, $f$ as a function of $\lambda$ has $l$ vertical asymptotes in $\lambda = a_i$, and for $\lambda \neq a_i$ one has $\partial f/\partial \lambda > 0$.

If **x** belongs to the ellipsoid (11.91), necessarily $\lambda_1 = 0$ and the variables $\lambda_2, \ldots, \lambda_l$ yield a system of orthogonal coordinates on the ellipsoid.

It is not difficult to show that for every $i = 1, \ldots, l$ we have

$$x_i^2 = \frac{\prod_{j=1}^{l}(a_i - \lambda_j)}{\prod_{j=1, j\neq i}^{l}(a_i - a_j)}, \tag{11.93}$$

from which it follows that[1]

$$\sum_{i=1}^{l} \dot{x}_i^{\,2} = \frac{1}{4}\sum_{i=2}^{l} M_i \dot{\lambda}_i^2, \tag{11.94}$$

where

$$M_i = \frac{\prod_{j\neq i}(\lambda_j - \lambda_i)}{\prod_{j=1}^{l}(a_j - \lambda_i)}, \quad i = 2, \ldots, l. \tag{11.95}$$

The variables $\mu_i$ canonically conjugate to the $\lambda_i$ are

$$\mu_i = \frac{m M_i \dot{\lambda}_i}{4}, \quad i = 2, \ldots, l, \tag{11.96}$$

and the Hamiltonian of our problem is given by

$$H(\mu_2, \ldots, \mu_l, \lambda_2, \ldots, \lambda_l) = \frac{2}{m}\sum_{i=2}^{l} \frac{\mu_i^2}{M_i(\lambda_2, \ldots, \lambda_l)}. \tag{11.97}$$

A set of independent first integrals is constructed by means of a remarkable formula due to Jacobi:

$$\sum_{i=1}^{l} \frac{\lambda_i^n}{\prod_{j\neq i}(\lambda_i - \lambda_j)} = \begin{cases} 0, & \text{if } n < l-1, \\ 1, & \text{if } n = l-1. \end{cases} \tag{11.98}$$

We leave its verification as an exercise. From equation (11.98), and substituting the definition (11.95) of $M_i$ into (11.97), we find that the following identity holds:

$$\sum_{i=1}^{l} \frac{\sum_{n=0}^{l-1} F_n \lambda_i^n}{\prod_{j\neq i}(\lambda_i - \lambda_j)} = \frac{2}{m}\sum_{i=1}^{l} \frac{\mu_i^2 \prod_{k=1}^{l}(\lambda_i - a_k)}{\prod_{j\neq i}(\lambda_i - \lambda_j)}. \tag{11.99}$$

---

[1] An easy proof is provided by the computation of the residues of (11.92) considered as a rational function of $\lambda$.

In this formula $F_{l-1} = H$, while for the moment, $F_0, F_1, \ldots, F_{l-2}$ are arbitrary. However, if we set

$$\sum_{n=0}^{l-1} F_n \lambda_i^n = \frac{2}{m}\mu_i^2 \prod_{k=1}^{l}(\lambda_i - a_k), \quad i = 2, \ldots l, \quad (11.100)$$

from this system of equations we find $F_0, F_1, \ldots, F_{l-2}$ as functions of $\lambda$ and $\mu$. These, together with $F_{l-1}$, yield a set of $l$ independent integrals of the motion which can be seen to be in involution. ∎

## 11.5 Invariant $l$-dimensional tori. The theorem of Arnol'd

Liouville's theorem implies that if an autonomous Hamiltonian system with $l$ degrees of freedom has $l$ integrals that are independent and in involution, then the Hamilton–Jacobi equation has a complete integral and the equations of motion are integrable by quadratures.

In this section we intend to study the geometry of invariant manifolds of integrable systems with several degrees of freedom. In particular, we prove the following theorem, which clarifies in which cases it is possible to give a *global* parametric representation of the manifold $M_\mathbf{a}$ using $l$ angular coordinates (in which case $M_\mathbf{a}$ is diffeomorphic to a torus $\mathbf{T}^l$).

THEOREM 11.5 (Arnol'd) *Let $H(\mathbf{p}, \mathbf{q})$ be a given autonomous Hamiltonian system with $l$ degrees of freedom and which has $l$ first integrals of the motion $f_1(\mathbf{p}, \mathbf{q}), \ldots, f_l(\mathbf{p}, \mathbf{q})$ that are independent and in involution. If the level manifold $M_\mathbf{a}$ of the first integrals is compact and connected, then it is diffeomorphic to an $l$-dimensional torus.* ∎

*Remark* 11.14
Sometimes $M_\mathbf{a}$ has several connected components. In this case, Theorem 11.5 applies separately to each connected component. ∎

*Remark* 11.15
There exist Hamiltonian systems such that the level manifold $M_\mathbf{a}$ is not compact and/or not connected. These systems satisfy the hypotheses of Liouville's theorem, but not of the theorem of Arnol'd above. An important example is the case of linearised equations of a system with two degrees of freedom in a neighbourhood of a saddle point of the potential energy:

$$H(\mathbf{p}, \mathbf{q}) = \frac{1}{2}[(p_1^2 + \omega_1^2 q_1^2) + (p_2^2 - \omega_2^2 q_2^2)].$$

Setting $f_1 = (p_1^2 + \omega_1^2 q_1^2/2)$, $f_2 = (p_2^2 - \omega_2^2 q_2^2/2)$, $M_\mathbf{a}$ is the Cartesian product of an ellipse (corresponding to the curve $f_1 = a_1$ in the $(p_1, q_1)$ plane) with two branches of the hyperbola (corresponding to $f_2 = a_2$ in the $(p_2, q_2)$ plane), and hence it is neither compact nor connected. ∎

## 11.5  Analytic mechanics: Hamilton–Jacobi theory and integrability

Theorem 11.5 is a non-trivial extension of a very simple property, which we observed when $l = 1$ (the manifold $M_\mathbf{a}$ reduces to the phase curve $\gamma_E$). The proof of Theorem 11.5 can be omitted at a first reading. It is possible to skip directly to the following section, after reading the statement of Proposition 5.1 and the subsequent remarks.

We devote this section to the proof of the theorem of Arnol'd and to its consequences.

We have already remarked (see Remark 10.30) that the integrals $f_1, \ldots, f_l$ induce $l$ Hamiltonian phase flows $g_1^t, \ldots, g_l^t$ that leave $M_\mathbf{a}$ invariant. The idea of the proof of Theorem 11.5 is to use these flows to construct an atlas of the manifold $M_\mathbf{a}$, and then to prove that this atlas is compatible with the definition of the $l$-dimensional torus (see Examples 1.38 and 1.39).

Choose $\mathbf{t} = (t_1, \ldots, t_l) \in \mathbf{R}^l$, and consider the composition $g^\mathbf{t}$ of the flows $g_i^t$:

$$g^\mathbf{t} = g_1^{t_1} \circ \cdots \circ g_l^{t_l}. \tag{11.101}$$

Since $\{f_i, f_j\} = 0$, the flows commute, and $g^\mathbf{t}$ does not depend on the order in which the individual flows are applied, but only on $\mathbf{t}$. We hence define an $l$-parameter family of transformations from $M_\mathbf{a}$ to itself, i.e. a map $g : \mathbf{R}^l \times M_\mathbf{a} \to M_\mathbf{a}$, defined by $g(\mathbf{t}, \mathbf{x}) = g^\mathbf{t}(\mathbf{x})$, satisfying the group conditions required by Definition 1.33. We then say that $g^\mathbf{t}$ is an $l$-parameter group of transformations of $M_\mathbf{a}$, and that $\mathbf{R}^l$ acts on $M_\mathbf{a}$ through $g^\mathbf{t}$, and hence that $g^\mathbf{t}$ defines an action of $\mathbf{R}^l$ on $M_\mathbf{a}$.

**LEMMA 11.1**  *Let $\mathbf{x}_0$ be any point of $M_\mathbf{a}$. The map $g_{\mathbf{x}_0} : \mathbf{R}^l \to M_\mathbf{a}$, $g_{\mathbf{x}_0}(\mathbf{t}) = g^\mathbf{t}(\mathbf{x}_0)$ is a local diffeomorphism (Section 1.7), and hence there exist an open neighbourhood $U$ of $\mathbf{t} = \mathbf{0}$ in $\mathbf{R}^l$ and an open neighbourhood $V$ of $\mathbf{x}_0$ in $M_\mathbf{a}$ such that $g_{\mathbf{x}_0}(U) = V$, and $g_{\mathbf{x}_0}$ restricted to $U$ is a diffeomorphism (Fig. 11.7).*

*Proof*
Since the integrals $f_1, \ldots, f_l$ are independent, for every $\mathbf{x} \in M_\mathbf{a}$ the vectors $\mathcal{I}\nabla_\mathbf{x} f_i(\mathbf{x}) \in T_\mathbf{x} M_\mathbf{a}$ are linearly independent and are a basis of $T_\mathbf{x} M_\mathbf{a}$. Integrating along the directions of these vectors, it is possible to parametrise every point $\mathbf{y} \in V$ of a neighbourhood of $\mathbf{x}_0 \in M_\mathbf{a}$ through $\mathbf{t}$:

$$\mathbf{y} = \mathbf{y}(t_1, \ldots, t_l) = g^\mathbf{t}(\mathbf{x}_0), \tag{11.102}$$

where $\mathbf{t} = (t_1, \ldots, t_l)$ belongs to a neighbourhood $U$ of $\mathbf{0}$ (note that $\mathbf{x}_0 = g^\mathbf{0}(\mathbf{x}_0)$).

The invertibility of the transformation is a consequence of the independence of the first integrals, which ensures that the determinant of the Jacobian matrix of the parametrisation (11.102) is non-zero. Indeed if, for example, $M_\mathbf{a}$ is parametrisable through the variables $(q_1, \ldots, q_l)$ (hence $\mathbf{p} = \hat{\mathbf{p}}(\mathbf{a}, \mathbf{q})$) in the neighbourhood $V$ of $\mathbf{x}_0 = (\mathbf{p}_0, \mathbf{q}_0) = (\hat{\mathbf{p}}(\mathbf{a}, \mathbf{q}_0), \mathbf{q}_0)$, equation (11.102) can be written as

$$\mathbf{q} = \mathbf{q}(t_1, \ldots, t_l) = g^\mathbf{t}(\hat{\mathbf{p}}(\mathbf{a}, \mathbf{q}_0), \mathbf{q}_0). \tag{11.103}$$

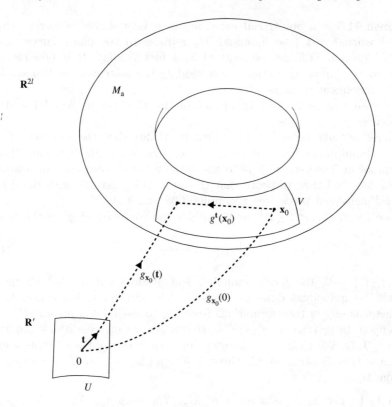

Fig. 11.7

Since the flows $g_i^{t_j}$ are canonical, $\partial \mathbf{q}/\partial t_j = \nabla_{\mathbf{p}} f_j$, and hence

$$\frac{\partial(q_1, \ldots, q_l)}{\partial(t_1, \ldots, t_l)} = \left(\frac{\partial(f_1, \ldots, f_l)}{\partial(p_1, \ldots, p_l)}\right)^T, \tag{11.104}$$

which is clearly non-singular. ∎

*Remark* 11.16

Evidently the map $g_{\mathbf{x}_0}$ cannot be a global diffeomorphism, because $M_{\mathbf{a}}$ is assumed to be a compact manifold, while $\mathbf{R}^l$ is not compact. It is worth noting that, because of the local character of this lemma, we made no use of the compactness assumption in the proof. ∎

LEMMA 11.2  *The action of $\mathbf{R}^l$ on $M_{\mathbf{a}}$ defined by $g^{\mathbf{t}}$ is transitive, and hence for each pair of points $\mathbf{x}_1$, $\mathbf{x}_2$ belonging to $M_{\mathbf{a}}$ there exists $\mathbf{t} \in \mathbf{R}^l$ such that $g^{\mathbf{t}}(\mathbf{x}_1) = \mathbf{x}_2$.*

*Proof*

Since $M_{\mathbf{a}}$ is a connected manifold, there exists a regular curve $\gamma \colon [0, 1] \to M_{\mathbf{a}}$ joining $\mathbf{x}_1$ and $\mathbf{x}_2$: $\gamma(0) = \mathbf{x}_1$, $\gamma(1) = \mathbf{x}_2$. By Lemma 11.1 every point $\gamma(\tau)$ of

## 11.5 Analytic mechanics: Hamilton–Jacobi theory and integrability

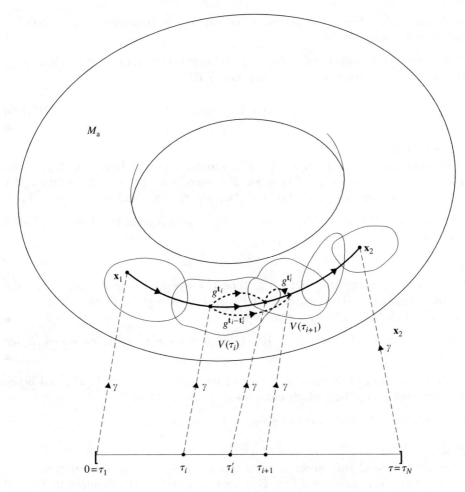

**Fig. 11.8**

the curve, $0 \leq \tau \leq 1$, has an open neighbourhood $V(\tau)$ restricted to which $g^{\mathbf{t}}$ acts as a local diffeomorphism. The family $\{V(\tau)\}_{\tau \in [0,1]}$ is an open covering of the curve $\gamma$. By compactness, there exists a finite subcovering $\{V(\tau_i)\}_{i=1}^{N}$, with $\tau_1 = 0$ and $\tau_N = 1$. Consider any sequence of points $\gamma(\tau_i')$ of the curve defined by the conditions $\gamma(\tau_i') = V(\tau_i) \cap V(\tau_{i+1}) \cap \gamma([0,1])$, $\tau_i' > \tau_i$ (Fig. 11.8), as $i = 1, \ldots, N-1$ varies. Since $g^{\mathbf{t}}$ is a local diffeomorphism between an open set of $\mathbf{R}^l$ and every open set $V(\tau_i)$, there exist $\mathbf{t}_i$ and $\mathbf{t}_i'$ such that $g^{\mathbf{t}_i}\gamma(\tau_i) = \gamma(\tau_i')$ and $g^{\mathbf{t}_i'}\gamma(\tau_{i+1}) = \gamma(\tau_i')$. It follows that $\gamma(\tau_{i+1}) = g^{\mathbf{t}_i - \mathbf{t}_i'}\gamma(\tau_i)$, and therefore $\mathbf{x}_2 = g^{\mathbf{t}}\mathbf{x}_1$ where $\mathbf{t} = \sum_{i=1}^{N-1}(\mathbf{t}_i - \mathbf{t}_i')$. ∎

The two previous lemmas show that the action of $\mathbf{R}^l$ on $M_{\mathbf{a}}$ yields a way to construct an atlas of $M_{\mathbf{a}}$ whose elements are the local parametrisations defined by Lemma 11.1. Since the action of $\mathbf{R}^l$ on $M_\mathbf{a}$ is transitive, $M_\mathbf{a}$ is called a *homogeneous space* of $\mathbf{R}^l$.

DEFINITION 11.3  Given $\mathbf{x}_0 \in M_\mathbf{a}$, the stationary subgroup *of the action* $g^\mathbf{t}$ *of* $\mathbf{R}^l$ *on* $M_\mathbf{a}$ *at the point* $\mathbf{x}_0$ *is the subgroup of* $\mathbf{R}^l$:

$$\Gamma_{\mathbf{x}_0} = \{\mathbf{t} \in \mathbf{R}^l | g^\mathbf{t}\mathbf{x}_0 = \mathbf{x}_0\}. \tag{11.105}$$

∎

*Remark* 11.17
It is immediate to verify that $\Gamma_{\mathbf{x}_0}$ is a subgroup of $\mathbf{R}^l$. Indeed $\mathbf{0} \in \Gamma_{\mathbf{x}_0}$, and if $\mathbf{t} \in \Gamma_{\mathbf{x}_0}$ then $g^{-\mathbf{t}}\mathbf{x}_0 = g^{-\mathbf{t}}g^\mathbf{t}\mathbf{x}_0 = \mathbf{x}_0$, and therefore $-\mathbf{t} \in \Gamma_{\mathbf{x}_0}$. In addition, if $\mathbf{t}$ and $\mathbf{s}$ both belong to $\Gamma_{\mathbf{x}_0}$, $g^{\mathbf{t}+\mathbf{s}}\mathbf{x}_0 = g^\mathbf{t}g^\mathbf{s}\mathbf{x}_0 = g^\mathbf{t}\mathbf{x}_0 = \mathbf{x}_0$, and hence $\mathbf{t}+\mathbf{s} \in \Gamma_{\mathbf{x}_0}$.∎

LEMMA 11.3  *The stationary subgroup* $\Gamma_{\mathbf{x}_0}$ *is independent of* $\mathbf{x}_0$ *(we shall henceforth denote it simply by* $\Gamma$).

*Proof*
It is enough to prove that if $\mathbf{t} \in \Gamma_{\mathbf{x}_0}$ then $g^\mathbf{t}\mathbf{x} = \mathbf{x}$ for every $\mathbf{x} \in M_\mathbf{a}$. Since the action is transitive, there exists $\mathbf{s} \in \mathbf{R}^l$ such that $\mathbf{x} = g^\mathbf{s}\mathbf{x}_0$. From this it follows that $g^\mathbf{t}\mathbf{x} = g^\mathbf{t}g^\mathbf{s}\mathbf{x}_0 = g^\mathbf{s}g^\mathbf{t}\mathbf{x}_0 = g^\mathbf{s}\mathbf{x}_0 = \mathbf{x}$.
∎

DEFINITION 11.4  *A subgroup of* $\mathbf{R}^l$ *is called* discrete *if it has no accumulation point*.
∎

Thus a subgroup $\Gamma$ is discrete if each of its points is isolated in $\mathbf{R}^l$, and hence it is the centre of a ball which contains no other point of $\Gamma$.

LEMMA 11.4  *The stationary subgroup* $\Gamma$ *is discrete*.

*Proof*
Since $g^\mathbf{t}$ is a local diffeomorphism, the origin $\mathbf{0} \in \Gamma$ is an isolated point, and hence it has a neighbourhood $U \subset \mathbf{R}^l$ such that $\Gamma \cap U = \{\mathbf{0}\}$. Suppose that $\mathbf{t} \neq \mathbf{0}$ is an accumulation point of $\Gamma$. Then

$$\mathbf{t} + U = \{\mathbf{s} + \mathbf{t} | \mathbf{s} \in U\}$$

is a neighbourhood of $\mathbf{t}$, and hence there exists $\mathbf{s} \in (\mathbf{t}+U) \cap \Gamma$, $\mathbf{s} \neq \mathbf{t}$. But then $\mathbf{s}-\mathbf{t} \neq \mathbf{0}$ and $\mathbf{s}-\mathbf{t} \in \Gamma \cap U$, contradicting the hypothesis $\Gamma \cap U = \{\mathbf{0}\}$. ∎

The following lemma yields a classification of all discrete subgroups of $\mathbf{R}^l$.

LEMMA 11.5  *Every discrete subgroup* $G$ *of* $\mathbf{R}^l$ *is isomorphic to* $\mathbf{Z}^k$, *where* $k \in \{0,\ldots,l\}$. *Hence there exist* $k$ *linearly independent vectors* $\mathbf{e}_1,\ldots,\mathbf{e}_k$ *in* $\mathbf{R}^l$ *such that*

$$G = \{m_1\mathbf{e}_1 + \cdots + m_k\mathbf{e}_k | \mathbf{m} = (m_1,\ldots,m_k) \in \mathbf{Z}^k\}. \tag{11.106}$$

*The vectors* $\mathbf{e}_1,\ldots,\mathbf{e}_k$ *are called* generators *(or* periods *or* bases*) of* $G$.

*Proof*
If $l = 1$ every discrete subgroup $G$ of $\mathbf{R}$ is either trivial, $G = \{0\}$, or else it is of the form $G = \{m\mathbf{e}_1, m \in \mathbf{Z}\}$, where $\mathbf{e}_1 = \min_{x \in G\setminus\{0\}} |x|$. Indeed, since $G$ is discrete, $\mathbf{e}_1$ is a non-zero element of $G$ and every other element $x$ of $G$ must be an integer multiple of it, otherwise the remainder $r$ of the division of $|x|$ by $\mathbf{e}_1$ would give another element of $G$, $0 < r < \mathbf{e}_1$, which contradicts the definition of $\mathbf{e}_1$.

For $l \geq 2$ the proof that $G$ is isomorphic to $\mathbf{Z}^k$ with $0 \leq k \leq l$ can be obtained by induction on $l$, by projecting $G$ onto $\mathbf{R}^{l-1}$ orthogonally to any element $\mathbf{e}_1$ of $G\setminus\{0\}$ of minimum norm. Since the projection of $G$ is again a discrete subgroup, this yields the proof. ∎

The representation (11.106) of the discrete subgroup $G$ is not unique. If $\mathbf{e}_1, \ldots, \mathbf{e}_k$ generates $G$, evidently it also true that $\mathbf{e}_1 + \mathbf{e}_2, \mathbf{e}_2, \ldots, \mathbf{e}_k$ generate $G$, and so on. However, it is possible to characterise uniquely all possible choices of the generators of a discrete subgroup, as shown by the following.

LEMMA 11.6  $\mathbf{e}_1, \ldots, \mathbf{e}_k$ and $\mathbf{e}'_1, \ldots, \mathbf{e}'_k$ *are two $k$-tuples of generators of the same discrete subgroup $G$ of $\mathbf{R}^l$ if and only if there exists a $k \times k$ matrix $A$, with integer coefficients and with determinant equal to $\pm 1$ (i.e. $A \in \mathrm{GL}(k, \mathbf{Z})$), such that for every $i = 1, \ldots, k$ we have*

$$\mathbf{e}'_i = \sum_{j=1}^{k} A_{ij}\mathbf{e}_i. \tag{11.107}$$

*Proof*
Evidently if $\mathbf{e}_1, \ldots, \mathbf{e}_k$ generates $G$, and $A \in \mathrm{GL}(k, \mathbf{Z})$, the $k$-tuple $\mathbf{e}'_1, \ldots, \mathbf{e}'_k$ defined by (11.107) generates a discrete subgroup $G'$ of $\mathbf{R}^l$. In addition $G' \subset G$, for if $\mathbf{t}' \in G'$, $\mathbf{t}' = \sum_{i=1}^{k} m'_i \mathbf{e}'_i = \sum_{i=1}^{k}\sum_{j=1}^{k} m'_i A_{ij} \mathbf{e}_j = \sum_{j=1}^{k} m_j \mathbf{e}_j$, where $m_j = \sum_{i=1}^{k} m'_i A_{ij} \in \mathbf{Z}$. Since $\det A = \pm 1$ the inverse matrix $A^{-1}$ also has integer coefficients and $\det A^{-1} = \pm 1$; therefore $A^{-1} \in \mathrm{GL}(k, \mathbf{Z})$ and it can be shown immediately that $G \subset G'$. Conversely, if $\mathbf{e}'_1, \ldots, \mathbf{e}'_k$ also generates $G$, let $A$ be the $k \times k$ matrix defined by (11.107) which transforms $\mathbf{e}_1, \ldots, \mathbf{e}_k$ in $\mathbf{e}'_1, \ldots, \mathbf{e}'_k$. The coefficients of $A$ are integers, as $\mathbf{e}'_i \in G$ for every $i = 1, \ldots, k$. Applying the same reasoning to $A^{-1}$ we see that the latter must also have integer coefficients. It follows that there exist two integers $m$ and $n$ such that $\det(A) = m$, $\det(A^{-1}) = n$. But $\det(A)\det(A^{-1}) = 1$, and therefore $m = n = \pm 1$ and $A \in \mathrm{GL}(k, \mathbf{Z})$. ∎

We can finally prove the theorem of Arnol'd.

*Proof of Theorem 11.5*
Since the stationary subgroup $\Gamma$ of the action of $\mathbf{R}^l$ on $M_\mathbf{a}$ is discrete, there exists $k \in \{0, \ldots, l\}$ such that $\Gamma$ is isomorphic to $\mathbf{Z}^k$. Therefore, there exist $k$ linearly independent vectors $\mathbf{e}_1, \ldots, \mathbf{e}_k$ in $\mathbf{R}^l$ that generate $\Gamma$. Let $\tilde{\mathbf{e}}_{k+1}, \ldots, \tilde{\mathbf{e}}_l$ be $l - k$ vectors in $\mathbf{R}^l$ chosen arbitrarily in such a way that $\mathbf{e}_1, \ldots, \mathbf{e}_k, \tilde{\mathbf{e}}_{k+1}, \ldots, \tilde{\mathbf{e}}_l$ is a basis of $\mathbf{R}^l$. Setting then

$$\mathbf{t} = \frac{\psi_1}{2\pi}\mathbf{e}_1 + \cdots + \frac{\psi_k}{2\pi}\mathbf{e}_k + t_{k+1}\tilde{\mathbf{e}}_{k+1} + \cdots + t_l \tilde{\mathbf{e}}_l, \tag{11.108}$$

when $(\psi_1, \ldots, \psi_k, t_{k+1}, \ldots, t_l) \in \mathbf{R}^l$ vary, the action $g^{\mathbf{t}}$ defines a parametrisation of $M_{\mathbf{a}}$. By Lemmas 11.1, 11.2, and 11.4 (recalling also Example 1.39) the manifold $M_{\mathbf{a}}$ is diffeomorphic to $\mathbf{T}^k \times \mathbf{R}^{l-k}$. But $M_{\mathbf{a}}$ is compact, and hence $k = l$. ∎

We conclude this section by proving the following.

PROPOSITION 11.1  *Under the hypotheses of the theorem of Arnol'd, there exists a neighbourhood* $\mathcal{U} \subset \mathbf{R}^{2l}$ *of* $M_{\mathbf{a}}$ *that is diffeomorphic to the direct product of a neighbourhood of an open set* $V \subset \mathbf{R}^l$ *with a torus* $\mathbf{T}^l$: $\mathcal{U} \approx V \times \mathbf{T}^l$.

*Proof*
The idea is to prove that the functions $f_1, \ldots, f_l$ and the angles $\psi_1, \ldots, \psi_l$ constructed in the proof of Theorem 11.5 give a regular parametrisation of a neighbourhood $\mathcal{U}$ of $M_{\mathbf{a}}$. Indeed, in a neighbourhood of any point $P$ of $M_{\mathbf{a}}$ the functions $f(\mathbf{p}, \mathbf{q})$ can be inverted with respect to $(x_{i_1}, \ldots, x_{i_l})$, and hence to $l$ of the variables $\mathbf{x} = (\mathbf{p}, \mathbf{q})$. This is due to the independence of the $f_i$ which ensures that the condition (11.7) is always satisfied. Hence we determine a regular submanifold $N$ of dimension $l$, implicitly defined by

$$x_{i_k} = \hat{x}_{i_k}(\mathbf{f}), \tag{11.109}$$

where $k = 1, \ldots, l$ and $\mathbf{f}$ varies in a neighbourhood $V$ of $\mathbf{a}$ in $\mathbf{R}^l$.

At each point of $N$, determined by fixing the values of $\mathbf{f}$, we can apply Lemma 11.1 and construct a local parametrisation of the corresponding manifold $M_{\mathbf{f}}$:

$$\mathbf{x} = \hat{\mathbf{x}}(\mathbf{f}, \mathbf{t}). \tag{11.110}$$

The parametrisation (11.110) is by construction differentiable, and invertible with respect to both $\mathbf{f}$ and $\mathbf{t}$. It follows that we have a local diffeomorphism between a neighbourhood $\mathcal{U}_P \subset \mathbf{R}^{2l}$ of any point of $M_{\mathbf{a}}$ and a domain $V \times W$, where $W$ is a neighbourhood of $\mathbf{0}$ in $\mathbf{R}^l$.

Since by Lemma 11.2 the action of $\mathbf{R}^l$ on $M_{\mathbf{a}}$ is transitive, considering any other point $P'$ in $M_{\mathbf{a}}$, there exists $\mathbf{t}' \in \mathbf{R}^l$ such that $g^{\mathbf{t}'} P = P'$. It is immediate to verify that $g^{\mathbf{t}'}(\mathcal{U}_P)$ is a neighbourhood of $P'$ that is diffeomorphic to $\mathcal{U}_P$ and hence also to $V \times W$ via the parametrisation $\mathbf{x} = \hat{\mathbf{x}}(\mathbf{f}, \mathbf{t} + \mathbf{t}')$. Since $P'$ is arbitrary, we conclude that there exists a neighbourhood $\mathcal{U}$ of $M_{\mathbf{a}}$ that can be parametrised by coordinates $(\mathbf{f}, \mathbf{t}) \in V \times \mathbf{R}^l$ through a differentiable function $\mathbf{x} = \hat{\mathbf{x}}(\mathbf{f}, \mathbf{t})$. However this function is not invertible, because the stationary subgroup $\Gamma$ of the action of $\mathbf{R}^l$ used to construct it is isomorphic to $\mathbf{Z}^l$. If $\mathbf{e}_1, \ldots, \mathbf{e}_l$ generates $\Gamma$, although $g^{\mathbf{e}_i} P = P$, the map from $g^{\mathbf{e}_i}$ to $N$ generates a new submanifold $N' = g^{\mathbf{e}_i}(N)$ containing $P$ but distinct from $N$.

The existence of a local parametrisation $\hat{\mathbf{x}}(\mathbf{f}, \mathbf{t})$ in a neighbourhood of $M_{\mathbf{a}}$ ensures that for every point $P' \in N'$ (determined uniquely by the corresponding value $\mathbf{f}$) there exist $l$ differentiable functions $\tau_i(\mathbf{f})$, $i = 1, \ldots, l$, and a point $P'' \in N$ such that $\tau_i(\mathbf{a}) = 0$ and $g^{\tau_i(\mathbf{f})} P' = P''$ for every $i = 1, \ldots, l$. Hence $g^{\mathbf{e}_i + \tau_i(\mathbf{f})} P'' = P''$, and in a neighbourhood of $P$ we can construct generators

$\mathbf{e}_i(\mathbf{f})$, with regular dependence on $\mathbf{f}$ and which on every level manifold $M_{\mathbf{f}}$ determine the stationary subgroup of the action of $\mathbf{R}^l$. On each manifold $M_{\mathbf{f}}$ we can finally consider $l$ angles $\psi_1,\ldots,\psi_l$ providing a global parametrisation, and thus obtain a regular parametrisation

$$\mathbf{x} = \tilde{\mathbf{x}}(\mathbf{f},\boldsymbol{\psi}), \tag{11.111}$$

of a neighbourhood of $M_{\mathbf{a}}$ through coordinates $(\mathbf{f},\boldsymbol{\psi}) \in V \times \mathbf{T}^l$. ∎

*Remark* 11.18
In general, the coordinates $(\mathbf{f},\boldsymbol{\psi})$ constructed in the course of the previous proof *are not canonical* (recall the analogous discussion for the case $l=1$, Section 11.3). Liouville's theorem guarantees the existence of $l$ coordinates $\beta_1,\ldots,\beta_l$, canonically conjugate to $f_1,\ldots,f_l$, but the variables $\beta_1,\ldots,\beta_l$ *are not angles*, as required by the previous proposition. In the next section we show how to overcome this difficulty, by introducing the action-angle variables. ∎

*Remark* 11.19
The previous proposition is sufficient to prove that the phase space of an autonomous Hamiltonian system having as many first integrals independent and in involution as degrees of freedom is *foliated in invariant tori*, provided all trajectories are bounded. In that case the invariant tori in the family $\{M_{\mathbf{a}}\}_{\mathbf{a}\in V}$ depend regularly on $\mathbf{a}$. This is an important geometric characterisation of integrable Hamiltonian systems, which will be discussed in depth in the next section. ∎

## 11.6 Integrable systems with several degrees of freedom: action-angle variables

In Section 11.3 we introduced action-angle variables for one-dimensional systems. We started from the observation that, for example for oscillatory motions, every phase curve is diffeomorphic to a circle enclosing the same area.

In the case of an autonomous Hamiltonian system with $l$ degrees of freedom, which admits $l$ integrals that are independent and in involution, the analogous observation is that the level manifold of the first integrals, $M_{\mathbf{a}}$, when is compact, it is diffeomorphic to an $l$-dimensional torus (Theorem 11.5).

Starting from this, we try to extend the construction of the action-angle variables to systems with several degrees of freedom.

DEFINITION 11.5 *An autonomous Hamiltonian system, with Hamiltonian* $H(\mathbf{p},\mathbf{q})$ *having $l$ degrees of freedom, is called* completely canonically integrable *if there exists a completely canonical transformation*

$$\begin{aligned} \mathbf{p} &= \hat{\mathbf{p}}(\mathbf{J},\boldsymbol{\chi}), \\ \mathbf{q} &= \hat{\mathbf{q}}(\mathbf{J},\boldsymbol{\chi}) \end{aligned} \tag{11.112}$$

*(where the dependence of $\hat{\mathbf{p}}$ and $\hat{\mathbf{q}}$ on each variable $\chi_i$ is $2\pi$-periodic) to new variables* $(\mathbf{J},\boldsymbol{\chi}) \in \mathbf{R}^l \times \mathbf{T}^l$, *called* action-angle variables, *such that the new Hamiltonian*

$K$ is only a function of the actions $\mathbf{J}$:

$$K = H(\hat{\mathbf{p}}(\mathbf{J},\boldsymbol{\chi}),\hat{\mathbf{q}}(\mathbf{J},\boldsymbol{\chi})) = K(\mathbf{J}). \tag{11.113}$$

■

If a system is completely canonically integrable, from (11.113) it follows that Hamilton's equations can be written as

$$\begin{aligned}\dot{\mathbf{J}} &= -\nabla_{\boldsymbol{\chi}} K = 0, \\ \dot{\boldsymbol{\chi}} &= \nabla_{\mathbf{J}} K \equiv \boldsymbol{\omega}(\mathbf{J}).\end{aligned} \tag{11.114}$$

The system (11.114) can be immediately integrated:

$$\begin{aligned}\mathbf{J}(t) &= \mathbf{J}(0), \\ \boldsymbol{\chi}(t) &= \boldsymbol{\chi}(0) + \boldsymbol{\omega}(\mathbf{J}(0))t,\end{aligned} \tag{11.115}$$

for every $t \in \mathbf{R}$. The actions are therefore a system of $l$ integrals that are independent and in involution, while each angle variable $\chi_i$, by hypothesis defined mod $2\pi$, has a time period

$$T_i = \frac{2\pi}{\omega_i(\mathbf{J})}. \tag{11.116}$$

Since the dependence of $(\mathbf{p},\mathbf{q})$ on $(\mathbf{J},\boldsymbol{\chi})$ is regular and $2\pi$-periodic with respect to each angle, it follows that the motions of a completely canonically integrable system are *bounded* and *quasi-periodic* (see Section 11.7).

We aim to prove the following theorem.

THEOREM 11.6 *Let $H(\mathbf{p},\mathbf{q})$ be a Hamiltonian system with $l$ degrees of freedom which admits $l$ first integrals $f_1(\mathbf{p},\mathbf{q}),\ldots,f_l(\mathbf{p},\mathbf{q})$ that are independent and in involution. Assume that for a certain fixed value $\mathbf{a} \in \mathbf{R}^l$ the level manifold $M_\mathbf{a}$ of the integrals is compact and connected. Then there exists a canonical transformation of the variables $(\mathbf{p},\mathbf{q}) \in U$ to action-angle variables $(\mathbf{J},\boldsymbol{\chi}) \in V \times \mathbf{T}^l$ (where $V$ is an open subset of $\mathbf{R}^l$). The system is therefore completely canonically integrable.*

■

Theorem 11.5 implies that it is possible to parametrise $M_\mathbf{a}$ through $l$ angles $(\psi_1,\ldots,\psi_l)$. This fact is essential in the proof of Theorem 11.6. More precisely, we refer to the conclusion of Proposition 11.1, that in a neighbourhood of $M_\mathbf{a}$ in $\mathbf{R}^{2l}$ one can introduce the generalised coordinates (not canonical) $(f_1,\ldots,f_l,\psi_1,\ldots,\psi_l)$. For fixed $\mathbf{f} = \mathbf{a}$ and varying between 0 and $2\pi$ only one of the angles $\psi_i$, we obtain a *cycle* $\gamma_i \subset M_\mathbf{a}$ (corresponding to one of the generators of its fundamental group). Hence we can construct $l$ cycles $\gamma_1,\ldots,\gamma_l$ that are and not continuously reducible to one another (hence not *homotopic*). It is now possible to introduce the action variables, in analogy with Definition 11.2.

## 11.6  Analytic mechanics: Hamilton–Jacobi theory and integrability

DEFINITION 11.6   The action variables *are the variables* $(J_1, \ldots, J_l)$ *defined as*

$$J_i = \frac{1}{2\pi} \oint_{\gamma_i} \sum_{j=1}^{l} p_j \, dq_j, \qquad (11.117)$$

*where* $i = 1, \ldots, l$. ■

Apparently the definition we have just given of action variables has some degree of arbitrariness, due to the indetermination of the cycles $\gamma_i$, i.e. the arbitrariness in the choice of the variables $\psi_j$, $j \neq i$. However, the invariant manifolds $M_\mathbf{a}$ are Lagrangian (see Remark 11.9), and this can be used to show that the above definition determines the action variables uniquely. More precisely, one has the following.

PROPOSITION 11.2   *The action variables* $J_i$ *do not depend on the choice of the cycles* $\gamma_i$ *inside the same class of homotopy: if* $\gamma'_i$ *is a new cycle obtained by a continuous deformation of* $\gamma_i$ *we have*

$$\oint_{\gamma_i} \sum_{j=1}^{l} p_j \, dq_j = \oint_{\gamma'_i} \sum_{j=1}^{l} p'_j \, dq_j. \qquad (11.118)$$

*The action variables depend only on the integrals* $f_1(\mathbf{p}, \mathbf{q}), \ldots, f_l(\mathbf{p}, \mathbf{q})$, *and are independent and in involution.*

*Proof*
The independence of the choice of $\gamma_i$ in the same class of homotopy is an immediate consequence of (11.69) and of Stokes' theorem (see Appendix 4).

On the other hand, by (11.117) every action variable $J_i$ is independent of $\psi_i$ and cannot depend on the other angles $\psi_j$, $j \neq i$, either, since as $\psi_j$ varies, the cycle $\gamma_i$ is continuously deformed and the integral (11.117) does not change. Hence the actions are only functions of the integrals $f_1, \ldots, f_l$. They are also in involution, as

$$\{J_i, J_k\} = \sum_{m,n=1}^{l} \frac{\partial J_i}{\partial f_m} \frac{\partial J_k}{\partial f_n} \{f_m, f_n\} = 0.$$

The independence of the actions can be proved by showing that

$$\det\left(\frac{\partial J_i}{\partial f_j}\right) \neq 0, \qquad (11.119)$$

and then using the independence of the integrals $f_i$. The proof is simplified when the variables are separable, which is the most interesting case in practice. Indeed, in this case the set $M_\mathbf{a}$ is the Cartesian product of curves in each subspace $(p_i, q_i)$, identifiable with the cycles. Following the procedure of separation of variables we obtain that $J_i$ depends only on $f_1, \ldots, f_i$ so that the Jacobian matrix is triangular. Therefore (11.119) amounts to showing that $\partial J_i/\partial f_i \neq 0$, $i = 1, \ldots, l$, which follows by repeating in each subspace the same argument used in the one-dimensional case. For simplicity, we limit the proof to the separable case. ∎

*Proof of Theorem 11.6*
By Proposition 11.2 the action variables are a set of independent integrals which are also in involution. By Liouville's theorem (see in particular (11.78) and (11.80)) the function

$$W(\mathbf{J}, \mathbf{q}) = \int_{\mathbf{q}_0}^{\mathbf{q}} \sum_{j=1}^{l} p_j \, \mathrm{d}q_j \bigg|_{M_\mathbf{J}} \tag{11.120}$$

is the generating function of a completely canonical transformation to new variables $(\mathbf{J}, \boldsymbol{\chi})$, and the new Hamiltonian $K$ is a function only of the action variables $\mathbf{J}$. To complete the proof it is then sufficient to show that the new coordinates $\boldsymbol{\chi}$ are angles defined mod $2\pi$. By (11.117), the increment of the function $W(\mathbf{J}, \mathbf{q})$ when integrating along a cycle $\gamma_i$ is

$$\Delta_i W = 2\pi J_i, \tag{11.121}$$

and hence the increment of each variable $\chi_k$ along the same cycle is

$$\Delta_i \chi_k = \Delta_i \frac{\partial W}{\partial J_k} = \frac{\partial}{\partial J_k} \Delta_i W = 2\pi \delta_{ik}. \tag{11.122}$$

It follows that $\boldsymbol{\chi} \in \mathbf{T}^l$. ∎

*Remark 11.20*
The action-angle variables are evidently not unique. The construction of action-angle variables depends on the choice of the homotopy classes of the cycles $\gamma_1, \ldots, \gamma_l$ generating the fundamental group of the torus (see Singer and Thorpe 1980, and Dubrovin et al. 1991b), and a different choice (of other cycles not homotopic) produces a different determination of the variables (Fig. 11.9). Because of this arbitrariness it is possible to have completely canonical transformations to new action-angle variables. ∎

PROPOSITION 11.3 *Let $\mathbf{J}$, $\boldsymbol{\chi}$ be action-angle variables. The variables $\tilde{\mathbf{J}}$, $\tilde{\boldsymbol{\chi}}$ obtained through any of the following completely canonical transformations are still action-angle variables.*

(1) *Translations of the actions: for fixed $\mathbf{a} \in \mathbf{R}^l$ we have*

$$\tilde{\mathbf{J}} = \mathbf{J} + \mathbf{a}, \quad \tilde{\boldsymbol{\chi}} = \boldsymbol{\chi}. \tag{11.123}$$

(2) *Translation of the origin of the angles on each torus: let $\delta : \mathbf{R}^l \to \mathbf{R}$ be an arbitrary regular function, then*

$$\tilde{\mathbf{J}} = \mathbf{J}, \quad \tilde{\boldsymbol{\chi}} = \boldsymbol{\chi} + \nabla_\mathbf{J} \delta(\mathbf{J}). \tag{11.124}$$

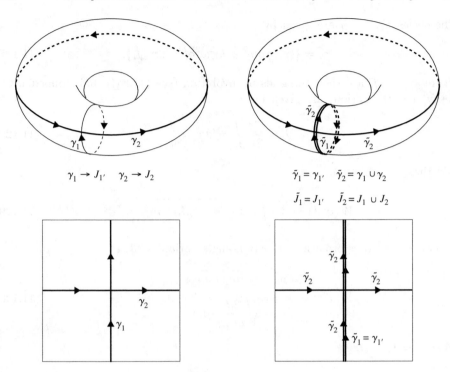

Fig. 11.9

*(3) Linear transformations of the torus onto itself:* let $A$ be a matrix in $\mathrm{GL}(l, \mathbf{Z})$ (hence an $l \times l$ matrix with integer entries and $|\det A| = 1$), then

$$\tilde{\mathbf{J}} = (A^T)^{-1}\mathbf{J}, \quad \tilde{\chi} = A\chi. \tag{11.125}$$

*Proof*
The verification that these transformations are completely canonical is left to the reader. Evidently the transformations (11.123) and (11.124) are canonical and leave invariant the property of being action-angle variables. We remark that the function $W(\mathbf{J}, q)$ in (11.120) is defined up to an arbitrary function $\delta(\mathbf{J})$, which reflects precisely the transformation (11.124). As for (11.125) it is sufficient to note that $\tilde{K} = K(A^T \tilde{\mathbf{J}})$ and the variables $\tilde{\chi}$ are still defined mod $2\pi$. We observe that $A^{-1}$ also has integer entries, thanks to $|\det A| = 1$, which also preserves the measure of the torus. ∎

*Example 11.14*
Consider a system of $l$ harmonic oscillators:

$$H(\mathbf{p}, \mathbf{q}) = \sum_{i=1}^{l} \frac{p_i^2 + m_i^2 \omega_i^2 q_i^2}{2m_i}. \tag{11.126}$$

The cycles $\gamma_1, \ldots, \gamma_l$ are given by

$$\gamma_i = \{(p_i, q_i) | p_i^2 + m_i^2 \omega_i^2 q_i^2 = 2 m_i f_i\}, \qquad (11.127)$$

where $f_1, \ldots, f_l$ are the $l$ integrals in involution (see (11.84)). It is immediate to verify that the actions are given by

$$J_i = \frac{1}{2\pi} \oint_{\gamma_i} p_i \, dq_i = \frac{f_i}{\omega_i}, \qquad (11.128)$$

and that the function

$$W(\mathbf{q}, \mathbf{J}) = \sum_{i=1}^{l} \int_0^{q_i} \pm \sqrt{2 m_i \omega_i J_i - m_i^2 \omega_i^2 \xi_i^2} \, d\xi_i \qquad (11.129)$$

generates the transformation to action-angle variables $(\mathbf{J}, \boldsymbol{\chi})$:

$$\begin{aligned} p_i &= \sqrt{2 m_i \omega_i J_i} \cos \chi_i, \\ q_i &= \sqrt{\frac{2 J_i}{m_i \omega_i}} \sin \chi_i, \end{aligned} \qquad (11.130)$$

where $i = 1, \ldots, l$. ∎

## 11.7 Quasi-periodic motions and functions

The analysis of the previous sections yields the conclusion that the integrable completely canonical Hamiltonian systems are characterised by the fact that they admit $l$ independent integrals in involution, and the phase space is foliated in invariant tori. On these tori, the motion is governed by the equations

$$\dot{\boldsymbol{\chi}} = \boldsymbol{\omega}. \qquad (11.131)$$

In what follows we ignore the trivial case $\boldsymbol{\omega} = \mathbf{0}$. If $l = 1$ the motions are periodic. In the more general case that $l \geq 2$, the motions are not necessarily periodic. Before starting a more detailed analysis, we consider the case $l = 2$. In this case, the solution of equation (11.131) can be written as

$$\chi_1(t) = \chi_1(0) + \omega_1 t, \quad \chi_2(t) = \chi_2(0) + \omega_2 t. \qquad (11.132)$$

Eliminating time $t$, the orbit is given by the line

$$\omega_2 (\chi_1 - \chi_1(0)) - \omega_1 (\chi_2 - \chi_2(0)) = 0. \qquad (11.133)$$

We can therefore assume without loss of generality that $\chi_1(0) = \chi_2(0) = 0$, so that the line passes through the origin (otherwise, it is sufficient to translate the origin to $(\chi_1(0), \chi_2(0))$). Since $(\chi_1, \chi_2) \in \mathbf{T}^2$, and $\mathbf{T}^2 = \mathbf{R}^2/(2\pi\mathbf{Z})^2$, it is

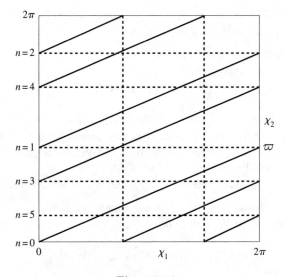

Fig. 11.10

clear that the line must be represented in the square $[0, 2\pi]^2$ with opposite sides identified with each other, according to the rule $\chi' \equiv \chi + 2\pi \mathbf{m}$, where $\mathbf{m} \in \mathbf{Z}^2$. The segments obtained are necessarily parallel (Fig. 11.10).

If $\omega_1 \neq 0$, the sequence of intersections of the orbit with the vertical segment $[0, 2\pi]$ on the $\chi_2$-axis is given by $\{n\varpi \pmod{2\pi}\}_{n=0}^{\infty}$, where

$$\varpi = 2\pi \frac{\omega_2}{\omega_1}, \qquad (11.134)$$

while if $\omega_1 = 0$ all trajectories are clearly periodic.

We thus obtain a map of $\mathbf{T}^1$ onto itself defined by a rotation of angle $\varpi$.

THEOREM 11.7 *The sequence $\{n\varpi \pmod{2\pi}\}_{n=0}^{\infty}$ on the circle $\mathbf{T}^1$ is periodic if and only if $\varpi/2\pi \in \mathbf{Q}$. Else if $\varpi/2\pi$ is irrational, the sequence is dense in $\mathbf{T}^1$.*

*Proof*
A necessary and sufficient condition for the sequence to be periodic is that there exists an integer $s > 0$ such that $s\varpi \pmod{2\pi} = 0$, and hence that there exists an integer $r$ such that $s\varpi = 2r\pi$, from which it follows that $\varpi/2\pi = r/s$. If $\varpi/2\pi$ is irrational, all points of the sequence are distinct. Since the circle is compact, for every $\varepsilon > 0$ there exist integers $r$, $s$ such that $|(r\varpi - s\varpi) \pmod{2\pi}| < \varepsilon$. Setting $j = |r - s|$, the subsequence $\{nj\varpi \pmod{2\pi}\}_{n=0}^{[2\pi/\varepsilon]}$ subdivides the circle into adjacent intervals of length less than $\varepsilon$, and hence every point of the circle is at a distance less than $\varepsilon$ from a point of the sequence. Since $\varepsilon$ is arbitrary, the sequence is dense. ∎

An obvious corollary of this proposition is the following.

COROLLARY 11.1 *The orbit (11.32) on the torus $\mathbf{T}^2$ is periodic if and only if $\omega_2/\omega_1$ is rational or $\omega_1 = 0$, otherwise it is dense on the torus.* ∎

If $l > 2$, the solution of (11.131) is still given by

$$\chi_i(t) = \chi_i(0) + \omega_i t, \qquad (11.135)$$

where $i = 1, \ldots, l$. Eliminating time, we find that the orbit is still represented by a line in $\mathbf{T}^l = \mathbf{R}^l/(2\pi\mathbf{Z})^l$, and hence with the points $\boldsymbol{\chi}' \equiv \boldsymbol{\chi} + 2\pi\mathbf{m}$ identified, where $\mathbf{m} \in \mathbf{Z}^l$, and can therefore be represented in the hypercube $[0, 2\pi]^l$ with opposite faces identified.

To study the periodicity of the orbit it will be useful to introduce the following.

DEFINITION 11.7 *Choose $\boldsymbol{\omega} \in \mathbf{R}^l$. The resonance module $\mathcal{M}_\omega$ of the frequencies vector $\boldsymbol{\omega}$ is the subset[2] of $\mathbf{Z}^l$ given by*

$$\mathcal{M}_\omega = \{\mathbf{m} \in \mathbf{Z}^l | \mathbf{m} \cdot \boldsymbol{\omega} = 0\}. \qquad (11.136)$$

∎

The dimension of the resonance module $\mathcal{M}_\omega$ represents the number of independent resonance relations $\mathbf{m} \cdot \boldsymbol{\omega} = 0$ satisfied by $\boldsymbol{\omega}$. It is also called *resonance multiplicity*. Since we excluded the case $\boldsymbol{\omega} = \mathbf{0}$ we have[3] $0 \leq \dim \mathcal{M}_\omega \leq l - 1$. If $l = 2$ only the extreme cases $\dim \mathcal{M}_\omega = 0$, and hence $\mathcal{M}_\omega = \{\mathbf{0}\}$, and $\dim \mathcal{M}_\omega = l - 1$ are possible, and are called, respectively, *non-resonance* and *complete resonance*. Corollary 11.1 implies that in this case the orbit (11.132) is periodic for complete resonance and dense for non-resonance. We can indeed prove the following generalisation of Theorem 11.7.

THEOREM 11.8 *Let $\mathcal{M}_\omega$ be the resonance module associated with the frequency vector $\boldsymbol{\omega}$ of the motions (11.135). Then*

*(1) the orbit is periodic if and only if $\dim \mathcal{M}_\omega = l - 1$ (complete resonance);*
*(2) if $\dim \mathcal{M}_\omega = 0$ the orbit is dense on the whole torus $\mathbf{T}^l$;*
*(3) if $0 < d < l - 1$, with $d = \dim \mathcal{M}_\omega$, the orbit is dense on a torus of dimension $l - d$ embedded into $\mathbf{T}^l$.*

*The motions corresponding to the cases (2) and (3) are called* quasi-periodic. ∎

We subdivide the proof of Theorem 11.8 into a series of partial results, of some interest by themselves.

Consider an arbitrary invertible linear transformation of coordinates of the torus $\mathbf{T}^l$ which preserves orientation. By Lemma 11.6 its general form is

$$\boldsymbol{\chi}' = M\boldsymbol{\chi}, \qquad (11.137)$$

---

[2] Evidently $\mathcal{M}_\omega$ is a module of $\mathbf{Z}^l$.
[3] Indeed if $\boldsymbol{\omega} = \mathbf{0}$ we have $\dim \mathcal{M}_\omega = l$, but then all the points of the torus $\mathbf{T}^l$ are fixed.

where $M \in \mathrm{SG}(l, \mathbf{Z})$. The system of equations (11.131) is transformed by (11.137) into

$$\dot{\chi}' = \omega', \tag{11.138}$$

where

$$\omega' = M\omega. \tag{11.139}$$

LEMMA 11.7  Let $\mathcal{M}_\omega$ be the resonance module corresponding to $\omega$. There exists a coordinate transformation (11.137) of the torus $\mathbf{T}^l$ such that $\omega'_{l-d+1} = \cdots = \omega'_l = 0$, where $d = \dim \mathcal{M}_\omega$.

*Proof*
First of all we note that a collection of $l$ vectors of $\mathbf{Z}^l$ ($\mathbf{e}_1, \ldots, \mathbf{e}_l$) is a basis of $\mathbf{Z}^l$ if and only if the parallelepiped with sides $\mathbf{e}_1, \ldots, \mathbf{e}_l$ has volume 1. Indeed, the canonical basis $\mathbf{e}_1 = (1, 0, \ldots, 0)$, $\mathbf{e}_2 = (0, 1, 0, \ldots, 0), \ldots$ generates a cube of side 1, and by Lemma 11.6 every other basis in $\mathbf{Z}^l$ is related to the canonical one by a volume-preserving linear transformation.

We now try to complete an arbitrary basis $(\mathbf{m}_1, \ldots, \mathbf{m}_d)$ of $\mathcal{M}_\omega$ with $l - d$ linearly independent vectors of $\mathbf{Z}^l$, $(\boldsymbol{\mu}_1, \ldots, \boldsymbol{\mu}_{l-d})$, in such a way that $(\boldsymbol{\mu}_1, \ldots, \boldsymbol{\mu}_{l-d}, \mathbf{m}_1, \ldots, \mathbf{m}_d)$ is a basis of $\mathbf{Z}^l$. If this is possible, then the lemma is proved by constructing the matrix $M$ whose rows are the components of the vectors $(\boldsymbol{\mu}_1, \ldots, \boldsymbol{\mu}_{l-d}, \mathbf{m}_1, \ldots, \mathbf{m}_d)$; indeed $\omega'_{l-d+j} = \omega \cdot \mathbf{m}_j = 0$ for every $j = 1, \ldots, d$. The matrix $M$ has integer components and determinant equal to $\pm 1$ by the previous remark, and hence it induces an invertible coordinate transformation on the torus $\mathbf{T}^l$ which satisfies the statement.

On the other hand, it is immediate to prove that such a choice of $(\boldsymbol{\mu}_1, \ldots, \boldsymbol{\mu}_{l-d})$ is possible. Let $(\boldsymbol{\mu}_1, \ldots, \boldsymbol{\mu}_{l-d})$ be linearly independent vectors of

$$\mathcal{M}_\omega^\perp = \{\boldsymbol{\mu} \in \mathbf{Z}^l | \boldsymbol{\mu} \cdot \mathbf{m} = 0 \text{ for every } \mathbf{m} \in \mathcal{M}_\omega\}. \tag{11.140}$$

Evidently $(\boldsymbol{\mu}_1, \ldots, \boldsymbol{\mu}_{l-d}, \mathbf{m}_1, \ldots, \mathbf{m}_d)$ is a basis of $\mathbf{R}^l$. If the volume of the parallelepiped they generate is equal to 1, it is also a basis of $\mathbf{Z}^l$ and the proof is finished. Otherwise, since the volume is a positive integer, there exists a non-zero vector $\mathbf{v} \in \mathbf{Z}^l$ inside the parallelepiped:

$$\mathbf{v} = \lambda_1 \boldsymbol{\mu}_1 + \cdots + \lambda_{l-d} \boldsymbol{\mu}_{l-d} + \lambda_{l-d+1} \mathbf{m}_1 + \cdots + \lambda_l \mathbf{m}_d, \tag{11.141}$$

with $0 \leq \lambda_j < 1$ and $\lambda_j$ a suitable rational, for every $j = 1, \ldots, l$. Since the subspace of $\mathbf{R}^l$ generated by $\mathcal{M}_\omega$ does not contain any point of $\mathbf{Z}^l$ different from those of $\mathcal{M}_\omega$, the vector $\mathbf{v}$ cannot belong to $\mathcal{M}_\omega$ (which has no vectors inside the parallelepiped), and therefore it is not restrictive to assume that $\lambda_1 \neq 0$. Hence replacing $\mathbf{v}$ by $\boldsymbol{\mu}_1$, we find a new $l$-tuple of linearly independent vectors of $\mathbf{R}^l$

such that

$$\det\begin{pmatrix}\mathbf{v}\\ \boldsymbol{\mu}_2\\ \vdots\\ \boldsymbol{\mu}_{l-d}\\ \mathbf{m}_1\\ \vdots\\ \mathbf{m}_d\end{pmatrix} \leq \det\begin{pmatrix}\boldsymbol{\mu}_1\\ \boldsymbol{\mu}_2\\ \vdots\\ \boldsymbol{\mu}_{l-d}\\ \mathbf{m}_1\\ \vdots\\ \mathbf{m}_d\end{pmatrix} - 1.$$

The volume of the parallelepiped generated by the basis is therefore diminished by at least one unit. If it is not equal to 1, by repeating this procedure a sufficient number of times, we find the basis sought. ∎

*Example* 11.15
Consider a system of three independent harmonic oscillators; the Hamiltonian of the system in action-angle variables is given by (see Example 11.14)

$$K(\mathbf{J}) = \omega_1 J_1 + \omega_2 J_2 + \omega_3 J_3. \tag{11.142}$$

Suppose that the frequencies satisfy the resonance relations

$$\omega_1 + 2\omega_2 - 4\omega_3 = 0, \quad \omega_1 - \omega_2 = 0, \tag{11.143}$$

so that $\mathcal{M}_\omega$ has dimension 2 and a basis for it is clearly given by $\mathbf{m}_1 = (1, 2, -4)$, $\mathbf{m}_2 = (1, -1, 0)$. In this case the canonical linear transformation

$$\tilde{\mathbf{J}} = (M^T)^{-1}\mathbf{J}, \quad \tilde{\boldsymbol{\chi}} = M\boldsymbol{\chi}, \tag{11.144}$$

where $M$ is the following matrix of $\mathrm{SL}(3, \mathbf{Z})$:

$$M = \begin{pmatrix} 1 & 2 & -4 \\ 1 & -1 & 0 \\ 0 & -1 & 1 \end{pmatrix}, \tag{11.145}$$

transforms the Hamiltonian (11.142) into

$$\tilde{K}(\tilde{\mathbf{J}}) = \tilde{\mathbf{J}} \cdot M\boldsymbol{\omega} = (\omega_3 - \omega_2)\tilde{J}_3, \tag{11.146}$$

and hence in the new variables two frequencies vanish ($\tilde{\omega}_1 = \tilde{\omega}_2 = 0$, $\tilde{\omega}_3 = \omega_3 - \omega_2$). ∎

DEFINITION 11.8 *A continuous function $\phi: \mathbf{R} \to \mathbf{R}$ is called* quasi-periodic *if there exist a continuous function $f: \mathbf{T}^l \to \mathbf{R}$ and a vector $\boldsymbol{\omega} \in \mathbf{R}^l$ such that*

$$\phi(t) = f(\omega_1 t, \ldots, \omega_l t). \tag{11.147}$$

The time average $\langle \phi \rangle_T$ of a quasi-periodic function is given by

$$\langle \phi \rangle_T = \lim_{T \to \infty} \frac{1}{T} \int_0^T \phi(t)\,\mathrm{d}t = \lim_{T \to \infty} \frac{1}{T} \int_0^T f(\omega_1 t, \ldots, \omega_l t)\,\mathrm{d}t, \qquad (11.148)$$

as long as the limit exists. ∎

Evidently, the kinetic momenta and the coordinates $(\mathbf{p}, \mathbf{q})$ of a completely integrable Hamiltonian system are examples of quasi-periodic functions. More generally, if $f$ is any continuous function defined on the torus $\mathbf{T}^l$, then if we consider the values $f(\boldsymbol{\chi}(0) + \boldsymbol{\omega} t)$ that the function takes along the flow (11.135) we find a quasi-periodic function, for which it is meaningful to consider the time average (11.148) (this is a function of the orbit considered, parametrised by the initial data $\boldsymbol{\chi}(0)$) and also the *phase average*, i.e. the average on the torus $\mathbf{T}^l$:

$$\langle f \rangle = \frac{1}{(2\pi)^l} \int_{\mathbf{T}^l} f(\boldsymbol{\chi})\,\mathrm{d}^l \boldsymbol{\chi}. \qquad (11.149)$$

The comparison of the time average with the phase average allows us to establish whether the motion on $\mathbf{T}^l$ is dense. Indeed, we have the following.

THEOREM 11.9 *Let $f: \mathbf{T}^l \to \mathbf{R}$ be a continuous function, and consider the quasi-periodic function obtained by composing $f$ with the flow (11.35): $\phi(t) = f(\boldsymbol{\chi}(0) + \boldsymbol{\omega} t)$. If the frequencies $\boldsymbol{\omega}$ are not resonant, i.e. if $\dim \mathcal{M}_{\boldsymbol{\omega}} = 0$, the time average $\langle \phi \rangle_T(\boldsymbol{\chi}(0))$ exists everywhere, it is constant on $\mathbf{T}^l$ and coincides with the phase average (11.49).*

*Proof*
First of all we prove the theorem in the special case that $f$ is a trigonometric polynomial and hence can be written as

$$f(\boldsymbol{\chi}) = \sum_{\mathbf{m} \in \mathcal{F}} \hat{f}_{\mathbf{m}} e^{i\mathbf{m}\cdot\boldsymbol{\chi}}, \qquad (11.150)$$

where $\mathcal{F} \subset \mathbf{Z}^l$ is a finite set of indices.

If $\mathcal{F}$ is made of only one index $\mathbf{m}$, then if $\mathbf{m} = \mathbf{0}$ the function is constant and $\langle \phi \rangle_T = \hat{f}_{\mathbf{0}} = \langle f \rangle$. Otherwise, if $\mathbf{m} \neq \mathbf{0}$ it is immediate to check that the phase average is zero and the time average is given by

$$\langle \phi \rangle_T = e^{i\mathbf{m}\cdot\boldsymbol{\chi}(0)} \lim_{T \to \infty} \frac{1}{T} \int_0^T e^{i\mathbf{m}\cdot\boldsymbol{\omega} t}\,\mathrm{d}t = \frac{e^{i\mathbf{m}\cdot\boldsymbol{\chi}(0)}}{i\mathbf{m}\cdot\boldsymbol{\omega}} \lim_{T \to \infty} \frac{e^{i\mathbf{m}\cdot\boldsymbol{\omega} T} - 1}{T} = 0,$$

for any $\boldsymbol{\chi}(0) \in \mathbf{T}^l$. If $\mathcal{F}$ has a finite number of indices, one can use the linearity of the time average and phase average operators to show that the averages in phase and time coincide.

Now let $f$ be a generic continuous function. By Weierstrass's theorem (see Giusti 1989) for every $\varepsilon > 0$ there exists a trigonometric polynomial $P_\varepsilon$

approximating $f$ uniformly on $\mathbf{T}^l$ up to $\varepsilon/2$:

$$\max_{\chi \in \mathbf{T}^l} |f(\chi) - P_\varepsilon(\chi)| \leq \varepsilon/2. \tag{11.151}$$

Setting $P_- = P_\varepsilon - \varepsilon/2$ and $P_+ = P_\varepsilon + \varepsilon/2$, we have $P_- \leq f \leq P_+$ and

$$\frac{1}{(2\pi)^l} \int_{\mathbf{T}^l} (P_+(\chi) - P_-(\chi)) \, d^l\chi \leq \varepsilon.$$

Therefore for every $\varepsilon > 0$ there exist two trigonometric polynomials $P_-$ and $P_+$ such that

$$\langle f \rangle - \langle P_- \rangle \leq \varepsilon, \quad \langle P_+ \rangle - \langle f \rangle \leq \varepsilon, \tag{11.152}$$

and for every $T > 0$ we have

$$\frac{1}{T} \int_0^T P_-(\chi(0) + \omega t) \, dt \leq \frac{1}{T} \int_0^T f(\chi(0) + \omega t) \, dt \leq \frac{1}{T} \int_0^T P_+(\chi(0) + \omega t) \, dt. \tag{11.153}$$

However, by the previous remarks, for every $\varepsilon > 0$ there exists $\overline{T}(\varepsilon) > 0$ such that for every $T > \overline{T}(\varepsilon)$ one has

$$\left| \langle P_\pm \rangle - \frac{1}{T} \int_0^T P_\pm(\chi(0) + \omega t) \, dt \right| \leq \varepsilon. \tag{11.154}$$

Combining (11.152)–(11.154) we find that for every $\varepsilon > 0$ and for every $T > \overline{T}(\varepsilon)$ we have

$$\left| \langle f \rangle - \frac{1}{T} \int_0^T f(\chi(0) + \omega t) \, dt \right| \leq 2\varepsilon, \tag{11.155}$$

and the theorem is proved. ∎

It is not difficult now to prove Theorem 11.8.

*Proof of Theorem 11.8*

Statement (1) is of immediate verification, and it is left to the reader.

Suppose now that $\dim \mathcal{M}_\omega = 0$. If there exist a point $\overline{\chi} \in \mathbf{T}^l$ and an open neighbourhood $U$ not visited by the orbit, take any continuous function $f \colon \mathbf{T}^l \to \mathbf{R}$ with the following properties:

(a) $\langle f \rangle = 1$;
(b) $f(\chi) = 0$ for every $\chi \notin U$.

The function $f$ would then have zero time average, different from the phase average, contradicting Theorem 11.9.

Finally, if $\dim \mathcal{M}_\omega = d$ and $0 < d < l-1$, by Lemma 11.7 there exists a coordinate transformation on $\mathbf{T}^l$ which annihilates the last $d$ frequencies. It is therefore sufficient to repeat the previous argument restricted to the torus $\mathbf{T}^{l-d}$ with points $(\chi_1, \ldots, \chi_{l-d}, \chi_{l-d+1}(0), \ldots, \chi_l(0))$. ∎

*Example* 11.16
Apply Theorem 11.9 to solve a celebrated problem proposed by Arnol'd. Consider the sequence constructed by taking the first digit of $2^n$ for $n \geq 0$: $1, 2, 4, 8, 1, 3, 6, 1, 2, 5, 1, \ldots$ and compute the frequency with which each integer $i$ appears in the sequence.

The first digit of $2^n$ is equal to $i$ if and only if

$$\log_{10} i \leq \{n \log_{10} 2\} < \log_{10}(i+1), \quad 1 \leq i \leq 9,$$

where $\{x\}$ denotes the fractional part of $x$: $\{x\} = x \pmod 1$. On the other hand, $\log_{10} 2$ is irrational and, by Theorem 11.7, the sequence $\{n \log_{10} 2\}$ is dense on the interval $[0, 1]$.

The frequency $\nu_i$ with which the integer $i$ appears in the sequence is given by

$$\nu_i = \lim_{N \to +\infty} \frac{\operatorname{card}(\{\{n \log_{10} 2\} \in [\log_{10} i, \log_{10}(i+1)) | 0 \leq n \leq N - 1\})}{N}, \quad (11.156)$$

where $\operatorname{card}(A)$ indicates the cardinality of the set $A$.

Evidently (11.156) coincides with the time average $\langle \chi_i \rangle_T$ of the function $\chi_i : [0, 1] \to \mathbf{R}$ given by

$$\chi_i(x) = \begin{cases} 1, & \text{if } x \in [\log_{10} i, \log_{10}(i+1)), \\ 0, & \text{otherwise,} \end{cases} \quad (11.157)$$

computed for the sequence $\{n \log_{10} 2\}$:

$$\nu_i = \langle \chi_i \rangle_T = \lim_{N \to +\infty} \frac{1}{N} \sum_{j=0}^{N-1} \chi_i(\{j \log_{10} 2\}). \quad (11.158)$$

It is not difficult to prove, by adapting the proof of Theorem 11.9,[4] that, although the function $\chi_i$ is not continuous, the conclusions of the theorem still hold, and in particular, that the average $\langle \chi_i \rangle_T$ is constant and equal to the average of $\chi_i$ on the interval $[0, 1]$:

$$\nu_i = \langle \chi_i \rangle_T = \int_0^1 \chi_i(x) \, \mathrm{d}x = \log_{10}(i+1) - \log_{10} i. \quad (11.159)$$

Hence the frequency of $1, 2, \ldots, 9$ in the sequence of the first digit of $2^n$ is approximately equal to $0.301, 0.176, 0.125, 0.097, 0.079, 0.067, 0.058, 0.051, 0.046$, respectively.

---

[4] Note that the function $\chi_i$ can be approximated by trigonometric polynomials, although the convergence occurs only pointwise.

Considering only the first 40 terms of the sequence, it would appear that the sequence is periodic, with period 10: $1, 2, 4, 8, 1, 3, 6, 1, 2, 5$, etc. The number 7 appears for the first time for $n = 46$, and 9 for $n = 53$. This behaviour illustrates how the convergence to the limit (11.158) is possibly very slow, and in our case it can be explained by observing that $\log_{10} 2 = 0.301029996\ldots$, while an irrational number is very close to $3/10$, which would produce the sequence $\{3n/10\}$ that is periodic with period 10.

It is interesting to compare the behaviour of the sequence $\{n \log_{10} 2\}$ with $\{n(\sqrt{5} - 1)/2\}$. ∎

## 11.8 Action-angle variables for the Kepler problem. Canonical elements, Delaunay and Poincaré variables

The Hamiltonian of the Kepler problem in spherical coordinates is given by

$$H(p_r, p_\theta, p_\varphi, r, \theta, \varphi) = \frac{1}{2m}\left(p_r^2 + \frac{p_\theta^2}{r^2} + \frac{p_\varphi^2}{r^2 \sin^2\theta}\right) - \frac{k}{r}. \tag{11.160}$$

The moment $p_\varphi$ canonically conjugate to the azimuthal angle $\varphi$ coincides with the component along the $z$-axis (normal to the ecliptic) of the angular momentum, as can be immediately verified from the definitions. In addition we have

$$|\mathbf{L}|^2 = |m\mathbf{r} \times \dot{\mathbf{r}}|^2 = m^2 r^4 (\dot\theta^2 + \sin^2\theta \dot\varphi^2) = p_\theta^2 + \frac{p_\varphi^2}{\sin^2\theta}, \tag{11.161}$$

and $|\mathbf{L}| =$ constant (areas constant).

If $i$ indicates the angle of inclination of the orbit with respect to the ecliptic $z = 0$, evidently

$$p_\varphi = |\mathbf{L}| \cos i. \tag{11.162}$$

The angle $\varphi$ is cyclic, and hence $L_z = p_\varphi$ is a first integral of the motion. It is very easy to check that all assumptions of Arnol'd's theorem are satisfied. Two angular coordinates $\varphi, \theta$ are immediately available to obtain the respective cycles $\gamma_\varphi, \gamma_\theta$. The first action variable for the Kepler problem therefore coincides with $p_\varphi$:

$$J_\varphi = \frac{1}{2\pi} \oint_{\gamma_\varphi} p_\varphi \, d\varphi = p_\varphi \tag{11.163}$$

(here $\gamma_\varphi$ is the cycle obtained by varying $\varphi \in \mathbf{S}^1$ and keeping $r, \theta, p_r, p_\theta, p_\varphi$ constant).

The second action variable is given by

$$J_\theta = \frac{1}{2\pi} \oint_{\gamma_\theta} p_\theta \, d\theta. \tag{11.164}$$

## 11.8  Analytic mechanics: Hamilton–Jacobi theory and integrability

The equation of the cycle $\gamma_\theta$ is indeed (11.161) from which

$$J_\theta = \frac{1}{2\pi} \oint_{\gamma_\theta} \sqrt{|\mathbf{L}|^2 - \frac{J_\varphi^2}{\sin^2 \theta}}\, d\theta. \tag{11.165}$$

On the other hand, $J_\varphi = |\mathbf{L}|\cos i$, $\theta$ on a cycle varies between $\pi/2 - i$ and $\pi/2 + i$ while $r$ and $\varphi$ remain constant, and hence we have

$$J_\theta = -\frac{4|\mathbf{L}|}{2\pi} \int_{\pi/2}^{\pi/2-i} \frac{1}{\sin\theta} \sqrt{\sin^2 i - \cos^2 \theta}\, d\theta = \frac{2|\mathbf{L}|}{\pi} \sin^2 i \int_0^{\pi/2} \frac{\cos^2 \psi}{1 - \sin^2 i \sin^2 \psi}\, d\psi,$$

where we have substituted $\cos\theta = \sin i \sin\psi$. Setting now $u = \tan\psi$ we finally find

$$\begin{aligned} J_\theta &= \frac{2|\mathbf{L}|}{\pi} \int_0^{+\infty} \left[\frac{du}{1+u^2} - \cos^2 i \frac{du}{1 + u^2 \cos^2 i}\right] \\ &= \frac{2|\mathbf{L}|}{\pi} \left(\frac{\pi}{2} - \frac{\pi}{2} \cos i\right) = |\mathbf{L}|(1 - \cos i), \end{aligned}$$

from which, since $|\mathbf{L}|\cos i = J_\varphi$ we can deduce that $|\mathbf{L}| = J_\theta + J_\varphi$. The third and last action variable $J_r$ is given by

$$J_r = \frac{1}{2\pi} \oint_{\gamma_r} p_r\, dr = \frac{1}{2\pi} \oint_{\gamma_r} \sqrt{2m\left(E + \frac{k}{r}\right) - \frac{(J_\theta + J_\varphi)^2}{r^2}}\, dr. \tag{11.166}$$

Note that, because of equations (11.160), (11.161), the cycle $\gamma_r$ in the plane $(p_r, r)$ has precisely the equation

$$p_r^2 = 2m\left(E + \frac{k}{r}\right) - \frac{|\mathbf{L}|^2}{r^2}, \tag{11.167}$$

from which we immediately find the extreme values of $r$:

$$r_\pm = a\left[1 \pm \sqrt{1 - \frac{|\mathbf{L}|^2}{mka}}\right],$$

with $a = -k/2E > 0$. The interpretation of $J_\varphi$, $J_\theta$ and $J_r$ is clear in terms of the areas of the cycles depicted in Fig. 11.11.

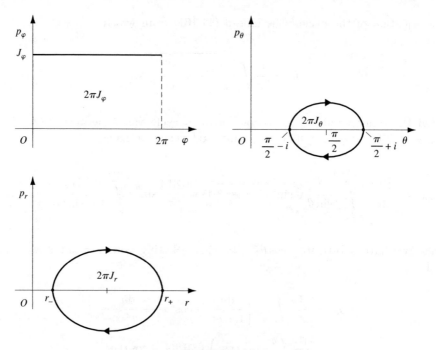

**Fig. 11.11**

The integral (11.166) can be computed by elementary means and the final result is

$$J_r = -(J_\theta + J_\varphi) + k\sqrt{\frac{m}{-2E}},$$

from which

$$H = E = -\frac{mk^2}{2(J_r + J_\theta + J_\varphi)^2}. \tag{11.168}$$

Differentiating with respect to the action variables we find the frequencies

$$\omega = \frac{\partial H}{\partial J_r} = \frac{\partial H}{\partial J_\theta} = \frac{\partial H}{\partial J_\varphi} = \frac{mk^2}{(J_r + J_\theta + J_\varphi)^3} = mk^2 \left(\frac{-2E}{mk^2}\right)^{3/2}. \tag{11.169}$$

Since the frequencies are all equal, the problem is completely resonant and all orbits are periodic with period

$$T = \frac{2\pi}{\omega} = \frac{2\pi}{mk^2}\left(\frac{mk^2}{-2E}\right)^{3/2}. \tag{11.170}$$

## 11.8  Analytic mechanics: Hamilton–Jacobi theory and integrability

From the relation $a = -k/2E$ linking the major semi-axis with the energy, one derives Kepler's third law (see (5.43) and recall that $k/m$ is independent of $m$):

$$\frac{a^3}{T^2} = \frac{k}{4\pi^2 m}.$$

The so-called *Delaunay elements*, which can be interpreted as orbital elements, are defined through the linear canonical transformation of the kind (11.125), naturally suggested by the physical meaning of $J_\varphi$, $J_\varphi + J_\theta$, $J_\varphi + J_\theta + J_r$:

$$\begin{aligned}
\mathcal{L} &= J_\theta + J_\varphi + J_r, \\
\mathcal{G} &= J_\varphi + J_\theta, \\
\mathcal{H} &= J_\varphi; \\
l &= \chi_r, \\
g &= \chi_\theta - \chi_r, \\
h &= \chi_\varphi - \chi_\theta,
\end{aligned} \qquad (11.171)$$

where $(\chi_r, \chi_\theta, \chi_\varphi)$ are the angle variables conjugate to $(J_r, J_\theta, J_\varphi)$. Relation (11.171) annihilates two frequencies (see Lemma 11.7) and the Hamiltonian in the new variables is written

$$H = -\frac{mk^2}{2\mathcal{L}^2}. \qquad (11.172)$$

It follows that the only non-constant element is $l$. On the other hand, the first three elements are combinations of constants, while the constancy of $g$ and $h$ is a consequence of complete resonance. It is not difficult to see that $l$ is the mean anomaly, $g$ is the perihelion argument and $h$ is the ascending node longitude (Fig. 11.12). Here $\mathcal{L}$, $\mathcal{G}$ and $\mathcal{H}$ are related to the semi-major axis $a$, the eccentricity $e$ and the inclination $i$ of the orbit by

$$\begin{aligned}
\mathcal{L} &= \sqrt{mka}, \\
\mathcal{G} &= |\mathbf{L}| = \mathcal{L}\sqrt{1 - e^2}, \\
\mathcal{H} &= |\mathbf{L}| \cos i.
\end{aligned} \qquad (11.173)$$

Although appropriate to the complete resonance of the Kepler problem, the Delaunay variables are not particularly convenient to describe the orbits of the planets of the Solar System. This is due to the fact that these variables become singular in correspondence to circular orbits ($e = 0$, therefore $\mathcal{L} = \mathcal{G}$ and the argument of the perihelion $g$ is not defined) and to horizontal orbits ($i = 0$ or $i = \pi$, therefore $\mathcal{G} = \mathcal{H}$ and the ascending node longitude $h$ is not defined). All the planets of the Solar System have almost circular orbits (except Mercury, Mars and Pluto) and small inclinations (see Table 11.1, taken from Danby (1988)).

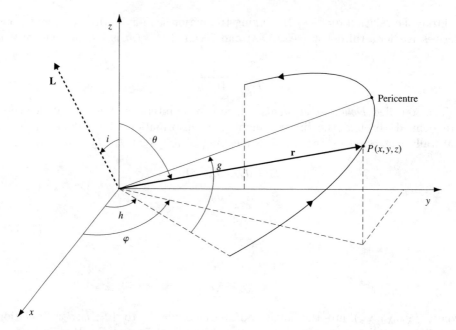

Fig. 11.12

**Table 11.1** Orbital elements of the planets of the Solar System. Here $\varpi, h, i$ and $\lambda$ are expressed in degrees, $a$ is expressed in astronomical units (1 A.U. $= 1.5 \times 10^8$ km), $\varpi$ is the perihelion longitude: $\varpi = g+h$, $\lambda = \varpi + l$ is the average longitude (at a fixed time)

| Planet | $\varpi$ | $h$ | $i$ | $e$ | $\lambda$ | $a$ |
|---|---|---|---|---|---|---|
| Mercury | 77.4561 | 48.3309 | 7.0050 | 0.205632 | 252.2509 | 0.387104 |
| Venus | 131.5637 | 76.6799 | 3.3947 | 0.006772 | 181.9798 | 0.723307 |
| Earth | 102.9373 | | | 0.016709 | 1000.4664 | 1.000012 |
| Mars | 336.0602 | 49.5581 | 1.8497 | 0.093401 | 355.4333 | 1.523711 |
| Jupiter | 374.3313 | 100.4644 | 1.3033 | 0.048495 | 34.3515 | 5.210216 |
| Saturn | 93.0568 | 113.6655 | 2.4889 | 0.055509 | 50.0775 | 9.538070 |
| Uranus | 173.0052 | 74.0060 | 0.7732 | 0.046296 | 314.0550 | 19.183302 |
| Neptune | 48.1237 | 131.7841 | 1.7700 | 0.008989 | 304.3487 | 30.055144 |
| Pluto | 224.6148 | 110.4065 | 17.1323 | 0.250877 | 218.8874 | 39.537580 |

This difficulty can be resolved by introducing a new set of action-angle variables $(\Lambda, Z_1, Z_2, \lambda, \zeta_1, \zeta_2) \in \mathbf{R}^3 \times \mathbf{T}^3$: $\Lambda = \mathcal{L}$, $Z_1 = \mathcal{L} - \mathcal{G}$, $Z_2 = \mathcal{G} - \mathcal{H}$, $\lambda = h+g+l$, $\zeta_1 = -g-h$, $\zeta_2 = -h$ ($\lambda$ is called the mean longitude, $-\zeta_1$ is the perihelion longitude). Hence considering the pairs $(Z_1, \zeta_1)$ and $(Z_2, \zeta_2)$ as polar coordinates we find

$$\xi_1 = \sqrt{2Z_1} \cos \zeta_1, \quad \eta_1 = \sqrt{2Z_1} \sin \zeta_1, \quad \xi_2 = \sqrt{2Z_2} \cos \zeta_2, \quad \eta_2 = \sqrt{2Z_2} \sin \zeta_2.$$
(11.174)

The canonical variables $(\Lambda, \xi_1, \xi_2, \lambda, \eta_1, \eta_2) \in \mathbf{R}_+ \times \mathbf{R}^2 \times \mathbf{T}^1 \times \mathbf{R}^2$ are called *Poincaré variables* and are also well defined in the case of a circular orbit ($Z_1 = 0$) or a horizontal orbit ($Z_2 = 0$). In these new variables the Hamiltonian of Kepler's problem is $H = -mk^2/2\Lambda^2$; therefore $\Lambda, \xi_1, \xi_2, \eta_1, \eta_2$ are constants of the motion. The advantage of the Poincaré variables over the Delaunay ones is that the former are then suitable for both the complete resonance of the Kepler problem and for the study of the planets of the Solar System. The relation between the Poincaré and the original variables momentum and position $(\mathbf{p}, \mathbf{q})$ is more complicated and will not be discussed here (the interested reader can refer to Poincaré (1905, chapter III) or Laskar (1989a)). Note however that $\Lambda$ is proportional to $\sqrt{a}$,

$$\sqrt{\xi_1^2 + \eta_1^2} \simeq \sqrt{\Lambda}e(1 + \mathcal{O}(e^2)), \quad \sqrt{\xi_2^2 + \eta_2^2} \simeq \sqrt{\Lambda}i(1 + \mathcal{O}(i^2) + \mathcal{O}(e^2)).$$

In applications one often uses the orbital elements as (non-canonical) coordinates.

## 11.9 Wave interpretation of mechanics

In this section we intend to illustrate how through the Hamilton–Jacobi equation, we can associate a *wave front* to a Hamiltonian system. What follows is a seemingly abstract analysis of classical mechanics, which however comes surprisingly close to the fundamental concepts of quantum mechanics.

Consider an autonomous system with Hamiltonian $H(\mathbf{p}, \mathbf{q})$ having $l \geq 2$ degrees of freedom, and assume that the Hamilton principal function is known. This function can be written in the form

$$S(\mathbf{q}, \boldsymbol{\alpha}, t) = W(\mathbf{q}, \boldsymbol{\alpha}) - E(\boldsymbol{\alpha})t, \qquad (11.175)$$

up to an inessential additive constant. The constants $\alpha_1, \ldots, \alpha_l$ are determined by the initial conditions. For $t = 0$ we have

$$S = W_0 = W(\mathbf{q}(0), \boldsymbol{\alpha}). \qquad (11.176)$$

For every $t > 0$ the equation

$$S(\mathbf{q}, \boldsymbol{\alpha}, t) = W_0 \qquad (11.177)$$

defines a regular $(l-1)$-dimensional manifold $M(t)$ in the space of configurations $\mathbf{R}^l$. This manifold is identified with the level set

$$W(\mathbf{q}, \boldsymbol{\alpha}) = W_0 + E(\boldsymbol{\alpha})t. \qquad (11.178)$$

At time $t = 0$ equation (11.178) selects a family of initial states, determined by the pairs of vectors $(\mathbf{q}(0), \boldsymbol{\alpha})$, such that $W(\mathbf{q}(0), \boldsymbol{\alpha}) = W_0$. As $t$ varies, $M(t)$ moves within the family $\Sigma$ of manifolds $W(\mathbf{q}, \boldsymbol{\alpha}) = \text{constant}$, according to (11.178).

Hence the dynamics of the system becomes associated with the propagation of a 'front' $M(t)$. There exist interesting relations between the motion of the system and the motion of $M(t)$ in $\Sigma$.

**Proposition 11.4** *If $T = \frac{1}{2}\sum_{i,j=1}^{l} a_{ij}\dot{q}_i\dot{q}_j$ is the kinetic energy of the system and if $\mathbf{R}^l$ is endowed with the metric*

$$\mathrm{d}s_T^2 = \sum_{i,j=1}^{l} a_{ij}\,\mathrm{d}q_i\,\mathrm{d}q_j, \tag{11.179}$$

*then the trajectories of the system in the configuration space are orthogonal to each manifold in the family $\Sigma$.*

*Proof*
For a fixed time $t_0$ and a point $\mathbf{q}_0 \in M(t_0)$, for every vector $\delta\mathbf{q} \in T_{\mathbf{q}_0} M(t_0)$ we have by (11.178),

$$\nabla_{\mathbf{q}} W(\mathbf{q}_0, \boldsymbol{\alpha}) \cdot \delta\mathbf{q} = 0 \tag{11.180}$$

to first order in $|\delta\mathbf{q}|$. We know that $(\mathbf{q}_0, \boldsymbol{\alpha})$ determines uniquely a vector $\mathbf{p}_0$ through the relation $\mathbf{p}_0 = \nabla_{\mathbf{q}} W(\mathbf{q}_0, \boldsymbol{\alpha})$ and that in addition, by definition, $p_{0i} = \sum_{j=1}^{l} a_{ij}(\mathbf{q}_0)\dot{q}_j$. Equation (11.180) can be interpreted as $(\dot{\mathbf{q}}, \delta\mathbf{q})_T = 0$, where we denote by $(\mathbf{x}, \mathbf{y})_T = \sum_{i,j=1}^{l} a_{ij}x_iy_j$ the scalar product induced by the metric (11.179). ∎

We can now deduce information on the velocity of the points of $M(t)$. Consider the family of the trajectories of the system issuing from the points of $M(0)$ and the family $\Sigma$ of the manifolds $M(t)$ (Fig. 11.13). For every fixed trajectory $\gamma$ we define the vector $\mathbf{q}_\gamma(t) = \gamma \cap M(t)$.

**Proposition 11.5** *According to the metric (11.179) we have*

$$|\dot{\mathbf{q}}_\gamma|_T = \frac{|E|}{\sqrt{2(E - V(\mathbf{q}_\gamma))}}. \tag{11.181}$$

*Proof*
From the identity

$$W(\mathbf{q}_\gamma(t), \boldsymbol{\alpha}) = W_0 + Et \tag{11.182}$$

we deduce

$$\nabla_{\mathbf{q}} W(\mathbf{q}_\gamma(t), \boldsymbol{\alpha}) \cdot \dot{\mathbf{q}}_\gamma = E, \tag{11.183}$$

which can be interpreted as $(\dot{\mathbf{q}}, \dot{\mathbf{q}}_\gamma)_T = E$, where $\dot{\mathbf{q}}$ is defined through the vector $\mathbf{p} = \nabla_{\mathbf{q}} W(\mathbf{q}_\gamma(t), \boldsymbol{\alpha})$. By construction $\dot{\mathbf{q}}_\gamma$ and $\dot{\mathbf{q}}$ are proportional at every instant, and therefore $(\dot{\mathbf{q}}, \dot{\mathbf{q}}_\gamma)_T = |\dot{\mathbf{q}}|_T |\dot{\mathbf{q}}_\gamma|_T = |E|$.

Since $|\dot{\mathbf{q}}|_T^2 = 2T = 2(E - V)$, we can deduce equation (11.181). ∎

In the case of a single free point particle, the space of configurations coincides with the physical space, the manifold $M(t)$ is a surface, and the metric (11.179) can be identified with the Euclidean metric.

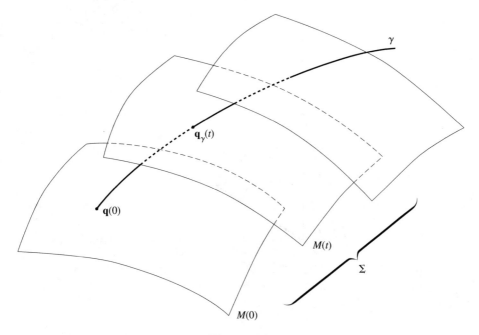

Fig. 11.13

*Example* 11.17
For a free point particle we have $\dot{\mathbf{q}} = \mathbf{c}_0$, a constant, $\mathbf{p} = m\mathbf{c}_0 = \boldsymbol{\alpha}$, and hence $W = \boldsymbol{\alpha}\cdot\mathbf{q}$. In this case, $W$ is the generating function of the identity transformation. The surfaces $W = $ constant are the planes orthogonal to $\boldsymbol{\alpha}$ (Fig. 11.14) and the 'front' $M(t)$ behaves as the phase of a plane wave:

$$S = \boldsymbol{\alpha} \cdot \mathbf{q} - Et. \qquad (11.184)$$

■

This simple example leads us to reinterpret the propagation of the front $M(t)$ in the context of a field theory analogous to the theory describing the propagation of light in a non-uniform optical medium.

We start from the observation that a plane wave with velocity $\mathbf{c}_0$ can be represented in the form

$$\phi = \phi_0 e^{i(\mathbf{k}\cdot\mathbf{q}-\omega t)}, \qquad (11.185)$$

with $\phi_0$ a constant, $\mathbf{k}$ a vector parallel to $\mathbf{c}_0$ and

$$\omega = kc_0. \qquad (11.186)$$

The absolute value $k$ is the *wave number* which defines the *wavelength* $\lambda = 2\pi/k$. Equation (11.185), with $\omega$ given by (11.186), is a solution of the wave

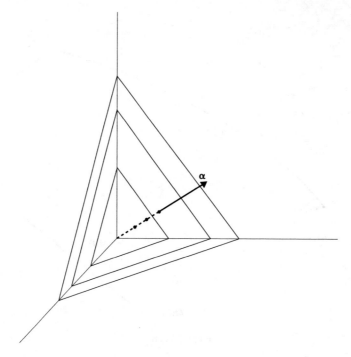

**Fig. 11.14**

equation

$$\Delta \phi - \frac{1}{c_0^2} \frac{\partial^2 \phi}{\partial t^2} = 0, \tag{11.187}$$

describing the propagation of the electromagnetic field in a medium with *refractive index*

$$n_0 = \frac{c}{c_0}, \tag{11.188}$$

where $c$ is the speed of light in the void. The *wave phase* can also be written in the form

$$k_0(n_0 \mathbf{e} \cdot \mathbf{q} - ct) = 2\pi \left( \frac{n_0}{\lambda_0} \mathbf{e} \cdot \mathbf{q} - \nu t \right), \tag{11.189}$$

where $\mathbf{e} = \mathbf{k}/n$, $k_0 = kc_0/c$, $\lambda_0 = 2\pi/k_0$, $\nu = \omega/2\pi$. Up to here, the analogy with Example 11.17 is evident.

We now consider the modifications which must be introduced in order to describe the motion of a more general mechanical system by means of this optical model. For simplicity we shall deal only with the case of a single free point particle, subject to a field with potential energy $V(\mathbf{q})$. We know then

## 11.9    Analytic mechanics: Hamilton–Jacobi theory and integrability

that at every point of the associated moving surface $M(t)$ we can define the propagation velocity

$$u = \frac{E}{\sqrt{2m(E-V)}}, \tag{11.190}$$

and we try to reproduce this behaviour in the case of a wave of the form

$$\phi = \phi_0 e^{A(\mathbf{q}) + ik_0(L(\mathbf{q}) - ct)}, \tag{11.191}$$

by imposing the validity of an equation of the kind (11.187). Note that in contrast to the case of a plane wave, we now have a variable amplitude $\phi_0 e^{A(\mathbf{q})}$ and that the function $L(\mathbf{q})$, called *eikonal*, replaces the linear function $n_0 \mathbf{e} \cdot \mathbf{q}$. If we insert the function (11.191) into the modified equation

$$\Delta \phi - \frac{1}{u^2(\mathbf{q})} \frac{\partial^2 \phi}{\partial t^2} = 0, \tag{11.192}$$

with $u(\mathbf{q})$ given by (11.190), separating real and imaginary parts we find

$$|\nabla L|^2 = n^2(\mathbf{q}) + \left(\frac{\lambda_0}{2\pi}\right)^2 [\Delta A + |\nabla A|^2], \tag{11.193}$$

$$\Delta L + 2 \nabla L \cdot \nabla A = 0, \tag{11.194}$$

where $n(\mathbf{q}) = c/u(\mathbf{q})$.

If $\lambda_0$ tends to zero in (11.193) (*geometrical optic limit*) we find for $L(\mathbf{q})$ the *equation of geometrical optics*:

$$|\nabla L|^2 = n^2(\mathbf{q}), \tag{11.195}$$

which is structurally a Hamilton–Jacobi equation. The analogy with the equation for the function $W$ can be made closer by noting that the front with constant phase:

$$2\pi \left(\frac{1}{\lambda_0} L(\mathbf{q}) - \nu t\right) = \text{constant} \tag{11.196}$$

varies in the family $\Sigma$ of surfaces $L(\mathbf{q}) = \text{constant}$.

We can impose proportionality between the phase (11.196) and the function $S = W - Et$. This yields:

(a) a proportionality relation between $|E|$ and $\nu$:

$$|E| = h\nu \tag{11.197}$$

($h$ is the *Planck constant*),

(b) the proportionality $W = h/\lambda_0 L$, so that equation (11.193) coincides with the Hamilton–Jacobi equation for $W$ if

$$n = \frac{\lambda_0}{h}\sqrt{2m(E-V)}, \tag{11.198}$$

in agreement with the definition

$$n = c/u = \frac{c\sqrt{2m(E-V)}}{E} = \frac{c}{h\nu}\sqrt{2m(E-V)}, \quad c/\nu = \lambda_0.$$

The hypothesis that allowed us to develop this analogy is that we assumed the validity of the *geometrical optics approximation*, corresponding to neglecting the terms in $\lambda_0^2$ in equation (11.193).

We can also introduce the wave number of the equivalent field:

$$k = \frac{2\pi}{u} = \frac{1}{\hbar}\sqrt{2m(E-V)}, \quad \hbar = \frac{h}{2\pi}. \tag{11.199}$$

We can easily check that if we rewrite equation (11.191) in the form

$$\phi = \psi(\mathbf{q})e^{-i\omega t} \tag{11.200}$$

$\psi$ must satisfy the equation

$$\Delta \psi + k^2 \psi = 0, \tag{11.201}$$

and hence

$$\Delta \psi + \frac{2}{\hbar^2} m(E-V)\psi = 0, \tag{11.202}$$

called the *Schrödinger equation of wave mechanics*.

We note that this can also be derived from the Schrödinger equation for quantum mechanics:

$$\frac{1}{2m}\Delta\phi - \frac{1}{\hbar^2}V\phi = -\frac{i}{\hbar}\frac{\partial\phi}{\partial t}, \tag{11.203}$$

by substituting into it the expression (11.200).

If we now return to the parallel idea of considering a wave with phase $S/\hbar$, and hence $\phi = \phi_0 e^{iS/\hbar}$, equation (11.203) yields the equation

$$\frac{1}{2m}|\nabla S|^2 + V + \frac{\partial S}{\partial t} = \frac{i}{2m}\hbar \Delta S, \tag{11.204}$$

which reduces to the Hamilton–Jacobi equation if we consider the limit $h \to 0$ (*classical limit*). It is interesting to note that $h \to 0$ is equivalent to $\nu \to \infty$ (by (11.197)), and hence to $\lambda_0 \to 0$: the classical limit of the Schrödinger equation is equivalent to the limit of geometrical optics in the context of wave theory.

## 11.10 Problems

**1.** Solve the Hamilton–Jacobi equation for the motion in space of a point particle of mass $m$ subject to weight.

**2.** Separate variables in the Hamilton–Jacobi equation for the motion of a point particle of mass $m$ subject to the action of weight and constrained to move on a surface of rotation around the $z$-axis.

**3.** A point particle of unit mass moves without any external forces on a surface whose first fundamental form is $(ds)^2 = (U(u) + V(v))((du)^2 + (dv)^2)$, where $(u, v) \in \mathbf{R}^2$ (*Liouville surface*).
(a) Write down the Hamiltonian of the system.
(b) Write down the Hamilton–Jacobi equation and separate variables.
(c) Solve the Hamilton–Jacobi equation in the case that $U(u) = u^2$, $V(v) = v$, where $u > 0$, $v > 0$.
(Answer: (a) $H = (p_u^2 + p_v^2)/[2(U(u) + V(v))]$; (b) setting $S = W_1(u) + W_2(v) - Et$, we have $(W_1'(u))^2 - 2EU(u) = \alpha = (W_1'(v))^2 - 2EV(v)$; (c) $W_1(u) = 1/2[u\sqrt{\alpha + 2Eu^2} + (\alpha/\sqrt{2E})\mathrm{arcsinh}(\sqrt{2E/\alpha}u)]$, and $W_2(v) = 1/3E(\alpha + 2Ev)^{3/2}$.)

**4.** Use the Hamilton–Jacobi method to solve Problems 23, 24 and 28 of Section 1.13.

**5.** Write down the Hamiltonian, solve the Hamilton–Jacobi equation and find the action variables for the systems described in Problems 11 and 15 of Section 3.7 and Problems 8, 9 and 12 of Section 4.12.

**6.** Consider a point particle of unit mass freely moving on the surface of a tri-axial ellipsoid:

$$\frac{x^2}{a^2} + \frac{y^2}{b^2} + \frac{z^2}{c^2} = 1,$$

where $a < b < c$. Prove that in the variables $(u, v) \in [b, c] \times [a, b]$ defined by

$$x = \sqrt{a\frac{(u-a)(v-a)}{(b-a)(c-a)}},$$

$$y = \sqrt{b\frac{(u-b)(v-b)}{(c-b)(a-b)}},$$

$$z = \sqrt{c\frac{(u-c)(v-c)}{(a-c)(b-c)}},$$

the Hamiltonian is given by

$$H(u, v, p_u, p_v) = \frac{1}{2}\left[\frac{p_u^2}{(u-v)A(u)} + \frac{p_v^2}{(v-u)A(v)}\right],$$

where

$$A(\lambda) = \frac{1}{4} \frac{\lambda}{(a-\lambda)(b-\lambda)(c-\lambda)}.$$

Write down the Hamilton–Jacobi equation and separate variables.

**7.** Two point particles of equal mass move along a line interacting through a force field with potential energy $V(x_1 - x_2)$, where $x_1$ and $x_2$ are the coordinates of the two points.
(a) Write down the Hamiltonian.
(b) Write down the Hamilton–Jacobi equation and separate variables (hint: introduce as new coordinates $x = (x_1 + x_2)/\sqrt{2}$ and $y = (x_1 - x_2)/\sqrt{2}$).
(c) Setting $V(x_1 - x_2) = V_0 e^{(x_1 - x_2)/d}$, where $V_0$ and $d$ are two prescribed positive constants, compute the solution of the Hamilton–Jacobi equation.

**8.** Write down and solve the Hamilton–Jacobi equation for a point particle of mass $m$ moving in space under the action of a central field $V(r) = k/r^2 + hr^2$, where $k$ and $h$ are two positive constants. Find the action variables and express the energy as a function of them. Compute the frequencies of the motions. Find the resonance conditions. Do periodic orbits exist?

**9.** Consider the Hamiltonian (see Problem 9 of Section 3.7)

$$H(p,q) = \frac{p^2}{2m} + V_0 \left(\frac{q}{d}\right)^{2n},$$

where $m$, $V_0$ and $d$ are prescribed positive constants and $n$ is an integer greater than or equal to 2. Let $E$ be the fixed value of energy. Prove that if $J$ indicates the action variable, then

$$E = \left(\frac{\pi n J}{d B_n}\right)^{2n/(n+1)} \left(\frac{1}{2m}\right)^{n/(n+1)} V_0^{1/(n+1)},$$

where $B_n = \int_0^1 \sqrt{1-u}\, u^{1/2n - 1}\, du$. Prove that the period $T$ of the motion is equal to

$$T = d \sqrt{\frac{2m}{E}} \left(\frac{E}{V_0}\right)^{1/2n} \frac{n+1}{n^2} B_n.$$

**10.** Given a system of Hamiltonian (see Problem 14 of Section 3.7)

$$H(p,q) = \frac{p^2}{2m} + V_0 (e^{-2q/d} - 2e^{-q/d}),$$

where $V_0, d$ are prescribed positive constants, compute the action variable $J$ and check that the energy expressed in terms of the action variable is

$$E = -V_0 \left(1 - \frac{J}{d\sqrt{2mV_0}}\right)^2.$$

Compute the period of the motion.

11. Given a system of Hamiltonian (see Problem 13 of Section 3.7)

$$H(p,q) = \frac{p^2}{2m} - \frac{V_0}{\cosh^2(q/d)},$$

where $V_0, d$ are prescribed positive constants, compute the action variable corresponding to librations, and verify that the energy expressed in terms of the action variable is

$$E = -V_0 \left(1 - \frac{J}{d\sqrt{2mV_0}}\right).$$

Compute the period of the motions.

12. Given a system with Hamiltonian

$$H(p,q) = \frac{p^2}{2m} + V_0|q|,$$

where $V_0$ is a prescribed positive constant, find the transformation to action-angle variables and determine the frequency of the motion.

13. Given a system with Hamiltonian

$$H(p,q) = \frac{p^2}{2m} + V(q),$$

where $V$ is periodic with period $2\pi$ and

$$V(q) = \begin{cases} -V_0 q, & \text{if } -\pi \leq q \leq 0, \\ +V_0 q, & \text{if } 0 \leq q \leq \pi, \end{cases}$$

with $V_0$ a prescribed positive constant (see Problem 12 of Section 3.7), find the transformation to action-angle variables and determine the frequency of the motion for librations and for rotations.

14. Consider the Hamiltonian system (see Problem 10 of Section 3.7)

$$H(p,q) = \frac{p^2}{2m} + V_0 \tan^2(q/d),$$

where $V_0, d$ are two fixed positive constants. Compute the action variable $J$ and prove that the energy $E$ expressed in terms of $J$ is given by

$$E = \frac{1}{2m}\frac{J}{d^2}(J + 2d\sqrt{2mV_0}).$$

Verify that the period of the motion is

$$T = \frac{2\pi m d^2}{J + 2d\sqrt{2mV_0}},$$

and compute the angle variable.

**15.** A point particle of mass $m$ moves with velocity $v$ along a segment and is reflected elastically at the endpoints of the segment. Prove that the action $J$ for the system is $J = pl/\pi$, where $p = mv$ and $l$ is the length of the segment. Prove that the energy is $E = 1/2m\,(\pi J/l)^2$.

**16.** Use separability to express the function $W$ for the Kepler problem as the sum of three functions related to the three cycles in Fig. 11.11. Then compute the angle variables, showing that one of them coincides with the mean anomaly.

## 11.11 Additional remarks and bibliographical notes

In this chapter we discussed the Hamilton–Jacobi method for solving the equations of motion and *a* notion of integrability[5] for the Hamilton's equations, corresponding to the existence of *bounded and quasi-periodic* orbits, and hence of a foliation of the phase space in invariant tori. The presentation of the Hamilton–Jacobi equations and of the method of separation of variables for its solution follows the text of Landau and Lifschitz (1976).

The book by Levi-Civita and Amaldi (1927) contains a detailed discussion of the theory presented in the first two sections, and of the examples we chose to illustrate it.

General references for this chapter are the book of Whittaker (1936), which contains many of the classical results obtained by the Italian school[6] at the beginning of the twentieth century on the classification of the cases when the Hamilton–Jacobi equation is solvable by separation of variables, and the treatise of Agostinelli and Pignedoli (1989).

For a more complete treatment of the notion of a completely canonically integrable system, and of action-angle variables, we recommend the article of Nekhoroshev (1972), and the review of Gallavotti (1984). The latter is also an excellent basis for a more advanced study of the topics discussed in the next chapter. We partially followed it in our proof of the theorem of Liouville given in Section 11.4.

The lecture notes of Giorgilli (1990) have been particularly useful in preparing Section 11.5, in particular for the proof of Proposition 11.1.

A more detailed discussion of the action-angle variables for the problem of Kepler, and their relation with the orbital elements, can be found in the first

---

[5] There indeed exist various other notions of integrability, just as there exist methods different from the Hamilton–Jacobi one for solving the equations of motion. The reader interested in a more detailed treatment of these themes can start by reading the review article of Kozlov (1983).

[6] Morera, Levi-Civita, Burgatti, Dall'Acqua and many more.

chapter of Beletski (1986). In the second chapter of this very pleasant and informal introduction to celestial mechanics one can find a study of the problem of two centres of force (see Example 11.6), with the completion of its solution via separation of variables in the Hamilton–Jacobi equation, and the explicit computation of the trajectory of a polar satellite using elliptic functions. In the following third chapter, parabolic coordinates are used for the study of the pressure of solar radiation upon an artificial satellite orbiting around the Earth, and there is a complete classification of the trajectories corresponding to the planar case.

We stress that for the sake of brevity we did not illustrate the mechanics of rigid systems as an example of completely canonically integrable systems. The book of Gallavotti (1980) contains this illustration fully.

In addition, we did not develop very extensively the theme, touched upon in Section 11.9, of the relations existing between the Hamilton–Jacobi equation, geometrical optics and the semiclassical approximation in quantum mechanics. The text of Arnol'd (1978a), already referred to, also contains these topics (sections 46 and 47 and appendices 11 and 12). The complex relation between the notion of action variable and the so-called 'old quantum' of Bohr and Sommerfeld, described in Graffi (1993), is also very interesting. An excellent reading on this is the treatise of Born (1927), which can also be read as an introduction to the study of the canonical theory of perturbations, studied in the next chapter.

## 11.12 Additional solved problems

*Problem 1*
A point particle of unit mass moves along a plane curve of equation $y = P(x)$, where $P$ is a polynomial of degree $n \geq 1$, and it is subject to a conservative force field with potential energy $V(x) = ax^2 + bx + c$, $a \neq 0$. Use the method of Hamilton–Jacobi to determine the travelling time $t = \hat{t}(x, x(0), \dot{x}(0))$ of a solution.

*Solution*
The Lagrangian of the particle is

$$L(x, \dot{x}) = \frac{1}{2}\dot{x}^2(1 + P'(x)^2) - V(x),$$

from which it follows that $p = \partial L/\partial \dot{x} = \dot{x}(1 + P'(x)^2)$ and

$$H(p, x) = \frac{p^2}{2(1 + P'(x)^2)} + V(x). \tag{11.205}$$

The Hamilton–Jacobi equation for Hamilton's characteristic function $W = W(x, E)$ is

$$\frac{1}{2}\left(\frac{\partial W}{\partial x}\right)^2 (1 + P'(x)^2)^{-1} + V(x) = E, \tag{11.206}$$

and hence
$$W(x, E) = \pm \int_{x(0)}^{x} (1 + P'(\xi)^2)\sqrt{2(E - V(\xi))}\,d\xi, \qquad (11.207)$$

where the sign is determined by the sign of $\dot{x}(0)$. Differentiating (11.207) with respect to $E$ and recalling Remark 11.5, we find

$$t = \frac{\partial W}{\partial E} = \pm \int_{x(0)}^{x} \frac{1 + P'(\xi)^2}{\sqrt{2[E - V(\xi)]}}\,d\xi. \qquad (11.208)$$

The integrand is of the form $Q(\xi)/\sqrt{\alpha\xi^2 + \beta\xi + \gamma}$, where $Q(\xi) = \sum_{j=0}^{k} q_j \xi^j$ (in our case $q_0 = 1$, $k = 2n - 2$, $\alpha = -2a$, $\beta = -2b$, $\gamma = 2(E - c)$). Its primitives are of the form

$$\left(\sum_{j=0}^{k-1} \tilde{q}_j \xi^j\right) \sqrt{\alpha\xi^2 + \beta\xi + \gamma} + \tilde{q}_k \int \frac{d\xi}{\sqrt{\alpha\xi^2 + \beta\xi + \gamma}}, \qquad (11.209)$$

where the $k$ coefficients $\tilde{q}_0, \tilde{q}_1, \ldots, \tilde{q}_k$ can be determined starting from the known coefficients $q_0, \ldots, q_k$, $\alpha, \beta, \gamma$ and multiplying the relation

$$\frac{\sum_{j=0}^{k} q_j \xi^j}{\sqrt{\alpha\xi^2 + \beta\xi + \gamma}} = \frac{d}{d\xi}\left[\left(\sum_{j=0}^{k-1} \tilde{q}_j \xi^j\right)\sqrt{\alpha\xi^2 + \beta\xi + \gamma}\right] + \frac{\tilde{q}_k}{\sqrt{\alpha\xi^2 + \beta\xi + \gamma}} \qquad (11.210)$$

by the square root and identifying the polynomials obtained on the two sides of the identity. Recall that, setting $\Delta = \beta^2 - 4\alpha\gamma$, we have, up to additive constants,

$$\int \frac{d\xi}{\sqrt{\alpha\xi^2 + \beta\xi + \gamma}} = \frac{1}{\sqrt{\alpha}} \ln|2\sqrt{\alpha\xi^2 + \beta\xi + \gamma} + 2\alpha\xi + \beta| \quad \text{(if } \alpha > 0\text{)}$$

$$= \begin{cases} \frac{1}{\sqrt{\alpha}} \text{arcsinh}\frac{2\alpha\xi + \beta}{\sqrt{-\Delta}}, & \text{if } \Delta < 0, \alpha > 0, \\ \frac{1}{\sqrt{\alpha}} \ln|2\alpha\xi + \beta|, & \text{if } \Delta = 0, \alpha > 0, \\ -\frac{1}{\sqrt{-\alpha}} \arcsin\frac{2\alpha\xi + \beta}{\sqrt{\Delta}}, & \text{if } \Delta > 0, \alpha < 0. \end{cases} \qquad (11.211)$$

From (11.209)–(11.211) it follows that it is possible to compute explicitly the integral (11.208) and hence obtain $t = \hat{t}(x, x(0), \dot{x}(0))$.

*Problem 2*
Consider the following canonical transformation of $\mathbf{R}^+ \times \mathbf{R}^3$:

$$Q_1 = -e^t(1 + p_1)\sqrt{q_1},$$

$$Q_2 = \arcsin\frac{q_2}{\sqrt{p_2^2 + q_2^2}},$$

$$P_1 = e^{-t}(1 - p_1)\sqrt{q_1},$$

$$P_2 = \frac{p_2^2 + q_2^2}{2}.$$

$$(11.212)$$

How does the Hamiltonian $H(p_1,p_2,q_1,q_2) = q_1(p_1^2 - 1) + (p_2^2 + q_2^2)/2$ transform? Use the result obtained to completely solve Hamilton's equations associated with $H$. For the associated system determine whether the hypotheses of the theorem of Liouville and of the theorem of Arnol'd are satisfied.

*Solution*
The transformation (11.212) admits the generating function

$$F_2(q_1,q_2,P_1,P_2,t) = -2(e^t P_1)\sqrt{q_1} + \frac{1}{2}(e^t P_1)^2 + q_1$$

$$+ \frac{q_2}{2}\sqrt{2P_2 - q_2^2} + P_2 \arcsin\frac{q_2}{\sqrt{2P_2}}.$$

The Hamiltonian $H$ is transformed into $K(P_1,P_2,Q_1,Q_2,t) = H + \partial F_2/\partial t$:

$$K = P_1 Q_1 + P_2 + \frac{\partial F_2}{\partial t} = P_1 Q_1 + P_2 - 2e^t P_1 \sqrt{q_1} + P_1^2 e^{2t} = 2P_1 Q_1 + P_2.$$
(11.213)

Hamilton's equations associated with $K$ can be solved immediately:

$$P_1(t) = P_1(0)e^{-2t}, \quad Q_1(t) = Q_1(0)e^{2t},$$
$$P_2(t) = P_2(0), \quad Q_2(t) = Q_2(0) + t.$$
(11.214)

Then from the inverse of (11.122):

$$q_1 = \frac{e^{2t}}{4}(P_1 - e^{-2t}Q_1)^2, \quad p_1 = -\frac{P_1 + e^{-2t}Q_1}{P_1 - e^{-2t}Q_1},$$

$$q_2 = \sqrt{2P_2}\sin Q_2, \quad p_2 = \sqrt{2P_2}\cos Q_2,$$
(11.215)

we deduce the solution of Hamilton's equations associated with $H$ by substituting (11.214) into (11.215) and using the relations

$$P_1(0) = (1 - p_1(0))\sqrt{q_1(0)}, \quad P_2(0) = (p_2^2(0) + q_2^2(0))/2,$$

$$Q_1(0) = -(1 + p_1(0))\sqrt{q_1(0)}, \quad Q_2(0) = \arcsin\frac{q_2(0)}{\sqrt{p_2^2(0) + q_2^2(0)}}.$$

The two functions

$$f_1(p_1,p_2,q_1,q_2) = -q_1(1 - p_1^2),$$

$$f_2(p_1,p_2,q_1,q_2) = \frac{p_2^2 + q_2^2}{2}$$

are first integrals for $H = f_1 + f_2$, in involution and independent except in three planes $\pi_1, \pi_2, \pi_3$ of equations $q_1 = 0$, $p_1 = 1$; $q_1 = 0$, $p_1 = -1$; $q_2 = p_2 = 0$, respectively. The hypotheses of the theorem of Liouville are therefore satisfied on $\mathbf{R}^4 \setminus (\pi_1 \cup \pi_2 \cup \pi_3)$ while those of the theorem of Arnol'd are not, because the level sets of $f_1$ are not compact.

*Problem 3*
Consider the system described by the Hamiltonian

$$H : \mathbf{R}^2 \times (\mathbf{R} \setminus \{-1, 1\}) \times \mathbf{R} \to \mathbf{R}, \quad H(p_1, p_2, q_1, q_2) = \frac{p_1^2}{q_1^2 - 1}(1 + p_2^2 + q_2^2).$$
(11.216)

(i) Write down Hamilton's equations and determine all constant solutions.
(ii) Linearise the equations around $p_1 = p_2 = q_1 = q_2 = 0$ and solve the linearised equations.
(iii) Determine two first integrals of the motion, independent and in involution, and express the Hamiltonian through these first integrals.
(iv) Write down the Hamilton–Jacobi equation associated with $H$ and solve it by separation of variables.
(v) Construct when it is possible the action-angle variables, write the Hamiltonian as a function of the actions only and determine the frequencies. For what initial conditions are the motions periodic?

*Solution*
Hamilton's equations are

$$\dot{p}_1 = \frac{2q_1 p_1^2}{(q_1^2 - 1)^2}(1 + p_2^2 + q_2^2), \quad \dot{q}_1 = \frac{2p_1}{q_1^2 - 1}(1 + p_2^2 + q_2^2),$$
$$\dot{p}_2 = -\frac{2q_2 p_1^2}{q_1^2 - 1}, \quad \dot{q}_2 = \frac{2p_2 p_1^2}{q_1^2 - 1},$$
(11.217)

from which we see immediately that the constant solutions are given by $p_1 = 0$, for any $(p_2, q_1, q_2)$. The equations linearised around the origin are

$$\dot{p}_1 = 0, \quad \dot{p}_2 = 0, \quad \dot{q}_1 = -2p_1, \quad \dot{q}_2 = 0.$$

Denoting by $(P_1, P_2, Q_1, Q_2)$ the initial conditions, the corresponding solution is clearly

$$p_1(t) \equiv P_1, \quad p_2(t) \equiv P_2, \quad q_1(t) = Q_1 - 2P_1 t, \quad q_2(t) \equiv Q_2.$$

Since the Hamiltonian is of the form $H(p_1, p_2, q_1, q_2) = f_1(p_1, q_1) f_2(p_2, q_2)$, with $f_1(p_1, q_1) = p_1^2/(q_1^2 - 1)$, $f_2(p_2, q_2) = 1 + p_2^2 + q_2^2$, we deduce immediately that $f_1$ and $f_2$ are two first integrals that are independent and in involution. Indeed, the involution is guaranteed by the fact that $f_1$ and $f_2$ are functions of distinct pairs of canonically conjugate variables. Moreover $\{f_1, H\} = \{f_1, f_1 f_2\} = \{f_1, f_2\} f_1 +$

$\{f_1, f_1\} f_2 = 0$ and similarly $\{f_2, H\} = 0$. The independence holds on the whole of $\mathbf{R}^2 \times (\mathbf{R} \setminus \{-1, 1\}) \times \mathbf{R}$ except where

$$\nabla f_1 = \left( \frac{2p_1}{q_1^2 - 1}, 0, -\frac{2q_1 p_1^2}{(q_1^2 - 1)^2}, 0 \right) = (0, 0, 0, 0),$$

or

$$\nabla f_2 = (0, 2p_2, 0, 2q_2) = (0, 0, 0, 0),$$

and hence $\{p_1 = 0\}$ and $\{p_2 = q_2 = 0\}$, respectively. Since $H$ is independent of time, if $E$ denotes the energy, the Hamilton–Jacobi equation for the characteristic function $W(q_1, q_2, \alpha_1, \alpha_2)$ can be written as

$$\frac{1}{q_1^2 - 1} \left( \frac{\partial W}{\partial q_1} \right)^2 \left[ 1 + \left( \frac{\partial W}{\partial q_2} \right)^2 + q_2^2 \right] = E$$

and can be solved by separation of variables: $W(\mathbf{q}, \boldsymbol{\alpha}) = W_1(q_1, \alpha_1) + W_2(q_2, \alpha_2)$, with

$$\left( \frac{\partial W_1}{\partial q_1} \right)^2 = \alpha_1 (q_1^2 - 1),$$

$$\left( \frac{\partial W_2}{\partial q_2} \right)^2 = \alpha_2 - 1 - q_2^2, \quad \alpha_2 \geq 1$$

$$E = \alpha_1 \alpha_2,$$

from which it follows that, up to additive constants,

$$W_1(q_1, \alpha_1) = \pm \begin{cases} \frac{q_1}{2} \sqrt{\alpha_1(q_1^2 - 1)} - \frac{1}{2\sqrt{\alpha_1}} \ln \left( \sqrt{\alpha_1} q_1 + \sqrt{\alpha_1(q_1^2 - 1)} \right), & \text{if } \alpha_1 > 0, \\ 0, & \text{if } \alpha_1 = 0, \\ \frac{q_1}{2} \sqrt{-\alpha_1(q_1^2 - 1)} + \frac{1}{2\sqrt{-\alpha_1}} \arcsin q_1, & \text{if } \alpha_1 < 0 \end{cases}$$

$$W_2(q_2, \alpha_2) = \pm \left[ \frac{q_2}{2} \sqrt{\alpha_2 - 1 - q_2^2} + \frac{1}{2}(\alpha_2 - 1) \arcsin \frac{q_2}{\sqrt{\alpha_2 - 1}} \right].$$

Following the steps outlined in Section 11.1 it is then possible to compute explicitly the Hamiltonian flow associated with (11.216). For a completely canonical transformation to action-angle variables to exist, the level set $M_{\alpha_1, \alpha_2} = \{(p_1, p_2, q_1, q_2) \mid f_1(p_1, q_1) = \alpha_1, \ f_2(p_2, q_2) = \alpha_2\}$ must be compact and connected. This is the case only if $(\alpha_1, \alpha_2) \in (-\infty, 0) \times (1, +\infty)$. In this case, the two equations

$$p_1^2 - \alpha_1 q_1^2 = -\alpha_1, \quad p_2^2 + q_2^2 = \alpha_2 - 1$$

determine two ellipses in the planes $(p_1, q_1)$, $(p_2, q_2)$ and therefore $M_{\alpha_1,\alpha_2}$ is evidently diffeomorphic to a two-dimensional torus $\mathbf{T}^2$. From equation (11.61) it follows that

$$J_1 = \frac{1}{2\pi}\pi\sqrt{-\alpha_1} = \frac{\sqrt{-\alpha_1}}{2}, \quad J_2 = \frac{1}{2\pi}\pi(\alpha_2 - 1) = \frac{\alpha_2 - 1}{2},$$

from which $\alpha_1 = -4J_1^2$, $\alpha_2 = 2J_2+1$, $E = -4J_1^2(2J_2+1) = K(J_1, J_2)$. Substituting the latter into $W_1$ and $W_2$ we find the generating function of the transformation to action-angle variables.

The frequencies are $\omega_1 = -8J_1(2J_2+1)$, $\omega_2 = -8J_1^2$; therefore the motions are periodic only if $\omega_1/\omega_2$ is rational, i.e. for initial conditions such that $(2J_2(0) + 1)/J_1(0) \in \mathbf{Q}$ (or if $J_1(0) = 0$).

# 12 ANALYTICAL MECHANICS: CANONICAL PERTURBATION THEORY

## 12.1 Introduction to canonical perturbation theory

The so-called 'perturbation methods' for studying differential equations are traditionally of great importance for their applications to celestial mechanics (indeed, this field of study initially motivated their development). In spite of the efforts of generations of celebrated mathematicians (Lagrange, Laplace, Weierstrass and above all Poincaré, who can be considered the father of the modern theory) until recently the majority of techniques used did not have a rigorous mathematical justification. Proving the convergence (or divergence) of the perturbation series is not just an abstract goal, of secondary interest for physicists. On the contrary, it stems from the need to understand in depth the domains of applicability to physical problems, and the limitations, of perturbation techniques.

On the other hand, the number of problems that can be treated with these techniques justifies a more detailed analysis, even if most of the modern developments go beyond the scope of the present text.

The central question we want to consider is the study of a system whose Hamiltonian is a 'small' perturbation of the Hamiltonian of a completely canonically integrable system. According to Poincaré (1893) this is to be considered the 'fundamental problem of classical mechanics'.

In what follows we assume systematically that the Hamiltonian functions we consider are sufficiently regular.

DEFINITION 12.1 *A Hamiltonian system is called* quasi-integrable *if its Hamilton function is of the form*

$$h(\mathbf{p},\mathbf{q},\varepsilon) = h_0(\mathbf{p},\mathbf{q}) + \varepsilon f(\mathbf{p},\mathbf{q}), \tag{12.1}$$

*where* $(\mathbf{p},\mathbf{q}) \in \mathbf{R}^{2l}$, $\varepsilon$ *is a* small parameter *($0 \leq |\varepsilon| \ll 1$) and $h_0$ is the Hamiltonian of a completely canonically integrable system.* ∎

*Remark* 12.1
The previous definition is not completely satisfactory, unless we make more precise in what sense $\varepsilon$ is a small parameter. As $(\mathbf{p},\mathbf{q})$ varies in a compact subset $K$ of $\mathbf{R}^{2l}$ (in which we want to study the motions and where $h_0$ does not have singularities) there exists a constant $M_K > 0$ such that

$$\max_{(\mathbf{p},\mathbf{q})\in K} |f(\mathbf{p},\mathbf{q})| \leq M_K \max_{(\mathbf{p},\mathbf{q})\in K} |h_0(\mathbf{p},\mathbf{q})|. \tag{12.2}$$

The requirement that the perturbation be small can be expressed through the condition

$$|\varepsilon M_K| \ll 1. \qquad (12.3)$$

(Naturally, we suppose that neither $h_0$ nor $f$ contain terms that are independent of $\mathbf{q},\mathbf{p}$.) ∎

Since the system with Hamiltonian $h_0$ is completely canonically integrable, there exists a completely canonical variable transformation from $(\mathbf{p},\mathbf{q})$ to action-angle variables $(\mathbf{J},\boldsymbol{\chi})$ with respect to which the Hamiltonian $h_0$ is expressed through a function $H_0$ that depends only on the action variables. After this transformation, the Hamiltonian (12.1) in the new coordinates is

$$H(\mathbf{J},\boldsymbol{\chi},\varepsilon) = H_0(\mathbf{J}) + \varepsilon F(\mathbf{J},\boldsymbol{\chi}), \qquad (12.4)$$

where $H_0$ and $F$ are the functions $h_0$ and $f$ expressed in the new variables, respectively. The action variables $\mathbf{J}$ are defined on some open subset of $\mathbf{R}^l$, while the angle variables are by their nature variables on a torus $\mathbf{T}^l$ of dimension $l$. In other words, the function $F$ is periodic separately in each of its variables $\chi_1, \ldots, \chi_l$ with fixed periodicity, for example equal to $2\pi$.

We also assume that the functions $H$, $H_0$ and $F$ are regular (of class $\mathcal{C}^\infty$ or analytic when needed) in each argument.

*Example* 12.1
The so-called *Fermi–Pasta–Ulam model* (Fermi et al. 1954) consists of a chain of $l+2$ equal particles linked by non-linear springs. The two particles at the extremes of the chain are fixed. If $(\mathbf{p},\mathbf{q}) = (p_1, \ldots, p_l, q_1, \ldots, q_l)$ are the kinetic moments and the coordinates of the $l$ moving particles, setting $q_0 = q_{l+1} = 0$, the Hamiltonian of the model is

$$H(\mathbf{p},\mathbf{q},\varepsilon) = h_0(\mathbf{p},\mathbf{q}) + \varepsilon f(\mathbf{q})$$

where

$$h_0(\mathbf{p},\mathbf{q}) = \sum_{i=1}^{l} \frac{p_i^2}{2m} + \frac{k}{2} \sum_{i=0}^{l} (q_{i+1} - q_i)^2,$$

corresponds to the integrable part, and

$$f(\mathbf{q}) = \frac{\lambda}{r} \sum_{i=0}^{l} (q_{i+1} - q_i)^r,$$

where $r = 3$ or $r = 4$, defines the non-linearity of the springs of the chain, and $\lambda$ is a constant. Fermi, Pasta and Ulam introduced this model to study numerically the ergodic hypothesis and the equipartition theorem of statistical mechanics (cf. Section 15.2). ∎

*Example* 12.2
Consider a system of $l$ identical particles, each performing a uniform rotation on a fixed circle. Let $q_i$ be the angular coordinate identifying the $i$th particle and

we examine a weak perturbation with potential energy

$$V(q_i, q_{i+1}) = -\varepsilon V_0 \cos(q_{i+1} - q_i), \qquad V_0 > 0,$$

where $0 \leq |\varepsilon| \ll 1$ measures the intensity of the coupling. If we assume that the last particle is coupled to the penultimate and the first, setting $q_0 = q_l$, the Hamiltonian of the system is

$$H(\mathbf{p}, \mathbf{q}, \varepsilon) = h_0(\mathbf{p}) + \varepsilon f(\mathbf{q}),$$

with

$$h_0(\mathbf{p}) = \sum_{i=1}^{l} \frac{p_i^2}{2}, \qquad f(\mathbf{q}) = -V_0 \sum_{i=1}^{l} \cos(q_i - q_{i-1}).$$

The action variables for the unperturbed system coincide with the kinetic moments $\mathbf{p}$ and the angle variables correspond to the angles $\mathbf{q} \in \mathbf{T}^l$. This system can also be considered as a classic model for the study of the so-called 'spin systems' of statistical mechanics. ∎

In the following two examples the perturbation is periodic in time.

*Example* 12.3
The Hamiltonian of the *restricted three-body problem* is quasi-integrable. Consider the effect of the attraction of Jupiter on the revolution around the Sun of a minor planet of the Solar System (the Earth, or Mercury) or of an asteroid. As a first approximation we can consider the orbit of Jupiter as circular and fixed (hence neglecting the effect of the attraction of the minor body on Jupiter). The resulting problem has three degrees of freedom, but the Hamiltonian depends on time $t$ periodically (the period is equal to the period of revolution of Jupiter around the Sun). The motion of the minor body is then described with respect to a reference system with origin at the Sun, and axes moving with respect to an inertial reference system (Fig. 12.1). Note that this system is however not inertial, because the Sun has an acceleration (due to the attraction of Jupiter):

$$\mathbf{a}_S = K\varepsilon \frac{\mathbf{r}_G}{|\mathbf{r}_G|^3},$$

where $K$ is a constant proportional to the mass of the Sun, $\varepsilon$ is the ratio $M_G/M_S$ between the mass of Jupiter and of the Sun (so that $\varepsilon \approx 10^{-3}$), and $\mathbf{r}_G$ is the position vector of Jupiter. The Hamiltonian of the system is then equal to

$$H(\mathbf{p}, \mathbf{q}, t, \varepsilon) = \frac{|\mathbf{p}|^2}{2} - \frac{K}{|\mathbf{q}|} - \frac{K\varepsilon}{|\mathbf{q} - \mathbf{r}_G(t)|} + K\varepsilon \frac{\mathbf{r}_G(t) \cdot \mathbf{q}}{|\mathbf{r}_G(t)|^3},$$

where the mass of the minor body is equal to 1, its position vector is denoted by $\mathbf{q} \in \mathbf{R}^3$ and $\mathbf{p}$ denotes the relative conjugate kinetic momentum. The last term in $H$ is the (generalised) potential energy of the inertia force responsible for the

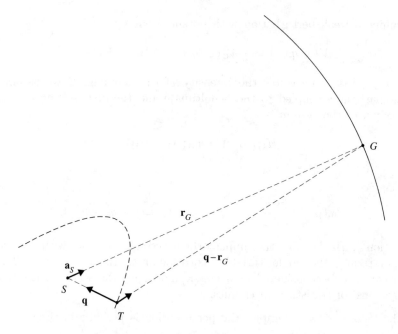

Fig. 12.1

acceleration $\mathbf{a}_S$ of the origin of the reference system considered. Evidently the system is quasi-integrable, with

$$h_0(\mathbf{p},\mathbf{q}) = \frac{|\mathbf{p}|^2}{2} - \frac{K}{|\mathbf{q}|},$$

$$f(\mathbf{p},\mathbf{q},t) = -\frac{K}{|\mathbf{q}-\mathbf{r}_G(t)|} + K\frac{\mathbf{r}_G(t)\cdot\mathbf{q}}{|\mathbf{r}_G(t)|^3}.$$ ∎

*Example 12.4: the spin–orbit problem*
Consider a satellite $S$ in orbit around a planet $P$. Suppose that the satellite is a rigid body with the form of a tri-axial homogeneous ellipsoid. The three axes $A_1 > A_2 > A_3$ of the ellipsoid coincide with the principal axes of inertia. Since the ellipsoid is homogeneous, the corresponding principal moments of inertia are $\mathcal{I}_1 < \mathcal{I}_2 < \mathcal{I}_3$, and hence the maximum momentum $\mathcal{I}_3$ is associated with the shortest axis of the ellipsoid. Suppose also that the orbit of the satellite is a fixed Keplerian ellipse with $P$ at one of the foci. We denote by $e$ the eccentricity of the orbit. We also assume that the axis of rotation of the satellite coincides with the $x_3$-axis and is directed orthogonally to the plane of the orbit. Since the orientation of the satellite is completely determined by the angle between the major axis of the ellipsoid and the direction of the pericentre of the orbit, the problem has only one degree of freedom. We also neglect dissipative forces which may be acting on the system and all perturbations due to other bodies (which

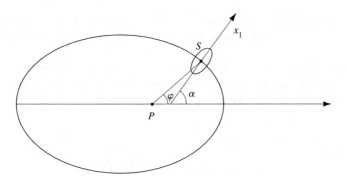

**Fig. 12.2**

may for example be responsible for changes in the orbital parameters, cf. Laskar and Robutel (1993)). The orientation of the satellite varies only under the effect of the torque of the gravitational attraction of $P$ on the ellipsoid $S$.

If $\alpha$ is the angle between the $x_1$-axis of the ellipsoid and the direction of the pericentre of the orbit, $\varphi$ is the polar angle, $a$ is the semi-major axis and $r$ is the instantaneous orbital radius (Fig. 11.2) the equation of the motion can be written as (cf. Goldreich and Peale 1966, Danby 1988, section 14.3)

$$\ddot{\alpha} + \frac{3}{2}\frac{\mathcal{I}_2 - \mathcal{I}_1}{\mathcal{I}_3}\left(\frac{a}{r(t)}\right)^3 \sin(2\alpha - 2\varphi(t)) = 0.$$

Note that if the ellipsoid is a surface of revolution, then $\mathcal{I}_1 = \mathcal{I}_2$ and the equation is trivially integrable. In addition, since $r$ and $\varphi$ are periodic functions of $t$ (with period equal to the period $T$ of revolution of $S$ around $P$), by choosing the unit of time appropriately we can assume that $r$ and $\varphi$ are $2\pi$-periodic functions. Finally, setting

$$x = 2\alpha, \qquad \varepsilon = 3\frac{\mathcal{I}_2 - \mathcal{I}_1}{\mathcal{I}_3},$$

and expanding $(a/r(t))^3 \sin(x - 2\varphi(t))$ in Fourier series we find

$$\ddot{x} + \varepsilon \sum_{\substack{m \in \mathbb{Z} \\ m \neq 0}} \hat{s}_m(e) \sin(x - mt) = 0.$$

This equation corresponds to a quasi-integrable Hamiltonian system (depending on time) with one degree of freedom:

$$H(p, x, t, \varepsilon) = \frac{p^2}{2} - \varepsilon \sum_{\substack{m \in \mathbb{Z} \\ m \neq 0}} \hat{s}_m(e) \cos(x - mt),$$

which will be the object of a more detailed study in the next section ($\varepsilon$ is a small parameter because, in the majority of cases arising in celestial mechanics, $\varepsilon \approx 10^{-3}$–$10^{-4}$).

The computations of the coefficients $\hat{s}_m(e)$ is somewhat laborious (see Cayley 1861). They can be expressed as a power series in $e$ and at the lowest order they are proportional to $e^{|m-2|}$. As an example, we have

$$\hat{s}_{-2}(e) = \frac{e^4}{24} + \frac{7e^6}{240} + \mathcal{O}(e^8),$$

$$\hat{s}_{-1}(e) = \frac{e^3}{48} + \frac{11e^5}{768} + \mathcal{O}(e^7),$$

$$\hat{s}_1(e) = -\frac{e}{2} + \frac{e^3}{16} - \frac{5}{384}e^5 + \mathcal{O}(e^7),$$

$$\hat{s}_2(e) = 1 - \frac{5e^2}{2} + \frac{13e^4}{16} - \frac{35e^6}{288} + \mathcal{O}(e^8),$$

$$\hat{s}_3(e) = \frac{7e}{2} - \frac{123e^3}{16} + \frac{489e^5}{128} + \mathcal{O}(e^7),$$

$$\hat{s}_4(e) = \frac{17e^2}{2} - \frac{115e^4}{6} + \frac{601e^6}{48} + \mathcal{O}(e^8),$$

$$\hat{s}_5(e) = \frac{845}{48}e^3 - \frac{32525}{768}e^5 + \mathcal{O}(e^7).$$

In the Earth–Moon system (cf. Celletti 1990) the orbital eccentricity is $e = 0.0549$, while $\varepsilon = 7 \times 10^{-4}$. If we neglect the terms which give a contribution to the Hamiltonian of less than $10^{-6}$ we find

$$H(p,x,t) = \frac{p^2}{2} - \varepsilon \left[ -\frac{e}{2}\cos(x-t) + \left(1 - \frac{5}{2}e^2\right)\cos(x-2t) + \frac{7e}{2}\cos(x-3t) \right.$$
$$\left. + \frac{17}{2}e^2\cos(x-4t) + \frac{845}{48}e^3\cos(x-5t) \right].$$

∎

If $\varepsilon = 0$ the system (12.4) is integrable and Hamilton's equations

$$\dot{J}_i = 0, \qquad \dot{\chi}_i = \frac{\partial H_0}{\partial J_i}(\mathbf{J}) \qquad (12.5)$$

are trivially integrable: the actions are first integrals of the motion for the system, i.e. $J_i(t) = J_i(0)$ for every $i = 1, \ldots, l$, while each angle has a period $2\pi/w_i$, where

$$w_i = w_i(\mathbf{J}(0)) = \frac{\partial H_0}{\partial J_i}(\mathbf{J}(0))$$

is the frequency of the angular motion, depending on the initial conditions for the action variables. All motions are therefore *bounded* and *quasi-periodic* and the system admits as many independent first integrals as the number of degrees of freedom. The phase space is foliated into invariant tori of dimension $l$ (cf. Remark 11.19) for the Hamiltonian flow and each torus is identified by the constant values of the actions $\mathbf{J}$.

When $\varepsilon \neq 0$ the motion equations change; in particular for the action variables we have

$$\dot{J}_i = -\varepsilon \frac{\partial F}{\partial \chi_i}(\mathbf{J}, \boldsymbol{\chi}), \qquad i = 1, \ldots, l, \tag{12.6}$$

and they are no longer constants of the motion. From the regularity of $F$ there follows the possibility of estimating the time difference of the action from its initial value:

$$|J_i(t) - J_i(0)| \leq \left\| \frac{\partial F}{\partial \chi_i} \right\| \varepsilon t, \tag{12.7}$$

where $\|\cdot\|$ indicates the maximum norm on a compact subset $K$ of $\mathbf{R}^l$ to which $\mathbf{J}(0)$ belongs, and on $\mathbf{T}^l$ for the angles. The estimate (12.7), while significant for times $t$ of order $\mathcal{O}(1)$, may yield little information for longer times. This is shown by the following trivial example.

*Example* 12.5
Let $l = 1$ and $H(J, \chi, \varepsilon) = J + \varepsilon \cos \chi$. In this case, Hamilton's equations are

$$\dot{J} = \varepsilon \sin \chi, \qquad \dot{\chi} = 1,$$

and hence

$$J(t) = J(0) + \varepsilon[\cos \chi(0) - \cos(\chi(0) + t)], \qquad \chi(t) = \chi(0) + t.$$

It follows that

$$|J(t) - J(0)| \leq 2\varepsilon$$

for all times $t$ and not only for times $t = \mathcal{O}(1)$ as predicted by (12.7). ∎

This drawback of (12.7) can be attributed to the fact that in deriving this inequality we did not take into account the sign variations in $\partial F / \partial \chi_i$. These variations can yield some compensations which extend the validity of the estimate. The perturbation $\partial F / \partial \chi_i$ is not generally constant (except when its arguments are constant), and it does not have a constant sign. Indeed, the function $\partial F / \partial \chi_i$ is periodic but has zero mean, and therefore it cannot have a constant sign unless it is identically zero.

The perturbation method for Hamiltonian systems of type (12.4) consists of solving the following problem.

*Problem*
*Find a completely canonical transformation which eliminates the dependence of the Hamiltonian on the angular variables, to first order in $\varepsilon$. Then iterate this*

*procedure until the dependence on $\chi$ to all orders in $\varepsilon$, or at least to a prescribed order, is eliminated.*

Hence we seek the generating function $W(\mathbf{J}', \chi, \varepsilon)$ of a canonical transformation from the action-angle variables $(\mathbf{J}, \chi)$ corresponding to the integrable system with Hamiltonian $H_0$ to new variables $(\mathbf{J}', \chi')$, with respect to which the Hamiltonian (12.4) has an expression $H'(\mathbf{J}', \chi', \varepsilon)$ that is independent of the angular variables, at least in the terms up to order $\mathcal{O}(\varepsilon^2)$:

$$H'(\mathbf{J}', \chi', \varepsilon) = H'_0(\mathbf{J}') + \varepsilon H'_1(\mathbf{J}') + \varepsilon^2 F'(\mathbf{J}', \chi', \varepsilon). \tag{12.8}$$

Here $F'$ is a *remainder* depending on $\varepsilon$, but that may fail to tend to 0 when $\varepsilon \to 0$ (however we assume it to be *bounded* together with its first derivatives).

When $\varepsilon = 0$ the starting Hamiltonian is independent of the angle variables. Hence the transformation sought is $\varepsilon$-near the identity and we can try to expand the generating function $W$ into a power series in $\varepsilon$ whose zero-order term is the generating function of the identity transformation. We therefore write

$$W(\mathbf{J}', \chi, \varepsilon) = \mathbf{J}' \cdot \chi + \varepsilon W^{(1)}(\mathbf{J}', \chi) + \mathcal{O}(\varepsilon^2), \tag{12.9}$$

with $W^{(1)}(\mathbf{J}', \chi)$ unknown. The transformation generated by (12.9) is

$$J_i = J'_i + \varepsilon \frac{\partial W^{(1)}}{\partial \chi_i}(\mathbf{J}', \chi) + \mathcal{O}(\varepsilon^2), \quad i = 1, \ldots, l,$$

$$\chi'_i = \chi_i + \varepsilon \frac{\partial W^{(1)}}{\partial J'_i}(\mathbf{J}', \chi) + \mathcal{O}(\varepsilon^2), \quad i = 1, \ldots, l. \tag{12.10}$$

Substituting the first of equations (12.10) into (12.4) and requiring that the transformed Hamiltonian has the form (12.8), we find the equation

$$H_0(\mathbf{J}' + \varepsilon \nabla_\chi W^{(1)}) + \varepsilon F(\mathbf{J}', \chi) + \mathcal{O}(\varepsilon^2) = H'_0(\mathbf{J}') + \varepsilon H'_1(\mathbf{J}') + \mathcal{O}(\varepsilon^2), \tag{12.11}$$

where the functions $H'_0$, $H'_1$ are to be determined. Expanding $H_0$ to first order and equating the corresponding powers of $\varepsilon$ we find for the term of zero order in $\varepsilon$:

$$H'_0(\mathbf{J}') = H_0(\mathbf{J}'). \tag{12.12}$$

This ensures—as was obvious from the previous considerations—that to zero order in $\varepsilon$ the new Hamiltonian coincides with the starting one (expressed in the new action variables).

At the first order in $\varepsilon$ we find the equation

$$\boldsymbol{\omega}(\mathbf{J}') \cdot \nabla_\chi W^{(1)}(\mathbf{J}', \chi) + F(\mathbf{J}', \chi) = H'_1(\mathbf{J}'), \tag{12.13}$$

for the unknowns $W^{(1)}(\mathbf{J}', \chi)$ and $H'_1(\mathbf{J}')$, where $\boldsymbol{\omega}(\mathbf{J}') = \nabla_{\mathbf{J}'} H'_0$ is the vector of frequencies of the new Hamiltonian. For fixed actions $\mathbf{J}'$, equation (12.13) is a linear partial differential equation of first order on the torus $\mathbf{T}^l$ whose solution will be studied in Sections 12.3 and 12.4.

We shall see that the iteration to higher order terms of the perturbation method always leads to solving equations of the type (12.13). For this reason, the latter is called the *fundamental equation of classical perturbation theory*.

If equation (12.13) admits a solution, i.e. if there exist two functions $H_1'(\mathbf{J}')$ and $W^{(1)}(\mathbf{J}', \boldsymbol{\chi})$ (the second $2\pi$-periodic with respect to $\boldsymbol{\chi}$) which satisfy (12.13), the equations of motion for the new action variables are

$$\dot{J}_i' = -\frac{\partial H'}{\partial \chi_i}(\mathbf{J}', \varepsilon) = \mathcal{O}(\varepsilon^2),$$

where $i = 1, \ldots, l$. Therefore, for all times $t$ in the interval $[0, 1/\varepsilon]$ we have

$$|\mathbf{J}'(t) - \mathbf{J}'(0)| = \mathcal{O}(\varepsilon).$$

The new action variables are approximately (up to $\mathcal{O}(\varepsilon)$ terms) constant over a time interval of length $1/\varepsilon$. One arrives at the same conclusion for the action variables $\mathbf{J}$, exploiting the fact that the transformation (12.10) is near the identity. Indeed,

$$\mathbf{J}(t) - \mathbf{J}(0) = (\mathbf{J}(t) - \mathbf{J}'(t)) + (\mathbf{J}'(t) - \mathbf{J}'(0)) + (\mathbf{J}'(0) - \mathbf{J}(0)),$$

and given that the first and last terms are also $\mathcal{O}(\varepsilon)$ (uniformly with respect to time $t$) we have

$$|\mathbf{J}(t) - \mathbf{J}(0)| = \mathcal{O}(\varepsilon),$$

for every $t \in [0, 1/\varepsilon]$.

*Remark* 12.2

Equation (12.11) is simply the Hamilton–Jacobi equation approximated up to terms of order $\varepsilon^2$ for the Hamiltonian (12.4). Indeed, the Hamilton–Jacobi equation for the Hamiltonian (12.4) can be written as

$$H(\nabla_\chi W, \boldsymbol{\chi}, \varepsilon) = H_0(\nabla_\chi W) + \varepsilon F(\nabla_\chi W, \boldsymbol{\chi}) = H'(\mathbf{J}', \varepsilon), \qquad (12.14)$$

and equation (12.11) is then obtained by substituting the expansion (12.9) into equation (12.14) and neglecting all terms of order $\mathcal{O}(\varepsilon^2)$. ∎

Before starting a more detailed study of equation (12.13) when $l \geq 2$, we consider the case $l = 1$. If the system has only one degree of freedom, then as we saw (cf. Section 11.3), it is completely canonically integrable, as long as the motions are periodic (hence outside the separatrix curves in phase space). Therefore the following theorem should not come as a surprise to the reader.

**Theorem 12.1** *If $l = 1$ and $\omega(J') \neq 0$, equation (12.13) has solution*

$$H_1'(J') = \frac{1}{2\pi} \int_0^{2\pi} F(J', \chi) \, d\chi, \tag{12.15}$$

$$W^{(1)}(J', \chi) = \frac{1}{\omega(J')} \int_0^{\chi} [H_1'(J') - F(J', x)] \, dx. \tag{12.16}$$

*This solution is unique, if we require that the mean value of $W^{(1)}$ on $\mathbf{S}^1$ be zero, and hence that*

$$\frac{1}{2\pi} \int_0^{2\pi} W^{(1)}(J', \chi) \, d\chi = 0. \tag{12.17}$$

*Proof*
Expression (12.15) is the only possible choice for $H_1'(J')$, because the $\chi$-average of $\omega(J') \partial W^{(1)} / \partial \chi$ vanishes due to the periodicity of $W^{(1)}$. Therefore $H_1'(J')$ must be the mean of $F(J', \chi)$ with respect to $\chi$. After this, it is immediate to check that (12.16) actually satisfies (12.13). The uniqueness of the solution follows in a similar way. Let $\widetilde{W}^{(1)}, \widetilde{H}_1'$ be a second solution of (12.13). Then

$$\omega(J') \frac{\partial}{\partial \chi} (\widetilde{W}^{(1)} - W^{(1)})(J', \chi) = \widetilde{H}_1'(J') - H_1'(J'). \tag{12.18}$$

However

$$\int_0^{2\pi} \frac{\partial}{\partial \chi} (\widetilde{W}^{(1)} - W^{(1)})(J', \chi) \, d\chi$$
$$= \widetilde{W}^{(1)}(J', 2\pi) - W^{(1)}(J', 2\pi) - (\widetilde{W}^{(1)}(J', 0) - W^{(1)}(J', 0)) = 0,$$

by the periodicity of $\widetilde{W}^{(1)}$ and $W^{(1)}$. Hence integrating both sides of equation (12.18) we find that $\widetilde{H}_1'(J') = H_1'(J')$. Therefore

$$\omega(J') \frac{\partial}{\partial \chi} (\widetilde{W}^{(1)} - W^{(1)})(J', \chi) = 0,$$

from which it follows that $\widetilde{W}^{(1)}(J', \chi) = W^{(1)}(J', \chi) + g(J')$. If we impose that $\widetilde{W}^{(1)}$ has zero average, then necessarily $g \equiv 0$. ■

*Example* 12.6
Consider the following quasi-integrable system with one degree of freedom (dimensionless variables):

$$H(J, \chi, \varepsilon) = J^2 + \varepsilon J^3 \sin^2 \chi.$$

## 12.1  Analytical mechanics: canonical perturbation theory

The generating function
$$W(J',\chi,\varepsilon) = J'\chi + \frac{\varepsilon J'^2}{8}\sin 2\chi$$
transforms the Hamiltonian $H$ to
$$H'(J',\varepsilon) = J'^2 + \frac{\varepsilon}{2}J'^3 + \mathcal{O}(\varepsilon^2).$$

The frequency of the motions corresponding to $H'$ is $\omega'(J',\varepsilon) = 2J' + \frac{3}{2}\varepsilon J'^2$. ∎

In the case of one degree of freedom, it is possible to formally solve the Hamilton–Jacobi equation (12.14) to all orders in $\varepsilon$ (neglecting the question of the convergence of the series), assuming that the frequency of the motions is not zero. Canonical perturbation theory thus yields (at least formally) the complete integrability of these systems.

**THEOREM 12.2**  *If $l = 1$ and $\omega(J') \neq 0$, the Hamilton–Jacobi equation (12.14) admits a formal solution:*

$$H'(J',\varepsilon) = \sum_{n=0}^{\infty} \varepsilon^n H'_n(J'), \tag{12.19}$$

$$W(J',\chi,\varepsilon) = J'\chi + \sum_{n=1}^{\infty} \varepsilon^n W^{(n)}(J',\chi). \tag{12.20}$$

*The solution is unique if we require that $W^{(n)}$ has zero average with respect to $\chi$ for every $n \geq 1$.*

*Proof*
Substituting (12.19) and (12.20) into equation (12.14) we have

$$H_0\left(J' + \sum_{n=1}^{\infty} \varepsilon^n \frac{\partial W^{(n)}}{\partial \chi}\right) + \varepsilon F\left(J' + \sum_{n=1}^{\infty} \varepsilon^n \frac{\partial W^{(n)}}{\partial \chi}, \chi\right) = \sum_{k=0}^{\infty} \varepsilon^k H'_k(J'),$$

and expanding $H_0$ in Taylor series around $J'$ we find

$$H_0\left(J' + \sum_{n=1}^{\infty} \varepsilon^n \frac{\partial W^{(n)}}{\partial \chi}\right)$$
$$= H_0(J') + \omega(J')\sum_{n=1}^{\infty} \varepsilon^n \frac{\partial W^{(n)}}{\partial \chi}$$
$$+ \frac{1}{2}\frac{d^2 H_0}{dJ^2}\sum_{n=2}^{\infty} \varepsilon^n \sum_{n_1+n_2=n} \frac{\partial W^{(n_1)}}{\partial \chi}\frac{\partial W^{(n_2)}}{\partial \chi} + \cdots$$
$$+ \frac{1}{k!}\frac{d^k H_0}{dJ^k}\sum_{n=k}^{\infty} \varepsilon^n \sum_{n_1+n_2+\cdots+n_k=n} \frac{\partial W^{(n_1)}}{\partial \chi}\frac{\partial W^{(n_2)}}{\partial \chi}\cdots\frac{\partial W^{(n_k)}}{\partial \chi} + \cdots, \tag{12.21}$$

where $\omega = dH_0/dJ$. Similarly, expanding $F$ we find

$$F\left(J' + \sum_{n=1}^{\infty} \varepsilon^n \frac{\partial W^{(n)}}{\partial \chi}, \chi\right)$$

$$= F(J', \chi) + \frac{\partial F}{\partial J} \sum_{n=1}^{\infty} \varepsilon^n \frac{\partial W^{(n)}}{\partial \chi}$$

$$+ \frac{1}{2} \frac{\partial^2 F}{\partial J^2} \sum_{n=2}^{\infty} \varepsilon^n \sum_{n_1+n_2=n} \frac{\partial W^{(n_1)}}{\partial \chi} \frac{\partial W^{(n_2)}}{\partial \chi} + \cdots$$ (12.22)

$$+ \frac{1}{k!} \frac{\partial^k F}{\partial J^k} \sum_{n=k}^{\infty} \varepsilon^n \sum_{n_1+n_2+\cdots+n_k=n} \frac{\partial W^{(n_1)}}{\partial \chi} \frac{\partial W^{(n_2)}}{\partial \chi} \cdots \frac{\partial W^{(n_k)}}{\partial \chi} + \cdots.$$

Therefore to order $k \geq 2$ we must solve the equation

$$\omega(J') \frac{\partial W^{(k)}}{\partial \chi}(J', \chi) + F^{(k)}(J', \chi) = H'_k(J'),$$ (12.23)

where the functions $F^{(k)}$ are given by

$$F^{(k)} = \sum_{i=2}^{k} \frac{1}{i!} \frac{d^i H_0}{dJ^i} \sum_{n_1+\cdots+n_i=k} \frac{\partial W^{(n_1)}}{\partial \chi} \cdots \frac{\partial W^{(n_i)}}{\partial \chi}$$

$$+ \sum_{i=1}^{k-1} \frac{1}{i!} \frac{\partial^i F}{\partial J^i} \sum_{n_1+\cdots+n_i=k-1} \frac{\partial W^{(n_1)}}{\partial \chi} \cdots \frac{\partial W^{(n_i)}}{\partial \chi}$$ (12.24)

and thus contain $W^{(n)}$ only with $n < k$. Equation (12.23) is exactly of the type (12.13). It then follows from Theorem 12.1 that

$$H'_k(J') = \frac{1}{2\pi} \int_0^{2\pi} F^{(k)}(J', \chi) \, d\chi,$$ (12.25)

$$W^{(k)}(J', \chi) = \frac{1}{\omega(J')} \int_0^{\chi} [H'_k(J') - F^{(k)}(J', x)] \, dx.$$ (12.26)

The uniqueness of the solution follows from the uniqueness of the Taylor series expansion in $\varepsilon$ of $H'$ and $W$, and from Theorem 12.1. ∎

*Remark* 12.3
It is not difficult to prove the uniform convergence of the series expansions (12.19) and (12.20) under the assumption that $H_0$ and $F$ are analytic functions of all their arguments, but the proof goes beyond the scope of this introduction as it requires some knowledge of the theory of analytic functions of one or more complex variables. ∎

*Example* 12.7
Consider the following quasi-integrable system with one degree of freedom:

$$H(J, \chi, \varepsilon) = \cos J + \varepsilon \frac{J^2}{2} \sin^2 \chi,$$

and solve the Hamilton–Jacobi equation up to terms of order $\mathcal{O}(\varepsilon^3)$.

Substituting the expansions (12.19) and (12.20) into
$$\cos\frac{\partial W}{\partial \chi} + \frac{\varepsilon}{2}\left(\frac{\partial W}{\partial \chi}\right)^2 \sin^2\chi = H'(J',\varepsilon)$$
and neglecting terms of order higher than 3 in $\varepsilon$ we find
$$\cos\left(J' + \varepsilon\frac{\partial W^{(1)}}{\partial \chi} + \varepsilon^2\frac{\partial W^{(2)}}{\partial \chi}\right) + \frac{\varepsilon}{2}\left(J' + \varepsilon\frac{\partial W^{(1)}}{\partial \chi}\right)^2 \sin^2\chi = H'_0 + \varepsilon H'_1 + \varepsilon^2 H'_2.$$

Then it follows that
$$\cos J' - \sin J' \left[\varepsilon\frac{\partial W^{(1)}}{\partial \chi} + \varepsilon^2\frac{\partial W^{(2)}}{\partial \chi}\right] - \frac{1}{2}\varepsilon^2 \cos J'\left(\frac{\partial W^{(1)}}{\partial \chi}\right)^2$$
$$+ \frac{\varepsilon}{2}\left(J'^2 + 2\varepsilon J'\frac{\partial W^{(1)}}{\partial \chi}\right)\sin^2\chi = H'_0 + \varepsilon H'_1 + \varepsilon^2 H'_2.$$

Equating the terms corresponding to the same power of $\varepsilon$ in the expansion and solving the resulting equations by using equations (12.25), (12.26), we find
$$H'_0(J') = \cos J',$$
$$H'_1(J') = \frac{1}{4}J'^2,$$
$$H'_2(J') = -\frac{1}{64}\frac{(\cos J')J'^4}{\sin^2 J'}$$

and
$$W^{(1)}(J',\chi) = -\frac{J'^2}{8\sin J'}\sin(2\chi).$$

We leave the computation of $W^{(2)}$ as an exercise for the reader. ∎

## 12.2 Time periodic perturbations of one-dimensional uniform motions

Consider a point particle of unit mass moving along a line under the action of a weak force field depending periodically on the position $x$ of the particle and on time $t$. For simplicity of exposition, we systematically use dimensionless variables. The Lagrangian and the Hamiltonian of the system can then be written as follows:
$$L(x,\dot{x},t,\varepsilon) = \frac{1}{2}\dot{x}^2 - \varepsilon V(x,t), \qquad H(p,x,t,\varepsilon) = \frac{p^2}{2} + \varepsilon V(x,t), \qquad (12.27)$$

where $\varepsilon$ is a small parameter, $0 \leq \varepsilon \ll 1$, and $V$ is the (generalised) potential of the applied force (which we assume to be non-constant and of class $\mathcal{C}^\infty$).

The periodicity assumption implies that $V(x+2\pi,t) = V(x,t+2\pi) = V(x,t)$, for appropriately normalised units of space and time. The periodicity in space of the force field acting on the particle yields that the $x$-coordinates of the particle that differ by an integer multiple of $2\pi$ are identifiable; the phase space of the system is therefore a cylinder $(x,\dot{x}) \in \mathbf{S}^1 \times \mathbf{R}$.

An example of a field that satisfies these assumptions is the case discussed in Example 2.4.

If $\varepsilon = 0$, the velocity $\dot{x} = \omega$ of the particle is constant. Since $\varepsilon$ is small, it is reasonable to expect that there exists an invertible coordinate transformation, depending on time, which transforms the equation of motion

$$\ddot{x} + \varepsilon V_x(x,t) = 0, \tag{12.28}$$

where $V_x = \partial V/\partial x$, into

$$\ddot{\xi} = 0, \tag{12.29}$$

and for which the velocity $\dot{\xi} = \omega$ is a conserved quantity. Such a transformation certainly exists—if $\omega \neq 0$ and $\varepsilon$ is sufficiently small—when $V$ does not depend on time (it is easy to conclude this in view of the complete integrability of the associated Hamiltonian system; $\xi$ is the angle variable corresponding to rotations).

We therefore seek a transformation of the type

$$x = \xi + u(\xi,t;\varepsilon), \tag{12.30}$$

with $u \in C^\infty$, where $(\xi,t) \in \mathbf{T}^2$ (hence the function $u$ is $2\pi$-periodic with respect to $\xi$ and $t$), transforming equation (12.28) into (12.29). Since for $\varepsilon = 0$ the two equations coincide, we must impose that $u(\xi,t;0) = 0$, and hence $u(\xi,t,\varepsilon) = \mathcal{O}(\varepsilon)$. If we also require that

$$\left|\frac{\partial u}{\partial \xi}\right| < 1, \tag{12.31}$$

the local invertibility of the transformation is guaranteed.

Condition (12.31) is satisfied as long as $\varepsilon$ is chosen sufficiently small.

Differentiating equation (12.30) once with respect to $t$, and recalling that $\dot{\xi} = \omega$ is constant, we find

$$\dot{x} = \omega + (D_\omega u)(\xi,t;\varepsilon), \tag{12.32}$$

where $D_\omega$ denotes the linear partial differential operator of first order:

$$(D_\omega u)(\xi,t;\varepsilon) = \omega \frac{\partial u}{\partial \xi}(\xi,t;\varepsilon) + \frac{\partial u}{\partial t}(\xi,t;\varepsilon). \tag{12.33}$$

Differentiating (12.32) again with respect to time, we find

$$\ddot{x} = D_\omega^2 u = \omega^2 \frac{\partial^2 u}{\partial \xi^2} + 2\omega \frac{\partial^2 u}{\partial \xi \partial t} + \frac{\partial^2 u}{\partial t^2}. \tag{12.34}$$

## 12.2 Analytical mechanics: canonical perturbation theory

The equation of motion (12.28) then becomes

$$(D_\omega^2 u)(\xi, t; \varepsilon) + \varepsilon V_x(\xi + u(\xi, t; \varepsilon), t) = 0. \tag{12.35}$$

Equation (12.35) is a partial differential equation of second order, which is *non-linear* because of the term $V_x(\xi + u, t)$. We try to solve equation (12.35) starting from the remark that this equation is identically satisfied by $u \equiv 0$ if $\varepsilon = 0$. Then we expand the function $u$ in power series of $\varepsilon$:

$$u(\xi, t; \varepsilon) = \sum_{n=1}^{\infty} \varepsilon^n u^{(n)}(\xi, t). \tag{12.36}$$

Each function $u^{(n)}(\xi, t)$ is periodic with period $2\pi$, both in the space coordinate $\xi$ and in time $t$. If we substitute the expansion (12.36) into the term $V_x(\xi + u, t)$ of equation (12.35) we find

$$
\begin{aligned}
V_x(\xi + u, t) &= V_x(\xi, t) + \sum_{m=1}^{\infty} \frac{1}{m!} \left(\frac{\partial^m}{\partial x^m} V_x\right)(\xi, t)(u(\xi, t; \varepsilon))^m \\
&= V_x(\xi, t) + \sum_{m=1}^{\infty} \frac{1}{m!} \left(\frac{\partial^m}{\partial x^m} V_x\right)(\xi, t) \left(\sum_{n=1}^{\infty} \varepsilon^n u^{(n)}(\xi, t)\right)^m \\
&= V_x(\xi, t) + \sum_{m=1}^{\infty} \frac{1}{m!} \left(\frac{\partial^m}{\partial x^m} V_x\right)(\xi, t) \\
&\quad \times \sum_{n_1, \ldots, n_m} \varepsilon^{n_1 + \cdots + n_m} u^{(n_1)}(\xi, t) \ldots u^{(n_m)}(\xi, t).
\end{aligned}
\tag{12.37}
$$

Hence, reordering the second sum in increasing powers $n_1 + \cdots + n_m = n$ of $\varepsilon$ we have

$$
\begin{aligned}
&V_x(\xi + u, t) \\
&= V_x(\xi, t) + \sum_{n=1}^{\infty} \varepsilon^n \sum_{m=1}^{n} \frac{1}{m!} \left(\frac{\partial^m}{\partial x^m} V_x\right)(\xi, t) \sum_{n_1 + \cdots + n_m = n} u^{(n_1)}(\xi, t) \ldots u^{(n_m)}(\xi, t).
\end{aligned}
\tag{12.38}
$$

Substituting (12.38) and (12.36) into equation (12.35) we find

$$
\begin{aligned}
&\sum_{n=1}^{\infty} \varepsilon^n (D_\omega^2 u^{(n)})(\xi, t) + \varepsilon V_x(\xi, t) \\
&+ \sum_{n=2}^{\infty} \varepsilon^n \sum_{m=1}^{n-1} \frac{1}{m!} \left(\frac{\partial^m}{\partial x^m} V_x\right)(\xi, t) \sum_{n_1 + \cdots + n_m = n-1} u^{(n_1)}(\xi, t) \ldots u^{(n_m)}(\xi, t) = 0.
\end{aligned}
\tag{12.39}
$$

Therefore, equation (12.35) has a solution if the following infinite system of *linear* equations admits a solution:

$$D_\omega^2 u^{(1)}(\xi,t) + V_x(\xi,t) = 0,$$
$$D_\omega^2 u^{(2)}(\xi,t) + V_{xx}(\xi,t) u^{(1)}(\xi,t) = 0,$$
$$D_\omega^2 u^{(3)}(\xi,t) + V_{xx}(\xi,t) u^{(2)}(\xi,t) + \frac{1}{2} V_{xxx}(\xi,t)(u^{(1)}(\xi,t))^2 = 0, \quad (12.40)$$
$$\ldots,$$
$$D_\omega^2 u^{(n)}(\xi,t) + P_n(\xi,t) = 0,$$
$$\ldots,$$

where $V_{xx} = \partial^2 V/\partial x^2$, $V_{xxx} = \partial^3 V/\partial x^3$ and $P_n$ is a function depending only on $V$, on its derivatives (up to order $n$) and on the functions $u^{(1)}, u^{(2)}, \ldots, u^{(n-1)}$. We must hence study the linear equation

$$(D_\omega^2 u)(\xi,t) = v(\xi,t), \quad (12.41)$$

where $v$ is a known function, periodic in $x$ and in $t$. Equation (12.41) is a partial differential equation with constant coefficients on the torus $\mathbf{T}^2$, analogous to equation (12.39).

Evidently, the existence of a solution of equation (12.41) is a consequence of the invertibility of the linear operator $D_\omega$ which we will now discuss.

## 12.3 The equation $D_\omega u = v$. Conclusion of the previous analysis

We remark first of all that the eigenvalues $\lambda$ and the eigenvectors $u_\lambda$ of the linear operator

$$D_\omega = \omega \frac{\partial}{\partial \xi} + \frac{\partial}{\partial t} \quad (12.42)$$

are given by

$$D_\omega u_\lambda = \lambda u_\lambda, \quad (12.43)$$

and take the form

$$u_\lambda = e^{i(m\xi + nt)}, \qquad \lambda = i(m\omega + n), \quad (12.44)$$

where $(m,n) \in \mathbf{Z}^2$. Hence if $\omega$ is an irrational number, the eigenvalue $\lambda = 0$ corresponds to the choice $m = n = 0$, and hence has multiplicity one. If, on the other hand, $\omega$ is rational, $\omega = j/k$, the eigenvalue $\lambda = 0$ corresponds to the choice of $(m,n) \in \mathbf{Z}^2$ such that $mj + nk = 0$ and therefore has infinite multiplicity. We shall now see that if $\omega \in \mathbf{R}\backslash\mathbf{Q}$, it is sufficient to impose the *zero mean condition*

$$\hat{v}_{0,0} = \frac{1}{(2\pi)^2} \int_0^{2\pi} d\xi \int_0^{2\pi} v(\xi,t) dt = 0 \qquad (12.45)$$

to ensure the existence of a formal solution $u$ of the equation

$$D_\omega u = v. \qquad (12.46)$$

This means that it is possible to determine the coefficients of the Fourier series expansion of $u$, neglecting the question of its convergence.

If, on the other hand, $\omega$ is rational, then it is necessary to impose infinitely many conditions, corresponding to the vanishing of all the coefficients $\hat{v}_{m,n}$ of the Fourier expansion of $v$ with $m,n$ such that $mj + nk = 0$. We will not study further the latter case, which would lead to the study of the so-called *resonant normal forms*.

PROPOSITION 12.1  *If $\omega \in \mathbf{R}\backslash\mathbf{Q}$ and if $v(\xi,t)$ has zero mean (hence if it satisfies the condition (12.45)) there exists a formal solution $u$ of the equation (12.46). The solution is unique if we impose that the mean of $u$ be zero: $\hat{u}_{0,0} = 0$.*

*Proof*
Expanding both $v$ and $u$ in Fourier series, and substituting these series into (12.46), we find

$$\sum_{(m,n)\in\mathbf{Z}^2} \hat{u}_{m,n} i(m\omega + n) e^{i(m\xi+nt)} = \sum_{(m,n)\in\mathbf{Z}^2} \hat{v}_{m,n} e^{i(m\xi+nt)}.$$

Hence, by the uniqueness of Fourier expansions, it follows that for every $(m,n) \in \mathbf{Z}^2$ we have

$$\hat{u}_{m,n} i(m\omega + n) = \hat{v}_{m,n}$$

(yielding $\hat{v}_{0,0} = 0$), and therefore

$$\hat{u}_{m,n} = \frac{\hat{v}_{m,n}}{i(m\omega + n)}$$

if $(m,n) \neq (0,0)$, while $\hat{u}_{0,0}$ is undetermined. ∎

It follows that if $\omega$ is irrational and $v$ has zero mean, there exists a unique formal solution

$$u(\xi,t) = \sum_{(m,n)\in\mathbf{Z}^2\backslash\{(0,0)\}} \frac{\hat{v}_{m,n}}{i(m\omega+n)} e^{i(m\xi+nt)} \qquad (12.47)$$

of $D_\omega u = v$, and similarly, a unique formal solution

$$w(\xi,t) = \sum_{(m,n)\in\mathbf{Z}^2\backslash\{(0,0)\}} \frac{\hat{v}_{m,n}}{-(m\omega+n)^2} e^{i(m\xi+nt)} \qquad (12.48)$$

of $D_\omega^2 w = v$. We now discuss the convergence of the series (12.47) and (12.48). We start by remarking that the most serious difficulty is the need to control the denominators $|m\omega + n|$ which can become arbitrarily small even if $\omega$ is irrational.

THEOREM 12.3 (Dirichlet) *Let $\omega$ be irrational; there exist infinitely many distinct pairs $(m,n) \in \mathbf{Z}^2 \setminus \{(0,0)\}$, $m > 0$, such that $|m\omega + n| < 1/|m|$.*

*Proof*
Let $M$ be a fixed integer, and consider the sequence $\{|m\omega + \overline{n}|, m = 0, \ldots, M\}$, where $\overline{n} \in \mathbf{Z}$ is prescribed for every $m$ in such a way that $|m\omega + \overline{n}| \leq 1$ (such a choice is always possible). The points of the sequence then belong to the interval [0,1] and are necessarily distinct, as $\omega$ is irrational. If we consider the decomposition of

$$[0,1] = \bigcup_{j=0}^{M-1} \left[\frac{j}{M}, \frac{j+1}{M}\right]$$

into $M$ intervals, it follows that at least two points in the sequence $|m\omega + \overline{n}|$ must belong to the same subinterval $[j/M, (j+1)/M]$ (indeed, there are $M+1$ points and only $M$ intervals). Denote these two points by $m'\omega + \overline{n}'$ and $m''\omega + \overline{n}''$, and note that it is not restrictive to assume that $0 \leq m'' - m' \leq M$. Therefore

$$|(m'' - m')\omega + \overline{n}'' - \overline{n}'| \leq \frac{1}{M} < \frac{1}{m'' - m'},$$

and we have found one pair satisfying the claim. The existence of infinitely many such pairs $(m,n)$ follows from a simple proof by contradiction. Suppose that $(m_1, n_1), (m_2, n_2), \ldots, (m_k, n_k)$ are all the solutions of $|m\omega + n| < 1/|m|$. Then if $M$ is an integer such that

$$|m_j \omega + n_j| > \frac{1}{M}, \quad j = 1, \ldots, k,$$

there exists a pair $(m,n) \in \mathbf{Z}^2 \setminus \{0,0\}$ such that

$$|m\omega + n| \leq \frac{1}{M} < \frac{1}{|m|},$$

which is a contradiction. ∎

From this theorem it also follows that

$$\inf_{(m,n)\in \mathbf{Z}^2\setminus\{(0,0)\}} |m\omega + n| = 0,$$

and therefore $1/|m\omega + n|$ is not bounded from above. This fact yields serious difficulties in the proof of the convergence of the series (12.47) and (12.48). This problem is called the *problem of small divisors* and was already known to Poincaré and the astronomers of the nineteenth century.

We shall now see that it is possible to make some hypotheses on $\omega$—verified by almost any $\omega$ with respect to the Lebesgue measure—under which it is possible to prove the convergence of the series (12.47) and (12.48).

**DEFINITION 12.2** *We say that an irrational number $\omega$ satisfies a diophantine condition (with constant $\gamma > 0$ and exponent $\mu \geq 1$), and we denote it by $\omega \in C_{\gamma,\mu}$, if for every $(m,n) \in \mathbf{Z}^2 \setminus \{(0,0)\}$ we have*

$$|m\omega + n| \geq \gamma(|m| + |n|)^{-\mu}. \tag{12.49}$$

∎

*Remark* 12.4
The need for the condition $\mu \geq 1$ in the previous definition is an immediate consequence of the theorem of Dirichlet, which guarantees that there do not exist diophantine irrationals with exponent $\mu < 1$.
∎

**PROPOSITION 12.2** *Let $\mu > 1$ be fixed. The Lebesgue measure $|\cdot|$ of the set $C_{\gamma,\mu} \cap (0,1)$ satisfies the inequality*

$$|C_{\gamma,\mu} \cap (0,1)| \geq 1 - 4\zeta(\mu)\gamma, \tag{12.50}$$

*where $\zeta(\mu) = \sum_{k=1}^{\infty} k^{-\mu}$ is the Riemann zeta function computed at $\mu$.*

*Proof*
Let $R_{\gamma,\mu} = (0,1) \setminus (C_{\gamma,\mu} \cap (0,1))$ be the complement in $(0,1)$ of $C_{\gamma,\mu}$. From the definition of $C_{\gamma,\mu}$ it immediately follows that

$$R_{\gamma,\mu} = \bigcup_{j,k} \left\{ x \in (0,1) \,\Big|\, |kx - j| < \frac{\gamma}{(k+j)^\mu} \right\}$$

$$= \bigcup_{j,k} \left\{ x \in (0,1) \,\Big|\, \left|x - \frac{j}{k}\right| < \frac{\gamma}{k(k+j)^\mu} \right\},$$

with the conditions $k \geq 1$, $0 \leq j \leq k$. This yields

$$|R_{\gamma,\mu}| \leq \sum_{k=1}^{\infty} \sum_{j=0}^{k} \frac{2\gamma}{k(k+j)^\mu} \leq \sum_{k=1}^{\infty} \sum_{j=0}^{k} 2\gamma k^{-1-\mu} \leq 4\gamma \sum_{k=1}^{\infty} \frac{1}{k^\mu}.$$

Evidently $\zeta(\mu) = \sum_{k=1}^{\infty} 1/k^\mu < +\infty$, since $\mu > 1$. Equation (12.50) follows by observing that $|C_{\gamma,\mu} \cap (0,1)| = 1 - |R_{\gamma,\mu}|$.
∎

From the previous result it follows immediately that

$$\left| \bigcup_{\gamma > 0} C_{\gamma,\mu} \cap (0,1) \right| = 1, \tag{12.51}$$

and therefore for almost every $\omega \in (0,1)$ there exists a constant $\gamma > 0$ such that $\omega \in C_{\gamma,\mu}$. Note that if $\gamma' < \gamma$, then $C_{\gamma,\mu} \subseteq C_{\gamma',\mu}$.

*Remark 12.5*
It is not difficult to prove that if $\omega$ is an algebraic number of degree $d \geq 2$, i.e. if $\omega \in \mathbf{R} \setminus \mathbf{Q}$ is a zero of an irreducible polynomial with rational coefficients and of degree $d$, then $\omega$ is diophantine with exponent $\mu = d - 1$ (Liouville's theorem). It is possible to prove that in fact all algebraic numbers are diophantine for every exponent $\mu > 1$, independent of their degree (by a theorem of Roth, cf. Schmidt 1980). ∎

If $\omega$ satisfies a diophantine condition, and $v$ is of class $\mathcal{C}^\infty$ or analytic, the series (12.47) and (12.48) converge uniformly and define a function of class $\mathcal{C}^\infty$ and analytic, respectively.

THEOREM 12.4 *Let $\omega \in C_{\gamma,\mu}$ and $v \in \mathcal{C}^\infty$. Then the series (12.47) and (12.48) converge uniformly.*

*Proof*
Consider the series (12.47) (a similar argument applies to the series (12.48)):

$$\left| \sum_{(m,n)\in\mathbf{Z}^2\setminus\{(0,0)\}} \frac{\hat{v}_{m,n}}{i(m\omega+n)} e^{i(m\xi+nt)} \right| \leq \sum_{(m,n)\in\mathbf{Z}^2\setminus\{(0,0)\}} \frac{|\hat{v}_{m,n}|}{|m\omega+n|}$$

$$\leq \sum_{(m,n)\in\mathbf{Z}^2\setminus\{(0,0)\}} \frac{|\hat{v}_{m,n}|}{\gamma}(|m|+|n|)^\mu.$$

Since $v \in \mathcal{C}^\infty$, for $r > \mu + 2$ we have the inequality (cf. Appendix 7)

$$\sum_{(m,n)\in\mathbf{Z}^2\setminus\{(0,0)\}} \frac{|\hat{v}_{m,n}|}{\gamma}(|m|+|n|)^\mu \leq \frac{M}{\gamma} \sum_{(m,n)\in\mathbf{Z}^2\setminus\{(0,0)\}} \frac{1}{(|m|+|n|)^{r-\mu}} < +\infty.$$

To prove that $\sum_{(m,n)\in\mathbf{Z}^2\setminus\{(0,0)\}} \frac{1}{(|m|+|n|)^{r-\mu}} < +\infty$ compare the series with the integral $\int_1^\infty \int_1^\infty \frac{dx\, dy}{(x+y)^{r-\mu}}$. ∎

We can now conclude our discussion of the example from which we started. Indeed, we have reduced the solution of equation (12.35) to the system of linear equations (12.40). Each of the equations in (12.40) has the form (12.41) and the previous theorem guarantees that if $\omega$ satisfies a diophantine condition with constant $\gamma$ and exponent $\mu$, and if $V(x,t)$ is of class $\mathcal{C}^\infty$ (or analytic), the system (12.40) admits a solution of class $\mathcal{C}^\infty$ (or analytic), so that the functions $u^{(1)}, \ldots, u^{(k)}, \ldots$ that are solutions of (12.40) exist and are functions of class $\mathcal{C}^\infty$ (or analytic) of $(\xi, t)$. In fact, from the first equation of the system (12.40):

$$D_\omega^2 u^{(1)}(\xi, t) + V_x(\xi, t) = 0, \tag{12.52}$$

and expanding $u^{(1)}$ and $V$ in Fourier series:

$$u^{(1)}(\xi,t) = \sum_{(m,n)\in \mathbf{Z}^2\setminus\{(0,0)\}} \hat{u}^{(1)}_{m,n} e^{i(m\xi+nt)},$$
$$V(\xi,t) = \sum_{(m,n)\in \mathbf{Z}^2\setminus\{(0,0)\}} \hat{V}_{m,n} e^{i(m\xi+nt)},$$
(12.53)

we find

$$-(m\omega+n)^2 \hat{u}^{(1)}_{m,n} + im\hat{V}_{m,n} = 0,$$

for every $(m,n)\in \mathbf{Z}^2\setminus\{(0,0)\}$. Hence

$$u^{(1)}(\xi,t) = \sum_{(m,n)\in \mathbf{Z}^2\setminus\{(0,0)\}} \frac{im\hat{V}_{m,n}}{(m\omega+n)^2} e^{i(m\xi+nt)}. \qquad (12.54)$$

The regularity of $u^{(1)}$ follows from the regularity of $V$ and from the assumption that $\omega$ satisfies a diophantine condition.

Since $u^{(1)}$ is $\mathcal{C}^\infty$, we can substitute this into the second equation of the system (12.40):

$$D_\omega^2 u^{(2)}(\xi,t) + V_{xx}(\xi,t) u^{(1)}(\xi,t) = 0. \qquad (12.55)$$

One can check that $V_{xx}(\xi,t) u^{(1)}(\xi,t)$ has zero mean, and thus we can compute $u^{(2)}$, which is then of class $\mathcal{C}^\infty$, and so on. We are still left with the more difficult problem of the convergence of the series (12.36). What we have seen so far only guarantees that each term in the series is well defined. The convergence of (12.36) under our assumptions (regularity of $V$ and $\omega$ satisfying a diophantine condition) is guaranteed by the following theorem, whose proof is beyond the scope of this introduction (cf. Salomon and Zehnder 1989).

THEOREM 12.5  Let $\omega \in C_{\gamma,\mu}$ and suppose that $V$ is analytic. Then there exists a unique solution $u(\xi,t;\varepsilon)$ of (12.35) that is analytic in $(\xi,t;\varepsilon)$. Moreover there exists a constant $\varepsilon_0 > 0$ such that the series expansion (12.36) of $u(\xi,t;\varepsilon)$ converges uniformly with respect to $(\xi,t)$ for all $\varepsilon$ such that $|\varepsilon| < \varepsilon_0$. ∎

The constant $\varepsilon_0$ of the previous theorem depends only on $V$ and on $\omega$. If $\omega = (\sqrt{5}-1)/2$ and $V = -\cos\xi - \cos(\xi-t)$, $\varepsilon_0$ has a value of approximately 0.03. The computation of $\varepsilon_0$—and its physical significance—have been discussed, e.g. in Escande (1985).

## 12.4 Discussion of the fundamental equation of canonical perturbation theory. Theorem of Poincaré on the non-existence of first integrals of the motion

We consider again the fundamental equation of canonical perturbation theory (12.13), and we show how the discussion of equation (12.46) extends to the more general case.

Since the mean on the torus $\mathbf{T}^l$ of the term $\boldsymbol{\omega}(\mathbf{J}') \cdot \nabla_\chi W^{(1)}$ is equal to zero because of the periodicity of $W$, a *necessary* condition (which clearly is not sufficient) for (12.13) to have a solution is

$$\frac{1}{(2\pi)^l} \oint_{\mathbf{T}^l} (H'_1(\mathbf{J}') - F(\mathbf{J}', \chi)) \, d\chi_1 \ldots d\chi_l = 0, \tag{12.56}$$

which allows the determination of $H'_1$ as the mean of the perturbation:

$$H'_1(\mathbf{J}') = \frac{1}{(2\pi)^l} \oint_{\mathbf{T}^l} F(\mathbf{J}', \chi) \, d\chi_1 \ldots d\chi_l = F_0(\mathbf{J}'), \tag{12.57}$$

as we have already seen when $l = 1$ (cf. (12.15)).

Fixing the values of the actions $\mathbf{J}'$, the linear operator

$$D_\omega = \boldsymbol{\omega} \cdot \nabla_\chi \tag{12.58}$$

has constant coefficients. Its eigenvalues $\lambda$ and eigenfunctions $u_\lambda(\chi)$ are of the form

$$\lambda = i\mathbf{m} \cdot \boldsymbol{\omega}, \qquad u_\lambda = e^{i\mathbf{m} \cdot \chi}, \tag{12.59}$$

where $\mathbf{m} \in \mathbf{Z}^l$ and $\boldsymbol{\omega} = \boldsymbol{\omega}(\mathbf{J}')$ is the vector of frequencies.

DEFINITION 12.3  *The frequencies $\boldsymbol{\omega} \in \mathbf{R}^l$ are called* non-resonant *if for every $\mathbf{m} \in \mathbf{Z}^l, \mathbf{m} \neq \mathbf{0}$,*

$$\mathbf{m} \cdot \boldsymbol{\omega} \neq 0. \tag{12.60}$$

*Otherwise (hence if there exists $\mathbf{m} \in \mathbf{Z}^l$, $\mathbf{m} \neq \mathbf{0}$, such that $\mathbf{m} \cdot \boldsymbol{\omega} = 0$) the frequencies $\boldsymbol{\omega}$ are said to be* resonant. ∎

*Example* 12.8
The vector $(1, \sqrt{2}, \sqrt{3}) \in \mathbf{R}^3$ is non-resonant, while $(1, \sqrt{2}, 1/\sqrt{2})$ is resonant (for example consider $\mathbf{m} = (0, 1, -2)$). ∎

*Remark* 12.6
We could naturally examine the various possible kinds of resonance, and consider the associated modules of resonance (cf. Definition 11.7). This would lead us to the study of resonant normal forms, which goes beyond the scope of this introduction. ∎

If $\boldsymbol{\omega}$ is non-resonant, the eigenvalue $\lambda = 0$ of $D_\omega$ corresponds to the choice $\mathbf{m} = \mathbf{0}$ and has multiplicity one. The fundamental equation of the canonical theory of perturbations is therefore *formally* solvable (neglecting the question of the convergence of the series arising when considering the Fourier expansions of $W^{(1)}$ and $F$).

THEOREM 12.6  *If $\boldsymbol{\omega}$ is non-resonant, there exists a formal solution $W^{(1)}$ of equation (12.13). The solution is unique if we require that the mean of $W^{(1)}$ on the torus $\mathbf{T}^l$ is zero: $\hat{W}^{(1)}_0 = 0$.*

*Proof*
Expanding both $F$ and $W^{(1)}$ in Fourier series (see Appendix 7):

$$F(\mathbf{J}', \boldsymbol{\chi}) = \sum_{\mathbf{m} \in \mathbf{Z}^l} \hat{F}_{\mathbf{m}}(\mathbf{J}')e^{i\mathbf{m}\cdot\boldsymbol{\chi}},$$
$$W^{(1)}(\mathbf{J}', \boldsymbol{\chi}) = \sum_{\mathbf{m} \in \mathbf{Z}^l} \hat{W}_{\mathbf{m}}^{(1)}(\mathbf{J}')e^{i\mathbf{m}\cdot\boldsymbol{\chi}}, \quad (12.61)$$

and substituting these expansions into (12.13) we find

$$i\mathbf{m} \cdot \boldsymbol{\omega}(\mathbf{J}')\hat{W}_{\mathbf{m}}^{(1)}(\mathbf{J}') + \hat{F}_{\mathbf{m}}(\mathbf{J}') = 0, \quad (12.62)$$

for every $\mathbf{m} \in \mathbf{Z}^l \setminus \{\mathbf{0}\}$, from which it follows immediately that

$$\hat{W}_{\mathbf{m}}^{(1)}(\mathbf{J}') = \frac{\hat{F}_{\mathbf{m}}(\mathbf{J}')}{-i\mathbf{m} \cdot \boldsymbol{\omega}(\mathbf{J}')}. \quad (12.63)$$

The non-resonance hypothesis (12.60) guarantees that the denominators in (12.63) never vanish. ∎

When the Hamiltonian $H_0$ is linear in the action variables (harmonic oscillators)

$$H_0(\mathbf{J}) = \mathbf{J} \cdot \boldsymbol{\omega} = \sum_{k=1}^{l} \omega_k J_k, \quad (12.64)$$

the non-resonance condition is a hypothesis on the unperturbed system, and *not* on the values of the action variables, as the frequencies do not depend on the actions. However, in general the frequencies $\boldsymbol{\omega}$ depend on the action variables, and hence contrary to the case of (12.64), the function $\boldsymbol{\omega}(\mathbf{J}')$ is not constant, and the non-resonance condition will only hold on a subset of the phase space.

DEFINITION 12.4  *A Hamiltonian integrable system $H_0(\mathbf{J})$ is non-degenerate (in an open subset $A \subset \mathbf{R}^l$) if there exists a constant $c > 0$ such that for every $\mathbf{J} \in A$,*

$$\left| \det \left( \frac{\partial^2 H_0}{\partial J_i \partial J_k} \right)(\mathbf{J}) \right| \geq c. \quad (12.65)$$

∎

If a system is non-degenerate, by the local invertibility theorem the map $\boldsymbol{\omega} : A \to \mathbf{R}^l$,

$$\mathbf{J} \to \boldsymbol{\omega}(\mathbf{J}) = \nabla_\mathbf{J} H_0(\mathbf{J}),$$

is a local diffeomorphism. In this case, the hypothesis of non-resonance (12.60) selects some values of the action variables, and disregards others. Since the set of vectors of $\mathbf{R}^l$ orthogonal to vectors of $\mathbf{Z}^l$ is dense in $\mathbf{R}^l$, the resonance condition $\boldsymbol{\omega} \cdot \mathbf{m} = 0$ is satisfied for any $\mathbf{m} \in \mathbf{Z}^l$, $\mathbf{m} \neq \mathbf{0}$ in a dense subset $\Omega_r$ of $\mathbf{R}^l$:

$$\Omega_r = \bigcup_{\substack{\mathbf{m} \in \mathbf{Z}^l \\ \mathbf{m} \neq \mathbf{0}}} \{\boldsymbol{\omega} \in \mathbf{R}^l | \boldsymbol{\omega} \cdot \mathbf{m} = 0\}.$$

However, since the frequencies $\boldsymbol{\omega}$ are in continuous one-to-one correspondence with the action variables, the resonance condition is satisfied by values of the actions $\mathbf{J}$ which belong to a dense subset $A_r$ of $A$:

$$A_r = \{\mathbf{J} \in A | \boldsymbol{\omega}(\mathbf{J}) \in \Omega_r\} = \bigcup_{\substack{\mathbf{m} \in \mathbf{Z}^l \\ \mathbf{m} \neq \mathbf{0}}} \{\mathbf{J} \in A | \boldsymbol{\omega}(\mathbf{J}) \cdot \mathbf{m} = 0\}.$$

We shall see shortly (cf. Theorem 12.7) that the density of $A_r$ in $A$ makes it impossible to define the canonical transformation generated by $\mathbf{J}' \cdot \boldsymbol{\chi} + \varepsilon W^{(1)}$ as a regular transformation on an open susbset of the phase space, and it precludes the existence of analytic first integrals of the motion, independent of the Hamiltonian, in quasi-integrable systems (cf. Theorem 12.8). This was proved by Poincaré in 1893.

DEFINITION 12.5  *A function $F : A \times \mathbf{T}^l \to \mathbf{R}$, $F = F(\mathbf{J}, \boldsymbol{\chi})$ has a generic Fourier series expansion if for every $\mathbf{J} \in A$ and every $\mathbf{m} \in \mathbf{Z}^l$ there exists $\mathbf{m}' \in \mathbf{Z}^l$ parallel to $\mathbf{m}$ such that $\hat{F}_{\mathbf{m}'}(\mathbf{J}) \neq 0$.* ∎

THEOREM 12.7 (Poincaré)  *If the integrable part of the Hamiltonian (12.4) is non-degenerate in an open set $A$ and the perturbation $F$ has a generic Fourier series expansion, the fundamental equation of perturbation theory (12.13) does not admit a solution $W^{(1)}(\mathbf{J}', \boldsymbol{\chi})$ which is regular as the action variables vary in the open set $A$.*

*Proof*
The proof is by contradiction. Suppose that the fundamental equation of perturbation theory (12.13) admits a solution $W^{(1)}$ regular with respect to the actions. The non-degeneracy of the Hamiltonian $H_0$ guarantees the invertibility of the relation between the actions $\mathbf{J}'$ and the frequencies $\boldsymbol{\omega}$, as well as the continuity of both transformations (from actions to frequencies and vice versa). The set $\Omega_r$ of resonant frequencies is dense in every open subset of $\mathbf{R}^l$. It follows that the set $A_r$ of the $\mathbf{J}'$ resonant actions, to which there corresponds a resonant frequency $\boldsymbol{\omega}(\mathbf{J}')$, is dense in $A$. Therefore, for every $\mathbf{J}' \in A$ there exists an action $\overline{\mathbf{J}} \in A$, arbitrarily close to $\mathbf{J}'$, and a vector $\overline{\mathbf{m}} \in \mathbf{Z}$, $\overline{\mathbf{m}} \neq \mathbf{0}$, such that $\mathbf{m} \cdot \boldsymbol{\omega}(\overline{\mathbf{J}}) = 0$ for $\mathbf{m} = \overline{\mathbf{m}}$ and for all vectors $\mathbf{m} = \overline{\mathbf{m}}'$ parallel to it. From (12.62) it then follows that necessarily $F_{\overline{\mathbf{m}}}(\overline{\mathbf{J}}) = 0$ and by continuity also that $F_{\overline{\mathbf{m}}}(\mathbf{J}) = 0$, and hence $F_{\overline{\mathbf{m}}'}(\mathbf{J}) = 0$ for every $\overline{\mathbf{m}}'$ parallel to $\overline{\mathbf{m}}$, contradicting the hypothesis that $F$ has a generic Fourier series expansion. ∎

The density of the set $A_r$ of the actions corresponding to resonant values of the frequencies has significant consequences for the problem of the existence of analytic first integrals, independent of the Hamiltonian.

## 12.4 Analytical mechanics: canonical perturbation theory

Consider the Hamiltonian quasi-integrable system (12.4) and seek a solution for the equation of the first integrals

$$\{I, H\} = 0 \tag{12.66}$$

in the form of a power series in $\varepsilon$:

$$I(\mathbf{J}, \boldsymbol{\chi}, \varepsilon) = \sum_{n=0}^{\infty} \varepsilon^n I^{(n)}(\mathbf{J}, \boldsymbol{\chi}). \tag{12.67}$$

Substituting equation (12.67) into (12.66), taking into account the form (12.4) of $H$ and equating terms of the same order in $\varepsilon$, we obtain an infinite system of equations for the (unknown) coefficients of the expansion (12.67) of the first integral sought:

$$\begin{aligned} \{I^{(0)}, H_0\} &= 0, \\ \{I^{(1)}, H_0\} &= \{F, I^{(0)}\}, \\ &\cdots\cdots \\ \{I^{(n)}, H_0\} &= \{F, I^{(n-1)}\}. \end{aligned} \tag{12.68}$$

We remark first of all that the Poisson bracket with $H_0$ is an operator of the form

$$\{\cdot, H_0\} = \omega(\mathbf{J}) \cdot \nabla_{\boldsymbol{\chi}}, \tag{12.69}$$

and hence it coincides with the operator $D_\omega$ (12.58). Each equation of the infinite system (12.68) therefore has the form of the fundamental equation of canonical perturbation theory (12.13). We start by proving that the first of equations (12.68) implies that $I^{(0)}$ does not depend on the angles $\boldsymbol{\chi}$.

**PROPOSITION 12.3** *If the Hamiltonian $H_0(\mathbf{J})$ is non-degenerate and $I^{(0)}$ is a first integral that is regular for the Hamiltonian flow associated with $H_0$, i.e. a regular solution of the equation*

$$\{H_0, I^{(0)}\} = 0, \tag{12.70}$$

*then $I^{(0)}$ does not depend on the angles $\boldsymbol{\chi}$, and hence $I^{(0)} = I^{(0)}(\mathbf{J})$.*

*Proof*
Assume that $I^{(0)}(\mathbf{J}, \boldsymbol{\chi})$ is a solution of (12.70). Substituting the equation into the Fourier series expansion of $I^{(0)}$:

$$I^{(0)}(\mathbf{J}, \boldsymbol{\chi}) = \sum_{\mathbf{m} \in \mathbb{Z}^l} \hat{I}_\mathbf{m}^{(0)}(\mathbf{J}) e^{i\mathbf{m}\cdot\boldsymbol{\chi}},$$

we find

$$i \sum_{\mathbf{m} \in \mathbb{Z}^l} (\mathbf{m} \cdot \omega(\mathbf{J})) \hat{I}_\mathbf{m}^{(0)}(\mathbf{J}) e^{i\mathbf{m}\cdot\boldsymbol{\chi}} = 0,$$

and hence it follows that for every $\mathbf{m} \in \mathbf{Z}^l$ we have

$$\hat{I}_{\mathbf{m}}^{(0)}(\mathbf{J}) \equiv 0 \quad \text{or} \quad \mathbf{m} \cdot \boldsymbol{\omega}(\mathbf{J}) \equiv 0.$$

Differentiating the latter relation with respect to the actions, we find

$$\sum_{i=1}^{l} m_i \frac{\partial \omega_i}{\partial J_k} = 0,$$

for every $k = 1, \ldots, l$, which, when $\mathbf{m} \neq \mathbf{0}$, is satisfied only if

$$\det\left(\frac{\partial \omega_i}{\partial J_k}\right) = \det\left(\frac{\partial^2 H_0}{\partial J_i \partial J_k}\right) = 0,$$

contradicting the hypothesis of non-degeneracy (12.65). It follows that the only non-zero Fourier coefficient is the one corresponding to $\mathbf{m} = \mathbf{0}$ and the solutions of the first of equations (12.68) are necessarily of the form

$$I^{(0)} = I^{(0)}(\mathbf{J}).$$

∎

We now use induction, and assume that we have solved equations (12.68) for $I^{(1)}, \ldots, I^{(n-1)}$. Consider then the equation

$$\{I^{(n)}, H_0\} = \{F, I^{(n-1)}\}. \tag{12.71}$$

Indicating by $F^{(n)}$ the term, known by the inductive hypothesis, which appears on the right-hand side, by expanding in Fourier series both $I^{(n)}$ and $F^{(n)}$ we find the relation

$$i\mathbf{m} \cdot \boldsymbol{\omega}(\mathbf{J}) \hat{I}_{\mathbf{m}}^{(n)}(\mathbf{J}) = \hat{F}_{\mathbf{m}}^{(n)}(\mathbf{J}), \tag{12.72}$$

which must hold for every $\mathbf{m} \in \mathbf{Z}^l$.

There are therefore two problems to be solved in order to prove the existence of a solution of (12.72).

(a) We must prove that $\hat{F}_{\mathbf{0}}^{(n)}(\mathbf{J}) \equiv 0$, and hence that $\{F, I^{(n-1)}\}$ has zero mean value. This is immediate for $n = 1$ (since $\hat{I}^{(0)}$ is independent of $\boldsymbol{\chi}$ and $F$ is periodic in $\boldsymbol{\chi}$) but it is non-trivial for $n \geq 2$.

(b) We again need a non-resonance condition for $\boldsymbol{\omega}(\mathbf{J})$ (unless $\hat{F}_{\mathbf{m}}^{(n)}(\mathbf{J})$ vanishes when $\mathbf{m} \cdot \boldsymbol{\omega}(\mathbf{J}) = 0$) to guarantee at least the existence of a *formal* solution (still neglecting the problem of the convergence of the series).

While the first problem can be solved generally by a more in-depth study of the series (cf. Cherry 1924a,b; Whittaker 1936, chapter 16; Diana et al., 1975), the second is at the heart of the non-existence theorem of Poincaré (Poincaré 1892, sections 81–3).

DEFINITION 12.6 *An analytic first integral of the motion I depends only on H if there exists a non-constant analytic function g of one variable such that $I = g(H)$. Otherwise, I is independent of H.* ∎

THEOREM 12.8 (Poincaré) *If $H(\mathbf{J}, \boldsymbol{\chi}, \varepsilon)$ is a Hamiltonian quasi-integrable system satisfying the same hypotheses as Theorem 12.6 (non-degeneracy and genericity), there does not exist an analytic first integral of the motion $I(\mathbf{J}, \boldsymbol{\chi}, \varepsilon)$ (for which the expansion (12.67) is therefore well defined and convergent if $\varepsilon$ is sufficiently small, uniformly with respect to $\mathbf{J} \in A$ and $\boldsymbol{\chi} \in \mathbf{T}^l$) which is independent of H.* ∎

The proof of the theorem of Poincaré uses the following.

LEMMA 12.1 *An analytic first integral I, such that $I^{(0)}$ is independent of $H_0$, is also independent of H. Conversely, if I is an analytic first integral that is independent of H, one can associate with it an analytic first integral $\tilde{I}$ with $\tilde{I}^{(0)}$ independent of $H_0$.*

*Proof*
If $I$ depends on $H$, $I^{(0)}$ necessarily depends on $H_0$. Indeed since $I = g(H) = g(H_0 + \varepsilon F)$, expanding in Taylor series it follows that $I = g(H_0) + \varepsilon g'(H_0) F + \cdots$.
Comparing with (12.67) we find $I^{(0)} = g(H_0)$, proving the first part of the proposition.

Now let $I_0$ be an analytic first integral that is independent of $H$ and consider the power series expansion in $\varepsilon$:

$$I_0 = I_0^{(0)} + \varepsilon I_0^{(1)} + \varepsilon^2 I_0^{(2)} + \cdots. \tag{12.73}$$

We want to prove that if $I_0^{(0)}$ is not independent of $H_0$, starting from $I_0$ one can construct another first integral $\tilde{I}$, analytic and independent of $H$, for which $\tilde{I}^{(0)}$ is independent of $H_0$.

Indeed, if $I^{(0)}$ depends on $H_0$, i.e.

$$I^{(0)} = g_0(H_0), \tag{12.74}$$

then if $I_0$ is a first integral $I_0 - g_0(H)$ is a first integral too. Moreover from equations (12.73) and (12.74) it follows that

$$\begin{aligned} I_0 - g_0(H) &= I_0^{(0)} + \varepsilon I_0^{(1)} + \mathcal{O}(\varepsilon^2) - g_0(H_0 + \varepsilon F) \\ &= I_0^{(0)} + \varepsilon I_0^{(1)} - g_0(H_0) - \varepsilon g_0'(H_0) F + \mathcal{O}(\varepsilon^2) \\ &= \varepsilon [I_0^{(1)} - g_0'(H_0) F] + \mathcal{O}(\varepsilon^2); \end{aligned}$$

hence setting

$$I_1 = \frac{I_0 - g_0(H)}{\varepsilon}, \tag{12.75}$$

$I_1$ is a new analytic first integral that is independent of $H$ (because by hypothesis $I_0$ cannot be expressed as a function of $H$) and

$$I_1 = I_1^{(0)} + \varepsilon I_1^{(1)} + \varepsilon^2 I_1^{(2)} + \cdots \tag{12.76}$$

is its expansion in powers of $\varepsilon$. The coefficients $I_1^{(j)}$ are obtained starting from the coefficients $I_0^{(k)}$ and from the Taylor series expansion of $g_0(H)$:

$$\sum_{j=0}^{\infty} I_1^{(j)} \varepsilon^j = \frac{1}{\varepsilon} \left[ I_0^{(0)} + \sum_{k=1}^{\infty} I_0^{(k)} \varepsilon^k - g_0(H_0 + \varepsilon F) \right]$$

$$= \frac{1}{\varepsilon} \left[ I_0^{(0)} + \sum_{k=1}^{\infty} I_0^{(k)} \varepsilon^k - g_0(H_0) - \sum_{k=1}^{\infty} \frac{\varepsilon^k}{k!} g_0^{(k)}(H_0) F^k \right]$$

$$= \sum_{k=1}^{\infty} \left[ I_0^{(k)} - \frac{1}{k!} g_0^{(k)}(H_0) F^k \right] \varepsilon^{k-1},$$

where $g_0^{(k)}$ is the $k$th-order derivative of $g_0$. Therefore we have

$$I_1^{(j)} = I_0^{(j+1)} - \frac{1}{(j+1)!} g_0^{(j+1)}(H_0)(H - H_0)^{j+1}, \tag{12.77}$$

for every $j \geq 0$. We can again ask if $I_1^{(0)}$ is independent of $H_0$ or not. In the first case, the proof is finished: $\widetilde{I} = I_1$. If however

$$I_1^{(0)} = g_1(H_0),$$

by repeating the previous argument, setting

$$I_2 = \frac{I_1 - g_1(H)}{\varepsilon},$$

$I_2$ is a new analytic first integral that is independent of $H$. If $I_2^{(0)}$ depends again on $H_0$, we must iterate this procedure. But after a *finite number* $n$ of iterations we necessarily obtain an integral $I_n$ for which $I_n^{(0)}$ does not depend on $H_0$. Indeed, if otherwise $I_n^{(0)} = g_n(H_0)$ for every $n \geq 0$, since (cf. (12.77))

$$g_n(H_0) = I_n^{(0)} = I_{n-1}^{(1)} - g'_{n-1}(H_0)(H - H_0)$$

$$= I_{n-2}^{(2)} - \frac{1}{2!} g''_{n-2}(H_0)(H - H_0)^2 - g'_{n-1}(H_0)(H - H_0) = \cdots$$

$$= I_0^{(n)} - \sum_{k=0}^{n-1} \frac{1}{(n-k)!} g_k^{(n-k)}(H_0)(H - H_0)^{n-k},$$

we would find

$$I_0^{(n)} = g_n(H_0) + \sum_{k=0}^{n-1} \frac{1}{(n-k)!} g_k^{(n-k)}(H_0)(H-H_0)^{n-k},$$

for every $n$, and hence $I_0$ would depend on $H$, contradicting the hypothesis. ∎

We now prove the theorem of Poincaré.

*Proof of Theorem 12.8*
Let $I$ be an analytic first integral of the motion. By Proposition 12.3, $I^{(0)}$ is only a function of the action variables. Expanding in Fourier series $I^{(1)}$ and $F$ in equation (12.71) for $n = 1$ we therefore find the equation

$$i\mathbf{m} \cdot \boldsymbol{\omega}(\mathbf{J})\hat{I}_\mathbf{m}^{(1)}(\mathbf{J}) = i(\mathbf{m} \cdot \nabla_\mathbf{J} I^{(0)}(\mathbf{J}))\hat{F}_\mathbf{m}(\mathbf{J}),$$

for every $\mathbf{m} \in \mathbf{Z}^l$. Hence, for $\hat{I}_\mathbf{m}^{(1)}(\mathbf{J})$ to be well defined, $\mathbf{m} \cdot \boldsymbol{\omega}(\mathbf{J})$ must vanish for every value of $\mathbf{J}$ annihilating the right-hand side. By the hypothesis of genericity of the Fourier series expansion of $F$, there is no loss of generality in assuming that $\hat{F}_\mathbf{m}(\mathbf{J}) \neq 0$ (otherwise, there certainly exists a vector $\mathbf{m}'$ parallel to $\mathbf{m}$ for which $\hat{F}_{\mathbf{m}'}(\mathbf{J}) \neq 0$; but $\mathbf{m}'$ is parallel to $\mathbf{m}$ only if there exists an integer $k$ such that $\mathbf{m}' = k\mathbf{m}$). Therefore $\mathbf{m} \cdot \nabla_\mathbf{J} I^{(0)}(\mathbf{J})$ must vanish every time that $\mathbf{m} \cdot \boldsymbol{\omega}(\mathbf{J})$ is zero (and vice versa).

For a fixed resonant vector $\boldsymbol{\omega} \in \Omega_r$, consider the associated resonance module $\mathcal{M}_\omega$ (see Definition 11.7). The condition that $\mathbf{m} \cdot \nabla_\mathbf{J} I^{(0)}(\mathbf{J})$ and $\mathbf{m} \cdot \boldsymbol{\omega}(\mathbf{J})$ are both zero is equivalent to imposing that $\boldsymbol{\omega}$ and $\nabla_\mathbf{J} I^{(0)}$ are both orthogonal to $\mathcal{M}_\omega$. Hence, if the dimension of $\mathcal{M}_\omega$ is equal to $l-1$, the orthogonal complement of $\mathcal{M}_\omega$ has dimension 1 and $\boldsymbol{\omega}$ and $\nabla_\mathbf{J} I^{(0)}$ are parallel. By the non-degeneracy hypothesis, the correspondence between $\boldsymbol{\omega}$ and $\mathbf{J}$ is bijective and continuous, and hence the set $A_r$ of the values of $\mathbf{J} \in A$ corresponding to resonant frequencies $\boldsymbol{\omega}(\mathbf{J})$ is dense.[1] It follows that $\boldsymbol{\omega}(\mathbf{J}) = \nabla_\mathbf{J} H_0(\mathbf{J})$ and $\nabla_\mathbf{J} I^{(0)}(\mathbf{J})$ must be parallel as $\mathbf{J}$ varies in a dense set in $\mathbf{R}^l$. By continuity there must then exist a scalar function $\alpha(\mathbf{J})$ such that

$$\nabla_\mathbf{J} H_0(\mathbf{J}) = \alpha(\mathbf{J})\nabla_\mathbf{J} I^{(0)}(\mathbf{J}),$$

for every $\mathbf{J} \in \mathbf{R}^l$. Hence there exists a function $A : \mathbf{R} \to \mathbf{R}$ such that $\alpha(\mathbf{J}) = (A'(H_0(\mathbf{J})))^{-1}$ and $I^{(0)}(\mathbf{J}) = A(H_0(\mathbf{J}))$. By Lemma 4.1 the integral $I$ is then a function of $H$. ∎

The 'negative' results proved in this section apparently leave only two possibilities of establishing the existence of a regular solution of equation (12.13):

(a) consider only degenerate Hamiltonian systems—for example systems that are linear in the action variables, as in the case of harmonic oscillators;
(b) admit that the solution does not have a regular dependence on the actions.

---

[1] It is not difficult to check that the subset of $\Omega_r$ made of the vectors $\omega$ whose resonance module $\mathcal{M}_\omega$ has dimension $l-1$ is dense in $\mathbf{R}^l$: it is enough to observe that $\dim \mathcal{M}_\omega = l-1$ if and only if there exists $\nu \in \mathbf{R}$ and $\mathbf{m} \in \mathbf{Z}^l$, such that $\omega = \nu\mathbf{m}$.

Both cases are possible and lead to physically significant results. In the next section, we examine briefly the first possibility, then we survey the important developments related to the second.

## 12.5 Birkhoff series: perturbations of harmonic oscillators

While not eternity, this is a considerable slice of it.[2]

In the previous section we showed (Theorem 12.6) that it is not possible to find a regular solution of the fundamental equation of the canonical theory of perturbations for non-degenerate Hamiltonians. In this section we restrict our analysis only to *degenerate* quasi-integrable Hamiltonian systems:

$$H(\mathbf{J}, \boldsymbol{\chi}, \varepsilon) = \boldsymbol{\omega} \cdot \mathbf{J} + \varepsilon F(\mathbf{J}, \boldsymbol{\chi}). \tag{12.78}$$

In this case, the frequencies $\boldsymbol{\omega}$ are fixed constants that are *independent of the actions*. The condition of non-resonance for the frequencies does not imply any restriction on the action variables, as opposed to what happens in the case of non-degenerate systems, and Theorem 12.7 ensures the existence of a *formal* solution of equation (12.13) for all $\mathbf{J}' \in A$ and $\boldsymbol{\chi} \in \mathbf{T}^l$. We shall indeed show that it is possible to prove a result analogous to Theorem 12.4: if the frequencies satisfy a diophantine condition, the formal solution (12.63) gives rise to a convergent Fourier series and the fundamental equation of the canonical theory of perturbations admits a regular solution for $\mathbf{J}' \in A$ and $\boldsymbol{\chi} \in \mathbf{T}^l$.

DEFINITION 12.7 *Fix $l > 1$. A vector $\boldsymbol{\omega} \in \mathbf{R}^l$ satisfies a diophantine condition (of constant $\gamma > 0$ and exponent $\mu \geq l - 1$), and we write $\boldsymbol{\omega} \in C_{\gamma,\mu}$, if for every $\mathbf{m} \in \mathbf{Z}^l$, $\mathbf{m} \neq \mathbf{0}$, we have*

$$|\mathbf{m} \cdot \boldsymbol{\omega}| \geq \gamma |\mathbf{m}|^{-\mu}, \tag{12.79}$$

*where $|\mathbf{m}| = |m_1| + \cdots + |m_l|$.* ∎

*Remark* 12.7
It is not difficult to show, generalising Theorem 12.3 (of Dirichlet), that the condition $\mu \geq l - 1$ is necessary, and hence that if $\mu < l - 1$ there does not exist a vector $\boldsymbol{\omega} \in \mathbf{R}^l$ that satisfies (12.79) for every $\mathbf{m} \neq \mathbf{0}$. In addition, it can be proved—using an argument slightly more sophisticated than the one used in Section 12.3—that for every fixed $\mu > l - 1$ the Lebesgue measure of $C_{\gamma,\mu} \cap [0, 1]^l$ satisfies the inequality

$$|C_{\gamma,\mu} \cap [0, 1]^l| \geq 1 - a\gamma\zeta(\mu + 2 - l) > 0, \tag{12.80}$$

where $a$ is a constant depending only on $l$. Note that if $l = 2$ we again find (12.50) (and $a = 4$). Hence for almost every $\boldsymbol{\omega} \in [0, 1]^l$ there exists $\gamma > 0$ such that $\boldsymbol{\omega} \in C_{\gamma,\mu}$, for fixed $\mu > l - 1$. ∎

---

[2] Littlewood (1959a, p. 343).

*Remark* 12.8
It is possible to prove the following generalisation of the theorem of Liouville referred to in Remark 12.5. Suppose that $(\omega_1, \ldots, \omega_l)$ is a basis on $\mathbf{Q}$ of a field of algebraic real numbers. Then $\boldsymbol{\omega} = (\omega_1, \ldots, \omega_l)$ satisfies the diophantine condition (12.79) with $\mu = l - 1$ (see Meyer 1972, proposition 2, p. 16). Hence, for example, $(1, \sqrt{2}, \sqrt{3}, \sqrt{6})$ and $(1, 2^{1/3}, 2^{2/3})$ satisfy equation (12.79) with $\mu = l - 1$ and $l = 4, l = 3$, respectively. There also exists a generalisation of the theorem of Roth known as the subspace theorem, see Schmidt (1991). ∎

THEOREM 12.9  *Consider* $\boldsymbol{\omega} \in C_{\gamma,\mu}$, *and let* $A$ *be an open subset of* $\mathbf{R}^l$ *and* $F : A \times \mathbf{T}^l \to \mathbf{R}$, $F = F(\mathbf{J}, \boldsymbol{\chi})$, *a function of class* $\mathcal{C}^\infty$. *The Fourier series*

$$W^{(1)}(\mathbf{J}', \boldsymbol{\chi}) = \sum_{\substack{\mathbf{m} \in \mathbf{Z}^l \\ \mathbf{m} \neq 0}} -\frac{\hat{F}_{\mathbf{m}}(\mathbf{J}')}{i\mathbf{m} \cdot \boldsymbol{\omega}} e^{i\mathbf{m} \cdot \boldsymbol{\chi}} \quad (12.81)$$

*converges uniformly for* $(\mathbf{J}', \boldsymbol{\chi}) \in K \times \mathbf{T}^l$, *where* $K$ *is any compact subset of* $A$.

*Proof*
The proof is analogous to that of Theorem 12.4. Indeed, exploiting the diophantine condition on $\boldsymbol{\omega}$, we find

$$\left| \sum_{\substack{\mathbf{m} \in \mathbf{Z}^l \\ \mathbf{m} \neq 0}} -\frac{\hat{F}_{\mathbf{m}}(\mathbf{J}')}{i\mathbf{m} \cdot \boldsymbol{\omega}} e^{i\mathbf{m} \cdot \boldsymbol{\chi}} \right| \leq \sum_{\substack{\mathbf{m} \in \mathbf{Z}^l \\ \mathbf{m} \neq 0}} \frac{\|\hat{F}_{\mathbf{m}}\|}{\gamma} |\mathbf{m}|^\mu,$$

where

$$\|\hat{F}_{\mathbf{m}}\| = \max_{\mathbf{J}' \in K} |\hat{F}_{\mathbf{m}}(\mathbf{J}')|.$$

Since $F$ is of class $\mathcal{C}^\infty$, for any $r > \mu + l$, there exists a constant $M > 0$ depending only on $r$ and $K$ such that

$$\|\hat{F}_{\mathbf{m}}\| \leq M |\mathbf{m}|^{-r}$$

(see Appendix 7), and therefore

$$\sum_{\substack{\mathbf{m} \in \mathbf{Z}^l \\ \mathbf{m} \neq 0}} \frac{\|\hat{F}_{\mathbf{m}}\|}{\gamma} |\mathbf{m}|^\mu \leq \frac{M}{\gamma} \sum_{\substack{\mathbf{m} \in \mathbf{Z}^l \\ \mathbf{m} \neq 0}} |\mathbf{m}|^{\mu-r} < +\infty.$$
∎

Consider now the Hamiltonian systems

$$h(\mathbf{p}, \mathbf{q}) = \sum_{j=1}^{l} \left( \frac{p_j^2}{2} + \frac{\omega_j^2 q_j^2}{2} \right) + \sum_{r=3}^{\infty} f_r(\mathbf{p}, \mathbf{q}), \quad (12.82)$$

where $f_r$ is a homogeneous trigonometric polynomial of degree $r$ in the variables $(\mathbf{q}, \mathbf{p})$. The Hamiltonian (12.82) represents the perturbation of a system of $l$

harmonic oscillators. In this problem the perturbation parameter $\varepsilon$ does not appear explicitly, as in (12.4), but an analogous role is played by the distance in phase space from the linearly stable equilibrium position corresponding to the origin $(\mathbf{p}, \mathbf{q}) = (\mathbf{0}, \mathbf{0})$. Indeed, consider the set

$$B_\varepsilon = \left\{ \mathbf{J} \in \mathbf{R}^l \mid J_i = \frac{p_i^2 + \omega_i^2 q_i^2}{2\omega_i} < \varepsilon, \text{ for every } i = 1, \ldots, l \right\}, \qquad (12.83)$$

and suppose for simplicity that the sum in (12.82) is extended only to odd indices $r$ (this assumption guarantees that only integer powers of $\varepsilon$ will appear in the series expansion (12.84)). Then if we perform a change of scale of the actions $\mathbf{J} \to \mathbf{J}/\varepsilon$ such that $B_\varepsilon$ is transformed to $B_1$, and a change of time scale $t \to \varepsilon t$ and of the Hamiltonian $H \to H/\varepsilon$ (recall that $t$ and $H$ are canonically conjugate variables, see Remark 10.21), we find

$$H(\mathbf{J}, \boldsymbol{\chi}) = \boldsymbol{\omega} \cdot \mathbf{J} + \sum_{r=1}^{\infty} \varepsilon^r F_r(\mathbf{J}, \boldsymbol{\chi}), \qquad (12.84)$$

where we have introduced the action-angle variables $(\mathbf{J}, \boldsymbol{\chi})$ of the unperturbed harmonic oscillators, and the functions $F_r$ are homogeneous trigonometric polynomials of degree $2(r+1)$:

$$F_r(\mathbf{J}, \boldsymbol{\chi}) = \sum_{\substack{\mathbf{m} \in \mathbf{Z}^l \\ |\mathbf{m}| = 2(r+1)}} \hat{F}_\mathbf{m}^{(r)}(\mathbf{J}) e^{i \mathbf{m} \cdot \boldsymbol{\chi}}. \qquad (12.85)$$

Suppose that the frequency $\boldsymbol{\omega}$ is not resonant (hence that the condition (12.60) is satisfied). In the series expansion in powers of $\varepsilon$ of the perturbation, the corresponding term $F_r$ has to all orders a finite number of Fourier components.

We now show how it is possible, at least formally, to construct the series of the canonical theory of perturbations, to all orders $\varepsilon^r$, $r \geq 1$.

Denote by $W$ the generating function of the canonical transformation near the identity that transforms the Hamiltonian (12.84) into a new Hamiltonian $H'$, depending only on the new action variables $\mathbf{J}'$:

$$H'(\mathbf{J}', \varepsilon) = \sum_{r=0}^{\infty} \varepsilon^r H_r'(\mathbf{J}'). \qquad (12.86)$$

Expanding $W = W(\mathbf{J}', \boldsymbol{\chi}, \varepsilon)$ in a series of powers of $\varepsilon$:

$$W(\mathbf{J}', \boldsymbol{\chi}, \varepsilon) = \mathbf{J}' \cdot \boldsymbol{\chi} + \sum_{r=1}^{\infty} \varepsilon^r W^{(r)}(\mathbf{J}', \boldsymbol{\chi}), \qquad (12.87)$$

and substituting the transformation induced by equation (12.87)

$$\mathbf{J} = \mathbf{J}' + \sum_{r=1}^{\infty} \varepsilon^r \nabla_{\boldsymbol{\chi}} W^{(r)}(\mathbf{J}', \boldsymbol{\chi}) \qquad (12.88)$$

into the Hamilton–Jacobi equation for the Hamiltonian (12.87), we find the equation

$$\boldsymbol{\omega} \cdot \left[ \mathbf{J}' + \sum_{r=1}^{\infty} \varepsilon^r \nabla_{\boldsymbol{\chi}} W^{(r)} \right] + \sum_{r=1}^{\infty} \varepsilon^r F_r \left( \mathbf{J}' + \sum_{r=1}^{\infty} \varepsilon^r \nabla_{\boldsymbol{\chi}} W^{(r)}, \boldsymbol{\chi} \right) = \sum_{r=0}^{\infty} \varepsilon^r H'_r(\mathbf{J}'). \tag{12.89}$$

Expanding in Taylor series the second term:

$$F_r \left( \mathbf{J}' + \sum_{r=1}^{\infty} \varepsilon^r \nabla_{\boldsymbol{\chi}} W^{(r)}, \boldsymbol{\chi} \right) = F_r + \nabla_{\mathbf{J}} F_r \cdot \sum_{r=1}^{\infty} \varepsilon^r \nabla_{\boldsymbol{\chi}} W^{(r)} + \cdots$$

$$+ \frac{1}{k!} \sum_{m_1,\ldots,m_k=1}^{l} \frac{\partial^k F_r}{\partial J_{m_1} \cdots \partial J_{m_k}} \sum_{n=k}^{\infty} \varepsilon^n \tag{12.90}$$

$$\times \sum_{j_1+\cdots+j_k=n} \frac{\partial W^{(j_1)}}{\partial \chi_{m_1}} \cdots \frac{\partial W^{(j_k)}}{\partial \chi_{m_k}} + \cdots,$$

equation (12.89) can be written as

$$(\boldsymbol{\omega} \cdot \mathbf{J}' - H'_0) + \varepsilon(\boldsymbol{\omega} \cdot \nabla_{\boldsymbol{\chi}} W^{(1)} + F_1 - H'_1)$$
$$+ \varepsilon^2(\boldsymbol{\omega} \cdot \nabla_{\boldsymbol{\chi}} W^{(2)} + F_2 + \nabla_{\mathbf{J}} F_1 \cdot \nabla_{\boldsymbol{\chi}} W^{(1)} - H'_2) + \cdots \tag{12.91}$$
$$+ \varepsilon^r(\boldsymbol{\omega} \cdot \nabla_{\boldsymbol{\chi}} W^{(r)} + F_r + \nabla_{\mathbf{J}} F_{r-1} \cdot \nabla_{\boldsymbol{\chi}} W^{(1)} + \cdots - H'_r) + \cdots = 0.$$

To all orders in $\varepsilon$ we must solve the fundamental equation of the theory of perturbations:

$$\boldsymbol{\omega} \cdot \nabla_{\boldsymbol{\chi}} W^{(r)}(\mathbf{J}', \boldsymbol{\chi}) + \mathcal{F}^{(r)}(\mathbf{J}', \boldsymbol{\chi}) = H'_r(\mathbf{J}'), \tag{12.92}$$

where

$$\mathcal{F}^{(r)} = F_r + \sum_{n=1}^{r-1} \sum_{k=1}^{n} \frac{1}{k!} \sum_{m_1,\ldots,m_k=1}^{l} \frac{\partial^k F_{r-n}}{\partial J_{m_1} \cdots \partial J_{m_k}} \sum_{j_1+\cdots+j_k=n} \frac{\partial W^{(j_1)}}{\partial \chi_{m_1}} \cdots \frac{\partial W^{(j_k)}}{\partial \chi_{m_k}} \tag{12.93}$$

depends only on $F_1, \ldots, F_r$ and on $W^{(1)}, \ldots, W^{(r-1)}$. Therefore $H'_r$ is determined by the average of $\mathcal{F}^{(r)}$ on $\mathbf{T}^l$:

$$H'_r(\mathbf{J}') = \hat{\mathcal{F}}_0^{(r)}(\mathbf{J}'), \tag{12.94}$$

while $W^{(r)}$ is a homogenous trigonometric polynomial of degree $2(r+1)$.

If the series (12.86) and (12.87), called *Birkhoff series*, converge for $|\varepsilon| < \varepsilon_0$ in the domain $A \times \mathbf{T}^l$, where $A$ is an open set of $\mathbf{R}^l$, the Hamiltonian (12.80) would be completely canonically integrable. Indeed, we would have a perturbative solution of the Hamilton–Jacobi equation (12.89); $W$ would generate a completely canonical transformation transforming (12.82) to a Hamiltonian that is independent of the new angle variables.

In general the series (12.86) and (12.87) diverge, and hence perturbations of harmonic oscillators do not give rise to integrable problems. The divergence of the Birkhoff series can be easily illustrated by an example, as is shown in Problem 7 of Section 12.8. In addition, there holds a theorem analogous to Theorem 12.7 (see Siegel 1941, 1954), which we simply state.

Consider the set $\mathcal{H}$ of the Hamiltonians $h : \mathbf{R}^{2l} \to \mathbf{R}$ which are analytic and of the form (12.82). We can associate with every Hamiltonian $h$ its power series expansion

$$h(\mathbf{p}, \mathbf{q}) = \sum_{\mathbf{k}, \mathbf{n} \in \mathbf{N}^l} h_{\mathbf{k},\mathbf{n}} p_1^{k_1} \ldots p_l^{k_l} q_1^{n_1} \ldots q_l^{n_l}. \tag{12.95}$$

Comparing with (12.82) it follows that for every $r \geq 3$ we have

$$f_r(\mathbf{p}, \mathbf{q}) = \sum_{|\mathbf{k}|+|\mathbf{n}|=r} h_{\mathbf{k},\mathbf{n}} p_1^{k_1} \ldots p_l^{k_l} q_1^{n_1} \ldots q_l^{n_l}, \tag{12.96}$$

where $|\mathbf{k}| = k_1 + \cdots + k_l$.

DEFINITION 12.8 *Let $h^* \in \mathcal{H}$. A neighbourhood of $h^*$ in $\mathcal{H}$ is given by the set of all Hamiltonians $h \in \mathcal{H}$ such that for every $\mathbf{k}, \mathbf{n} \in \mathbf{N}^l$ we have*

$$|h_{\mathbf{k},\mathbf{n}} - h^*_{\mathbf{k},\mathbf{n}}| < \varepsilon_{\mathbf{k},\mathbf{n}}, \tag{12.97}$$

*where $\{\varepsilon_{\mathbf{k},\mathbf{n}}\}_{\mathbf{k},\mathbf{n} \in \mathbf{N}^l}$ is an arbitrary fixed sequence of positive numbers such that $\varepsilon_{\mathbf{k},\mathbf{n}} \to 0$ for $|\mathbf{k}| + |\mathbf{n}| \to \infty$.* ∎

Two Hamiltonians are therefore close if all the coefficients of the corresponding power series expansions are close.

THEOREM 12.10 (Siegel) *In every neighbourhood of a Hamiltonian $h^* \in \mathcal{H}$ there exists a Hamiltonian $h$ such that the corresponding flow does not admit a first integral of the motion which is analytic and independent of $h$.* ∎

Systems which are not (completely canonically) integrable are therefore *dense* in $\mathcal{H}$, and hence the set of Hamiltonians for which the Birkhoff series diverge is also dense.

*Remark* 12.9
Siegel's theorem also shows how, in general, the Hamilton–Jacobi equation does not admit a complete integral near a point of linearly stable equilibrium (see Remark 11.2). Indeed, moving the equilibrium point into the origin, the Hamiltonian has the form (12.82), and therefore it belongs to $\mathcal{H}$. If the Hamilton–Jacobi equation admitted a complete integral, the system would have $l$ first integrals of the motion, independent of $h$. By Theorem 12.10 this is not the case for any $h$ in a dense subset of $\mathcal{H}$. ∎

Birkhoff series, although divergent, are very important *in practice*, for the qualitative study of degenerate Hamiltonian systems, and for the study of the stability of the Hamiltonian flow for finite but long time. Indeed, there holds the following remarkable result (see Nekhoroshev 1977, Gallavotti 1984).

THEOREM 12.11  *Consider a Hamiltonian quasi-integrable system, degenerate and of the form (12.78), and assume that:*

*(1) the Hamiltonian (12.78) is analytic with respect to $\mathbf{J}$, $\boldsymbol{\chi}$ and $\varepsilon$ for $|\varepsilon| \leq 1$;*
*(2) the frequency vector $\omega$ satisfies a diophantine condition (12.79).*

*Then there exist two constants $\varepsilon_0 > 0$ and $\rho_0 > 0$ and a completely canonical transformation, analytic and near the identity:*

$$\begin{aligned} \mathbf{J} &= \mathbf{J}' + \varepsilon \mathbf{A}(\mathbf{J}', \boldsymbol{\chi}', \varepsilon), \\ \boldsymbol{\chi} &= \boldsymbol{\chi}' + \varepsilon \mathbf{B}(\mathbf{J}', \boldsymbol{\chi}', \varepsilon), \end{aligned} \tag{12.98}$$

*defined for $|\varepsilon| \leq \varepsilon_0$ and $\|\mathbf{J}'\| \leq \rho_0$, such that the transformed Hamiltonian $H'(\mathbf{J}', \boldsymbol{\chi}', \varepsilon)$ is of the form*

$$H'(\mathbf{J}', \boldsymbol{\chi}', \varepsilon) = \omega \cdot \mathbf{J}' + \varepsilon K'(\mathbf{J}', \varepsilon) + \frac{\varepsilon}{\varepsilon_0} \exp\left[-(l+3) \left(\frac{\varepsilon}{\varepsilon_0}\right)^{1/(l+3)}\right] R(\mathbf{J}', \boldsymbol{\chi}', \varepsilon), \tag{12.99}$$

*where $K'$ and $R$ are analytic functions of their arguments and $K'(\mathbf{J}', 0) = 0$, $R(\mathbf{J}', \boldsymbol{\chi}', 0) = 0$.* ∎

An interesting consequence is the following.

COROLLARY 12.1  *There exist two constants $C_1 > 0$ and $C_2 > 0$ such that if $(\mathbf{J}(t), \boldsymbol{\chi}(t))$ is the solution of Hamilton's equations for the Hamiltonian (12.78) with initial data $(\mathbf{J}(0), \boldsymbol{\chi}(0))$, for every time $t$ such that*

$$|t| \leq C_1 \exp\left[(l+3)\left(\frac{\varepsilon}{\varepsilon_0}\right)^{1/(l+3)}\right], \tag{12.100}$$

*we have*

$$|\mathbf{J}(t) - \mathbf{J}(0)| \leq C_2 \frac{\varepsilon}{\varepsilon_0}. \tag{12.101}$$

*Proof (sketch)*
From equation (12.99) it follows that

$$\dot{\mathbf{J}}' = -\frac{\varepsilon}{\varepsilon_0} \exp\left[-(l+3)\left(\frac{\varepsilon}{\varepsilon_0}\right)^{1/(l+3)}\right] \nabla_{\boldsymbol{\chi}} R(\mathbf{J}', \boldsymbol{\chi}', \varepsilon), \tag{12.102}$$

and therefore, if $t$ is chosen as in (12.100), then

$$|\mathbf{J}'(t) - \mathbf{J}'(0)| \leq C_3 \frac{\varepsilon}{\varepsilon_0}, \tag{12.103}$$

where $\mathbf{J}'(0)$ is the initial condition corresponding to $\mathbf{J}(0)$,

$$C_3 = \max |\nabla_\chi R(\mathbf{J}', \chi', \varepsilon)|, \qquad (12.104)$$

and the maximum is taken as $\mathbf{J}'$ varies on the sphere of radius $\rho_0$, while $\chi \in \mathbf{T}^l$ and $\varepsilon \in [-\varepsilon_0, \varepsilon_0]$. The inequality (12.101) follows from the remark that the canonical transformation (12.98) is near the identity and from the inequality

$$|\mathbf{J}(t) - \mathbf{J}(0)| \le |\mathbf{J}(t) - \mathbf{J}'(t)| + |\mathbf{J}'(t) - \mathbf{J}'(0)| + |\mathbf{J}'(0) - \mathbf{J}(0)|. \qquad (12.105)$$

∎

It is not difficult to convince oneself, by a careful inspection of (12.100) as the ratio $\varepsilon/\varepsilon_0$ varies in $[-1, 1]$, that the order of magnitude of the time over which the previous corollary ensures the validity of (12.101) can be very large. As an example, in the applications to celestial mechanics (see Giorgilli et al. 1989) one can obtain stability results for the restricted three-body problem for times of the order of billions of years, and hence comparable with the age of the Solar System.

Littlewood (1959a,b), who first thought of a 'rigorous' application of Birkhoff series to the three-body problem, wrote that, 'while not eternity, this is a considerable slice of it.'

## 12.6 The Kolmogorov–Arnol'd–Moser theorem

In Section 12.4 we saw that, under fairly general hypotheses, the fundamental equation of perturbation theory does not admit regular solutions. In Section 12.5 we studied a special case, which does not satisfy the assumptions of Theorems 12.7 and 12.8 of Poincaré. Under appropriate hypotheses of non-resonance, for these systems it is possible to write formally the series of the canonical theory of perturbations to all orders. However, these series are in general divergent (see Theorem 12.10).

It would therefore seem impossible to prove the existence of quasi-periodic motions for Hamiltonian quasi-integrable systems, and the theory of perturbations seems, from this point of view, bound to fail. (It can still yield interesting information about the stability problem, though. This is shown by Theorem 12.11.)

Consider a quasi-integrable Hamiltonian system. If $\varepsilon = 0$ the system is integrable and *all* motions are bounded and quasi-periodic. When $\varepsilon \ne 0$, instead of requiring that this property is preserved, and hence that the system is still integrable, as we did so far, we can ask if at least *some* of these quasi-periodic unperturbed motions persist in the perturbed version. We shall not therefore seek a regular foliation of the phase space in invariant tori, but simply try to prove the existence, for values $\varepsilon \ne 0$, of 'some' invariant tori, without requiring that their dependence on the action $\mathbf{J}$ is regular as $\mathbf{J}$ varies in an open subset $A$ of $\mathbf{R}^l$.

The Kolmogorov–Arnol'd–Moser (KAM) theorem gives a positive answer to this question: for sufficiently small values of $\varepsilon$ the 'majority' (in a sense to be clarified shortly) of invariant tori corresponding to diophantine frequencies $\omega$ are conserved, and are slightly deformed by the perturbation. The motions on these tori are quasi-periodic with the same frequency $\omega$ which characterises them for $\varepsilon = 0$.

To be able to state the KAM theorem precisely, we must first give a meaning to the statement that 'the invariant tori are slightly deformed' under the action of a perturbation.

Let

$$H(\mathbf{J}, \boldsymbol{\chi}, \varepsilon) = H_0(\mathbf{J}) + \varepsilon F(\mathbf{J}, \boldsymbol{\chi}) \tag{12.106}$$

be a quasi-integrable Hamiltonian system. Suppose that, for fixed $\varepsilon_0 > 0$, $H : A \times \mathbf{T}^l \times (-\varepsilon_0, \varepsilon_0) \to \mathbf{R}$ is an analytic function and that $H_0$ is non-degenerate (cf. Definition 12.4). Every invariant $l$-dimensional unperturbed torus $\mathcal{T}_0 = \{\mathbf{J}_0\} \times \mathbf{T}^l \subset A \times \mathbf{T}^l$ is uniquely characterised by the vector $\boldsymbol{\omega}_0 = \boldsymbol{\omega}(\mathbf{J}_0)$ of the frequencies of the quasi-periodic motions that stay on it.

DEFINITION 12.9  Let $\varepsilon_0 > 0$ be fixed. A one-parameter family $\{\mathcal{T}_\varepsilon\}_{\varepsilon \in (-\varepsilon_0, \varepsilon_0)}$ of $l$-dimensional submanifolds of $\mathbf{R}^{2l}$ is an analytic deformation of a torus $\mathcal{T}_0 = \{\mathbf{J}_0\} \times \mathbf{T}^l$ if, for every $\varepsilon \in (-\varepsilon_0, \varepsilon_0)$, $\mathcal{T}_\varepsilon$ has parametric equations

$$\begin{aligned}\mathbf{J} &= \mathbf{J}_0 + \varepsilon \mathbf{A}(\psi, \varepsilon), \\ \boldsymbol{\chi} &= \psi + \varepsilon \mathbf{B}(\psi, \varepsilon),\end{aligned} \tag{12.107}$$

where $\psi \in \mathbf{T}^l$, $\mathbf{A} : \mathbf{T}^l \times [-\varepsilon_0, \varepsilon_0] \to \mathbf{R}^l$ and $\mathbf{B} : \mathbf{T}^l \times [-\varepsilon_0, \varepsilon_0] \to \mathbf{T}^l$ are analytic functions. ∎

Note that setting $\varepsilon = 0$ in (12.107) we again find the torus $\mathcal{T}_0 = \{\mathbf{J}_0\} \times \mathbf{T}^l$.

*Remark* 12.10

The function $\mathbf{B}$ in (12.107) has the additional property that its average $\hat{\mathbf{B}}_0$ on the torus $\mathbf{T}^l$ is zero. Indeed, $\varepsilon \mathbf{B} = \boldsymbol{\chi} - \psi$, and since $\boldsymbol{\chi}$ and $\psi$ are both coordinates on a torus $\mathbf{T}^l$ we have

$$\frac{\varepsilon}{(2\pi)^l} \int_{\mathbf{T}^l} \mathbf{B}(\psi, \varepsilon) \, \mathrm{d}^l \psi = \frac{1}{(2\pi)^l} \left[ \int_{\mathbf{T}^l} \boldsymbol{\chi} \, \mathrm{d}^l \boldsymbol{\chi} - \int_{\mathbf{T}^l} \psi \, \mathrm{d}^l \psi \right] = 0.$$

∎

For fixed $\varepsilon \in (-\varepsilon_0, \varepsilon_0)$, equations (12.107) establish a correspondence of every point $\psi_0 \in \mathbf{T}^l$ with the point $\mathcal{T}_\varepsilon$ of coordinates

$$\begin{aligned}\mathbf{J} &= \mathbf{J}_0 + \varepsilon \mathbf{A}(\psi_0, \varepsilon), \\ \boldsymbol{\chi} &= \psi_0 + \varepsilon \mathbf{B}(\psi_0, \varepsilon).\end{aligned} \tag{12.108}$$

Denote by $(\mathbf{J}(t, \psi_0), \boldsymbol{\chi}(t, \psi_0))$ the solution of the Hamilton equations associated with (12.106) and passing through the point of coordinates (12.108) at time $t = 0$.

DEFINITION 12.10  A deformation $\{\mathcal{T}_\varepsilon\}_{\varepsilon\in(-\varepsilon_0,\varepsilon_0)}$ of $\mathcal{T}_0$ is a deformation of $\mathcal{T}_0$ into invariant tori for the quasi-integrable system (12.106) if, for fixed $\varepsilon \in (-\varepsilon_0,\varepsilon_0)$, and every choice of $\psi_0 \in \mathbf{T}^l$ the Hamiltonian flow $(\mathbf{J}(t,\psi_0),\boldsymbol{\chi}(t,\psi_0))$ can be obtained from equations (12.107) by setting $\psi = \psi_0 + \boldsymbol{\omega}(\mathbf{J}_0)t$:

$$\mathbf{J}(t,\psi_0) = \mathbf{J}_0 + \varepsilon \mathbf{A}(\psi_0 + \boldsymbol{\omega}(\mathbf{J}_0)t, \varepsilon),$$
$$\boldsymbol{\chi}(t,\psi_0) = \psi_0 + \boldsymbol{\omega}(\mathbf{J}_0)t + \varepsilon \mathbf{B}(\psi_0 + \boldsymbol{\omega}(\mathbf{J}_0)t, \varepsilon). \tag{12.109}$$

It follows that $(\mathbf{J}(t,\psi_0),\boldsymbol{\chi}(t,\psi_0))$ belongs to $\mathcal{T}_\varepsilon$ for every $t \in \mathbf{R}$. ∎

*Remark* 12.11
The motions on $\mathcal{T}_\varepsilon$ are quasi-periodic with the same frequency vector $\boldsymbol{\omega}_0$ of the motions on $\mathcal{T}_0$. ∎

We now show how it is possible to carry out, by means of a perturbative approach, the computation of the functions $\mathbf{A}$ and $\mathbf{B}$.

Setting $\boldsymbol{\omega}_0 = \boldsymbol{\omega}(\mathbf{J}_0)$ and $\psi = \psi_0 + \boldsymbol{\omega}_0 t$, from equation (12.109) it follows that

$$\dot{\mathbf{J}} = \varepsilon \frac{d\mathbf{A}}{dt}(\psi,\varepsilon) = \varepsilon \boldsymbol{\omega}_0 \cdot \nabla_\psi \mathbf{A}(\psi,\varepsilon),$$
$$\dot{\boldsymbol{\chi}} = \boldsymbol{\omega}_0 + \varepsilon \frac{d\mathbf{B}}{dt}(\psi,\varepsilon) = \boldsymbol{\omega}_0 + \varepsilon \boldsymbol{\omega}_0 \cdot \nabla_\psi \mathbf{B}(\psi,\varepsilon), \tag{12.110}$$

to be compared with Hamilton's equations associated with (12.106) and computed along the flow (12.109):

$$\dot{\mathbf{J}} = -\varepsilon \nabla_{\boldsymbol{\chi}} F(\mathbf{J}(t,\psi_0),\boldsymbol{\chi}(t,\psi_0)) = -\varepsilon \nabla_{\boldsymbol{\chi}} F(\mathbf{J}_0 + \varepsilon \mathbf{A}(\psi,\varepsilon), \psi + \varepsilon \mathbf{B}(\psi,\varepsilon)),$$
$$\dot{\boldsymbol{\chi}} = \boldsymbol{\omega}(\mathbf{J}(t,\psi_0)) + \varepsilon \nabla_{\mathbf{J}} F(\mathbf{J}(t,\psi_0),\boldsymbol{\chi}(t,\psi_0)) \tag{12.111}$$
$$= \boldsymbol{\omega}(\mathbf{J}_0 + \varepsilon \mathbf{A}(\psi,\varepsilon)) + \varepsilon \nabla_{\mathbf{J}} F(\mathbf{J}_0 + \varepsilon \mathbf{A}(\psi,\varepsilon), \psi + \varepsilon \mathbf{B}(\psi,\varepsilon)).$$

Expanding $\mathbf{A}$ and $\mathbf{B}$ in power series in $\varepsilon$ (the so-called *Lindstedt series*):

$$\mathbf{A}(\psi,\varepsilon) = \sum_{k=0}^{\infty} \varepsilon^k \mathbf{A}^{(k)}(\psi) = \mathbf{A}^{(0)}(\psi) + \varepsilon \mathbf{A}^{(1)}(\psi) + \cdots,$$
$$\mathbf{B}(\psi,\varepsilon) = \sum_{k=0}^{\infty} \varepsilon^k \mathbf{B}^{(k)}(\psi) = \mathbf{B}^{(0)}(\psi) + \varepsilon \mathbf{B}^{(1)}(\psi) + \cdots, \tag{12.112}$$

and $\boldsymbol{\omega}$ in Taylor series around $\mathbf{J}_0$:

$$\boldsymbol{\omega}(\mathbf{J}_0 + \varepsilon \mathbf{A}(\psi,\varepsilon)) = \boldsymbol{\omega}_0 + \varepsilon \nabla_{\mathbf{J}} \boldsymbol{\omega}(\mathbf{J}_0) \cdot \mathbf{A}(\psi,\varepsilon) + \cdots, \tag{12.113}$$

and then comparing (12.111) and (12.110) to first order in $\varepsilon$, we find

$$\varepsilon \boldsymbol{\omega}_0 \cdot \nabla_\psi \mathbf{A}^{(0)}(\psi) = -\varepsilon \nabla_{\boldsymbol{\chi}} F(\mathbf{J}_0,\psi), \tag{12.114}$$
$$\varepsilon \boldsymbol{\omega}_0 \cdot \nabla_\psi \mathbf{B}^{(0)}(\psi) = \varepsilon \nabla_{\mathbf{J}} \boldsymbol{\omega}(\mathbf{J}_0) \cdot \mathbf{A}^{(0)}(\psi) + \varepsilon \nabla_{\mathbf{J}} F(\mathbf{J}_0,\psi). \tag{12.115}$$

Equation (12.114) can be solved immediately by expanding $\mathbf{A}^{(0)}$ and $F$ in Fourier series: setting

$$\mathbf{A}^{(0)}(\psi) = \sum_{\mathbf{m} \in \mathbf{Z}^l} \hat{\mathbf{A}}_{\mathbf{m}}^{(0)} e^{i\mathbf{m}\cdot\psi}, \tag{12.116}$$

since

$$\nabla_{\chi} F(\mathbf{J}_0, \psi) = \sum_{\mathbf{m} \in \mathbf{Z}^l} i\mathbf{m} \hat{F}_{\mathbf{m}}(\mathbf{J}_0) e^{i\mathbf{m}\cdot\psi}, \tag{12.117}$$

by the uniqueness of Fourier series we have for all $\mathbf{m} \in \mathbf{Z}^l$ that

$$i\mathbf{m}\cdot\boldsymbol{\omega}_0 \hat{\mathbf{A}}_{\mathbf{m}}^{(0)} = -i\mathbf{m}\hat{F}_{\mathbf{m}}(\mathbf{J}_0). \tag{12.118}$$

The solution, if $\boldsymbol{\omega}_0 = \boldsymbol{\omega}(\mathbf{J}_0)$ is non-resonant, is given by

$$\hat{\mathbf{A}}_{\mathbf{m}}^{(0)} = -\frac{\mathbf{m}\hat{F}_{\mathbf{m}}(\mathbf{J}_0)}{\mathbf{m}\cdot\boldsymbol{\omega}_0}, \tag{12.119}$$

for $\mathbf{m} \neq \mathbf{0}$, while for the time being the average $\hat{\mathbf{A}}_{\mathbf{0}}^{(0)}$ of $\mathbf{A}$ on the torus $\mathbf{T}^l$ is undetermined.

Substituting the solution (12.116), (12.119) into the expression (12.115), and expanding in turn $\mathbf{B}^{(0)}$ in Fourier series, we similarly find the coefficients $\hat{\mathbf{B}}_{\mathbf{m}}^{(0)}$ for $\mathbf{m} \neq \mathbf{0}$. Note that integrating both sides of (12.115) on $\mathbf{T}^l$, and taking into account the periodicity of $\mathbf{B}$ with respect to $\psi$, we find

$$\nabla_{\mathbf{J}}\boldsymbol{\omega}(\mathbf{J}_0) \cdot \hat{\mathbf{A}}_{\mathbf{0}}^{(0)} + \nabla_{\mathbf{J}}\hat{F}_0(\mathbf{J}_0) = 0. \tag{12.120}$$

Since $\nabla_{\mathbf{J}}\hat{F}_0(\mathbf{J}_0)$ can be non-zero, for equation (12.120) (hence also (12.115)) to have a solution we must require that the matrix

$$\nabla_{\mathbf{J}}\boldsymbol{\omega}(\mathbf{J}_0) = \left(\frac{\partial^2 H_0}{\partial J_i \partial J_k}(\mathbf{J}_0)\right) \tag{12.121}$$

be invertible, and hence that the unperturbed Hamiltonian $H_0$ be *non-degenerate* in a neighbourhood of $\mathbf{J}_0 \in A$. In this case

$$\hat{\mathbf{A}}_{\mathbf{0}}^{(0)} = -(\nabla_{\mathbf{J}}\boldsymbol{\omega}(\mathbf{J}_0))^{-1}\nabla_{\mathbf{J}}\hat{F}_0(\mathbf{J}_0), \tag{12.122}$$

and this determines the average of $\mathbf{A}^{(0)}$ on $\mathbf{T}^l$.

This discussion can be summarised in the following proposition.

PROPOSITION 12.4 *If the Hamiltonian $H_0$ is non-degenerate on the open set $A$, for fixed $\mathbf{J}_0 \in A$ such that $\boldsymbol{\omega}_0 = \boldsymbol{\omega}(\mathbf{J}_0)$ is non-resonant, the system (12.114), (12.115) admits a formal solution.* ∎

We can in fact prove that the argument we have just presented to obtain functions $\mathbf{A}$ and $\mathbf{B}$ as first-order perturbations can be iterated to all orders.

Under the hypotheses of non-degeneracy for $H_0$ and of non-resonance for $\omega_0$ as in Proposition 12.2 it is possible to define for every $k \geq 0$ the functions $\mathbf{A}^{(k)}$ and $\mathbf{B}^{(k)}$ in (12.112) through their Fourier series expansions:

$$\mathbf{A}^{(k)}(\psi) = \sum_{\mathbf{m} \in \mathbf{Z}^l} \hat{\mathbf{A}}_{\mathbf{m}}^{(k)} e^{i \mathbf{m} \cdot \psi},$$
$$\mathbf{B}^{(k)}(\psi) = \sum_{\mathbf{m} \in \mathbf{Z}^l, \mathbf{m} \neq 0} \hat{\mathbf{B}}_{\mathbf{m}}^{(k)} e^{i \mathbf{m} \cdot \psi}, \tag{12.123}$$

at least *formally*, and hence neglecting the problem of the convergence of the series (12.123). The coefficients $\hat{\mathbf{A}}_{\mathbf{m}}^{(k)}$ and $\hat{\mathbf{B}}_{\mathbf{m}}^{(k)}$ of the series expansions (12.123) can be computed from the solution of a system of the form

$$\omega_0 \cdot \nabla_\psi \mathbf{A}^{(k)}(\psi) = \mathcal{A}^{(k)}(\mathbf{J}_0, \psi), \tag{12.124}$$
$$\omega_0 \cdot \nabla_\psi \mathbf{B}^{(k)}(\psi) = \mathcal{B}^{(k)}(\mathbf{J}_0, \psi), \tag{12.125}$$

where $\mathcal{A}^{(k)}$ and $\mathcal{B}^{(k)}$ depend on $\mathbf{A}^{(0)}, \ldots, \mathbf{A}^{(k-1)}$, $\mathbf{B}^{(0)}, \ldots, \mathbf{B}^{(k-1)}$ and on the derivatives of $F$ with respect to $\mathbf{J}$ and $\chi$ up to order $k+1$. Here $\mathcal{B}^{(k)}$ also depends on $\mathbf{A}^{(k)}$ and on the derivatives of $\omega$ with respect to $\mathbf{J}$ up to order $k+1$ (hence on the derivatives of $H_0$ with respect to $\mathbf{J}$ up to order $k+2$).

Note that the structure of equations (12.124) and (12.125) is the same as that of the fundamental equation of perturbation theory (12.13), and it constitutes the natural generalisation of equation (12.46) to the case $l > 2$.

Indeed, Poincaré proved in chapter IX of his *Méthodes Nouvelles* (second volume, 1893) that the functions $\mathcal{A}^k$ and $\mathcal{B}^k$ appearing on the right-hand side of (12.124) and (12.125) have zero mean on the torus $\mathbf{T}^l$, and therefore the formal solvability of the two equations is guaranteed.

It follows that we have the following significant extension of Proposition 12.2.

PROPOSITION 12.5 *If the Hamiltonian $H_0$ is non-degenerate in the open set $A$, for any fixed $\mathbf{J}_0 \in A$ such that $\omega_0 = \omega(\mathbf{J}_0)$ is non-resonant, it is always possible to determine formally the functions $\mathbf{A}$ and $\mathbf{B}$ (parametrising the deformation to invariant tori of the torus of frequency $\omega_0$) via the series expansions (12.112) and (12.123).* ∎

Two problems are still open:

(a) the question of the *convergence of the Fourier series expansions* (12.123) of the functions $\mathbf{A}^{(k)}(\psi)$ and $\mathbf{B}^{(k)}(\psi)$;
(b) the question of the *convergence of the power series* (12.112).

The first question has an easy solution. For fixed $\mu > l - 1$, consider the set of diophantine frequencies $\omega_0$ (see Definition 12.7) of constant $\gamma > 0$ (and exponent $\mu$):

$$C_{\gamma,\mu} = \{\omega_0 \in \mathbf{R}^l \,|\, |\omega_0 \cdot \mathbf{m}| \geq \gamma |\mathbf{m}|^{-\mu}\}. \tag{12.126}$$

Since the Hamiltonian $H_0$ is non-degenerate, to every $\boldsymbol{\omega}_0 \in C_{\gamma,\mu}$ there corresponds a unique vector $\mathbf{J}_0$ of the actions for which $\boldsymbol{\omega}(\mathbf{J}_0) = \boldsymbol{\omega}_0$. Let

$$A_{\gamma,\mu} = \{\mathbf{J}_0 \in A | \boldsymbol{\omega}(\mathbf{J}_0) \in C_{\gamma,\mu}\}. \tag{12.127}$$

Evidently

$$A_{\gamma,\mu} = \boldsymbol{\omega}^{-1}(C_{\gamma,\mu}) \tag{12.128}$$

(recall that the hypothesis of non-degeneracy of $H_0$ guarantees that the map $\mathbf{J} \to \boldsymbol{\omega}(\mathbf{J})$ is a local diffeomorphism).

Now fix a value of the actions $\mathbf{J}_0 \in A_{\gamma,\mu}$, so that the corresponding frequency is $\boldsymbol{\omega}_0 = \boldsymbol{\omega}(\mathbf{J}_0) \in C_{\gamma,\mu}$. Then we can extend the arguments considered in the proofs of Theorems 12.4 and 12.9 to the equations (12.124) and (12.125) for the Fourier series expansions of $\mathbf{A}^{(k)}(\psi)$ and of $\mathbf{B}^{(k)}(\psi)$, and prove their convergence.

PROPOSITION 12.6  *If the Hamiltonian $H_0$ is non-degenerate in the open set $A$, for fixed $\mathbf{J}_0 \in A_{\gamma,\mu}$ the functions $\mathbf{A}^{(k)} : \mathbf{T}^l \to \mathbf{R}^l$ and $\mathbf{B}^{(k)} : \mathbf{T}^l \to \mathbf{R}^l$ which solve the system (12.124), (12.125) have a convergent Fourier series expansion, for every $k \geq 0$.*  ∎

Problem (a) is therefore solved. The solution of problem (b) is much more difficult. However, it is necessary to give this question an affirmative answer if the existence of deformations of a torus into invariant tori is to be proven.

Poincaré was sceptical of the possibility of proving the convergence of the Lindstedt series, and in *Méthodes Nouvelles*, volume II, p. 104 he comments that

Supposons pour simplifier qu'il y ait deux degrés de liberté; les séries ne pourraient-elles pas, par exemple, converger quand $x_1^0$ et $x_2^0$ ont été choisis de telle sorte que le rapport $\dfrac{n_1}{n_2}$ soit incommensurable, et que son carré soit au contraire commensurable (ou quand le rapport $\dfrac{n_1}{n_2}$ est assujetti à une autre condition analogue à celle que je viens d'énoncer un peu au hasard)? Les raisonnements de ce chapitre ne me permettent pas d'affirmer que ce fait ne se présentera pas. Tout ce qu'il m'est permis de dire, c'est qu'il est fort invraisemblable.

[Suppose for simplicity there are two degrees of freedom; would it be possible for the series to converge when, for example, $x_1^0$ and $x_2^0$—the initial conditions—are chosen in such a way that the ratio $n_1/n_2$ of the frequencies—in our notation $\omega_0 = (n_1, n_2)$—is irrational, however such that its square is rational (or when the ratio $n_1/n_2$ satisfies some other condition analogous to the one I just stated a bit randomly)? The arguments in this chapter do not allow me to rule out this case, although it appears to me rather unrealistic.]

Weierstrass, as opposed to Poincaré, was convinced of the possibility that the Lindstedt series could converge (see Barrow-Green 1997).

It is nevertheless surprising that the condition referred to as 'a bit randomly' by Poincaré—implying that $\boldsymbol{\omega}_0$ satisfies a diophantine condition, see Remark 12.8—is correct.

The Kolmogorov–Arnol'd–Moser theorem (see Kolmogorov 1954, Arnol'd 1961, 1963a, Moser 1962, 1967), whose proof goes beyond the scope of the present introduction, guarantees in practice the convergence of the power series (12.112) as long as the frequency $\omega_0$ satisfies a diophantine condition.

THEOREM 12.12 (KAM) *Consider a quasi-integrable Hamiltonian system (12.106) and assume that the Hamiltonian H is analytic and non-degenerate. Let $\mu > l - 1$ and $\gamma > 0$ be fixed. There exists a constant $\varepsilon_c > 0$, depending on $\gamma$, such that for every $\mathbf{J}_0 \in A_{\gamma,\mu}$ there exists a deformation $\{\mathcal{T}_\varepsilon\}_{\varepsilon \in (-\varepsilon_c, \varepsilon_c)}$ of the torus $\mathcal{T}_0 = \mathbf{J}_0 \times \mathbf{T}^l$ into invariant tori for the quasi-integrable system (12.106).* ∎

*Remark 12.12*
It is possible to prove that $\varepsilon_c = \mathcal{O}(\gamma^2)$ (see Pöschel 1982, Arnol'd et al. 1983). ∎

*Remark 12.13*
Since we assume that $H_0$ is non-degenerate, the correspondence between actions $\mathbf{J}$ and frequencies $\boldsymbol{\omega}$ is a diffeomorphism, and there therefore exists the inverse function $\mathbf{J} = \mathbf{J}(\boldsymbol{\omega})$ of $\boldsymbol{\omega} = \boldsymbol{\omega}(\mathbf{J}) = \nabla_\mathbf{J} H_0(\mathbf{J})$. Hence, thanks to (12.65),

$$|A \setminus A_{\gamma,\mu}| = \int_{A \setminus A_{\gamma,\mu}} \mathrm{d}^l \mathbf{J}$$
$$= \int_{\boldsymbol{\omega}(A) \setminus C_{\gamma,\mu}} \left| \det \left( \frac{\partial^2 H_0}{\partial J_i \partial J_k}(\mathbf{J}(\boldsymbol{\omega})) \right)^{-1} \right| \mathrm{d}^l \boldsymbol{\omega} \leq c^{-1} |\boldsymbol{\omega}(A) \setminus C_{\gamma,\mu}|. \quad (12.129)$$

Assume for simplicity that the open set $A$ of $\mathbf{R}^l$ is obtained as the preimage of $(0,1)^l$ via the map $\boldsymbol{\omega} \to \mathbf{J}(\boldsymbol{\omega})$. Then $\boldsymbol{\omega}(A) = (0,1)^l$, and from (12.129), taking into account (12.80), it follows that

$$|A \setminus A_{\gamma,\mu}| \leq c^{-1}(|(0,1)^l| - |C_{\gamma,\mu} \cap (0,1)^l|) \leq c^{-1} a\gamma \zeta(\mu + 2 - l). \quad (12.130)$$

By Remark 12.12 $\gamma = \mathcal{O}(\sqrt{\varepsilon})$, and hence *the Lebesgue measure of the complement, in the phase space, of the set of invariant tori is $\mathcal{O}(\sqrt{\varepsilon})$; therefore it tends to 0 for $\varepsilon \to 0$.* ∎

*Remark 12.14*
The set $A_{\gamma,\mu}$ has a rather complex structure: it is closed but totally disconnected, and it is a *Cantor set*.[3] Because of the density in $\mathbf{R}^l$ of resonant frequencies, the complement of $A_{\gamma,\mu}$ is dense. ∎

*Remark 12.15*
In practice, in the proof of the KAM theorem one constructs a canonical transformation near the identity of the variables $(\mathbf{J}, \boldsymbol{\chi})$ to new variables $(\widetilde{\mathbf{J}}, \widetilde{\boldsymbol{\chi}})$ with generating function $\boldsymbol{\chi} \cdot \widetilde{\mathbf{J}} + \varepsilon \widetilde{W}(\boldsymbol{\chi}, \widetilde{\mathbf{J}}, \varepsilon)$ and a new Hamiltonian $K(\widetilde{\mathbf{J}}, \varepsilon)$, satisfying

$$H_0(\widetilde{\mathbf{J}} + \varepsilon \nabla_{\boldsymbol{\chi}} \widetilde{W}) + \varepsilon F(\widetilde{\mathbf{J}} + \varepsilon \nabla_{\boldsymbol{\chi}} \widetilde{W}, \boldsymbol{\chi}) = K(\widetilde{\mathbf{J}}, \varepsilon)$$

---

[3] A closed set is a Cantor set if it is totally disconnected and has no isolated points.

every time that $\mathbf{J} \in A_{\gamma,\mu}$. The Hamilton–Jacobi equation therefore admits a solution in the set of invariant tori $A_{\gamma,\mu}$ (see Chierchia and Gallavotti 1982, Pöschel 1982). Hence to the system (12.106) there are associated $l$ first integrals of the motions (the new actions). However, these integrals are not defined everywhere, but only on $A_{\gamma,\mu}$; hence, although the dependence on $\chi$ and on $\varepsilon$ is regular, they do not have a regular dependence on $\mathbf{J}$, and the result is not in contradiction with Theorem 12.8. ∎

For more details on this topic, which we had no pretension to treat exhaustively, we recommend reading chapter 5 of Arnol'd et al. (1983).

## 12.7 Adiabatic invariants

Consider a Hamiltonian system with one degree of freedom, depending on one parameter $r$, so that its Hamilton function has the form

$$H = H(p, q, r). \tag{12.131}$$

As an example, we can consider a pendulum (see Example 11.11) and take as parameter the length $l$, or a harmonic oscillator (see (11.28)) and treat the frequency $\omega$ as a parameter.

If for every fixed value of the parameter $r$ the system admits motions of rotation or of libration, the Hamiltonian (12.131) is completely canonically integrable and there exists a canonical transformation depending on the parameter $r$ to action-angle variables $(J, \chi)$. Let $W(q, J, r)$ be the generating function of this canonical transformation, where we emphasise the dependence on the parameter $r$.

We denote by $K_0(J, r)$ the Hamiltonian corresponding to the new variables, and by $\omega_0(J, r) = (\partial K_0/\partial J)(J, r)$ the frequency of the motion. Note that the action $J$ is a function of $(p, q, r)$.

Suppose that the system is subject to an external influence, expressed as a time dependence $r = r(t)$ of the parameter $r$. If the rate of change of the parameter is comparable with the frequency $\omega_0(J, r)$ of the motion of the system corresponding to a fixed value of $r$, in general the system is no longer integrable, because of the overwhelming effect of the external influence, and it is not possible to find a first integral—not even in an 'approximate' sense (note that the energy is not conserved, because $\mathrm{d}H/\mathrm{d}t = \partial H/\partial t = \partial H/\partial r \dot{r}$). The situation is however substantially different if the variation of the parameter in time is *slow*, and hence if $|\dot{r}| \leq \varepsilon \ll 1$, where $r$ and $t$ are dimensionless with respect to two respective 'natural' scales.[4]

---

[4] It is however possible to introduce the notion of a smooth function on a Cantor set (Whitney smoothness) and prove that in this wider sense the dependence of $W$ on $\widetilde{W}$ is smooth; see Pöschel (1982) for details.

In this case, the dependence on time of the parameter can be expressed through the so-called *slow time*:

$$r = r(\tau), \qquad \tau = \varepsilon t, \qquad (12.132)$$

and it is possible to find a constant of the motion in an approximate sense that we now clarify.

DEFINITION 12.11   *A function $A(p,q,r)$ is an* adiabatic invariant *of the system (12.131) subject to a slow variation (12.132) of the parameter $r$, if for every $\delta > 0$ there exists $\varepsilon_0 > 0$ such that for every fixed $\varepsilon \in (0, \varepsilon_0)$ and for every $t \in [0, 1/\varepsilon]$ we have*

$$|A(p(t), q(t), r(\varepsilon t)) - A(p(0), q(0), r(0))| < \delta, \qquad (12.133)$$

*where $(p(t), q(t))$ is the solution of the system of Hamilton's equations corresponding to $H(p,q,r(\varepsilon t))$:*

$$\begin{aligned} \dot{p} &= -\frac{\partial H}{\partial q}(p, q, r(\varepsilon t)), \\ \dot{q} &= \frac{\partial H}{\partial p}(p, q, r(\varepsilon t)), \end{aligned} \qquad (12.134)$$

*with initial conditions $(p(0), q(0))$.*   ∎

*Remark 12.16*
An adiabatic invariant is an approximate constant of the motion of the Hamiltonian flow associated with (12.134) for a bounded time interval of length $1/\varepsilon$, which grows indefinitely if the rate of change of the parameter $\varepsilon \to 0$. If for a fixed value of $\varepsilon > 0$ a function $A(p,q,r)$ satisfies equation (12.133) *for all times $t \geq 0$*, then $A$ is a *perpetual* adiabatic invariant.   ∎

*Remark 12.17*
It is immediate to realise that the energy is not, in general, an adiabatic invariant. Consider, for example, a point particle in the absence of forces, whose mass changes slowly with time, so that its Hamiltonian is $H = p^2/2m(\varepsilon t)$. If $m = m_0(2 - \sin(\pi \varepsilon t/2))$, since $p(t) = p(0)$, we have $E(1/\varepsilon) = p^2(0)/2m_0 = 2E(0)$.   ∎

THEOREM 12.13   *Assume that the Hamiltonian (12.131) is of class $\mathcal{C}^3$ and that the dependence $r(\tau)$ of the parameter on the slow time has the same regularity. If there exists a $\delta > 0$ such that for all $\tau \in [0, 1]$ we have*

$$\omega_0(J, r(\tau)) > \delta, \qquad (12.135)$$

*the action $J(p, q, r)$ is an adiabatic invariant.*

*Proof*
Since the parameter depends on time, $r = r(\varepsilon t)$, the function $W(q, J, r(\varepsilon t))$ generates a canonical transformation depending on time, and the new Hamiltonian is

$$K(J, \chi, \varepsilon t) = K_0(J, r(\varepsilon t)) + \frac{\partial W}{\partial t}(q(J, \chi, r(\varepsilon t)), J, r(\varepsilon t)) \qquad (12.136)$$
$$= K_0(J, r(\varepsilon t)) + \varepsilon f(J, \chi, \varepsilon t),$$

where $f(J, \chi, \varepsilon t) = r'(\varepsilon t) \partial W / \partial r$. The corresponding Hamilton equations are

$$\dot{J} = -\varepsilon \frac{\partial f}{\partial \chi}(J, \chi, \varepsilon t), \qquad \dot{\chi} = \omega_0(J, r(\varepsilon t)) + \varepsilon \frac{\partial f}{\partial J}(J, \chi, \varepsilon t). \qquad (12.137)$$

We now seek the generating function $\widetilde{W}(\chi, \tilde{J}, \varepsilon t)$ of a canonical transformation near the identity that would eliminate the dependence on the angle in the Hamiltonian, to first order in $\varepsilon$, and hence a solution of

$$K_0\left(\frac{\partial \widetilde{W}}{\partial \chi}, r(\varepsilon t)\right) + \varepsilon f\left(\frac{\partial \widetilde{W}}{\partial \chi}, \chi, \varepsilon t\right) + \frac{\partial \widetilde{W}}{\partial t} = \widetilde{K}_0(\tilde{J}, \varepsilon t) + \varepsilon \widetilde{K}_1(\tilde{J}, \varepsilon t) + \mathcal{O}(\varepsilon^2). \qquad (12.138)$$

Setting $\widetilde{W} = \chi \tilde{J} + \varepsilon \widetilde{W}^{(1)}(\chi, \tilde{J}, \varepsilon t)$, substituting and equating the corresponding terms in the expansion in $\varepsilon$ we find:

$$\widetilde{K}_0(\tilde{J}, \varepsilon t) = K_0(\tilde{J}, r(\varepsilon t)). \qquad (12.139)$$

To first order we therefore have

$$\omega_0(\tilde{J}, r(\varepsilon t)) \frac{\partial \widetilde{W}^{(1)}}{\partial \chi}(\chi, \tilde{J}, \varepsilon t) + f(\tilde{J}, \chi, \varepsilon t) = \widetilde{K}_1(\tilde{J}, \varepsilon t), \qquad (12.140)$$

since

$$\frac{\partial \widetilde{W}}{\partial t} = \varepsilon \frac{\partial \widetilde{W}^{(1)}}{\partial t} = \varepsilon^2 \frac{\partial \widetilde{W}^{(1)}}{\partial \tau} = \mathcal{O}(\varepsilon^2). \qquad (12.141)$$

Condition (12.135) guarantees that the solution of (12.140) exists and (recall Theorem 12.1) is given by

$$\widetilde{K}_1(\tilde{J}, \varepsilon t) = \frac{1}{2\pi} \int_0^{2\pi} f(\tilde{J}, \chi, \varepsilon t) \, d\chi,$$
$$\widetilde{W}^{(1)}(\chi, \tilde{J}, \varepsilon t) = \frac{1}{\omega_0(\tilde{J}, r(\varepsilon t))} \int_0^\chi [K_1(\tilde{J}, \varepsilon t) - f(\tilde{J}, \xi, \varepsilon t)] \, d\xi. \qquad (12.142)$$

The hypothesis that $H$ is of class $\mathcal{C}^3$ ensures that $\widetilde{W}^{(1)}$ is of class $\mathcal{C}^2$ and generates a canonical transformation. From

$$\dot{\tilde{J}} = -\frac{\partial \widetilde{K}}{\partial \chi} = \mathcal{O}(\varepsilon^2), \tag{12.143}$$

it follows that, for every $t \in [0, 1/\varepsilon]$,

$$|\tilde{J}(t) - \tilde{J}(0)| = \mathcal{O}(\varepsilon), \tag{12.144}$$

and therefore our claim holds, as

$$|J(t) - J(0)| \leq |J(t) - \tilde{J}(t)| + |\tilde{J}(t) - \tilde{J}(0)| + |\tilde{J}(0) - J(0)|, \tag{12.145}$$

and the transformation from $J$ to $\tilde{J}$ is near the identity. ∎

*Remark 12.18*
Arnol'd (1963b) proved that the KAM theorem guarantees the perpetual adiabatic invariance of the action if the dependence of the parameter $r$ on the slow time $\tau$ is periodic, and hence if there exists a $T > 0$ such that $r(\tau) = r(\tau + T)$ for every $\tau$. It is however necessary to impose the condition of non-degeneracy:

$$\frac{\partial^2 K_0}{\partial J^2} = \frac{\partial \omega_0}{\partial J} \neq 0, \tag{12.146}$$

to assume that the Hamiltonian is an analytic function of $(p, q, r)$, and that the dependence of $r$ on $\tau$ is also analytic. ∎

*Remark 12.19*
It is possible to extend Theorem 12.13 to the case of more degrees of freedom, but the proof is much more complicated (see Neishtadt 1976, Golin et al. 1989), because one must overcome the difficulties generated by the presence of small denominators and by the dependence of the frequencies (and of the non-resonance condition) on the parameter. The proof is much simpler, and similar to that of Theorem 12.13, if the frequencies do not depend explicitly on the parameter (see Golin and Marmi 1990).

## 12.8 Problems

1. Compute the first order of the canonical perturbation theory for the Hamiltonian

$$H = \left(\frac{p^2 + x^2}{2\sqrt{2}}\right)^2 + \frac{\varepsilon x}{\sqrt{2}}.$$

Write down explicitly the generating function $W$ and the new action and angle variables $J'$ and $\chi'$. (Solution: $J' = J + (\varepsilon \sin \chi)/\sqrt{J}$, $\chi' = \chi - (\varepsilon \cos \chi)/2J^{3/2}$, $W = J'\chi + (\varepsilon \cos \chi)/\sqrt{J'}$.)

2. If $V = -\cos x - \cos(x-t)$ compute $u^{(1)}$ in the expansion (12.36).
(Solution: $u^{(1)} = (1/\omega^2)\sin\xi + [1/(\omega-1)^2]\sin(\xi-t)$.)

3. If $V = -\sum_{k=1}^{\infty} e^{-k}\cos(x-kt)$ compute $u^{(1)}$ in the expansion (12.36).

(Solution: $u^{(1)} = \sum_{k=1}^{\infty}[e^{-k}/(\omega-k)^2]\sin(\xi-kt)$.)

4. Prove that if $V$ is a trigonometric polynomial of degree $r$, then $u^{(n)}$ in the expansion (12.36) is a trigonometric polynomial of degree $nr$ for every $n \geq 1$.

5. Check directly that $\mathcal{F}^{(1)}$ and $\mathcal{F}^{(2)}$ in equation (12.92) are homogeneous trigonometric polynomials of degree 4 and 6, respectively. Prove that $\mathcal{F}^{(r)}$ is a homogeneous trigonometric polynomial of degree $2(r+1)$.

6. Given the Hamiltonian

$$H = J_1 + \omega J_2 + \frac{4\varepsilon}{\omega}J_1 J_2 \cos^2\chi_1 \cos^2\chi_2,$$

prove that the Birkhoff series (12.86) to third order is given by

$$H' = J_1 + \omega J_2 + \varepsilon\frac{J_1 J_2}{\omega} - \varepsilon^2\frac{J_1 J_2}{\omega^2}\left[\frac{J_1}{2\omega} + \frac{J_2}{2} + \frac{J_2 - J_1}{8(1-\omega)} + \frac{J_1 + J_2}{8(1+\omega)}\right]$$

$$+ \varepsilon^3\left[\frac{4J_1^2 J_2^2}{\omega^3}\left(\frac{1}{\omega^2} + \frac{2}{\omega} - \frac{2\omega}{(1-\omega^2)^2}\right)\right.$$

$$+ \frac{J_1 + \omega J_2}{\omega^2}\left(\frac{J_1 + J_2}{8(1+\omega)} + \frac{J_2 - J_1}{8(1-\omega)} + \frac{8J_1}{\omega} + 8J_2\right)\Bigg]$$

$$+ \varepsilon^3\left[\frac{J_2 - J_1}{\omega(1-\omega)}\left(\frac{J_2 - J_1}{8(1-\omega)} + 4J_2 + \frac{4J_1}{\omega}\right)\right.$$

$$+ \frac{J_1 + J_2}{\omega(1+\omega)}\left(\frac{J_1 + J_2}{8(1+\omega)} + 4J_2 + \frac{4J_1}{\omega}\right)\Bigg].$$

(The first two orders are computed quickly but the third order requires more work.)

7. Given the Hamiltonian

$$H = J_1 + \omega J_2 + \varepsilon[J_2 + F(\chi_1, \chi_2)],$$

where $F(\chi_1, \chi_2) = \sum_{\mathbf{m}\in\mathbb{Z}^2\setminus 0} e^{-|m_1|-|m_2|}e^{i(m_1\chi_1+m_2\chi_2)}$ and $\omega$ is an irrational number, prove that the formal solution of the Hamilton–Jacobi equation (12.14) for $H$ is given by

$$H' = J_1' + (\omega + \varepsilon)J_2',$$

$$W = J_1'\chi_1 + J_2'\chi_2 + i\varepsilon\sum_{\mathbf{m}\in\mathbb{Z}^2\setminus 0}\frac{e^{-|m_1|-|m_2|}e^{i(m_1\chi_1+m_2\chi_2)}}{m_1 + m_2(\omega + \varepsilon)}.$$

Note that for every irrational $\omega$ there exists a sequence $\varepsilon_n \to 0$ such that $\omega + \varepsilon_n$ is rational. Deduce from this fact the divergence of the series expansion of $W$.

**8.** Solve Hamilton's equations for the Hamiltonian of the previous problem. Prove that if $\omega + \varepsilon$ is rational, $J_1(t)$ and $J_2(t)$ are proportional to $t$.

**9.** Compute the first order of canonical perturbation theory for the Hamiltonian

$$H = \left(\frac{p_1^2 + \omega_1^2 q_1^2}{2}\right)^3 + \left(\frac{p_2^2 + \omega_2^2 q_2^2}{2}\right)^3 + \varepsilon q_1^2 q_2^2 p_1^2 p_2^2.$$

Under which hypotheses is this procedure justified? (Solution: $H' = (\omega_1 J_1')^3 + (\omega_2 J_2')^3 + (\varepsilon/4)(J_1' J_2')^2$ and the hypothesis is that $k(\omega_1 J_1)^3 - j(\omega_2 J_2)^3 \neq 0$ for all integers $j$ and $k$ with $|k| \leq 2$.)

**10.** Consider a quasi-integrable system with two degrees of freedom described by the Hamiltonian $H(q_1, q_2, p_1, p_2, \varepsilon) = \frac{1}{2}p_1^2 + q_1^2 + \frac{1}{2}p_2^2 + \varepsilon q_1^2 p_2^2 \cos^2 q_2$.

(a) Introduce the action-angle variables $\chi_1, \chi_2, J_1, J_2$ for the integrable system obtained by setting $\varepsilon = 0$ and express the Hamiltonian $H$ in these variables.
(b) Compute the Hamiltonian $K(\tilde{J}_1, \tilde{J}_2, \varepsilon)$ obtained through the use of first-order canonical perturbation theory, and the frequencies $\omega_1(\tilde{J}_1, \tilde{J}_2)$, $\omega_2(\tilde{J}_1, \tilde{J}_2)$ of the motions.
(c) Under which conditions on $\tilde{J}_1, \tilde{J}_2$ is this procedure justified?

**11.** Consider the quasi-integrable system with two degrees of freedom described by the Hamiltonian

$$H(p_1, p_2, q_1, q_2, \varepsilon) = \frac{p_1^2 + p_2^2}{2} + \frac{3q_1^2 + q_2^2}{2} + \varepsilon q_1^2 q_2^2.$$

(a) Introduce the action-angle variables $(J_1, J_2, \chi_1, \chi_2)$ for the integrable system obtained by setting $\varepsilon = 0$ and express the Hamiltonian $H$ in these variables.
(b) Compute the generating function $W(\chi_1, \chi_2, \tilde{J}_1, \tilde{J}_2, \varepsilon)$ of the canonical transformation near the identity to new variables $\tilde{\chi}_1, \tilde{\chi}_2, \tilde{J}_1, \tilde{J}_2$, transforming the Hamiltonian $H$ into a new Hamiltonian $K$ which depends (up to terms of order $\mathcal{O}(\varepsilon^2)$) only on the new action variables. Compute the new Hamiltonian $K(\tilde{J}_1, \tilde{J}_2, \varepsilon)$ and the frequencies of the corresponding motions.
(c) Under which conditions on $(\tilde{J}_1, \tilde{J}_2)$ is this procedure justified?
(d) Compute the new Hamiltonian $K$ which depends only on the new action variables up to terms of order $\mathcal{O}(\varepsilon^3)$.

## 12.9 Additional remarks and bibliographical notes

In this chapter we briefly introduced a few perturbation methods for studying the motion of quasi-integrable Hamiltonian systems, and in particular we considered the problem of the existence of (approximate) first integrals of the motion

(Sections 12.4 and 12.5), of bounded and quasi-periodic motions (Section 12.6), and of the existence of adiabatic invariants (Section 12.7).

In the last twenty years, the study of canonical perturbation theory saw a very significant development, justified both by the new theoretical results on the problem of convergence of the series expansions (in particular, the KAM theorem), and because of the appearance of new fields of application (plasma physics, elementary particle accelerators, physical chemistry, dynamics of galaxies, etc.) which complement the classical domain of application of this theory, celestial mechanics.

Although this is traditionally considered a 'difficult' subject, too difficult to enter the syllabus of an undergraduate course, we thought that it was necessary, from the point of view of general scientific culture, to present, if only briefly, the general lines of the modern theory, skipping many mathematical details.

Chapters 5 and 6 of Arnol'd *et al.* (1983) contain a more detailed exposition, of exceptional clarity, of the material we summarised and of many more results, including a large bibliography.

The textbook by Gallavotti (1980) is also a good source for further study, as is the excellent review article by the same author (Gallavotti 1984).

The Birkhoff series and their applications are discussed in detail in an article by Moser (1968). See also Moser (1986), especially for the study (here omitted) of discrete Hamiltonian dynamical systems which are quasi-integrable (see Arrowsmith and Place (1990) for a short introduction, Moser (1973) and Siegel and Moser (1971) for a more detailed exposition).

A very readable proof of the KAM theorem, developing the original argument due to Kolmogorov (1954), is given by Benettin *et al.* (1984). More recent developments of the KAM theory are discussed in Bost (1986) and Yoccoz (1992).

Finally, we must recommend the reading of the vast original work of Poincaré on the subject (Poincaré 1892, 1893, 1899), which remains, after a century, a constant source of inspiration for research in the field. It is not possible to even hint at the richness of the topics considered, or to illustrate the depth of Poincaré's reasoning. The reader interested in the personality of one of the founders of modern mathematics can find interesting material in Boutroux (1914).

## 12.10 Additional solved problems

*Problem 1*
Consider a harmonic oscillator with Hamiltonian

$$h(p, q, \varepsilon) = \frac{p^2}{2m} + \frac{1}{2}m\omega^2 q^2 + \varepsilon a q^3,$$

where $\varepsilon$ is a small parameter. Compute, using the perturbation method, the variation in the frequency of the motion to the first significant order in $\varepsilon$. Compare the result obtained with the direct computation of the action and of the frequencies of the motions associated with the completely canonically integrable Hamiltonian $h(p, q, \varepsilon)$.

*Solution*

The action-angle variables associated with the unperturbed motion ($\varepsilon = 0$) are

$$p = \sqrt{2m\omega J} \cos \chi, \quad q = \sqrt{\frac{2J}{m\omega}} \sin \chi.$$

Substituting in $h$, we obtain from this the expression for the Hamiltonian $H(J, \chi, \varepsilon)$:

$$H(J, \chi, \varepsilon) = \omega J + \varepsilon a \left(\frac{2J}{m\omega}\right)^{3/2} \sin^3 \chi.$$

We seek a generating function $W(J', \chi, \varepsilon) = J'\chi + \varepsilon W^{(1)}(J', \chi) + \varepsilon^2 W^{(2)}(J', \chi) + \cdots$ which transforms $H$ to $K(J', \varepsilon) = \omega J' + \varepsilon K_1(J') + \varepsilon^2 K_2(J') + \cdots$. Following the procedure described in Section 12.1 we find the equations

$$\omega \frac{\partial W^{(1)}}{\partial \chi}(J', \chi) + F(J', \chi) = K_1(J'),$$

$$\omega \frac{\partial W^{(2)}}{\partial \chi}(J', \chi) + F^{(2)}(J', \chi) = K_2(J'),$$

where

$$F(J', \chi) = a \left(\frac{2J}{m\omega}\right)^{3/2} \sin^3 \chi, \quad F^{(2)}(J', \chi) = \frac{\partial F}{\partial J'}(J', \chi) \frac{\partial W^{(1)}}{\partial \chi}(J', \chi).$$

Since we are seeking the variation in the frequency of the motions and $F^{(2)}$ depends only on $\partial W^{(1)}/\partial \chi$, it is not necessary to compute explicitly $W^{(1)}$ and $W^{(2)}$ and it is sufficient to compute $K_1$ and $K_2$.

From the first equation we obtain

$$K_1(J') = \frac{1}{2\pi} \int_0^{2\pi} F(J', \chi) \, d\chi = 0,$$

and therefore

$$\frac{\partial W^{(1)}}{\partial \chi}(J', \chi) = -\frac{F(J', \chi)}{\omega},$$

from which it follows that

$$K_2(J') = \frac{1}{2\pi} \int_0^{2\pi} F^{(2)}(J', \chi) \, d\chi = -\frac{1}{2\pi} \int_0^{2\pi} \frac{3a^2}{2\omega} \left(\frac{2}{m\omega}\right)^3 (J')^2 \sin^6 \chi \, d\chi$$

$$= -\frac{3a^2}{2\omega} \left(\frac{2}{m\omega}\right)^3 (J')^2 \frac{5}{16} = -\frac{15}{4} \frac{a^2 (J')^2}{m^3 \omega^4}.$$

Hence the first significant variation in the frequency of the motions happens to the second order in $\varepsilon$ and has value

$$\omega(J',\varepsilon) = \omega + \varepsilon^2 \frac{\partial K_2}{\partial J'} + \mathcal{O}(\varepsilon^3) = \omega - \frac{15}{2} \frac{a^2 J'}{m^3 \omega^4} \varepsilon^2 + \mathcal{O}(\varepsilon^3).$$

Let us now see how, thanks to the integrability of the one-dimensional motions, it is possible to arrive at the same conclusion by computing directly the dependence of the energy on the action $\tilde{J}$ of the completely integrable Hamiltonian $h(p,q,\varepsilon)$.

Indeed, from the equation

$$\frac{p^2}{2m} + \frac{1}{2} m\omega^2 q^2 + \varepsilon a q^3 = E$$

one can deduce the relation between the action $\tilde{J}$ associated with the oscillations (near $q = p = 0$) and the energy $E$ and the parameter $\varepsilon$, through an elliptic integral. Since we are only interested in small values of the parameter $\varepsilon$, we can compute the variation of $\tilde{J}$ and $E$ in the form of an expansion in series of powers of $\varepsilon$.

Indeed, we have

$$p = \sqrt{2m\varepsilon a f(q)},$$

where

$$f(q) = -q^3 - \frac{m\omega^2}{2\varepsilon a} q^2 + \frac{E}{\varepsilon a} = (e_1 - q)(q - e_2)(q - e_3). \tag{12.147}$$

For small values of $\varepsilon$ two roots of the polynomial $f(q)$, which we indicated by $e_1, e_2$, are in a neighbourhood of the points $\pm\sqrt{2E/m\omega^2}$ and the oscillation takes place in the interval $e_1 \le q \le e_2$. The third root $e_3$ is of the order of $-m\omega^2/2\varepsilon a$. We can therefore expand

$$\sqrt{f(q)} = \sqrt{-e_3(e_1 - q)(q - e_2)\left(1 - \frac{q}{e_3}\right)}$$

$$= \sqrt{-e_3(e_1 - q)(q - e_2)} \left(1 - \frac{q}{2e_3} - \frac{q^2}{8e_3^2} + \cdots\right),$$

from which it follows that

$$\tilde{J} = \frac{1}{\pi} \int_{e_1}^{e_2} \sqrt{2m\varepsilon a f(q)}\, dq = \sqrt{-2m\varepsilon a e_3} \left(\tilde{J}_0 - \frac{1}{2e_3} \tilde{J}_1 - \frac{1}{8e_3^2} \tilde{J}_2 + \cdots\right),$$

where

$$\tilde{J}_k = \frac{1}{\pi} \int_{e_1}^{e_2} q^k \sqrt{(e_1 - q)(q - e_2)}\, dq, \qquad k = 0, 1, 2, \ldots.$$

Using the substitution $[2q - (e_1 + e_2)]/(e_1 - e_2) = \sin\psi$ which transforms the integration interval $[e_1, e_2]$ into $\left[\frac{1}{2}\pi, \frac{3}{2}\pi\right]$ and explicitly computing we find

$$\tilde{J}_0 = \frac{(e_1 - e_2)^2}{8}, \quad \tilde{J}_1 = \frac{(e_1 + e_2)^2}{2}\tilde{J}_0, \quad \tilde{J}_2 = \frac{\tilde{J}_0}{16}[5(e_1 + e_2)^2 - 4e_1 e_2].$$

We must now determine $e_1, e_2, e_3$ as functions of $\varepsilon$. Identifying the coefficients in (12.147) and setting $\theta = \varepsilon a$, we find

$$e_1 + e_2 + e_3 = -\frac{m\omega^2}{2\theta}, \quad (e_1 + e_2)e_3 = -e_1 e_2, \quad e_1 e_2 e_3 = \frac{E}{\theta}. \tag{12.148}$$

Writing $e_1 = -\sqrt{2E/m\omega^2} + \bar{e}_1$, $e_2 = \sqrt{2E/m\omega^2} + \bar{e}_2$, $e_3 = \xi - m\omega^2/2\theta$ and defining $\eta = \bar{e}_1 + \bar{e}_2$, $\zeta = \bar{e}_1 - \bar{e}_2$, from equations (12.148) we arrive at the system

$$\eta + \xi = 0, \tag{12.149}$$

$$\eta\left(\eta\theta + \frac{m\omega^2}{2}\right)^2 = -E\theta, \tag{12.150}$$

$$\eta\left(\eta\theta + \frac{m\omega^2}{2}\right) = -\frac{2E}{m\omega^2}\theta + \sqrt{\frac{2E}{m\omega^2}}\zeta\theta + \frac{\eta^2 - \zeta^2}{4}\theta. \tag{12.151}$$

From (12.150) one computes the values taken by $\eta', \eta'', \eta'''$ at $\theta = 0$:

$$\eta'_0 = -E\left(\frac{m\omega^2}{2}\right)^{-2}, \quad \eta''_0 = 0, \quad \eta'''_0 = -12E^2\left(\frac{m\omega^2}{2}\right)^{-5}. \tag{12.152}$$

From (12.151), by differentiating three times, we obtain the values of $\zeta', \zeta''$:

$$\zeta'_0 = 0, \quad \zeta''_0 = -\frac{5}{2}E^{3/2}\left(\frac{m\omega^2}{2}\right)^{-7/2}. \tag{12.153}$$

We now only need to use that

$$\bar{e}_1 = \frac{1}{2}\left(\eta'_0\theta + \frac{1}{2}\zeta''_0\theta^2\right) + \mathcal{O}(\theta^3), \quad \bar{e}_2 = \frac{1}{2}\left(\eta'_0\theta - \frac{1}{2}\zeta''_0\theta^2\right) + \mathcal{O}(\theta^3),$$

$$\xi = -\eta'_0\theta + \mathcal{O}(\theta^3)$$

to obtain

$$e_1 = -\sqrt{\frac{2E}{m\omega^2}} - \frac{1}{2}E\left(\frac{m\omega^2}{2}\right)^{-2}\theta - \frac{5}{8}E^{3/2}\left(\frac{m\omega^2}{2}\right)^{-7/2}\theta^2 + \mathcal{O}(\theta^3),$$

$$e_2 = \sqrt{\frac{2E}{m\omega^2}} - \frac{1}{2}E\left(\frac{m\omega^2}{2}\right)^{-2}\theta + \frac{5}{8}E^{3/2}\left(\frac{m\omega^2}{2}\right)^{-7/2}\theta^2 + \mathcal{O}(\theta^3),$$

$$e_3 = -\frac{m\omega^2}{2\theta} + E\left(\frac{m\omega^2}{2}\right)^{-2}\theta + \mathcal{O}(\theta^3).$$

Substituting the expansions of $e_1, e_2, e_3$ into the expressions for $\tilde{J}_0, \tilde{J}_1$ and $\tilde{J}_2$ we finally find

$$\tilde{J} = \frac{E}{\omega}\left(1 + \frac{15}{4}\frac{\varepsilon^2 a^2 E}{m^3 \omega^6} + \cdots\right).$$

Inverting these relations, we have

$$E = \omega \tilde{J} - \frac{15\varepsilon^2 a^2}{4m^3 \omega^4}\tilde{J}^2 + \cdots,$$

and finally the frequency

$$\omega(\tilde{J}) = \frac{\mathrm{d}E}{\mathrm{d}\tilde{J}} = \omega - \frac{15}{2}\frac{\varepsilon^2 a^2}{m^3 \omega^4}\tilde{J} + \mathcal{O}(\varepsilon^3).$$

*Problem 2*
Consider a harmonic oscillator with two degrees of freedom and Hamiltonian

$$h(p_1, p_2, q_1, q_2) = \frac{p_1^2 + p_2^2}{2m} + \frac{1}{2}m(\omega_1^2 q_1^2 + \omega_2^2 q_2^2) + a_{30} q_1^3 + a_{21} q_1^2 q_2 + a_{12} q_1 q_2^2 + a_{03} q_2^3.$$

Introduce the action-angle variables $(J_1, J_2, \chi_1, \chi_2)$ associated with the harmonic part

$$h_0(p_1, p_2, q_1, q_2) = \frac{p_1^2 + p_2^2}{2m} + \frac{1}{2}m(\omega_1^2 q_1^2 + \omega_2^2 q_2^2)$$

of the Hamiltonian $h$ and determine an approximate first integral $I$ in the form $I(J_1, J_2, \chi_1, \chi_2) = \omega_1 J_1 - \omega_2 J_2 + I^{(3)}(J_1, J_2, \chi_1, \chi_2)$, where $I^{(3)}(J_1, J_2, \chi_1, \chi_2) = \sum_{j=0}^{3} J_1^{(3-j)/2} J_2^{j/2} P_j(\chi_1, \chi_2)$ and $P_j$ is a trigonometric polynomial of degree 3 (the adelphic integral of Whittaker, cf. Whittaker (1936, chapter XVI)).

*Solution*
With respect to the action-angle variables

$$p_i = \sqrt{2m\omega_i J_i}\cos\chi_i, \quad q_i = \sqrt{\frac{2J_i}{m\omega_i}}\sin\chi_i, \quad i = 1, 2,$$

the Hamiltonian $h$ becomes

$$H(\mathbf{J}, \boldsymbol{\chi}) = \omega_1 J_1 + \omega_2 J_2 + F(J_1, J_2, \chi_1, \chi_2),$$

$$F(J_1, J_2, \chi_1, \chi_2) = J_1^{3/2}(F_1 \sin\chi_1 + F_2 \sin 3\chi_1) + J_1 J_2^{1/2}[F_3 \sin\chi_2 + F_4 \sin(2\chi_1 + \chi_2)$$
$$+ F_5 \sin(2\chi_1 - \chi_2)] + J_1^{1/2} J_2[F_6 \sin\chi_1 + F_7 \sin(2\chi_2 + \chi_1)$$
$$+ F_8 \sin(2\chi_2 - \chi_1)] + J_2^{3/2}[F_9 \sin\chi_2 + F_{10} \sin 3\chi_2],$$

where $F_1, \ldots, F_{10}$ are constants depending on $m, \omega_1, \omega_2, a_{30}, a_{21}, a_{12}, a_{03}$ (for example: $F_1 = \frac{3}{4}(2/m\omega_1)^{3/2} a_{30}$, $F_2 = -\frac{1}{4}(2/m\omega_1)^{3/2} a_{30}$, etc.).
Setting $H_0 = \omega_1 J_1 + \omega_2 J_2$ and $I^{(2)} = \omega_1 J_1 - \omega_2 J_2$ we must impose the condition

$$\{H, I\} = \{H_0 + F, I^{(2)} + I^{(3)}\}$$
$$= \{H_0, I^{(2)}\} + \{H_0, I^{(3)}\} + \{F, I^{(2)}\} + \{F, I^{(3)}\} = 0,$$

neglecting terms of degree $\geq 4$ in $J_1^{1/2}, J_2^{1/2}$ and in their products. Since $\{H_0, I^{(2)}\} = 0$ and $\{F, I^{(3)}\}$ is of degree 4 we arrive at the equation

$$\{H_0, I^{(3)}\} = -\{F, I^{(2)}\};$$

hence

$$\omega_1 \frac{\partial I^{(3)}}{\partial \chi_1} + \omega_2 \frac{\partial I^{(3)}}{\partial \chi_2} = \omega_1 \frac{\partial F}{\partial \chi_1} - \omega_2 \frac{\partial F}{\partial \chi_2},$$

from which it follows that to each term $A\sin(m\chi_1 + n\chi_2)$ appearing in $F$ there corresponds a term

$$\frac{\omega_1 m - \omega_2 n}{\omega_1 m + \omega_2 n} A \sin(m\chi_1 + n\chi_2)$$

in $I^{(3)}$. Therefore the required integral is

$$I^{(3)}(J, \chi) = J_1^{3/2}[F_1 \sin \chi_1 + F_2 \sin 3\chi_1] + J_1 J_2^{1/2}\left[-F_3 \sin \chi_2\right.$$
$$\left. + \frac{2\omega_1 - \omega_2}{2\omega_1 + \omega_2} F_4 \sin(2\chi_1 + \chi_2) + \frac{2\omega_1 + \omega_2}{2\omega_1 - \omega_2} F_5 \sin(2\chi_1 - \chi_2)\right]$$
$$+ J_1^{1/2} J_2 \left[F_6 \sin \chi_1 + \frac{\omega_1 - 2\omega_2}{\omega_1 + 2\omega_2} F_7 \sin(2\chi_2 + \chi_1)\right.$$
$$\left.+ \frac{\omega_1 + 2\omega_2}{\omega_1 - 2\omega_2} F_8 \sin(2\chi_2 - \chi_1)\right] + J_2^{3/2}[-F_9 \sin \chi_2 - F_{10} \sin 3\chi_2].$$

The procedure followed is justified as long as

$$2\omega_1 \pm \omega_2 \neq 0, \quad \omega_1 \pm 2\omega_2 \neq 0.$$

*Problem 3*
Consider the motion of a ball of mass $m$ bouncing elastically between two walls that are slowly moving towards one another, and prove that the action is an adiabatic invariant.

*Solution*
Consider the motion of a ball of mass $m$ bouncing elastically between two fixed walls at a distance $d$ (see Percival and Richards 1986). Although this system cannot be described by a regular Hamiltonian (because the speed $v$ of the ball varies discontinuously from $v \to -v$ at each hit), the motion can be studied simply and the system is integrable. Let $E = \tfrac{1}{2}mv^2$ be the energy of the ball, $q$ be its position and $p = mv$ be the momentum. The phase curves are rectangles (Fig. 12.3), the action $J$ is given by

$$ J = \frac{1}{2\pi} \text{ (area of the rectangle)} = \frac{1}{2\pi}(2mvd) = \frac{d}{\pi}\sqrt{2mE}, $$

and the energy, as a function of the action, is given by

$$ E = \frac{1}{2m}\left(\frac{\pi J}{d}\right)^2. $$

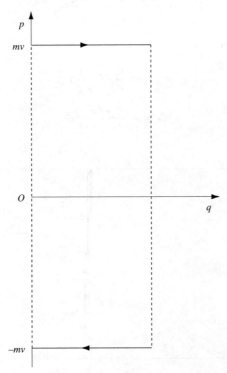

Fig. 12.3

Suppose now that one of the two walls moves towards the other with a velocity $\varepsilon$, such that at time $t$ the distance between the walls is $x(t) = d - \varepsilon t$. Let

$$\varepsilon \ll \omega_0 = \frac{\pi^2}{md^2} J = \frac{\pi}{d} v,$$

and indicate by $v_n$ the velocity of the ball just before the $n$th collision with the moving wall, and by $v_{n+1}$ the velocity immediately after the collision (Fig. 12.4). Evidently

$$v_{n+1} = v_n + 2\varepsilon,$$

from which it follows that

$$v_n = v_0 + 2n\varepsilon.$$

If $x_n$ is the distance between the planes at the moment corresponding to the $n$th collision, and $\Delta t_n$ measures the time interval between the $(n+1)$th and the $n$th collisions, we have

$$\Delta t_n = \frac{x_{n+1} + x_n}{v_{n+1}} = \frac{x_n - x_{n+1}}{\varepsilon},$$

from which we deduce

$$x_{n+1} = \frac{v_{n+1} - \varepsilon}{v_{n+1} + \varepsilon} x_n,$$

$$\Delta t_n = \frac{2 x_n}{v_{n+1} + \varepsilon}.$$
(12.154)

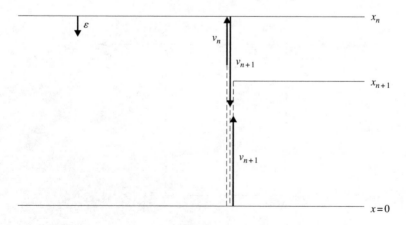

Fig. 12.4

**Table 12.1** $x_0 = 1, v_0 = 1, m = 1, \varepsilon = 0.01$.

| $n$ | $x_n$ | $v_n$ | $t_n$ | $E_n$ | $\pi \cdot J_n$ |
|---|---|---|---|---|---|
| 0 | 1 | 1.0 | 0 | 0.5 | 1 |
| 10 | 0.835 | 1.2 | 16.5 | 0.72 | 1.002 |
| 50 | 0.502 | 2.0 | 49.8 | 2.0 | 1.005 |
| 100 | 0.336 | 3.0 | 66.4 | 4.5 | 1.007 |
| 500 | 0.092 | 11.0 | 90.8 | 60.5 | 1.009 |
| 1000 | 0.048 | 21.0 | 95.2 | 220.5 | 1.010 |

**Table 12.2** $x_0 = 1, v_0 = 1, m = 1, \varepsilon = 0.1$.

| $n$ | $x_n$ | $v_n$ | $t_n$ | $E_n$ | $\pi \cdot J_n$ |
|---|---|---|---|---|---|
| 0 | 1 | 1 | 0 | 0.5 | 1 |
| 10 | 0.355 | 3 | 6.45 | 4.5 | 1.065 |
| 50 | 0.099 | 11 | 9.01 | 60.5 | 1.089 |
| 100 | 0.052 | 21 | 9.48 | 220.5 | 1.092 |
| 500 | 0.01088 | 101 | 9.89 | 5100.5 | 1.099 |
| 1000 | 0.0055 | 201 | 9.95 | 20200.5 | 1.099 |

Since $v_{j+1} - \varepsilon = v_j + \varepsilon$, iterating equation (12.154) we find

$$x_{n+1} = x_0 \prod_{j=0}^{n} \frac{v_j + \varepsilon}{v_{j+1} + \varepsilon} = \frac{v_0 + \varepsilon}{v_{n+1} + \varepsilon} x_0 = \frac{v_0 + \varepsilon}{v_0 + (2n+3)\varepsilon} x_0, \qquad (12.155)$$

and hence the $(n+1)$th collision happens at the instant

$$t_{n+1} - t_0 = \sum_{j=0}^{n} \Delta t_j = \frac{x_0 - x_{n+1}}{\varepsilon} = \frac{2(n+1)\varepsilon}{v_0 + (2n+3)\varepsilon}.$$

The action between two successive collisions is given by

$$J_n = \frac{m}{\pi} v_n x_n$$

(the system keeps the memory of the last hit); therefore by equation (12.155) we have

$$J_n = \frac{m}{\pi}(v_0 + 2n\varepsilon)\frac{v_0 + \varepsilon}{v_0 + (2n+1)\varepsilon} x_0 = J_0 + \varepsilon x_0 \frac{m}{\pi} \frac{2n\varepsilon}{v_0 + (2n+1)\varepsilon}, \qquad (12.156)$$

while

$$E_n = \frac{1}{2}mv_n^2 = \frac{1}{2}m(v_0 + 2n\varepsilon)^2 = E_0 + 2\varepsilon v_0 nm + 2mn^2\varepsilon^2.$$

It is immediate to check that from (12.156) it follows that
$$|J_n - J_0| \le \frac{m\varepsilon x_0}{\pi},$$
for all $n \in \mathbb{N}$. The action is therefore an adiabatic invariant, while the energy is not, because
$$E_n - E_0 = 2m\varepsilon n(v_0 + n\varepsilon),$$
and therefore $E_n - E_0 = \mathcal{O}(1)$ if $n = \mathcal{O}(1/\varepsilon)$. Tables 11.1 and 11.2 show the values of $E, J, x$ and $t$ for $n = 10, 50, 100, 500$ and $1000$, corresponding to $x_0 = 1$, $v_0 = 1$ and $\varepsilon = 0.01$ and $\varepsilon = 0.1$, respectively. ∎

# 13 ANALYTICAL MECHANICS: AN INTRODUCTION TO ERGODIC THEORY AND TO CHAOTIC MOTION

The completely canonically integrable systems are the mechanical model for the study of systems with an *orderly and regular* behaviour. The main idea in all studies in the nineteenth century has been to reduce the study of mechanical systems to the study of integrable systems, both exactly by using canonical transformations and the Hamilton–Jacobi equations, and approximately using the canonical theory of perturbations.

Poincaré proved however that this is not always possible, and that Hamiltonian systems may exhibit a behaviour that is totally different from the behaviour of integrable systems, exhibiting *disorderly and chaotic* orbits. The appropriate language for the study of these systems connects the study of dynamical systems to probability theory (of which we recall the main introductory notions). This is the point of view underlying *ergodic theory*, which we introduce in this chapter.

We start by introducing the notions of *measure* and a *measurable dynamical system*.

## 13.1 The concept of measure

DEFINITION 13.1  Let $X$ be a non-empty set. A non-empty family $\mathcal{A}$ of subsets of $X$ is a $\sigma$-algebra on $X$ if it satisfies the following properties:

(1) $A \in \mathcal{A} \Rightarrow A^c \in \mathcal{A}$;
(2) for every sequence $\{A_i\}$, $i \in \mathbf{N}$ of elements in $\mathcal{A}$ we have $\cup_{i \in \mathbf{N}} A_i \in \mathcal{A}$. ∎

Any family of subsets of $X$ for which (1) and (2) are valid for *finite* sequences is called an *algebra*. It is immediate to verify that any $\sigma$-algebra is also an algebra. In particular if $A, B \in \mathcal{A}$ then $A \cup B \in \mathcal{A}$.

*Example* 13.1
For a given set $X$ we can obtain trivial examples of $\sigma$-algebras by choosing

(a) the family of all subsets of $X$;
(b) the pair $\{\emptyset, X\}$. ∎

*Remark* 13.1
If $\mathcal{A}$ is a $\sigma$-algebra on $X$ it is easy to prove that the following properties hold.

(i) Imposing (2) is equivalent to imposing that $\cap_{i \in \mathbf{N}} A_i \in \mathcal{A}$ (it is enough to note that $\cap_{i \in \mathbf{N}} A_i = (\cup_{i \in \mathbf{N}} A_i^c)^c$ and $A_i^c \in \mathcal{A}$); in particular $A \cap B \in \mathcal{A}$ for every $A, B \in \mathcal{A}$.
(ii) $\emptyset \in \mathcal{A}$, $X \in \mathcal{A}$ (indeed $A \in \mathcal{A} \Rightarrow X = A \cup A^c \in \mathcal{A}$, $\emptyset = A \cap A^c \in \mathcal{A}$).
(iii) $A, B \in \mathcal{A} \Rightarrow A \setminus B \in \mathcal{A}$ (indeed $A \setminus B = A \cap B^c$).
(iv) The intersection of $\sigma$-algebras on $X$ is a $\sigma$-algebra (if $\mathcal{I}$ denotes the intersection, $\mathcal{I} \ni X$ and hence it is non-empty, the properties (1) and (2) of Definition 13.1 are easily proved). ∎

The latter property allows us to generate the smallest $\sigma$-algebra on $X$ containing a prescribed family $\mathcal{F}$ of subsets of $X$.

DEFINITION 13.2 *Given a family $\mathcal{F}$ of subsets of $X$ the $\sigma$-algebra on $X$ generated by $\mathcal{F}$ is the intersection of all $\sigma$-algebras $\mathcal{A}$ such that $\mathcal{A} \supset \mathcal{F}$.* ∎

The definition is meaningful because there exists at least one $\sigma$-algebra $\mathcal{A}$ such that $\mathcal{A} \supset \mathcal{F}$ (the $\sigma$-algebra of all subsets of $X$). An important case is the following.

DEFINITION 13.3 *Let $X = \mathbf{R}^l$. We call a Borel $\sigma$-algebra on $\mathbf{R}^l$ (denoted by $\mathcal{B}(\mathbf{R}^l)$) the one generated by the family of open subsets of $\mathbf{R}^l$. The elements of $\mathcal{B}(\mathbf{R}^l)$ are called* Borelian sets *of $\mathbf{R}^l$. More generally, if $X$ is any topological space, the Borel $\sigma$-algebra of $X$ is the $\sigma$-algebra generated by the open subsets of $X$.* ∎

We can now define the concept of measure.

DEFINITION 13.4 *Given a set $X$ and a $\sigma$-algebra $\mathcal{A}$ on $X$, a measure is a function $\mu : \mathcal{A} \to [0, +\infty]$ such that*

(1) $\mu(\emptyset) = 0$,

(2) $\mu\left(\bigcup_{i \in \mathbf{N}} A_i\right) = \sum_{i \in \mathbf{N}} \mu(A_i)$

*for every sequence $\{A_i\}$ of disjoint elements of $\mathcal{A}$.* ∎

Note that the function $\mu$ is allowed to take the value $+\infty$.

DEFINITION 13.5 *A triple $(X, \mathcal{A}, \mu)$ of a set $X$, a $\sigma$-algebra $\mathcal{A}$ on $X$ and a measure $\mu$ are called a* measure space. ∎

A set $A \subset X$ has *zero measure* if there exists $A_1 \in \mathcal{A}$ such that $A \subset A_1$ and $\mu(A_1) = 0$.

Two sets $A_1, A_2$ *coincide (mod 0)* and we write $A_1 = A_2 \pmod{0}$ if the symmetric difference $A_1 \triangle A_2$ has zero measure.

If a property is valid for all points of $A \subset X$ except for those in a set of measure zero, we say that the property is true for $\mu$-almost all $x \in A$ (written as $\mu$-a.a. $x \in A$).

An important case is the case of $\mathbf{R}$ and of the *Lebesgue measure* on $\mathcal{B}(\mathbf{R})$ which associates with intervals their lengths, and at the same time the case of the Lebesgue measure on $\mathcal{B}(\mathbf{R}^l)$. It can be shown that the Lebesgue measure $\lambda : \mathcal{B}(\mathbf{R}^l) \to [0, +\infty]$ is the only measure with the property that for every $A = (a_1, b_1) \times \cdots \times (a_l, b_l)$, we have

$$\lambda(A) = (b_1 - a_1)(b_2 - a_2) \cdots (b_l - a_l).$$

*Example* 13.2
A simple example of measure space is given by a finite set $X = \{x_1, \ldots, x_N\}$ with the $\sigma$-algebra $\mathcal{A} = \mathcal{P}(X)$, the set of parts of $X$. A measure is defined by assigning to every element $x_i \in X$ a real number $p_i \geq 0$. The measure of the subset $\{x_{i_1}, \ldots, x_{i_k}\} \subset X$ is therefore $p_{i_1} + \ldots + p_{i_k}$. If $\sum_{i=1}^N p_i = 1$ the measure is called a *probability measure*. Interesting examples are given by $X = \{0, 1\}$ or $X = \{1, 2, 3, 4, 5, 6\}$ with probabilities $p_1 = p_2 = \frac{1}{2}$ and $p_1 = p_2 = \ldots = p_6 = \frac{1}{6}$, respectively, which can be chosen to represent the probability spaces associated with the toss of a coin or the roll of a die. ■

*Example* 13.3
Let $(X_i, \mathcal{A}_i, \mu_i)$, $i = 1, \ldots, l$, be measure spaces. The Cartesian product $X = X_1 \times \ldots \times X_l$ has a natural structure of a measure space, whose $\sigma$-algebra $\mathcal{A}$ is the smallest $\sigma$-algebra of subsets of $X$ containing the subsets of the form $A_1 \times \ldots \times A_l$, where $A_i \in \mathcal{A}_i$, $i = 1, \ldots, l$. On these subsets the measure $\mu$ is defined by

$$\mu(A_1 \times \ldots \times A_l) = \mu_1(A_1) \ldots \mu_l(A_l). \tag{13.1}$$

It can be proved (see Lasota and Mackey 1985, theorem 2.2.2, p. 24) that there exists a unique extension of the measure $\mu$ defined by (13.1) to the $\sigma$-algebra $\mathcal{A}$ of $X$. The space $(X, \mathcal{A}, \mu)$ thus obtained is called the *product space* and the measure $\mu$ is called the *product measure*.

If $X_1 = \ldots = X_l = \{0, 1\}$ or $\{1, 2, 3, 4, 5, 6\}$ and the measures $\mu_i$ coincide with the measure defined in the previous example, the product space coincides with the space of finite sequences of tosses of a coin or rolls of a die, and the product measure with the probability associated with each sequence. ■

DEFINITION 13.6 *If $\mu(X) = 1$, a measure $\mu$ is called a* probability measure *and the triple $(X, \mathcal{A}, \mu)$ is a* probability space. ■

In what follows we sometimes denote by $\mathcal{M}(X)$ the set of probability measures on a measure space $(X, \mathcal{A}, \mu)$.

## 13.2 Measurable functions. Integrability

The theory of Lebesgue measurable functions (see Giusti 1989), with its most significant results (the theory of integration, Fatou's theorems on monotone and dominated convergence, the absolute continuity of the integral, and so on), can be easily extended to the functions $f : X \to \mathbf{R}$, where $(X, \mathcal{A}, \mu)$ is an arbitrary measure space (see Rudin 1974).

We recall first of all the notion of an integral of a measurable function.

DEFINITION 13.7  Let $f : A \to [-\infty, +\infty]$ be defined on $A \subset X$ belonging to a $\sigma$-algebra $\mathcal{A}$ on $X$. The function $f$ is called measurable (with respect to $\mathcal{A}$) if $\{x \in A \mid f(x) < t\} \in \mathcal{A}, \; \forall \, t \in \mathbf{R}$. ∎

It is possible to prove that the inequality $f(x) < t$ can be replaced by one of the following: $f(x) \leq t$, $f(x) > t$, $f(x) \geq t$.

To define the integral on a measure space $(X, \mathcal{A}, \mu)$ consider first the so-called *simple* functions, of the form

$$g = \sum_{i=1}^{n} \alpha_i \chi_{A_i}, \tag{13.2}$$

with $n$ finite, $\alpha_i \geq 0$, $A_i \in \mathcal{A}$ disjoint and $\chi_{A_i}$ the characteristic function of $A_i$, and hence

$$\chi_{A_i}(x) = \begin{cases} 1, & \text{if } x \in A_i, \\ 0, & \text{if } x \in A_i^c. \end{cases}$$

In this case we define

$$\int_X g \, \mathrm{d}\mu = \sum_{i=1}^{n} \alpha_i \mu(A_i). \tag{13.3}$$

In particular $\int_X \chi_A \, \mathrm{d}\mu = \mu(A)$, $\forall A \in \mathcal{A}$.

If $f : X \to [0, +\infty]$, we set

$$\int_X f \, \mathrm{d}\mu = \sup_{g \in \mathcal{G}(f)} \int_X g \, \mathrm{d}\mu, \tag{13.4}$$

where $\mathcal{G}(f)$ is the set of simple functions such that $g \leq f$.

Finally for a generic $f : X \to [-\infty, +\infty]$ we define

$$\int_X f \, \mathrm{d}\mu = \int_X f^+ \, \mathrm{d}\mu - \int_X f^- \, \mathrm{d}\mu, \tag{13.5}$$

where $f^+(x) = \max(0, f(x))$, $f^-(x) = \max(0, -f(x))$, if at least one of these integrals is finite. In this case $f$ is called $\mu$-summable. If $\int_X |f| \, \mathrm{d}\mu < +\infty$ we say that $f$ is $\mu$-integrable. Let $A \in \mathcal{A}$; $f$ is said to be $\mu$-integrable on $A$ if the function $f\chi_A$ is $\mu$-integrable on $X$.

We set
$$\int_A f \, d\mu = \int_X f \chi_A \, d\mu. \tag{13.6}$$

The space of $\mu$-integrable functions on $X$ is denoted by $L^1(X, \mathcal{A}, \mu)$.

Consider $0 < p < +\infty$. The space of functions $f$ such that $|f|^p$ is $\mu$-integrable on $X$ is denoted by $L^p(X, \mathcal{A}, \mu)$.

A particular and well-known case is that of the functions on $\mathbf{R}^l$ which are Lebesgue integrable.

*Remark* 13.2

Assume that $X$ is a compact metric space and $\mathcal{A}$ is the $\sigma$-algebra of Borel sets $X$. In this case one can define the *support* of a measure $\mu$ as the smallest compact set $K \subset X$ such that $\mu(A) = 0$ for all $A \subset X \setminus K$. Moreover, it is possible to endow $\mathcal{M}(X)$ with a topological structure by defining at every point $\mu \in \mathcal{M}(X)$ a basis of neighbourhoods

$$V_{\varphi,\varepsilon}(\mu) := \left\{ \nu \in \mathcal{M}(X) \mid \left| \int_X \varphi \, d\nu - \int_X \varphi \, d\mu \right| \leq \varepsilon \right\}, \tag{13.7}$$

where $\varepsilon > 0$ and $\varphi : X \to \mathbf{R}$ is continuous. In this topology a sequence of measures $(\mu_n)_{n \in \mathbf{N}} \subseteq \mathcal{M}(X)$ converges to $\mu \in \mathcal{M}(X)$ if for every $\varphi : X \to \mathbf{R}$ continuous we have

$$\int_X \varphi \, d\mu_n \to \int_X \varphi \, d\mu. \tag{13.8}$$

∎

*Remark* 13.3

In what follows we always assume that $X$ is *totally $\sigma$-finite*, and hence that $X = \bigcup_{i=1}^\infty A_i$, where the sets $A_i \in \mathcal{A}$ have measure $\mu(A_i) < +\infty$ for every $i \in \mathbf{N}$. ∎

DEFINITION 13.8 *Let $X$ be a set, and $\mathcal{A}$ be a $\sigma$-algebra of subsets of $X$. If $\mu, \nu : \mathcal{A} \to [0, +\infty]$ are two measures, we say that $\mu$ is absolutely continuous with respect to $\nu$ if for every $A \in \mathcal{A}$ such that $\nu(A) = 0$ we have $\mu(A) = 0$. If $\mu$ is not absolutely continuous with respect to $\nu$ then it is said to be* singular *(with respect to $\nu$).* ∎

An important characterisation of measures which are absolutely continuous with respect to another measure is given by the following theorem. The proof, that goes beyond the scope of this limited introduction, can be found in the book of Rudin, already cited.

THEOREM 13.1 (Radon–Nikodym) *A measure $\mu : \mathcal{A} \to [0, +\infty]$ is absolutely continuous with respect to another measure $\nu : \mathcal{A} \to [0, +\infty]$ if and only if there exists a function $\rho : X \to \mathbf{R}$, integrable with respect to $\nu$ on every subset $A \in \mathcal{A}$ such that $\nu(A) < +\infty$, and such that for every $A \in \mathcal{A}$ we have*

$$\mu(A) = \int_A \rho \, d\nu. \tag{13.9}$$

The function $\rho$ is unique (if we identify any two functions which only differ on a set of $\nu$ measure zero), and it is called the Radon–Nikodym derivative of $\mu$ with respect to $\nu$, or density of $\mu$ with respect to $\nu$, and it is denoted by $\rho = \mathrm{d}\mu/\mathrm{d}\nu$. ∎

*Remark* 13.4
We have that $g \in L^1(X, \mathcal{A}, \mu)$ if and only if $g\rho \in L^1(X, \mathcal{A}, \nu)$. In this case

$$\int_X g \, \mathrm{d}\mu = \int_X g\rho \, \mathrm{d}\nu. \tag{13.10}$$

∎

## 13.3 Measurable dynamical systems

The objects of study of ergodic theory are the dynamical systems that preserve a measure, in a sense that we now make precise.

DEFINITION 13.9 *Let $(X, \mathcal{A}, \mu)$ be a measure space. A transformation $S : X \to X$ is said to be* measurable *if for every $A \in \mathcal{A}$, we have $S^{-1}(A) \in \mathcal{A}$.*

*A measurable transformation is* non-singular *if $\mu(S^{-1}(A)) = 0$ for all $A \in \mathcal{A}$ such that $\mu(A) = 0$.* ∎

Obviously $S^{-1}(A) = \{x \in X \mid S(x) \in A\}$ and $S$ is not necessarily invertible.

For example, if $X$ is a topological space, $\mathcal{A}$ is the $\sigma$-algebra of Borel sets and $S$ is a homeomorphism, then $S$ is measurable and non-singular if and only if the inverse map is measurable and non-singular.

DEFINITION 13.10 *Let $(X, \mathcal{A}, \mu)$ be a measure space. A measurable non-singular transformation $S : X \to X$* preserves the measure *(i.e. the measure $\mu$ is* invariant with respect to the transformation $S$*) if for every $A \in \mathcal{A}$, we have $\mu(S^{-1}(A)) = \mu(A)$.* ∎

If $S$ is invertible with a measurable non-singular inverse and if it preserves the measure, then clearly $\mu(S^{-1}(A)) = \mu(A) = \mu(S(A))$, $\forall A \in \mathcal{A}$.

If however $S$ is not invertible, the following simple example highlights the need to use the condition $\mu(S^{-1}(A)) = \mu(A)$ in the previous definition. Choose $X = (0, 1)$ and the $\sigma$-algebra $\mathcal{A}$ of Borel sets on $(0, 1)$; the transformation $S(x) = 2x \,(\mathrm{mod}\, 1)$ preserves the Lebesgue measure $\lambda$, while if we take an interval $(a, b) \subset (0, 1)$ then $\lambda(S(a, b)) = 2\lambda((a, b))$.

*Remark* 13.5
Let $f$ be $\mu$-integrable and assume that $S$ preserves the measure $\mu$. Then

$$\int_X f(x) \, \mathrm{d}\mu = \int_X f(S(x)) \, \mathrm{d}\mu.$$

Conversely, if this property holds for every $f : X \to \mathbf{R}$ continuous, then $S$ preserves the measure $\mu$. ∎

DEFINITION 13.11 *A* measurable dynamical system *$(X, \mathcal{A}, \mu, S)$ is constituted by a probability space $(X, \mathcal{A}, \mu)$ and by a transformation $S : X \to X$ which preserves* the measure $\mu$. The orbit of a point $x \in X$ is the infinite sequence of points $x, S(x), S^2(x) = S(S(x)), \ldots, S^{n+1}(x) = S(S^n(x)), \ldots$ obtained by iterating $S$. ∎

**Remark** 13.6
The recurrence theorem of Poincaré (Theorem 8.4) can be extended without difficulty to the case of measurable dynamical systems $(X, \mathcal{A}, \mu, S)$. We state it, and leave the proof as an exercise: for every $A \in \mathcal{A}$ the subset $A_0$ of all points $x \in A$ such that $S^n(x) \in A$ for infinitely many values of $n \in \mathbf{N}$ belongs to $\mathcal{A}$ and $\mu(A) = \mu(A_0)$.

A 'topological' version of the recurrence theorem of Poincaré is presented in Problem 13.15. ∎

A particularly interesting case arises when $X$ is a subset of $\mathbf{R}^l$ (or, more generally, of a differentiable Riemannian manifold $M$ of dimension $l$) and $\mu_\rho$ is a probability measure which is absolutely continuous with respect to the Lebesgue measure

$$d\mu_\rho(\mathbf{x}) = \rho(\mathbf{x})\, d\mathbf{x} \tag{13.11}$$

(or $d\mu_\rho = \rho\, dV_g$, where $dV_g = \sqrt{\det(g_{ij})}\, d^l\mathbf{x}$ is the volume element associated with the metric $g$ on the manifold $M$). The definition of a measure that is invariant with respect to a non-singular transformation $S : X \to X$ is therefore equivalent to

$$\int_{S^{-1}(A)} \rho(\mathbf{x})\, d\mathbf{x} = \int_A \rho(\mathbf{x})\, d\mathbf{x}, \tag{13.12}$$

for every $A \in \mathcal{A}$. A very important problem in ergodic theory is the problem of determining all measures that are invariant for a given transformation. A case when this is possible is given by the following systems.

Let $X = [0, 1]$, and $S : X \to X$ be non-singular. Assume that $S$ is piecewise monotone and of class $\mathcal{C}^1$, and hence that there exists a finite or countable decomposition of the interval $[0, 1]$ into intervals $[a_i, a_{i+1}]$, $i \in \mathcal{I}$, on which $S$ is monotone (and $\mathcal{C}^1$ in the interior). On each of these subintervals the inverse $S_i^{-1}$ of $S$ is well defined. Let $A = [0, x]$. Equation (13.12) becomes

$$\int_0^x \rho(s)\, ds = \sum_{i \in \mathcal{I}} \int_{S_i^{-1}([0,x])} \rho(s)\, ds,$$

from which, by differentiating with respect to $x$, we obtain

$$\rho(x) = \sum_{i \in \mathcal{I}_x} \frac{\rho(S_i^{-1}(x))}{|S'(S_i^{-1}(x))|}, \tag{13.13}$$

where $\mathcal{I}_x$ indicates the subset of $\mathcal{I}$ corresponding to the indices $i$ such that $S_i^{-1}(x) \neq \emptyset$. Equation (13.13) is therefore a condition (necessary and sufficient) for the density $\rho$ for a measure that is absolutely continuous with respect to the Lebesgue measure to be invariant with respect to $S$.

*Example* 13.4 (Ulam and von Neumann 1947)
Consider $X = [0,1]$, $S(x) = 4x(1-x)$. The probability measure $d\mu(x) = dx/\pi\sqrt{x(1-x)}$ is invariant.

Indeed, to every point $x \in X$ there correspond two preimages $S_1^{-1}(x) = \frac{1}{2}(1-\sqrt{1-x}) \in [0,1/2]$ and $S_2^{-1}(x) = \frac{1}{2}(1+\sqrt{1-x}) \in [1/2,1]$. Therefore, equation (13.13) becomes

$$\frac{1}{\sqrt{x(1-x)}} = \frac{1}{\sqrt{S_1^{-1}(x)(1-S_1^{-1}(x))}\,|4-8S_1^{-1}(x)|}$$
$$+ \frac{1}{\sqrt{S_2^{-1}(x)(1-S_2^{-1}(x))}\,|4-8S_2^{-1}(x)|},$$

which is immediately verified. ∎

*Example* 13.5: *the p-adic transformation*
Consider $X = [0,1]$, and $p \in \mathbf{N}$, $S(x) = px \pmod{1}$, and hence $S(x) = px - m$ if $m/p \le x < (m+1)/p$, $m = 0, \ldots, p-1$, $S(1) = 1$. The $p$-adic transformation preserves the Lebesgue measure. ∎

*Example* 13.6: *the Gauss transformation*
Consider $X = [0,1)$, $S(x) = 1/x - [1/x]$ if $x \ne 0$, $S(0) = 0$, where $[\cdot]$ denotes the integer part of a number. The probability measure $d\mu(x) = dx/(1+x)\log 2$ is invariant. Indeed, $S$ is invertible on the intervals $[1/(n+1), 1/n]$, $n \in \mathbf{N}$, with inverse $S_n^{-1}(x) = 1/(n+x)$, and

$$\sum_{n=1}^{\infty} \frac{1}{1+S_n^{-1}(x)} \frac{1}{|S'(S_n^{-1}(x))|} = \sum_{n=1}^{\infty} \frac{1}{1+1/(n+x)} \frac{1}{(n+x)^2}$$
$$= \sum_{n=1}^{\infty} \frac{1}{(n+x+1)(n+x)}$$
$$= \sum_{n=1}^{\infty} \left(\frac{1}{n+x} - \frac{1}{n+x+1}\right) = \frac{1}{x+1}. \quad ∎$$

*Example* 13.7: *the 'baker's transformation'*
If $X = [0,1] \times [0,1]$, then

$$S(x,y) = \begin{cases} \left(2x, \dfrac{1}{2}y\right), & \text{if } 0 \le x < \dfrac{1}{2}, \\ \left(2x-1, \dfrac{1}{2}y + \dfrac{1}{2}\right), & \text{if } \dfrac{1}{2} \le x \le 1 \end{cases}$$

(Fig. 13.1) preserves the Lebesgue measure. From a geometrical point of view, $S$ transforms the square $[0,1] \times [0,1]$ in the rectangle $[0,2] \times [0,\frac{1}{2}]$, cuts out the right half of this rectangle and translates it on top of the left half. ∎

*Example* 13.8
Let $X$ be a Riemannian manifold, and $S : X \to X$ be an isometry. The measure $dV_g$ is invariant with respect to $S$. ∎

*Example* 13.9
Let $X = \mathbf{R}^{2l}$, and $S : X \to X$ be a completely canonical transformation. By the Liouville theorem, the Lebesgue measure is invariant with respect to $S$. ∎

*Example* 13.10: 'Arnol'd's cat' (Arnol'd and Avez 1967)
If $X = \mathbf{T}^2$, then $S(x_1, x_2) = (x_1 + x_2, x_1 + 2x_2) \pmod 1$ preserves the Lebesgue measure. ∎

Assume that $X$ is a Riemannian manifold and that $S : X \to X$ is a diffeomorphism of $M$ such that, $\forall\, x \in X$, $|\det(DS(x))| < 1$. Then it can be shown that there exists an *attractor* $\Omega \subset X$ and a *basin of attraction* $U$, i.e. a neighbourhood of $\Omega$ such that $S(\overline{U}) \subset U$ and $\cap_{n \geq 0} S^n(U) = \Omega$. In addition, the volume of $\Omega$ (with respect to the volume form induced by the Riemannian structure on $X$) vanishes and all the probability measures that are invariant for $S$ have support contained in the attractor $\Omega$.

Some obvious examples are $X = \mathbf{R}$, $S(x) = x/2$, $\Omega = \{0\}$, $U = \mathbf{R}$, where the only invariant measure is the Dirac measure $\delta_0(x)$ at the point $x = 0$:

$$\delta_0(A) = \begin{cases} 0, & \text{if } 0 \notin A, \\ 1, & \text{if } 0 \in A. \end{cases}$$

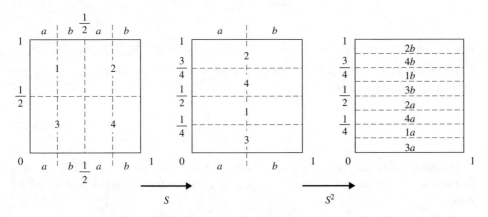

**Fig. 13.1**

In general

$$\delta_y(A) = \begin{cases} 0, & \text{if } y \notin A, \\ 1, & \text{if } y \in A. \end{cases}$$

Another example is given by $X = \mathbf{R}^2$, where $S(x,y)$ is the flow at time $t=1$ of the following system of ordinary differential equations:

$$\dot{x} = x(1 - x^2 - y^2) - y,$$
$$\dot{y} = y(1 - x^2 - y^2) + x.$$

Introducing polar coordinates $x = r\cos\theta$, $y = r\sin\theta$ it is immediate to verify that $\dot{r} = r(1 - r^2)$, $\dot{\theta} = 1$, and therefore the circle $r = 1$ is an attractor. Note that $\dot{r} > 0$ if $0 < r < 1$, while $\dot{r} < 0$ if $1 < r < +\infty$.

In this case $\Omega = \mathbf{S}^1$, $U = \mathbf{R}^2 \setminus \{0\}$ and the invariant measure is $d\mu(r,\theta) = \delta_{r=1}d\theta/2\pi$. The support of this measure is precisely the limit cycle $\mathbf{S}^1 = \{r = 1\}$.

## 13.4 Ergodicity and frequency of visits

Consider a measurable dynamical system $(X, \mathcal{A}, \mu, S)$. The first fundamental notion associated with such a system is its 'statistics', which is the frequency with which the orbit $\{S^j x\}_{j \in \mathbf{N}}$ of a point $x \in X$ visits a prescribed measurable set $A \in \mathcal{A}$.

To this end, we define for every $n \in \mathbf{N}$ the *number of visits* $T(x, A, n)$ of $A$ by the orbit of $x$:

$$T(x, A, n) := \sum_{j=0}^{n-1} \chi_A(S^j x), \tag{13.14}$$

where $\chi_A$ indicates the characteristic function of the set $A$.

DEFINITION 13.12 We call the frequency of visits $\nu(x, A)$ of the set $A$ by the orbit of $x$ the limit (when it exists)

$$\nu(x, A) = \lim_{n \to +\infty} \frac{1}{n} T(x, A, n). \tag{13.15}$$

∎

The first important result for the study of the orbit statistics of a dynamical system is the existence of the frequency of visits for $\mu$-almost every initial condition.

## 13.4 Analytical mechanics

**THEOREM 13.2** *For $\mu$-almost every $x \in X$ the frequency of visits $\nu(x, A)$ exists.*

*Proof*
Let $n \in \mathbf{N}$ be fixed, and set

$$\nu(x, A, n) := \frac{1}{n} T(x, A, n) \quad \text{(average frequency after } n \text{ steps)},$$
$$\overline{\nu}(x, A) := \limsup\nolimits_{n \to +\infty} \nu(x, A, n).$$

Obviously $0 \leq \overline{\nu}(x, A) \leq 1$ and $\forall\, k \in \mathbf{N}$ we have $\overline{\nu}(S^k x, A) = \overline{\nu}(x, A)$.
Analogous properties hold for $\underline{\nu}(x, A) := \liminf\limits_{n \to +\infty} \nu(x, A, n)$.
We want to prove that for $\mu$-almost every $x \in X$, we have

$$\overline{\nu}(x, A) = \underline{\nu}(x, A).$$

To this end, introduce $\varepsilon > 0$ and the function

$$\overline{\tau}_A(x, \varepsilon) := \min\{n \in \mathbf{N} \text{ such that } \nu(x, A, n) \geq \overline{\nu}(x, A) - \varepsilon\}$$

so that $\nu(x, A, \overline{\tau}_A(x, \varepsilon)) \geq \overline{\nu}(x, A) - \varepsilon$. Suppose that there exists $M > 0$ such that $\overline{\tau}_A(x, \varepsilon) \leq M$ for every $x \in X$. In this case we can decompose the orbit of $x$ up to time $n$, i.e. the finite sequence $(S^j x)_{0 \leq j < n}$, in parts on each of which the average frequency of visits of $A$ is at least $\overline{\nu}(x, A) - \varepsilon$. Indeed, consider the points $x_0 = x$, $x_1 = S^{\overline{\tau}_A(x_0, \varepsilon)} x_0$, $x_2 = S^{\overline{\tau}_A(x_1, \varepsilon)} x_1 = S^{\overline{\tau}_A(x_1, \varepsilon) + \overline{\tau}_A(x_0, \varepsilon)} x_0$, and so on.
We then have

$$\begin{aligned} T(x_j, A, \overline{\tau}_A(x_j, \varepsilon)) &= \overline{\tau}_A(x_j, \varepsilon) \nu(x_j, A, \overline{\tau}_A(x_j, \varepsilon)) \\ &\geq \overline{\tau}_A(x_j, \varepsilon)(\overline{\nu}(x_j, A) - \varepsilon) \\ &= \overline{\tau}_A(x_j, \varepsilon)(\overline{\nu}(x, A) - \varepsilon), \end{aligned}$$

where we have used the previous remark $\overline{\nu}(S^k x, A) = \overline{\nu}(x, A)\ \forall\, k \in \mathbf{N}$.
For fixed $n > M$, proceed in this way until a point $x_J = S^{\overline{\tau}_J} x$, with $\overline{\tau}_J = \sum_{j=0}^{J-1} \overline{\tau}_A(x_j, \varepsilon) < n$ and $\overline{\tau}_J + \overline{\tau}_A(x_J, \varepsilon) \geq n$, is reached. We then have

$$T(x, A, n) = \sum_{j=0}^{J-1} T(x_j, A, \overline{\tau}_A(x_j, \varepsilon)) + T(x_J, A, n - \overline{\tau}_J)$$

$$\geq \left( \sum_{j=0}^{J-1} \overline{\tau}_A(x_j, \varepsilon) \right) (\overline{\nu}(x, A) - \varepsilon) = \overline{\tau}_J (\overline{\nu}(x, A) - \varepsilon).$$

On the other hand, $\overline{\tau}_J \geq n - \overline{\tau}_A(x_J, \varepsilon) \geq n - M$, and hence we found that $\forall\, n > M$,

$$T(x, A, n) \geq (n - M)(\overline{\nu}(x, A) - \varepsilon).$$

Integrating this inequality over all the set $X$, since $\int_X T(x, A, n)\, d\mu = \sum_{j=0}^{n-1} \int_X \chi_A(S^j x)\, d\mu = \sum_{j=0}^{n-1} \int_X \chi_A(x)\, d\mu = n\mu(A)$, we find

$$n\mu(A) \geq (n - M) \int_X [\overline{\nu}(x, A) - \varepsilon]\, d\mu;$$

hence we find that $\forall\, \varepsilon > 0$,

$$\mu(A) \geq \int_X \overline{\nu}(x, A)\, d\mu - \varepsilon$$

or $\mu(A) \geq \int_X \overline{\nu}(x, A)\, d\mu$.

It is now possible to repeat this procedure considering $\underline{\nu}(x, A)$ in place of $\overline{\nu}(x, A)$, defining

$$\underline{\tau}_A(x, \varepsilon) := \min\{n \in \mathbf{N} \text{ such that } \nu(x, A, n) \leq \underline{\nu}(x, A) + \varepsilon\}$$

and supposing that this is also bounded, to arrive at the conclusion that

$$\int_X \underline{\nu}(x, A)\, d\mu \geq \mu(A) \geq \int_X \overline{\nu}(x, A)\, d\mu.$$

From this, taking into account that $\underline{\nu}$ and $\overline{\nu}$ are non-negative, it obviously follows that $\underline{\nu}(x, A) \geq \overline{\nu}(x, A)$ $\mu$-almost everywhere, and therefore that

$$\nu(x, A) \quad \text{exists } \mu\text{-almost everywhere.}$$

We still need to consider the case that $\overline{\tau}_A$ (or $\underline{\tau}_A$) is not a bounded function. In this case, remembering the definition, we choose $M > 0$ sufficiently large, so that we have

$$\mu(\{x \in X \mid \overline{\tau}_A(x, \varepsilon) > M\}) < \varepsilon.$$

From this choice it follows that

$$\overline{A} = A \cup \{x \in X \mid \overline{\tau}_A(x,\varepsilon) > M\},$$

so we have $\mu(\overline{A}) \leq \mu(A) + \varepsilon$. Considering now the number of visits $T(x, \overline{A}, n)$ relative to $\overline{A}$, setting $\overline{\nu}(x, \overline{A})$ as before, the function $\overline{\tau}_{\overline{A}}(x, \varepsilon)$ relative to $\overline{A}$ is now bounded. Hence proceeding as above, we arrive at the inequality

$$\mu(\overline{A}) > \int_X \overline{\nu}(x, \overline{A}) \, d\mu - \varepsilon,$$

from which, taking into account that $\mu(\overline{A}) \leq \mu(A) + \varepsilon$ and that $\overline{\nu}(x, \overline{A}) \geq \overline{\nu}(x, A)$, we deduce

$$\mu(A) \geq \int_X \overline{\nu}(x, A) \, d\mu - 2\varepsilon.$$

Since $\varepsilon$ is arbitrary, $\mu(A) \geq \int_X \overline{\nu}(x, A) \, d\mu$, exactly as in the case that $\overline{\tau}_A$ is bounded. ∎

The frequency of visits describes the 'statistics' of an orbit and can depend essentially upon it.

*Example* 13.11: *the square billiard*
Consider a point particle of unit mass moving freely in a square of side $2\pi$, and reflected elastically by the walls.

To study the motion, instead of reflecting the trajectory when it meets a wall, we can reflect the square with respect to the wall and consider the motion as undisturbed (note that this argument shows how to extend the motion of the particle to the case in which the trajectory meets one of the vertices of the square). In this way each trajectory of the billiard corresponds to a geodesic of the flat torus (recall the results of Sections I.7 and I.8), and hence we can apply the results of Section 11.7. In particular we find that if $\alpha$ denotes the angle of incidence (constant) of the trajectory on a wall, then the latter is periodic if $\tan \alpha$ is a rational number, and it is dense on the torus if $\tan \alpha$ is irrational. Given any measurable subset $A$ of the torus $\mathbf{T}^2$ it is evident that the frequency of visits of $A$ by one of the billiard's orbits depends essentially on the initial condition $(s, \alpha)$. However, it is possible to compute it exactly thanks to Theorem 11.9 and the following theorem of Lusin (see Rudin 1974, pp. 69–70).

There exists a sequence of continuous functions $\chi_j : \mathbf{T}^2 \to [0,1]$ such that $\chi_j \to \chi_A$ for $j \to \infty$ almost everywhere.

Applying Theorem 11.9 to the sequence $\chi_j$ we find that if $\tan\alpha$ is irrational, we have

$$\nu(s,\alpha,A) = \lim_{T\to\infty} \frac{1}{T} \int_0^T \chi_A(s+t\cos\alpha, t\sin\alpha)\, dt$$

$$= \lim_{j\to\infty} \lim_{T\to\infty} \frac{1}{T} \int_0^T \chi_j(s+t\cos\alpha, t\sin\alpha)\, dt$$

$$= \lim_{j\to\infty} \int_0^{2\pi} dx_1 \int_0^{2\pi} dx_2\, \chi_j(x_1,x_2) = \mu(A),$$

where $\mu$ denotes $1/(2\pi)^2$ multiplied by the Lebesgue measure on $\mathbf{T}^2$.

Therefore, *for almost every* initial condition, the frequency of visits of a measurable set $A$ by the corresponding trajectory of the billiard in the square is simply equal to the measure of $A$, and is hence *independent of the initial condition*. ∎

What we have just discussed is a first example of an *ergodic* system.

DEFINITION 13.13 *A measurable dynamical system* $(X,\mathcal{A},\mu,S)$ *is called* ergodic *if for every choice of* $A \in \mathcal{A}$ *it holds that* $\nu(x,A) = \mu(A)$ *for $\mu$-almost every* $x \in X$. ∎

We now turn to the study of ergodic systems and their properties.

We start with a remark: consider $A \in \mathcal{A}$ and let $\chi_A$ be its characteristic function. Since $\mu(A) = \int_X \chi_A(x)\, d\mu$, the ergodicity is equivalent to the statement that $\forall\, A \in \mathcal{A}$ and for $\mu$-almost every $x \in X$ one has

$$\lim_{n\to\infty} \frac{1}{n} \sum_{j=0}^{n-1} \chi_A(S^j x) = \int_X \chi_A(x)\, d\mu. \tag{13.16}$$

If instead of the characteristic function of a set we consider arbitrary integrable functions $f \in L^1(X,\mathcal{A},\mu)$, the following corresponding generalisation of Theorem 13.2 is called *Birkhoff's theorem*.

THEOREM 13.3 *Let* $(X,\mathcal{A},\mu,S)$ *be a measurable dynamical system, and let* $f \in L^1(X,\mathcal{A},\mu)$. *For $\mu$-almost every* $x \in X$ *the limit*

$$\hat{f}(x) := \lim_{n\to\infty} \frac{1}{n} \sum_{j=0}^{n-1} f(S^j x) \tag{13.17}$$

*exists and it is called* the time average *of* $f$ *along the orbit of the point* $x \in X$. ∎

For the proof of this theorem see Gallavotti (1981) and Cornfeld et al. (1982). We remark however that from the proof of Theorem 4.1 it follows in fact that the time average exists whenever $f$ is a finite linear combination of characteristic functions of measurable sets (hence every time that $f$ is a *simple* function):

$$f = \sum_{k=1}^m a_j \chi_{A_j}, \quad a_j \in \mathbf{R},\ A_j \in \mathcal{A},\ \forall\, j = 1,\ldots,m. \tag{13.18}$$

## 13.4 Analytical mechanics

Recall that every function $f \in L^1(X, \mathcal{A}, \mu)$ is the limit a.e. of a sequence of simple functions.

*Remark 13.7*
It is obvious that $\hat{f}(Sx) = \hat{f}(x)$, and hence that the time average depends on the orbit and not on the initial point chosen along the orbit. In addition, since $\mu$ is $S$-invariant, by Remark 13.5 we have

$$\int_X f(x)\,d\mu = \int_X f(Sx)\,d\mu,$$

from which it follows, by an application of the theorem of Lebesgue on dominated convergence to (13.17):

$$\langle f \rangle_\mu := \int_X f(x)\,d\mu = \int_X \hat{f}(x)\,d\mu. \tag{13.19}$$

The quantity $\langle f \rangle_\mu = \int_X f\,d\mu$ is called the *phase average* of $f$ (or *expectation* of $f$) and equation (13.19) implies that $f$ and its time average $\hat{f}$ have the same expectation value. ∎

The ergodicity of a dynamical system has as the important consequence that the phase and time averages are equal almost everywhere, as the following theorem shows (see property 4).

**THEOREM 13.4** *Let $(X, \mathcal{A}, \mu, S)$ be a measurable dynamical system. The following properties are equivalent.*

(1) *The system is ergodic.*
(2) *The system is metrically indecomposable: every invariant set $A \in \mathcal{A}$ (i.e. every set such that $S^{-1}(A) = A$) has measure $\mu(A)$ either zero or equal to $\mu(X) = 1$.*
(3) *If $f \in L^1(X, \mathcal{A}, \mu)$ is invariant (i.e. $f \circ S = f$ $\mu$-almost everywhere) then $f$ is constant $\mu$-almost everywhere.*
(4) *If $f \in L^1(X, \mathcal{A}, \mu)$ then $\langle f \rangle_\mu = \hat{f}(x)$ for $\mu$-almost every $x \in X$.*
(5) *$\forall\, A, B \in \mathcal{A}$ then*

$$\lim_{n \to +\infty} \frac{1}{n} \sum_{j=0}^{n-1} \mu(S^{-j}(A) \cap B) = \mu(A)\mu(B). \tag{13.20}$$

*Proof*
(1) $\Rightarrow$ (2) Suppose that there exists an invariant set $A \in \mathcal{A}$ with measure $\mu(A) > 0$. Since $A$ is invariant for every choice of $x \in A$ the frequency of visits of $A$ is precisely $\nu(x, A) = 1$. But since the system is ergodic for $\mu$-almost every $x$, then $\nu(x, A) = \mu(A)$. The hypothesis that $\mu(A) > 0$ then yields $\mu(A) = 1$.

(2) $\Rightarrow$ (3) If $f \in L^1(X, \mathcal{A}, \mu)$ is invariant, for every choice of $\gamma \in \mathbf{R}$ the set $A_\gamma = \{x \in X \mid f(x) \leq \gamma\}$ is invariant. Since the system is metrically indecomposable it follows that either $\mu(A_\gamma) = 0$ or $\mu(A_\gamma) = 1$. On the other

hand, if $\gamma_1 < \gamma_2$ clearly $A_{\gamma_1} \subset A_{\gamma_2}$. Therefore setting $\gamma_f = \inf\{\gamma \in \mathbf{R} \mid \mu(A_\gamma) = 1\}$ it follows that $f(x) = \gamma_f$ for $\mu$-almost every $x$.

(3) $\Rightarrow$ (4) Since the time average $\hat{f}$ is invariant we have that $\hat{f}$ is constant $\mu$-almost everywhere. From equation (13.19) it then follows that $\hat{f}(x) = \langle f \rangle_\mu$ for $\mu$-almost every $x \in X$.

(4) $\Rightarrow$ (1) It suffices to apply hypothesis (4) to the characteristic function $\chi_A$ of the set $A$.

(4) $\Rightarrow$ (5) Let $f = \chi_A$. For $\mu$-a.e. $x \in X$ we have

$$\mu(A) = \int_X \chi_A \, d\mu = \hat{\chi}_A(x) = \lim_{n \to \infty} \frac{1}{n} \sum_{j=0}^{n-1} \chi_A(S^j(x)).$$

By the dominated convergence theorem, we have

$$\mu(A)\mu(B) = \int_X \lim_{n \to \infty} \frac{1}{n} \sum_{j=0}^{n-1} \chi_A(S^j(x)) \chi_B(x) \, d\mu$$

$$= \lim_{n \to \infty} \frac{1}{n} \int_X \sum_{j=0}^{n-1} \chi_A(S^j(x)) \chi_B(x) \, d\mu$$

$$= \lim_{n \to \infty} \frac{1}{n} \sum_{j=0}^{n-1} \mu(S^{-j}(A) \cap B).$$

(5) $\Rightarrow$ (2) Let $A$ be invariant. Setting $B = A^c$ we have by (5) that

$$\mu(A)\mu(A^c) = \lim_{n \to \infty} \frac{1}{n} \sum_{j=0}^{n-1} \mu(S^{-j}(A) \cap A^c) = 0$$

because of the invariance of $A$. Hence $\mu(A) = 0$ or $\mu(A^c) = 0$. ∎

In general a dynamical system has more than just one invariant measure. For example if it has a periodic orbit $\{x_i\}_{i=1}^n$, $x_{i+1} = S(x_i)$ $i = 0, \ldots, n-1$, $x_0 = S(x_n)$, the measure

$$\mu(x) = \frac{1}{n} \sum_{i=1}^n \delta_{x_i}(x) \tag{13.21}$$

is invariant, where $\delta_y(x)$ denotes the Dirac measure at the point $y$:

$$\delta_y(A) = \begin{cases} 0, & \text{if } y \notin A, \\ 1, & \text{if } y \in A. \end{cases} \tag{13.22}$$

Given that a system often has many periodic orbits it follows that it also has many distinct invariant measures, and (13.21) clearly implies that they are not absolutely continuous with respect to one another. For ergodic transformations, the distinct invariant measures are necessarily singular.

**THEOREM 13.5** *Assume that $(X, \mathcal{A}, \mu, S)$ is ergodic and that $\mu_1 : \mathcal{A} \to [0,1]$ is another S-invariant probability measure. The following statements are then* *equivalent:*

*(1)* $\mu_1 \neq \mu$;
*(2)* $\mu_1$ *is not absolutely continuous with respect to* $\mu$;
*(3)* *there exists an invariant set* $A \in \mathcal{A}$ *such that* $\mu(A) = 0$ *and* $\mu_1(A) \neq 0$.

*Proof*
(1) $\Rightarrow$ (2) If $\mu_1$ were absolutely continuous with respect to $\mu$, the Radon–Nykodim derivative $d\mu_1/d\mu$ would be an invariant function in $L^1(X, \mathcal{A}, \mu)$. Since the system $(X, \mathcal{A}, \mu, S)$ is ergodic it follows that $(d\mu_1/d\mu)(x)$ is constant $\mu$-almost everywhere and therefore it is necessarily equal to 1 as both $\mu$ and $\mu_1$ are probability measures. It follows that $\mu_1 = \mu$, a contradiction.

(2) $\Rightarrow$ (3) Since $\mu_1$ is not absolutely continuous with respect to $\mu$ there exists $B \in \mathcal{A}$ such that $\mu(B) = 0$, while $\mu_1(B) \neq 0$. Setting $A = \bigcup_{i=0}^{\infty} S^j(B)$, it is immediate to verify that $A \in \mathcal{A}$, $\mu(A) = 0$, $\mu_1(A) \neq 0$. ∎

(3) $\Rightarrow$ (1) Obvious.

Suppose that $X$ is a compact metric space and $\mathcal{A}$ is the $\sigma$-algebra of the Borel sets of $X$.

In some exceptional cases, a dynamical system $(X, \mathcal{A}, \mu, S)$ can have a *unique invariant measure*. In this case the system is called *uniquely ergodic*. This has the following motivation.

THEOREM 13.6  *Let* $(X, \mathcal{A}, \mu, S)$ *be a uniquely ergodic system. Then the system is ergodic and for every choice of* $f : X \to \mathbf{R}$ *continuous and* $x \in X$ *the sequence* $(1/n)\sum_{j=0}^{n-1} f(S^j(x))$ *converges uniformly to a constant that is independent of* $x$. *Therefore the time average exists for every* $x \in X$ *and has value* $\int_X f \, d\mu$.

*Proof*
If the system were metrically decomposable, there would exist an invariant subset $A \subset X$ such that $0 < \mu(A) < 1$. The measure $d\nu = \chi_A(d\mu/\mu(A))$ is an invariant probability measure distinct from $\mu$: $\mu(A^c) = 1 - \mu(A)$ while $\nu(A^c) = 0$. This contradicts the hypothesis that the system is uniquely ergodic, and hence the system cannot be metrically decomposable.

Suppose then that there exists a continuous function $f : X \to \mathbf{R}$ for which the sequence of functions $\left((1/n)\sum_{j=0}^{n-1} f \circ S^j\right)_{n \in \mathbf{N}}$ does not converge uniformly to $\int_X f \, d\mu$ (because of ergodicity, this is the limit of the sequence $\mu$-almost everywhere). There then exist $\varepsilon > 0$, and two sequences $(n_i)_{i \in \mathbf{N}} \subset \mathbf{N}$, $n_i \to \infty$ and $(x_i)_{i \in \mathbf{N}} \subset X$ such that for every $i \in \mathbf{N}$ we have

$$\left| \frac{1}{n_i} \sum_{j=0}^{n_i - 1} f(S^j(x_i)) - \int_X f \, d\mu \right| \geq \varepsilon. \tag{13.23}$$

Consider the sequence of probability measures on $X$:

$$\nu_i := \frac{1}{n_i} \sum_{j=0}^{n_i-1} \delta_{S^j(x_i)}. \qquad (13.24)$$

By the compactness of the space of probability measures on $X$ (see Problem 1 of Section 13.13 for a proof) there is no loss of generality in assuming that the sequence $\nu_i$ converges to a probability measure $\nu$. We show that $\nu$ is invariant; to this end, thanks to Remark 13.5, it is sufficient to show that for every continuous $g: X \to \mathbf{R}$ we have $\int_X g(S(x))\,\mathrm{d}\nu = \int_X g(x)\,\mathrm{d}\nu$. On the other hand

$$\begin{aligned}\int_X g(S(x))\,\mathrm{d}\nu &= \lim_{i\to\infty} \int_X g(S(x))\,\mathrm{d}\nu_i = \lim_{i\to\infty} \frac{1}{n_i} \sum_{j=0}^{n_i-1} g(S^{j+1}(x_i)) \\ &= \lim_{i\to\infty} \left[ \int_X g(x)\,\mathrm{d}\nu_i - \frac{1}{n_i} g(x_i) + \frac{1}{n_i} g(S^{n_i+1}(x_i)) \right]. \end{aligned} \qquad (13.25)$$

Since $X$ is compact and $g$ is continuous, the second and third terms of the sum in (13.25) have limits of zero. It follows that the measure $\nu$ is invariant. Recalling equations (13.23) and (13.24) we have

$$\begin{aligned}\left| \int_X f\,\mathrm{d}\nu - \int_X f\,\mathrm{d}\mu \right| &= \lim_{i\to\infty} \left| \int_X f\,\mathrm{d}\nu_i - \int_X f\,\mathrm{d}\mu \right| \\ &= \lim_{i\to\infty} \left| \frac{1}{n_i} \sum_{j=0}^{n_i-1} f(S^j(x_i)) - \int_X f\,\mathrm{d}\mu \right| \geq \varepsilon,\end{aligned}$$

which shows that $\nu \neq \mu$ and contradicts the hypothesis that $\mu$ is the only invariant measure of the system. ∎

**Remark 13.8**
It is not difficult to prove that if for every continuous function $f: X \to \mathbf{R}$ the limit $\lim_{n\to\infty}(1/n)\sum_{j=0}^{n-1} f(S^j(x))$ exists for every fixed point $x$, independently of $x$, then the system is uniquely ergodic. ∎

**Example 13.12**
Let $\omega \in \mathbf{R}^l$ be such that $\omega \cdot \mathbf{k} + p \neq 0$ for every $p \in \mathbf{Z}$ and for every $\mathbf{k} \in \mathbf{Z}^l\setminus\{0\}$. Consider the measurable dynamical system determined by $X = \mathbf{T}^l$, $\mathcal{A}$ is the $\sigma$-algebra of Borel sets on $\mathbf{T}^l$, $\chi \in \mathbf{T}^l$, $\mathrm{d}\mu(\chi) = 1/(2\pi)^l \mathrm{d}^l\chi$ is the Haar measure on $\mathbf{T}^l$ and $S\chi = \chi + \omega \pmod{2\pi\mathbf{Z}^l}$. Theorem 11.9 guarantees that the time average exists $\forall \chi \in \mathbf{T}^l$ and it is independent of the choice of the initial point $\chi \in \mathbf{T}^l$. Therefore, by the previous remark, the system is uniquely ergodic. ∎

## 13.5 Mixing

One of the equivalent characterisations of ergodicity for a measurable dynamical system $(X, \mathcal{A}, \mu, S)$ is the fact that *on average* the measure of the preimages $S^{-j}(A)$ of any set $A \in \mathcal{A}$ is distributed uniformly on the whole support of the measure $\mu$ in the sense described by (13.20) (see Theorem 13.4). However, the existence of the limit in (13.20) does not guarantee that the limit of the sequence $\mu(S^{-j}(A) \cap B)$ exists, but it guarantees that if this sequence converges, then its limit is $\mu(A)\mu(B)$.[1]

It is therefore natural to consider the dynamical systems satisfying the following definition.

DEFINITION 13.14  A *measurable dynamical system* $(X, \mathcal{A}, \mu, S)$ *is* mixing *if* $\forall\ A, B \in \mathcal{A}$ *one has*

$$\lim_{n \to +\infty} \mu(S^{-n}(A) \cap B) = \mu(A)\mu(B). \tag{13.26}$$

■

Since equation (13.26) implies (13.20) *every mixing system is ergodic*. An independent verification of this fact can be obtained assuming that $A$ is invariant, in which case from (13.26) it follows that

$$\mu(A)\mu(A^c) = \lim_{n \to \infty} \mu(S^{-n}(A) \cap A^c) = \mu(A \cap A^c) = 0,$$

and therefore either $\mu(A) = 0$ or $\mu(A^c) = 0$, and the system is metrically indecomposable.

The converse is false: the irrational translations on tori (see Example 13.12) are uniquely ergodic but not a mixing (see Problem 9 of Section 13.12). A simple example of a mixing dynamical system is given by the so-called 'baker's transformation' of Example 3.4, as we shall see below (see Problem 2 of Section 13.13).

Just as ergodicity has an equivalent formulation in terms of the behaviour of the time average of integrable functions, mixing can be characterised by studying the functions $f : X \to \mathbf{R}$ which are measurable and square integrable.

DEFINITION 13.15  *Let the measurable dynamical system* $(X, \mathcal{A}, \mu, S)$ *be given. The linear operator* $U_S : L^2(X, \mathcal{A}, \mu) \to L^2(X, \mathcal{A}, S)$ *defined by*

$$U_S f = f \circ S \tag{13.27}$$

*is called* Koopman's operator.

■

Recalling the definition of the scalar product of two functions $f, g \in L^2(X, \mathcal{A}, \mu)$:

$$\langle f, g \rangle := \int_X fg \, d\mu, \tag{13.28}$$

---

[1] Recall that in a probability space $(X, \mathcal{A}, \mu, S)$, two sets (or events) $A, B \in \mathcal{A}$ are *independent* if $\mu(A \cap B) = \mu(A)\mu(B)$.

it is immediate to verify that since $S$ preserves the measure $\mu$, $U_S$ is an isometry:

$$\langle U_S f, U_S g \rangle = \langle f, g \rangle, \qquad \forall\ f, g \in L^2(X, \mathcal{A}, \mu). \tag{13.29}$$

THEOREM 13.7 *A necessary and sufficient condition for the measurable dynamical system $(X, \mathcal{A}, \mu, S)$ to be mixing is that*

$$\lim_{n \to \infty} \langle U_S^n f, g \rangle = \langle f, 1 \rangle \langle 1, g \rangle \tag{13.30}$$

*for every $f, g \in L^2(X, \mathcal{A}, \mu)$.* ∎

*Remark 13.9*
The quantity

$$\langle U_S^n f, g \rangle - \langle f, 1 \rangle \langle 1, g \rangle = \int_X f \circ S^n g \, \mathrm{d}\mu - \int_X f \, \mathrm{d}\mu \int_X g \, \mathrm{d}\mu$$

is also called the *correlation between $f$ and $g$ at time $n$*. Theorem 13.7 therefore states that a system is mixing if and only if the correlation between any two functions tends to zero as $n \to \infty$. ∎

*Proof of Theorem 13.7*
It is immediate to verify that (13.30) implies that the system is mixing; it is enough to apply it to $f = \chi_A$ and $g = \chi_B$, $A, B \in \mathcal{A}$.

Conversely, assuming that $(X, \mathcal{A}, \mu, S)$ is mixing, then (13.26) implies that equation (13.30) holds when $f$ and $g$ are two characteristic functions of sets belonging to $\mathcal{A}$. By linearity, we therefore find that equation (13.30) is valid when $f$ and $g$ are two simple functions.

Recall now that simple functions are dense in $L^2(X, \mathcal{A}, \mu)$ (see Rudin 1974); hence it follows that $\forall\ f, g \in L^2(X, \mathcal{A}, \mu)$ and $\forall\ \varepsilon > 0$ there exist two simple functions $f_0, g_0 \in L^2(X, \mathcal{A}, \mu)$ such that

$$\|f - f_0\| = \sqrt{\langle f - f_0, f - f_0 \rangle} \leq \varepsilon$$
$$\|g - g_0\| = \sqrt{\langle g - g_0, g - g_0 \rangle} \leq \varepsilon$$
$$\lim_{n \to \infty} \langle U_S^n f_0, g_0 \rangle = \langle f_0, 1 \rangle \langle 1, g_0 \rangle.$$

Writing

$$\langle U_S^n f, g \rangle = \langle U_S^n f_0, g_0 \rangle + \langle U_S^n f, g - g_0 \rangle + \langle U_S^n (f - f_0), g_0 \rangle$$

since $U_S$ is an isometry, using the Schwarz inequality $|\langle f, g \rangle| \geq \|f\| \|g\|$ one has

$$|\langle U_S^n f, g \rangle - \langle f, 1 \rangle \langle 1, g \rangle| \leq |\langle U_S^n f_0, g_0 \rangle - \langle f_0, 1 \rangle \langle 1, g_0 \rangle|$$
$$+ \|f\| \|g - g_0\| + \|f - f_0\| \|g_0\| + \varepsilon \|f\| + \varepsilon \|g_0\|.$$

There then exists a constant $c > 0$ such that if $n$ is sufficiently large

$$|\langle U_S^n f, g \rangle - \langle f, 1 \rangle \langle 1, g \rangle| \leq c\varepsilon,$$

and hence (13.30) follows. ∎

*Example 13.13: linear automorphisms of the torus* $\mathbf{T}^2$

Consider the flat two-dimensional torus with the $\sigma$-algebra of Borel sets, and the Haar measure $d\mu(\chi) = (1/4\pi^2) \, d\chi_1 \, d\chi_2$. A linear automorphism of the torus is given by

$$S(\chi_1, \chi_2) = (a\chi_1 + b\chi_2, c\chi_1 + d\chi_2) \bmod 2\pi \mathbf{Z}^2, \tag{13.31}$$

where $a, b, c, d \in \mathbf{Z}$ and $|ad - bc| = 1$. It is easy to verify that the Haar measure is $S$-invariant.

We now prove that if the matrix $\sigma = \begin{pmatrix} a & b \\ c & d \end{pmatrix}$ has no eigenvalue with unit modulus, then the system is mixing.

To this end, we check that (13.30) is satisfied by the functions $f_{\mathbf{k}}(\chi) = e^{i\mathbf{k}\cdot\chi}$, $\mathbf{k} \in \mathbf{Z}^2$, which form a basis of $L^2(\mathbf{T}^2)$. We want to show therefore that for every pair $\mathbf{k}, \mathbf{k}' \in \mathbf{Z}^2$ we have

$$\lim_{n \to \infty} \int_{\mathbf{T}^2} f_{\mathbf{k}}(S^n(\chi)) \overline{f_{\mathbf{k}'}}(\chi) \, d\mu(\chi) = \int_{\mathbf{T}^2} f_{\mathbf{k}} \chi \, d\mu(\chi) \int_{\mathbf{T}^2} \overline{f_{\mathbf{k}'}} \chi \, d\mu(\chi). \tag{13.32}$$

If $\mathbf{k} = \mathbf{k}' = \mathbf{0}$ the two sides are constant and equal to 1 for every $n \in \mathbf{N}$. It is not restrictive to assume that $\mathbf{k} \neq \mathbf{0}$ which yields immediately that the right-hand side is equal to 0. On the other hand, we have $f_{\mathbf{k}}(S(\chi)) = f_{\sigma^T \mathbf{k}}(\chi)$ hence $f_{\mathbf{k}}(S^n(\chi)) = f_{(\sigma^T)^n \mathbf{k}}(\chi)$ and since $\sigma$ has an eigenvalue with absolute value $> 1$ the norm is $|(\sigma^T)^n \mathbf{k}| \to \infty$.[2] It follows that if $n$ is sufficiently large we necessarily have $(\sigma^T)^n \mathbf{k} \neq \mathbf{k}'$ and as the basis $(f_{\mathbf{k}})_{\mathbf{k} \in \mathbf{Z}^2}$ is orthonormal, then the left-hand side of (13.32) vanishes. This concludes the proof that the system is mixing. ∎

## 13.6 Entropy

Let $(X, \mathcal{A}, \mu, S)$ be a measurable dynamical system. Ergodicity and mixing give two qualitative indications of the degree of randomness (or stochasticity) of the system. An indication of quantitative type is given by the notion of *entropy* which we shall soon introduce.

We start by considering the following situation. Let $\alpha$ be an experiment with $m \in \mathbf{N}$ possible mutually exclusive outcomes $A_1, \ldots, A_m$ (for example the toss of

---

[2] Since $\sigma$ transforms the vectors of $\mathbf{Z}^2$ into vectors of $\mathbf{Z}^2$ and is invertible, no non-zero vector with integer components can be entirely contained in the eigenspace corresponding to the eigenvalue less than 1, because this would imply that by iterating $\sigma$ a finite number of times the vector has norm less than 1, contradicting the hypothesis that it belongs to $\mathbf{Z}^2$.

a coin $m = 2$ or the roll of a die $m = 6$). Assume that each outcome $A_i$ happens with probability $p_i \in [0,1]$: $\sum_{i=1}^{m} p_i = 1$.

In a probability space $(X, \mathcal{A}, \mu, S)$ this situation is described by assigning a finite *partition* of $X = A_1 \cup \ldots \cup A_m$ (mod 0), $A_i \in \mathcal{A}$, $A_i \cap A_j = \emptyset$ if $i \neq j$, $\mu(A_i) = p_i$.

The following definition describes the properties which must hold for a function measuring the *uncertainty* of the prediction of an outcome of the experiment (equivalently, the *information* acquired from the execution of the experiment $\alpha$).

Let $\Delta^{(m)}$ be the $(m-1)$-dimensional standard symplex of $\mathbf{R}^m$, given by

$$\Delta^{(m)} = \left\{ (x_1, \ldots, x_m) \in \mathbf{R}^m \mid x_i \in [0,1], \sum_{i=1}^{m} x_i = 1 \right\}.$$

DEFINITION 13.16  *A family of continuous functions $H^{(m)} : \Delta^{(m)} \to [0, +\infty]$, where $m \in \mathbf{N}$, is called an* entropy *if the following properties hold:*

(1) *symmetry:* $\forall\ i, j \in \{1, \ldots, m\}$ *we have*

$$H^{(m)}(p_1, \ldots, p_i, \ldots, p_j, \ldots, p_m) = H(p_1, \ldots, p_j, \ldots, p_i, \ldots, p_m);$$

(2) $H^{(m)}(1, 0, \ldots, 0) = 0$;
(3) $H^{(m)}(0, p_2, \ldots, p_m) = H^{(m-1)}(p_2, \ldots, p_m)$, $\forall\ m \geq 2, \forall\ (p_2, \ldots, p_m) \in \Delta^{(m-1)}$;
(4) $\forall\ (p_1, \ldots, p_m) \in \Delta^{(m)}$ *we have* $H^{(m)}(p_1, \ldots, p_m) \leq H^{(m)}(1/m, \ldots, 1/m)$ *and the equality holds if and only if $p_i = 1/m$ for every $i = 1, \ldots, m$;*
(5) *consider* $(\pi_{11}, \ldots, \pi_{1l}, \pi_{21}, \ldots, \pi_{2l}, \ldots, \pi_{m1}, \ldots, \pi_{ml}) \in \Delta^{(ml)}$; *then for every* $(p_1, \ldots, p_m) \in \Delta^{(m)}$ *we have*

$$H^{(ml)}(\pi_{11}, \ldots, \pi_{1l}, \pi_{21}, \ldots, \pi_{ml})$$
$$= H^{(m)}(p_1, \ldots, p_m) + \sum_{i=1}^{m} p_i H^{(l)}\left(\frac{\pi_{i1}}{p_i}, \ldots, \frac{\pi_{il}}{p_i}\right). \blacksquare$$

Property (2) expresses the absence of uncertainty of a certain event. Property (3) means that no information is gained by impossible outcomes and (4) means that the maximal uncertainty is attained when all outcomes are equally probable.

Property (5) describes the behaviour of the entropy when distinct experiments are compared. Let $\beta$ be another experiment with possible outcomes $B_1, \ldots, B_l$ (i.e. another partition of $(X, \mathcal{A}, \mu, S)$). Let $\pi_{ij}$ be the probability of $A_i$ and $B_j$ together. The probability of $B_j$ conditional on the fact that the outcome of $\alpha$ is $A_i$ is prob $(B_j \mid A_i) = \pi_{ij}/p_i (= \mu(A_i \cap B_j))$. Clearly the uncertainty in the prediction of the outcome of the experiment $\beta$ when the outcome of $\alpha$ is $A_i$ is measured by $H^{(l)}(\pi_{i1}/p_i, \ldots, \pi_{il}/p_i)$. From this fact stems the requirement that (5) be satisfied. In the following, we use the simpler notation $H(p_1, \ldots, p_m)$.

THEOREM 13.8  *The function*

$$H(p_1, \ldots, p_m) = -\sum_{i=1}^{m} p_i \log p_i \qquad (13.33)$$

(with the convention $0\log 0 = 0$) is, up to a constant positive multiplier, the only function satisfying (1)–(5).

*Proof* (see Khinchin 1957, pp. 10–13).
Let $H(p_1,\ldots,p_m)$ be an entropy function, and for any $m$ set $K(m) = H(1/m,\ldots,1/m)$. We show first of all that $K(m) = +c\log m$, where $c$ is a positive constant.

Properties (3) and (4) imply that $K$ is a non-decreasing function. Indeed,

$$K(m) = H\left(0, \frac{1}{m}, \ldots, \frac{1}{m}\right) \leq H\left(\frac{1}{m+1}, \ldots, \frac{1}{m+1}\right) = K(m+1).$$

Consider now any two positive integers $m$ and $l$. The property (5) applied to the case $\pi_{ij} \equiv 1/ml$, $p_i \equiv 1/m$ yields

$$K(lm) = K(m) + \sum_{i=1}^{m} \frac{1}{m} K(l) = K(m) + K(l),$$

from which it follows that

$$K(l^m) = mK(l).$$

Given any three integers $r, n, l$ let $m$ be such that $l^m \leq r^n \leq l^{m+1}$, i.e.

$$\frac{m}{n} \leq \frac{\log r}{\log l} \leq \frac{m}{n} + \frac{1}{n}.$$

We know that

$$mK(l) = K(l^m) \leq K(r^n) = nK(r) \leq K(l^{m+1}) = (m+1)K(l),$$

from which it follows that

$$\frac{m}{n} \leq \frac{K(r)}{K(l)} \leq \frac{m}{n} + \frac{1}{n}, \quad \text{i.e.} \quad \left|\frac{K(r)}{K(l)} - \frac{\log r}{\log l}\right| \leq \frac{1}{n}.$$

Because of the arbitrariness of $n$ we deduce that $K(r)/\log r = K(l)/\log l$ and therefore $K(m) = c\log m$, $c > 0$.

Assume now that $p_1,\ldots,p_m$ are rational numbers. Setting the least common multiple of the denominators equal to $s$, we have $p_i = r_i/s$, with $\sum_{i=1}^{m} r_i = s$. In addition to the experiment $\alpha$ with outcomes $A_1,\ldots,A_m$ with respective probabilities $p_1,\ldots,p_m$ we consider an experiment $\beta$ constituted by $s$ outcomes $B_1,\ldots,B_s$ divided into $m$ groups, each containing, respectively, $r_1,\ldots,r_m$ outcomes.

We now set $\pi_{ij} = p_i/r_i = 1/s$, $i = 1,\ldots,m$, $j = 1,\ldots,r_i$.

Given any outcome $A_i$ of $\alpha$, we therefore have that the outcome $\beta$ is the outcome of an experiment with $r_i$ equally probable outcomes, and hence

$$H\left(\frac{\pi_{i1}}{p_i}, \ldots, \frac{\pi_{ir_i}}{p_i}\right) = c\log r_i$$

and

$$\sum_{i=1}^{m} p_i H\left(\frac{\pi_{i1}}{p_i}, \ldots, \frac{\pi_{ir_i}}{p_i}\right) = c\sum_{i=1}^{m} p_i \log r_i = c\sum_{i=1}^{m} p_i \log p_i + c\log s.$$

On the other hand, $H(\pi_{11}, \ldots, \pi_{mr_m}) = c\log s$ and by property (5) we have

$$H(p_1, \ldots, p_m) = H(\pi_{11}, \ldots, \pi_{mr_m}) - \sum_{i=1}^{m} p_i H\left(\frac{\pi_{i1}}{p_i}, \ldots, \frac{\pi_{ir_i}}{p_i}\right) = -c\sum_{i=1}^{m} p_i \log p_i.$$

The continuity of $H$ ensures that the formula (13.33), proved so far when $p_i \in \mathbf{Q}$, is also valid when $p_i$ is a real number. ∎

*Remark* 13.10
$H$ can be characterised as the $(-1/N) \times$ logarithm of the probability of a 'typical' outcome of the experiment $\alpha$ repeated $N$ times. Indeed, if $N$ is large, repeating the experiment $\alpha$ $N$ times one expects to observe each outcome $A_i$ approximately $p_i N$ times (this is a formulation of the so-called *law of large numbers*).

The probability of a typical outcome containing $p_1 N$ times $A_1$, $p_2 N$ times $A_2$, etc. is therefore

$$p_1^{p_1 N} p_2^{p_2 N} \ldots p_m^{p_m N}.$$

From this it follows precisely that

$$H(p_1, \ldots, p_m) = -\frac{1}{N}\log\left[p_1^{p_1 N} \ldots p_m^{p_m N}\right] = -\sum_{i=1}^{m} p_i \log p_i. \quad \blacksquare$$

*Remark* 13.11
The maximum value of $H$ is attained when $p_i = 1/m$, $i = 1, \ldots, m$ (as required by property (4)) and has value $H(1/m, \ldots, 1/m) = \log m$. ∎

We now consider how to extend the notion of entropy to measurable dynamical systems $(X, \mathcal{A}, \mu, S)$.

We introduce some notation. If $\alpha$ and $\beta$ are two partitions of $\mathcal{A}$, the *joined partition* $\alpha \vee \beta$ of $\alpha$ and $\beta$ is defined by the subsets $\{A \cap B,\ A \in \alpha,\ B \in \beta\}$. If $\alpha_1, \ldots, \alpha_n$ are partitions, we write $\bigvee_{i=1}^{n} \alpha_i$ for the joined partition of $\alpha_1, \ldots, \alpha_n$. If $S$ is measurable and non-singular, and $\alpha$ is a partition, $S^{-1}\alpha$ is the partition defined by the subsets $\{S^{-1}A,\ A \in \alpha\}$. Finally, we say that a partition $\beta$ is *finer* than $\alpha$, which we denote by $\alpha < \beta$, if $\forall\ B \in \beta$ there exists $A \in \alpha$ such that $B \subset A$. Obviously, the joined partitions are finer than the starting ones. The entropy $H(\alpha)$ of a partition $\alpha = \{A_1, \ldots, A_m\}$ is given by $H(\alpha) = -\sum_{i=1}^{m} \mu(A_i) \log \mu(A_i)$.

DEFINITION 13.17 *Let $(X, \mathcal{A}, \mu, S)$ be a measurable dynamical system, and let $\alpha$ be a partition. The entropy of $S$ relative to the partition $\alpha$ is defined by*

$$h(S, \alpha) := \lim_{n \to \infty} \frac{1}{n} H \left( \bigvee_{i=0}^{n-1} S^{-i} \alpha \right). \tag{13.34}$$

*The entropy of $S$ is*

$$h(S) := \sup\{h(S, \alpha), \ \alpha \text{ is a finite partition of } X\}. \tag{13.35}$$

∎

*Remark* 13.12
It is possible to prove, exploiting the strict convexity of the function $x \log x$ on $\mathbf{R}_+$, that the limit (13.34) exists. Indeed, the sequence $(1/n) H \left( \bigvee_{i=0}^{n-1} S^{-i} \alpha \right)$ is monotone non-increasing and non-negative. Hence $h(S, \alpha) \geq 0$ for every $\alpha$. ∎

*Remark* 13.13
The entropy of a partition $\alpha$ measures the quantity of information acquired by observing the system using an instrument that distinguishes between the points of $X$ with the resolution given by the sets of the partition $\{A_1, \ldots, A_m\} = \alpha$.
For $x \in X$, consider the orbit of $x$ up to time $n - 1$:

$$x, \ Sx, \ S^2 x, \ \ldots, \ S^{n-1} x.$$

Since $\alpha$ is a partition of $X$, the points $S^i x$, $0 \leq i \leq n - 1$, belong to precisely one of the sets of the partition $\alpha$: setting $x_0 = x$, $x_i = S^i x$, we have $x_i \in A_{k_i}$ with $k_i \in \{1, \ldots, m\}$ for every $i = 0, \ldots, n - 1$.
$H \left( \bigvee_{i=0}^{n-1} S^{-i} \alpha \right)$ measures the quantity of information deduced from the knowledge of the distribution with respect to the partition $\alpha$ of a segment of 'duration' $n$ of the orbit. Therefore $(1/n) H \left( \bigvee_{i=0}^{n-1} S^{-i} \alpha \right)$ is the average quantity of information per unit of time and $h(S, \alpha)$ is the the quantity of information acquired (asymptotically) at each iteration of the dynamical system, knowing how the orbit of a point is distributed with respect to the partition $\alpha$. ∎

This remark is made rigorous by the following theorem. The proof can be found in Mañe (1987).

THEOREM 13.9 (Shannon–Breiman–McMillan) *Let $(X, \mathcal{A}, \mu, S)$ be a measurable ergodic dynamical system, and $\alpha$ be a finite partition of $X$. Given $x \in X$ let $\alpha^n(x)$ be the element of $\bigvee_{i=0}^{n-1} S^{-i} \alpha$ which contains $x$. Then for $\mu$-almost every $x \in X$ we have*

$$h(S, \alpha) = \lim_{n \to \infty} -\frac{1}{n} \log \mu(\alpha^n(x)). \tag{13.36}$$

∎

An interpretation of the Shannon–Breiman–McMillan theorem is the following. For an ergodic system there exists a number $h$ such that $\forall\ \varepsilon > 0$, if $\alpha$ is a sufficiently fine partition of $X$, then there exists a positive integer $N$ such that for every $n \geq N$ there exists a subset $X_n$ of $X$ of measure $\mu(X_n) > 1-\varepsilon$ made of approximately $e^{nh}$ elements of $\bigvee_{i=0}^{n-1} S^{-i}\alpha$, each of measure approximately $e^{-nh}$.

If $X$ is a compact metric space and $\mathcal{A}$ is the $\sigma$-algebra of Borel sets $X$, Brin and Katok (1983) have given an interesting topological version of the Shannon–Breiman–McMillan theorem. Let $B(x,\varepsilon)$ be the ball of centre $x \in X$ and radius $\varepsilon$. Assume that $S: X \to X$ is continuous and preserves the probability measure $\mu: \mathcal{A} \to [0,1]$. Consider

$$B(x,\varepsilon,n) := \{y \in X \mid d(S^i x, S^i y) \leq \varepsilon \text{ for every } i = 0,\ldots,n-1\},$$

i.e. $B(x,\varepsilon,n)$ is the set of points $y \in X$ whose orbit remains at a distance less than $\varepsilon$ from that of $x$ for at least $n-1$ iterations. It is possible to prove the following.

**THEOREM 13.10 (Brin–Katok)** *Assume that $(X, \mathcal{A}, \mu, S)$ is ergodic. For $\mu$-almost every $x \in X$ we have*

$$\sup_{\varepsilon > 0} \limsup_{n \to \infty} -\frac{1}{n} \log \mu(B(x,\varepsilon,n)) = h(S). \tag{13.37}$$

∎

An interesting corollary of the previous theorem is that the entropy of the translations over tori $\mathbf{T}^l$ is zero. Indeed, in this case $d(Sx, Sy) = d(x,y)$ and therefore $\forall\ n \in \mathbf{N}$ and $\forall\ \varepsilon > 0$ we have $B(x,\varepsilon,n) = B(x,\varepsilon)$, from which it follows that $h(S) = 0$.

The same is true, more generally, if $S$ is an isometry of the metric space $(X,d)$.

The notion of entropy allows one to distinguish between systems in terms of the 'predictability' of their observables. When the entropy is positive, at least part of the observables cannot be computed from the knowledge of the past history.

*Chaotic* systems are therefore the systems that have *positive entropy*. Taking into account the Brin–Katok theorem and the recurrence theorem of Poincaré, one sees how in chaotic systems the orbits are subject to two constraints, apparently contradicting each other. On the one hand, almost every orbit is recurrent and in the future will pass infinitely many times near the starting-point. On the other hand, the probability that two orbits remain close for a given time interval $n$ decays exponentially as $n$ grows.

Since two orbits, that were originally close to each other, must return infinitely many times near the starting-point, they must be entirely uncorrelated if the entropy is positive, and hence they must go far and come back in different times.

This complexity of motions is called chaos, and it clearly shows how difficult (or impossible) it is to compute the future values of an observable (corresponding to a function $f: X \to \mathbf{R}$) simply from the knowledge of its past history.

## 13.7 Computation of the entropy. Bernoulli schemes. Isomorphism of dynamical systems

In the definition of entropy $h(S)$ of a measurable dynamical system, it is necessary to compute the supremum of $h(S,\alpha)$ as $\alpha$ varies among all finite partitions of $X$. This seems to exclude the *practical* possibility of computing $h(S)$. In reality, this is not the case, and one can proceed in a much simpler way.

In this section we identify a partition $\alpha$ with the $\sigma$-algebra generated by $\alpha$, and $\bigvee_{i=0}^{\infty} S^{-i}\alpha$ with the smallest $\sigma$-algebra containing all the partitions $\bigvee_{i=0}^{n-1} S^{-i}\alpha$ for every $n \in \mathbf{N}$.

Recall that two $\sigma$-algebras $\mathcal{A}$ and $\mathcal{B}$ are *equal* (mod 0), denoted by $\mathcal{A} = \mathcal{B}$ (mod 0), if $\forall\ A \in \mathcal{A}$ there exists $B \in \mathcal{B}$ such that $\mu(A \triangle B) = 0$, and vice versa.

The discovery of Kolmogorov and Sinai, which makes possible the computation of the entropy overcoming the need to compute the supremum in (13.35), is that it suffices to consider the finite partitions $\alpha$ that *generate* the $\sigma$-algebra $\mathcal{A}$, and hence such that $\bigvee_{-\infty}^{+\infty} S^{-i}\alpha = \mathcal{A}$ (mod 0) if $S$ is invertible, or $\bigvee_{i=0}^{\infty} S^{-i}\alpha = \mathcal{A}$ (mod 0) if $S$ is not invertible. Indeed one can prove the following.

THEOREM 13.11 (Kolmogorov–Sinai) *If $\alpha$ is a partition of $X$ generating the $\sigma$-algebra $\mathcal{A}$, the entropy of the measurable dynamical system $(X, \mathcal{A}, \mu, S)$ is given by*

$$h(S) = h(S, \alpha). \qquad (13.38)$$

∎

The proof of this theorem does not present special difficulties but it is tedious and will be omitted (see Mañe 1987).

Among the measurable dynamical systems for which it is possible to compute the entropy, the *Bernoulli schemes*, which we now introduce, constitute the fundamental example of systems with strong stochastic properties.

Consider the space $X$ of infinite sequences $x = (x_i)_{i \in \mathbf{N}}$, where the variable $x_i$ can only take a finite number of values which, for simplicity, we assume to be the integers $\{0, \ldots, N-1\}$ (we sometimes use the notation $\mathbf{Z}_N$ to denote the integers $\{0, 1, \ldots, N-1\}$). The space of sequences $X$ is often denoted by $\mathbf{Z}_N^{\mathbf{N}}$.[3] When we want to model an infinite sequence of outcomes of the toss of a coin (or the roll of a die) we fix $N = 2$ (respectively $N = 6$) and each possible value of $x_i$ is equally probable.

Consider on $X$ the transformation $S : X \to X$ defined by

$$(S(x))_i = x_{i+1}, \qquad \forall\ i \in \mathbf{N}, \qquad (13.39)$$

usually known as a *shift*.

---

[3] If instead of one-sided sequences $(x_i)_{i \in \mathbf{N}} \in \mathbf{Z}_N^{\mathbf{N}}$ the space $X$ is made of two-sided (doubly infinite) sequences $(x_i)_{i \in \mathbf{Z}} \in \mathbf{Z}_N^{\mathbf{Z}}$ we have a so-called bilateral Bernoulli scheme. All considerations to be developed trivially extend to the case of bilateral Bernoulli schemes.

We proceed as in Example 13.2, associating with $\mathbf{Z}_N$ a probability measure and assigning to the value $j \in \mathbf{Z}_N$ a probability equal to $p_j > 0$, with the condition $\sum_{j=0}^{N-1} p_j = 1$.

This choice induces a probability measure on the space of sequences $X$ that we now describe.

Consider first of all the $\sigma$-algebras $\mathcal{A}$ on $X$ generated by the *cylinders*, i.e. the subsets of $X$ corresponding to sequences for which a finite number of values is fixed. Given $k \geq 1$ elements $j_1, \ldots, j_k \in \mathbf{Z}_N$, not necessarily distinct, and $k$ distinct positions $i_1 < i_2 < \ldots < i_k \in \mathbf{N}$, the corresponding cylinder is

$$C = C\begin{pmatrix} j_1, \ldots, j_k \\ i_1, \ldots, i_k \end{pmatrix} = \{x \in X \mid x_{i_1} = j_1,\ x_{i_2} = j_2, \ldots,\ x_{i_k} = j_k\}. \tag{13.40}$$

Therefore all sequences in $X$ which take the prescribed values in the positions corresponding to the indices $i_1, \ldots, i_k$ belong to $C$.

We therefore define the measure $\mu$ on $\mathcal{A}$ by prescribing its value on cylinders:

$$\mu\left(C\begin{pmatrix} j_1, \ldots, j_k \\ i_1, \ldots, i_k \end{pmatrix}\right) = p_{j_1} \cdots p_{j_k}. \tag{13.41}$$

Note that in (13.41) the positions $i_1, \ldots, i_k$ do not play any role. Hence it is immediate to deduce that if $C$ is a cylinder, then $\mu(S^{-1}(C)) = \mu(C)$, and recalling that the $\sigma$-algebra $\mathcal{A}$ is generated by cylinders we conclude that $(X, \mathcal{A}, \mu, S)$ is a measurable dynamical system (hence that $S$ preserves the measure $\mu$). This system is known as a *Bernoulli scheme* with probability $(p_0, \ldots, p_{N-1})$ and it is denoted by $SB(p_0, \ldots, p_{N-1})$.

We leave as an exercise the verification that a Bernoulli scheme is mixing (see Problem 10 of Section 13.12) but we show that the entropy of $SB(p_0, \ldots, p_{N-1})$ is $-\sum_{i=0}^{N-1} p_i \log p_i$.

The partition $\alpha$ into the cylinders $\{C\binom{j}{0}\}_{j=0,\ldots,N-1}$ generates the $\sigma$-algebra $\mathcal{A}$. Indeed we have

$$\alpha \vee S^{-1}\alpha = \left\{C\begin{pmatrix} j_0 & j_1 \\ 0 & 1 \end{pmatrix}\right\}_{j_0,\ j_1 = 0, \ldots, N-1},$$

$$\alpha \vee S^{-1}\alpha \vee S^{-2}\alpha = \left\{C\begin{pmatrix} j_0 & j_1 & j_2 \\ 0 & 1 & 2 \end{pmatrix}\right\}_{\substack{j_i = 0, \ldots, N-1 \\ i = 0, 1, 2}},$$

and so on. The corresponding entropies are (use (13.41)):

$$H(\alpha) = -\sum_{j=0}^{N-1} p_j \log p_j,$$

$$H(\alpha \vee S^{-1}\alpha) = -\sum_{j_0=0}^{N-1}\sum_{j_1=0}^{N-1} p_{j_0} p_{j_1} \log p_{j_0} p_{j_1}$$

$$= -\sum_{j_0=0}^{N-1} (p_{j_0} \log p_{j_0}) \sum_{j_1=0}^{N-1} p_{j_1} - \sum_{j_1=0}^{N-1} (p_{j_1} \log p_{j_1}) \sum_{j_0=0}^{N-1} p_{j_0}$$

$$= -2 \sum_{j=0}^{N-1} p_j \log p_j,$$

$$H(\alpha \vee S_\alpha^{-1} \vee S_\alpha^{-2}) = -\sum_{j_0,j_1,j_2} p_{j_0} p_{j_1} p_{j_2} \log p_{j_0} p_{j_1} p_{j_2} = -3 \sum_{j=0}^{N-1} p_j \log p_j,$$

and so on. From this it follows that $h(S, \alpha) = -\sum_{j=0}^{N-1} p_j \log p_j$ and thus the entropy of $SB(p_0, \ldots, p_{N-1})$ also follows by the Kolmogorov–Sinai theorem.

We examine again the $p$-adic transformation $S$ of Example 13.5 and consider the partition $\alpha = \{(j/p, (j+1)/p)\}_{j=0,\ldots,p-1}$. Using the fact that $\vee_{i=0}^n S^{-i}\alpha = \{(j/(p^{n+1}), (j+1)/(p^{n+1}))\}_{j=0,\ldots,p^{n+1}-1}$ it is not difficult to verify that $\alpha$ is a generating partition and therefore $h(S) = h(S, \alpha)$.

On the other hand,

$$H\left(\bigvee_{i=0}^n S^{-i}\alpha\right) = p^{n+1} \cdot p^{-(n+1)} \log p^{-(n+1)} = -(n+1) \log p,$$

from which it follows that $h(S, \alpha) = \log p$.

Note that $SB(1/p, \ldots, 1/p)$ has the same entropy. It is indeed possible to pass from one system to the other by a very easy construction.

With every point $\xi \in (0,1)$ we associate the sequence $x \in \mathbf{Z}_p^N$ defined as follows: for every $i = 0, 1, \ldots$ we set

$$x_i = j \Leftrightarrow S^i(\xi) \in \left(\frac{j}{p}, \frac{j+1}{p}\right). \tag{13.42}$$

Denote by $(X, \mathcal{A}, \mu, S)$ and by $(X', \mathcal{A}', \mu', S')$, respectively, the two 4-tuples: ((0,1), $\sigma$-algebra of Borel sets of (0,1), Lebesgue measure, $p$-adic transformation) and ($\mathbf{Z}_p^N$, the $\sigma$-algebra generated by the cylinders, the measure corresponding to $SB(1/p, \ldots, 1/p)$, and the shift).

In addition, denote by $T: X \to X'$ the transformation defined in (13.42).

The following facts are of immediate verification:
(a) $T$ is measurable;
(b) $\forall\ A' \in \mathcal{A}'$, $\mu(T^{-1}A') = \mu'(A')$;
(c) for $\mu$-a.e. $x \in X$, $T(S(x)) = S'(T(x))$;
(d) $T$ is invertible (mod 0), i.e. there exists a measurable transformation $T' : X' \to X$, which preserves the measures (so that $\forall\ A \in \mathcal{A}$, $\mu'(T'^{-1}A) = \mu(A)$), such that $T'(T(x)) = x$ for $\mu$-a.e. $x \in X$ and $T(T'(x')) = x'$ for $\mu'$-a.e. $x' \in X'$.

In general, we have the following.

DEFINITION 13.18  Let $(X, \mathcal{A}, \mu, S)$, $(X', \mathcal{A}', \mu', S')$ be two measurable dynamical systems. A transformation $T : X \to X'$ satisfying the conditions (a), (b), (c), (d) is called an isomorphism of measurable dynamical systems and the two systems are then isomorphic. ∎

Ergodic theory does not distinguish between isomorphic systems: two isomorphic systems have the same 'stochastic' properties.

It is an exercise to prove the following.

THEOREM 13.12  Two isomorphic systems have the same entropy. If one system is mixing, then the other is also mixing. If one system is ergodic, then the other is also ergodic. ∎

In the particular case of the Bernoulli schemes the equality of entropy is not only a necessary condition but it is also sufficient for two schemes to be isomorphic.

THEOREM 13.13 (Ornstein)  Two Bernoulli schemes with the same entropy are isomorphic. ∎

The proof of this result goes beyond the scope of this book. Besides the original article of Ornstein (1970), see also Cornfeld et al. (1982, section 7, chapter 10).

A consequence of this theorem of Ornstein is that the Bernoulli schemes are completely classified (up to isomorphism) by their entropy. The last result we quote in this section shows how the entropy also classifies the hyperbolic isomorphisms of tori (see Example 13.13): these are given by matrices $\sigma \in \mathrm{GL}(l, \mathbf{Z})$ with no eigenvalue of absolute value $= 1$.

THEOREM 13.14 (Katznelson)  Every linear hyperbolic automorphism of $\mathbf{T}^l$ is isomorphic to a Bernoulli scheme. ∎

Due to the theorem of Ornstein the classification of the ergodic properties of the automorphisms of $\mathbf{T}^l$ is given by the entropy.

It can be proved (see Walters 1982, sections 8.4 and 8.10) that if $\nu_1, \ldots, \nu_l$ are the eigenvalues of the automorphism $\sigma$ then

$$h(\sigma) = \sum_{\{i\,|\,|\nu_i|>1\}} \log |\nu_i|. \tag{13.43}$$

We conclude with the definition of Bernoulli systems.

DEFINITION 13.19 *A measurable dynamical system $(X, \mathcal{A}, \mu, S)$ is a Bernoulli system if it is isomorphic to a Bernoulli scheme.* ∎

Bernoulli systems exhibit the most significant stochastic properties. Their equivalence classes up to isomorphism, due to the theorem of Ornstein, are completely classified by only one invariant, the entropy.

## 13.8 Dispersive billiards

Many important models of classical statistical mechanics are systems of point particles or rigid spheres moving freely except for the effect of elastic collisions, either with fixed obstacles or among themselves. To study the behaviour of electron gases in metals, Lorentz introduced in 1905 the following model: a point particle moves in $\mathbf{R}^l$ subject only to elastic collisions with a distribution of infinitely many fixed rigid spheres (see Fig. 13.2).

Another important model is the *hard spheres gas*: a system of spheres which move freely in a domain $V \subset \mathbf{R}^l$ interacting through elastic collisions between them and with the boundary $\partial V$ of the domain (see Fig. 13.3).

In all these cases, the main element of the model is the condition that the collision be *elastic*. This is the characterising feature of all dynamical systems of billiard type.

DEFINITION 13.20 *A billiard is a dynamical system constituted by the motion of a point particle with constant velocity inside a bounded open subset $V \subseteq \mathbf{R}^d$ with piecewise smooth boundary ($\mathcal{C}^\infty$), and with a finite number of smooth components intersecting transversally. The particle is subject to elastic reflections when it collides with $\partial V$ (see Fig. 13.3): the incidence angle is equal to the reflection angle and the energy is conserved.* ∎

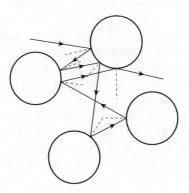

**Fig. 13.2** Lorentz gas.

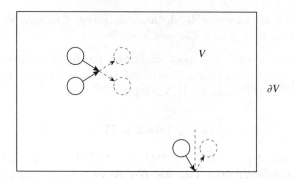

**Fig. 13.3** Hard spheres gas.

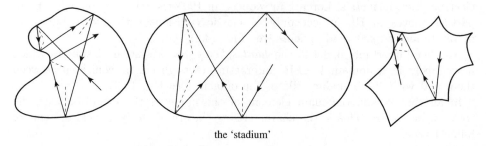

the 'stadium'

**Fig. 13.4** Examples of plane billiards.

In our short introduction to the study of billiards we shall restrict ourselves to the plane case, when $l = 2$ (See Fig. 13.4 for some examples of plane billiards). This is the only case whose stochastic properties are sufficiently understood.

Since the absolute value of the velocity is constant, it is possible to describe the motion using a system with discrete time.

We parametrise $\partial V$ using the natural parameter $s$ and suppose that the length of $\partial V$ is equal to $2\pi$; we can then characterise $\mathbf{x} \in \partial V$ by choosing arbitrarily the origin corresponding to $s = 0$ and via the application $\mathbf{S}^1 \ni s \mapsto \mathbf{x}(s)$ (note that $\mathbf{x}(s+2\pi) = \mathbf{x}(s)$) (see Fig. 13.5). The elastic collision with $\partial V$ is completely described by assigning the pair $(s, \alpha) \in \mathbf{S}^1 \times (0, \pi)$, where $\mathbf{x}(s)$ is the collision point in $\partial V$ and $\alpha$ is the angle formed by the reflected velocity (i.e. the velocity immediately after the collision) and the unit vector tangent to $\partial V$.

Consider the phase space $X = \mathbf{S}^1 \times (0, \pi)$ with the $\sigma$-algebra $\mathcal{A}$ of Borel sets, and the transformation $S : X \to X$ which associates with $(s, \alpha)$ the next collision point and reflection angle $(s', \alpha')$.

PROPOSITION 13.1  *S preserves the probability measure* $\mathrm{d}\mu(s, \alpha) = 1/4\pi \sin \alpha \, \mathrm{d}s \, \mathrm{d}\alpha$.

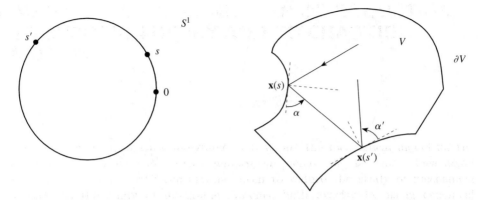

**Fig. 13.5** Parametrisation of the billiard.

*Proof*
Let $l(s, s')$ be the length of the segment $[\mathbf{x}(s), \mathbf{x}(s')] \subset V$. It is immediate to verify that

$$\frac{\partial l}{\partial s}(s, s') = -\cos \alpha, \quad \frac{\partial l}{\partial s'}(s, s') = \cos \alpha',$$

from which it follows that

$$dl = -\cos \alpha \, ds + \cos \alpha' \, ds'.$$

Since $d^2 l = 0$ we deduce

$$\sin \alpha \, d\alpha \wedge ds = \sin \alpha' \, d\alpha' \wedge ds. \qquad \blacksquare$$

*Remark* 13.14
If the boundary of $V$ is not smooth, but it is in fact constituted by a finite number of smooth arcs that intersect transversally, the transformation $S$ is not defined in correspondence to the values $s_1, \ldots, s_N$ associated with the vertices of the billiard. This set has $\mu$ measure equal to zero. $\blacksquare$

The measurable dynamical system $(X, \mathcal{A}, \mu, S)$ is in general not ergodic. An important class of ergodic billiards was discovered by Sinai (1970). These billiards have a piecewise smooth boundary $\partial V$ whose smooth components are internally strictly convex (see Fig. 13.6) and intersect transversally.

A beam made of parallel rays, after reflection on one side of the Sinai billiard, becomes *dispersive* (see Fig. 13.6c). Each consecutive reflection forces the beam to diverge further. This property is at the origin of the stochastic behaviour of the orbits in dispersive billiards. Indeed, we have the following two results.

THEOREM 13.15 (Sinai 1970) *Dispersive billiards are ergodic.* $\blacksquare$

THEOREM 13.16 (Gallavotti and Ornstein 1974) *Dispersive billiards are Bernoulli systems.* $\blacksquare$

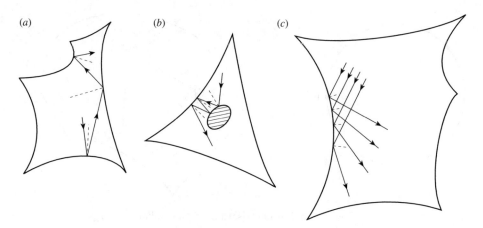

**Fig. 13.6** Billiards of Sinai. Dispersion.

The proofs are very technical and go beyond the scope of this book.

A good introduction to the study of billiards can be found in the monograph by Tabachnikov (1995).

## 13.9 Characteristic exponents of Lyapunov. The theorem of Oseledec

A necessary condition for a measurable dynamical system $(X, \mathcal{A}, \mu, S)$ to be strongly stochastic (e.g. a Bernoulli system) is that orbits corresponding to initial conditions that are close will quickly get away from each other (hence are unstable). For example, in the case of the $p$-adic map $S$ (Example 13.5) consider two initial conditions $x_1, x_2 \in (0, 1)$ and write the corresponding expansions in base $p$: $x_i = \sum_{j=1}^{\infty} x_{i,j} p^{-j}$, where $x_{i,j} \in \mathbf{Z}_p$ for every $i = 1, 2$ and $j \in \mathbf{N}$. Two initial conditions can be made arbitrarily close to each other by making the first digits of the corresponding expansions in base $p$ coincide up to a sufficiently high order: $x_{1,j} = x_{2,j}$ for every $j = 1, 2, \ldots, j_0$, while $x_{1,j_0+1} \neq x_{2,j_0+1}$, in which case $(x_1 - x_2) < p^{-j_0}$.

Recall that $S$ acts as a shift on the expansions in base $p$. Hence we immediately find that if $0 < k < j_0$ we have

$$|S^k(x_1) - S^k(x_2)| = p^k |x_1 - x_2|.$$

In this case, the exponential rate at which the two orbits distance themselves from one another is given by $1/k \log(|S^k(x_1) - S^k(x_2)|)/|x_1 - x_2| = \log p$, and hence it is the entropy of the map $S$. This is far more than a coincidence, as we shall discuss in the next section, but it is useful to introduce quantities that measure the exponential rate of divergence of orbits corresponding to nearby initial conditions: the *Lyapunov characteristic exponents*.

Before considering the most interesting case, when a measurable dynamical system $(X, \mathcal{A}, \mu, S)$ also has the property that the transformation $S$ and the space $X$ are regular in some sense (e.g. $X$ is a smooth differentiable manifold and $S$ is a piecewise $\mathcal{C}^1$ map), we introduce Lyapunov's characteristic exponents through a more abstract procedure.

The fundamental result on which our construction is based, and which we do not prove, is the following.

THEOREM 13.17 (Multiplicative ergodic theorem, Oseledec 1968) *Let $(X, \mathcal{A}, \mu, S)$ be an ergodic system. Let $T : X \to GL(m, \mathbf{R})$ be a measurable map such that*

$$\int_X \log^+ \|T(x)\| \, \mathrm{d}\mu < +\infty, \tag{13.44}$$

*where* $\log^+ u = \max(0, \log u)$. *Set*

$$T_x^n := T(S^{n-1}(x))T(S^{n-2}(x)) \cdots T(x) = \prod_{j=1}^n T(S^{n-j}(x)) \tag{13.45}$$

*for $\mu$-almost every $x \in X$. Then the limit*

$$\lim_{n \to \infty} ((T_x^n)^T T_x^n)^{1/2n} = \Lambda_x \tag{13.46}$$

*exists (where $(T_x^n)^T$ denotes the transpose matrix of $T_x^n$) and it is a symmetric positive semidefinite matrix.* ∎

DEFINITION 13.21 *The logarithms of the eigenvalues of the matrix $\Lambda_x$ are called Lyapunov's characteristic exponents of the system $(X, \mathcal{A}, \mu, S, T)$.* ∎

In what follows the characteristic exponents are ordered in a decreasing sequence $\lambda_1(x) \geq \lambda_2(x) \geq \cdots$. Note that for ergodic systems, they are constant $\mu$-almost everywhere. Now let $\lambda^{(1)} > \lambda^{(2)} > \cdots$ be the characteristic exponents again, but now not repeated according to their multiplicity, and let $m^{(i)}$ be the multiplicity of $\lambda^{(i)}$. Let $E_x^{(i)}$ be the vector subspace of $\mathbf{R}^m$ corresponding to the eigenvalues $\leq \exp \lambda^{(i)}$ of $\Lambda_x$. We thus obtain a 'filtration' of $\mathbf{R}^m$ in subspaces:

$$\mathbf{R}^m = E_x^{(1)} \supset E_x^{(2)} \supset \cdots, \tag{13.47}$$

and moreover the following refinement of Theorem 13.17 holds.

THEOREM 13.18 *Let $(X, \mathcal{A}, \mu, S)$ be as in Theorem 13.17. For $\mu$-almost every $x \in X$, if $\mathbf{v} \in E_x^{(i)} \setminus E_x^{(i+1)}$, we have*

$$\exists \lim_{n \to \infty} \frac{1}{n} \log \|T_x^n \mathbf{v}\| = \lambda^{(i)}. \tag{13.48}$$

*In particular for all vectors $\mathbf{v} \in \mathbf{R}^m \setminus E_x^{(2)}$ (hence for almost every vector $\mathbf{v} \in \mathbf{R}^m$ with respect to the Lebesgue measure) the limit (13.48) is the highest characteristic exponent $\lambda^{(1)}$.* ∎

*Remark* 13.15

For the case $m = 1$ the multiplicative ergodic Theorem 13.18 reduces to the Birkhoff Theorem 13.3 (with the restriction that the functions $f$ in (13.17) are the logarithm of a measurable positive function). Oseledec overcame the additional difficulty that the products of matrices are non-commutative for $m > 1$. ∎

Suppose now that $X$ is $\mathbf{R}^l$ or a compact Riemannian manifold, $\mathcal{A}$ is the $\sigma$-algebra of Borel sets, $S : X \to X$ is a piecewise differentiable transformation and $\mu$ is an invariant ergodic probability measure (if $X = \mathbf{R}^l$ we assume that the support of $\mu$ is compact).

Choose

$$T(\mathbf{x}) = \left(\frac{\partial S_i}{\partial x_j}(\mathbf{x})\right)_{i,j=1,\ldots,l} \in \mathrm{GL}(l, \mathbf{R}), \quad (13.49)$$

the Jacobian matrix of $S$. The hypotheses of the theorem of Oseledec are satisfied and Lyapunov's characteristic exponents are defined for the system $(X, \mathcal{A}, \mu, S)$. From the chain rule it follows that

$$\left(\frac{\partial (S^n)_i}{\partial x_j}(\mathbf{x})\right) = \prod_{k=1}^{n} T(S^{n-k}(\mathbf{x})) = T_x^n, \quad (13.50)$$

and therefore if we consider an infinitesimal change $\delta\mathbf{x}(0) \in \mathbf{R}^l$ in the initial condition, after $n$ iterations of $S$ the latter becomes

$$\delta\mathbf{x}(n) = T_x^n \delta\mathbf{x}(0). \quad (13.51)$$

By Theorem 13.18, for almost every choice of $\delta\mathbf{x}(0)$ we have

$$\delta\mathbf{x}(n) \sim e^{n\lambda^{(1)}} \delta\mathbf{x}(0) \quad (13.52)$$

and the (exponential) instability of the trajectories corresponds to $\lambda^{(1)} > 0$, where $\lambda^{(1)}$ is the largest Lyapunov characteristic exponent.

In the one-dimensional case ($l = 1$) it is possible to compute Lyapunov's characteristic exponent by using the Birkhoff theorem; indeed for $\mu$-a.e. $x \in X$ we have

$$\lambda = \lim_{n\to\infty} \frac{1}{n} \log |T_x^n| = \lim_{n\to\infty} \frac{1}{n} \log \prod_{j=1}^{n} S'(S^{n-j}(x))$$

$$= \lim_{n\to\infty} \frac{1}{n} \sum_{j=0}^{n-1} \log |S'(S^j(x))| = \int_X \log |S'(x)| \, \mathrm{d}\mu, \quad (13.53)$$

where $S'$ denotes the derivative of $S$.

*Example* 13.14

Consider the transformation of Example 13.4: $X = [0, 1]$, $\mathcal{A}$ = Borel sets, $S(x) = 4x(1-x)$, $\mathrm{d}\mu(x) = \mathrm{d}x/[\pi\sqrt{x(1-x)}]$ and assume known that it is ergodic.

We now apply the ergodic theorem (of Birkhoff or of Oseledec; in a one-dimensional situation there is no difference) and set $T_x^n = \prod_{j=1}^{n} S'(S^{n-j}(x))$:

$$\lim_{n\to\infty} \frac{1}{n} \log |T_x^n| = \frac{1}{\pi} \int_0^1 \frac{\log|4(1-2x)|}{\sqrt{x(1-x)}} \, dx$$

$$= \frac{1}{\pi} \left[ 2\arcsin\sqrt{x} \log|4(1-2x)| \right]_0^1 - \frac{2}{\pi} \int_0^1 \frac{\arcsin\sqrt{x}}{4(1-2x)} \, dx$$

$$= \log 2.$$

It follows that the characteristic exponent of $S$ is $\lambda = \log 2$. Since the isomorphism $\Phi : [0,1] \to [0,1]$, $\Phi(x) = 2/\pi \arcsin\sqrt{x}$ transforms $S$ in the diadic map $x \mapsto 2x$ (mod 1) which is also isomorphic to the Bernoulli scheme $SB(1/2, 1/2)$ it follows that $S$ is ergodic and also that $h(S) = \log 2 = \lambda$. ∎

In the general case $l > 1$ there are no formulae that allow the explicit computation (in general) of the characteristic exponents of Lyapunov.

## 13.10 Characteristic exponents and entropy

In the previous section, we saw that Lyapunov's characteristic exponents measure the exponential rate of divergence of two orbits which are initially close. Therefore these exponents give a 'geometric' measure of the complexity of a measurable dynamical system.

On the other hand, the entropy is a purely probabilistic notion, and it measures the complexity of a transformation in the sense of information theory.

These seem at first to be two completely different approaches. However Theorem 12.10 (Brin–Katok) shows how the entropy is also created by the exponential divergence of close orbits, measured by the rate of exponential decrease of the sets

$$B(x, \varepsilon, n) = \{y \in X \mid d(S^i x, S^i y) \leq \varepsilon, \ \forall \ i = 0, 1, \ldots, n-1\}.$$

Just as the rate of exponential growth of an infinitesimal vector $\delta\mathbf{x}(n)$ is given by $e^{n\lambda^{(1)}}$, where $\lambda^{(1)}$ is the largest Lyapunov exponent, the rate of growth of the $k$th element of volume $\delta_1\mathbf{x}(n) \wedge \ldots \wedge \delta_k\mathbf{x}(n)$ is given by $\exp[n(\lambda_1 + \ldots + \lambda_k)]$.

These heuristic remarks suggest that there exists a relation between the *positive* characteristic exponents of Lyapunov and the entropy.

In what follows we assume that $X$ is a compact Riemannian manifold, that $S: X \to X$ is a diffeomorphism of $X$ of class $C^2$, $\mathcal{A}$ is the $\sigma$-algebra of the Borel sets $X$ and $\mu$ is an ergodic invariant probability measure for $S$.

We denote by $\lambda^{(1)} > \lambda^{(2)} > \ldots$ Lyapunov's characteristic exponents of $(X, \mathcal{A}, \mu, S, S')$ and by $m^{(i)}$ the multiplicity of $\lambda^{(i)}$. Finally we set $u^+ = \max(0, u)$, so that $\{\lambda^{(i)+}\}$ is the set of positive characteristic exponents.

The following are the two fundamental results linking entropy and characteristic exponents.

THEOREM 13.19 (Ruelle's inequality)
$$h(S) \leq \sum \lambda^{(i)+} m_i. \tag{13.54}$$
∎

THEOREM 13.20 (Pesin's formula) *If the invariant measure $\mu$ is equivalent to the volume associated with the Riemannian metric on $X$ then*
$$h(S) = \sum \lambda^{(i)+} m_i. \tag{13.55}$$
∎

For a proof of these results, besides the original articles of Pesin (1977) and Ruelle (1978), we recommend Mañe (1987) and Young (1995).

*Example* 13.15
Take $X = \mathbf{T}^l$ with the flat metric, $\mu$ the Haar measure ($= [1/(2\pi)^l] \times$ Lebesgue measure), and $S$ a hyperbolic automorphism. In this case $S' = S$ and if $\nu_1 \geq \nu_2 \geq \ldots \geq \nu_l$ are the eigenvalues of $S$ the characteristic exponents are $\lambda_i = \log|\nu_i|$. Since the Haar measure (Example 13.12) is equivalent to the Lebesgue measure (differing from it only in the choice of normalising factor), the hypotheses of Pesin's formula hold and
$$h(S) = \sum \lambda^{(i)+} m_i = \sum_{|\nu_j|>1} \log|\nu_j|,$$
i.e. formula (13.43). ∎

## 13.11 Chaotic behaviour of the orbits of planets in the Solar System

The problem of the long-term behaviour of the planets in the Solar System has been central to the investigation of astronomers and mathematicians. Newton was convinced that the Solar System is unstable: he believed that perturbations between the planets are sufficiently strong to destroy in the long term the Keplerian orbits. Newton even conjectured that from time to time God intervened directly to 'reorder things' so that the Solar System could survive. In the *Principia* we find:

Planetae sex principales revolvuntur circum solem in circulis soli concentricis, eadem motus directione, in eodem plano quamproxime. Lunae decem revolvuntur circum terram, jovem et saturnum in circulis concentricis, eadem motus directione, in planis orbitum planetarum quamproxime. Et hi omnes motus regulares originem non habent ex causis mechanicis (...). Elegantissima haecce solis, planetarum et cometarum compage non nisi consilio et dominio entis intelligentis et potentis oriri potuit.[4]

---

[4] Newton I., *Principia Mathematica Philosophiae Naturalis, Liber Tertius:* De Mundi Systemate. Pars II Scholium Generale 672–3 ('The six primary planets revolve about the sun in

Already in the seventeenth century the stability of the orbits of the planets in the Solar System was considered as a concrete problem: Halley, analysing Chaldean observations reported in Ptolemy's work, proved that Saturn was distancing itself from the Sun, while Jupiter was approaching it. An extrapolation of those data leads to a possible collision between the two planets in 6 million years.

From a mathematical point of view, arguments in favour of the stability of the orbits of planets were advocated by Lagrange, Laplace and Poisson in the eighteenth century. Using the theory of perturbations, they could prove the absence of 'secular terms' (hence terms with polynomial dependence on time) in the time evolution of the semi-major axes of the planets, up to errors of third order in the planetary masses. The extrapolation just mentioned is therefore not justified.

On the contrary, the research of Poincaré and Birkhoff showed the possibility of strong instability in the planets' dynamics and found that the phase space must have a very complex structure.

Modern theoretical research, mostly based on the KAM theorem, suggests that the situation could have two aspects: the majority of the orbits in the sense of measure theory (hence corresponding to the majority of initial conditions with respect to the Lebesgue measure) would be stable, but in any neighbourhood of them there exist unstable orbits. 'Therefore, although the motion of a planet or of an asteroid is regular, an arbitrarily small perturbation of the initial conditions is sufficient to transform the orbit in a chaotic orbit' (Arnol'd 1990, p. 82).

It is a delicate issue, even if one neglects the actual physical data of the problem (masses and orbital data of the planets of the Solar System), to consider just idealised and simplified problems. For example, at a recent International Congress of Mathematicians, in one of the plenary talks the following question was posed, whose answer appears to be very difficult:

Consider the $n$-body problem ($n \geq 3$) in which one of the masses is much greater than all others, and a solution with circular orbits around the principal mass, which lie in the same plane and are traced in the same direction. Do there exist wandering domains[5] *in every neighbourhood of it?*[6]

circles concentric with the sun, with the same direction of motion, and very nearly in the same plane. Ten moons revolve about the earth, Jupiter, and Saturn in concentric circles, with the same direction of motion, very nearly in the planes of the orbits of the planets. And all these regular motions do not have their origin in mechanical causes (...) This most elegant system of the sun, planets, and comets could not have arisen without the design and dominion of an intelligent and powerful being.' (Translated by I. Bernard Cohen and Anne Whitman, University of California Press.)

[5] An open set $V$ is called wandering in the Hamiltonian flow $f^t$ if there exists a time $t_0 > 0$ such that $f^t(V) \cap V = \emptyset$ for every $t > t_0$.
[6] Herman M. R., *Some open problems in dynamical systems*, International Congress of Mathematicians, Berlin, 1998.

The problem of the stability of the orbits of the planets has also been studied by the numerical integration of the Newton equations. A severe limitation of this approach is the small size of the time-step necessary (from about 40 days for Jupiter down to 12 hours for Mercury). Hence, until 1991 the only numerical integration of a realistic model of the Solar System could simulate its evolution only for 44 centuries.

This limitation forces, even for numerical studies, an analytical approach using the appropriate variables and ideas from the canonical theory of perturbations. Therefore one can replace Newton's equations by the so-called *secular system* introduced by Lagrange, where the rapidly varying angular parameters, i.e. the mean anomalies, are eliminated, together with the corresponding canonically conjugate variables, i.e. the action variables (proportional to the semi-major axes of the orbits). The system thus obtained describes the slow deformation of the orbits of the planets since the remaining variables are proportional to the eccentricity, to the inclination of the orbit, to the longitude of the ascending node and to the argument of the perihelion. Considering the eight principal planets, we obtain in this way a system with 16 degrees of freedom.

Laskar integrated numerically a model of a secular system for the Solar System (Laskar 1989b, 1990), accurate to second order in masses and to fifth order in the eccentricities and inclinations. The result is a system containing approximately 150 000 polynomial terms.

The main result of this numerical study is that the inner Solar System (Mercury, Venus, Earth and Mars) is chaotic, with a Lyapunov exponent of the order of $1/5$ (million years)$^{-1}$. This result indicates that it is impossible in practice to predict exactly the motion of the planets for a period longer than 100 million years. This sensitivity to initial conditions leads to a total lack of determination for the orientation of the orbit (hence to the impossibility of predicting the time evolution of the longitude of the ascending node and of the perihelion). The variations in the eccentricity and in the inclination are much slower, and become relevant only on a time-scale of the order of a billion years.

Additional numerical studies have shown that in a time of the order of 4 billion years the eccentricity of Mercury might increase to a value 0.5, which would bring it to intersect the orbit of Venus. In this case, the expulsion of Mercury from the Solar System cannot be excluded.

## 13.12 Problems

**1.** Prove that the $\sigma$-algebra $\mathcal{B}(\mathbf{R})$ of Borel sets is generated not only by the open sets of $\mathbf{R}$ but also by each of the following families: the closed sets of $\mathbf{R}$; the intervals of the type $(a, b]$; the intervals of the type $(-\infty, b]$.

**2.** Consider a measurable dynamical system $(X, \mathcal{A}, \mu, S)$. Prove that if there exists a set $\mathcal{F} \subseteq L^1(X, \mathcal{A}, \mu)$ dense in $L^1(X, \mathcal{A}, \mu)$ and such that for every $f \in \mathcal{F}$ then $\hat{f}(x) = \langle f \rangle_\mu$ for $\mu$-almost every $x$, the system is ergodic.

**3.** Let $1 < p < \infty$ and $(X, \mathcal{A}, \mu, S)$ be a measurable dynamical system. Prove that the system is ergodic if and only if every $S$-invariant function $f \in L^p(X, \mathcal{A}, \mu)$ is constant $\mu$-almost everywhere.

**4.** Let $X$ be a compact metric space, $S : X \to X$ a continuous map, and $\mathcal{B}$ the $\sigma$-algebra of Borel sets on $X$. Prove that there exists at least one probability measure on $X$ which is invariant for $S$.
(Hint: associate with $S$ the continuous transformation $S^* : \mathcal{M}(X) \to \mathcal{M}(X)$ defined by $(S^*\mu)(A) = \mu(S^{-1}(A))$ for every $A \in \mathcal{B}$. An invariant probability measure $\mu$ satisfies $S^*\mu = \mu$. Given any measure $\mu_0 \in \mathcal{M}(X)$ consider the sequence $\mu_m = 1/m \sum_{j=0}^{m-1} S^{*m}\mu_0$ and use the compactness of $\mathcal{M}(X)$ (see Problem 1).)

**5.** Let $X$ be a topological space (locally compact, separable and metrisable) and let $S : X \to X$ be continuous. $S$ is *topologically transitive* if for every pair of non-empty open sets $U, V \subset X$ there exists an integer $N = N(U, V)$ such that $S^N(U) \cap V \neq \emptyset$. $S$ is *topologically mixing* if for every pair $U, V$ as above there exists $N = N(U, V)$ such that $S^n(U) \cap V \neq \emptyset$ for every $n \geq N$.

(1) Prove that if $S$ is topologically transitive, then there exists $x \in X$ whose orbit $(S^n(x))_{n \in \mathbf{N}}$ is dense in $X$.

(2) If $S$ is topologically transitive, the only continuous functions $f : X \to \mathbf{R}$ which are $S$-invariant are the constant functions.

(3) Prove that irrational translations on the tori (Example 13.12) are not topologically mixing but they are topologically transitive.

(4) Prove that for every integer $m \geq 2$ the transformation $S : \mathbf{S}^1 \to \mathbf{S}^1$, $\chi \mapsto m\chi$ (mod $2\pi \mathbf{Z}$) is topologically mixing.

**6.** Let $X$ be a topological space, and $S : X \to X$ measurable with respect to the $\sigma$-algebra $\mathcal{A}$ of Borel sets in $X$ preserving the measure $\mu$. If $S$ is mixing and $\mu(A) > 0$ for every open set $A \in \mathcal{A}$ then $S$ is topologically mixing.

**7.** Prove that if $(X, \mathcal{A}, \mu, S)$ is mixing, equation (13.30) is valid also $\forall\, f \in L^\infty(X, \mathcal{A}, \mu)$ and $\forall\, g \in L^1(X, \mathcal{A}, \mu)$.

**8.** Let $(X, \mathcal{A}, \mu, S)$ be a mixing dynamical system. Assume that $\lambda : \mathcal{A} \to [0, 1]$ is another probability measure not necessarily preserved by $S$ but absolutely continuous with respect to $\mu$. Prove that $\lim_{n \to +\infty} \lambda(S^{-n}(A)) = \mu(A)$ for every $A \in \mathcal{A}$.

**9.** Prove that the irrational translations on the tori (described in Example 13.12) are not mixing.

**10.** Prove that a Bernoulli scheme is mixing. (Hint: prove first that equation (13.27) is satisfied if $A$ and $B$ are cylindrical sets.)

**11.** Let $(X, \mathcal{A}, \mu, S)$ be a measurable dynamical system. Prove that, for every $m \in \mathbf{N}$, $h(S^m) = mh(S)$. Show also that if $S$ is invertible then $h(S^m) = |m|h(S)$ for every $m \in \mathbf{Z}$ (equivalently, $S$ and its inverse have the same entropy).

**12.** Let $(X_1, \mathcal{A}_1, \mu_1, S_1)$ and $(X_2, \mathcal{A}_2, \mu_2, S_2)$ be two measurable dynamical systems. Consider $X = X_1 \times X_2$ with the measure space structure induced by the product (see Example 13.3). Prove that $S : X \to X$, defined by setting $S(x_1, x_2) = (S_1(x_1), S_2(x_2))$, preserves the product measure and that $h(S) = h(S_1) + h(S_2)$.

**13.** Prove Theorem 13.12.

**14.** Prove that the transformation of Gauss (Example 13.6) is exact (see Problem 3 of Section 13.13), and therefore ergodic. Then prove that the Lyapunov exponent of the transformation is $\pi^2/6 \log 2$. (Hint: expanding $1/1+x = \sum_{n=0}^{\infty}(-1)^n x^n$ show that $\int_0^1 \log x \, dx/(1+x) = \sum_{n=1}^{\infty}(-1)^n/n^2$. To see that $\sum_{n=1}^{\infty}(-1)^n/n^2 = -\pi^2/12$ compute the Fourier series expansion of the $2\pi$-periodic function which takes value $-x^2/4$ in the interval $(-\pi, \pi)$ and evaluate it at $x = 0$.)

**15.** Let $X$ be a separable metric space, $d$ the metric, $\mathcal{A}$ the associated $\sigma$-algebra of Borel sets, $\mu$ a probability measure and $S : X \to X$ a map preserving the measure $\mu$. With every point $x \in X$ we associate the $\omega$-*limit* set

$$\omega(x) := \{y \in X \mid \liminf_{n \to \infty} d(S^n(x), y) = 0\}.$$

From the theorem of Poincaré (Remark 13.6) we deduce that $\mu(\{x \in X \mid x \notin \omega(x)\}) = 0$. Since $\omega(x)$ is the set of accumulation points of the orbit $x, S(x), S^2(x), \ldots$, the previous statement shows that $\mu$-a.e. point $x \in X$ is an accumulation point for its own orbit.

## 13.13 Additional solved problems

*Problem 1*
Let $X$ be a compact metric space and let $\mathcal{M}(X)$ be the set of invariant measures on $X$ with the usual topology. Prove that $\mathcal{M}(X)$ is a compact metric space (see Mañe 1987).

*Solution*
Consider the Banach space $\mathcal{C}(X)$ of continuous functions $f : X \to \mathbf{R}$ with the usual norm

$$\|f\| = \sup_{x \in X} |f(x)|. \tag{13.56}$$

Since $X$ is metric and compact it is also separable, and therefore there exists a countable set $(g_i)_{i \in \mathbf{N}} \subset \mathcal{C}(X)$ that is dense in the unit ball $B = \{f \in \mathcal{C}(X) \mid \|f\| \leq 1\}$.

Using the functions $(g_i)_{i \in \mathbf{N}}$ it is possible to define a metric on $\mathcal{M}(X)$: if $\mu$ and $\nu$ are two probability measures on $X$ we define

$$d(\mu, \nu) = \sum_{j=1}^{\infty} 2^{-j} \left| \int_X g_j \, d\mu - \int_X g_j \, d\nu \right|. \tag{13.57}$$

It is trivial to verify that $d$ satisfies the triangle inequality and moreover $\forall\, i \in \mathbf{N}$ we obviously have

$$\left| \int_X g_i \, d\mu - \int_X g_i \, d\nu \right| \leq 2^i \, d(\mu, \nu).$$

This shows that if $d(\mu_n, \mu) \to 0$ for $n \to \infty$ then for every $i \in \mathbf{N}$ it follows that $\int_X g_i \, d\mu_n \to \int_X g_i \, d\mu$. Using the density of the functions $(g_i)_{i \in \mathbf{N}}$ in $B$ we can conclude that for every function $g \in \mathcal{C}(X)$ we have

$$\lim_{n \to \infty} d(\mu_n, \mu) = 0 \Leftrightarrow \lim_{n \to \infty} \left| \int_X g \, d\mu_n - \int_X g \, d\mu \right| = 0.$$

Therefore the topology induced by the metric (13.57) is the same as that defined by (13.7) (or (13.8)).

Since $\mathcal{M}(X)$ is a metric space its compactness is equivalent to compactness for sequences, and hence we only need to show that every sequence $(\mu_n)_{n \in \mathbf{N}} \subset \mathcal{M}(X)$ has a convergent subsequence. The fundamental ingredient in the proof is given by the Riesz theorem (see Rudin 1974) given as follows.

Let $\Phi : \mathcal{C}(X) \to \mathbf{R}$ be a positive linear functional (hence such that $\Phi(f) \geq 0$ if $f \geq 0$). There exists a unique probability measure $\mu \in \mathcal{M}(X)$ such that

$$\int_X f \, d\mu = \Phi(f) \qquad (13.58)$$

for every $f \in \mathcal{C}(X)$.

Let there be given a bounded sequence $(\mu_n)_{n \in \mathbf{N}} \subset \mathcal{M}(X)$. With every measure $\mu_n$ we associate the sequence $(\tilde{\mu}_{n,i})_{i \in \mathbf{N}} \subset [-1, 1]$ defined by setting

$$\tilde{\mu}_{n,i} = \int_X g_i \, d\mu_n.$$

By the compactness in the space of sequences in $[-1, 1]$ there exists a subsequence $(\mu_{n_m})_{m \in \mathbf{N}}$ such that for every $i \in \mathbf{N}$ the sequence $(\tilde{\mu}_{n_m, i})_{m \in \mathbf{N}} \subset [-1, 1]$ is convergent, i.e. for every $i \in \mathbf{N}$ the sequence in $m$ given by $\int_X g_i \, d\mu_{n_m}$ converges. Using again the density of the sequence of functions $(g_i)_{i \in \mathbf{N}}$ it follows that for every $g \in \mathcal{C}(X)$ the sequence $\left( \int_X g \, d\mu_{n_m} \right)_{m \in \mathbf{N}} \subset \mathbf{R}$ is convergent. Now let $\Phi : \mathcal{C}(X) \to \mathbf{R}$ be defined by

$$\Phi(f) = \lim_{m \to \infty} \int_X f \, d\mu_{n_m}. \qquad (13.59)$$

It is immediate to verify that $\Phi$ is a positive linear functional, and therefore by Riesz's theorem there exists $\mu \in \mathcal{M}(X)$ such that for every $f \in \mathcal{C}(X)$ we have

$$\Phi(f) = \int_X f \, d\mu. \qquad (13.60)$$

Comparing (13.59) with (13.60) shows that $\mu_{n_m} \to \mu$; hence the subsequence $\mu_{n_m}$ is convergent, and the proof is finished.

*Problem 2*
Prove that the baker's transformation (Example 13.7) is a Bernoulli system, and compute its entropy.

*Solution*
We note first of all that the baker's transformation $S$ is invertible: its inverse is

$$S^{-1}(x,y) = \begin{cases} \left(\dfrac{x}{2}, 2y\right), & \text{if } y \in (0, \tfrac{1}{2}), \\ \left(\dfrac{x+1}{2}, 2y-1\right), & \text{if } y \in (\tfrac{1}{2}, 1). \end{cases} \qquad (13.61)$$

We can then construct an isomorphism between $S$ and a bilateral Bernoulli scheme, namely $SB(1/2, 1/2)$. From this fact it immediately follows that $h(S) = \log 2$.

Consider the map $T : \mathbf{Z}_2^{\mathbf{Z}} \to [0,1] \times [0,1]$ defined as follows: if $\xi = (\xi_i)_{i \in \mathbf{Z}} \in \mathbf{Z}_2^{\mathbf{Z}}$ set

$$(x, y) = T(\xi) = \left( \sum_{i=0}^{+\infty} \xi_i 2^{-i-1}, \sum_{i=-1}^{-\infty} \xi_i 2^i \right). \qquad (13.62)$$

The map $T$ therefore associates with a doubly infinite sequence $\xi$ the point in the square whose base 2 expansion of the $x$ and $y$ coordinates is given, respectively, by $(\xi_i)_{i \geq 0}$ and $(\xi_i)_{i<0}$. It is immediate to verify that the properties (a) and (b) of Definition 13.18 are satisfied. In addition we have

$$T(\sigma(\xi)) = T((\xi_{i+1})_{i \in \mathbf{Z}}) = \left( \sum_{i=0}^{+\infty} \xi_{i+1} 2^{-i-1}, \sum_{i=-1}^{-\infty} \xi_{i+1} 2^i \right)$$

$$= \left( 2 \sum_{i=1}^{+\infty} \xi_i 2^{-i-1}, \frac{1}{2} \sum_{i=0}^{-\infty} \xi_i 2^i \right) \qquad (13.63)$$

$$= \left( 2x - \xi_0, \frac{y + \xi_0}{2} \right)$$

$$= S(x, y)$$

since $\xi_0 = 1$ if $x \geq \tfrac{1}{2}$ and zero otherwise. It follows that (c) is also fulfilled.

To conclude the proof, it is enough to construct the inverse map $T'$ (mod 0) of $T$. Taking into account the interpretation of $T$ in terms of the expansions $x = \sum_{i=1}^{\infty} x_i 2^{-i}$, $y = \sum_{i=1}^{\infty} y_i 2^{-i}$, it is immediate to check that

$$\xi = (\xi_i)_{i \in \mathbf{Z}} = T'(x, y) = \begin{cases} x_{i+1}, & i \geq 0, \\ y_{-i}, & i < 0 \end{cases} \qquad (13.64)$$

is the sought transformation and satisfies all the conditions of (d).

*Problem 3*
Let $(X, \mathcal{A}, \mu, S)$ be a measurable dynamical system and let $S$ be non-invertible (mod 0). The system is *exact* if

$$\bigcap_{n=0}^{+\infty} S^{-n}\mathcal{A} = \mathcal{N}, \qquad (13.65)$$

where $\mathcal{N}$ is the trivial $\sigma$-algebra of measurable sets $A \in \mathcal{A}$ such that (modifying $A$ mod 0 if necessary) $A = S^{-n}(S^n(A))$. Prove that:

(a) $S$ is exact if and only if $\forall\, A \in \mathcal{A}$ such that $\mu(A) > 0$ and $S^j A \in \mathcal{A},\ \forall\, j \geq 0$ we have

$$\lim_{j \to +\infty} \mu(S^j(A)) = 1; \qquad (13.66)$$

(b) every exact system is ergodic.[7]

*Solution*
Let $A$ be as in (a) and let us show that if $S$ is exact, $\lim_{j\to\infty} \mu(S^j(A)) = 1$. Since the sequence $A, S^{-1}(S(A)), S^{-2}(S^2(A)), \ldots$ is increasing, the union $B = \cup_{k=0}^{+\infty} S^{-k}(S^k(A))$ satisfies

$$B = \bigcup_{k=n}^{+\infty} S^{-k}(S^k(A)) = S^{-n}(S^n(B))$$

for every $n \in \mathbf{N}$. Hence $B \in \cap_{n=0}^{\infty} S^{-n}\mathcal{A}$ and since $\mu(B) > \mu(A) > 0$ it necessarily follows that $\mu(B) = 1$, and therefore that

$$\lim_{j \to \infty} \mu(S^j(A)) = \lim_{j \to \infty} \mu(S^{-j}(S^j(A))) = \mu(B) = 1. \qquad (13.67)$$

Conversely, let us assume (13.66) holds and show how to deduce (13.65). Let $A \in \mathcal{A}$ be such that $S^{-n}(S^n(A)) = A$ for every $n \in \mathbf{N}$. Clearly $\mu(S^n(A)) = \mu(A)$ and $\lim_{n \to \infty} \mu(S^n(A)) = \mu(A)$. Then if $\mu(A) > 0$ necessarily $\mu(A) = 1$. This ends the proof of (a).

We now show that an exact system is metrically indecomposable. Let $\mathcal{A}_S$ be the sub-$\sigma$-algebra of $\mathcal{A}$ of all $S$-invariant sets. The fact that the system is metrically indecomposable is equivalent to the condition that $\mathcal{A}_S \subset \mathcal{N}$, and hence every $S$-invariant set has measure zero or one.

It is clear that $\mathcal{A}_S \subset S^{-n}\mathcal{A}$ for every $n \in \mathbf{N}$. Therefore

$$\mathcal{A}_S \subset \bigcap_{n=0}^{+\infty} S^{-n}\mathcal{A} = \mathcal{N}.$$

---

[7] It can be proved (see Rohlin 1964) that exact systems are mixing.

## 13.14 Additional remarks and bibliographical notes

Our brief introduction to ergodic theory has been strongly influenced by the beautiful monograph of Mañe (1987) and by the excellent article of Young (1995).

The relation with the more physical aspects of the theory and in particular with 'strange' attractors and turbulence is discussed in the review by Eckmann and Ruelle (1985), where it is possible to also find an interesting discussion of the various notions of fractal dimensions and of how to compute them experimentally using time series.

A great impulse to the development of ergodic theory came also from the problem of the foundations of classical statistical mechanics. In addition to reference works (Khinchin 1949, Krylov 1979), now slightly dated, for an introduction to a modern point of view we recommend Gallavotti and Ruelle (1997) and Gallavotti (1998) for their originality.

To read more about the chaotic behaviour of the orbits of the planets of the solar system we recommend Laskar (1992) and Marmi (2000).

The collection of articles by Bedford *et al.* (1991) can be useful to the reader looking for an introduction to the study of hyperbolic dynamical systems (see Yoccoz 1995), of which an important example is given by geodesic flows on manifolds with constant negative curvature (Hadamard 1898; Anosov 1963, 1967).

# 14 STATISTICAL MECHANICS: KINETIC THEORY

## 14.1 Distribution functions

In this chapter we present a brief introduction to the statistical approach to mechanics, developed by Ludwig Boltzmann. The great importance and immense bearing of the ideas of Boltzmann deserves ampler space, but this is not feasible within the context of the present book. We recommend the monographs of Cercignani (1988, 1997) and the deep analysis of Gallavotti (1995), in addition to the treatise of Cercignani et al. (1997).

Consider a gas of $N$ particles, which for simplicity we assume to be identical. The gas is contained in a volume $V$. The typical values of $N$ and $V$, at standard conditions of temperature and pressure ($T = 300\,\mathrm{K}$, $P = 1\,\mathrm{atm}$) are $N = 6.02 \times 10^{23}$ (*Avogadro's number*) and $V = 22.7\,\mathrm{l}$. We assume from now on that all collisions with the walls of the container are non-dissipative.

It is clearly impractical to follow the motion of the single particles taking into account their mutual interactions and possible external forces. In fact this is impossible, for example because we cannot know the initial conditions of all particles. Statistics proves to be a more appropriate tool. Thus the methodology of kinetic theory to study the evolution of a system and the achievement of an equilibrium state is the following. We introduce a six-dimensional space, which we use as phase space with momentum and position coordinates $(\mathbf{p}, \mathbf{q})$, and we plot in this space the representative points of each particle. This space is traditionally called the *space $\mu$*.

We neglect the internal degrees of freedom of the particles, treating them effectively as points. In what follows we always use this simplification to avoid a heavily technical exposition, but this is only a reasonable assumption for *monatomic gases*.

Consider in the space $\mu$ a cell of volume $\Delta$ and count at a given time $t$ the number $\nu(\Delta, t)$ of representative points contained in this cell. If the ratio $N/V$ is, e.g. of order $10^{18}\,\mathrm{cm}^{-3}$, we note that the ratio $\nu(\Delta)/\Delta$ stabilises, as the diameter of the cell becomes sufficiently (but not excessively) small, to a value depending on the centre of the cell $(\mathbf{p}, \mathbf{q})$ and on the time $t$ considered. The value thus obtained defines a function $f(\mathbf{p}, \mathbf{q}, t)$ called the *distribution function*. This procedure is analogous to the procedure defining the density of a system in the mathematical model adopted by the mechanics of continuous systems.

Thus the set of representative points in the space $\mu$ is treated as a continuous distribution. Therefore the number of particles $\nu(\Omega, t)$, whose kinematic state at time $t$ is described by a point that belongs to a given measurable subset $\Omega$ of the space $\mu$, is given by the integral

$$\nu(\Omega, t) = \int_\Omega f(\mathbf{p}, \mathbf{q}, t) \, \mathrm{d}\mathbf{p} \, \mathrm{d}\mathbf{q}. \tag{14.1}$$

Hence

$$N = \int f(\mathbf{p}, \mathbf{q}, t) \, \mathrm{d}\mathbf{p} \, \mathrm{d}\mathbf{q}, \tag{14.2}$$

where the domain of integration is the whole space $\mu$.

If the spatial distribution of the particles is uniform, the distribution function is independent of the space vector $\mathbf{q}$ inside the container (and it is zero outside) and the integration with respect to the $\mathbf{q}$ in (14.2) simply leads to factorisation of the volume $V$ occupied by the system. In this case, we obtain the following expression for the number $n$ of particles per unit of volume, relative to the whole system:

$$n = \frac{N}{V} = \int f(\mathbf{p}, t) \, \mathrm{d}\mathbf{p}, \tag{14.3}$$

where the domain of integration is $\mathbf{R}^3$.

The states of the system are described by the distribution function $f$, and therefore it must in principle be possible to derive from this function the thermodynamical properties of the system.

## 14.2 The Boltzmann equation

In this section we want to describe the line of thought that led Boltzmann to deduce the equation governing the distribution function. Maxwell had assumed the system to be in equilibrium (hence a distribution function independent of time) and had looked for the conditions on $f$ such that the equilibrium would be stable. On the other hand, Boltzmann was interested by the problem, logically very important, of how such equilibrium—whose experimental evidence is given by the success of classical thermodynamics—can be achieved through the collisions between the molecules.

The rate at which the distribution function $f$ varies in time is given by

$$\frac{\mathrm{d}f}{\mathrm{d}t} = \frac{\partial f}{\partial t} + \nabla_\mathbf{q} f \cdot \frac{\mathbf{p}}{m} + \nabla_\mathbf{p} f \cdot \mathbf{F},$$

where we take into account that $\dot{\mathbf{q}} = \mathbf{p}/m$ and $\dot{\mathbf{p}} = \mathbf{F}$, where $\mathbf{F}$ is any external force acting on the system.

If the dilution of the gas were so strong that we could neglect the interaction between the molecules, we would have that $\mathrm{d}f/\mathrm{d}t = 0$. This can be proved starting from the conservation of the volume occupied by each set of representative points in the space $\mu$ (Liouville's theorem, Theorem 8.3). The variations of $f$ can therefore be attributed to the 'collisions' between molecules, where the term 'collision' is used in the generic sense of a *short-range interaction*. We mean therefore that the molecules interact only when they arrive at a mutual distance comparable to their diameters.

In the simplest model we make the following assumptions.

(1) *Hard spheres*; we assume that the molecules are identical hard spheres, of radius $R$ and mass $m$.[1]
(2) *Strong dilution*; if $n = N/V$, we assume that

$$nR^3 \ll 1,$$

and therefore the probability that two molecules are at a distance of order $R$ (hence 'colliding') is very small.

(3) *Perfectly elastic binary collisions*; we exclude all situations where three or more molecules collide at the same time. From a physical point of view, this assumption is reasonable if the gas is strongly diluted, because the *mean free path* of a molecule (the average distance between between two consecutive collisions) is then much larger than the average diameter of the molecules.

(4) *Molecular chaos (Stosszahlansatz)*;[2] the distribution function of a pair of colliding molecules—hence the probability that at time $t$ we can determine a binary collision at a position $\mathbf{q}$ between two molecules with momenta $\mathbf{p}_1$ and $\mathbf{p}_2$—is proportional to the product

$$f(\mathbf{q}, \mathbf{p}_1, t) f(\mathbf{q}, \mathbf{p}_2, t). \tag{14.4}$$

The statistical significance of (14.4) is the weak correlation between the motion of the two colliding particles before the collision. Hence we neglect the possibility that the two particles have already collided with each other or separately with the same particles.

From the assumption that the collisions are non-dissipative, it follows that the two colliding molecules with initial momenta $\mathbf{p}_1, \mathbf{p}_2$ emerge from the collision with new momenta $\mathbf{p}'_1, \mathbf{p}'_2$, which must satisfy the fundamental laws of conservation of momentum and energy:

$$\mathbf{p}_1 + \mathbf{p}_2 = \mathbf{p}'_1 + \mathbf{p}'_2 = \mathbf{P}, \tag{14.5}$$

$$\mathbf{p}_1^2 + \mathbf{p}_2^2 = \mathbf{p}'^2_1 + \mathbf{p}'^2_2 = 2mE. \tag{14.6}$$

---

[1] The typical order of magnitude of $R$ is $10^{-7} - 10^{-8}$ cm, and the order of magnitude of $m$ is $10^{-22} - 10^{-24}$ g.

[2] This assumption is still discussed today, and it is essentially statistical, as opposed to the assumption that the collisions are only binary. The rigorous deduction of the assumption of molecular chaos for appropriate initial conditions $f_0$ for the distribution function (in the so-called Grad–Boltzmann limit $R \to 0$ and $n \to \infty$, so that $nR^2 \to$ constant, corresponding to fixing the mean free path, as we shall see in Section 14.6) is an important success of modern mathematical physics, due to Lanford (1975).

In reality, the following considerations apply to any interaction model satisfying (14.5), (14.6).

The transitions of the pair $(\mathbf{p}_1, \mathbf{p}_2)$ to the admissible pairs $(\mathbf{p}'_1, \mathbf{p}'_2)$ do not, in general, have equal probability, but they are described by a *transition kernel* $\tau(\mathbf{p}_1, \mathbf{p}_2, \mathbf{p}'_1, \mathbf{p}'_2)$ which must be symmetric with respect to the interchange of the pairs $(\mathbf{p}_1, \mathbf{p}_2)$ and $(\mathbf{p}'_1, \mathbf{p}'_2)$, because the inverse transition has the same probability, due to the reversibility of the microscopic evolution equations (the equations of Hamilton). The kernel is also symmetric separately for the interchange of $\mathbf{p}_1$ and $\mathbf{p}_2$ and of $\mathbf{p}'_1$ and $\mathbf{p}'_2$, since we assumed that the particles are identical.

Finally, it is reasonable to assume that $\tau$ depends on the *modulus* of the relative velocity of the colliding particles, in addition to the angular coordinates of the collision, for reasons of *isotropy*.

If we now consider the function

$$f_1 = f(\mathbf{p}_1, \mathbf{q}, t), \tag{14.7}$$

we see that its total derivative with respect to time is the sum of a negative term due to the transitions $(\mathbf{p}_1, \mathbf{p}_2) \to (\mathbf{p}'_1, \mathbf{p}'_2)$ for any $\mathbf{p}_2$, and of a positive term due to the inverse transitions. For fixed $\mathbf{p}_1$, we must consider all the possible vectors $\mathbf{p}_2$ and all the possible pairs $(\mathbf{p}'_1, \mathbf{p}'_2)$ that are compatible with the conservation laws (14.5) and (14.6).

Because of the assumption (14.4) the frequency of the transitions $(\mathbf{p}_1, \mathbf{p}_2) \to (\mathbf{p}'_1, \mathbf{p}'_2)$ and the frequency of the inverse ones are proportional to the products $f_1 f_2$ and $f'_1 f'_2$, respectively, by where analogy with (14.7) we have used the symbols $f_i = f(\mathbf{p}_i, \mathbf{q}, t)$, $f'_i = f(\mathbf{p}'_i, \mathbf{q}, t)$, $i = 1, 2$. The transition kernel weighs such products to obtain the respective frequencies. Hence at every point $\mathbf{q}$, for fixed $\mathbf{p}_1$ and $\mathbf{p}_2$, the frequency of the collisions that make a particle leave the class described by the function $f_1$ is $\tau(\mathbf{p}_1, \mathbf{p}_2, \mathbf{p}'_1, \mathbf{p}'_2) f_1 f_2$, while the frequency of the collisions that enrich this class is $\tau(\mathbf{p}_1, \mathbf{p}_2, \mathbf{p}'_1, \mathbf{p}'_2) f'_1 f'_2$. To obtain the collision term that equates with $df_1/dt$ we must therefore integrate the expression $\tau(f'_1 f'_2 - f_1 f_2)$ over all the momenta $\mathbf{p}_2$ and on the regular two-dimensional submanifold of $\mathbf{R}^6$ made of the pairs $(\mathbf{p}'_1, \mathbf{p}'_2)$ subject to the constraints (14.5), (14.6), where the invariants $\mathbf{P}, E$ are fixed in correspondence to $\mathbf{p}_1, \mathbf{p}_2$. Denoting by $\Sigma(\mathbf{P}, E)$ this manifold, we can finally write the balance equation for $f_1$ in the form

$$\left(\frac{\partial}{\partial t} + \frac{\mathbf{p}_1}{m} \cdot \nabla_{\mathbf{q}} + \mathbf{F} \cdot \nabla_{\mathbf{p}}\right) f_1 = \int_{\mathbf{R}^3} d\mathbf{p}_2 \int_{\Sigma(\mathbf{P},E)} \tau(\mathbf{p}_1, \mathbf{p}_2, \mathbf{p}'_1, \mathbf{p}'_2)(f'_1 f'_2 - f_1 f_2) \, d\Sigma \tag{14.8}$$

(*Boltzmann equation*). The surface $\Sigma(\mathbf{P}, E)$ is a sphere (see Problem 1 of Section 14.9) with radius $p_r/2$, where $\mathbf{p}_r = \mathbf{p}_1 - \mathbf{p}_2$ is the relative momentum. Hence the integral on the right-hand side of the Boltzmann equation (14.8) can be written, in angular coordinates (colatitude $\theta$ and longitude $\varphi$) with respect to the polar axis $\mathbf{p}_r$:

$$\int_{R^3} d\mathbf{p}_2 \int_0^{2\pi} d\varphi \int_0^{\pi} d\theta\, \tilde{\tau}(p_r,\theta,\varphi)(f_1'f_2' - f_1 f_2) \tag{14.9}$$

where $\tilde{\tau}(p_r,\theta,\varphi) = p_r^2/4\sin\theta\, \tau(\mathbf{p}_1,\mathbf{p}_2,\mathbf{p}_1',\mathbf{p}_2')$ has dimensions $[\tilde{\tau}] = [l^3 t^{-1}]$. The kernel $\tilde{\tau}$ can be interpreted, for the transitions described by $(p_r,\theta,\varphi)$, as an 'effective volume' traced in the unit of time by the incident particle. Making the dependence on the modulus of the velocity of the latter, i.e. $p_r/m$, explicit we obtain a particularly transparent form of $\tilde{\tau}$:

$$\tilde{\tau}(p_r,\theta,\varphi) = \frac{p_r}{m}\sigma(p_r,\theta,\varphi), \tag{14.10}$$

where $[\sigma] = [l^2]$, so that $\sigma(p_r,\theta,\varphi)$ is the area of an ideal disc, with centre in the incident particle and normal to its velocity, which traces the effective volume for collisions. In particular, since the product $f_1 f_2$ is independent of $\theta$ and $\varphi$, it makes sense to consider the integral

$$\Sigma_{\text{TOT}}(p_r) = \int_0^{2\pi} d\varphi \int_0^{\pi} \sigma(p_r,\theta,\varphi)\, d\theta, \tag{14.11}$$

called the *total cross-section* (irrespective of the outcome of the collision). The role of the partial cross-section $\sigma$ can be clarified by considering the classical example of when the particles are modelled as hard spheres (see the next section). In this simple case, $\sigma$ depends only on $\theta$.

The Boltzmann equation is a fundamental tool for the study of systems with many particles, whose evolution is due to the interactions between the particles. There exists a great variety of situations, each requiring the correct description of the collision term. There are systems of charged particles (plasma), heterogeneous systems, systems of particles which collide with the molecules of a fixed structure with possible absorption. A relevant example is a neutron gas in a nuclear reactor, where it is known that the cross-section necessary to capture a neutron by a uranium isotope $U^{235}$ depends on the energy of the incident particle. A classical reference is the treatise of Cercignani (1988).

*Remark* 14.1
Integrating the right-hand side of (14.8) with respect to $\mathbf{p}_1$ we find zero. Indeed the substitution of $(\mathbf{p}_1,\mathbf{p}_2)$ with $(\mathbf{p}_1',\mathbf{p}_2')$ formally changes the integral into its opposite. However the integral is symmetric in the four momenta, and hence it is itself invariant. This fact has a simple interpretation. If for example we consider $f$ to be independent of $\mathbf{q}$ and assume $f(\mathbf{p},t)$ is zero for $|\mathbf{p}| \to \infty$ then the integral

of the left-hand side of (14.8) reduces to $dn/dt$ and its vanishing corresponds to the conservation of the density of the particles. ∎

*Remark 14.2*
The mathematical literature on the Boltzmann equation is very extensive. An existence and uniqueness theorem for a model of a gas with hard spheres with perfectly elastic collisions was proved by Carleman (1957).

The initial value problem turns out to be of extreme complexity. While several results have been obtained under particular assumptions, in its generality the problem has only recently been solved by Di Perna and Lions (1990). ∎

## 14.3 The hard spheres model

We compute the cross-section for a hard spheres gas of radius $R$, interacting via elastic collisions, neglecting as usual the energy associated with the rotations of the spheres. In addition to the reference frame in which the particle with momentum $\mathbf{p}_2$ (in the laboratory system) is at rest, it is convenient to also consider the centre of mass frame. In the latter frame the momenta $\hat{\mathbf{p}}_i, \hat{\mathbf{p}}'_i$ ($i = 1, 2$) are obtained by subtracting from $\mathbf{p}_i, \mathbf{p}'_i$ the momentum of the centre of mass $\mathbf{p}_0 = \frac{1}{2}(\mathbf{p}_1 + \mathbf{p}_2)$. It follows that

$$\hat{\mathbf{p}}_1 = \frac{1}{2}\mathbf{p}_r, \qquad \hat{\mathbf{p}}_2 = -\frac{1}{2}\mathbf{p}_r.$$

The outgoing momenta must be opposite and with the same magnitude as the incoming momenta:

$$\hat{\mathbf{p}}'_1 = -\hat{\mathbf{p}}'_2, \qquad \hat{p}'_1 = \hat{p}'_2 = \frac{1}{2}p_r.$$

To establish the direction of $\hat{\mathbf{p}}'_1, \hat{\mathbf{p}}'_2$ we say that in the centre of mass frame the collision between two spheres follows the optical reflection law: the angles between $\hat{\mathbf{p}}_1$ and $\hat{\mathbf{p}}'_1$ with the line $O_1O_2$ joining the centres of the two spheres are equal, and the four momenta all lie in the same plane (Fig. 14.1a).

Adding to all the momenta $\frac{1}{2}\mathbf{p}_r$ we return to the frame where $\tilde{\mathbf{p}}_2 = 0$; hence we deduce that, in the latter frame,

$$\tilde{\mathbf{p}}'_i = \hat{\mathbf{p}}'_i + \frac{1}{2}\mathbf{p}_r.$$

This has the following interpretation: the momenta $\tilde{\mathbf{p}}'_1, \tilde{\mathbf{p}}'_2$ are the diagonals of the parallelograms with sides $\pm\hat{\mathbf{p}}'_1$ and $\frac{1}{2}\mathbf{p}_r$ (Fig. 14.1b).

### 14.3 Statistical mechanics: kinetic theory

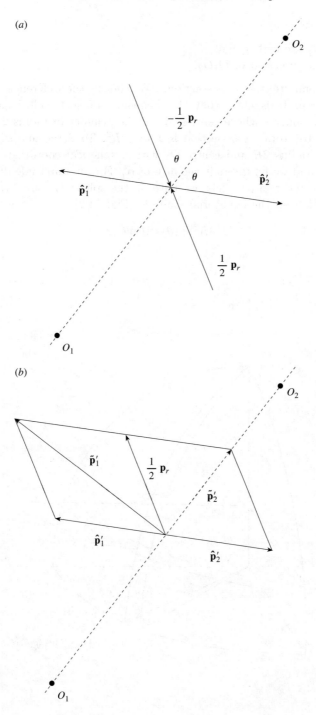

Fig. 14.1

Therefore

(a) $\widetilde{\mathbf{p}}'_1$ and $\widetilde{\mathbf{p}}'_2$ are orthogonal;
(b) $\widetilde{\mathbf{p}}'_2$ has the direction of $O_1 O_2$.

We now compute the cross-section. We choose the reference frame where $\widetilde{\mathbf{p}}_1 = \mathbf{p}_r$, $\widetilde{\mathbf{p}}_2 = 0$. It is clear that the frequency of such collisions is equal to the number of spheres whose centres are in the cylinder of radius $2R$ and height $p_r/m$. Hence the total cross-section is $\Sigma = 4\pi R^2$. To determine $\sigma(\theta)$ we endow the sphere of radius $2R$ and centre $O_1$ with a spherical coordinate system with polar axis $\mathbf{p}_r$ and we fix the unit vector $\mathbf{e}$ of $\widetilde{\mathbf{p}}'_2$. For an amplitude $d\theta, d\varphi$ between two meridians and two parallels, we have on the sphere the area $4R^2 \sin\theta \, d\theta \, d\varphi$, whose projection on the equatorial plane is (Fig. 14.2)

$$4R^2 \sin\theta \cos\theta \, d\theta \, d\varphi.$$

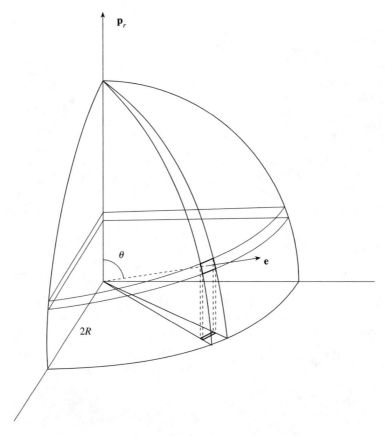

Fig. 14.2

Integrating with respect to $d\varphi$ we find

$$\sigma(\theta) = 8\pi R^2 \sin\theta \cos\theta$$

(integrating in $d\theta$ between 0 and $\pi/2$ gives naturally $\Sigma_{TOT} = 4\pi R^2$).

Do not confuse the present coordinate $\theta$ (the incidence angle, varying between 0 and $\pi/2$) with the colatitude used in (14.10), which is twice the incidence angle.

## 14.4 The Maxwell–Boltzmann distribution

The equilibrium states of a system governed by equation (14.8) are described by the stationary solutions. We seek such solutions assuming that $\mathbf{F} = 0$ and that the distribution function $f$ does not depend on the position coordinates $\mathbf{q}$. In other words, we look for an equilibrium solution of the kind of $f_0(\mathbf{p})$. A sufficient condition for $f_0(\mathbf{p})$ to be a stationary solution of the Boltzmann equation is that it satisfies the equality

$$f_0(\mathbf{p}_1) f_0(\mathbf{p}_2) = f_0(\mathbf{p}'_1) f_0(\mathbf{p}'_2) \tag{14.12}$$

for every pair of states $(\mathbf{p}_1, \mathbf{p}_2)$, $(\mathbf{p}'_1, \mathbf{p}'_2)$ satisfying (14.5) and (14.6). We shall see in what follows that this condition is also necessary ('theorem H' of Boltzmann).

Equation (14.12) expresses a conservation law for the product $f_0(\mathbf{p}_1) f_0(\mathbf{p}_2)$. However our hypotheses (in particular the absence of internal structure in the molecules) imply that the only conserved quantities in the collision are the kinetic energy and the total momentum. Therefore the function $f_0(\mathbf{p})$ must be such that the product $f_0(\mathbf{p}_1) f_0(\mathbf{p}_2)$ depends only on the invariants $\mathbf{P}$ and $E$. Note that for an arbitrary vector $\mathbf{p}_0$ we have

$$(\mathbf{p}_1 - \mathbf{p}_0)^2 + (\mathbf{p}_2 - \mathbf{p}_0)^2 = 2mE - 2\mathbf{P} \cdot \mathbf{p}_0 + 2\mathbf{p}_0^2,$$

and hence a possible choice of $f_0$ satisfying (14.12) (and in addition such that $f_0(\mathbf{p}) \to 0$ for $|\mathbf{p}| \to \infty$) is

$$f_0(\mathbf{p}) = C e^{-A(\mathbf{p} - \mathbf{p}_0)^2}, \tag{14.13}$$

with $A$ and $C$ positive constants, whose meaning will be elucidated.

We now define the *mean value* of a quantity $G(\mathbf{p})$ relative to the distribution (14.13) by the formula

$$\langle G \rangle = \frac{\int G(\mathbf{p}) f_0(\mathbf{p}) \, d\mathbf{p}}{\int f_0(\mathbf{p}) \, d\mathbf{p}}. \tag{14.14}$$

Recall that by the definition of the distribution function, as we saw in (14.3), the denominator in (14.14) represents the density $n = N/V$ of particles.

We can therefore easily compute that the mean value of the momentum $\mathbf{p}$ is given by

$$\langle \mathbf{p} \rangle = \frac{\int \mathbf{p} f_0(\mathbf{p}) \, d\mathbf{p}}{\int f_0(\mathbf{p}) \, d\mathbf{p}} = \mathbf{p}_0, \qquad (14.15)$$

since

$$\int \mathbf{p} f_0(\mathbf{p}) \, d\mathbf{p} = C \int (\mathbf{p} + \mathbf{p}_0) e^{-Ap^2} \, d\mathbf{p} = \mathbf{p}_0 \int f_0(\mathbf{p}) \, d\mathbf{p}.$$

Hence $\mathbf{p}_0$ expresses a uniform translation of the whole frame. It is always possible to choose a reference frame moving with this translation, so that in it we have $\mathbf{p}_0 = 0$.

The normalising condition

$$\int f_0(\mathbf{p}) \, d\mathbf{p} = n \qquad (14.16)$$

fixes the constant $C$ in terms of

$$n = \int f_0(\mathbf{p}) \, d\mathbf{p} = 4\pi C \int_0^\infty p^2 e^{-Ap^2} \, dp = C \left(\frac{\pi}{A}\right)^{3/2}$$

(see Appendix 8), and therefore

$$C = n \left(\frac{A}{\pi}\right)^{3/2}. \qquad (14.17)$$

The constant $A$ is in turn linked to the average kinetic energy $\varepsilon$ of a molecule:

$$\varepsilon = \frac{\int p^2/2m \, f_0(\mathbf{p}) \, d\mathbf{p}}{\int f_0(\mathbf{p}) \, d\mathbf{p}}. \qquad (14.18)$$

Indeed from (14.13) and (14.17) it follows that

$$\varepsilon = \frac{2\pi}{m} \left(\frac{A}{\pi}\right)^{3/2} \int_0^\infty p^4 e^{-Ap^2} \, dp = \frac{3}{4Am},$$

and hence

$$A = \frac{3}{4\varepsilon m}. \qquad (14.19)$$

This yields the following expression for the equilibrium distribution, called the *Maxwell–Boltzmann distribution*:

$$f_0(\mathbf{p}) = n \left(\frac{3}{4\pi \varepsilon m}\right)^{3/2} \exp\left[\frac{-p^2}{2m} \bigg/ \left(\frac{2\varepsilon}{3}\right)\right]. \qquad (14.20)$$

Equation (14.20) was deduced by Maxwell in the essay *On the Dynamical Theory of Gases*, assuming the statistical independence of the velocities of two colliding molecules, and using the conservation of the total kinetic energy during an elastic collision. These are the same assumptions that we adopted in the previous section to derive the Boltzmann equation.

If the gas is subject to an external conservative force,

$$\mathbf{F} = -\nabla_q \Phi(\mathbf{q}), \tag{14.21}$$

and occupies a bounded region $V$, we can show that the Boltzmann equation admits the stationary solution

$$f(\mathbf{p}, \mathbf{q}) = f_0(\mathbf{p}) \left[ \frac{1}{|V|} \int_V e^{-\Phi(q)/(2\varepsilon/3)} d\mathbf{q} \right]^{-1} e^{-\Phi(q)/(2\varepsilon/3)}. \tag{14.22}$$

Indeed, we note that equation (14.12) is still obviously satisfied. Therefore, if we seek $f(\mathbf{p}, \mathbf{q})$ in the form $f = f_0(\mathbf{p})g(\mathbf{q})$, we have on the left-hand side of (14.8) that

$$f_0 \nabla_\mathbf{q} g \cdot \frac{\mathbf{p}}{m} + g \nabla_\mathbf{p} f_0 \cdot (-\nabla_\mathbf{q} \Phi) = 0,$$

which yields the equation for $g$:

$$\nabla_\mathbf{q} g + g \nabla_\mathbf{q} \Phi \Big/ \left(\frac{2\varepsilon}{3}\right) = 0,$$

with solution $g(\mathbf{q}) = c\exp[-\Phi(\mathbf{q})/(2\varepsilon/3)]$. The constant $c$ is a result of the normalisation $\int f_0(\mathbf{p})\, d\mathbf{p} \int_V g(\mathbf{q})\, d\mathbf{q} = N$.

Note that the equilibrium distributions (14.20) and (14.22) *are independent of the function* $\tau$ appearing in the Boltzmann equation, and hence of the kind of two-body interaction between the molecules of the gas.

It is interesting to note, in view of future developments, that once the kernel $\tau$ is defined, the mechanics of the collision do not depend on the identification of the particles. Indeed, the indices of the outgoing particles are assigned for convenience, but the symmetry properties of the kernel allow them to be interchanged, so that the outgoing particles are not only identical, but also indistinguishable.

## 14.5 Absolute pressure and absolute temperature in an ideal monatomic gas

Consider a surface exposed to the action of the gas molecules, and assume that it is perfectly reflecting.

By definition, the force acting (on average) on any of its infinitesimal elements $d\sigma$ is in magnitude equal to $P\,d\sigma$, where $P$ is the pressure. This force can be computed by observing that every molecule colliding with $d\sigma$ is subject to a variation of its momentum in the direction normal to $d\sigma$ and equal to twice the normal component $p_n$ of its momentum preceding the collision. The force exerted on $d\sigma$ is obtained by multiplying $2p_n$ by the number of collisions experienced in one unit of time by particles with momentum component $p_n$, and integrating on the space of momenta which produce collisions ($p_n > 0$). We compute the expression for $P$ corresponding to the distribution (14.20).

Since $1/m\, p_n f_0(\mathbf{p})\, d\sigma\, d\mathbf{p}$ is the number of collisions per unit time due to the particles with momentum in the cell $d\mathbf{p}$ centred at $\mathbf{p}$, we find the expression

$$P = \frac{1}{m} \int_{p_n > 0} 2p_n^2 f_0(\mathbf{p})\, d\mathbf{p} = \frac{1}{m} \int p_n^2 f_0(\mathbf{p})\, d\mathbf{p}, \tag{14.23}$$

which is proportional to the average $\langle p_n^2 \rangle$.

Because of the symmetry of $f_0(\mathbf{p})$, it follows that $\langle p^2 \rangle$ is equal to the sum of the averages $\langle p_i^2 \rangle$, where $p_i$ are the projections in three mutually orthogonal directions, which are all equal. It follows that $\langle p_n^2 \rangle = \frac{1}{3} \langle p^2 \rangle$ and therefore we can substitute $\frac{1}{3} p^2$ for $p_n^2$ in (14.23).

Hence we find

$$P = \frac{4\pi}{3m} \int_0^\infty p^4 f_0(p)\, dp,$$

and we arrive at the so-called *state equation*:

$$P = \frac{2}{3} n\varepsilon. \tag{14.24}$$

Equation (14.24) expresses a relation between two macroscopic quantities, which we can make more explicit by introducing the absolute temperature in the following way.

DEFINITION 14.1  *The absolute temperature $T$ is related to the average kinetic energy $\varepsilon$ of the gas by*

$$\varepsilon = \frac{3}{2} kT, \tag{14.25}$$

*where $k$ is the Boltzmann constant ($1.380 \times 10^{-16}$ erg/K).* ∎

This definition may appear rather abstract, and can be reformulated differently. What is important is that it is consistent with classical thermodynamics.

Considering (14.24) and (14.25) together we obtain the well-known relation

$$P = nkT \tag{14.26}$$

(which could have been used as the definition of $T$).[3] In addition, equation (14.25) yields the following alternative form for (14.20):

$$f_0(\mathbf{p}) = n(2\pi mkT)^{-3/2} \exp\left[-\frac{p^2}{2mkT}\right]. \tag{14.27}$$

*Remark* 14.3
With reference to the more general case, when there is also the action of an external field, we note that the equilibrium distribution (14.22) contains the factor $e^{-\beta h(\mathbf{p},\mathbf{q})}$, where $\beta = 1/kT$ and $h(\mathbf{p},\mathbf{q}) = (p^2/2m) + \Phi(\mathbf{q})$ is the Hamiltonian of each particle, but where the internal forces do not contribute (confirming the fact that in our assumptions these do not change the structure of the equilibrium, although they play a determining role in leading the system towards it). ■

For a prescribed value of the mean kinetic energy of the molecules, the following definition appears natural, and links the total kinetic energy to the state of molecular motion, under the usual assumptions (monatomic gas, non-dissipative collisions, etc.).

DEFINITION 14.2 *We call the* internal energy *of the system the quantity*

$$U(T) = N\varepsilon = \frac{3}{2} NkT. \tag{14.28}$$

■

The definition of the internal energy allows us to complete the logical path from the microscopic model to the thermodynamics of the system.

In an infinitesimal thermodynamical transformation the work done by the system for a variation $dV$ of its volume is clearly $P\,dV$. If the transformation is adiabatic the work is done entirely at the expense (or in favour) of the internal energy, i.e. $dU + P\,dV = 0$. If the transformation is not adiabatic the energy balance is achieved by writing

$$dQ = dU + P\,dV. \tag{14.29}$$

The identification of $dQ$ with the *quantity of heat* exchanged with the exterior leads to the first principle of thermodynamics. We can now use $dQ$, defined

---

[3] Since $N = \nu N_A$ we again find the well-known law $PV = \nu RT$, where the universal gas constant is $R = kN_A = 8.31 \times 10^7$ erg/mole K.

by equation (14.29), to introduce the *thermal capacity* $C$ (relative to a generic transformation):

$$C\,dT = dQ. \tag{14.30}$$

Since $dU = \frac{3}{2} Nk\,dT$, we easily find the expression for the thermal capacity at *constant volume* of a monatomic gas:

$$C_V = \frac{3}{2} Nk. \tag{14.31}$$

## 14.6 Mean free path

We can now obtain the expression for the mean free path in a hard sphere gas following the Maxwell–Boltzmann distribution. Recall that if $\delta$ is the diameter of the spheres, the *cross-section* is measured by $\pi\delta^2$.

If we consider the pairs of molecules with momenta $\mathbf{p}_1$ and $\mathbf{p}_2$ and we fix a reference frame translating with one of the particles, the magnitude of the velocity of one with respect to the other particle is $1/m\,|\mathbf{p}_1 - \mathbf{p}_2|$. In a time $dt$ only the particles within a volume $\pi\delta^2/m\,|\mathbf{p}_1 - \mathbf{p}_2|\,dt$ can collide. To find the number of collisions per unit volume, we must multiply the latter volume by the functions $f_0(\mathbf{p}_1)$ and $f_0(\mathbf{p}_2)$ (in agreement with (14.4)) and then integrate on $\mathbf{p}_1$ and $\mathbf{p}_2$. Dividing by $dt$ we find the frequency of the collisions per unit volume as

$$\nu_u = \frac{\pi\delta^2}{m} \int |\mathbf{p}_1 - \mathbf{p}_2| f_0(\mathbf{p}_1) f_0(\mathbf{p}_2)\,d\mathbf{p}_1 d\mathbf{p}_2. \tag{14.32}$$

Since every collision involves two and only two particles, the total number of collisions to which a molecule is subject per unit time can be found by dividing $2\nu_u$ by the density $n$ of molecules.

The mean free path is then obtained by dividing the average velocity by the number of collisions found above:

$$\lambda = \frac{n\langle v \rangle}{2\nu_u}. \tag{14.33}$$

It is not difficult to compute that $\langle v \rangle = 2\sqrt{2kT/\pi m}$ (of the order of magnitude of $10^5\,\mathrm{cm\,s}^{-1}$ at $T = 300\,\mathrm{K}$ and $m \sim 10^{-23}\,\mathrm{g}$), so that

$$\lambda = \frac{n}{\nu_u} \sqrt{\frac{2kT}{\pi m}}. \tag{14.34}$$

The computation of $\nu_u$ can be easily achieved recalling that (see (14.27))

$$f_0(\mathbf{p}_1) f_0(\mathbf{p}_2) = \frac{n^2}{(2\pi mkT)^3} \exp\left[-\frac{p_1^2 + p_2^2}{2mkT}\right].$$

It is convenient to change variables to

$$\mathbf{P} = \mathbf{p}_1 + \mathbf{p}_2, \qquad \boldsymbol{\eta} = \mathbf{p}_1 - \mathbf{p}_2,$$

thus expressing the integral in (14.32) in the form

$$\nu_u = \frac{1}{8m} \left(\frac{n\delta}{\pi}\right)^2 \frac{1}{(2mkT)^3} \int |\boldsymbol{\eta}| \exp\left(-\frac{P^2}{4mkT}\right) \exp\left(-\frac{\eta^2}{4mkT}\right) \mathrm{d}\mathbf{P}\,\mathrm{d}\boldsymbol{\eta}$$

$$= \frac{2^5}{m}(n\delta)^2 \sqrt{mkT} \int_0^\infty \xi^2 e^{-\xi^2}\,\mathrm{d}\xi \int_0^\infty \xi^3 e^{-\xi^2}\,\mathrm{d}\xi = 4\sqrt{\pi}\,(n\delta)^2 \sqrt{\frac{kT}{m}}.$$

Finally, from this it follows that

$$\lambda = \frac{1}{2\sqrt{2}} \frac{1}{\pi \delta^2 n}, \tag{14.35}$$

independent of the temperature ($n \sim 10^{18}$ cm$^{-3}$, $\delta \sim 10^{-7}$ cm yields $\lambda \sim 10^{-5}$ cm). We remark that equation (14.35) justifies our previous statement that the product $n\delta^2$ determines the mean free path.

## 14.7 The 'H theorem' of Boltzmann. Entropy

We now examine again the Boltzmann equation (14.8) to show that the condition (14.12) (from which we deduced the Maxwell–Boltzmann distribution (14.20)) is not only sufficient but also necessary for the distribution $f_0$ to be an equilibrium distribution. This is a consequence of the 'H theorem', which we state below. Its implications are far more relevant, as they yield the concept of entropy.

Assume for simplicity that the molecular distribution is spatially uniform (hence that $f$ does not depend on the coordinates $\mathbf{q}$) and that the gas is not subject to external forces. The distribution function $f(\mathbf{p}, t)$ then satisfies the equation

$$\frac{\partial f}{\partial t}(\mathbf{p}_1, t) = \int \mathrm{d}\mathbf{p}_2 \int_{\Sigma(\mathbf{P}, E)} \tau(\mathbf{p}_1, \mathbf{p}_2, \mathbf{p}'_1, \mathbf{p}'_2)[f(\mathbf{p}'_1, t)f(\mathbf{p}'_2, t) - f(\mathbf{p}_1, t)f(\mathbf{p}_2, t)]\,\mathrm{d}\Sigma,$$
$$\tag{14.36}$$

where the manifold $\Sigma$ has been described in Section 14.2.

We now want to use equation (14.36) to describe the time evolution of the $H$ *functional* of Boltzmann, defined by

$$H(t) = \int f(\mathbf{p}, t) \log f(\mathbf{p}, t)\,\mathrm{d}\mathbf{p}. \tag{14.37}$$

Obviously when writing equation (14.37) one must only consider the functions $f(\mathbf{p}, t)$ whose integral is convergent; we assume that this is the case in what follows.

*Remark* 14.4

Considering that $f/n$ plays the role of a probability density, we note the analogy of (14.37) with the definition of entropy given in the study of ergodic theory (see (13.33)). ∎

We have the following theorem.

THEOREM 14.1 (Boltzmann's H theorem) *If the distribution $f(\mathbf{p},t)$ appearing in the definition (14.37) of $H(t)$ is a solution of equation (14.36), then*

$$\frac{\mathrm{d}H}{\mathrm{d}t} \leq 0. \tag{14.38}$$

*In expression (14.38) equality holds if and only if $f_1 f_2 = f_1' f_2'$.*

*Proof*

Substituting (14.36) into the expression

$$\frac{\mathrm{d}H}{\mathrm{d}t} = \int_{\mathbf{R}^3} \frac{\partial f}{\partial t} [1 + \log f(\mathbf{p},t)] \, \mathrm{d}\mathbf{p}$$

we find (setting $\mathbf{p} = \mathbf{p}_1$)

$$\frac{\mathrm{d}H}{\mathrm{d}t} = \int_{\mathbf{R}^3} \mathrm{d}\mathbf{p}_1 \int_{\mathbf{R}^3} \mathrm{d}\mathbf{p}_2 \int_{\Sigma(\mathbf{P},E)} \tau(\mathbf{p}_1, \mathbf{p}_2, \mathbf{p}_1', \mathbf{p}_2') \tag{14.39}$$

$$\times [f(\mathbf{p}_1', t)f(\mathbf{p}_2', t) - f(\mathbf{p}_1, t)f(\mathbf{p}_2, t)][1 + \log f(\mathbf{p}_1, t)] \, \mathrm{d}\Sigma.$$

In view of future developments, it is preferable to treat symmetrically the four momentum vectors $\mathbf{p}_1, \mathbf{p}_2, \mathbf{p}_1', \mathbf{p}_2'$ and to define the manifold $\Omega$ of 4-tuples $(\mathbf{p}_1, \mathbf{p}_2, \mathbf{p}_1', \mathbf{p}_2')$ satisfying (14.5) and (14.6).

By the symmetry of the kernel $\tau$ with respect to the interchange of $\mathbf{p}_1$ with $\mathbf{p}_2$ we find an equation analogous to (14.39), i.e.

$$\frac{\mathrm{d}H}{\mathrm{d}t} = \int_{\Omega} \tau(\mathbf{p}_1, \mathbf{p}_2, \mathbf{p}_1', \mathbf{p}_2')[f(\mathbf{p}_1', t)f(\mathbf{p}_2', t) - f(\mathbf{p}_1, t)f(\mathbf{p}_2, t)][1 + \log f(\mathbf{p}_2, t)] \, \mathrm{d}\Omega, \tag{14.40}$$

where $f(\mathbf{p}_2, t)$ has simply replaced $f(\mathbf{p}_1, t)$ in the last term. Adding equations (14.39) and (14.40), we find

$$\frac{\mathrm{d}H}{\mathrm{d}t} = \frac{1}{2} \int_{\Omega} \tau(\mathbf{p}_1, \mathbf{p}_2, \mathbf{p}_1', \mathbf{p}_2') \cdot [f(\mathbf{p}_1', t)f(\mathbf{p}_2', t) - f(\mathbf{p}_1, t)f(\mathbf{p}_2, t)] \tag{14.41}$$

$$\times [2 + \log(f(\mathbf{p}_1, t)f(\mathbf{p}_2, t))] \, \mathrm{d}\Omega.$$

Recalling the symmetry of the kernel $\tau$ with respect to the interchange of the pairs $(\mathbf{p}_1, \mathbf{p}_2)$ and $(\mathbf{p}'_1, \mathbf{p}'_2)$, we also have

$$\frac{dH}{dt} = -\frac{1}{2} \int_\Omega \tau(\mathbf{p}_1, \mathbf{p}_2, \mathbf{p}'_1, \mathbf{p}'_2) \cdot [f(\mathbf{p}'_1, t)f(\mathbf{p}'_2, t) - f(\mathbf{p}_1, t)f(\mathbf{p}_2, t)] \qquad (14.42)$$
$$\times [2 + \log(f(\mathbf{p}'_1, t)f(\mathbf{p}'_2, t))]\, d\Omega.$$

Adding (14.41) and (14.42), we finally find the expression

$$\frac{dH}{dt} = \frac{1}{4} \int_\Omega \tau(\mathbf{p}_1, \mathbf{p}_2, \mathbf{p}'_1, \mathbf{p}'_2)[f(\mathbf{p}'_1, t)f(\mathbf{p}'_2, t) - f(\mathbf{p}_1, t)f(\mathbf{p}_2, t)] \qquad (14.43)$$
$$\times [\log(f(\mathbf{p}_1, t)f(\mathbf{p}_2, t)) - \log(f(\mathbf{p}'_1, t)f(\mathbf{p}'_2, t))]\, d\Omega,$$

which is clearly non-positive, since for each pair of positive real numbers $(x, y)$ we have

$$(y - x)(\log x - \log y) \leq 0,$$

with equality only if $x = y$. ∎

We can also deduce from the proof of the H theorem the following corollaries.

COROLLARY 14.1 *The condition (14.12) for a distribution to be in equilibrium is not only sufficient but also necessary.*

*Proof*
For a stationary solution we have $dH/dt = 0$ that necessarily—from equation (14.43)—yields (14.12). ∎

The monotonicity of $H$ finally yields the following.

COROLLARY 14.2 *For any initial distribution $f(\mathbf{p}, 0)$ the system converges asymptotically towards the stationary solution.* ∎

The H theorem plays a fundamental role in the kinetic theory of gases, as it allows the introduction of entropy and the deduction of the second law of thermodynamics. Indeed, it is enough to define the entropy so that it is proportional to $-H(t)$ and also that it is *extensive* (i.e. increasing proportionally with the volume, when the average density $n$ is fixed).

DEFINITION 14.3 *If $V$ indicates the volume occupied by the gas, we call* entropy *the extensive quantity*

$$S = -kVH + \text{constant}. \qquad (14.44)$$

∎

*Remark* 14.5

In the definition (14.37) of $H$ we assume that the argument of the logarithm is dimensionless (and that modifying it we modify $H$ by a constant proportional to $n$). It follows that $H$ has the dimension of $V^{-1}$ and in equation (14.44) $S$ has the same dimensions as the Boltzmann constant $k$. ∎

The relation between the H theorem and the second law of thermodynamics is an immediate consequence of Definition 14.3 of entropy: the entropy of a system grows until equilibrium is achieved.

The $H$ functional computed corresponding to the Maxwell–Boltzmann distribution (14.27) is

$$H_0 = n\left\{\log\left[\lambda^{-1}n\left(\frac{1}{2\pi mkT}\right)^{3/2}\right] - \frac{3}{2}\right\}, \qquad (14.45)$$

where $\lambda > 0$ is a factor yielding a dimensionless quantity, and therefore

$$S_0(E,V) = kN\left\{\log\left[\hat{\lambda}\frac{V}{N}\left(\frac{E}{N}\right)^{3/2}\right] + \frac{3}{2}\right\}, \quad \hat{\lambda} = \lambda\left(\frac{4}{3}\pi m\right)^{3/2}. \qquad (14.46)$$

This formula emphasises the additivity of $S_0$.

The computation of (14.45) is simple, since when we set $f = f_0(\mathbf{p})$ in (14.37) the integrand depends on $p^2$. Hence

$$H_0 = \int_0^\infty 4\pi p^2 f_0(p) \log[\lambda f_0(p)]\,dp.$$

From (14.46) it is immediate to check that

$$\frac{\partial S_0}{\partial E} = \frac{3}{2}k\frac{N}{E} = \frac{1}{T},$$

which is simply the usual definition of absolute temperature (note that we could avoid expressing $\varepsilon$ through equation (14.25) and introduce the temperature at this point). Indeed, setting in (14.29) $U = E$ and $dQ = T\,dS(E,V)$, we find precisely

$$\frac{\partial S}{\partial E} = \frac{1}{T}$$

and

$$\frac{\partial S}{\partial V} = \frac{P}{T}.$$

This last relation is easily verified for (14.46).

*Remark* 14.6

The equation $T(\partial S/\partial V) = P$ can in general be deduced from (14.44). Indeed, setting $f(\mathbf{p}) = n\varphi(\mathbf{p})$ with $\int_{\mathbf{R}^3} \varphi(\mathbf{p})\,\mathrm{d}\mathbf{p} = 1$, we can write

$$H = \frac{N}{V}\int_{\mathbf{R}^3} \varphi \log\left(\frac{N}{V}\varphi\right) \mathrm{d}\mathbf{p},$$

yielding

$$\frac{\partial H}{\partial V} = -\frac{1}{V}(H + n)$$

and eventually $T(\partial S/\partial V) = nkT = P$. ∎

*Remark* 14.7

We cannot discuss here the many 'paradoxes' stemming from the interpretation of the H theorem as the manifestation of the *irreversibility* of the process achieving *macroscopic* equilibrium, as opposed to the *reversible* and *recurrent* behaviour (see Theorem 5.1) of the Hamiltonian flow governing the *microscopic* dynamics of the system. For a discussion of these important problems, we refer the reader to the texts of Uhlenbeck and Ford (1963), Thompson (1972) and Huang (1987). We also note the pleasant article by Cercignani (1988). ∎

## 14.8 Problems

1. The *Ehrenfest model* (1912). Consider a gas of $N$ molecules 'P' non interacting and moving in the plane. We also introduce the obstacles 'Q' modelled by squares of side $a$ with diagonals parallel to the axes $x$ and $y$. The obstacles Q are fixed, uniformly but randomly distributed, and they model a strongly diluted gas (the average distance between any two of them is much larger than $a$). The molecules P are moving at constant speed $c$, equal for all of them, uniquely in the directions of the axes $x$ or $y$ (positive or negative); when they meet the obstacles Q they undergo an elastic collision. We denote by $f_1(t)$, $f_2(t)$, $f_3(t)$ and $f_4(t)$ the number of molecules P which at time $t$ move, respectively, in the positive $x$ direction (direction 1), the positive $y$ direction (2), the negative $x$ direction (3) and the negative $y$ direction (4). Clearly $f_1 + f_2 + f_3 + f_4 = N$. The functions $f_i$ play the same role as the distribution function. Let $N_{12}\Delta t$ be the number of molecules P which, after collision with an obstacle, in the time interval $\Delta t$, pass from moving in the direction 1 to motion in the direction 2. The assumption of molecular chaos (*Stosszahlansatz*) can be formulated for this model as follows: $N_{12}\Delta t = \alpha f_1 \Delta t$, where $\alpha = nca/\sqrt{2}$ and $n$ is the density of obstacles Q in the plane; analogously for the other transitions. Note that $\alpha \Delta t$ is the ratio of the total area occupied by the strips $S_{ij}$ which are parallelograms of length $c\Delta t$ and basis resting on each of the obstacles Q on the side where the collision occurs,

changing the direction of the motion of the molecules P from $i$ to $j$. Prove that the average number of collisions in the interval $\Delta t$ is given by $2N\alpha\,\Delta t$ and that the average time interval between any two collisions is $T = 1/\sqrt{2}acnN$. Prove that the equation modelling the evolution of the distribution functions (Boltzmann equation) is given by the system of ordinary differential equations

$$\frac{df_1}{dt} = \alpha(f_2 + f_4 - 2f_1),$$

$$\frac{df_2}{dt} = \alpha(f_3 + f_1 - 2f_2),$$

$$\frac{df_3}{dt} = \alpha(f_4 + f_2 - 2f_3),$$

$$\frac{df_4}{dt} = \alpha(f_1 + f_3 - 2f_4).$$

Verify that the equilibrium distribution (stationary) is given by $f_1 = f_2 = f_3 = f_4 = N/4$. Prove that an arbitrary initial distribution converges to the equilibrium distribution and that the time $\tau$ of relaxation is of the order of $1/\alpha$, and therefore much larger than $T$. Finally, if $H(t) = f_1(t)\log f_1(t) + f_2(t)\log f_2(t) + f_3(t)\log f_3(t) + f_4(t)\log f_4(t)$, prove that $dH/dt \leq 0$, and that the derivative vanishes only for the equilibrium distribution. (Hint: show that $dH/dt$ as a function of $f_1, f_2, f_3, f_4$ subject to the constraint $\sum_{i=1}^{4} f_i = N$ has an absolute maximum equal to zero in correspondence with $f_1 = f_2 = f_3 = f_4 = N/4$.)

## 14.9 Additional solved problems

*Problem 1*
Prove that the surface $\Sigma(\mathbf{P}, E)$ is a sphere and deduce the expression (14.9) for the integral on the right-hand side of the Boltzmann equation (14.8).

*Solution*
In the reference frame in which the particle with momentum $\mathbf{p}_2$ is at rest (which is uniformly translating with respect to the laboratory frame), the new momenta are

$$\tilde{\mathbf{p}}_1 = \mathbf{p}_r\,, \quad \tilde{\mathbf{p}}_2 = 0\,, \quad \tilde{\mathbf{p}}_1' = \mathbf{p}_1' - \mathbf{p}_2\,, \quad \tilde{\mathbf{p}}_2' = \mathbf{p}_2' - \mathbf{p}_2,$$

and equations (14.5), (14.6) become

$$\tilde{\mathbf{p}}_1' + \tilde{\mathbf{p}}_2' = \mathbf{p}_r, \quad \tilde{p}_1'^2 + \tilde{p}_2'^2 = p_r^2.$$

Therefore the vectors $\tilde{\mathbf{p}}_1'$ and $\tilde{\mathbf{p}}_2'$ are the sides of a right-angled triangle with hypotenuse $\mathbf{p}_r$ and $\Sigma(\mathbf{P}, E)$ is the sphere of diameter $\mathbf{p}_r$. The form (14.9) of the integral on the right-hand side of (14.8) can be deduced immediately after introducing angular coordinates (colatitude and longitude), choosing $\mathbf{p}_r$ as the polar axis of the sphere $\Sigma(\mathbf{P}, E)$.

*Problem 2*

Let $F(\mathbf{p}, \mathbf{q})$ be some observable quantity, associated with the molecules at $\mathbf{q}$ with momentum $\mathbf{p}$, and preserved by binary collisions; hence such that

$$F(\mathbf{p}_1, \mathbf{q}) + F(\mathbf{p}_2, \mathbf{q}) = F(\mathbf{p}'_1, \mathbf{q}) + F(\mathbf{p}'_2, \mathbf{q}). \tag{14.47}$$

Prove that its expectation $\langle F \rangle$ does not vary with time.

*Solution*

From equation (14.8) we find

$$\frac{\mathrm{d}\langle F \rangle}{\mathrm{d}t} = \int \mathrm{d}\mathbf{q} \int_{\mathbb{R}^3} \mathrm{d}\mathbf{p}_1 \, F(\mathbf{p}_1, \mathbf{q}) \int \mathrm{d}\mathbf{p}_2 \int_{\Sigma(P,E)} \mathrm{d}\sum \tau(\mathbf{p}_1, \mathbf{p}_2, \mathbf{p}'_1, \mathbf{p}'_2)(f'_1 f'_2 - f_1 f_2).$$

Using the same kind of argument as used to prove the H theorem, considering the possible exchanges of variables ($\mathbf{p}_1$ with $\mathbf{p}_2$; $\mathbf{p}_1$ with $\mathbf{p}'_1$ and $\mathbf{p}_2$ with $\mathbf{p}'_2$; $\mathbf{p}_1$ with $\mathbf{p}'_2$ and $\mathbf{p}_2$ with $\mathbf{p}'_1$) and adding all contributions thus obtained, we find

$$4\frac{\mathrm{d}\langle F \rangle}{\mathrm{d}t} = \int \mathrm{d}\mathbf{q} \int_{\mathbb{R}^3} \mathrm{d}\mathbf{p}_1 \int \mathrm{d}\mathbf{p}_2 \int_{\Sigma(P,E)} \mathrm{d}\sum \tau(\mathbf{p}_1, \mathbf{p}_2, \mathbf{p}'_1, \mathbf{p}'_2)$$
$$\times (f'_1 f'_2 - f_1 f_2)(F_1 + F_2 - F'_1 - F'_2), \tag{14.48}$$

where we set $F_i = F(\mathbf{p}_i, \mathbf{q})$, $F'_i = F(\mathbf{p}'_i, \mathbf{q})$. Thanks to the conservation law (14.47) the right-hand side of (14.82) vanishes, and the proof follows.

## 14.10 Additional remarks and bibliographical notes

Kinetic theory is a field with many applications to a variety of different physical situations (fluid dynamics, plasma physics, many-body dynamics, etc.). In addition to the mentioned treatise of Cercignani (1988) the reader interested in physical applications can refer to Bertin (2000).

In our brief introduction we have deliberately avoided the discussion of the problem of irreversibility; for an introduction to the most recent developments, see Sinai (1979).

The statistical mechanics of equilibria, to be discussed in the next chapter, in addition to being extremely successful, has many connections with the ergodic theory of dynamical systems. Recently, newly-opened research directions aim to describe the statistical mechanics of non-equilibrium states through the introduction of stationary states described by probability measures invariant for the microscopic description. In the presence of a thermostat the stationary states correspond to the SRB measures (after Sinai, Ruelle, Bowen) of ergodic theory. In particular, the recent proof given by Gallavotti and Cohen (1995) of a fluctuation theorem for the production of entropy (Ruelle 1996, 1997) is significant progress towards a dynamical approach to the statistical mechanics of non-equilibrium.

The reader interested in learning more about this fascinating subject can refer to the review work of Gallavotti (1998) and Ruelle (1999).

# 15 STATISTICAL MECHANICS: GIBBS SETS

## 15.1 The concept of a statistical set

In the previous chapter we considered the study of the evolution of a diluted gas, disregarding the (impossible) task of describing the motion of each molecule, and referring instead to a quantity, the distribution function, with an extrapolation to the continuous setting in the space $\mu$. We then related the distribution function to thermodynamical quantities through averaging, and to entropy through the $H$ functional.

The procedure we followed was based on rather restrictive assumptions on the structure and the kind of interaction between particles, for example the assumption that the particles are elastic spheres. In other words, we used repeatedly the laws governing particle collisions in the construction of the evolution equation for the distribution function. At the same time, we concluded that, within the same approximation, the way in which binary interactions between particles take place is not essential (as long as it is of collisional type) for determining the equilibrium distribution. Such a distribution contains the factor $\mathrm{e}^{-\beta h}$, where $h$ is the Hamiltonian without the interaction potential.

The statistical mechanics in the treatment of Gibbs, presented in the famous treatise of 1902, focuses on the states of equilibrium of systems with many degrees of freedom, with the aim of deducing their thermodynamical behaviour starting from their mechanical nature, and hence from the Hamiltonian. On the one hand, if this aim may seem more restrictive, one should recall that Gibbs' studies led to the creation of statistical mechanics as an independent discipline, and yielded a great number of applications and discoveries. We must state that it would be wrong, historically and scientifically, to contrast the ideas of Boltzmann and Gibbs, not only because Gibbs' work is based on the work of Boltzmann, but also because many of the basic points in Gibbs' theory had already been stated by Boltzmann, within a different formalism. It is therefore not surprising to find many contact points between the two theories, the one presented in the previous chapter and the one that we are about to discuss.

Consider a system of $N$ identical particles, with fixed total mechanical energy $E$, contained in a bounded region of the space $\mathbf{R}^3$ of volume $V$ (the walls of the container are assumed to be perfectly reflecting). The evolution of such a system in the $6N$-dimensional phase space with coordinates $(\mathbf{P}, \mathbf{Q}) = (\mathbf{p}_1, \ldots, \mathbf{p}_N, \mathbf{q}_1, \ldots, \mathbf{q}_N)$, the so-called *space* $\Gamma$, is governed by a Hamiltonian $H$ which for simplicity we assume to have the following form:

$$H(\mathbf{P}, \mathbf{Q}) = \sum_{i=1}^{N} \frac{\mathbf{p}_i^2}{2m} + \sum_{1 \leq i < j \leq N} \Phi(\mathbf{q}_i - \mathbf{q}_j) + \sum_{i=1}^{N} \Phi_e(\mathbf{q}_i). \tag{15.1}$$

Naturally $(\mathbf{p}_i, \mathbf{q}_i)$ are the momentum and position coordinates of the $i$th particle, $\Phi$ is the interaction potential energy between pairs of molecules and $\Phi_e$ is the potential energy of possible external fields. Writing the expression (15.1) we tacitly assume that $H = +\infty$ outside the accessible region, according to the discussion in Section 2.6. As in the previous chapter, we neglect the internal degrees of freedom of the particles and the associated energy. In addition we can possibly consider that the system is subject to external random perturbations, in a sense to be made precise.

The objective of statistical mechanics is evidently not to follow the trajectories in the space $\Gamma$ (as impossible as following the trajectories in the space $\mu$), but rather deriving the macroscopic properties of the system starting from its Hamiltonian (15.1). These macroscopic properties are determined by a few thermodynamical quantities which are *experimentally observable*, whose values determine *macroscopic states*. The macroscopic variables must be derived from the microscopic ones (position and momentum of each molecule) through certain averaging operations. We then confront two fundamental problems: the justification for the interpretation of averages as physical macroscopic quantities, and the development of methods to compute such averages, typically via asymptotic expressions reproducing the thermodynamical quantities in the limit that the number of degrees of freedom tends to infinity.

Note that to any given macroscopic state there corresponds a set of representative points in the space $\Gamma$, associated with different microscopic states which reproduce the given macroscopic state. For example, interchanging two molecules we obtain a new point in the space $\Gamma$, but this evidently does not change the macroscopic state (as the distribution function in the space $\mu$ is unaffected). These considerations justify the introduction of the set of points $\mathcal{E}$ in the space $\Gamma$ with which it is possible to associate a prescribed macroscopic state. At the same time we need to define a procedure to compute the macroscopic quantities. To solve this problem, and to visualise the set $\mathcal{E}$, Gibbs considered a family (he called it an *ensemble*) constituted by a large number of copies of the system. A point in the set of representative points corresponding to the thermodynamical equilibrium considered is associated with each such copy. Therefore it is possible to consider the Gibbs ensemble as being produced by an extremely numerous sampling of kinematic states of the system in the same situation of thermodynamical equilibrium.

It is reasonable to expect that the points of $\mathcal{E}$ are not uniformly distributed in $\Gamma$ and therefore that they contribute differently to the average of any prescribed quantity. Using a limiting procedure analogous to the one adopted with the distribution function in the space $\mu$, we can treat $\mathcal{E}$ as a continuous set, endowed with a *density function* $\rho(\mathbf{X}) \geq 0$, integrable on $\mathcal{E}$. Hence the number $\nu(\Omega)$ of the states of $\mathcal{E}$ contained in a region $\Omega$ of the space $\Gamma$ is given by

$$\nu(\Omega) = \int_\Omega \rho(\mathbf{X}) \, d\mathbf{X}. \tag{15.2}$$

We can therefore define the *average* of a quantity $F(\mathbf{X})$ over a statistical set $\mathcal{E}$ with density $\rho$:

$$\langle F \rangle_\rho = \frac{\int_\mathcal{E} F(\mathbf{X})\rho(\mathbf{X})\,\mathrm{d}\mathbf{X}}{\int_\mathcal{E} \rho(\mathbf{X})\,\mathrm{d}\mathbf{X}}, \qquad (15.3)$$

where we clearly mean that the product $F\rho$ is integrable over $\mathcal{E}$. In Gibbs' interpretation, this value corresponds to the value attained by the corresponding macroscopic quantity at the *equilibrium state* described by the density $\rho$.

These considerations justify the following.

DEFINITION 15.1   *A statistical set according to Gibbs is described by a density $\rho(\mathbf{X}) \geq 0$ in the space $\Gamma$. The set $\mathcal{E}$ of points where $\rho > 0$ is called the* support *of the density $\rho$. If the density $\rho$ is normalised in such a way that*

$$\int_\mathcal{E} \rho(\mathbf{X})\,\mathrm{d}\mathbf{X} = 1,$$

*then it is called a* probability density. *We denote a statistical set by the symbol $(\mathcal{E}, \rho)$.*   ∎

The fundamental problem of statistical mechanics is the quest for statistical sets on which it is possible to define, through averages of the type (15.3), the macroscopic quantities satisfying the known laws of thermodynamics. A statistical set which constitutes a good model of thermodynamics is called (following Boltzmann) *orthodic*.

The theory of statistical sets presents three important questions:

(1) existence and description of orthodic statistical sets;
(2) equivalence of the thermodynamics described by these sets;
(3) comparison between experimental data and the predictions of the state equations derived starting from such statistical sets.

Before considering these questions, it is useful to briefly discuss the justification for the interpretation of observable quantities as averages, i.e. the so-called *ergodic hypothesis*. We shall present here only brief introductory remarks, and refer to Chapter 13 for a more detailed study of this question. However the present chapter can be read independently of Chapter 13.

## 15.2 The ergodic hypothesis: averages and measurements of observable quantities

Firstly, we need to make precise the fact that, assuming the number of particles to be constant, we confront two clearly distinct situations:

(a) the system is isolated, in the sense that the value of the Hamiltonian (15.1) is prescribed;
(b) the system is subject to external random perturbations (in a precise thermodynamical context) which make its energy fluctuate.

It is intuitively clear that the structure of the statistical set $(\mathcal{E}, \rho)$ is different in the two cases. From the physical point of view, we can state that what distinguishes (a) and (b) is that in the first case the value of the energy is fixed, while in the second case, the average energy, and hence the temperature, is fixed.

In the case (a) we know that the Hamiltonian flow defines a group of one-parameter transformations $S^t$ (the parameter is time) of the space $\Gamma$ into itself. The set $\mathcal{E}$ is a $(6N-1)$-dimensional manifold $H(\mathbf{P}, \mathbf{Q}) = E$ (we shall see in the following how to define density on it). In addition if $\mathbf{X}$ and $\mathbf{X}_0$ belong to the same trajectory, and hence if $\mathbf{X} = S^t \mathbf{X}_0$ for some $t$, there exists between them a deterministic correspondence, and therefore we must attribute to the two points the same probability density (because the volume of a cell containing $\mathbf{X}_0$ is not modified by the Hamiltonian flow). We can then state the following.

THEOREM 15.1 *If the Hamiltonian $H(\mathbf{X})$ is a first integral, then the same is true of the density $\rho(\mathbf{X})$.* ∎

The case (b) presents a different picture. The typical realisation that we consider is the one where the system of Hamiltonian (15.1), which we denote now by $H_1$, is in contact with a second 'much larger' system, of Hamiltonian $H_2$. The resulting system has Hamiltonian $H_{\text{tot}} = H_1 + H_2 + H_{\text{int}}$ (the last is the coupling term) and is isolated, in the sense that $H_{\text{tot}} = E_{\text{tot}}$, a constant. In the corresponding space $\Gamma_{\text{tot}}$ we could apply the considerations just discussed. If, however, we restrict our observation to the projection $\Gamma_1$ of $\Gamma_{\text{tot}}$, which is the phase space of the first system, then the Hamiltonian $H_1$ is not constant along the trajectories in $\Gamma_1$, but instead fluctuates because of the action of $H_{\text{int}}$, which is perceived as a random perturbation. This explains why $\rho$ is also not constant along the trajectories in $\Gamma_1$, which do not establish between their points a deterministic correspondence. As we shall see, the presence of the second system (the so-called thermostat) is needed to fix the temperature, in the sense that the energy of the first system must fluctuate near a prescribed average.

We can now deduce a simple but very useful result.

If $M \subset \Gamma$ is a measurable subset of the phase space, and we denote by $M_t = S^t M$ the image of $M$ according to the Hamiltonian flow at time $t$, for every integrable function $f$ we have

$$\int_M f(\mathbf{X})\,\mathrm{d}\mathbf{X} = \int_{M_t} f(S^{-t}\mathbf{Y})\,\mathrm{d}\mathbf{Y}, \qquad (15.4)$$

where $\mathbf{X} = (\mathbf{P},\mathbf{Q}) \in \Gamma$ indicates a generic point in the phase space and $\mathbf{Y} = S^t\mathbf{X}$.

DEFINITION 15.2  *A set $M$ is called* invariant *if $S^t M = M$ for every $t \in \mathbf{R}$* ∎

Clearly if $M$ is invariant, equation (15.4) yields

$$\int_M f(S^t\mathbf{X})\,\mathrm{d}\mathbf{X} = \text{constant}. \qquad (15.5)$$

DEFINITION 15.3  *To every statistical set $(\mathcal{E},\rho)$ one can associate a measure $|\cdot|_\rho$ in the space $\Gamma$ defined by*

$$|M|_\rho = \int_M \rho(\mathbf{X})\,\mathrm{d}\mathbf{X}, \qquad (15.6)$$

*where $M$ is any subset of $\Gamma$ measurable with respect to the Lebesgue measure. Any property that is satisfied everywhere except than in a set $A$ of measure $|A|_\rho = 0$ is said to hold $\rho$-almost everywhere. A function $f : \mathcal{E} \to \mathbf{R}$ is $\rho$-integrable if and only if $\int_\mathcal{E} |f(\mathbf{X})|\rho(\mathbf{X})\,\mathrm{d}\mathbf{X} < +\infty$.* ∎

For an introduction to measure theory, see Sections 13.1 and 13.2.

*Remark* 15.1
Clearly $|\Gamma|_\rho = |\mathcal{E}|_\rho$. If $\rho$ is an integrable function and a set $A$ has Lebesgue measure $|A| = 0$ then $|A|_\rho = 0$. ∎

If we apply equation (15.1) to the density $\rho(\mathbf{X})$ and take into account Theorem 15.1, we arrive at the following conclusion.

COROLLARY 15.1  *In the case that $H = $ constant the measure $|\cdot|_\rho$ is invariant with respect to the one-parameter group of transformations $S^t$: for every measurable subset $M$ of $\Gamma$ we have*

$$|M_t|_\rho = |M|_\rho, \qquad (15.7)$$

*for every time $t \in \mathbf{R}$.* ∎

*Remark* 15.2
Consider the map $S = S^1$ and denote by $\mathcal{B}(\Gamma)$ the $\sigma$-algebra of Borel sets on $\Gamma$. The system $(\mathcal{E}, \mathcal{B}(\Gamma), \rho, S)$ is an example of a measurable dynamical system (see Section 13.3 and, in particular, Example 13.9). ∎

*Remark* 15.3
From what we have just seen, the measure $|M|_\rho$ is proportional (equal if $\rho$ is a probability density) to the probability that the system is in a microscopic state described by a point in the space $\Gamma$ belonging to $M$. ∎

It is not obvious, and it is indeed a much debated issue in classical statistical mechanics, that one can interpret the average $\langle f \rangle_\rho$ as the value to attribute to the quantity $f$ in correspondence to the equilibrium described by the statistical set $(\mathcal{E}, \rho)$.

In an experimental measurement process on a system made up of a large number of particles, the system interacts with the instrumentation *for a certain time*, which—although short on a macroscopic scale—is typically very long with respect to the characteristic times involved at the microscopic level. We mean that the observation of the quantity is not done by picking up a precise microscopic state, and hence a point of the space $\Gamma$, but rather it refers to an arc of the trajectory of a point in the space $\Gamma$ (even neglecting the non-trivial fact that the system itself is perturbed by the observation—this point is crucial in quantum statistical mechanics).

Thus it seems closer to the reality of the measurement process to consider the time average of $f$ on arcs of the trajectory of the system. The first problem we face is then to prove the existence of the time average of $f$ along the Hamiltonian flow $S^t$. This is guaranteed by an important theorem due to Birkhoff (see Theorem 13.2).

THEOREM 15.2 *Let $M$ be an invariant subset with finite Lebesgue measure $|M|$ in the phase space $\Gamma$, and let $f$ be an integrable function on $M$. The limit*

$$\hat{f}(\mathbf{X}) = \lim_{T \to +\infty} \frac{1}{T} \int_0^T f(S^t \mathbf{X}) \, dt \tag{15.8}$$

*exists for almost every point $\mathbf{X} \in M$ with respect to the Lebesgue measure. The same conclusion holds if $+\infty$ is replaced by $-\infty$ in (15.8). In addition, it is immediate to verify that for every $t \in \mathbf{R}$ we have*

$$\hat{f}(S^t \mathbf{X}) = \hat{f}(\mathbf{X}). \tag{15.9}$$
∎

The limit (15.8) defines the *time average* of a function $f$. The time average of a given quantity along an arc of a trajectory (corresponding to the time interval during which the measurement is taken) can take—in general—very different values on different intervals. The theorem of Birkhoff guarantees the existence, for almost every trajectory, of the time average, and it establishes that the averages over *sufficiently long* intervals are approximately equal (as they must all tend to $\hat{f}(\mathbf{X})$ for $T \to \infty$).

However, as we have already stated many times, the computation of averages is only a hypothetical operation, as it is not practically possible to determine a Hamiltonian flow of such complexity nor know its initial conditions. This question is at the heart of Gibbs' approach: if the Hamiltonian flow is such that it visits every subset of $\mathcal{E}$ with positive measure, then we can expect that the time average can be identified with the ensemble average (15.3), a quantity that can actually be computed. To make this intuition precise we introduce the concept of *metric indecomposability*.

DEFINITION 15.4 *An invariant subset $M$ of $\Gamma$ is called* metrically indecomposable (with respect to the measure $|\cdot|_\rho$) *if it cannot be decomposed into the union of disjoint measurable subsets $M_1$ and $M_2$, each invariant and of positive measure. Equivalently, if $M = M_1 \cup M_2$, with $M_1$ and $M_2$ measurable, invariant and disjoint, then $|M_1|_\rho = |M|_\rho$ and $|M_2|_\rho = 0$, or vice versa. A statistical set is* metrically indecomposable *if $\mathcal{E}$ is metrically indecomposable with respect to the measure $|\cdot|_\rho$.* ∎

If a set is metrically indecomposable, necessarily its time average is constant almost everywhere, and vice versa, as the following theorem states.

THEOREM 15.3 *Let $(\mathcal{E}, \rho)$ be metrically indecomposable with respect to the measure $|\cdot|_\rho$. Then for any $\rho$-integrable function $f$ on $\mathcal{E}$, the time average $\hat{f}(\mathbf{X})$ is constant $\rho$-almost everywhere. Conversely, if for all integrable functions the time average is constant $\rho$-almost everywhere, then $(\mathcal{E}, \rho)$ is metrically indecomposable.* ∎

The proof of this theorem is the same as the proof of the equivalence of (2) and (4) in Theorem 13.4. The importance of the notion of metric indecomposability in the context of statistical mechanics of equilibrium is due to the following fundamental result.

THEOREM 15.4 *If $(\mathcal{E}, \rho)$ is metrically indecomposable and $f$ is $\rho$-integrable, then*

$$\hat{f}(\mathbf{X}) = \frac{1}{|\mathcal{E}|_\rho} \int_\mathcal{E} f(\mathbf{X}) \rho(\mathbf{X}) \, d\mathbf{X} = \langle f \rangle_\rho \tag{15.10}$$

*for almost every $\mathbf{X} \in \mathcal{E}$.* ∎

Once again, for the proof see Section 13.4.

Metric indecomposability therefore implies the possibility of interpreting the set average (15.3) as the result of the measurement of $f$.

The hypothesis that the support of a Gibbs statistical set is metrically indecomposable is known as the *ergodic hypothesis*. We saw that this hypothesis is equivalent to the condition (15.10) that the time average is equal to the set average. This fact justifies the following definition.

DEFINITION 15.5 *A statistical set $(\mathcal{E}, \rho)$ is* ergodic *if and only if condition (15.10) is satisfied for every $\rho$-integrable $f$ (hence the time average is equal to the set average). If a Hamiltonian system admits an ergodic statistical set, then we say that it satisfies the* ergodic hypothesis. ∎

*Remark* 15.4

We have deliberately neglected so far a critical discussion of the identification of the result of a measurement with the time average. We would then face the following problem: how much time must pass (hence how large must $T$ be in (15.8)) for the difference between the average of a quantity $f$ on the interval $[0,T]$ and the time average $\hat{f}$ (hence the set average $\langle f \rangle_\rho$) to be less than a prescribed tolerance? This problem is known as the problem of *relaxation times* at the equilibrium value for an observable quantity. It is a problem of central importance in classical statistical mechanics, and it is still the object of intense research (see Krylov (1979) for a detailed study of this problem). ∎

## 15.3 Fluctuations around the average

In order to understand what is the degree of confidence we may attach to $\langle f \rangle_\rho$ as the equilibrium value of an observable it is convenient to analyse the *quadratic dispersion* $(f^2 - \langle f \rangle_\rho)^2$. Weighing this with the density $\rho$, we obtain the *variance*: $\langle (f - \langle f \rangle_\rho)^2 \rangle_\rho = \langle f^2 \rangle_\rho - \langle f \rangle_\rho^2$. The ratio between the latter and $\langle f^2 \rangle_\rho$ (or $\langle f \rangle_\rho^2$) is the *mean quadratic fluctuation*:

$$\eta = \frac{\langle f^2 \rangle_\rho - \langle f \rangle_\rho^2}{\langle f^2 \rangle_\rho}. \tag{15.11}$$

Usually we consider extensive quantities, for which $\langle f^2 \rangle_\rho$ and $\langle f \rangle_\rho^2 \sim \mathcal{O}(N^2)$. Hence what is required for $\langle f \rangle_\rho$ to be a significant value is that $\eta \ll 1$ for $N \gg 1$ (typically $\eta \sim \mathcal{O}(1/N)$). Hence instead of (15.11) it is equivalent to consider (as we shall do in what follows)

$$\eta = \frac{\langle f^2 \rangle_\rho - \langle f \rangle_\rho^2}{\langle f \rangle_\rho^2}. \tag{15.12}$$

In the same spirit, we can interpret $\langle f \rangle_\rho$ as the *by far most probable* value of $f$ if the contribution of the average comes 'mainly' from a 'very thin' region of $\Gamma$, centred at the level set $A(\langle f \rangle_\rho)$, where $A(\varphi) = \{\mathbf{X} \in \Gamma \mid f(\mathbf{X}) = \varphi\}$. We refer here to $C^1$ functions. To make this concept more precise, we consider the set $\Omega_\delta$ defined by $\Omega_\delta = \{\mathbf{X} \in \Gamma \mid |f - \langle f \rangle_\rho| < \delta/2\}$. We say that $\Omega_\delta$ is 'thin' if $\varepsilon = \delta/\langle f \rangle_\rho \ll 1$ for $N \gg 1$ (we still refer to the case that $\langle f \rangle_\rho = \mathcal{O}(N)$). We say that $\langle f \rangle_\rho$ is the by far most probable value of $f$ if for some $\delta$ satisfying the condition above, we have

$$\langle f \rangle_\rho \simeq \frac{1}{|\mathcal{E}|_\rho} \int_{\Omega_\delta} \rho(\mathbf{X}) \, f(\mathbf{X}) \, \mathrm{d}\mathbf{X} \tag{15.13}$$

up to $\mathcal{O}(\delta)$.

In typical cases, $\nabla_{\mathbf{X}} f \neq 0$ on $A(\langle f \rangle_\rho)$ and to the same order of approximation we can write

$$\int_{\Omega_\delta} \rho f \, d\mathbf{X} \simeq \delta \langle f \rangle_\rho \int_{A(\langle f \rangle_\rho)} \frac{\rho}{|\nabla_{\mathbf{X}} f|} \, d\Sigma, \qquad (15.14)$$

and hence (15.13) is equivalent to

$$\frac{\delta}{|\mathcal{E}|_\rho} \int_{A(\langle f \rangle_\rho)} \frac{\rho}{|\nabla_{\mathbf{X}} f|} \, d\Sigma = 1 + \mathcal{O}(\varepsilon). \qquad (15.15)$$

The meaning of (15.15) is that when this condition is valid with $\delta/\langle f \rangle_\rho \ll 1$, the 'overwhelming majority' of the states contributing to the average $\langle f \rangle_\rho$ is concentrated (in the sense of the density $\rho$) close to $\Omega_\delta$.

Equation (15.14) suggests that the value $\overline{f}$ of $f$ which naturally takes the role of *most probable value* is the value maximising the function

$$F(\varphi) = \varphi \int_{A(\varphi)} \frac{\rho}{|\nabla_{\mathbf{X}} f|} \, d\Sigma. \qquad (15.16)$$

If $F(\varphi)$ decays rapidly in a neighbourhood of $\varphi = \overline{f}$ then we expect that $\overline{f} \simeq \langle f \rangle_\rho$.

We conclude by observing that if $\Omega_\delta$ gives the main contribution to the averages $\langle f \rangle_\rho$ and $\langle f^2 \rangle_\rho$, then we can write

$$\eta \simeq \frac{1}{|\mathcal{E}|_\rho} \int_{\Omega_\delta} \rho \frac{f^2 - \langle f \rangle_\rho^2}{\langle f \rangle_\rho^2} \, d\mathbf{X}$$

up to order $\mathcal{O}(\delta^2/N^2)$. Since in $\Omega_\delta$ we have $|(f - \langle f \rangle_\rho)(f + \langle f \rangle_\rho)| \leq \frac{1}{2}\delta(2|\langle f \rangle_\rho| + \frac{1}{2}\delta)$, implying $\eta \leq \mathcal{O}(\varepsilon)$, the same conditions guaranteeing that $\langle f \rangle_\rho$ is the most probable value also ensure that the mean quadratic fluctuation is small.

## 15.4 The ergodic problem and the existence of first integrals

We saw how the ergodic hypothesis is the basis of the formalism of statistical sets, and allows one to interpret the averages of observable thermodynamical quantities as their equilibrium values.

A condition equivalent to ergodicity, which highlights even more clearly the connection with the dynamics associated with the Hamiltonian (15.1) when the latter is constant, is given by the following theorem.

THEOREM 15.5  *Consider a system described by the Hamiltonian (15.1) and isolated (in the sense that $H = $ constant). The corresponding statistical set $(\mathcal{E}, \rho)$ is ergodic if and only if every first integral is constant almost everywhere on $\mathcal{E}$.* ∎

For the proof we refer to Section 13.4.

*Remark* 15.5

In the previous statement, a first integral is any *measurable* function $f(\mathbf{X})$, invariant along the orbits of the Hamiltonian flow: for any $\mathbf{X}$ in the domain of $f$, $f(S^t\mathbf{X}) = f(\mathbf{X})$ for every time $t \in \mathbf{R}$. ∎

At this point, it is appropriate to insert a few general remarks on the ergodic hypothesis, connected with the results of the canonical theory of perturbations considered in Chapter 12. These remarks can be omitted in a first reading of this chapter.

For systems which are typically studied by statistical mechanics, it is possible in general to recognise in the Hamiltonian a part corresponding to a completely canonically integrable system. The difference between the Hamiltonian (15.1) and this integrable part is 'small', and the system is therefore in the form (12.4) of quasi-integrable systems which are the object of study of the canonical theory of perturbations:

$$H = H_0(\mathbf{J}) + \varepsilon F(\mathbf{J}, \chi), \tag{15.17}$$

where $(\mathbf{J}, \chi)$ are the action-angle variables associated with the completely canonically integrable system described by the Hamiltonian $H_0$ and $\varepsilon$ is a small parameter, $0 \leq |\varepsilon| \ll 1$.

As an example, for a sufficiently diluted particle gas (where the particles do not necessarily all have the same mass), the integrable part of the Hamiltonian (15.1) corresponds to the total kinetic energy

$$T = \sum_{j=1}^{N} \frac{\mathbf{p}_j^2}{2m_j}, \tag{15.18}$$

and the interaction potential $V$ can be considered almost always as a 'small perturbation', because it can always be neglected except during collisions, and can then be expressed in the form $V = \varepsilon F$.

*Remark* 15.6

The possibility that the quasi-integrable system (15.17) is ergodic is encoded in the presence of the perturbation (the foliation in invariant tori implies metric decomposability). Nevertheless, in the course of the computation of thermodynamical quantities, in the formalism of statistical sets the contribution of $\varepsilon F$ is usually neglected. ∎

On the other hand, in Section 12.4, we discussed and proved the non-existence theorem of first integrals, due to Poincaré (Theorem 12.8). The latter states that, under appropriate regularity, genericity and non-degeneracy assumptions, actually satisfied by many systems of interest for statistical mechanics, there do not exist first integrals regular in $\varepsilon$, $\mathbf{J}$, $\chi$ and independent of the Hamiltonian (15.17).

In a series of interesting papers, Fermi (1923a,b,c, 1924) discussed the consequences of the theorem of Poincaré for the ergodic problem of statistical mechanics, and proved the following theorem.

THEOREM 15.6 (Fermi)  *Under the assumptions of the theorem of Poincaré (Theorem 12.8) a quasi-integrable Hamiltonian system (15.17) with $l > 2$ degrees of freedom does not have $(2l - 1)$-dimensional manifolds which depend regularly on $\varepsilon$ and are invariant for the Hamiltonian flow, with the exception of the manifold with constant energy.* ■

The proof of Fermi's theorem is evidently obtained by showing that there does not exist a regular function $f(\mathbf{J}, \chi, \varepsilon)$ (whose zero level set $M_{f,0}$ defines the invariant manifold) which is at the same time regular in its arguments, a solution of $\{f, H\} = 0$ and independent of $H$ (in the sense that at every point of $M_{f,0}$ the gradients of $f$ and of $H$ are linearly independent). Fermi's proof is very similar to the proof of the theorem of Poincaré. The interested reader is referred to the original paper of Fermi (1923b) or to the recent, excellent exposition of Benettin et al. (1982).

It is interesting to remark how Fermi tried to deduce from this result the (wrong) conclusion that generally, quasi-integrable systems with at least three degrees of freedom are ergodic, and in particular the metric indecomposability of the constant energy surface. Fermi's argument (1923a,c) is roughly the following: if the manifold of constant energy

$$M_E = \{(\mathbf{J}, \chi) | H(\mathbf{J}, \chi, \varepsilon) = E\}$$

were metrically decomposable into two parts with positive measure, the set separating these two parts, and hence their common boundary, could be interpreted as (a part of) an invariant manifold distinct from the manifold of constant energy $M_E$, contradicting the previous theorem.

As was immediately remarked by Urbanski (1924) and recognised by Fermi himself (1924), Fermi's theorem only excludes the possibility that the manifold of constant energy is decomposable into two parts with a *regular* interface, while it is possible for the boundary to be irregular, i.e. not locally expressible as the graph of a differentiable function but at most a measurable one. This is in fact the general situation. The Kolmogorov–Arnol'd–Moser theorem (see Section 12.6) ensures, for sufficiently small values of $\varepsilon$, the existence of an invariant subset of the constant energy surface (which is the union of the invariant tori corresponding to diophantine frequencies) and of positive measure, whose boundary is not regular, but only measurable. We may therefore end up in the paradoxical situation that we can 'prove' that quasi-integrable Hamiltonian systems are not ergodic for 'small' values of $\varepsilon$. The situation is, however, much more complicated, especially as the maximum values $\varepsilon_c$ of $\varepsilon$ admitted under the assumptions of the theorem depend heavily on the number of degrees of freedom of the system,[1] for example

---

[1] In Remark 6.3 we did not stress the dependence of $\varepsilon_c$ on $l$ but only on $\gamma$, since we considered $\mu > l - 1$ *fixed*.

through laws such as $|\varepsilon_c| \leq$ constant $l^{-l}$, which make the KAM theorem not of practical applicability to systems of statistical interest. On the other hand, we do not know any physical system that is both described by a Hamiltonian such as (15.1) (or (15.17)), where the potential energy is a regular function of its arguments (excluding therefore the possibility of situations such as that of a 'hard sphere gas with perfectly elastic collisions'), and for which the ergodic hypothesis has been proved. The problem of the ergodicity of Hamiltonian systems is therefore still fundamentally open, and is the object of intense research, both analytically and using numerical simulations (started by Fermi himself, see Fermi et al. 1954).

## 15.5 Closed isolated systems (prescribed energy). Microcanonical set

In Section 15.2 we anticipated that we would study two typical situations for closed systems (case (a) and case (b)). We now examine the first of these. Consider a system of $N$ particles described by the Hamiltonian (15.1) and occupying a bounded region of volume $V$ with perfectly reflecting walls. Assume that this system is *closed* (fixed number of particles) and *isolated*.

In this case, we saw how the support $\mathcal{E}$ of the density for the corresponding statistical set (if we want it to be ergodic) must coincide with the manifold of constant energy

$$\Sigma_E = \{\mathbf{X} \in \Gamma \mid H(\mathbf{X}) = E\}. \tag{15.19}$$

However the latter has (Lebesgue) measure zero in the space $\Gamma$, and hence the definition of density is non-trivial. To overcome this difficulty we introduce an approximation of the statistical set that we want to construct. Take as the set of states $\mathcal{E}_\Delta$ the accessible part of the space $\Gamma$ lying between the two manifolds $\Sigma_E$ and $\Sigma_{E+\Delta}$, where $\Delta$ is a fixed energy that later will go to zero, and we choose in this set the constant density. In this way we do not obtain a 'good' statistical set because this is not ergodic (since it is a collection of invariant sets). However what we obtain is a promising approximation to an ergodic set, because the energy variation $\Delta$ is very small, and the density (which is a first integral) is constant. To obtain a correct definition of a statistical set we must now 'condense' on the manifold $\Sigma_E$, by a limiting procedure, the information that can be gathered from the approximate set. To this end, we define a new quantity.

DEFINITION 15.6 *For fixed values of $E$ and $V$ the density of states of the system is the function*

$$\omega(E,V) = \lim_{\Delta \to 0} \frac{\Omega(E,V,\Delta)}{\Delta}, \tag{15.20}$$

*where $\Omega(E,V,\Delta)$ is the Lebesgue measure of the set $\mathcal{E}_\Delta$.* ∎

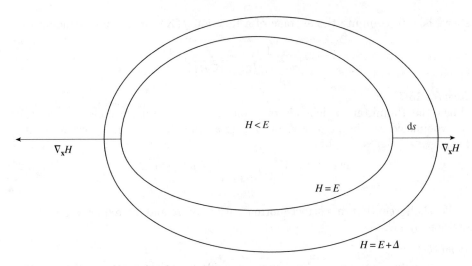

**Fig. 15.1**

In other words, $w(E,V)$ is the derivative with respect to $E$ of the measure of the region defined by the inequality $H \leq E$. It is easy to express $w(E,V)$ through the Hamiltonian. Indeed, at every point of $H = E$ the thickness $ds$ of the microcanonical set (Fig. 15.1) can be obtained from $\Delta = |\nabla_X H|\,ds$, to first order in $\Delta$. In this approximation we therefore have

$$\Omega(E,V,\Delta) = \Delta \int_{H=E} |\nabla_X H|^{-1}\,d\Sigma$$

(recall Remark 8.12), from which it follows that

$$w(E,V) = \int_{H=E} |\nabla_X H|^{-1}\,d\Sigma. \tag{15.21}$$

The next definition follows naturally from these considerations.

DEFINITION 15.7 *The microcanonical set is the ergodic statistical set corresponding to a closed isolated system, described by the manifold of constant energy $\mathcal{E} = \Sigma_E$ with measure*

$$d\rho(\mathbf{X}) = |\nabla_X H|^{-1}\,d\Sigma, \tag{15.22}$$

*where $d\Sigma$ denotes the volume element on the manifold $\Sigma_E$.* ∎

Note that the name 'density of states' assigned to the function $w$ can be misleading (a better name would be *microcanonical partition function*). In reality the role of density on $\mathcal{E}$ is played by the function $\rho$ defined by (15.22), which also shows how to compute the average of a function $f(\mathbf{X})$ on the microcanonical set:

$$\langle f \rangle = \frac{1}{\omega} \int_{H=E} \frac{f}{|\nabla H|} \, d\Sigma.$$

*Remark 15.7*
When the Hamiltonian depends only on the momentum variables, $H = H(\mathbf{P})$, the volume $V$ can be factorised in (15.21) as the result of the integration $d\mathbf{Q}$:

$$\omega(E, V) = V^N \int_{H(P)=E} |\nabla_P H|^{-1} \, d\Sigma_P. \tag{15.23}$$

∎

We shall see that a correct normalisation of $\rho$ and $\omega$ requires the further division by $N!$.

*Remark 15.8*
The procedure leading to the formula (15.22) is rather abstract. We can however deduce the same formula by reasoning only in mechanical terms, and this shows the natural relation between the Hamiltonian flow in the space $\Gamma$ of an isolated system and the density of the microcanonical set. This approach relies only on the condition of ergodicity. Assume that a partition of $\mathcal{E}$ in subsets of comparable measure and diameter has been defined. Following the Hamiltonian flow, if the system is ergodic, all cells are crossed. Thus if we perform a sampling $\mathbf{X}_1, \mathbf{X}_2, \ldots, \mathbf{X}_n, \ldots$ of the trajectory along a time sequence, the probability of finding points of the sample in a specified cell is proportional to the permanence time in the cell during the Hamiltonian flow. Evidently, the average time of permanence in a cell centred at $\mathbf{X}$ is proportional to $|\dot{\mathbf{X}}(\mathbf{X})|^{-1}$. Since $\dot{\mathbf{X}} = \mathbb{J}\nabla_{\mathbf{X}} H$, we find that the probability density associated with this cell is proportional to $|\nabla_{\mathbf{X}} H(\mathbf{X})|^{-1}$.

∎

*Remark 15.9*
Note that the elastic collisions with the walls and between the particles, which can be represented as singularities of the Hamiltonian, do not contribute to (15.21). However, we must stress that the ergodicity of the system is precisely due to the collisions. Indeed, in the absence of any collision the projection $\mathbf{P}$ of the point $\mathbf{X}$ onto the space of momenta would be constant.

∎

We have already remarked that the function $\omega(E, V)$ carries global, and hence macroscopic, information about the system. Its significance is due to the fact that it can be used to define the *entropy* of the system:

$$S(E, V) = k \log \left[ \frac{\omega(E, V)}{\omega_0} \right], \tag{15.24}$$

where $k$ is the Boltzmann constant and $\omega_0$ is an arbitrary constant making quantities dimensionless. Note that the logarithm makes the entropy an extensive quantity, as required by the thermodynamical formalism. Indeed, the partition function of a system obtained by the union of isoenergetic systems of identical particles is the product of the partition functions of the component systems.

Starting from (15.24) we can also define the temperature:

$$T = \left(\frac{\partial S}{\partial E}\right)^{-1} \tag{15.25}$$

and the pressure:

$$P = T\frac{\partial S}{\partial V} = \frac{\partial S}{\partial V}\left(\frac{\partial S}{\partial E}\right)^{-1} \tag{15.26}$$

in such a way that from the expression for the total differential of $S$:

$$\mathrm{d}S = \frac{\partial S}{\partial E}\mathrm{d}E + \frac{\partial S}{\partial V}\mathrm{d}V \tag{15.27}$$

we can deduce the first principle of thermodynamics for our system, by simply multiplying by the temperature $T$:

$$T\,\mathrm{d}S = \mathrm{d}E + P\,\mathrm{d}V. \tag{15.28}$$

It is a significant success of this theory to have been able to express the entropy $S(E,V)$ through the Hamiltonian $H(\mathbf{P},\mathbf{Q})$, and hence through the microscopic mechanical properties of the system.

*Remark* 15.10

We stress that from the physical point of view, the definition of thermodynamical quantities for a system with which the microcanonical set is associated has essentially theoretical interest, since by definition the system itself is not accessible to measurement. However, the definitions (15.24)–(15.26) are self-consistent and will be used in subsequent considerations.

## 15.6 Maxwell–Boltzmann distribution and fluctuations in the microcanonical set

Consider a system in equilibrium, closed and isolated, and described by the microcanonical set. We now want to prove two important facts, highlighting the relation between the formalism of the microcanonical set and the kinetic theory (Chapter 14).

PROPOSITION 15.1    *To every distribution function in the space $\mu$ it is possible to associate a volume in the microcanonical set. After the appropriate normalisation, this volume can be interpreted as the probability of the corresponding distribution.*    ∎

THEOREM 15.7    *The Maxwell–Boltzmann distribution in the space $\mu$ is the distribution to which there corresponds the maximum volume in the microcanonical set, and it is therefore the most probable.*    ∎

*Proof of Proposition 15.1*
Introduce a partition of the accessible region of the space $\mu$ in 'cells' (for example, cubic cells) of equal volume $\omega$. This volume must be small with respect to the total volume, but sufficiently large that we can find in each cell representative points of a large enough number of molecules. Since the accessible region in the space $\mu$ is bounded, the cells are a finite number: $\omega_1, \ldots, \omega_K$. Indicate by $n_i$ the number of representative points in the cell $\omega_i$ (*occupation number*). The numbers $n_i$ are subject to the characteristic conditions of the microcanonical set:

$$\sum_{i=1}^{K} n_i = N, \tag{15.29}$$

$$\sum_{i=1}^{K} \varepsilon_i n_i = E, \tag{15.30}$$

where $\varepsilon_i = p_i^2/2m$ and $\mathbf{p}_i$ is the momentum corresponding to the cell $\omega_i$ (because of the non-zero dimensions of the cell the momentum is not defined exactly, and therefore we have a finite variation in the energy, and hence the discretised microcanonical set in the space $\Gamma$ has non-zero thickness). A $K$-tuple $(n_1, \ldots, n_K)$ determines a (discretised) distribution function defined by $f(\mathbf{p}_i) = n_i/\omega$. To each prescribed distribution of the $N$ points in the cells $\omega_1, \ldots, \omega_K$ (hence to each microscopic state) there corresponds exactly one specific cell of volume $\omega^N$ in the space $\Gamma$. However to a $K$-tuple of occupation numbers there correspond more than one microscopic state, and hence a larger volume in the space $\Gamma$. Indeed, interchanging two particles with representative points in two distinct cells the occupation numbers do not change, but the representative point in the space $\Gamma$ does. On the other hand, nothing changes if we permute particles inside the same cell. Since $N!$ is the total number of permutations and $n_i!$ are those inside the cell $\omega_i$, which do not change the position of the representative volume element in the space $\Gamma$, we find that the total volume in $\Gamma$ space corresponding to a prescribed sequence of occupation numbers $(n_1, \ldots, n_K)$ is

$$\Omega(n_1, \ldots, n_K) = \frac{N!}{n_1! n_2! \cdots n_K!} \omega^N. \tag{15.31}$$

∎

*Proof of Theorem 15.7*
We seek the sequence $(\overline{n}_1, \ldots, \overline{n}_K)$ maximising $\Omega$ and therefore expressing the most probable macroscopic state (with respect to the microcanonical distribution).

Recall that $n_i \gg 1$. Using Stirling's formula, we obtain $\log n! \approx n \log n$, from which it follows that

$$\log \Omega(n_1, \ldots, n_K) = -\sum_{i=1}^{K} n_i \log n_i + \text{constant}. \tag{15.32}$$

Considering now the variables $n_i$ as continuous variables, we seek the maximum of the function (15.32) taking into account the constraints expressed by (15.29) and (15.30). Hence we seek to find the extrema of the function

$$\Lambda = -\sum_{i=1}^{K} n_i \log n_i - \lambda_1 \sum_{i=1}^{K} n_i - \lambda_2 \sum_{i=1}^{K} n_i \varepsilon_i,$$

where $\lambda_1, \lambda_2$ are the Lagrange multipliers. This procedure yields the $K$ equations

$$\log n_i + 1 + \lambda_1 + \varepsilon_i \lambda_2 = 0, \quad i = 1, \ldots, K, \qquad (15.33)$$

to which we again associate the conditions (15.29) and (15.30). Note that

$$\frac{\partial^2 \Lambda}{\partial n_i \partial n_j} = -\delta_{ij} \frac{1}{n_i},$$

and therefore the extremum is a maximum.

Redefining the parameters $\lambda_1, \lambda_2$ we can write the solutions of (15.33) in the form

$$\overline{n}_i = c e^{-\beta \varepsilon_i}, \quad i = 1, \ldots, K. \qquad (15.34)$$

Equation (15.34) is simply a discretised version of the Maxwell–Boltzmann distribution, setting $\varepsilon_i = \mathbf{p}_i^2/2m$ (or $\varepsilon_i = \mathbf{p}_i^2/2m + \Phi(\mathbf{q}_i)$ when there are external forces). It is easy to recognise that the continuous limit of the distribution associated with $(\overline{n}_1, \ldots, \overline{n}_K)$ is precisely the Maxwell–Boltzmann one, and this automatically leads to the determination of the constants $c$ and $\beta$ in (15.34). ∎

We have therefore proved that the Maxwell–Boltzmann distribution is the most probable in the macroscopic equilibrium state associated with the microcanonical set. It is important to realise that the Maxwell–Boltzmann distribution is also the distribution 'by far' most probable, in the sense that the fluctuations around it are very small. For this we must evaluate the relative differences

$$\frac{\langle n_i^2 \rangle - \overline{n}_i^2}{N^2}, \qquad (15.35)$$

where $\langle \cdot \rangle$ represents the average on the microcanonical set.

THEOREM 15.8  *The fluctuations (15.35) around the Maxwell–Boltzmann distribution tend to zero, for $N \to \infty$, at least as $N^{-1}$.*

*Proof*
Recall the discussion of Section 15.3 about the comparison between average and most probable value. It is then sufficient to compute $\langle n_i^2 \rangle - \langle n_i \rangle^2$. We follow a very elegant method, that can be found in Huang (1987).

If in place of (15.31) we consider the function

$$\widetilde{\Omega}(n_1, \ldots, n_K, \eta_1, \ldots, \eta_K) = N! \frac{\eta_1^{n_1} \cdots \eta_K^{n_K}}{n_1! n_2! \cdots n_K!} \omega^N, \qquad (15.36)$$

where $\eta_i$ are parameters varying between 0 and 1, using as before the Lagrange multipliers technique, we find that the maximum is attained for values of the occupation number $n_i = \tilde{n}_i(\eta_1, \ldots, \eta_K)$ given by

$$\tilde{n}_i = \eta_i \bar{n}_i = c\eta_i e^{-\beta n_i}, \quad i = 1, \ldots, K, \tag{15.37}$$

and that $\tilde{n}_i = \bar{n}_i$ and $\tilde{\Omega} = \Omega$ when $\eta_1 = \ldots = \eta_K = 1$. The parameters $\eta_i$ can therefore be considered as volume reduction factors for the cells $\omega_i$.

We note now that, for every $(n_1, \ldots, n_K)$ and $(\eta_1, \ldots, \eta_K)$, we have

$$\eta_i \frac{\partial}{\partial \eta_i} \tilde{\Omega}(n_1, \ldots, n_K, \eta_1, \ldots, \eta_K) = n_i \tilde{\Omega}(n_1, \ldots, n_K, \eta_1, \ldots, \eta_K).$$

Hence $\langle n_i \rangle$, which by definition is

$$\langle n_i \rangle = \frac{\sum_{\mathbf{n}} n_i \Omega(\mathbf{n})}{\sum_{\mathbf{n}} \Omega(\mathbf{n})}, \tag{15.38}$$

can also be expressed as

$$\langle n_i \rangle = \left. \frac{\sum_{\mathbf{n}} \eta_i \partial/\partial \eta_i \, \tilde{\Omega}(\mathbf{n}, \boldsymbol{\eta})}{\sum_{\mathbf{n}} \tilde{\Omega}(\mathbf{n}, \boldsymbol{\eta})} \right|_{\boldsymbol{\eta} = (1, \ldots, 1)}, \tag{15.39}$$

where $\mathbf{n} = (n_1, \ldots, n_K)$, $\boldsymbol{\eta} = (\eta_1, \ldots, \eta_K)$ and $\sum_{\mathbf{n}}$ denotes the sum over all sequences of occupation numbers. For the computation of $\langle n_i^2 \rangle$ we note that

$$\eta_i \frac{\partial}{\partial \eta_i} \left( \eta_i \frac{\partial}{\partial \eta_i} \tilde{\Omega} \right) = n_i^2 \tilde{\Omega}(\mathbf{n}, \boldsymbol{\eta}),$$

and therefore

$$\langle n_i^2 \rangle = \frac{\sum_{\mathbf{n}} n_i^2 \Omega(\mathbf{n})}{\sum_{\mathbf{n}} \Omega(\mathbf{n})} = \left[ \frac{\eta_i \partial/\partial \eta_i \left( \eta_i \partial/\partial \eta_i \sum_{\mathbf{n}} \tilde{\Omega}(\mathbf{n}, \boldsymbol{\eta}) \right)}{\sum_{\mathbf{n}} \tilde{\Omega}(\mathbf{n}, \boldsymbol{\eta})} \right]_{\eta_1 = \ldots = \eta_K = 1}. \tag{15.40}$$

The expression in square brackets can be rewritten and elaborated as follows:

$$\eta_i \frac{\partial}{\partial \eta_i} \left( \frac{1}{\sum_{\mathbf{n}} \tilde{\Omega}(\mathbf{n}, \boldsymbol{\eta})} \eta_i \frac{\partial}{\partial \eta_i} \sum_{\mathbf{n}} \tilde{\Omega}(\mathbf{n}, \boldsymbol{\eta}) \right)$$

$$- \left( \eta_i \frac{\partial}{\partial \eta_i} \frac{1}{\sum_{\mathbf{n}} \tilde{\Omega}(\mathbf{n}, \boldsymbol{\eta})} \right) \left( \eta_i \frac{\partial}{\partial \eta_i} \sum_{\mathbf{n}} \tilde{\Omega}(\mathbf{n}, \boldsymbol{\eta}) \right) \qquad (15.41)$$

$$= \eta_i \frac{\partial}{\partial \eta_i} \left( \frac{\sum_{\mathbf{n}} n_i \tilde{\Omega}(\mathbf{n}, \boldsymbol{\eta})}{\sum_{\mathbf{n}} \tilde{\Omega}(\mathbf{n}, \boldsymbol{\eta})} \right) + \left( \frac{1}{\sum_{\mathbf{n}} \tilde{\Omega}(\mathbf{n}, \boldsymbol{\eta})} \eta_i \frac{\partial}{\partial \eta_i} \sum_{\mathbf{n}} \tilde{\Omega}(\mathbf{n}, \boldsymbol{\eta}) \right)^2.$$

The last term is simply $\langle n_i \rangle^2$, after setting $\eta_1 = \ldots = \eta_K = 1$, while we identify the average of $n_i$ weighted by $\tilde{\Omega}(\mathbf{n}, \boldsymbol{\eta})$ with $\tilde{n}_i$, given by (15.37). Therefore the first term in (15.41) is simply $\eta_i (\partial/\partial \eta_i) \tilde{n}_i = \tilde{n}_i$ and reduces to $\bar{n}_i$ for $\eta_1 = \ldots = \eta_K = 1$.

Finally from (15.40) and (15.41) we find

$$\langle n_i^2 \rangle - \langle n_i \rangle^2 = \bar{n}_i. \qquad (15.42)$$

Recalling that $\bar{n}_i/N < 1$, from this it follows that

$$\left[ \frac{\langle n_i^2 \rangle}{N^2} - \frac{\langle n_i \rangle^2}{N^2} \right]^{1/2} < \frac{1}{\sqrt{N}}, \qquad (15.43)$$

proving the claim. ∎

Taking into account that $N$ can be of the order of $10^{23}$, equation (15.43) implies that the fluctuations around the Maxwell–Boltzmann distribution are in reality extremely small, and hence that the probability that the system takes a state very different from this particular distribution is very small.

## 15.7 Gibbs' paradox

We leave aside the study of the general properties of the function $S(E, V)$ defined by (15.24), and we only compute in a simple way the entropy of a perfect gas, hence assuming that the Hamiltonian of the system is of the form

$$H = \sum_{i=1}^{N} \frac{\mathbf{p}_i^2}{2m}. \qquad (15.44)$$

In this case, it is possible to use formula (15.23). In addition, $|\nabla_p H|^2 = \sum_{i=1}^{N} (\mathbf{p}_i/m)^2 = (2/m) H$. The integral (15.23) must be computed on the sphere

$\mathbf{P}^2 = 2mE$ in $\mathbf{R}^{3N}$. Denoting by $\chi_n$ the measure of the spherical surface of radius 1 in $\mathbf{R}^n$ $(\chi_n = 2\pi^{n/2}/\Gamma(n/2))$, we have

$$\omega(E,V) = V^N \left(\frac{m}{2}\right)^{1/2} E^{-1/2} \chi_{3N} (2mE)^{(3N-1)/2}. \tag{15.45}$$

Using the asymptotic expression

$$\Gamma(z+1) \approx \sqrt{2\pi} z^{z+(1/2)} e^{-z},$$

we can write $\chi_{3N} \approx \sqrt{2}(2\pi/3N)^{(3N-1)/2} e^{(3N/2)-1}$. Substituting into (15.45), taking the logarithm, *taking into account that* $N \gg 1$ and retaining only the terms in $E$, $V$, $N$, we obtain, because of (15.24), the following expression for the entropy:

$$S(E,V) = Nk \log\left[\tilde{\lambda} V \left(\frac{E}{N}\right)^{3/2}\right] + \frac{3}{2} kN, \tag{15.46}$$

where $\tilde{\lambda} > 0$ is the same constant appearing in expression (14.46) of the previous chapter.

The evident difference between the two expressions is that in (15.46) there appears $V$ instead of $V/N$. While this does not affect the validity of the expressions $\partial S/\partial E = 1/T$, $\partial S/\partial V = P/T$, it nevertheless makes (15.46) unacceptable as a state function of the system. Indeed, if we consider two systems with the same particle density ($n = N_1/V_1 = N_2/V_2$), and the same average energy per particle ($\varepsilon = E_1/N_1 = E_2/N_2$), we want the entropy of their union to be the sum of the entropies $S_1, S_2$. As written, equation (15.46) does not have this property, and yields the paradoxical consequence that it is not possible to partition the system into two or more parts with identical ratios $E_i/N_i, V_i/N_i$ and then reassemble it, again obtaining the starting entropy. This is *Gibbs' paradox*. This difficulty was immediately evident to Gibbs himself and he had no choice but to correct (15.46) by inserting $V/N$ in place of $V$:

$$S(E,V) = kN \log \frac{V}{NV_0} + \frac{3}{2} kN \log \frac{E}{NE_0} + C, \tag{15.47}$$

where $C$ is a constant and $E_0$, $V_0$ are constants that arise from making the variables dimensionless. Recalling the approximation $\log N! \approx N \log N$ for $N \gg 1$, we realise that this correction is equivalent to dividing the function $\omega(E,V)$ by $N!$. Therefore it amounts to a renormalisation of the density of states (*Boltzmann counting*), corresponding to considering the particles to be *indistinguishable* (the natural point of view of quantum mechanics), so that any permutation gives rise to the same state. The kinetic theory does not lead to the same paradox because, as we observed, it admits the interchange of particles.

Boltzmann renormalisation (which we henceforth adopt systematically) puts back on track classical statistical mechanics, by introducing in it the concept

## 15.7 Statistical mechanics: Gibbs sets

of non-individuality of the particles. The criticism aimed at classical statistical mechanics on the basis of Gibbs' paradox is, in our view, scarcely motivated, as the need for considering the particles as indistinguishable does not deprive the theory of its elegance and deep significance. However the theory presents more important limitations when one wants to study the contribution of each degree of freedom to the energy of the system. We shall discuss this topic in the next section.

To this end, it is useful to conclude this section with a few additional considerations on the definition of entropy.

Equation (15.24) is not the only possibility of defining the entropy starting from the structure of the microcanonical set.

If we assume that the set $\{H \leq E\}$ is bounded and denote by $B(E, V)$ its measure (recall that then $\omega(E, V) = (\partial/\partial E)\, B(E, V)$), we can show that another equivalent definition of entropy is given by

$$S(E, V) = k \log \left[\frac{B(E, V)}{B_0}\right]. \tag{15.48}$$

To verify this fact we denote by $S_\omega$ the entropy defined by (15.24) and by $S_B$ the entropy defined by (15.48). Define $T_\omega = (\partial S_\omega/\partial E)^{-1}$ and $T_B = (\partial S_B/\partial E)^{-1}$, and note that $T_B = (k\omega/B)^{-1}$. In addition,

$$S_\omega - S_B = k \log\left(\frac{c\omega}{B}\right) = k \log\left(\frac{c}{kT_B}\right)$$

($c$ constant with dimension of energy), and differentiating with respect to $E$ we have

$$\frac{1}{T_\omega} - \frac{1}{T_B} = -k\frac{1}{T_B}\frac{\partial T_B}{\partial E},$$

and hence

$$\frac{T_B}{T_\omega} = 1 - k\frac{\partial T_B}{\partial E}. \tag{15.49}$$

Now it is sufficient to recall that $k\, \partial T_B/\partial E = \mathcal{O}(1/N)$ (and precisely $2/3N$ in the case of a perfect gas) and conclude that in the thermodynamical limit (when $N \to \infty$) we can assume $T_\omega = T_B$, and that this equality is satisfied in practice for $N$ finite (of the characteristic order of magnitude $10^{23}$). In particular, we have obtained the following expression for the temperature:

$$T = \left(k\frac{\omega(E, V)}{B(E, V)}\right)^{-1}. \tag{15.50}$$

**Example 15.1**
It is immediate to verify the validity of equations (15.50) for the Hamiltonian (15.44). It is enough to recall that the volume of the sphere of

radius $R$ in $\mathbf{R}^n$ is $1/nR^n \chi_n$ ($\chi_n$ is the measure of the unit spherical surface). Therefore ($n = 3N$, $R = \sqrt{2mE}$) $B = \frac{1}{3}N(2mE)^{3N/2}\chi_{3N}$, and hence $\omega/B = \partial/\partial E \log B = \frac{3}{2}(N/E) = 1/kT$. ∎

*Example 15.2*
We now deduce the heat capacity at constant volume $C_V$ and that at constant pressure $C_P$ from the expression (15.47) for the entropy of a perfect monatomic gas, recalling that $dQ = T dS = dE + P dV$. For constant volume we then have $C_V = \partial E/\partial T = \frac{3}{2}Nk$. If, on the other hand, we impose $P = $ constant, we must use the substitution $V/N = kT/P$ in (15.47), and hence $S = kN\left(\log kT/P + \frac{3}{2}\log(E/N)\right)$, up to constants, from which it follows that $dS = kN\left(1/T + \frac{3}{2}(1/E)\frac{3}{2}Nk\right)dT$, hence $T dS = (kN + C_V)dT$, and finally $C_p = kN + C_V = \frac{5}{2}Nk$. ∎

## 15.8 Equipartition of the energy (prescribed total energy)

Recall that the average of a quantity $f$ on the microcanonical set,

$$\langle f \rangle = \lim_{\Delta \to 0} \frac{1}{\omega \Delta} \int_{E < H < E + \Delta} f \, d\mathbf{X}, \qquad (15.51)$$

is identified with

$$\langle f \rangle = \frac{1}{\omega} \int_{H=E} \frac{f}{|\nabla_{\mathbf{X}} H|} \, d\Sigma. \qquad (15.52)$$

Hence we compute $\langle X_i \, \partial H/\partial X_j \rangle$, $i, j = 1, \ldots, 6N$, assuming $\partial H/\partial X_j \neq 0$.
According to (15.52) we have

$$\left\langle X_i \frac{\partial H}{\partial X_j} \right\rangle = \frac{1}{\omega} \int_{H=E} X_i n_j \, d\Sigma, \qquad (15.53)$$

denoting by $n_j$ the $j$th component of the unit vector $\mathbf{n} = \nabla_{\mathbf{X}} H/|\nabla_{\mathbf{X}} H|$ normal to $H = E$. It is convenient to notice that if $H(\mathbf{X})$ does not depend on one of the coordinates $X_j$ the contribution of the latter to (15.53) is absent. Therefore, when some coordinates can be ignored, the set $\{H = E\}$ is implicitly replaced by its projection onto the subspace of coordinates $X_i$ for which $\partial H/\partial X_i \neq 0$. In addition we assume that the latter set is bounded. Denoting by $\mathbf{e}_j$ the unit vector in the $j$th direction in the space $\Gamma$, the integral in (15.53) is simply the flow through $H = E$ of the vector $X_i \mathbf{e}_j$. Therefore, if $n_j$ is non-zero, since $\mathbf{n}$ is oriented towards the exterior of the set $\{H < E\}$ we have

$$\left\langle X_i \frac{\partial H}{\partial X_j} \right\rangle = \frac{1}{\omega} \int_{H<E} \nabla_{\mathbf{X}} \cdot (X_i \mathbf{e}_j) \, d\mathbf{X}. \qquad (15.54)$$

Finally, since $\nabla_{\mathbf{X}} \cdot (X_i \mathbf{e}_j) = \delta_{ij}$, we arrive at

$$\left\langle X_i \frac{\partial H}{\partial X_j} \right\rangle = \delta_{ij} \frac{B(E,V)}{\omega(E,V)} = kT\delta_{ij} \tag{15.55}$$

(recall the hypothesis $\partial H/\partial X_j \neq 0$), where we used (15.50) in the last step. We have proved the so-called *equipartition theorem*.

**THEOREM 15.9** (Equipartition) *For a system described by the microcanonical set with Hamiltonian $H$, the average value of $X_i(\partial H/\partial X_j)$ is zero if $i \neq j$, and has value $kT$ if $i = j$, as long as $\partial H/\partial X_j \neq 0$.* ∎

**Remark 15.11**
If the Hamiltonian $H$ of the system contains terms of the kind $p_i^2/2m$, since $p_i(\partial/\partial p_i)(p_i^2/2m) = 2(p_i^2/2m)$, each one contributes to the average value of the energy with a factor $\frac{1}{2}kT$. Analogously, if the Hamiltonian $H$ contains terms of the kind $\frac{1}{2}m\omega_i^2 q_i^2$ we have that $q_i \partial/\partial q_i \frac{1}{2}m\omega_i^2 q_i^2 = m\omega_i^2 q_i^2$, and their contribution to the average value of the energy is equal to $\frac{1}{2}kT$. We consider more generally a system whose Hamiltonian is a quadratic form:

$$H = \frac{1}{2}\sum_{i=1}^{6N} A_i X_i^2 \geq 0, \quad A_i \geq 0, \tag{15.56}$$

or can be reduced to it via a canonical transformation.[2]

Note that by (15.56) we have

$$\sum_{i=1}^{6N} X_i \frac{\partial H}{\partial X_i} = 2H.$$

Then using equation (15.55), we can conclude that

$$\langle H \rangle = \frac{1}{2} lkT, \tag{15.57}$$

where $l$ is the number of non-zero coefficients in expression (15.56). The theorem of equipartition of energy states precisely that each (non-zero) term in the Hamiltonian (8.6) contributes to the average energy by the quantity $\frac{1}{2}kT$. ∎

**Remark 15.12**
For a perfect monatomic gas, the equipartition theorem implies that the average value $\varepsilon$ of the energy for each molecule is equal to $\frac{3}{2}kT$. Recall that in the kinetic theory, this same result was assumed as the definition of temperature. Here we deduced it from the definition of entropy and from the concept of the average of a set. The fact that the kinetic energy of the system is expressible as $\frac{3}{2}NkT$ is known as the *virial theorem*. ∎

---

[2] The restriction that $A_i \geq 0$ guarantees that the sets $\{H < E\}$ are bounded for every $E$ in the subspace of coordinates that cannot be ignored.

The equipartition theorem is the critical issue of classical statistical mechanics. This fact becomes clear when the internal degrees of freedom are increased, to include systems with great complexity, such as black bodies, to which infinitely many degrees of freedom (but not infinite energy) can be attributed. A profound and enlightening discussion of this problem is presented by Gallavotti (1995, pp. 65–85).

## 15.9 Closed systems with prescribed temperature. Canonical set

In Section 15.2 we announced that we would consider two typical situations for closed systems: systems with fixed energy (described by the microcanonical set) and systems with fluctuating energy, but with prescribed average. We anticipated that this corresponds to fixing the temperature of the system. Physically this is done by connecting the system with a thermostat. The statistical set describing the latter thermodynamical state was called by Gibbs the *canonical set*. Therefore, while for the microcanonical set the independent thermodynamical variables are $E, V$, for the canonical set they are $T, V$.

The union of the system under consideration, $S$ (with $N$ particles) and of the thermostat $\mathcal{T}$, must be assumed at constant energy, but because of the large size of the thermostat, the fluctuations of the energy in $S$ can in principle be as large as the energy of the entire system $S + \mathcal{T}$. Hence the set $\mathcal{E}$ associated with $S$ can span a region in the space $\Gamma$ of $S$ of significant $6N$-dimensional Lebesgue measure (where $\dim \Gamma = 6N$). The problem is now to define the density $\rho(\mathbf{X})$.

DEFINITION 15.8 *The* canonical set *is the statistical set corresponding to a closed system in thermal equilibrium with a thermostat at temperature $T$. It is obtained by setting $\mathcal{E} = \Gamma$ and assigning the density*

$$\rho(\mathbf{X}) = \frac{1}{N! h^{3N}} \exp\left(-\frac{H(\mathbf{X})}{kT}\right), \qquad (15.58)$$

*called the* canonical *or* Boltzmann distribution*, where $h$ is a constant with the same dimensions as the action $[h] = [p][q]$ and $H(\mathbf{X})$ is the Hamiltonian (15.1) of the system under consideration.* ∎

Typically, $h$ is identified with Planck's constant.

*Remark 15.13*
The introduction of $h^{-3N}$ is necessary for $\rho \, d\mathbf{X}$ to be dimensionless. The factor $1/N!$ is consistent with the rule of 'Boltzmann counting', already discussed. ∎

There exist in the literature derivations of (15.58) based on heuristic considerations (for example considering the canonical formalism that can be applied to the thermostat, when we can expect that the fluctuations of the energy are very small; this fact, not rigorously true, is largely justified a posteriori). We prefer to motivate (15.58) in a way that highlights the relation with the kinetic theory. To draw this parallel, we must limit ourselves to the case in which the interactions between particles are only of collisional type (short range). Since in the canonical set the fixed thermodynamical parameter is the temperature, it is convenient to start by recalling (Remark 14.3) that the equilibrium solution of the Boltzmann equation has the form

$$f(\mathbf{p},\mathbf{q}) = c e^{-\beta h(\mathbf{p},\mathbf{q})}, \quad \beta = \frac{1}{kT}, \qquad (15.59)$$

where $h(\mathbf{p},\mathbf{q}) = \mathbf{p}^2/2m + \Phi(\mathbf{q})$, and $\Phi$ is the potential energy of the external forces. Recall also that the internal forces influence the dynamics of the system in the space $\mu$ (through the interaction kernel), while the structure of the equilibrium depends only on the fact that the interactions are binary and preserve the energy (note that they can be modelled as singularities of the Hamiltonian and therefore they do not contribute to (15.59)).

We then examine again the partition in cells $\omega_1, \ldots, \omega_K$ (of volume $\omega$) of the space $\mu$, introduced in Section 15.6, and the corresponding discretisation of the space $\Gamma$ in cells of volume $\omega^N$. For a fixed cell $\Omega$ in the space $\Gamma$, by projection on the component subspaces $\mu_1, \mu_2, \ldots, \mu_N$ we can reconstruct the corresponding sequence of occupation numbers $(n_1, \ldots, n_K)$ in the cells $\omega_1, \ldots, \omega_K$. We know that this correspondence is not one to one, but now we are only interested in the map $\Omega \to (n_1, \ldots, n_K)$, and to obtain through this map information about the probability of finding a sampling of the representative point $\mathbf{X} \in \Gamma$ precisely in the cell $\Omega$. This probability is the product of the probabilities of finding, in the space $\mu$, $n_1$ points in the cell $\omega_1$, $n_2$ points in the cell $\omega_2$, and so on. Denoting by $h_i$ the values of the Hamiltonian $h(\mathbf{p},\mathbf{q})$ corresponding to the cells $\omega_i$, according to (15.59) such a product is proportional to

$$\exp\left[-\beta \sum_{i=1}^{K} n_i h_i\right] = \exp(-\beta H(\mathbf{X})), \qquad (15.60)$$

where $H = \sum_{i=1}^{K} n_i h_i$.

This result, and the procedure followed to obtain it, offers direction to remark on some particularly significant aspects.

(1) The distribution function $f(\mathbf{p},\mathbf{q})$ in the space $\mu$ can be normalised to a probability distribution by simply dividing it by $N$. Indeed, $\int (f/N) \, d\mathbf{p} \, d\mathbf{q} = 1$. Denoting by $f_i$ the value of $f$ in the cell $\omega_i$ of the partition of the space $\mu$, considering the product $f_1^{n_1} \ldots f_K^{n_K}$, to obtain the probability of finding $\mathbf{X} \in \Gamma$ in the cell $\Omega$ corresponding to the sequence $(n_1, \ldots, n_K)$, we should have divided it by $N^{n_1 + \cdots + n_K} = N^N$. For $N \gg 1$ this is in practice $N!$. This simple consideration shows that the factor $1/N!$ appears naturally in the expression for the density, multiplying the exponential $e^{-\beta H}$, and it confirms that in the formalism of Boltzmann it is implicitly the correct count of states, since taking the factors $f_i$ corresponds to labelling the cells, but not the particles.

(2) The reader will not have missed the rather puzzling fact that in Section 15.6, imposing the constraints of the microcanonical set ($\sum_{i=1}^{K} n_i = N$ $\sum_{i=1}^{K} n_i h_i = E$) we obtained as most probable distribution in the space $\mu$, using $N \gg 1$, precisely the same distribution we used to deduce the density of the canonical set! However, there is a subtle conceptual difference: in the microcanonical setting we must interpret $\beta$ as $\frac{3}{2}(E/N)^{-1}$, while in the canonical case, $\beta = 1/kT$. This does not have practical implications, since once we have correctly defined the temperature in the microcanonical case, the two expressions coincide. The essential coincidence of the equilibrium distributions in the space $\mu$ for the two cases carries significant physical information: for $N \gg 1$ the mean quadratic fluctuation of $H$ in the canonical set is so small that the important region of the space $\Gamma$ for the canonical set is a 'thin' shell (Section 15.3) around the manifold $H = \langle H \rangle_\rho$. This fact, which we prove below, implies that for $N \gg 1$, in spite of the significant formal differences, the two are essentially identifiable and the choice of one or the other formalism is only a question of computational convenience. We stress that this is only true for $N \gg 1$. Indeed, the canonical set has a unique characteristic: it is orthodic (hence it gives exact information about some averages) even when the number of particles is small (even for $N = 1$). However the fluctuations are then large. We shall return to this in an appropriate section.
(3) We finally remark that when there are internal forces, not of collisional type (hence acting at a distance), we must take their total potential energy into account in the expression for $H$.

DEFINITION 15.9 *The measure of the canonical set in the space $\Gamma$ is given by*

$$Z(V, N, T) = \int_\Gamma \frac{1}{N! h^{3N}} \exp(-\beta H(\mathbf{X})) \, d\mathbf{x} \qquad (15.61)$$

*and it is called the* (canonical) *partition function.* ∎

The extension of the integration to the whole space $\Gamma$ assumes the introduction of potential barriers simulating reflective walls and it requires implicitly the convergence of the integral.

It follows from the definition that the canonical partition function expresses the density of states. Normalising the canonical distribution, and considering $(1/Z)\rho(\mathbf{X})$ we obtain a probability distribution. If $f(\mathbf{X})$ is an observable quantity, its average value in the canonical set is given by

$$\langle f \rangle = \frac{1}{Z N! h^{3N}} \int_\Gamma f(\mathbf{X}) \exp(-\beta H(\mathbf{X})) \, d\mathbf{X}. \qquad (15.62)$$

## 15.9    Statistical mechanics: Gibbs sets

PROPOSITION 15.2   *The average value $\langle H \rangle$ of the energy of the system is equal to*

$$\langle H \rangle = -\frac{\partial}{\partial \beta} \log Z(V, N, T). \tag{15.63}$$

*Proof*
Since $Z = C \int_\Gamma e^{-\beta H} \, d\mathbf{X}$, where $C$ is independent of $\beta$, we have

$$\frac{\partial}{\partial \beta} \log Z = -\frac{\int_\Gamma H e^{-\beta H} \, d\mathbf{X}}{\int_\Gamma e^{-\beta H} \, d\mathbf{X}} = -\langle H \rangle. \qquad \blacksquare$$

*Example 15.3*
Consider the Hamiltonian of a perfect monatomic gas

$$H = \sum_{i=1}^{N} \frac{\mathbf{p}_i^2}{2m}.$$

The partition function of the system takes the value

$$Z(V, N, T) = \frac{V^N}{N! h^{3N}} \chi_{3N} \int_0^\infty p^{3N-1} e^{-\beta p^2/2m} \, dp,$$

where $\chi_{3N}$ denotes the measure of the spherical surface of radius 1 in $\mathbf{R}^{3N}$. Recalling the expression of Euler's gamma function (Appendix 8) we have

$$\int_0^\infty p^{3N-1} e^{-\beta p^2/2m} \, dp = \frac{1}{2}(2mkT)^{3N/2} \int_0^\infty t^{3N-2/2} e^{-t} \, dt = \frac{1}{2}(2mkT)^{3N/2} \Gamma(3N/2),$$

and hence

$$Z(V, N, T) = \frac{V^N}{N! h^{3N}} \chi_{3N} \frac{1}{2}(2mkT)^{3N/2} \Gamma\left(\frac{3N}{2}\right) = \frac{V^N}{N! h^{3N}} (2\pi mkT)^{3N/2}, \tag{15.64}$$

where we have used the fact that $\chi_{3N} = 2\pi^{3N/2}/\Gamma(3N/2)$. Therefore the average value of the energy is

$$\langle H \rangle = \frac{3}{2} NkT,$$

in agreement with our previous results.

Note also that this is a typical case where $\langle H \rangle$ can be interpreted as 'by far most probable value' of the energy, following the discussion of Section 15.3. Indeed, if for $N \gg 1$ we compute

$$\xi(\langle H \rangle) = \frac{1}{Z} \int_{H=\langle H \rangle} \frac{\rho}{|\nabla_\mathbf{X} H|} \, d\Sigma \approx \frac{\beta}{e} (3\pi N)^{-1/2},$$

we realise that equation (15.5) is satisfied with $\delta = kTe(3\pi N)^{1/2}$, and hence with $\delta/\langle H \rangle = \mathcal{O}(N^{-1/2})$. If we then compute (15.16), i.e. $F(E) = EZ\xi(E)$, we see that the function to be maximised in order to find the most probable value $\overline{E}$ is simply $E^{3N/2} e^{-\beta E}$, from which it follows that $\overline{E} = \frac{3}{2} NkT = \langle H \rangle$. ∎

## 15.10 Equipartition of the energy (prescribed temperature)

The equipartition theorem, proved for the microcanonical set, can be extended to the canonical set. We assume $H(\mathbf{X}) \to +\infty$ for $|\mathbf{X}| \to +\infty$.

THEOREM 15.10 (Equipartition) *The average of the product $X_i \, \partial H/\partial X_j$ in the canonical set is $\delta_{ij} kT$, as long as $\partial H/\partial X_j \neq 0$.*

*Proof*
It is enough to notice that

$$\int_\Gamma X_i \frac{\partial H}{\partial X_j} e^{-\beta H} \, d\mathbf{X} = -\frac{1}{\beta} \int_\Gamma X_i \frac{\partial}{\partial X_j} e^{-\beta H} \, d\mathbf{X}.$$

By integration by parts, since there are no boundary terms, we find

$$\langle X_i \frac{\partial H}{\partial X_j} \rangle = \left( \int_\Gamma e^{-\beta H} \, d\mathbf{X} \right)^{-1} \int_\Gamma X_i \frac{\partial H}{\partial X_j} e^{-\beta H} \, d\mathbf{X} = \delta_{ij} kT.$$ ∎

As a corollary we then obtain that in the canonical case as well, the quadratic Hamiltonians such as (15.56) have average $\frac{1}{2} kT$ times the number of non-zero coefficients of the form.

We have hinted already at the difficulties of the equipartition principle in the microcanonical set for more complex systems than those considered so far. Such difficulties also arise for the canonical set. However it is surprising that for the latter the equipartition principle is still valid (with some restrictions) for systems with few particles (even just one). These are systems with a 'simple' Hamiltonian, but that do not follow the associated Hamiltonian flow, but rather a motion subject to fluctuations in the energy at a prescribed temperature, which also determines the average energy.

## 15.10 Statistical mechanics: Gibbs sets

A coupling with a thermostat is therefore implicit, and with it the system, although it has few degrees of freedom, is in thermodynamical equilibrium. Speaking of a system with few particles 'at temperature $T$' we refer to the temperature of the thermostat. This concept is illustrated in the following example.

*Example* 15.4
Consider a system of $N \gg 1$ particles at temperature $T$ and the subset constituted by only one of the particles, in statistical equilibrium with the rest of the system. Consider the simplest possible case, when the Hamiltonian of the single particle is $H = \mathbf{p}^2/2m$ and let us try to apply the canonical formalism to this single particle. Define the density

$$\rho = \frac{1}{h^3} e^{-\beta \mathbf{p}^2/2m}$$

(note that we have used $N = 1$) and, assuming that the system is contained in a cube of side $L$, compute the partition function

$$Z = \frac{L^3}{h^3} \int_{\mathbf{R}^3} e^{-\beta \mathbf{p}^2/2m} \, d\mathbf{p} = \frac{L^3}{h^3} \int_0^\infty 4\pi p^2 e^{-\beta p^2/2m} \, dp = \frac{L^3}{h^3} \left(\frac{2\pi m}{\beta}\right)^{3/2}.$$

Formula (15.63) yields

$$\langle H \rangle = -\frac{\partial}{\partial \beta} \log Z = \frac{\partial}{\partial \beta}\left(\frac{3}{2} \log \beta\right) = \frac{3}{2} kT,$$

which is precisely the expected result.

What is essentially different from the case of systems with many particles is the fact that the mean quadratic fluctuation is not small: we find $\langle p^4/4m^2 \rangle = \frac{15}{4}(kT)^2$, so that $(\langle H^2 \rangle - \langle H \rangle^2)/\langle H \rangle^2 = \frac{2}{3}$.

Check that for a point in the plane inside a square of side $L$ the function $Z$ becomes $Z = L^2/h^2 (2\pi m/\beta)$ (hence $\langle H \rangle = kT$, $(\langle H^2 \rangle - \langle H \rangle^2)/\langle H \rangle^2 = 1$), while for a point on an interval of length $L$, $Z = L/h \, (2\pi m/\beta)^{1/2}$ (with $\langle H \rangle = \frac{1}{2} kT$ and $(\langle H^2 \rangle - \langle H \rangle^2)/\langle H \rangle^2 = 2$). ∎

*Example* 15.5
Consider the Hamiltonian of a harmonic oscillator with one degree of freedom:

$$H(q, p) = \frac{p^2}{2m} + \frac{1}{2} m\omega^2 q^2,$$

and compute the canonical partition function $Z$ and the average energy $\langle E \rangle$. Recalling the definition (15.61) of $Z$ and Appendix 8, we have

$$Z = \frac{1}{h} \int_{-\infty}^{+\infty} dq \int_{-\infty}^{+\infty} dp \, e^{-\beta H(q,p)} = \frac{2\pi}{h\omega\beta}.$$

We arrive at the same result by considering the Hamiltonian expressed in terms of the angle-action variables:

$$H(\chi, J) = \omega J.$$

In this case, recalling that $(\chi, J) \in \mathbf{T}^1 \times \mathbf{R}_+$, we have

$$Z = \frac{1}{h} \int_0^{2\pi} d\chi \int_0^{+\infty} e^{-\beta \omega J} \, dJ = \frac{2\pi}{h \omega \beta}.$$

The average value of the energy is equal to

$$\langle H \rangle = -\frac{d}{d\beta} \log Z = \frac{d}{d\beta} \log \beta = \frac{1}{\beta},$$

in agreement with what was predicted by the theorem of equipartition of the energy. It is also immediate to verify that

$$\langle q^2 \rangle = \frac{\int q^2 e^{-\beta H(q,p)} \, dp \, dq}{\int e^{-\beta H(q,p)} \, dp \, dq} = \frac{1}{m \omega^2 \beta}, \quad \langle p^2 \rangle = \frac{\int p^2 e^{-\beta H(q,p)} \, dp \, dq}{\int e^{-\beta H(q,p)} \, dp \, dq} = \frac{m}{\beta},$$

from which we arrive at the equipartition $\langle \frac{1}{2} m \omega^2 q^2 \rangle = \frac{1}{2} kT$, $\langle p^2/2m \rangle = \frac{1}{2} kT$.

It is interesting to compare these averages with the averages over one period of the deterministic motion. It is convenient to start from the transformation

$$p = \sqrt{2\omega m J} \cos \chi, \quad q = \sqrt{\frac{2J}{\omega m}} \sin \chi,$$

imposing for the energy the value $\omega J = kT$, and hence $J = kT/\omega$. Taking the average over a period of $\chi$ for these quantities, we find exactly the same results.

We now compute $\langle H^2 \rangle = (h \omega \beta / 2\pi)(1/h) \int_0^{2\pi} d\chi \int_0^{+\infty} \omega^2 J^2 e^{-\beta \omega J} \, dJ = 2(kT)^2$. Note that $(\langle H^2 \rangle - \langle H \rangle^2)/\langle H \rangle^2 = 1$.

It is elementary to observe that the latter result is necessarily different from the deterministic motion, for which the energy is constant. Hence the average over a period of any power of the energy coincides with the same power of the average (in other words $\overline{(H^2)} = (\overline{H})^2 = \omega^2 J^2$ and there is no fluctuation).

As an exercise, compare the average $\langle p^4/4m^2 \rangle$ in the statistical motion with the average $\overline{(p^4/4m^2)}$ over a period of the deterministic motion ($\frac{3}{4} k^2 T^2$ and $\frac{3}{8} k^2 T^2$, respectively). ∎

*Example 15.6*
Consider the Hamiltonian of a harmonic oscillator with two degrees of freedom:

$$H(q_1, q_2, p_1, p_2) = \frac{p_1^2 + p_2^2}{2m} + \frac{1}{2} m\omega^2 \left(q_1^2 + q_2^2\right),$$

and compute the canonical partition function $Z$ and the average energy $\langle E \rangle$.
Introducing planar polar coordinates, the Hamiltonian becomes

$$H(\varphi, r, p_\varphi, p_r) = \frac{1}{2m}\left(p_r^2 + \frac{p_\varphi^2}{r^2}\right) + \frac{1}{2} m\omega^2 r^2,$$

from which it follows that

$$Z = \frac{1}{h^2} \int_0^{2\pi} d\varphi \int_0^{+\infty} dr \int_{-\infty}^{+\infty} dp_\varphi \int_{-\infty}^{+\infty} e^{-\beta H(\varphi, r, p_\varphi, p_r)} dp_r$$

$$= \frac{2\pi}{h} \int_0^{+\infty} \frac{2\pi m}{\beta} r e^{-m\omega^2 \beta r^2 / 2} dr$$

$$= \frac{4\pi^2}{h^2 \omega^2 \beta^2} \int_0^{+\infty} e^{-m\omega^2 \beta r^2 / 2} d\left(\frac{m\omega^2 \beta r^2}{2}\right) = \frac{4\pi^2}{h^2 \omega^2 \beta^2},$$

and therefore

$$\langle H \rangle = -\frac{d}{d\beta} \log Z = \frac{d}{d\beta} 2 \log \beta = \frac{2}{\beta}.$$

We would arrive at the same result considering the Hamiltonian expressed in terms of the action-angle variables:

$$H(\chi_1, \chi_2, J_1, J_2) = \omega(J_1 + J_2),$$

by simply recalling that $(\chi_1, \chi_2, J_1, J_2) \in \mathbf{T}^2 \times \mathbf{R}_+^2$. ∎

*Example 15.7*
Consider the Hamiltonian of a point particle of mass $m$ constrained to move on the surface of a sphere of radius $R$ under the action of weight: in spherical coordinates we have

$$H(\vartheta, \varphi, p_\vartheta, p_\varphi) = \frac{1}{2mR^2}\left(p_\vartheta^2 + \frac{p_\varphi^2}{\sin^2 \vartheta}\right) + mgR \cos \vartheta,$$

where $0 \leq \vartheta \leq \pi$ and $0 \leq \varphi \leq 2\pi$. The canonical partition function is given by

$$Z = \frac{1}{h^2} \int_0^{2\pi} d\varphi \int_0^\pi d\vartheta \int_{-\infty}^{+\infty} dp_\varphi \int_{-\infty}^{+\infty} e^{-\beta H(\vartheta, \varphi, p_\vartheta, p_\varphi)} dp_\vartheta$$

$$= \frac{4\pi^2 m R^2}{h^2 \beta} \int_0^\pi e^{-mgR\beta \cos \vartheta} \sin \vartheta \, d\vartheta = \frac{8\pi^2 R}{gh^2} \frac{1}{\beta^2} \sinh(\beta mgR).$$

The average energy is therefore given by

$$\langle E \rangle = -\frac{\mathrm{d}}{\mathrm{d}\beta} \log Z = \frac{2}{\beta} - mgR \cotanh(\beta mgR),$$

from which it follows that

$$\langle E \rangle \approx \begin{cases} -mgR, & \text{for } \beta \to +\infty, \\ \dfrac{1}{\beta}, & \text{for } \beta \to 0. \end{cases}$$

Hence for low temperatures, the energy stabilises in correspondence with stable equilibrium, while at high temperatures the system tends to forget the gravitational potential energy and sees only the quadratic part of the Hamiltonian. ∎

*Example* 15.8
Consider a system of $N$ rigid segments of length $a$ and mass $m$, constrained to move on the semi-axis $x > 0$. Denoting by $(x_i, x_i + a)$ the interval occupied by the $i$th segment, the point $x_1$ is attracted with an elastic force of constant $\lambda$ by the point $x = 0$, while the adjacent endpoints $(x_{i-1} + a, x_i)$ of two consecutive segments are also subject to the same force. Defining $\xi_1 = x_1$ and $\xi_i = x_i - x_{i-1} - a$, $i = 2, \ldots, N$ and $\xi_i \geq 0$, we have the following expression for the Hamiltonian:

$$H(\xi_1, \ldots, \xi_N, p_1, \ldots, p_N) = \sum_{i=1}^{N} \left[ \frac{p_i^2}{2m} + \frac{\lambda}{2} \xi_i^2 \right].$$

If the system is kept at temperature $T$, its partition function has the value

$$Z = \frac{1}{N! h^N} \left( \int_{-\infty}^{+\infty} e^{-\beta p^2/2m} \, dp \int_{0}^{+\infty} e^{-\beta \lambda \xi^2/2} \, d\xi \right)^N$$

$$= \frac{1}{N! h^N} \left( \frac{2m\pi}{\beta} \right)^{N/2} \left( \frac{\pi}{2\beta\lambda} \right)^{N/2} = \frac{\pi^N}{N! h^N} \left( \frac{m}{\lambda} \right)^{N/2} \beta^{-N},$$

from which we deduce the average energy $\langle H \rangle = NkT$, in agreement with the equipartition theorem. We now want to compute the *average length* of the system, given by

$$\langle L \rangle = \langle x_N + a - \xi_1 \rangle = \left\langle Na + \sum_{i=2}^{N} \xi_i \right\rangle.$$

The averages $\xi_i$ are all equal to

$$\langle \xi \rangle = \int_{0}^{\infty} \xi e^{-\beta \lambda \xi^2/2} \, d\xi \bigg/ \left( \int_{0}^{\infty} e^{-\beta \lambda \xi^2/2} \, d\xi \right) = \sqrt{\frac{2}{\pi \beta \lambda}}.$$

Hence $\langle L \rangle = Na+(N-1)\sqrt{2/\pi\beta\lambda}$. The average $x$-coordinate of the last endpoint is $\langle x_N + a \rangle = Na + N\sqrt{2/\pi\beta\lambda}$. The result is consistent with physical intuition: when the intensity of the elastic interactions grows, $\langle L \rangle$ decreases; when the temperature rises, $\langle L \rangle$ increases. In addition, we can compute the *coefficient of thermal dilation* $\delta$ *of the system*, which we can define as the derivative with respect to $T$ of the average elongation $\langle : x_N + a - Na \rangle$, with respect to the extension $\langle x_N + a \rangle$:

$$\delta = \frac{1}{\langle x_N + a \rangle} \frac{\partial}{\partial T} \langle x_N + a - Na \rangle = \frac{k\beta}{a\sqrt{2\pi\beta\lambda} + 2}. \qquad \blacksquare$$

## 15.11 Helmholtz free energy and orthodicity of the canonical set

In the microcanonical set, given the parameters $E, V$ it is natural to choose the entropy $S(E,V)$ as the fundamental thermodynamical quantity. In the canonical set, the independent variables are rather $T, V$ and the connection with thermodynamics is established through the *Helmholtz free energy* $\Psi(V,T)$, defined by

$$T \frac{\partial \Psi}{\partial T} - \Psi = -U, \qquad (15.65)$$

where $U$ is the internal energy (hence $\langle H \rangle$). This expression highlights the fact that $-U$ is the Legendre transform of $\Psi$ (hence that $\Psi$ is the Legendre transform of $-U$). A more familiar expression is

$$U = \Psi + TS, \qquad (15.66)$$

from which, differentiating with respect to $T$, we find $S = -\partial \Psi / \partial T$; indeed we can write

$$\frac{\partial S}{\partial T} = \frac{\partial S}{\partial E} \frac{\partial E}{\partial T} = \frac{1}{T} \frac{\partial U}{\partial T}.$$

If $\langle H \rangle$ is independent of $V$, as for the equipartition case, by differentiating with respect to $V$ we deduce $P = -\partial \Psi / \partial V$ (it is enough to observe that $T\partial S/\partial V = P$).

We show now that a correct definition of the function $\Psi(V,T)$ starting from the canonical partition function is the following:

$$Z(V,T) = \exp[-\beta \Psi(V,T)]. \qquad (15.67)$$

Note that by altering the numerical factors in the definition of $Z$ the free energy is modified by the addition of a term linear in the temperature (which does not contribute to the expression $\Psi - T(\partial \Psi / \partial T)$ for the internal energy).

PROPOSITION 15.3 *The free energy* $\Psi(V,N,T) = -kT \log Z(V,N,T)$ *satisfies equation (15.65), and hence it coincides with the Helmholtz free energy.*

## Proof

Using the expression (15.67) in the computation of $\langle H \rangle$ through equation (15.63), we obtain

$$\langle H \rangle = \Psi - T \frac{\partial \Psi}{\partial T}. \tag{15.68}$$

Equation (15.68) reproduces the relation between the free energy and the internal energy (identified with the average value of the Hamiltonian). ∎

### Example 15.9

Recalling the expression (15.64) for the partition function $Z(V, T)$ for a monatomic perfect gas, we easily find

$$\Psi(V, T) = -NkT \log \left\{ \frac{V}{N} (2\pi mkT)^{3/2} h^{-3} \right\}. \tag{15.69}$$

From this it is immediate to check that $\partial \Psi / \partial T = -S$, $\partial \Psi / \partial V = -P$. Note also that equation (15.68) gives $\langle H \rangle = \frac{3}{2} NkT$. ∎

It is important to remark that the derivation of these results does not require the hypothesis $N \gg 1$ (explicitly used in the microcanonical set), i.e. *there is no need to invoke the thermodynamical limit* for their justification, and the canonical set is naturally orthodic.

## 15.12 Canonical set and energy fluctuations

In the discussion introducing the structure of the canonical set, we anticipated that for $N \gg 1$ this clusters around the microcanonical set, and hence the manifold $H = \langle H \rangle$. We can now verify this fact, studying the energy fluctuations.

**THEOREM 15.11** *In the limit $N \to \infty$, the mean quadratic energy fluctuation in the canonical set*

$$\frac{\langle H^2 \rangle - \langle H \rangle^2}{\langle H^2 \rangle} \tag{15.70}$$

*tends to zero. More precisely:*

$$\frac{\langle H^2 \rangle - \langle H \rangle^2}{\langle H^2 \rangle} \approx \frac{1}{N}, \quad for \quad N \to \infty. \tag{15.71}$$

## Proof

We start from the known relation $\langle H \rangle = -(\partial/\partial \beta) \log Z$. Differentiating with respect to $\beta$, we obtain

$$\frac{\partial \langle H \rangle}{\partial \beta} = \frac{1}{Z^2} \left( \frac{\partial Z}{\partial \beta} \right)^2 - \frac{1}{Z} \frac{\partial^2 Z}{\partial \beta^2}.$$

Since the first term on the right-hand side is just $\langle H \rangle^2$, while $\partial^2 Z/\partial \beta^2 = Z \langle H^2 \rangle$, we conclude that

$$\langle H^2 \rangle - \langle H \rangle^2 = -\frac{\partial \langle H \rangle}{\partial \beta}.$$

Note now that $-(\partial \langle H \rangle/\partial \beta) = kT^2 (\partial \langle H \rangle/\partial T) = kT^2 C_V$ is, like $\langle H \rangle$, proportional to $N$. Hence (15.71) is proved. ∎

*Example* 15.10
The explicit computation for a perfect monatomic gas gives the result

$$\frac{\langle H^2 \rangle - \langle H \rangle^2}{\langle H^2 \rangle} = \frac{2}{3N},$$

independent of the temperature. ∎

We can therefore confirm that the states of the canonical set are concentrated in a 'thin' region around the surface $H = \langle H \rangle$, and therefore the canonical set is not different from the microcanonical set in the limit $N \to \infty$, and the two formalisms give rise to thermodynamical descriptions which are essentially equivalent for sufficiently large values of $N$. Thanks to the results of the previous section we again find that the microcanonical set is orthodic in the limit $N \to \infty$.

## 15.13 Open systems with fixed temperature. Grand canonical set

Our discussion of statistical mechanics, and of Gibbs' statistical sets, has so far led us to consider two distributions, the microcanonical and the canonical. In the former, we consider closed and isolated systems, with a fixed number of particles $N$ in a region of fixed volume $V$ and with total energy $E$ fixed as well. In the latter, we studied closed but not isolated systems, and in contact with a thermostat at constant temperature $T$; hence with a fixed number of particles $N$ and volume $V$ but with variable energy $E$.

In many cases—for example when a chemical reaction is taking place—one needs to consider *open* systems in which the temperature $T$ and the volume $V$ are fixed, but the energy $E$ and the number of particles $N$ are variable. To model such a situation we can simply imagine eliminating the separation between the system and the thermostat, letting the two systems exchange their particles. We then have a fluctuation of the number of particles in the two compartments. The statistical set corresponding to this situation is the *grand canonical set*. In what follows we discuss briefly some properties of its density and its orthodicity in the thermodynamical limit.

Assume that $N_1, N_2$ are the number of particles in the system under consideration and in the thermostat, respectively, and that the sum $N_1 + N_2 = N$ is constant. Denote by $V_1, V_2$ ($V_1 + V_2 = V$ and $V_1 \ll V_2$) the respective volumes, and by $T$ the temperature. Naturally the two systems must have identical particles and we must assume that the respective Hamiltonians are subject to fluctuations also because of the exchange of particles. The space $\Gamma_1$ has the variable dimension $6N_1$, but for fixed $N_1$ we can consider the canonical partition function $Z_1(V_1, N_1, T)$ of the first system and simultaneously the one $Z_2(V_2, N_2, T)$ of the thermostat. In addition we can consider that the union of the two systems is in turn kept at temperature $T$ and therefore also consider the total partition function $Z(V, N, T)$. If $Z_1, Z_2, Z$ count the number of states in the respective systems, then the following relation must necessarily hold:

$$Z(V, N, T) = \sum_{N_1=0}^{N} Z_1(V_1, N_1, T) Z_2(V_2, N_2, T). \tag{15.72}$$

This suggests that we assume as density of the grand canonical set associated with system, the usual one of a canonical set with $N_1$ particles, corrected by the factor $Z_2/Z$, i.e.

$$\rho_G(\mathbf{X}_1, N_1, T) = \frac{Z_2}{Z} \frac{1}{N_1! h^{3N_1}} \exp[-\beta H(\mathbf{X}_1, N_1)], \tag{15.73}$$

and hence with integral

$$\int_{\Gamma_1} \rho_G(\mathbf{X}_1, N_1, T) \, d\mathbf{X}_1 = \frac{Z_1 Z_2}{Z}, \tag{15.74}$$

in such a way that $\rho_G$ is normalised to 1 when summing over all possible states of system 1:

$$\sum_{N_1=0}^{N} \int_{\Gamma_1} \rho_G(\mathbf{X}_1, N_1, T) \, d\mathbf{X}_1 = 1.$$

We now seek an expression for the correction factor $Z_2/Z$ that is valid in the situation in which not only $V_1 \ll V_2$, but also $N_1 \ll N_2$, as it is natural to expect. In this case we can consider $Z_2$ as a perturbation of $Z$. It is convenient to refer to the Helmholtz free energy and write

$$\frac{Z_2(V_2, N_2, T)}{Z(V, N, T)} = \exp[-\beta \Psi(V - V_1, N - N_1, T) + \beta \Psi(V, N, T)]. \tag{15.75}$$

Using the expansion to first order we obtain

$$\Psi(V - V_1, N - N_1, T) - \Psi(V, N, T) = -\frac{\partial \Psi}{\partial V} V_1 - \frac{\partial \Psi}{\partial N} N_1. \tag{15.76}$$

We already know that $-\partial \Psi / \partial V = P$. We now introduce two new quantities: the *chemical potential*:

$$\mu = \frac{\partial \Psi}{\partial N}, \tag{15.77}$$

and the *fugacity*:

$$z = e^{\beta\mu}, \tag{15.78}$$

through which we finally arrive at the expression

$$\frac{Z_2(V_2, N_2, T)}{Z(V, N, T)} = e^{-\beta P V_1} z^{N_1}. \tag{15.79}$$

The condition $N_1 \ll N_2$ can be verified *on average* a posteriori. Note that the pressure $P$ is an intensive quantity, defined in the global set, but also in each of its parts, and it is therefore admissible to interpret it as the pressure of the system with $N_1$ particles. The same can be argued of the chemical potential. Hence (15.79) is expressed only through variables referring to the latter system.

These considerations lead to defining the grand canonical set in the following way, dropping the index 1.

DEFINITION 15.10 *We call the density of the grand canonical set the function*

$$\rho_G(\mathbf{X}, N) = \frac{z^N}{h^{3N} N!} \exp[-\beta P V - \beta H(\mathbf{X}, N)], \tag{15.80}$$

*defined for every $N$ on the space $\Gamma(N)$ with $6N$ dimensions.* ∎

By integrating over the space $\Gamma(N)$ we obtain

$$\int_{\Gamma(N)} \rho_G(\mathbf{X}, N) \, d\mathbf{X} = z^N e^{-\beta P V} Z(V, N, T), \tag{15.81}$$

which is a reformulation of (15.74) taking into account (15.79), expressing the count of the states with $N$ particles. The probability of finding the system in any microscopic state with $N$ particles is found by dividing the right-hand side of (15.81) by the sum over $N$ of the same expression. We allow for $N$ to tend to infinity. We therefore conclude that the probability that the number of particles of the system is $N$ is given by

$$p(N) = \frac{z^N Z(V, N, T)}{Z_G(V, T, z)}, \tag{15.82}$$

where

$$Z_G(V, T, z) = \sum_{N=0}^{\infty} z^N Z(V, N, T) \tag{15.83}$$

is the *grand canonical partition function*.

We already know that, due to the normalisation of the function $\rho_G$, summing over $N$ expression (15.81) yields 1. Using the definition (15.83) we can deduce the equation

$$\beta P V = \log Z_G(V, T, z). \tag{15.84}$$

The average number of particles is given by

$$\langle N \rangle = \sum_{N=0}^{\infty} N p(N) = z \frac{\partial}{\partial z} \log Z_G(V,T,z). \tag{15.85}$$

Eliminating the variable $z$ between (15.84) and (15.85) we obtain the state equation.

*Example 15.11*
We compute the grand canonical partition function for a perfect monatomic gas. From (15.83), using the results of Example 15.3, we find immediately that

$$Z_G(V,T,z) = \exp[zV(2\pi mkT)^{3/2} h^{-3}], \tag{15.86}$$

from which it follows that for a perfect gas we have

$$z \frac{\partial}{\partial z} \log Z_G(V,T,z) = \langle N \rangle = \log Z_G(V,T,z), \tag{15.87}$$

$$\log Z_G(V,T,z) = zV(2\pi mkT)^{3/2} h^{-3} = \beta PV, \tag{15.88}$$

yielding the state equation

$$PV = \langle N \rangle kT. \tag{15.89}$$

In the expressions (15.86) and (15.88) there appears an arbitrary constant $h$, introduced by the function $Z$. It is clear that quantities such as $p(N)$, $\beta PV$, $\langle N \rangle$ given by (15.82), (15.84), (15.85) cannot depend on $h$. It is convenient to clarify this point using the explicit example of the perfect gas. First of all, note that the computation of $\mu$ and $z$ (see (15.77), (15.78)) is performed using the free energy $\Psi$ *of the thermostat*. The way we used the thermostat (see (15.83)) was by letting $N$ and $V$ tend to infinity, and implicitly choosing the limit value $V/N = v$ (hence the density of particles of the thermostat). Hence for the thermostat we must rewrite the expression (15.69) for the free energy in the form

$$\Psi(V,N,T,) = -NkT \log \left\{ v(2\pi mkT)^{3/2} h^{-3} \right\}. \tag{15.90}$$

As a consequence, we find

$$\beta \mu = -\log \left\{ v(2\pi mkT)^{3/2} h^{-3} \right\},$$
$$z = \left\{ v(2\pi mkT)^{3/2} h^{-3} \right\}^{-1}. \tag{15.91}$$

Note that $z$ is proportional to $h^3$, and hence the expression (15.86) is reduced to

$$Z_G = \exp\left(\frac{V}{v}\right) \tag{15.92}$$

and in addition

$$z^N Z(V,N,T,) = \frac{1}{N!}\left(\frac{V}{v}\right)^N, \qquad (15.93)$$

reproducing (15.92) when we sum over $N$. For the average value $\langle N\rangle$ we find precisely the value intuition would suggest, i.e.

$$\langle N\rangle = \frac{V}{v}. \qquad (15.94)$$

Introducing this value into the expression (15.89) we find $Pv = kT$, in agreement with the interpretation of $v$. Finally, note that if we substitute the values found for $z$ and $Z_G$ into (15.82), we find

$$p(N) = \frac{1}{N!}\frac{\langle N\rangle^N}{e^{\langle N\rangle}}. \qquad (15.95)$$

It is easy to verify that the average of a quantity $f(\mathbf{X},N)$ in the grand canonical set can be obtained from the formula

$$\langle f\rangle_G = \frac{\sum_{N=0}^{\infty} z^N \langle f\rangle_N Z(V,N,T)}{Z_G(V,T,z)}, \qquad (15.96)$$

where $\langle f\rangle_N$ denotes the average of $f$ in the canonical set with $N$ particles.

Applying this formula to the Hamiltonian of a perfect monatomic gas, we note that the numerator of (15.96) is (recall that $\langle H\rangle_N = \frac{3}{2}NkT$)

$$\frac{3}{2}kT \sum_{N=1}^{\infty} \frac{1}{(N-1)!}\left[\frac{zV}{h^3}(2\pi mkT)^{3/2}\right]^N,$$

and therefore

$$\langle H\rangle_G = \frac{3}{2}kT\frac{zV}{h^3}(2\pi mkT)^{3/2} = \frac{3}{2}\langle N\rangle_G kT, \qquad (15.97)$$

in agreement with physical intuition. ∎

## 15.14 Thermodynamical limit. Fluctuations in the grand canonical set

In the previous section, we used repeatedly the fact that the number $N$ of particles in the system in equilibrium with a thermostat is close to its average. This corresponds to interchanging $\langle N\rangle$ with its most probable value $\overline{N}$, characterised by the dominant term in the series expansion (15.83), in agreement with the definition (15.82) of $p(N)$.

DEFINITION 15.11 *A system admits the thermodynamical limit if, for fixed density* $\bar{n} = \overline{N}/V$ *the limit*

$$\lim_{V \to \infty} \frac{1}{\bar{n}V} \log Z(V, \bar{n}V, T) \stackrel{\text{def}}{=} -\beta\psi(\bar{n}, T) \qquad (15.98)$$

*exists.* ∎

*Remark 15.14*
The problem of finding sufficient conditions for the microscopic interactions (hence for the Hamiltonian) guaranteeing that the thermodynamical limit exists is one of the fundamental problems of statistical mechanics. For a discussion, we refer to Ruelle (1969) and Thompson (1972, 1988). Every time that the system admits the thermodynamical limit, we say that the system has the *property of being extensive*, since the thermodynamical quantities such as the entropy, the specific heat, etc. are asymptotically proportional to the size of the system. ∎

If we approximate $Z_G(V, T, z)$ by $z^{\overline{N}} Z(V, \overline{N}, T)$, assuming that this is the dominant term in the series expansion (15.83), from (15.98) it follows that

$$\lim_{V \to \infty} \frac{1}{V} \log Z_G(V, T, z) = \bar{n}(\log z - \beta\psi(\bar{n}, T)). \qquad (15.99)$$

The right-hand side defines the so-called *grand canonical potential* $\chi(T, z)$. Hence

$$\beta\psi(\bar{n}, T) = \log z - \frac{1}{\bar{n}} \chi(T, z). \qquad (15.100)$$

This conclusion should not come as a surprise, as by choosing only one term in the expansion of $Z_G$ we have identified the latter (up to a factor $z^{\overline{N}}$) with a canonical distribution (note that, with respect to (15.67), $\psi$ replaces $1/\overline{N}\, \Psi$ and $1/\bar{n}\, \chi$ replaces $1/\overline{N} \log Z$; the presence of $\log z$ in (15.100) is due to the different normalisation of $Z_G$).

Differentiating (15.100) with respect to $z$, we find the formula corresponding to the thermodynamical limit in (15.85):

$$\bar{n} = z \frac{\partial \chi}{\partial z}. \qquad (15.101)$$

We must finally prove that indeed the fluctuation of $N$ around $\langle N \rangle$ is small. We apply twice the operator $z\, \partial/\partial z$ to the function $\log Z_G$. We then easily see that

$$z \frac{\partial}{\partial z} z \frac{\partial}{\partial z} \log Z_G(V, T, z) = \langle N^2 \rangle - \langle N \rangle^2, \qquad (15.102)$$

by using the expression (15.82) for the probability $p(N)$. We note that

$$z \frac{\partial}{\partial z} = \frac{1}{\beta} \frac{\partial}{\partial \mu} \qquad (15.103)$$

and we can use (15.82) to rewrite (15.102) in the form

$$\langle N^2 \rangle - \langle N \rangle^2 = kTV \frac{\partial^2 P}{\partial \mu^2}. \tag{15.104}$$

Recalling the identities $\mu = \partial \Psi/\partial N$, $P = -\partial \Psi/\partial V$, setting $v = V/\overline{N} = 1/\overline{n}$ and, as is admissible under our assumptions, $\Psi = \overline{N}\psi(v,T)$, we obtain ($\partial/\partial N = \partial/\partial \overline{N}$, $\overline{N}\partial/\partial \overline{N} = -v\partial/\partial v$)

$$\mu = \psi + vP, \qquad P = -\frac{\partial \psi}{\partial v}. \tag{15.105}$$

From this, differentiating with respect to $v$ and $\mu$, it is easy to derive

$$\frac{\partial \mu}{\partial v} = -v\frac{\partial^2 \psi}{\partial v^2} = v\frac{\partial P}{\partial v}, \qquad \frac{\partial P}{\partial \mu} = \frac{1}{v}, \tag{15.106}$$

and

$$\frac{\partial^2 P}{\partial \mu^2} = \frac{\partial}{\partial v}\left(\frac{1}{v}\right)\frac{\partial v}{\partial \mu} = -\frac{1}{v^3}\left(\frac{\partial P}{\partial v}\right)^{-1}.$$

Define the factor of *isothermal compressibility* by

$$K_T = -\frac{1}{v}\left(\frac{\partial P}{\partial v}\right)^{-1}. \tag{15.107}$$

Then substituting this into (15.104), we find

$$\langle N^2 \rangle - \langle N \rangle^2 = \frac{\overline{N}kTK_T}{v}, \tag{15.108}$$

proving that

$$\frac{\langle N^2 \rangle - \langle N \rangle^2}{N^2} = \mathcal{O}\left(\frac{1}{N}\right). \tag{15.109}$$

This conclusion is correct except when $K_T \to \infty$, corresponding to the horizontal segments of the isothermal lines in the plane $(P,v)$ and to the triple point, i.e. to the phase transitions, discussed in the next section.

The smallness of the fluctuation of $N$ therefore confirms the equivalence in the thermodynamical limit of the descriptions given by the grand canonical and the canonical sets, and hence, following from what we have seen, the fact that the grand canonical set is also orthodic. In many practical applications, the latter behaves essentially like the canonical set corresponding to a system with $\langle N \rangle \approx \overline{N}$ particles.

## 15.15 Phase transitions

One of the most interesting problems of statistical mechanics concerns phase transitions. The latter are ubiquitous in the physical world: the boiling of a liquid, the melting of a solid, the spontaneous magnetisation of a magnetic material, up to the more exotic examples in superfluidity, superconductivity, and quantum chromodynamics. In its broadest sense, a phase transition happens any time a physical quantity, such as density or magnetisation, depends in a non-analytic (or non-differentiable, or discontinuous) way on some control parameter, such as temperature or magnetic field. An additional characteristic common to all phenomena of phase transitions is the generation (or destruction) in the macroscopic scale of ordered structures, starting from microscopic short-range interactions. Moreover, in the regions of the space of the parameters corresponding to critical phenomena (hence in a neighbourhood of a critical point), different systems have a similar behaviour even quantitatively. This fact generated the theory of the *universality* of critical behaviour.

Naturally a careful study of the theory of phase transitions and of critical phenomena goes beyond the scope of this book. Indeed, thanks to the impressive developments of the techniques for its solution, it constitutes one of the most significant achievements of modern theoretical physics. However, because of the physical (and mathematical) interest of the theory, and the extent of its applications, going beyond physics to biology and the theory of chaotic dynamical systems, we believe it appropriate to state the fundamental principles of classical statistical mechanics of equilibrium with a short reference to the theory of Lee and Yang (1952a,b) on phase transition, and their relation with the zeros of the grand canonical partition function in the thermodynamical limit.

As we mentioned, from a mathematical point of view, a phase transition is a singular point of the canonical partition function (see Section 15.9). It is however immediate to verify that for finite values of the volume $V$ and of the number $N$ of particles, the partition function depends analytically on the temperature $T$. At least for what concerns its mathematical description, the only way to determine if a phase transition is possible is to consider the thermodynamical limit.

DEFINITION 15.12 *A point in the phase diagram of a system (corresponding to real positive values of $T$, $v$ or $z$) is a phase transition point if at that point the free energy*

$$\psi(v,T) = - \lim_{N,V\to\infty} \frac{kT}{N} \log Z(N,V,T), \quad v = \frac{V}{N} \text{ given}, \qquad (15.110)$$

*or the grand canonical potential*

$$\chi(z,T) = \lim_{V\to\infty} \log Z_G(V,z,T) \qquad (15.111)$$

*are not analytic functions of their arguments.* ∎

*Remark* 15.15
Recall that the non-analyticity expresses the impossibility of representing a function as a Taylor series expansion converging to the function itself. It follows that it is not necessary for any of the derivatives of the function to diverge for a phase transition to be possible, though this is what frequently happens. ∎

The fundamental observation at the foundation of the theory of phase transitions of Lee and Yang is simple. Assume that the interacting particles have a 'hard core', and hence that they are impenetrable and of radius $r_0 > 0$. A volume $V$ can therefore fit at most $\nu(V) \approx V r_0^{-3}$ particles; hence the canonical partition function is

$$Z(V, N, T) = 0, \quad \text{if } N > \nu(V). \tag{15.112}$$

The grand canonical partition function (15.83) is then a polynomial in the fugacity of degree at most $\nu(V)$:

$$Z_G(V, z, T) = \sum_{N=0}^{\nu(V)} z^N Z(V, N, T) \tag{15.113}$$
$$= 1 + z Z(V, 1, T) + z^2 Z(V, 2, T) + \ldots + z^\nu Z(V, \nu, T).$$

Setting as a convention $Z(V, 0, T) = 1$ and denoting by $z_1, \ldots, z_{\nu(V)}$ the (complex) zeros of $Z_G(V, z, T)$, we have

$$Z_G(V, z, T) = \prod_{j=1}^{\nu(V)} \left(1 - \frac{z}{z_j}\right). \tag{15.114}$$

Note that, since all coefficients of the polynomial (15.113) are positive, it is not possible to have real positive zeros, and hence there can be no phase transitions for finite values of the volume $V$ (and of the number of particles $N$). Indeed, the parametric expression of the state equation of the system is (see (15.83) and (15.85))

$$\frac{P}{kT} = \frac{1}{V} \log Z(V, z, T),$$
$$\frac{1}{v} = \frac{1}{V} z \frac{\partial}{\partial z} \log Z(V, z, T). \tag{15.115}$$

For every finite value of $V$, from expressions (15.114) and (15.115) it follows that $P$ and $v$ are analytic functions of the fugacity $z$ in a region of the complex plane including the positive real axis. Therefore $P$ is an analytic function of $v$ for all physical values of $v$, and the thermodynamical functions are without singularities, and there cannot be phase transitions. For a phase transition to occur it is necessary to consider the thermodynamical limit.

Lee and Yang proved that phase transitions are controlled by the distribution of zeros of the grand canonical partition function in the plane $z \in \mathbf{C}$: a phase

## 15.16 Problems

**1.** A cylindrical container of radius $R$ and height $l$ contains a conducting cylinder of radius $r$, height $l$, electric charge $Q$ and axis coinciding with the axis of the container. The container is filled with a gas of $N$ point particles of mass $m$ and electric charge $q$. Assume that $l \gg R$ (so it is possible to neglect the axial component of the electric field) and do not take into account the electrostatic interaction between the particles. Assume also that the system is in thermal equilibrium with a thermostat at temperature $T$. Compute the canonical partition function and the average energy of the system.

**2.** A vertical cylindrical container has a base of area $S$ and height $l$. It contains an ideal gas made of $N$ molecules of mass $m$ and weight $mg$. Assume that the potential energy of a molecule on the lower base of the container is zero. The system is in thermal equilibrium with a thermostat at temperature $T$. Compute the canonical partition function, the average energy, the Helmholtz free energy, the entropy and the heat capacity of the system.

**3.** A spherical container of radius $R$ is filled with a perfect gas composed of $N$ point molecules of mass $m$ subject to the constant gravitational field $g$. Find the specific heat of the system as a function of the temperature.

**4.** A two-atom molecule is made of two ions both of mass $m$, with electric charges $q$ and $-q$, respectively, constrained to keep a fixed distance $d$ between them. The molecule is held in a container of volume $V$ and it is subject to a non-uniform electric field $\mathbf{E}(q)$. Write down the Hamiltonian of the molecule in the approximation in which the electric field is constant on the segment of length $d$ joining the two ions. Write down the canonical partition function for a gas of $N$ non-interacting molecules.

**5.** Compute $B(E, V)$ (see Section 15.7) when $H = \sum_{i=1}^{N} \mathbf{p}_i^2/2m$. Compare $S_B$ with $S_\omega$.

**6.** (Huang 1987). Consider a system of $N$ biatomic non-interacting molecules held in a container of volume $V$, in thermal equilibrium with a thermostat at temperature $T$. Each molecule has a Hamiltonian

$$H = \frac{1}{2m}\left(|\mathbf{p}_1|^2 + |\mathbf{p}_2|^2\right) + \frac{a}{2}|q_1 - q_2|^2,$$

where $(\mathbf{p}_1, \mathbf{p}_2, q_1, q_2)$ are the momenta and coordinates of the two atoms of the molecule. Compute the canonical partition function, the Helmholtz free energy, the specific heat at constant volume, and the mean square diameter $\langle |q_1 - q_2|^2 \rangle$.

**7.** A biatomic polar gas is made of $N$ molecules composed of two ions of mass $m$ and electric charges $q$ and $-q$, respectively, constrained to keep a fixed distance $d$ between them. The gas is held in two communicating containers $V_1$ and $V_2$ immersed in two electric fields of constant intensity $E_1$ and $E_2$, respectively. The system is in thermal equilibrium with a thermostat at temperature $T$. Neglecting the interactions between the molecules, determine the average number $N_1$ of molecules held in the first container, the free energy of the system, and the pressure on the walls of the containers.

**8.** A point particle of mass $m$ is constrained to move on a smooth circular paraboloid of equation $z = x^2 + y^2$ under the action of a conservative force with potential energy $V = V_0\sqrt{1+4z}$, where $V_0$ is a positive fixed constant. Introduce the Lagrangian coordinates $x = r\cos\varphi$, $y = r\sin\varphi$, $z = r^2$, where $r \in [0, +\infty)$, $\varphi \in [0, 2\pi]$.
(a) Write down the Hamiltonian of the problem.
(b) Assume that the system is in contact with a thermostat at temperature $T$, and compute the canonical partition function. Compute $\langle\varphi\rangle$ and $\langle H\rangle$.

**9.** Two point particles of mass $m$ move along the $x$-axis, subject to a potential

$$V(x_1, x_2) = \frac{1}{2}\left[ax_1^2 + ax_2^2 + b(x_1 - x_2)^2\right].$$

The system is in thermal equilibrium with a thermostat at temperature $T$. Compute the canonical partition function and the average value of the energy.

**10.** A one-dimensional system is composed of $N$ points of mass $m$ constrained to move along the $x$-axis and subject to the potential

$$V(x_1, \ldots, x_N) = v(x_1) + v(x_2 - x_1) + \ldots + v(x_N - x_{N-1}) + fx_N,$$

where $f$ is a prescribed positive constant and

$$v(x) = \begin{cases} +\infty, & \text{if } x < a, \\ b(x-a), & \text{if } x > a, \end{cases}$$

where $a$ and $b$ are two prescribed positive constants. Assume that the system is in thermal equilibrium with a thermostat at temperature $T$. Compute the canonical partition function, the heat capacity, the average length $\langle x_N \rangle$, the coefficient of thermal dilation and the elasticity module $(1/\langle x_N\rangle)(\partial\langle x_N\rangle/\partial f)$.

**11.** A point particle of mass $m$ is constrained to move along the $x$-axis under the action of a conservative force field with potential energy

$$V(x) = \begin{cases} \frac{1}{2}m\omega^2(x+a)^2, & \text{if } x \leq -a, \\ 0, & \text{if } -a \leq x \leq a, \\ \frac{1}{2}m\omega^2(x-a)^2, & \text{if } x \geq a, \end{cases}$$

where $a > 0$. Assuming that the system is in contact with a thermostat at temperature $T$, compute the canonical partition function. Compute $\langle H \rangle$ and show its graph as a function of $a > 0$.

**12.** Consider a system of $N$ point particles of mass $m$ moving along the $x$-axis under the action of a conservative force field of potential energy

$$V(x_1, \ldots, x_N) = v(x_1) + v(x_2 - x_1) + \ldots + v(x_N - x_{N-1}),$$

where

$$v(x) = \begin{cases} +\infty, & \text{if } x < a, \\ \frac{1}{2} m\omega^2 (x-a)^2, & \text{if } x \geq a \end{cases}$$

with $a > 0$. Assuming that the system is in contact with a thermostat at temperature $T$, compute the canonical partition function, $\langle H \rangle$, $\langle x_N \rangle$ (average length of the system) and $(1/\langle x_N \rangle)(\partial/\partial T) \langle x_N \rangle$ (coefficient of thermal dilation), and show their graphs as functions of the temperature $T$.

**13.** Prove that the grand canonical partition function for a system confined in a region of volume $V$ and with Hamiltonian $H = \sum_{i=1}^{N} (\mathbf{p}_i^2/2m + \phi(q_i))$ is given by $Z_G(V, T, z) = \exp(z \Phi(V, T))$, where $\Phi(V, T) = (2\pi m k T)^{3/2} \int_V \exp(-\beta \phi(\mathbf{Q})) \, d\mathbf{Q}$. Prove that for any external potential $\phi$ the state equation of this system is always the equation of a perfect gas.

**14.** (Thompson 1972) Prove that if the potential $\Phi(r)$ satisfies $\Phi(|q_l - q_m|) \leq C/|q_l - q_m|^{d+\varepsilon}$ if $|q_l - q_m| \geq R$, where $d$ is the dimension of the space of configurations of each particle, $C$, $\varepsilon$ are arbitrary positive constants, for every pair of subdomains $D_1$ and $D_2$, of volume $V_1$ and $V_2$ and containing $N_1$ and $N_2$ particles, respectively, of a domain $D$ of volume $V$ with a distance at least equal to $R$ between them, we have

$$Z(V, N_1 + N_2, T) \geq Z(V_1, N_1, T) Z(V_2, N_2, T) \exp(-N_1 N_2 \beta C/R^{d+\varepsilon}).$$

**15.** (Uhlenbeck and Ford 1963) Consider a system with partition function $Z(N, V) = \sum_{j=0}^{V} \binom{V}{N-k}$. Prove that the grand canonical partition function is $Z_G(V, z) = \sum_{N=0}^{\infty} z^N Z(N, V) = ((z^{V+1} - 1)/z - 1)(1 + z)^V$, and the grand canonical potential $\chi(z)$ in the thermodynamical limit is given by

$$\chi(z) = \begin{cases} \log(1+z), & \text{if } |z| \leq 1, \\ \log z(1+z), & \text{if } |z| > 1. \end{cases}$$

Deduce $p$ as a function of $v$ from the relations $\beta p = \chi(z)$ and $v^{-1} = z \partial \chi/\partial z$, eliminating the fugacity $z$. Prove that $\beta p = \log 2$ if $\frac{2}{3} \leq v \leq 2$.

**16.** (Huang 1987, Problem 9.5) Consider a system with grand canonical partition function

$$Z(z,V) = (1+z)^V(1+z^{\alpha V}),$$

where $\alpha > 0$ is fixed. Write the state equation (naturally, in the thermodynamical limit), eliminating the fugacity $z$ from the parametric form (15.115) and prove that there exists a phase transition. Find the specific volumes of the two phases. Find the zeros of the partition function $Z(z,V)$ in the $z \in \mathbf{C}$ plane (for fixed volume $V$) and prove that if $V \to \infty$, then the zeros approach the real axis at $z = 1$.

**17.** (One-dimensional Ising model) Consider a system made of a one-dimensional lattice such that to each site there corresponds a variable (*spin*) $s_i$ that can assume the values $\pm 1$. Each spin interacts with the two adjacent spins $s_{i\pm 1}$ and with an external magnetic field in such a way that the total energy of a configuration $\{s_i\}$ is given by

$$E(\{s_i\}) = -J \sum_{|i-j|=1} s_i s_j - H \sum_i s_i.$$

The case $J > 0$ corresponds to a ferromagnetic model, while $J < 0$ is associated with the antiferromagnetic case. Assume that $H = 0$ and that the total number of spins is $N$. In the case that the points $s_1$ and $s_N$ of the lattice are free, or the case when the lattice closes to form a ring, and hence that $s_1 = s_{N+1}$, prove that the canonical partition functions are given by

$$Z_f(x) = 2x^{-(N-1)/2}(x+1)^{N-1} \quad \text{in the free case,}$$
$$Z_r(x) = x^{-N/2}[(x+1)^N + (x-1)^N] \quad \text{in the ring case,}$$

where $x = e^{J/kT}$. Find the zeros $x_r$ and $x_f$ of the two partition functions (answer: $x_r = -1$ with multiplicity $N-1$; $x_f = i((2n+1)\pi)/2N$, where $i = \sqrt{-1}$ and $n = 0, 1, \ldots, N-1$) and check that, setting $x_f = is$, the density of the zeros $\mu(s) = (1/N)\,dn/ds$ in the thermodynamical limit $N \to \infty$ is given by

$$\mu(s) = \frac{1}{\pi} \frac{1}{1+s^2}.$$

Recalling that the physically significant region corresponds to $x$ real and positive (why?), do these models present phase transitions?

## 15.17 Additional remarks and bibliographical notes

In the classical literature, the fact that the ergodic hypothesis has been formulated by Boltzmann assuming the existence of a trajectory passing through all points in the phase space accessible to the system (hence corresponding to a fixed value of the energy) is often discussed. Clearly this condition would be sufficient to ensure that temporal averages and set averages are interchangeable, but at the same time its impossibility is evident. Indeed, the phase trajectory $S^t \mathbf{X}$ of a Hamiltonian flow is a regular curve of zero measure, and hence it can be *dense* at most on the constant energy surface.

The reasoning of Boltzmann is, however, much richer and more complex (and maybe this is the reason why it was not appreciated by his contemporaries) and it deserves a brief discussion. In addition, through a modern exposition of his ideas, we can criticise the tendency, which emerges in some texts (including the treatise of Huang 1987) to consider ergodic theory as a mathematical discipline without (almost) any physical relevance.

For the reader interested in going into more detail into the topics we are going to discuss below, we refer to the excellent article of Gallavotti *Classical statistical mechanics*, in the *Enciclopedia delle scienze fisiche*, published by the Istituto della Enciclopedia Italiana, from which we took most of the considerations that follow.

Consider a system of $N$ interacting particles described by the Hamiltonian (15.1) and contained in a finite volume $V$ with perfect walls (isolating and against which the particles collide elastically). Instead of assuming, as usual, that the system can take a continuum of states in the space $\Gamma$, we subdivide the latter into small cells $\Delta$, each determining the position and velocity of each particle with the uncertainty unavoidable in every measurement process. This approach is due to Boltzmann himself, and it is deeply innovative, anticipating in a sense (though not intentionally) the criticism to the determinism of classical mechanics which came much later with the uncertainty principle and the development of quantum mechanics.

If $h$ denotes the uncertainty in the measurements of position and velocity, and hence if

$$\delta q \delta p \approx h,$$

$h^{3N}$ is the volume of a cell. The *microscopic state space* is then the set of the cells $\Delta$ subdividing the space $\Gamma$.

The Hamiltonian flow $S^t$ associated with (15.1) induces, in this context, a transformation $\mathcal{S} = S^\tau$ which transforms the cells $\Delta$ into one another:

$$\mathcal{S} \Delta = \Delta'.$$

Here $\tau$ is a 'microscopic time', very short with respect to the duration $T$ of any macroscopic measurement of the system and on a scale in which the movement of the particles can be measured (accounting for the finite precision). Typical values for $\tau$ and $T$ are of the order of $10^{-12}$ seconds and one second, respectively. (A deeper discussion of this point is given by Gallavotti in the quoted article; the reader will notice how his arguments have many analogies with the typical arguments of kinetic theory.)

By the theorem of Liouville, the map $S$ is injective and surjective. Essentially $S$ is the canonical linear map obtained by solving over a time interval $\tau$ Hamilton's equations (15.1) linearised at the centre of the cell $\Delta$ considered. The effect of $S$ is therefore to *permute* the cells $\Delta$ among them.

Since the system under consideration is closed and isolated, its energy $E$ is macroscopically fixed (and lies between $E$ and $E + \Delta E$). Since the volume $V$ accessible to the particles is finite, the number $\mathcal{V}(\Delta)$ of cells representing the energetically possible states is very large, but finite. For example, if we assume for $N$, $V$ and $E$ the value of a mole of a perfect gas made by hydrogen molecules, with mass $m = 2 \times 10^{-24}$ g, in standard conditions of pressure and temperature, and hence $N = 6 \times 10^{23}$, $V = 22 l$ and $E = 4 \times 10^{10}$ erg, where $h$ is the Planck constant, $h = 6 \times 10^{-27}$ erg s and for $\Delta E$ the value $h/\tau = 6 \times 10^{-15}$ erg, we find that $\mathcal{V}(\Delta)$ is of the order of $10^{10^{25}}$. The cells can therefore be numbered: $\Delta_1, \ldots, \Delta_\mathcal{V}$.

The *temporal average* of a function $f$ becomes

$$\hat{f} = \lim_{j \to \infty} \frac{1}{j} \sum_{i=0}^{j-1} f(S^i \Delta), \qquad (15.116)$$

where $f(\Delta)$ is the value $f$ takes on the cell $\Delta$.

The *set average* is given by

$$\langle f \rangle = \frac{1}{\mathcal{V}(\Delta)} \sum_{i=1}^{\mathcal{V}(\Delta)} f(\Delta_i). \qquad (15.117)$$

The ergodic hypothesis becomes

$$\hat{f} = \langle f \rangle, \qquad (15.118)$$

and it is clearly equivalent to assume that $S$ acts as a *one-cycle permutation*: a given cell $\Delta$ evolves successively into different cells until it returns to the initial state in a number of steps *equal* to the number $\mathcal{V}(\Delta)$ of cells. It follows that, by numbering the cells appropriately, we have

$$S \Delta_i = \Delta_{i+1}, \quad i = 1, \ldots, \mathcal{V} - 1 \qquad (15.119)$$

and $S \Delta_\mathcal{V} = \Delta_1$.

The ergodic hypothesis is not necessary in the most general formulation. For statistical mechanics to be solidly based, it would be sufficient for (15.118) to be valid only for the few thermodynamical quantities of interest. It would be sufficient that, instead of satisfying (15.119), every cell in its evolution visited mainly those cells in which the observable quantities of interest take an approximately constant value, which as we saw (Section 15.6), in the Boltzmann interpretation are the majority of cells with fixed energy.

## 15.18  Additional solved problems

*Problem 1*
Consider a system of $N$ equal homogeneous plane plates, with a centre of symmetry and an axis $x$ through their centres and orthogonal to the plates. The centres of mass $G_1, \ldots, G_N$ are fixed (for example, they are equidistant) and the plates can rotate without friction around the $x$-axis. At rest, the plates occupy configurations that can be obtained by translating one into another along the $x$-axis. When not in equilibrium, there is a torsion energy $V = \sum_{i=1}^{N-1} \frac{1}{2}\gamma(\varphi_i - \varphi_{i-1})^2$ where $\varphi_i$ is the rotation angle of the $i$th plate with respect to equilibrium ($\varphi_0 = 0$) and $\gamma$ is a positive constant.

If the system is subject to energy fluctuations corresponding to the temperature $T$, find:

(i) the canonical partition function;
(ii) the average values of energy, kinetic energy, torsion energy, and the average value of the relative rotation angle between two contiguous plates and of its square.

*Solution*
If $I$ denotes the moment of inertia of the plates with respect to the $x$-axis, the Hamiltonian of the system is

$$H = \sum_{i=1}^{N} \frac{p_i^2}{2I} + \sum_{i=1}^{N-1} \frac{1}{2}\gamma(\varphi_{i+1} - \varphi_{i-1})^2.$$

The rotation angles vary between $-\infty$ and $+\infty$. To compute the partition function, we must compute the integral $\int_{\mathbf{R}^N} e^{-\frac{1}{2}\beta\gamma \sum_{i=1}^{N-1}(\varphi_{i+1}-\varphi_i)^2} d\varphi_1 \ldots d\varphi_N$. It is convenient to use the transformation $\varphi_i - \varphi_{i-1} = \eta_i$, $i = 1, \ldots, N$, whose Jacobian is equal to 1. The integral is thus reduced to $\left(\int_{-\infty}^{+\infty} e^{-\frac{1}{2}\beta\gamma\eta^2} d\eta\right)^N = (2\pi/\beta\gamma)^{N/2}$. It is immediately obvious that the integral in the space of momenta is factorised as in $\left(\int_{-\infty}^{+\infty} e^{(-\beta/2I)p^2} dp\right)^N = (2\pi I/\beta\gamma)^{N/2}$. Therefore the partition function is

$$Z = \frac{1}{N! h^N} \left(\frac{2\pi}{\beta}\right)^N \left(\frac{I}{\gamma}\right)^{N/2},$$

from which we find $\langle H \rangle = -\partial/\partial\beta \log Z = NkT$.
The average $\langle p_i^2/2I \rangle$ is given by

$$\left(\int_{-\infty}^{+\infty} \frac{p_i^2}{2I} e^{-\beta p_i^2/2I} dp_i\right) \left(\int_{-\infty}^{+\infty} e^{-\beta p_i^2/2I} dp_i\right)^{-1},$$

because all other factors cancel out, and hence we can write

$$\left\langle \frac{p_i^2}{2I} \right\rangle = -\frac{\partial}{\partial \beta} \log \left( \int_{-\infty}^{+\infty} e^{-\beta p_i^2/2I} \, dp_i \right) = -\frac{\partial}{\partial \beta} \log \left( \frac{2\pi I}{\beta} \right)^{1/2} = \frac{1}{2}kT.$$

It follows that the average kinetic energy is $\frac{1}{2}NkT$, from which we see that $\langle H \rangle$ is equipartitioned between the averages of the kinetic and torsion energy.

As for the angle of relative rotation, $\eta_i$, we clearly have $\langle \eta_i \rangle = 0$ and $\langle \eta_i^2 \rangle = kT/\gamma$, since $N/2\gamma\langle \eta_i^2 \rangle = \frac{1}{2}NkT$.

*Problem 2*
Consider the system of $N$ uncoupled harmonic oscillators with Hamiltonian $\frac{1}{2}(\mathbf{p}_i^2 + \omega^2 q_i^2)$ contained in the cube of side $2A$. Describe the corresponding canonical set at temperature $T$.

*Solution*
The partition function can be written as

$$Z = \frac{1}{N!h^{3N}} \left( \int_{-\infty}^{+\infty} e^{-\beta p^2/2} \, dp \right)^{3N} \left( \int_{-A}^{A} e^{-\beta \omega^2 q^2/2} \, dq \right)^{3N}.$$

Setting $f(\alpha) = \frac{1}{\sqrt{\pi}} \int_{-\alpha}^{\alpha} e^{-y^2} \, dy$, we have

$$Z = \frac{1}{N!h^{3N}} \left( \frac{2\pi}{\beta\omega} \right)^{3N} f^{3N}\left( \sqrt{\frac{\beta}{2}} \omega A \right).$$

Therefore we immediately find the average energy

$$\langle H \rangle = 3NkT - 3N \frac{\omega A}{\sqrt{2\pi\beta}} e^{-(\beta\omega^2/2)A^2} f^{-1}\left( \sqrt{\frac{\beta}{2}} \omega A \right)$$

(in agreement with the equipartition theorem for $A \to +\infty$). We can then compute the Helmholtz free energy

$$\Psi = -\frac{1}{\beta} \log Z$$

and the pressure $P = -\partial \Psi/\partial V$. Since $V = 8A^3$, we can write

$$P = 3NkT \frac{1}{24A^2} \frac{2e^{-\beta\omega^2 A^2/2}}{\int_{-A}^{A} e^{-\beta\omega^2 q^2/2} \, dq} = \frac{N}{V} kT \frac{2Ae^{-\beta\omega^2 A^2/2}}{\int_{-A}^{A} e^{-\beta\omega^2 q^2/2} \, dq}.$$

Note that for $\omega \to 0$ we obtain the perfect gas pressure $P = (N/V)kT$. The same happens, keeping the ratio $N/V$ fixed, for small $A$. However for $\omega$ large or $A$ large we see that $P$ tends to zero (the increase in the attractive force or moving away the walls have asymptotically the effect of suppressing the pressure).

## Problem 3

Model a system of $N$ biatomic particles by attributing to each pair the Hamiltonian

$$h_i = \frac{1}{2m}(\mathbf{p}_1^{(i)2} + \mathbf{p}_2^{(i)2}) + \frac{1}{2}a(q_1^{(i)} - q_2^{(i)})^2, \quad i = 1,\ldots,N.$$

The system is contained in the cube defined by $|q_{j,k}^{(i)}| \leq A$, with $k = 1,2$, $j = 1,2,3$, $i = 1,\ldots,N$. Using the Hamiltonian $H = \sum_{i=1}^N h_i$ and for prescribed temperature $T$, compute the canonical partition function, the average energy and the average square diameter $\langle|\mathbf{q}_1 - \mathbf{q}_2|^2\rangle$.

### Solution

We have a system of $2N$ particles in $\mathbf{R}^3$. Hence we write

$$Z = \frac{1}{(2N!)}\frac{1}{h^{6N}}\left(\int_{-\infty}^{+\infty} e^{-(\beta/2m)p^2}\,dp\right)^{6N}\left(\int_{-A}^{A}\int_{-A}^{A} e^{-(\beta a/2)(q_1-q_2)^2}\,dq_1\,dq_2\right)^{3N},$$

where we denote by $q_1, q_2$ two corresponding components of the vectors $\mathbf{q}_1^{(i)}, \mathbf{q}_2^{(i)}$, for generic $i$.

The first integral is simply $(2\pi m/\beta)^{3N}$. To evaluate the second integral it is convenient to use the transformation $q_1 + q_2 = \sqrt{2}\xi$, $-(q_1 - q_2) = \sqrt{2}\eta$, with unit Jacobian. The double integral then becomes $2\sqrt{2}A \int_{-\sqrt{2}A}^{\sqrt{2}A} e^{-\beta a\eta^2}\,d\eta = 8A^2 \frac{1}{\sqrt{2}A}\int_0^{\sqrt{2}A} e^{-\beta a\eta^2}\,d\eta$. In summary,

$$Z = \frac{V^{2N}}{(2N)!}\frac{1}{h^{6N}}\left(\frac{2\pi m}{\beta}\right)^{3N} 2^{3N}\left(\frac{1}{\sqrt{2}A}\int_0^{\sqrt{2}A} e^{-\beta a\eta^2}\,d\eta\right)^{3N}.$$

Taking the logarithm, the principal terms, up to factors that render the variables dimensionless, are ($N \gg 1$)

$$\log Z = 2N \log \frac{V}{2N} - 3N \log \beta + 3N \log \left(\frac{1}{\sqrt{2}A}\int_0^{\sqrt{2}A} e^{-\beta a\eta^2}\,d\eta\right).$$

We can now compute

$$\langle H \rangle = -\frac{\partial}{\partial \beta} \log Z = 3N \left( kT + kT \frac{\int_0^{A\sqrt{2\beta a}} y^2 e^{-y^2} dy}{\int_0^{A\sqrt{2\beta a}} e^{-y^2} dy} \right) \stackrel{\text{def}}{=} 3NkT(1 + \varphi(A\sqrt{2\beta a})).$$

In the case $A\sqrt{2\beta a} \gg 1$ we have $\langle H \rangle \approx \frac{9}{2}NkT$, since $\varphi(A\sqrt{2\beta a}) \approx \frac{1}{2}$.

Regarding the average $\langle 2\eta^2 \rangle$, we only need to compute

$$\langle 2\eta^2 \rangle = \frac{2\int_0^{\sqrt{2}A} \eta^2 e^{-\beta a \eta^2} d\eta}{\int_0^{\sqrt{2}A} e^{-\beta a \eta^2} d\eta} = \frac{2a}{\beta}\varphi(A\sqrt{2\beta a}),$$

and consequently we have

$$\langle |q_1 - q_2|^2 \rangle = 6\frac{a}{\beta}\varphi(A\sqrt{2\beta a}) \approx 3\frac{a}{\beta}, \quad \text{for } A\sqrt{2\beta a} \gg 1.$$

*Problem 4*

A cubic box of side $l$ resting on the horizontal plane $z = 0$ contains a system of $N$ particles subject to weight. Describe the canonical and microcanonical sets.

*Solution*

The Hamiltonian is

$$H = \frac{P^2}{2m} + mg \sum_{i=1}^{N} z_i,$$

where as usual $\mathbf{P}$ is the momentum vector in $\mathbf{R}^{3N}$.

We write the canonical partition function in the form ($V = l^3$)

$$Z = \frac{1}{N! h^{3N}} \int_{\mathbf{R}^{3N}} e^{-(\beta/2m)P^2} d\mathbf{P} \, V^N \left[ \frac{1}{l} \int_0^l e^{-\beta mgz} dz \right]^N.$$

Then

$$\int_{\mathbf{R}^{3N}} e^{-(\beta/2m P^2)} d\mathbf{P} = \left( \frac{2\pi m}{\beta} \right)^{3N/2},$$

from which it follows that

$$Z = \frac{V^N}{N! h^{3N}} \left( \frac{2\pi m}{\beta} \right)^{3N/2} \left( \frac{1 - e^{-\beta mgl}}{\beta mgl} \right)^N.$$

Taking $N \gg 1$, and neglecting $\beta$-independent terms we get

$$\log Z = -\frac{3}{2} N \log \beta - N \log(\beta mgl) + N \log(1 - e^{-\beta mgl}),$$

from which

$$\langle H \rangle = -\frac{\partial}{\partial \beta} \log Z = \frac{3}{2}\frac{N}{\beta} + \frac{N}{\beta} - \frac{N}{\beta}\frac{\beta mgl\, e^{-\beta mgl}}{1 - e^{-\beta mgl}} = NkT\left(\frac{5}{2} - \frac{xe^{-x}}{1-e^{-x}}\right),$$

with $x = \beta mgl$. For low values of $\beta$ (high temperatures) we have $\langle H \rangle \approx \frac{3}{2} NkT$ (the influence of gravity is not felt). At low temperatures, however, it remains $\langle H \rangle \approx \frac{5}{2} NkT$. We now evaluate the average height:

$$\langle z \rangle = \frac{\int_0^l z e^{-\beta mgz}\, dz}{\int_0^l e^{-\beta mgz}\, dz} = \frac{l}{x}\left(1 - \frac{xe^{-x}}{1-e^{-x}}\right), \quad x = \beta mgl.$$

At low temperatures ($x \to \infty$) we have $\langle z \rangle \approx 0$, while at high temperatures ($x \to 0$) we can easily verify that $\langle z \rangle \approx \frac{1}{2}$. Both results are consistent with physical intuition.

We can also compute the free energy:

$$-\Psi = \frac{1}{\beta}\log Z = \frac{1}{\beta} N \log \frac{V}{N} - \frac{3}{2}\frac{N}{\beta}\log \beta - \frac{N}{\beta}\log(\beta mgl) + \frac{N}{\beta}\log(1 - e^{-\beta mgl})$$

and the pressure:

$$P = -\frac{\partial \Psi}{\partial V} = \frac{N}{\beta V} + \frac{N}{\beta}\frac{xe^{-x}}{1-e^{-x}}\frac{1}{l}\frac{dl}{dV} - \frac{N}{\beta}\frac{1}{3V} = \frac{N}{V}kT\left(1 + \frac{1}{3}\frac{xe^{-x}}{1-e^{-x}} - \frac{1}{3}\right).$$

At high temperatures ($x \to 0$) we again find $P \approx N/VkT$, while at low temperatures ($x \to \infty$) the asymptotic value is $P \approx \frac{2}{3} N/VkT$.

We now try to describe the microcanonical set for the same system, considering energies $E > E_0 = Nmgl$ (which do not admit states with zero global momentum **P**). We study the set $\{H \leq E\}$ in the space $\Gamma$. For prescribed values of the heights $z_i$, we have for the norm of **P** the bound $P^2/2m \leq E - mg\sum_{i=1}^N z_i$. Setting $\sum_{i=1}^N z_i = N z_G$ the momentum **P** varies in the ball of $\mathbf{R}^{3N}$ with radius $\sqrt{2m(E-mgNz_G)}^{1/2}$ whose volume is $\chi_{3N} 1/3N[2m(E-mgNz_G)]^{3N/2} = v(E,z_G)$. Hence the measure $B(E,V)$ of the set $\{H \leq E\}$ is

$$B(E,V) = l^{2N}\int_0^l a(z_G) v(E, z_G)\, dz_G,$$

where $a(z_G)$ is the measure of the $(N-1)$-dimensional section of the cube in $\mathbf{R}^N$ of side $l$ with the hyperplane $\sum_{i=1}^N z_i = Nz_G$. Passing to the dimensionless variables $\xi_i = z_i/l \in (0,1)$ we can factorise in $B$ the coefficient $l^{3N} = V^N$, and hence in the entropy we can single out the term $N\log(V/N)$ (for $N \gg 1$). Since $\int_0^l a(z_G)\,dz_G = l^N$, when $E_0$ is negligible with respect to $E$ we find the same result obtained for a perfect gas.

## Problem 5

In a monatomic gas of $N$ particles confined in a square of side $l$ at temperature $T$ compute the probability that at least one particle has kinetic energy greater than:

(i) $\alpha \langle H \rangle / N = \alpha \varepsilon$, with $\alpha \in (0, \infty)$;
(ii) the sum of the energy of all other particles.

## Solution

These probabilities can be computed as ratios between the $\rho$-measure in the space $\Gamma$ in which the prescribed condition is verified and the $\rho$-measure of the whole space, i.e. the function $Z$.

In case (i) we must force a momentum $\mathbf{p}_i$ to be greater than the absolute value of $\sqrt{2m\alpha\varepsilon}$; hence we must compute the ratio

$$\frac{\int_{\sqrt{2m\alpha\varepsilon}}^{\infty} 2\pi p\, e^{-\beta p^2/2m}\, dp}{\int_0^{\infty} 2\pi p\, e^{-\beta p^2 2m}\, dp} = \int_{\sqrt{\alpha\beta\varepsilon}}^{\infty} y e^{-y^2}\, dy = e^{-\alpha\beta\varepsilon}.$$

Since $\varepsilon = 1/\beta$ (the system is plane), the sought probability is $\nu = e^{-\alpha}$ for a specific particle. Naturally $1 - \nu$ is the probability that a specific particle has energy less than $\alpha\varepsilon$. The probability that no particle has energy greater than $\alpha\varepsilon$ is $(1-\nu)^N$, and therefore the probability that at least one particle has energy greater than $\alpha\varepsilon$ is $1 - (1-\nu)^N$.

The probability that precisely $j$ particles have energy greater than $\alpha\varepsilon$ is $\binom{N}{N-j} \nu^j (1-\nu)^{N-j}$ (note that the sum over $j$ from 0 to $N$ yields 1). Considering $j$ as a continuous variable, the value maximising this probability is $j = \nu N = Ne^{-\alpha}$.

To answer the second question, we must compute the ratio

$$\int_{\mathbf{R}^{2(N-1)}} e^{-\beta P^{*2}/2m} \left( \int_{P>P^*} e^{-\beta P^2/2m}\, d\mathbf{P} \right) d\mathbf{P}^* \left[ \int_{\mathbf{R}^{2N}} e^{-\beta P^2/2m}\, d\mathbf{P} \right]^{-1},$$

which gives the probability $\nu$ that the event is verified for a specific particle. In the first integral $\mathbf{P}^*$ is a momentum in $\mathbf{R}^{2(N-1)}$. Hence we write

$$\nu = \frac{\chi_{2(N-1)} \int_0^\infty P^{*2N-3} e^{-\beta P^{*2}/2m} \left( \int_{P^*}^\infty 2\pi P e^{-\beta P^2/2m} \, dP \right) dP^*}{\chi_{2N} \int_0^\infty P^{2N-1} e^{-\beta P^2/2m} \, dP}$$

$$= \pi \frac{\chi_{2(N-1)}}{\chi_{2N}} \frac{\int_0^\infty X^{2N-3} e^{-2X^2} \, dX}{\int_0^\infty X^{2N-1} e^{-X^2} \, dX} = \left(\frac{1}{2}\right)^{N-1} \pi \frac{\chi_{2(N-1)}}{\chi_{2N}} \frac{\Gamma(N-1)}{\Gamma(N)} = \left(\frac{1}{2}\right)^{N-1}$$

(for example for $N=2$ we find trivially that one of the two particles has probability $\frac{1}{2}$ of having greater energy than the other). The probability that any particle has energy greater than the rest of the system is $N\left(\frac{1}{2}\right)^{N-1}$ (the events referring to a single particle are mutually exclusive).

Note that we could do all computations explicitly because we chose a two-dimensional system, but the same procedure applies in three dimensions. We suggest continuing the problem by finding the probability that sets of $2, 3, \ldots, \widehat{N} < N$ particles have globally energy greater than the energy of the complementary system.

### Problem 6
A cylinder of volume $V$ and cross-section $\Sigma$ contains $N$ particles and the system is at a temperature $T$.

(i) Find the average number of particles that pass through the generic section in unit time.
(ii) Find the pressure as an average of the momentum transfer rate per collision with the walls, proving that this average is equal to $\frac{2}{3}(\langle H \rangle / V)$.

### Solution
The average displacement of a molecule in unit time, in the direction orthogonal to $\Sigma$ ($x$-axis) with positive orientation, is

$$s = \int_0^\infty \frac{p_{1,1}}{m} e^{-\beta p_{1,1}^2/2m} \, dp_{1,1} \left( \int_0^\infty e^{-\beta p_{1,1}^2/2m} \, dp_{1,1} \right)^{-1} = \sqrt{\frac{2}{\pi}} \sqrt{\frac{1}{m\beta}}.$$

The global number of molecules which pass through $\Sigma$ in the unit of time (in both directions) is then

$$\nu = 2 \sqrt{\frac{2}{\pi}} \sqrt{\frac{kT}{m}} \frac{\Sigma}{V} N$$

(for $m \simeq 10^{-23}$ g, $T = 300$ K, $\Sigma = 1$ cm$^2$, $N/V = 10^{18}$ cm$^{-3}$, we find $\nu \simeq 10^{23}$ s$^{-1}$).

The momentum transfer per unit time on the unit surface normal to the $x$-axis is given by

$$\Pi_1^+ = \sum_{i=1}^N \frac{2}{m} \frac{1}{V} (p_{i,1})_+^2,$$

where the symbol $(\cdot)_+$ indicates that we only consider the positive components.

Without any computation we find $\langle \Pi_1^+ \rangle$. It is enough to note that $\sum_{j=1}^{3}(\Pi_j^+ + \Pi_j^-) = (4/V)H$ and passing to the averages (because of isotropy) we have

$$6\langle \Pi \rangle = \frac{4}{V}\langle H \rangle;$$

hence we can really identify $\langle \Pi \rangle$ with the pressure.

It should be noted that the results of the last problem can be obtained using the formalism of kinetic theory, and the Maxwell–Boltzmann distribution. The procedure is identical, even formally.

# 16 LAGRANGIAN FORMALISM IN CONTINUUM MECHANICS

## 16.1 Brief summary of the fundamental laws of continuum mechanics

The model of a continuum relies on the hypothesis that we can describe the distribution of mass through a *density function* $\rho(P,t)$, in such a way that the mass of each measurable part $D$ of the system under consideration is representable in the form

$$M(D) = \int_D \rho(\mathbf{x},t)\,d\mathbf{x}. \tag{16.1}$$

A continuum can be three-dimensional, two-dimensional (plates or membranes) or one-dimensional (strings and beams).

The following is a way to represent the configurations of a continuum with respect to a frame $S = (0, x_1, x_2, x_3)$. We choose a *reference configuration* $C^*$ and we denote by $x_1^*, x_2^*, x_3^*$ the coordinates of its points. Any other configuration $C$ is then described by a diffeomorphism:

$$\mathbf{x} = \mathbf{x}(\mathbf{x}^*), \qquad \mathbf{x}^* \in C^*. \tag{16.2}$$

The coordinates $x_1^*, x_2^*, x_3^*$ play the role of *Lagrangian coordinates*.

If the system is in motion, instead of (16.2) we have

$$\mathbf{x} = \mathbf{x}(\mathbf{x}^*, t), \qquad \mathbf{x}^* \in C^* \tag{16.3}$$

describing the motion of every single point (typically $\mathbf{x}^* = \mathbf{x}(\mathbf{x}^*, 0)$).

Expression (16.3) is the so-called *Lagrangian description* of the motion. Its inverse

$$\mathbf{x}^* = \mathbf{x}^*(\mathbf{x}, t) \tag{16.4}$$

provides, for every fixed $\mathbf{x}$ in the space, the Lagrangian coordinates of the points occupying the position $\mathbf{x}$ as time varies (*Eulerian description*).

The fundamental law of the kinematics of continua is *mass conservation*, expressed by the *continuity equation*

$$\frac{\partial \rho}{\partial t} + \mathrm{div}(\rho \mathbf{v}) = 0, \tag{16.5}$$

which is just a particular case of the balance equation

$$\frac{\partial \mathcal{G}}{\partial t} + \mathrm{div}\,\mathbf{j} = \gamma$$

for a scalar quantity $\mathcal{G}$, carried by a *current density* $\mathbf{j}$ (i.e. $\mathbf{j} \cdot \mathbf{n}$ = amount of $\mathcal{G}$ carried through the unit surface with normal $\mathbf{n}$ in unit time), where $\gamma$ represents the *source* or sink (rate of production or absorption of $\mathcal{G}$ per unit of volume). The proof is very simple. Equation (16.5) is written in Eulerian form. Since the derivative along the motion (Lagrangian derivative) is

$$\frac{d\mathcal{G}}{dt} = \frac{\partial \mathcal{G}}{\partial t} + v \cdot \nabla \mathcal{G},$$

the Lagrangian form of the continuity equation is

$$\frac{d\rho}{dt} + \rho \, \text{div} \, \mathbf{v} = 0. \tag{16.6}$$

The dynamics of continua require appropriate modelling of forces. We split the forces acting on a part $D$ of the continuum into two categories:

(a) *surface forces*: forces that are manifested through contact with the boundary of $D$;
(b) *body forces*: all other forces (a typical example is weight).

The model for body forces can be constructed using the simple hypothesis that they are proportional to the mass element $\rho(\mathbf{x},t)\,d\mathbf{x}$ on which they act, through a coefficient $\mathbf{f}(\mathbf{x},t)$, called *the specific mass force* (dimensionally, an acceleration: $\mathbf{g}$ in the case of weight).

To define surface forces, we consider an element $d\sigma$ of the boundary of $D$ with normal direction $\mathbf{n}$ external to $D$ and we say that the force that the complementary set exerts on $D$ through $d\sigma$ is expressed by $\Phi(\mathbf{x},t;\mathbf{n})\,d\sigma$, where $\Phi$ has the dimension of a pressure and it is called the *specific stress* ($\Phi \cdot \mathbf{n}$ is the *compression* stress if negative, and the *tension* stress if positive, while the component normal to $\mathbf{n}$ is called *shear stress*).

The basic theorem of the dynamics of continua is due to Cauchy.

THEOREM 16.1 (Cauchy) *For every unit vector* $\mathbf{n} = \sum_{i=1}^{3} \alpha_i \mathbf{e}_i$ *the specific stress has the following expression:*

$$\Phi(\mathbf{x},t;\mathbf{n}) = \sum_{i=1}^{3} \alpha_i \Phi(\mathbf{x},t;\mathbf{e}_i). \qquad \blacksquare$$

We omit the proof. Cauchy's theorem yields as a result that the products

$$T_{ij}(\mathbf{x},t) = \Phi(\mathbf{x},t;\mathbf{e}_i) \cdot \mathbf{e}_j, \tag{16.7}$$

with $\mathbf{e}_1, \mathbf{e}_2, \mathbf{e}_3$ an orthonormal triple in $\mathbf{x}$, are the elements of a tensor $T$ (the *stress tensor*) which defines the *stress state* in $(\mathbf{x},t)$.

Knowledge of $T_{ij}$ yields the reconstruction of the stress relative to every unit vector $\mathbf{n}$:

$$\Phi_j(\mathbf{x},t;\mathbf{n}) = \sum_{i=1}^{3} \alpha_i T_{ij}(\mathbf{x},t), \quad j = 1,2,3, \tag{16.8}$$

with $\alpha_1, \alpha_2, \alpha_3$ direction cosines of $\mathbf{n}$.

Using the theorem of Cauchy it is possible to deduce that the first and second cardinal equations applied to every subset of a continuous system yield, respectively, the following equations:

$$\rho(\mathbf{f} - \mathbf{a}) + \text{div } \mathbf{T} = 0, \tag{16.9}$$

$$T_{ij} = T_{ji}, \quad i \neq j. \tag{16.10}$$

In the former, by definition div $\mathbf{T}$ is the vector

$$\text{div}\,\mathbf{T} = \sum_{i,j=1}^{3} \frac{\partial T_{ij}}{\partial x_i} \mathbf{e}_j. \tag{16.11}$$

Equation (16.9) holds generically for all continua. The mechanical nature of the system must be specified through additional equations. Expression (16.10) represents the so-called *stress symmetry*.

A special case of great interest is the case of *fluids*.

DEFINITION 16.1 *A fluid is a continuum for which the shear stresses at equilibrium are zero. If this also happens in a dynamic situation, then the fluid is called perfect or ideal.* ∎

For fluids we have an additional simplification of the stress tensor, as the *diagonal elements are equal* (the proof is left as an exercise). Moreover, since the fluid resists only compression, the common value of the diagonal elements of the stress tensor must be negative: $T_{ij} = -p\delta_{ij}$, where $p > 0$ is the *pressure*. The equilibrium equation of a fluid can now be written as

$$\rho \mathbf{f} = \nabla p \tag{16.12}$$

(since div $T = -\nabla p$), and the equation of motion (for a perfect fluid) is

$$\rho(\mathbf{f} - \mathbf{a}) = \nabla p. \tag{16.13}$$

In both it is necessary to specify the relation between $\rho$ and $p$:

$$\rho = \rho(p), \quad \rho'(p) \geq 0 \tag{16.14}$$

(*state equation*). A fluid for which the relation (16.14) is known is called *barotropic*. For simplicity we only consider isothermal phenomena.

For a barotropic fluid we can introduce the function

$$\mathcal{P}(p) = \int \frac{dp}{\rho(p)} \tag{16.15}$$

(*potential energy of the pressure*), and hence $(1/\rho)\nabla_p = \nabla \mathcal{P}(p)$. If in addition $\mathbf{f} = \nabla u(\mathbf{x})$, equation (16.12) can be immediately integrated to give

$$\mathcal{P}(p(\mathbf{x})) = u(\mathbf{x}) + \text{constant}$$

(the constant can be determined using the *boundary condition* $p(x_0) = p_0$ at some given point $x_0$), while after some manipulations (16.13) can be written as

$$\frac{\partial \mathbf{v}}{\partial t} + \text{curl } \mathbf{v} \times \mathbf{v} = -\nabla B \tag{16.16}$$

(*Euler equation*), where

$$B = \frac{1}{2}v^2 - u + \mathcal{P} \tag{16.17}$$

is the *Bernoulli trinomial*. The Euler equation is invariant with respect to time reversal ($t \to -t$, $\mathbf{v} \to -\mathbf{v}$) and indeed it describes a non-dissipative phenomenon. While useful in many circumstances, it is not adequate to describe many phenomena of practical importance (for example the motion of objects in fluids). It is then necessary to construct a more sophisticated model of the fluid (the model of *viscous fluids*), which we do not discuss here. See for example Landau (1990).

*Example* 16.1: linear acoustics
Consider a perfect fluid in equilibrium (neglecting gravity) at uniform pressure $p_0$. Linearise the equation of state in a neighbourhood of $p_0$:

$$\rho = \rho_0 + \frac{p - p_0}{c^2},$$

with $c = [\rho'(p_0)]^{-1/2}$ having the dimension of a velocity ($\rho'(p_0)$ is assumed to be positive and the fluid is said to be *compressible*). In the small perturbations approximation (the linear approximation) we consider $p - p_0, \rho - \rho_0, \mathbf{v}$, etc. to be first-order perturbations, and we neglect higher-order terms such as curl $\mathbf{v} \times \mathbf{v}$, $\mathbf{v} \cdot \nabla \rho$, etc. Considering the linearised version of the Euler equation:

$$\frac{\partial \mathbf{v}}{\partial t} + \frac{1}{\rho_0} \nabla p = 0$$

and of the continuity equation:

$$\frac{d\rho}{dt} + \rho_0 \text{ div } \mathbf{v} = 0,$$

eliminating div $\mathbf{v}$ (after taking the divergence of the first equation) and writing $\partial \rho / \partial t = (1/c^2)(\partial p / \partial t)$, we find that the pressure satisfies the *wave equation* (or the *d'Alembert equation*)

$$\nabla^2 p - \frac{1}{c^2} \frac{\partial^2 p}{\partial t^2} = 0. \tag{16.18}$$

Particularly interesting solutions are the *plane waves* (depending on only one space coordinate, corresponding to the direction of propagation), which in the most common case can be represented as a superposition of *progressive waves*:

$$p(x,t) = f(x - ct) \tag{16.19}$$

and *regressive waves*:

$$p(x,t) = g(x+ct), \tag{16.20}$$

which highlight the role of $c$ as the *velocity of propagation* of the wave. If $\psi(x,t)$ is a plane wave then a *spherical wave* can be constructed via the transformation

$$\varphi(r,t) = \frac{1}{r}\psi(r,t), \tag{16.21}$$

where $r$ is the distance from the centre of the wave.

Plane waves can also appear as *stationary waves*:

$$\psi(x,t) = A \sin kx \cos \nu t, \tag{16.22}$$

where the *wave number* $k$ and the *frequency* $\nu$ must be related by $kc = \nu$. ∎

*Example 16.2: vibrating string*
Consider a perfectly flexible string kept straight with tension $T$ at the two endpoints. The equilibrium configuration is straight. If we perturb either the configuration (*plucked string*) or the velocity (*hammered string*), or both, in such a way that the string oscillates with velocity approximately orthogonal to the string at rest, we can easily prove the following facts regarding the linearised motions:

(i) the tension is constant along the string;
(ii) the equation of small *shear vibrations* is

$$\frac{\partial^2 u}{\partial x^2} - \frac{1}{c^2}\frac{\partial^2 u}{\partial t^2} = 0, \tag{16.23}$$

where $u(x,t)$ is the displacement from equilibrium and $c = \sqrt{T/\rho}$, where $\rho$ is the constant (linear) density of the string.

The *Cauchy problem* for equation (16.23) with initial values

$$u(x,0) = \varphi(x), \qquad -\infty < x < +\infty, \tag{16.24}$$

$$\left.\frac{\partial u}{\partial t}\right|_{t=0} = \psi(x), \qquad -\infty < x < +\infty \tag{16.25}$$

has the *d'Alembert solution*

$$u(x,t) = \frac{1}{2}\{\varphi(x-ct) + \varphi(x+ct)\} + \frac{1}{2c}\int_{x-ct}^{x+ct} \psi(\xi)\,d\xi, \tag{16.26}$$

where we recognise the progressive and regressive waves generated by the perturbation $\varphi$ and by the perturbation $\psi$. ∎

*Example 16.3: longitudinal vibrations of a rod*
Hooke's law for elastic materials applied to a homogeneous cylindrical rod subject to tension or compression $T_0(t)$ (for unit cross-section) implies that

$$T = E\frac{\Delta \ell}{\ell}, \tag{16.27}$$

where $\Delta \ell / \ell$ is the relative elongation and $E$ is *Young's modulus*.

Neglecting shear deformations, denoting by $u(x,t)$ the displacement from equilibrium and extrapolating Hooke's law to

$$T = E\frac{\partial u}{\partial x}, \tag{16.28}$$

the equation of motion can be written as

$$\frac{\partial^2 u}{\partial x^2} - \frac{1}{c^2}\frac{\partial^2 u}{\partial t^2} = 0, \tag{16.29}$$

where $c = \sqrt{E/\rho}$ ($\rho$ is the rod's density). ∎

Concerning the historical aspects of the theory of wave propagation, we suggest Truesdell (1968) or Manacorda (1991).

We refer the reader interested in the physics of musical instruments to Fletcher and Rossing (1991).

## 16.2 The passage from the discrete to the continuous model. The Lagrangian function

We consider again the problem of longitudinal vibrations of an elastic homogeneous rod (Example 16.3) and we aim to construct an approximation of this system using a discrete set of point particles. We denote by $\rho$ the density and by $S$ the area of the cross-section of the rod.

We subdivide the rod into $N$ equal parts and we replace them with a chain of point particles of mass $m/N$, where $m$ is the mass of the entire rod (Fig. 16.1). To model the internal forces we assume that two consecutive points are connected by springs of negligible mass, with an elastic constant $k$ which we specify later and with length at rest equal to $\varepsilon = \ell/N$ ($\ell$ is the length of the rod). We denote by $x_1^{(0)}, x_2^{(0)}, \ldots, x_n^{(0)}$ the $x$-coordinates of the point particles at equilibrium ($x_s^{(0)} = s\varepsilon$). Consider now the generic triple $(P_{i-1}, P_i, P_{i+1})$ and denote by $u_i$ the displacement of $P_i$ from equilibrium (Fig. 16.2). The stretching of the spring between $P_{i+1}$ and $P_i$ is $u_{i+1} - u_i$, and hence the global potential energy is

$$V = \frac{1}{2}k\sum_{i=1}^{N-1}(u_{i+1} - u_i)^2, \tag{16.30}$$

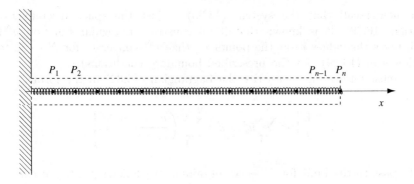

Fig. 16.1

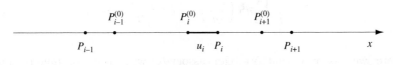

Fig. 16.2

while the kinetic energy is

$$T = \frac{1}{2}\frac{m}{N}\sum_{i=1}^{N}\dot{u}_i^2. \qquad (16.31)$$

We can then write the Lagrangian of the system

$$L(\mathbf{u},\dot{\mathbf{u}}) = \frac{1}{2}\left[\frac{m}{N}\sum_{i=1}^{N}\dot{u}_i^2 - k\sum_{i=1}^{N-1}(u_{i+1}-u_i)^2\right] \qquad (16.32)$$

and finally obtain the equations of motion

$$\frac{m}{N}\ddot{u}_i - k[u_{i+1} - 2u_i + u_{i-1}] = 0, \qquad i = 1, 2, \ldots, N$$

(with $u_0 = 0$), which we rewrite in the form

$$\frac{m}{\ell}\ddot{u}_i - k\varepsilon\frac{u_{i+1} - 2u_i + u_{i-1}}{\varepsilon^2} = 0, \qquad i = 1, \ldots, N, \qquad (16.33)$$

where there appears the discretisation of the second derivative with respect to $x$ and at the same time the product $k\varepsilon$. It is obvious that $m/l = \rho S$ is the linear density of the rod. We must make precise the choice of $k$. Since the elastic tension force between two contiguous points can be written in the form

$$k\varepsilon\frac{u_{i+1} - u_i}{\varepsilon}, \qquad (16.34)$$

following Hooke's law (16.27), we write $k\varepsilon = E$ (Young's modulus).

We now recall that the system (16.33) is just the space discretisation of equation (16.29). It is known that if we construct a regular function $u^N(x,t)$ which takes the values $u_i$ at the points $x_i$, then $u^N$ converges for $N \to \infty$ to the solution $u$ of (16.29) (for the prescribed boundary conditions).

It is most interesting to rewrite the Lagrangian (16.32) in the form

$$L = \frac{1}{2}\left[\rho S \sum_{i=1}^{N} \dot{u}_i^2 - ES \sum_{i=1}^{N-1} \left(\frac{u_{i+1} - u_i}{\varepsilon}\right)^2\right] \cdot \varepsilon$$

and to pass to the limit for $N \to \infty$, obtaining the following integral expression:

$$L = \int_0^\ell \left[\frac{1}{2}\rho S\left(\frac{\partial u}{\partial t}\right)^2 - ES\left(\frac{\partial u}{\partial x}\right)^2\right] dx. \tag{16.35}$$

Therefore we discover that we can associate with the continuous model a Lagrangian and also define a *Lagrangian density*:

$$\mathcal{L}\left(\frac{\partial u}{\partial x}, \frac{\partial u}{\partial t}\right) = \frac{1}{2}\rho\left(\frac{\partial u}{\partial t}\right)^2 - \frac{1}{2}E\left(\frac{\partial u}{\partial x}\right)^2 \tag{16.36}$$

(the factor $S$ can be replaced in (16.35) by a double integral over cross-sections) such that

$$L = \int_{\mathcal{C}} \mathcal{L}\, d\mathbf{x}, \tag{16.37}$$

where $\mathcal{C}$ is the configuration of the system under consideration.

## 16.3 Lagrangian formulation of continuum mechanics

It is now natural to consider whether the equation of motion (16.29) can be obtained by imposing the condition that an action-type functional related to the Lagrangian (16.37) is stationary with respect to certain classes of perturbations.

We consider the problem from a general point of view, assuming that with every continuum described by a *field function* $u(\mathbf{x}, t)$ (for example the displacement from the equilibrium configuration) we can associate a Lagrangian density with the necessary regularity with respect to its arguments:

$$\mathcal{L}\left(u, \frac{\partial u}{\partial x_1}, \frac{\partial u}{\partial x_2}, \frac{\partial u}{\partial x_3}, \frac{\partial u}{\partial t}, x_1, x_2, x_3, t\right) \tag{16.38}$$

## 16.3 Lagrangian formalism in continuum mechanics

and extend the validity of Hamilton's principle.

POSTULATE 16.1 *The natural motion of the system corresponds to a stationary point of the functional*

$$A = \int_{t_0}^{t_1} \int_{\mathcal{C}(t)} \mathcal{L} \, d\mathbf{x} \, dt \qquad (16.39)$$

*(where $\mathcal{C}(t)$ is the configuration of the system at time $t$) with respect to the (regular) perturbations $\delta u(\mathbf{x}, t)$ which vanish $\forall \mathbf{x} \in \mathcal{C}(t)$, when $t = t_0$, $t = t_1$, and on the boundary $\partial \mathcal{C}$, for every $t \in (t_0, t_1)$.* ■

Hence we seek the conditions for a point to be stationary for the functional (16.39) in the specified class. Denoting by $u^*(\mathbf{x}, t)$ the value of the field for the natural motion, we introduce the field

$$u = u^* + \delta u$$

and evaluate the first variation of $A$:

$$\delta A = \int_{t_0}^{t_1} \int_{\mathcal{C}(t)} \left[ \frac{\partial \mathcal{L}}{\partial u} \delta u + \sum_{i=1}^{3} \frac{\partial \mathcal{L}}{\partial \xi_i} \frac{\partial \delta u}{\partial x_i} + \frac{\partial \mathcal{L}}{\partial \zeta} \frac{\partial \delta u}{\partial t} \right] d\mathbf{x} \, dt,$$

where $\boldsymbol{\xi} = \nabla u$, $\zeta = \partial u / \partial t$. Given the assumptions on $\delta u$, the divergence theorem yields

$$\int_{t_0}^{t_1} \int_{\mathcal{C}} \sum_{i=1}^{3} \frac{\partial \mathcal{L}}{\partial \xi_i} \frac{\partial \delta u}{\partial x_i} \, d\mathbf{x} \, dt = - \int_{t_0}^{t_1} \int_{\mathcal{C}} \sum_{i=1}^{3} \frac{\partial}{\partial x_i} \frac{\partial \mathcal{L}}{\partial \xi_i} \delta u \, d\mathbf{x} \, dt.$$

In addition, using

$$\frac{d}{dt} \int_{\mathcal{C}(t)} \frac{\partial \mathcal{L}}{\partial \zeta} \delta u \, d\mathbf{x} = \int_{\mathcal{C}(t)} \frac{\partial}{\partial t} \left( \frac{\partial \mathcal{L}}{\partial \zeta} \delta u \right) d\mathbf{x} + \int_{\partial \mathcal{C}(t)} \frac{\partial \mathcal{L}}{\partial \zeta} \delta u \, v_n \, d\sigma,$$

with $v_n$ being the normal velocity of the points of $\partial \mathcal{C}(t)$, and remarking that the integral over $\partial \mathcal{C}(t)$ is zero, since $\delta u$ is zero on the boundary, we can rewrite the last term in $\delta A$ as

$$\int_{t_0}^{t_1} \int_{\mathcal{C}(t)} \frac{\partial \mathcal{L}}{\partial \zeta} \frac{\partial \delta u}{\partial t} \, d\mathbf{x} \, dt = - \int_{t_0}^{t_1} \int_{\mathcal{C}(t)} \frac{\partial}{\partial t} \frac{\partial \mathcal{L}}{\partial \zeta} \delta u \, d\mathbf{x} \, dt,$$

taking into account that $\delta u = 0$ on $\mathcal{C}(t_0)$ and $\mathcal{C}(t_1)$.

An argument similar to the one used to prove the analogous theorem in the discrete case yields the following conclusion.

THEOREM 16.2 *The characteristic condition for a point to be stationary for the functional (16.39) in the class of perturbations considered, is that*

$$\frac{\partial}{\partial t}\frac{\partial \mathcal{L}}{\partial \zeta} + \sum_{i=1}^{3} \frac{\partial}{\partial x_i}\frac{\partial \mathcal{L}}{\partial \xi_i} - \frac{\partial \mathcal{L}}{\partial u} = 0 \qquad (16.40)$$

*(recall that $\xi_i = \partial u/\partial x_i$, $\zeta = \partial u/\partial t$).* ■

Due to Postulate 16.1, equation (16.40) represents the equation of motion, naturally a partial differential equation.

*Remark* 16.1
From equation (16.40) we deduce that the terms linear in $\xi_i$ and in $\zeta$ (with constant coefficients) in the expression for $\mathcal{L}$ are not essential. ■

More generally, we can consider Lagrangian densities depending on $\ell$ scalar functions $u_1, u_2, \ldots, u_\ell$. The variational problem can be stated in a similar way, leading to an equation of the type (16.40) for every unknown function $u_k$.

## 16.4 Applications of the Lagrangian formalism to continuum mechanics

We now consider a few concrete examples illustrating the theory developed in the previous section.

(A) Longitudinal vibrations of an elastic rod

Using the Lagrangian density (16.36) in equation (16.40), we clearly find the d'Alembert equation (16.29).

(B) Linear acoustics

To determine the Lagrangian density for 'small perturbations' of a perfect gas, neglecting the effect of the body forces, we note that in the Lagrangian density there must appear two contributions, due to the specific kinetic energy, and to the specific potential energy (which must be subtracted from the former).

Denoting by $\mathbf{u}(\mathbf{x},t)$ the displacement vector, the kinetic energy of the unit of mass is $\frac{1}{2}(\partial \mathbf{u}/\partial t)^2$. To evaluate the potential energy $\mathcal{V}$ of the unit of mass we write the energy balance

$$d\mathcal{V} + p\,d\frac{1}{\rho} = 0, \qquad (16.41)$$

where $p\,d(1/\rho)$ is the work done by the unit of mass of the gas for the variation $d(1/\rho)$ of its volume. Recall that we are dealing with a barotropic fluid, and

## 16.4 Lagrangian formalism in continuum mechanics

hence that $\rho = \rho(p)$ (in the case of sound vibrations one must consider adiabatic transformations, hence $p\rho^{-\gamma} = \text{constant}$). We then obtain from (16.41) that

$$\mathcal{V} = -\int_{1/\rho_0}^{1/\rho} p \, d\eta,$$

where we have introduced the variable $\eta = 1/\rho$. Consider the linear approximation of $p$ as a function of $\eta$ around $\eta_0 = 1/\rho_0$ (henceforth, zero subscripts denote quantities at equilibrium):

$$p = p_0 + (\eta - \eta_0) \left( \frac{dp}{d\eta} \right)_{\eta=\eta_0}.$$

Computing the integral, we obtain

$$\mathcal{V} = -\left\{ p_0 \left( \frac{1}{\rho} - \frac{1}{\rho_0} \right) + \frac{1}{2} \left( \frac{dp}{d\eta} \right)_{\eta=\eta_0} \left( \frac{1}{\rho} - \frac{1}{\rho_0} \right)^2 \right\}. \tag{16.42}$$

We now compute $(dp/d\eta)_{\eta=\eta_0}$, writing $\eta = 1/\rho(p)$ and differentiating with respect to $\eta$:

$$1 = -\frac{\rho'(p)}{\rho^2(p)} \frac{dp}{d\eta},$$

from which

$$\left( \frac{dp}{d\eta} \right)_{\eta=\eta_0} = -\frac{\rho_0^2}{\rho_0'}. \tag{16.43}$$

We now express $(1/\rho) - (1/\rho_0)$ through the variation relative to $\rho$, i.e. $\delta = (\rho - \rho_0)/\rho$:

$$\frac{1}{\rho} - \frac{1}{\rho_0} = -\frac{1}{\rho_0} \delta, \tag{16.44}$$

and we substitute (16.43) and (16.44) into equation (16.42):

$$\mathcal{V} = \frac{p_0}{\rho_0} \delta + \frac{1}{2} \frac{1}{\rho_0'} \delta^2. \tag{16.45}$$

Recall that we set $1/\rho_0' = c^2$.

The last step to obtain the Lagrangian density consists of expressing $\delta$ in terms of the displacement $\mathbf{u}$. To this end, it is sufficient to write the linearised continuity equation

$$\frac{\partial \rho}{\partial t} + \rho_0 \, \text{div} \, \frac{\partial \mathbf{u}}{\partial t} = 0$$

in the form
$$\frac{\partial \delta}{\partial t} + \operatorname{div} \frac{\partial \mathbf{u}}{\partial t} = 0. \tag{16.46}$$

Integrating the latter expression (and denoting by $\gamma$ the value of $\delta + \operatorname{div} \mathbf{u}$ for $t = 0$) we find the relation
$$\delta = -\operatorname{div} \mathbf{u} + \gamma. \tag{16.47}$$

This yields the sought Lagrangian density
$$\mathcal{L} = \frac{1}{2}\left(\frac{\partial \mathbf{u}}{\partial t}\right)^2 + \left(\frac{p_0}{\rho_0} + c^2\gamma\right)\operatorname{div}\mathbf{u} - \frac{1}{2}c^2(\operatorname{div}\mathbf{u})^2,$$

and, recalling Remark 16.1, we can suppress the linear term in $\operatorname{div}\mathbf{u}$, and arrive at the expression
$$\mathcal{L} = \frac{1}{2}\left(\frac{\partial \mathbf{u}}{\partial t}\right)^2 - \frac{1}{2}c^2(\operatorname{div}\mathbf{u})^2. \tag{16.48}$$

It is a trivial exercise to check that (16.43) leads to
$$\nabla(\nabla \cdot \mathbf{u}) - \frac{1}{c^2}\frac{\partial^2 \mathbf{u}}{\partial t^2} = 0,$$

from which, by taking the divergence, we obtain the wave equation for $\delta$.

(C) Electromagnetic field

The idea of deducing the field equations from a Lagrangian density is entirely general and can be applied to fields other than mechanics as well, although outside the conceptual framework of mechanics, there does not exist a general criterion to deduce the Lagrangian density. We now consider an example and illustrate how it is possible to derive the Maxwell equations for the electromagnetic field in a vacuum from a Lagrangian density.

The unknown functions on which the Lagrangian density $\mathcal{L}$ depends are the scalar potential $\phi(\mathbf{x}, t)$ and the vector potential $\mathbf{A}(\mathbf{x}, t)$ (Section 4.7), through which we can express the electric field $\mathbf{E}$ and the magnetic induction field $\mathbf{B}$:
$$\mathbf{E} = -\nabla\phi - \frac{1}{c}\frac{\partial \mathbf{A}}{\partial t}, \tag{16.49}$$
$$\mathbf{B} = \operatorname{curl} \mathbf{A}. \tag{16.50}$$

The equations
$$\operatorname{div} \mathbf{B} = 0, \tag{16.51}$$
$$\operatorname{curl} \mathbf{E} + \frac{1}{c}\frac{\partial \mathbf{B}}{\partial t} = 0 \tag{16.52}$$

are automatically satisfied thanks to (16.49) and (16.50).

## 16.4 Lagrangian formalism in continuum mechanics

The equations to be deduced from the Lagrangian formulation are therefore the remaining Maxwell equations:

$$\text{div } \mathbf{E} = 4\pi\rho, \tag{16.53}$$

$$\text{curl } \mathbf{B} - \frac{1}{c}\frac{\partial \mathbf{E}}{\partial t} = \frac{1}{c}4\pi\mathbf{j}. \tag{16.54}$$

Let us check that a correct choice for $\mathcal{L}$ is

$$\mathcal{L} = \frac{1}{8\pi}(E^2 - B^2) + \frac{1}{c}\mathbf{j}\cdot\mathbf{A} - \rho\phi, \tag{16.55}$$

where $\mathbf{E}$ and $\mathbf{B}$ are given by (16.49), (16.50).

The equation for $\phi$ is

$$\sum_i \frac{\partial}{\partial x_i}\frac{\partial \mathcal{L}}{\partial(\partial\phi/\partial x_i)} - \frac{\partial \mathcal{L}}{\partial \phi} = 0. \tag{16.56}$$

Since

$$E^2 = \sum_i \left(\frac{\partial\phi}{\partial x_i}\right)^2 + \frac{2}{c}\sum_i \frac{\partial\phi}{\partial x_i}\frac{\partial A_i}{\partial t} + \frac{1}{c^2}\sum_i \left(\frac{\partial A_i}{\partial t}\right)^2,$$

equation (16.56) takes the form

$$\frac{1}{4\pi}\left(\Delta\phi + \frac{1}{c}\frac{\partial}{\partial t}\text{div}\mathbf{A}\right) + \rho = 0, \tag{16.57}$$

and hence it coincides with (16.53).

We now write

$$B^2 = \sum_{i<j}\left(\frac{\partial A_i}{\partial x_j} - \frac{\partial A_j}{\partial x_i}\right)^2$$

and compute for example

$$\frac{\partial}{\partial t}\frac{\partial \mathcal{L}}{\partial(\partial A_1/\partial t)} = \frac{1}{4\pi c^2}\frac{\partial^2 A_1}{\partial t^2} + \frac{1}{4\pi c}\frac{\partial^2 \phi}{\partial t\partial x_1} = -\frac{1}{4\pi c}\frac{\partial E_1}{\partial t},$$

$$-\sum_i \frac{\partial}{\partial x_i}\frac{\partial \mathcal{L}}{\partial(\partial A_1/\partial x_i)} = \frac{1}{4\pi}\frac{\partial}{\partial x_2}\left(\frac{\partial A_1}{\partial x_2} - \frac{\partial A_2}{\partial x_1}\right) + \frac{1}{4\pi}\frac{\partial}{\partial x_3}\left(\frac{\partial A_1}{\partial x_3} - \frac{\partial A_3}{\partial x_1}\right)$$

$$= -\frac{1}{4\pi}\left(\frac{\partial B_2}{\partial x_3} - \frac{\partial B_3}{\partial x_2}\right),$$

$$\frac{\partial \mathcal{L}}{\partial A_1} = \frac{1}{c}j_1.$$

It is now immediate to verify that the equation

$$\frac{\partial}{\partial t}\frac{\partial \mathcal{L}}{\partial(\partial A_1/\partial t)} + \sum_i \frac{\partial}{\partial x_i}\frac{\partial \mathcal{L}}{\partial(\partial A_1/\partial x_i)} - \frac{\partial \mathcal{L}}{\partial A_1} = 0$$

coincides with the first component of (16.54). The computation for the other components is similar.

## 16.5 Hamiltonian formalism

For the mechanics of continua and for field theory, just as for the mechanics of systems of point particles, it is possible to develop a Hamiltonian formalism parallel to the Lagrangian formalism.

If $\mathcal{L}(\mathbf{u}, \partial \mathbf{u}/\partial x_1, \partial \mathbf{u}/\partial x_2, \partial \mathbf{u}/\partial x_3, \partial \mathbf{u}/\partial t, \mathbf{x}, t)$ is a Lagrangian density depending on an unknown vector $\mathbf{u}(\mathbf{x}, t) \in \mathbf{R}^\ell$, we can define the vector $\boldsymbol{\pi}$ of the *densities of momenta*:

$$\pi_k = \frac{\partial \mathcal{L}}{\partial(\partial u_k/\partial t)}, \qquad k = 1, 2, \ldots, \ell. \tag{16.58}$$

To pass to the Hamiltonian formalism it is necessary for the system (16.58) to be invertible, and hence it must be possible to write $\partial u_k/\partial t$ as functions of the vector $\boldsymbol{\pi}$. We can then define the *Hamiltonian density*:

$$\mathcal{H} = \boldsymbol{\pi} \cdot \frac{\partial \mathbf{u}}{\partial t} - \mathcal{L}, \tag{16.59}$$

with

$$\mathcal{H} = \mathcal{H}\left(\mathbf{u}, \frac{\partial \mathbf{u}}{\partial x_1}, \frac{\partial \mathbf{u}}{\partial x_2}, \frac{\partial \mathbf{u}}{\partial x_3}, \boldsymbol{\pi}, \mathbf{x}, t\right). \tag{16.60}$$

Computing the derivatives $\partial \mathcal{H}/\partial u_k$, $\partial \mathcal{H}/\partial \pi_k$ using (16.58), (16.59) and (16.40) (written for each component $u_k$), we arrive at Hamilton's equations:

$$\begin{aligned}\frac{\partial \mathcal{H}}{\partial \pi_k} &= \frac{\partial u_k}{\partial t}, \\ \frac{\partial \mathcal{H}}{\partial u_k} &= -\frac{\partial \mathcal{L}}{\partial u_k} = -\frac{\partial \pi_k}{\partial t} - \sum_i \frac{\partial}{\partial x_i} \frac{\partial \mathcal{L}}{\partial(\partial u_k/\partial x_i)},\end{aligned} \tag{16.61}$$

and hence, noting that $\partial \mathcal{H}/[\partial(\partial u_k/\partial x_i)] = -\partial \mathcal{L}/[\partial(\partial u_k/\partial x_i)]$, we have

$$\frac{\partial \mathcal{H}}{\partial u_k} = -\frac{\partial \pi_k}{\partial t} + \sum_i \frac{\partial}{\partial x_i} \frac{\partial \mathcal{H}}{\partial(\partial u_k/\partial x_i)}. \tag{16.62}$$

The lack of formal symmetry between the two expressions (16.61) and (16.62) can be corrected by introducing the operators

$$\frac{\delta}{\delta u_k} = \frac{\partial}{\partial u_k} - \sum_i \frac{\partial}{\partial x_i} \frac{\partial}{\partial(\partial u_k/\partial x_i)},$$

$$\frac{\delta}{\delta \pi_k} = \frac{\partial}{\partial \pi_k} - \sum_i \frac{\partial}{\partial x_i} \frac{\partial}{\partial(\partial u_k/\partial \pi_i)} = \frac{\partial}{\partial \pi_k},$$

through which we find the more familiar expressions

$$\frac{\partial u_k}{\partial t} = \frac{\delta \mathcal{H}}{\delta \pi_k}, \tag{16.63}$$

$$\frac{\partial \pi_k}{\partial t} = -\frac{\delta \mathcal{H}}{\delta u_k}, \quad k = 1, 2, \ldots, \ell. \tag{16.64}$$

*Example* 16.4
Using the Lagrangian density (16.36) for the longitudinal vibrations of an elastic rod, we have

$$\pi = \rho \frac{\partial u}{\partial t},$$

from which it follows that

$$\mathcal{H} = \frac{\pi^2}{2\rho} + \frac{1}{2} E \left( \frac{\partial u}{\partial x} \right)^2,$$

and Hamilton's equations are

$$\frac{\partial u}{\partial t} = \frac{\pi}{\rho},$$

$$\frac{\partial \pi}{\partial t} = E \frac{\partial^2 u}{\partial x^2}.$$

The first is the definition of $\pi$; the combination of the two yields the d'Alembert equation. ∎

## 16.6 The equilibrium of continua as a variational problem. Suspended cables

The stationary points of the functional (16.39) with respect to the perturbations described in Postulate 16.1 define the motion of the system. Similarly equilibrium is characterised by the stationary points of the functional

$$V = \int_e \mathcal{V}\left(u, \frac{\partial u}{\partial x_1}, \frac{\partial u}{\partial x_2}, \frac{\partial u}{\partial x_3}, x_1, x_2, x_3\right) d\mathbf{x}, \tag{16.65}$$

where $\mathcal{V}$ is obtained by suppressing the kinematic terms in $\mathcal{L}$, with respect to the subclass of perturbations which vanish on the boundary. Clearly $V$ can be interpreted as the total potential energy. The Euler equation is therefore

$$\sum_i \frac{\partial}{\partial x_i} \frac{\partial \mathcal{V}}{\partial \xi_i} - \frac{\partial \mathcal{V}}{\partial u} = 0 \tag{16.66}$$

(as usual, $\xi_i = \partial u / \partial x_i$), with which one needs to associate possible constraints.

As an application, consider the two problems relating to the equilibrium of suspended cables, where a 'cable' is a homogeneous string, perfectly flexible and of constant length.

*(A) Cable subject only to weight (catenary, Fig. 16.3)*

If $y = f(x)$ is the equation for the generic configuration of the cable, the potential energy is proportional to the height of the centre of mass, defined by

$$\ell y_G(f) = \int_0^{x_B} f(x)\sqrt{1 + f'^2(x)}\,dx. \tag{16.67}$$

This must be minimised by imposing the constraint that the cable has fixed length $\ell$, and hence

$$\int_0^{x_B} \sqrt{1 + f'^2(x)}\,dx = \ell. \tag{16.68}$$

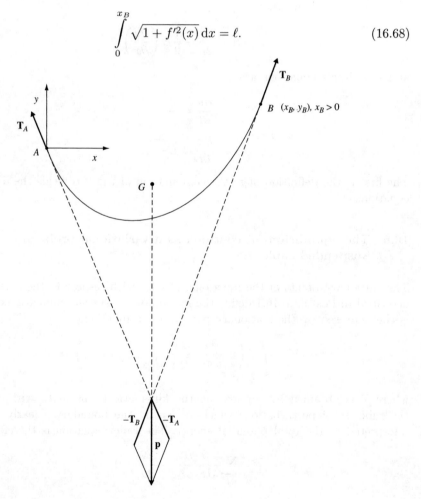

**Fig. 16.3** Suspended cable: catenary.

We now write the Euler equation for the functional

$$\int_0^{x_B} (f - \lambda)\sqrt{1 + f'^2}\, dx.$$

Setting $f - \lambda = g$, the latter becomes

$$F(g, g') = \int_0^{x_B} g\sqrt{1 + g'^2}\, dx, \qquad (16.69)$$

for which the Euler equation takes the form

$$gg' = 1 + g'^2. \qquad (16.70)$$

Recalling the identities $\cosh^2\alpha = 1 + \sinh^2\alpha$ and $(\cosh\alpha)'' = \cosh\alpha$, the general solution of (16.70) is

$$g(x) = \frac{1}{\mu}\cosh(\mu x + c),$$

from which

$$f(x) = \lambda + \frac{1}{\mu}\cosh(\mu x + c). \qquad (16.71)$$

The conditions determining the constants $\mu$ and $c$ and the multiplier $\lambda$ are the boundary conditions at $A$, $B$:

$$\lambda + \frac{1}{\mu}\cosh c = 0, \qquad (16.72)$$

$$\lambda + \frac{1}{\mu}\cosh(\mu x_B + c) = y_B \qquad (16.73)$$

and the constraint (16.68):

$$\ell = \int_0^{x_B} \sqrt{1 + \sinh^2(\mu x + c)}\, dx = \frac{1}{\mu}[\sinh(\mu x_B + c) - \sinh c]. \qquad (16.74)$$

Using the first two we eliminate $\lambda$:

$$y_B = \frac{1}{\mu}[\cosh(\mu x_B + c) - \cosh c], \qquad (16.75)$$

and between the latter and (16.74), isolating the hyperbolic sine and cosine of $\mu x_B + c$, taking the square, and subtracting, we find

$$\mu = \frac{e^{-c}}{\ell - y_B} - \frac{e^c}{\ell + y_B} \qquad (16.76)$$

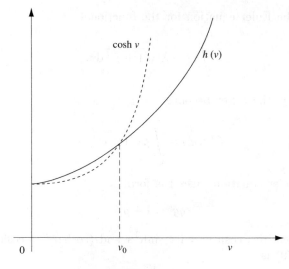

Fig. 16.4

(note that we must have $\ell > |y_B|$). In addition, adding term by term equations (16.77) and (16.75) we obtain

$$\mu(\ell + y_B) = e^c(e^{\mu x_B} - 1), \tag{16.77}$$

and we finally arrive at the equation for $\mu$:

$$\cosh \mu x_B = 1 + \frac{1}{2}\mu^2(\ell^2 - y_B^2). \tag{16.78}$$

Setting $\nu = \mu x_B$, we compare the functions $\cosh \nu$ and $h(\nu) = 1 + \frac{1}{2}\gamma\nu^2$, with $\gamma = (\ell^2 - y_B^2)/x_B^2 > 1$ (Fig. 16.4). Since $h''(\nu) = \gamma > 1$, we have $h(\nu) > \cosh \nu$ in a neighbourhood of the origin. Then the equation $\cosh \nu = h(\nu)$ has a first root $\nu_0$. Since also $\sinh(\bar{\nu}) = h'(\bar{\nu})$ for some $\bar{\nu} \in (0, \nu_0)$ and since $\sinh(\nu) > h'(\nu)$ for $\nu > \bar{\nu}$, there cannot be other roots. Hence $\mu$ is uniquely determined by (16.78) (the sign must be chosen compatibly with the orientation of the axes). The constant $c$ can then be found from (16.77), while $\lambda$ can be obtained from (16.72). The problem is therefore solved.

To determine the tensions at the two endpoints, note that for the second cardinal equation to be satisfied, the tensions act along lines (tangent to the cable) that must intersect on the vertical through the centre of mass of the cable (see Fig. 16.3). Denoting by **p** the weight of the cable, it is sufficient to decompose **p** along the two directions tangent at $A$ and $B$, to obtain $-\mathbf{T}_A$ and $-\mathbf{T}_B$.

The equilibrium profile of a suspended cable is a curve called a *catenary*.

*Remark 16.2*
It is a useful exercise to verify that for every arc of the catenary, the tangents through the endpoints intersect on the vertical through the centre of mass

of the arc. Imposing this condition it is in fact possible to obtain independently the equilibrium profile. ∎

PROPOSITION 16.1 *The horizontal component of the tension is constant along the cable, and the vertical component is equal to the weight of the arc of the catenary between the point considered and the vertex.*

*Proof*
Suppose first of all that the vertex $V$ belongs to the cable. Since the tension at the vertex is horizontal (Fig. 16.5), imposing the equilibrium conditions on each arc $\widehat{PV}$ we see that only the vertical component of $\mathbf{T}(P)$ varies to balance the weight of the arc $\widehat{PV}$. If $V$ does not belong to the cable, it is enough to consider the ideal extension of the cable along the same catenary and apply the same reasoning to the extended cable. ∎

*Remark* 16.3
The property that the horizontal component of the tension is constant depends only on the fact that the external forces distributed along the cable (in this case, gravity) are vertical. Indeed, if $\rho\mathbf{F}$ is the (linear) density of the external forces, the equilibrium equation of the cable is clearly

$$-\rho\mathbf{F}(s) = \frac{\mathrm{d}}{\mathrm{d}s}\mathbf{T}(s), \tag{16.79}$$

where $s$ is the curvilinear coordinate along the cable. ∎

*(B) Suspended bridge*

We now consider the problem of a *bridge suspended* by a cable through a series of hangers numerous enough that the weight can be considered to be distributed along the cable (Fig. 16.6).

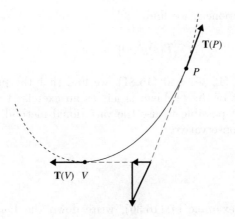

Fig. 16.5

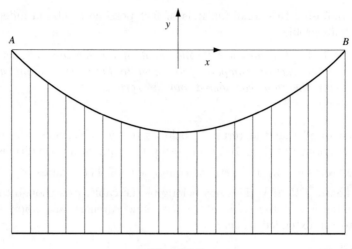

Fig. 16.6

Knowing the length of the cable and the weight of the load (such that we can neglect the weight of the cable), we want to determine the profile of the cable and the tension along it.

Since we neglect the weight of the cable, the force acting on each arc d$s$ is $\gamma$ ($\gamma$ = weight per unit length of the load). Therefore the specific force applied to the cable is $\gamma dx/ds$, and, if $y = f(x)$ describes the profile of the cable, this force has the expression $\gamma/\sqrt{1+f'^2}$. The dependence on $f'$ prevents the definition of a specific potential. The problem must therefore be solved by applying equation (16.79) directly.

We again find that the horizontal component of the tension is constant:

$$[T(s)x'(s)]' = 0. \tag{16.80}$$

For the vertical component we find

$$[T(s)y'(s)]' = \gamma x'(s). \tag{16.81}$$

Eliminating $T$ using (16.80) and (16.81), we find that the profile of the cable is a parabola. The rest of the problem is left as an exercise to the reader.

Note that it is not possible to use the variational method because in this case the forces are non-conservative.

## 16.7 Problems

**1.** Following the example of (16.36), write down the Lagrangian density for the vibrating string.

2. Extend the previous Lagrangian density to the case in which the function $u$ depends on two space variables $x_1$, $x_2$ and deduce the equation for the small vibrations of perfectly flexible elastic membranes.

3. Write down the Hamiltonian density corresponding to the Lagrangian density (16.48) for sound waves and show that Hamilton's equations reduce to wave equations.

## 16.8   Additional solved problems

*Problem 1*
Use equation (16.79) to solve the problem of a suspended cable, computing also the tension $\mathbf{T}(s)$.

*Solution*
Writing the two components of $\mathbf{T}$ in the form $Tdx/ds$, $Tdy/ds$, the equations to be integrated are

$$[T(s)x'(s)]' = 0,$$
$$[T(s)y'(s)]' = \rho g.$$

Since

$$\frac{d}{ds} = \left[1 + \left(\frac{dy}{dx}\right)^2\right]^{-1/2} \frac{d}{dx},$$

eliminating $T$ we arrive at a differential equation for the function $y(x)$:

$$\frac{d^2y}{dx^2} = c\left[1 + \left(\frac{dy}{dx}\right)^2\right]^{1/2},$$

where $c$ is a constant to be determined. This can be integrated by separation of variables, etc. Note that the vertical component of the tension is $\rho gs$ if the origin of the arcs is chosen at the vertex. This is therefore equal to the weight of the cable between the point $P(s)$ and the vertex (whether or not this belongs to the cable).

*Problem 2*
A heavy cable of linear density $\rho$ and length $\ell$ is fixed at the endpoint $A$, and runs without friction on a pulley $B$ at the same height as $A$. The cable is kept in tension by a weight $p$ applied to the other endpoint. Find the equilibrium configuration.

*Solution*
We know that the profile of the cable between the two suspending points $A$, $B$ is given by (16.71). Since $A$ and $B$ are at the same height, the catenary is

symmetric with respect to the $y$-axis, and hence we write (16.71) in the form

$$f(x) = \frac{1}{\mu}(\cosh \mu x - 1) - \beta. \tag{16.82}$$

Denoting by $2a$ the distance between the points $A$, $B$ (Fig. 16.7), we must impose the condition $f(\pm a) = 0$, and hence

$$\frac{1}{\mu}(\cosh \mu a - 1) = \beta. \tag{16.83}$$

We also know that the tension at the point $B$ is given by $p + \rho g(\ell - \lambda)$, where $\lambda$ is the length of the arc $\widehat{AB}$:

$$\lambda = 2 \int_0^a \sqrt{1 + f'^2}\, dx = \frac{2}{\mu} \sinh a\mu. \tag{16.84}$$

If $\varphi$ is the angle that the tension in $B$ forms with the $x$-axis, we have $\tan \varphi = f'(a) = \sinh a\mu$ and $\sin \varphi = \sinh a\mu / \cosh a\mu$.

We now use Proposition 16.1 to write

$$\frac{\rho g}{\mu} \sinh a\mu = \left[p + \rho g \left(\ell - \frac{2}{\mu} \sinh a\mu\right)\right] \frac{\sinh a\mu}{\cosh a\mu}, \tag{16.85}$$

which, setting $\nu = a\mu$, $\gamma = dx(p/\rho g)$, $\delta = dx(\gamma + l)/a$, reduces to

$$\cosh \nu + 2\sinh \nu = \delta \nu. \tag{16.86}$$

For $\nu > 0$ the left-hand side of (16.86) and all its derivatives are positive. Denote it by $\chi(\nu)$ and define $\nu_0$ such that $\chi'(\nu_0) = \delta$. This equation has only one positive solution $\nu_0(\delta)$, as long as $\delta > \chi'(0) = 2$. It is easy to verify that

$$e^{\nu_0} = \frac{1}{3}\left(\delta + \sqrt{\delta^2 - 3}\right). \tag{16.87}$$

We can now conclude that equation (16.86) is solvable if and only if $\delta \geq \delta_0$, with $\delta_0$ defined by

$$\delta_0 \nu_0(\delta_0) = \chi(\nu_0(\delta_0)). \tag{16.88}$$

We leave as an exercise the proof of the existence of a unique solution $\delta_0 > 2$ to (16.88). For fixed parameters $l$, $a$, $\rho$, the inequality $\delta \geq \delta_0$ becomes a condition on the weight $p$: if this is too small, there cannot exist a solution.

Equation (16.88) has the unique solution $\nu = \nu_0(\delta_0)$, if $\delta = \delta_0$, and has two solutions, $\nu_1(\delta)$, $\nu_2(\delta)$, such that $\nu_1(\delta) < \nu_0(\delta) < \nu_2(\delta)$, when $\delta > \delta_0$.

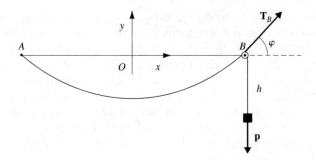

Fig. 16.7

We note that the solutions we obtained must be checked, to make sure they are compatible with the constraint $\lambda < l$, and hence, by (16.84), we have

$$\frac{l}{a}\nu > 2\sinh\nu, \qquad (16.89)$$

fixing a maximum admissible value $\nu^*$ for $\nu$. The solutions $\nu_i(\delta)$ are acceptable if they are in the interval $(0, \nu^*)$.

Note that for $\delta \to \infty$ we have $\nu_1 \to 0$ and $\nu_2 \to \infty$. Therefore for sufficiently large $p$ the problem always admits a solution, in correspondence to the root $\nu_1(\delta)$. The definition of the profile (16.82) of the catenary is completed by equation (16.83), which yields the value of $\beta$. It is possible in particular to study the case $p = 0$, characterising the conditions on $l$ that guarantee the existence of solutions.

# APPENDIX 1: SOME BASIC RESULTS ON ORDINARY DIFFERENTIAL EQUATIONS

## A1.1 General results

In this appendix we list some results of the theory of ordinary differential equations which are especially relevant for the aims of this book. For a more detailed exposition, or for the proofs that we omit, we refer the reader to Hirsch and Smale (1974) and Arnol'd (1978b).

Let $A \subset \mathbf{R}^l$ be an open set and $\mathbf{X} : A \to \mathbf{R}^l$ be a vector field of class $\mathcal{C}^1$. Consider the differential equation

$$\dot{\mathbf{x}} = \mathbf{X}(\mathbf{x}), \qquad (A1.1)$$

with the initial condition $\mathbf{x}(0) = \mathbf{x}_0 \in A$.

THEOREM A1.1 (Existence and uniqueness) *There exist $\delta > 0$ and a unique map $\mathbf{x} : (-\delta, \delta) \to A$, $\mathbf{x} = \mathbf{x}(t)$ of class $\mathcal{C}^1$, which is a solution of (A1.1) satisfying the initial condition $\mathbf{x}(0) = \mathbf{x}_0$.* ■

*Remark* A1.1
It is well known that our hypotheses are stronger than necessary. For the existence of a solution, it is sufficient for $\mathbf{X}$ to be continuous, while to guarantee uniqueness one must assume that $\mathbf{X}$ is locally Lipschitz: for every $\mathbf{x}_0 \in A$ there exist a neighbourhood $U_0 \subset A$ of $\mathbf{x}_0$ and a constant $K_0 > 0$ such that

$$|\mathbf{X}(\mathbf{x}) - \mathbf{X}(\mathbf{y})| \leq K_0 |\mathbf{x} - \mathbf{y}| \qquad (A1.2)$$

for every $\mathbf{x}, \mathbf{y} \in U_0$. ■

*Example* A1.1
The equation $\dot{x} = x^{2/3}$, $x \in \mathbf{R}$, has two distinct solutions such that $x(0) = 0$: $x(t) = 0$ and $x(t) = t^3/27$. ■

THEOREM A1.2 (Continuous dependence on the initial conditions) *Assume that $\mathbf{x}_1(t)$, $\mathbf{x}_2(t)$ are both solutions of (A1.1) in the interval $[0, \bar{t}]$ corresponding to the initial conditions $\mathbf{x}_1(0) = \mathbf{x}_{10}$, $\mathbf{x}_2(0) = \mathbf{x}_{20}$. There exists $K > 0$ such that for every $t \in [0, \bar{t}]$ we have*

$$|\mathbf{x}_1(t) - \mathbf{x}_2(t)| \leq |\mathbf{x}_{10} - \mathbf{x}_{20}| e^{Kt}. \qquad (A1.3)$$

■

*Remark* A1.2
$K$ can be chosen to be equal to the Lipschitz constant of $\mathbf{X}$. ■

*Remark A1.3*
The estimate (A1.3) is sharp. This can be verified by considering the equation $\dot{x} = kx$, whose solutions are $x(t) = x(0)e^{kt}$. ∎

In general the solutions of (A1.1) are not defined for every $t$. Theorem A1.1 guarantees existence of a solution only in an interval $(-\delta, \delta) \subseteq \mathbf{R}$. However it is not difficult to prove that for every $\mathbf{x}_0 \in A$ there exists a *maximal* open interval $(t_1, t_2)$, $0 \in (t_1, t_2)$, in which there exists a solution $\mathbf{x}(t)$ of (A1.1), satisfying the initial condition $\mathbf{x}(0) = \mathbf{x}_0$. Note that it is possible that $t_1 = -\infty$, or $t_2 = +\infty$, or both. If $t_1 = -\infty$, $t_2 = +\infty$ the solution $\mathbf{x}(t)$ is *global*.

The following theorem illustrates the behaviour of the solutions which are *not* global.

THEOREM A1.3  *If $\mathbf{x}(t)$ is a solution of (A1.1) whose maximal interval of definition is $(t_1, t_2)$ for bounded $t_2$, for every compact set $C \subseteq A$ there exists $t \in (t_1, t_2)$ such that $\mathbf{x}(t) \notin C$.* ∎

*Remark A1.4*
By Theorem A1.3, if $\mathbf{x}(t)$ is not a global solution, when $t \to t_2$ then either $\mathbf{x}(t) \to \partial A$ or $|\mathbf{x}(t)| \to +\infty$. ∎

*Example A1.2*
The equation $\dot{x} = -x + x^2$, $x \in \mathbf{R}$, has the solution $x(t) = -x(0)[e^t(x(0) - 1) - x(0)]^{-1}$. This solution is not global if $x(0) > 1$ or $x(0) < 0$. ∎

A result on the continuous dependence on the data, frequently used in Chapter 12, is the following.

LEMMA A1.1  *Let $\mathbf{X}$ be a vector field of class $\mathcal{C}^1$ and $A$ be an open subset of $\mathbf{R}^n$, such that*

$$\sup_{x \in A} |\mathbf{X}(\mathbf{x})| \leq \varepsilon.$$

*Let $\mathbf{x}(t)$ be a solution of $\dot{\mathbf{x}} = \mathbf{X}(\mathbf{x})$ with the initial condition $\mathbf{x}(0) = \mathbf{x}_0 \in A$, and let $(t_1, t_2)$ be the maximal interval of definition of $\mathbf{x}(t)$. Then*

$$|\mathbf{x}(t) - \mathbf{x}_0| \leq \varepsilon t$$

*for every $t \in (t_1, t_2)$.*

*Proof*
The function $f : (t_1, t_2) \to \mathbf{R}$, defined by $f(t) = |\mathbf{x}(t) - \mathbf{x}_0|$ satisfies $f(0) = 0$ and

$$\left| \frac{df}{dt}(t) \right| = \left| \frac{\mathbf{x}(t) - \mathbf{x}_0}{|\mathbf{x}(t) - \mathbf{x}_0|} \cdot \mathbf{X}(\mathbf{x}(t)) \right| \leq |\mathbf{X}(\mathbf{x}(t))| \leq \varepsilon$$

## A1.2 Systems of equations with constant coefficients

for every $t \in (t_1, t_2)$, from which it follows that

$$|f(t)| = \left| \int_0^t \frac{df}{dt}(t)\,dt \right| \le \int_0^t \left| \frac{df}{dt} \right| dt \le \varepsilon t. \qquad \blacksquare$$

### A1.2 Systems of equations with constant coefficients

In a neighbourhood of a singular point $\mathbf{x}_0$ one can obtain useful information about the solutions by considering the linearised equations, following a procedure analogous to the one described in Section 4.10. Setting $\mathbf{x} = \mathbf{x}_0 + \mathbf{y}$, substituting the latter into (A1.1) and expanding the result in Taylor series in $\mathbf{X}(\mathbf{x}_0 + \mathbf{y})$ to first order, *neglecting the remainder term* one arrives at the system

$$\dot{\mathbf{y}} = A\mathbf{y}, \qquad (A1.4)$$

where $A = \left( \dfrac{\partial X_i}{\partial x_j}(\mathbf{x}_0) \right)$.

The system of ordinary differential equations with constant coefficients (A1.4) can immediately be integrated: the solution corresponding to the initial condition $\mathbf{y}(0) = \mathbf{y}_0$ is given by $\mathbf{y}(t) = e^{tA}\mathbf{y}_0$, where

$$e^{tA} := \sum_{n=0}^{\infty} \frac{t^n}{n!} A^n \qquad (A1.5)$$

(cf. e.g. Arnol'd (1978b), section 14)).

*Example A1.3*

The matrix $A = \begin{pmatrix} 2 & 1 & 3 \\ 0 & 2 & 0 \\ 1 & 0 & 0 \end{pmatrix}$ has eigenvalues $\{-1, 2, 3\}$ to which there correspond

the eigenvectors $\begin{pmatrix} 1 \\ 0 \\ -1 \end{pmatrix}, \begin{pmatrix} 2 \\ -3 \\ 1 \end{pmatrix}, \begin{pmatrix} 3 \\ 0 \\ 1 \end{pmatrix}$.

Setting $M = \begin{pmatrix} 1 & 2 & 3 \\ 0 & -3 & 0 \\ -1 & 1 & 1 \end{pmatrix}$ one has $M^{-1} = \dfrac{1}{12}\begin{pmatrix} 3 & -1 & -9 \\ 0 & -4 & 0 \\ 3 & 3 & 3 \end{pmatrix}$ and

$A = M \begin{pmatrix} -1 & 0 & 0 \\ 0 & 2 & 0 \\ 0 & 0 & 3 \end{pmatrix} M^{-1}$, and therefore

$$e^{tA} = \sum_{n=0}^{\infty} \frac{t^n}{n!} A^n = \sum_{n=0}^{\infty} \frac{t^n}{n!} M \begin{pmatrix} -1 & 0 & 0 \\ 0 & 2 & 0 \\ 0 & 0 & 3 \end{pmatrix}^n M^{-1} = M \begin{pmatrix} e^{-t} & 0 & 0 \\ 0 & e^{2t} & 0 \\ 0 & 0 & e^{3t} \end{pmatrix} M^{-1}.$$

$\blacksquare$

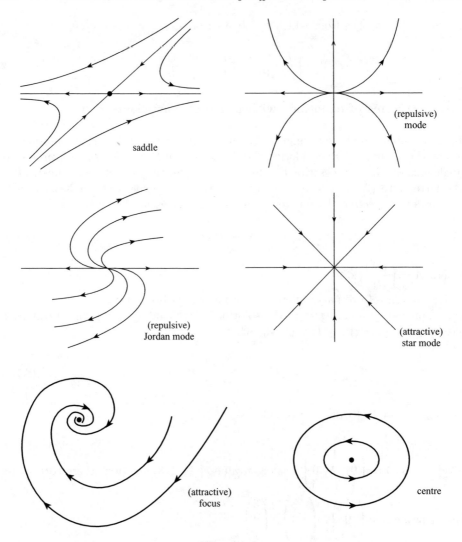

**Fig. A1.1**

We briefly summarise the behaviour of the solutions of (A1.4) in the case $\mathbf{y} \in \mathbf{R}^2$, with the help of Fig. A1.1.

Let $\lambda_1, \lambda_2 \in \mathbf{C}$ be the eigenvalues of $A$. We distinguish two cases:

*Case I*: the eigenvalues of $A$ are real;

*Case II*: the eigenvalues of $A$ are complex conjugates.

*Case I*: We need to distinguish various subcases. All eigenvalues of $A$ are real.

(I.1) $\lambda_1 < \lambda_2 < 0$ (*attracting node*)

Let $\mathbf{v}_1, \mathbf{v}_2$ be the eigenvectors corresponding to $\lambda_1, \lambda_2$. Setting $\mathbf{y}(t) = \eta_1(t)\mathbf{v}_1 + \eta_2(t)\mathbf{v}_2$ we find $\dot\eta_i = \lambda_i \eta_i$, $i = 1, 2$. Therefore $\mathbf{y}(t) = c_1 e^{t\lambda_1}\mathbf{v}_1 + c_2 e^{t\lambda_2}\mathbf{v}_2$

and the constants can be determined by decomposing the initial condition $y(0) = c_1 \mathbf{v} + c_2 \mathbf{v}_2$ in the basis of $\mathbf{R}^2$ given by $\mathbf{v}_1$ and $\mathbf{v}_2$. When $t \to \infty$ we have $\mathbf{y}(t) \to 0$ and the trajectory in the phase plane is tangent in $\mathbf{y} = 0$ to $\mathbf{v}_2$ (except if $\mathbf{y}(0) = c_1 \mathbf{v}_1$).

(I.2) $0 < \lambda_1 < \lambda_2$ (*repulsive node*)

The discussion for case (I.1) can be repeated for the limit $t \to -\infty$.

(I.3) $\lambda_1 < 0 < \lambda_2$ (*saddle*)

The solutions are $\mathbf{y}(t) = c_1 e^{t\lambda_1} \mathbf{v}_1 + c_2 e^{t\lambda_2} \mathbf{v}_2$, and hence are asymptotic to the direction $\mathbf{v}_1$ for $t \to +\infty$ and $\mathbf{v}_2$ for $t \to -\infty$.

(I.4) $\lambda_1 = \lambda_2$ and $A$ diagonalisable (*star node*)

In this case $A = \lambda_1 \begin{pmatrix} 1 & 0 \\ 0 & 1 \end{pmatrix}$, every vector of the plane is an eigenvector and the trajectories are rays of the form $\mathbf{y}(t) = \mathbf{y}(0) e^{\lambda_1 t}$.

(I.5) $\lambda_1 = \lambda_2$ and $A$ non-diagonalisable (*Jordan node*)

By an invertible linear transformation, $A$ can be reduced to a Jordan block $\begin{pmatrix} \lambda_1 & k \\ 0 & \lambda_1 \end{pmatrix}$, $k \neq 0$. If $\lambda_1 \neq 0$, the trajectories of the resulting system have equation $x_1 = (k/\lambda_1) x_2 \log|x_2/c|$.
The case $\lambda_1 = 0$ is trivial.

*Case II*: Since the eigenvalue equation is $\lambda^2 - \mathrm{Tr}(A)\lambda + \det(A) = 0$, setting $\theta = \frac{1}{2} \mathrm{Tr}(A)$ and $\omega = \sqrt{\det A - \theta^2}$ we obtain $\lambda_1 = \theta + i\omega$, $\lambda_2 = \theta - i\omega$. Note that in this case, $\det A > \theta^2$.

The matrix $A$ is diagonalisable in the complex field, and we can easily see that

$$S^{-1} \begin{pmatrix} \theta + i\omega & 0 \\ 0 & \theta - i\omega \end{pmatrix} S = \begin{pmatrix} \theta & -\omega \\ \omega & \theta \end{pmatrix},$$

with $S = \begin{pmatrix} 1 & i \\ i & 1 \end{pmatrix}$. We can therefore reduce to the case

$$A = \begin{pmatrix} \theta & -\omega \\ \omega & \theta \end{pmatrix},$$

where complex numbers do not appear. The corresponding differential system is

$$\dot{x}_1 = \theta x_1 - \omega x_2,$$
$$\dot{x}_2 = \omega x_1 + \theta x_2.$$

To study the trajectories in the plane $(x_1, x_2)$ it is convenient to change to polar coordinates $(r, \varphi)$ for which the equations decouple:

$$\dot{r} = \theta r, \quad \dot{\varphi} = \omega.$$

Hence we simply obtain $\dfrac{dr}{d\varphi} = \dfrac{\theta}{\omega} r$ and finally $r = r_0 e^{(\theta/\omega)\varphi}$.

Now the classification is evident.

(II.1) $\theta = 0$ (*centre*)
The trajectories are circles.

(II.2) $\theta \neq 0$ (*focus*)
The trajectories are spirals converging towards the centre if $\theta < 0$ (*attractive case*), and they move away from the centre if $\theta > 0$ (*repulsive case*).

*Remark* A1.5
A particularly interesting case for mechanics is when $A$ is Hamiltonian, i.e. a $2 \times 2$ matrix with trace zero:

$$A = \begin{pmatrix} a & b \\ c & -a \end{pmatrix}$$

which corresponds to the quadratic Hamiltonian $H = \frac{1}{2}(cx_1^2 - 2ax_1x_2 - bx_2^2)$.

The equation for the eigenvalues is $\lambda^2 + \det A = 0$. Therefore we have the following cases:

(1) $\det A = -(a^2 + bc) > 0$, the origin is a centre (this is the only stable case);
(2) $\det A < 0$, the eigenvalues are real and opposite (saddle point);
(3) $\det A = 0$, the eigenvalues are both zero (finish as an exercise). ∎

In the $n$-dimensional case, suppose that $\mathbf{u}_1, \ldots, \mathbf{u}_n$ is a basis of $\mathbf{R}^n$ of eigenvectors of $A$. Exploiting the invariance of the eigenspaces of $A$ for $e^{tA}$ we can better understand the behaviour of the solutions of (A1.4) by introducing

$$E^s = \left\{ \mathbf{v} \in \mathbf{R}^n, \ \mathbf{v} = \sum_{i=1}^{n_s} v_i \mathbf{u}_i \text{ where } A\mathbf{u}_i = \lambda_i \mathbf{u}_i, \ \Re\lambda_i < 0, \ i = 1, \ldots, n_s \right\}$$

$$E^u = \left\{ \mathbf{v} \in \mathbf{R}^n, \ \mathbf{v} = \sum_{i=n_s+1}^{n_u+n_s} v_i \mathbf{u}_i \text{ where } A\mathbf{u}_i = \lambda_i \mathbf{u}_i, \ \Re\lambda_i > 0, \ i = n_s+1, \ldots, n_u \right\}$$

$$E^c = \left\{ \mathbf{v} \in \mathbf{R}^n, \ \mathbf{v} = \sum_{i=n_s+n_u+1}^{n} v_i \mathbf{u}_i \right.$$

$$\left. \text{where } A\mathbf{u}_i = \lambda_i \mathbf{u}_i, \ \Re\lambda_i = 0, \ i = n_u + n_s + 1, \ldots, n \right\}.$$

These subspaces of $\mathbf{R}^n$ are invariant under $e^{tA}$ and are called, respectively, the *stable subspace* $E^s$, *unstable subspace* $E^u$ and *central subspace* $E^c$. Clearly $\mathbf{R}^n = E^s \oplus E^u \oplus E^c$.

*Example* A1.4

Assume $A = \begin{pmatrix} -1 & -1 & 0 \\ 1 & -1 & 0 \\ 0 & 0 & 2 \end{pmatrix}$. Then $E^u = \{(0, 0, y_3), y_3 \in \mathbf{R}\}$, corresponding to the eigenvalue $\lambda_3 = 2$; $E^s = \{(y_1, y_2, 0), (y_1, y_2) \in \mathbf{R}^2\}$, corresponding to the eigenvalues $\lambda_1 = -1 - i$, $\lambda_2 = -1 + i$. The restriction $A|_{E^s}$ has an attractive focus at the origin (see Fig. A1.2). ∎

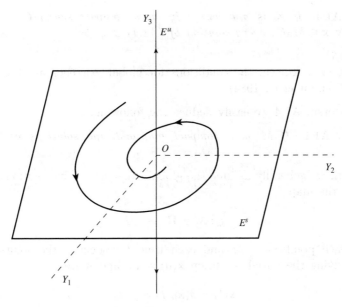

**Fig. A1.2**

DEFINITION A1.1  *A point $x_0$ is called* singular *if $X(x_0) = 0$.* ∎

THEOREM A1.4 (Rectification)  *If $x_0$ is not a singular point, there exists a neighbourhood $V_0$ of $x_0$ and an invertible coordinate transformation $y = y(x)$, defined on $V_0$ and of class $\mathcal{C}^1$ which transforms the equation (A1.1) into*

$$\dot{y}_1 = 1, \qquad \dot{y}_i = 0, \quad i = 2, \ldots, l. \tag{A1.6}$$

∎

*Remark* A1.6
If $X$ is of class $\mathcal{C}^r$, $1 \leq r \leq \infty$, the transformation $y$ is also of class $\mathcal{C}^r$. ∎

## A1.3 Dynamical systems on manifolds

The problem of the global existence of solutions of ordinary differential equations can be formulated in greatest generality in the context of differentiable manifolds. The existence, uniqueness, continuous dependence and rectification theorems are easily extended to the case of differential equations on manifolds.

Let $M$ be a differentiable manifold of dimension $l$, and $X : M \to TM$ be a $\mathcal{C}^1$ vector field. A curve $x : (t_1, t_2) \to M$ is a *solution* of the differential equation (A1.1) on the manifold $M$ if it is an *integral curve* of $X$, and hence if for every $t \in (t_1, t_2)$ the vector $\dot{x}(t) \in T_{x(t)}M$ satisfies $\dot{x}(t) = X(x(t))$ (note that by definition $X(x(t)) \in T_{x(t)}M$).

THEOREM A1.4  *If* **X** *is not zero only on a compact subset* $C \subseteq M$, *i.e. if* $\mathbf{X}(\mathbf{x}) = 0$, $\forall\ \mathbf{x} \in M \backslash C$, *every solution of (A1.1) is global.* ∎

*Remark A1.7*
If $M = \mathbf{R}^l$, as is known, the conditions for global existence are less restrictive (see, e.g. Piccinini et al. 1984). ∎

From Theorem A1.4 we easily deduce the following.

COROLLARY A1.1  *If $M$ is a compact manifold, the solutions of (A1.1) are global.* ∎

Henceforth we generally assume the global existence of the solutions of (A1.1). Consider the map

$$g : M \times \mathbf{R} \longrightarrow M,$$

which to each point $\mathbf{x}_0 \in M$ and each time $t$ associates the solution $\mathbf{x}(t)$ of (A1.1) satisfying the initial condition $\mathbf{x}(0) = \mathbf{x}_0$, and write

$$\mathbf{x}(t) = g(\mathbf{x}_0, t) = g^t \mathbf{x}_0.$$

Clearly

$$g(\mathbf{x}_0, 0) = g^0 \mathbf{x}_0 = \mathbf{x}_0, \qquad (A1.7)$$

for every $\mathbf{x}_0 \in M$, and from the uniqueness theorem it follows that $g^t$ is invertible:

$$\mathbf{x} = g^t \mathbf{x}_0 \Leftrightarrow \mathbf{x}_0 = g^{-t} \mathbf{x}. \qquad (A1.8)$$

Hence, for every $t \in \mathbf{R}$, $g^t$ is a *diffeomorphism* of $M$. In addition, for every $t$, $s \in \mathbf{R}$ and for every $\mathbf{x}_0 \in M$ we have

$$g^t(g^s \mathbf{x}_0) = g^{t+s} \mathbf{x}_0. \qquad (A1.9)$$

DEFINITION A1.2  *A one-parameter family $(g^t)_{t \in \mathbf{R}}$ of diffeomorphisms of $M$ satisfying the properties (A1.7)–(A1.9) is called a* one-parameter group of diffeomorphisms. ∎

*Remark A1.8*
A one-parameter group of diffeomorphisms of $M$ defines an *action* (cf. Section 1.8) of the additive group $\mathbf{R}$ on the manifold $M$. ∎

The manifold $M$ is called the *phase space* of the differential equation (A1.1), and the group $g^t$ is called the *phase flow* of the equation. The integral curve of the field **X** passing through $\mathbf{x}_0$ at time $t = 0$ is given by $\{\mathbf{x} \in M | \mathbf{x} = g^t \mathbf{x}_0\}$ and it is also called the *phase curve*.[1]

We can now give the *abstract* definition of a *dynamical system* on a manifold.

---

[1] The phase curves are therefore the orbits of the points of $M$ under the action of $\mathbf{R}$, determined by the phase flow.

DEFINITION A1.3  *A dynamical system on a manifold $M$ is an action of $\mathbf{R}$ on $M$.* ∎

Clearly the phase flow associated with a differential equation on a manifold is an example of a dynamical system on a manifold. Indeed, the two notions are equivalent.

THEOREM A1.5  *Every dynamical system on a manifold $M$ determines a differential equation on $M$.*

*Proof*
Let $g : \mathbf{R} \times M \to M$ be the given dynamical system; we denote by $g^t = g(t, \cdot)$ the associated one-parameter group of diffeomorphisms. The vector field

$$\mathbf{X}(\mathbf{x}) = \left.\frac{\partial g^t}{\partial t}(\mathbf{x})\right|_{t=0} \tag{A1.10}$$

is called the *infinitesimal generator* of $g^t$. Setting $\mathbf{x}(t) = g^t \mathbf{x}_0$, it is easy to verify that $\mathbf{x}(t)$ is the solution of (A1.1) with initial condition $\mathbf{x}(0) = \mathbf{x}_0$, where $\mathbf{X}$ is given by (A1.10). Indeed,

$$\begin{aligned}\dot{\mathbf{x}}(t) &= \lim_{\Delta t \to 0} \frac{g^{t+\Delta t}\mathbf{x}_0 - g^t \mathbf{x}_0}{\Delta t} \\ &= \lim_{\Delta t \to 0} \frac{g^{\Delta t}\mathbf{x}(t) - g^0 \mathbf{x}(t)}{\Delta t} = \mathbf{X}(\mathbf{x}(t)).\end{aligned} \tag{A1.11}$$

∎

*Remark A1.9*
An interesting notion connected to the ones just discussed is that of a *discrete dynamical system*, obtained by substituting $t \in \mathbf{R}$ with $t \in \mathbf{Z}$ in the definition of a one-parameter group of diffeomorphisms. For example, if $f : M \to M$ is a diffeomorphism, setting $f^0 = \mathrm{id}_M$, the identity on $M$, and $f^n = f \circ \cdots \circ f$ $n$ times, $f^{-n} = f^{-1} \circ \cdots \circ f^{-1}$ $n$ times, we see that $(f^n)_{n \in \mathbf{Z}}$ is a discrete dynamical system.

The study of discrete dynamical systems is as interesting as that of ordinary differential equations (see Hirsch and Smale 1974, Arrowsmith and Place 1990, and Giaquinta and Modica 1999). ∎

Besides the singular points, i.e. the fixed points of the infinitesimal generator, particularly important orbits of a dynamical system are the *periodic* orbits $\mathbf{x}(t) = g^t \mathbf{x}_0 = g^{t+T}\mathbf{x}_0 = \mathbf{x}(t+T)$ for every $t \in \mathbf{R}$. The *period* is $\min\{T \in \mathbf{R}$ such that $\mathbf{x}(t+T) = \mathbf{x}(t), \forall\, t \in \mathbf{R}\}$.

In the case of dynamical systems on the plane or on the sphere, the dynamics are described asymptotically by periodic orbits or by singular points. To make this idea more precise we introduce the $\omega$-*limit set* of a point $\mathbf{x}_0$ (cf. Problem 15 of Section 13.13, for the notion of an $\omega$-limit set in the discrete case): $\omega(\mathbf{x}_0) = \bigcap_{t_0 > 0} \overline{\{g^t \mathbf{x}_0,\ t \geq t_0\}}$. It is immediate to verify that $\mathbf{x} \in \omega(\mathbf{x}_0)$ if and only if there exists a sequence $t_n \to \infty$ such that $g^{t_n}\mathbf{x}_0 \to \mathbf{x}$ for $n \to \infty$.

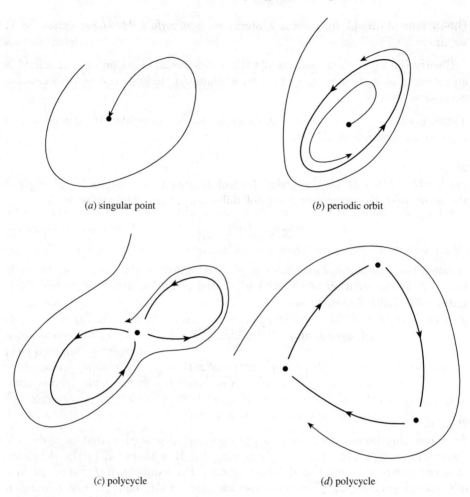

(a) singular point  (b) periodic orbit
(c) polycycle  (d) polycycle

Fig. A1.3

THEOREM A1.6 (Poincaré–Bendixon)  *Assume that the orbit $\{g^t \mathbf{x}_0,\ t \geq 0\}$ of a dynamical system on the plane (or on the two-dimensional sphere) is contained in a bounded open set. Then the $\omega$-limit set of $\mathbf{x}_0$ is necessarily a singular point or a periodic orbit or a polycycle, and hence the union of singular points and of phase curves each tending for $t \to \pm\infty$ to a singular point (not necessarily the same for all) (see Fig. A1.3).*

Only in dimension greater than two can the behaviour of a dynamical system be significantly more complex, including the possibility of chaotic motions, whose study employs the ideas of ergodic theory, introduced in Chapter 13.

# APPENDIX 2: ELLIPTIC INTEGRALS AND ELLIPTIC FUNCTIONS

The elliptic integrals owe their name to the fact that Wallis (in 1655) first introduced them in the calculation of the length of an arc of an ellipse. In their most general form they are given by

$$\int R(x,y)\,dx, \qquad (A2.1)$$

where $R$ is a rational function of its arguments and $y = \sqrt{P(x)}$, with $P$ a fourth-degree polynomial. Legendre showed in 1793 that every elliptic integral (A2.1) can be expressed as the sum of elementary functions plus a combination of integrals of the following *three kinds*:

(1) $$F(\varphi, k) = \int_0^\varphi \frac{d\psi}{\sqrt{1 - k^2 \sin^2 \psi}} = \int_0^z \frac{dx}{\sqrt{(1-x^2)(1-k^2x^2)}}, \qquad (A2.2)$$

(2) $$E(\varphi, k) = \int_0^\varphi \sqrt{1 - k^2 \sin^2 \psi}\, d\psi = \int_0^z \sqrt{\frac{1-k^2x^2}{1-x^2}}\, dx, \qquad (A2.3)$$

(3) $$\Pi(\varphi, k, n) = \int_0^\varphi \frac{d\psi}{(1+n \sin^2 \psi)\sqrt{1 - k^2 \sin^2 \psi}}$$

$$= \int_0^z \frac{dx}{(1+nx^2)\sqrt{(1-x^2)(1-k^2x^2)}}, \qquad (A2.4)$$

where $z = \sin \varphi$, $\varphi$ is called the *amplitude*, the number $k \in [0,1]$ is called the *modulus* and $n$ is the *parameter* (for elliptic integrals of the third kind). When $\varphi = \pi/2$, the elliptic integrals are called *complete*: we then have *the complete integral of the first kind*:

$$\mathbf{K}(k) = F\left(\frac{\pi}{2}, k\right) = \int_0^{\pi/2} \frac{d\psi}{\sqrt{1 - k^2 \sin^2 \psi}}. \qquad (A2.5)$$

It is easy to check that $\mathbf{K}(k)$ is a strictly increasing function of $k$, and $\mathbf{K}(0) = \pi/2$, while $\lim_{k \to 1^-} \mathbf{K}(k) = +\infty$. In addition, it admits the series expansion

$$\mathbf{K}(k) = \frac{\pi}{2}\left[1 + \sum_{n=1}^\infty \left(\frac{(2n-1)!!}{(2n)!!}\right)^2 k^{2n}\right]. \qquad (A2.6)$$

Indeed, expanding as a series $(1 - k^2 \sin^2 \psi)^{-1/2}$ we find

$$\int_0^{\pi/2} \frac{d\psi}{\sqrt{1 - k^2 \sin^2 \psi}} = \frac{\pi}{2}\left[1 + \frac{2}{\pi}\sum_{n=1}^{\infty} \frac{(2n-1)!!}{2^n n!} k^{2n} \int_0^{\pi/2} (\sin \psi)^{2n}\, d\psi\right],$$

from which equation (A2.6) follows, taking into account that

$$\int_0^{\pi/2} (\sin \psi)^{2n}\, d\psi = \frac{1}{2^{2n}} \binom{2n}{n} \frac{\pi}{2}$$

and the identity

$$\frac{(2n-1)!!}{2^{3n}(n!)^3}(2n)! = \left(\frac{(2n-1)!!}{(2n)!!}\right)^2,$$

that can be proved by induction.

Similarly we introduce *the complete integral of the second kind*:

$$\mathbf{E}(k) = E\left(\frac{\pi}{2}, k\right) = \int_0^{\pi/2} \sqrt{1 - k^2 \sin^2 \psi}\, d\psi. \tag{A2.7}$$

Setting $u = F(\varphi, k)$, the problem of the inversion of the elliptic integral consists of finding the unknown function $\varphi(u, k)$, and hence the amplitude as a function of $u$, for fixed $k$:

$$\varphi = \operatorname{am}(u). \tag{A2.8}$$

This is possible because $\partial F/\partial \varphi \neq 0$. The sine and cosine of $\varphi$ are called the *sine amplitude* and *cosine amplitude* of $u$ and are denoted sn and cn:

$$\begin{aligned}\operatorname{sn}(u) &= \sin \operatorname{am}(u),\\ \operatorname{cn}(u) &= \cos \operatorname{am}(u).\end{aligned} \tag{A2.9}$$

When it is necessary to stress the dependence on $k$ we write $\operatorname{sn}(k, u)$, etc.

We also set

$$\operatorname{dn}(u) = \sqrt{1 - k^2 \operatorname{sn}^2(u)}, \tag{A2.10}$$

and the function $\operatorname{dn}(u)$ is called the *delta amplitude*. The functions $\operatorname{sn}(u)$, $\operatorname{cn}(u)$ and $\operatorname{dn}(u)$ are the *Jacobi elliptic functions*, and as we have seen, they appear in the solution of the equation of motion in various problems of mechanics (see Chapters 3 and 7). The functions sn and cn are periodic of period $4\mathbf{K}(k)$, while dn is periodic of period $2\mathbf{K}(k)$.

In addition sn is odd, while cn and dn are even functions; sn and cn take values in the interval $[-1, 1]$, while dn takes values in the interval $[\sqrt{1 - k^2}, 1]$.

The following are important identities:

$$\operatorname{sn}^2(u) + \operatorname{cn}^2(u) = 1,$$
$$\operatorname{dn}^2(u) + k^2 \operatorname{sn}^2(u) = 1,$$
$$\operatorname{sn}(0) = \operatorname{sn}(2\mathbf{K}) = 0,$$
$$\operatorname{sn}(\mathbf{K}) = -\operatorname{sn}(3\mathbf{K}) = 1,$$

(A2.11)

and differentiation formulas:

$$\frac{d}{du}\operatorname{sn}(u) = \operatorname{cn}(u)\operatorname{dn}(u),$$
$$\frac{d}{du}\operatorname{cn}(u) = -\operatorname{sn}(u)\operatorname{dn}(u),$$
$$\frac{d}{du}\operatorname{dn}(u) = -k^2 \operatorname{sn}(u)\operatorname{cn}(u).$$

(A2.12)

If $k = 0$, the elliptic functions reduce to simple trigonometric functions (in this case $\varphi = u$), while for $k = 1$ the elliptic functions are no longer periodic and can be expressed through hyperbolic functions:

$$k = 1, \quad \operatorname{sn}(u) = \tanh(u), \quad \operatorname{cn}(u) = \operatorname{dn}(u) = \frac{1}{\cosh(u)}. \quad (A2.13)$$

The Jacobi elliptic functions, as functions of the complex variable $u \in \mathbf{C}$, have the following complex periods: $\operatorname{sn}(u+2\mathrm{i}\mathbf{K}(k')) = \operatorname{sn}(u)$, $\operatorname{cn}(u+2(\mathbf{K}(k)+\mathrm{i}\mathbf{K}(k'))) = \operatorname{cn}(u)$ and $\operatorname{dn}(u + 4\mathrm{i}\mathbf{K}(k')) = \operatorname{dn}(u)$, where $k' = \sqrt{1-k^2}$. They are therefore an example of *doubly periodic* functions.

For more information on elliptic functions and integrals, and for a more detailed study, we refer the reader to the beautiful classical books of Whittaker and Watson (1927) and Tricomi (1937). An interesting exposition of the history of elliptic functions can be found in Dieudonné (1978, chapter 7).

# APPENDIX 3: SECOND FUNDAMENTAL FORM OF A SURFACE

As seen in Section 1.6, the first fundamental form of a surface $S$ expresses in the tangent space the notion of a scalar product of the Euclidean space in which the surface is embedded, and allows one to measure lengths, angles and areas.

For planar curves the curvature measures how much the curve is far from being straight. To quantify how much a surface $S$ in three-dimensional Euclidean space deviates from the tangent plane at one of its points $P$, one can study the unit normal vector of $S$ in a neighbourhood of $P$.

The second fundamental form of a surface, which we discuss here, expresses precisely the rate of change of the normal to the surface $S$ for infinitesimal displacements on the surface. Since there exist two independent directions to move along the surface, the second fundamental form is a quadratic form.

Let $S$ be a regular surface, and $\mathbf{x}(u,v)$ be a local parametrisation. Let $\mathbf{n}$ be the normal unit vector

$$\mathbf{n} = \frac{\mathbf{x}_u \times \mathbf{x}_v}{|\mathbf{x}_u \times \mathbf{x}_v|} = \frac{\mathbf{x}_u \times \mathbf{x}_v}{\sqrt{EG - F^2}}. \tag{A3.1}$$

Consider a curve $s \to \mathbf{x}(s)$ on the surface $S$ parametrised by the arc length parameter $s$. Let $\mathbf{t}$ be the tangent unit vector to the curve and $k(s)$ its curvature. The *curvature vector* of the curve

$$\mathbf{k} = \frac{d\mathbf{t}}{ds}, \tag{A3.2}$$

whose modulus is the curvature $k(s)$, admits a unique decomposition

$$\mathbf{k} = \mathbf{k}_n + \mathbf{k}_g \tag{A3.3}$$

into two vectors: the *normal curvature vector*

$$\mathbf{k}_n = (\mathbf{k} \cdot \mathbf{n})\mathbf{n} \in (T_{\mathbf{x}(s)}S)^\perp, \tag{A3.4}$$

and the *geodesic curvature vector*

$$\mathbf{k}_g = \mathbf{k} - \mathbf{k}_n \in T_{\mathbf{x}(s)}S. \tag{A3.5}$$

The modulus $k_g = |\mathbf{k}_g|$ is called the *geodesic curvature* of the curve. We observe that if the curve is a geodesic, then its geodesic curvature is zero.

Since $\mathbf{t} \in T_{\mathbf{x}(s)}S$, it must be that $\mathbf{t} \cdot \mathbf{n} = 0$, and hence by differentiation it follows that the *normal curvature* $k_n$ has the expression

$$k_n = \mathbf{k} \cdot \mathbf{n} = -\frac{d\mathbf{x}}{ds} \cdot \frac{d\mathbf{n}}{ds}.$$

On the other hand $(ds)^2 = d\mathbf{x} \cdot d\mathbf{x}$, and therefore

$$k_n = -\frac{d\mathbf{x} \cdot d\mathbf{n}}{d\mathbf{x} \cdot d\mathbf{x}}. \qquad (A3.6)$$

Using the parametrisation of the surface, we have that

$$\begin{aligned} d\mathbf{x} &= \mathbf{x}_u\, du + \mathbf{x}_v\, dv, \\ d\mathbf{n} &= \mathbf{n}_u\, du + \mathbf{n}_v\, dv, \end{aligned} \qquad (A3.7)$$

where $\mathbf{n}_u = \partial\mathbf{n}/\partial u$ and $\mathbf{n}_v = \partial\mathbf{n}/\partial v$. Inserting the equations (A3.7) into the expression (A3.6) for the normal curvature we find

$$k_n = -\frac{(\mathbf{x}_u \cdot \mathbf{n}_u)(du)^2 + (\mathbf{x}_u \cdot \mathbf{n}_v + \mathbf{x}_v \cdot \mathbf{n}_u)(du)(dv) + (\mathbf{x}_v \cdot \mathbf{n}_v)(dv)^2}{E(du)^2 + 2F(du)(dv) + G(dv)^2}. \qquad (A3.8)$$

DEFINITION A3.1 *The numerator* $-d\mathbf{x} \cdot d\mathbf{n}$ *of (A3.8) is called the* second fundamental form *of the surface* $S$. *It is a quadratic form on the tangent space to the surface* $S$, *given by*

$$-d\mathbf{x} \cdot d\mathbf{n} = e(u,v)(du)^2 + 2f(u,v)(du)(dv) + g(u,v)(dv)^2, \qquad (A3.9)$$

*where*

$$\begin{aligned} e(u,v) &= -\mathbf{x}_u \cdot \mathbf{n}_u = \mathbf{x}_{uu} \cdot \mathbf{n}, \\ 2f(u,v) &= -(\mathbf{x}_u \cdot \mathbf{n}_v + \mathbf{x}_v \cdot \mathbf{n}_u) = 2\mathbf{x}_{uv} \cdot \mathbf{n}, \\ g(u,v) &= -\mathbf{x}_v \cdot \mathbf{n}_v = \mathbf{x}_{vv} \cdot \mathbf{n}. \end{aligned} \qquad (A3.10)$$

∎

It is immediate to check that the following relations hold:

$$\begin{aligned} e(u,v) &= \frac{\mathbf{x}_{uu} \cdot \mathbf{x}_u \times \mathbf{x}_v}{\sqrt{EG-F^2}} = \frac{1}{\sqrt{EG-F^2}} \begin{vmatrix} x_{uu} & y_{uu} & z_{uu} \\ x_u & y_u & z_u \\ x_v & y_v & z_v \end{vmatrix}, \\ f(u,v) &= \frac{\mathbf{x}_{uv} \cdot \mathbf{x}_u \times \mathbf{x}_v}{\sqrt{EG-F^2}} = \frac{1}{\sqrt{EG-F^2}} \begin{vmatrix} x_{uv} & y_{uv} & z_{uv} \\ x_u & y_u & z_u \\ x_v & y_v & z_v \end{vmatrix}, \\ g(u,v) &= \frac{\mathbf{x}_{vv} \cdot \mathbf{x}_u \times \mathbf{x}_v}{\sqrt{EG-F^2}} = \frac{1}{\sqrt{EG-F^2}} \begin{vmatrix} x_{vv} & y_{vv} & z_{vv} \\ x_u & y_u & z_u \\ x_v & y_v & z_v \end{vmatrix}. \end{aligned} \qquad (A3.11)$$

*Example* A3.1
Consider the sphere of radius $r$ with the parametrisation $\mathbf{x} = r(\cos u \cos v, \cos u \sin v, \sin u)$. The first fundamental form has value $(ds)^2 = r^2(du)^2 + r^2 \cos^2 u (dv)^2$, and hence $\sqrt{EG-F^2} = r^2 \cos u$. From the definition

of the normal unit vector it follows that $\mathbf{n} = -(\cos u \cos v, \cos u \sin v, \sin u)$, and it is immediate to check that the second fundamental form is given by

$$e = r, \quad f = 0, \quad g = r \cos^2 u.$$

∎

*Remark A3.1*

From (A3.8) it follows that the normal curvature $k_n$ depends only on the point $P$ (of coordinates $(u, v)$) on the surface and on the tangent space $T_P S$ (determined by $du/dv$ or by $dv/du$): *all curves through a point $P$ of the surface tangent to the same direction have the same normal curvature*. We can hence study how the normal curvature $k_n$ varies as the direction in a fixed point of the surface varies.

Since the first fundamental form is positive definite, the sign of the normal curvature $k_n$ depends only on the second fundamental form. There are three possible cases.

(1) If at a point $P$ of the surface $eg - f^2 > 0$, the second fundamental form applied to different directions always has the same sign, and the point is then called *elliptic*; the centres of curvature of all the normal sections to the surface passing through the point $P$ lie on the same side of the surface. This situation is satisfied, for example, at all points of a sphere or of an ellipsoid.

(2) If $eg - f^2 = 0$, there exists a direction in which the normal curvature vanishes. The point is then called *parabolic*. An example is given by any point of a cylinder.

(3) If $eg - f^2 < 0$, the second fundamental form changes sign as the direction varies: the surface $S$ crosses its tangent plane and the point is called *hyperbolic*. This is what happens if the point $P$ is a saddle point. ∎

We now look for the directions along which the normal curvature has a maximum or a minimum. A direction in the tangent space $T_P S$ to the surface at the point $P$ is determined by $\lambda = dv/du$, and the expression of the normal curvature $k_n$ in terms of $\lambda$ can be obtained immediately from equation (A3.8):

$$k_n = k_n(\lambda) = \frac{e + 2f\lambda + g\lambda^2}{E + 2F\lambda + G\lambda^2}. \tag{A3.12}$$

Hence the condition for a maximum or a minimum follows from requiring that

$$\frac{dk_n}{d\lambda}(\lambda) = 0,$$

i.e.

$$\frac{2(f + g\lambda)(E + 2F\lambda + G\lambda^2) - 2(F + G\lambda)(e + 2f\lambda + g\lambda^2)}{(E + 2F\lambda + G\lambda^2)^2} = 0. \tag{A3.13}$$

Since the first fundamental form is positive definite, the denominator of (A3.13) is never zero; the condition for the normal curvature to be stationary is

$$(E + F\lambda)(f + g\lambda) = (e + f\lambda)(F + G\lambda), \tag{A3.14}$$

which when substituted into (A3.12) gives

$$k_n = \frac{(e+f\lambda)+\lambda(f+g\lambda)}{(E+F\lambda)+\lambda(F+G\lambda)} = \frac{f+g\lambda}{F+G\lambda} = \frac{e+f\lambda}{E+F\lambda} \qquad (A3.15)$$
$$= \frac{e(du)+f(dv)}{E(du)+F(dv)} = \frac{f(du)+g(dv)}{F(du)+G(dv)}.$$

Hence we find that the maximum and minimum values of $k_n$ are solutions of the system

$$\begin{aligned}(e-k_nE)(du)+(f-k_nF)(dv) &= 0, \\ (f-k_nF)(du)+(g-k_nG)(dv) &= 0,\end{aligned} \qquad (A3.16)$$

and hence of the eigenvalue problem for the second fundamental form $\mathcal{F}_{II}$ relative to the first fundamental form $\mathcal{F}_I$:

$$\det(k_n\mathcal{F}_I - \mathcal{F}_{II}) = \begin{vmatrix} Ek_n - e & Fk_n - f \\ Fk_n - f & Gk_n - g \end{vmatrix} = 0. \qquad (A3.17)$$

The maximum and minimum values of $k_n$ are given by the roots of the characteristic polynomial

$$(EG - F^2)k_n^2 - (eG + Eg - 2fF)k_n + eg - f^2 = 0. \qquad (A3.18)$$

DEFINITION A3.2 *The two roots $k_1$ and $k_2$ of (A3.18) are called* the principal curvatures *of the surface $S$ at the point $P$. Moreover the mean curvature $M$ is the arithmetic mean of the principal curvatures:*

$$M = \frac{k_1 + k_2}{2} = \frac{Eg + eG - 2fF}{2(EG - F^2)}, \qquad (A3.19)$$

*while the* Gaussian curvature $K$ *is defined as the square of the geometric mean of the principal curvatures:*

$$K = k_1 k_2 = \frac{eg - f^2}{EG - F^2}. \qquad (A3.20)$$

■

Note that on the basis of the latter formula the classification given in Remark A3.1 can be reformulated in terms of the sign of $K$.

One can prove (cf. Dubrovin *et al.* 1991a) that the vanishing of the mean curvature characterises the *minimal* surfaces (i.e. the surfaces of minimal area).

The Gaussian curvature measures how far the metric of the surface is from the Euclidean metric. Indeed, we have the following.

THEOREM A3.1 *A necessary and sufficient condition for a surface to be isometric to an open set of a Euclidean plane is that the Gaussian curvature $K$ is identically zero.*

■

Clearly, the second fundamental form, and consequently the Gaussian curvature, are defined independently of the first. However, Gauss proved that $K$ is in fact determined by the first fundamental form.

**THEOREM A3.2** (*Egregium theorem of Gauss*) *The Gaussian curvature depends only on the first fundamental form and on its derivatives:*

$$K = \frac{1}{\sqrt{EG - F^2}} \left[ \frac{\partial}{\partial u} \left( \frac{FE_v - EG_u}{2E\sqrt{EG - F^2}} \right) + \frac{\partial}{\partial v} \left( \frac{2EF_u - FE_u - EE_v}{2E\sqrt{EG - F^2}} \right) \right], \tag{A3.21}$$

*where* $E_v = \partial E/\partial v$, $G_u = \partial G/\partial u$, *etc.* ■

**Remark A3.2**
If the coordinate system $u$, $v$ that parametrises the surface is orthogonal, and hence if $F = 0$, equation (A3.21) simplifies to

$$K = -\frac{1}{\sqrt{EG}} \left[ \frac{\partial}{\partial u} \frac{1}{\sqrt{E}} \frac{\partial}{\partial u} \sqrt{G} + \frac{\partial}{\partial v} \frac{1}{\sqrt{G}} \frac{\partial}{\partial v} \sqrt{E} \right].$$

If also $f = 0$, then (A3.18) becomes

$$EGk_n^2 - (gE + eG)k_n + eg = 0,$$

from which it follows that the principal curvatures are $k_1 = e/E$, $k_2 = g/G$ (corresponding in (A3.12) to the two cases $\lambda = 0$, $\lambda \to \infty$), and hence

$$M = \frac{eG + gE}{2EG},$$

$$K = \frac{eg}{EG}.$$
■

For a more detailed discussion of the theory of the curvature of a surface, and for its formulation on a Riemannian manifold, we refer the reader to the texts already cited. In addition, we recommend the survey article by Osserman (1990) which illustrates the various, fascinating developments of modern Riemannian geometry.

## Problems

1. Prove that the second fundamental form for surfaces of revolution, given by the parametrisation $\mathbf{x} = (u \cos v, u \sin v, \psi(u))$ has coefficients

$$e = \frac{\psi''(u)}{\sqrt{1 + (\psi'(u))^2}}, \quad f = 0, \quad g = \frac{u\psi'(u)}{\sqrt{1 + (\psi'(u))^2}}.$$

Along which directions do the principal curvatures lie?

**2.** Compute the second fundamental form for the ellipsoid with the parametrisation $\mathbf{x} = (a \cos u \cos v, b \cos u \sin v, c \sin u)$, where $a > b > c > 0$. Verify that in the case $a = b$ we again find the expression already derived for surfaces of revolution, and in the case $a = b = c$ the formula derived for the sphere.

**3.** Prove that the second fundamental form for the torus parametrised by $\mathbf{x} = (\cos v(1 + a \cos u), \sin v(1 + a \cos u), a \sin u)$, with $0 < a < 1$, has coefficients given by

$$e = a, \quad f = 0, \quad g = (1 + a \cos u) \cos u.$$

**4.** Compute the second fundamental form of the circular paraboloid $\mathbf{x} = (u \cos v, u \sin v, u^2)$.

**5.** Determine the elliptic, parabolic and hyperbolic points of the torus.

**6.** Compute the second fundamental form for a surface $S$ which is the graph of the function $\psi(x, y)$, and prove that its Gaussian curvature has value

$$K = \frac{\begin{vmatrix} \dfrac{\partial^2 \psi}{\partial x \partial x} & \dfrac{\partial^2 \psi}{\partial x \partial y} \\ \dfrac{\partial^2 \psi}{\partial y \partial x} & \dfrac{\partial^2 \psi}{\partial y \partial y} \end{vmatrix}}{\left(1 + \left(\dfrac{\partial \psi}{\partial x}\right)^2 + \left(\dfrac{\partial \psi}{\partial y}\right)^2\right)^2}.$$

**7.** Prove that the Gaussian curvature of an ellipsoid with semi-axes $a$, $b$, $c$ is

$$K = \frac{1}{a^2 b^2 c^2 \left(\dfrac{x^2}{a^4} + \dfrac{y^2}{b^4} + \dfrac{z^2}{c^4}\right)^2}.$$

**8.** Prove that the Gaussian curvature of a surface of revolution $\mathbf{x} = (u \cos v, u \sin v, \psi(u))$ is given by

$$K = \frac{\psi'(u) \psi''(u)}{u(1 + (\psi'(u))^2)^2}.$$

For example, for the circular paraboloid $\psi(u) = u^2$, we have

$$K = \frac{4}{(1 + 4u^2)^2},$$

which vanishes in the limit $u \to \infty$, in agreement with geometrical intuition.

**9.** Prove that the Gaussian curvature of the *catenary*

$$\mathbf{x} = \left(u \cos v, u \sin v, c \cosh^{-1}\left(\frac{u}{c}\right)\right),$$

where $c > 0$ is a fixed constant, is $K = -c^2/u^4$, and that the mean curvature is $M = 0$ (the catenary is an example of a 'minimal surface').

# APPENDIX 4: ALGEBRAIC FORMS, DIFFERENTIAL FORMS, TENSORS

The use of differential forms allows one to generalise to the case of manifolds of any dimension the ordinary concepts of work of a vector field along a path, of flow through a surface and in general the results of classical vector analysis.

The use of differential forms is important for a deeper understanding of Hamiltonian mechanics (see Abraham and Marsden 1978, Arnol'd 1979a, and Meyer and Hall 1992), although in the present text we have avoided their use (except for differential 1-forms).

In this appendix we limit ourselves to a brief introduction to the study of differential forms, and refer the interested reader to one of the numerous treatises on the subject (e.g. Flanders 1963, or the cited books of Abraham and Marsden and of Arnol'd) for a more detailed study and for the proofs we omit. In addition, we systematically adopt the repeated index summation convention (covariant and contravariant, below and above, respectively, following the classical notation).

## A4.1 Algebraic forms

Let $V$ be a real vector space of dimension $l$.

DEFINITION A4.1 *The dual space $V^*$ of $V$ is the space of all linear maps $\vartheta : V \to \mathbf{R}$. The elements $\vartheta \in V^*$ are called covectors or (algebraic) 1-forms.* ∎

It is immediate to check that $V^*$ is a real vector space, and that $\dim V^* = \dim V = l$. The sum of two covectors $\vartheta_1, \vartheta_2 \in V^*$ is defined by the formula

$$(\vartheta_1 + \vartheta_2)(\mathbf{v}) = \vartheta_1(\mathbf{v}) + \vartheta_2(\mathbf{v}), \tag{A4.1}$$

for every $\mathbf{v} \in V$, and the product with a real number $\lambda$ yields

$$(\lambda\vartheta)(\mathbf{v}) = \lambda\vartheta(\mathbf{v}). \tag{A4.2}$$

If $\mathbf{e}_1, \ldots, \mathbf{e}_l$ is any basis of $V$, we can associate with it the *dual basis* $\mathbf{e}^{1*}, \ldots, \mathbf{e}^{l*}$ of $V^*$, defined by the conditions

$$\mathbf{e}^{i*}(\mathbf{e}_j) = \delta^i_j = \begin{cases} 1, & \text{if } i = j, \\ 0, & \text{otherwise,} \end{cases} \tag{A4.3}$$

and every covector $\vartheta$ can be expressed through its components:

$$\vartheta = \vartheta_i \mathbf{e}^{i*}. \tag{A4.4}$$

It is not difficult to check that if $\mathbf{e}'_1, \ldots, \mathbf{e}'_l$ is a new basis of $V$, and $M$ is the $l \times l$ matrix whose entries $M_i^j$ are the components $e_i^{\prime j}$ of $\mathbf{e}'_i$ expressed in the basis $\mathbf{e}_1, \ldots, \mathbf{e}_l$, we have

$$\mathbf{e}^{\prime i*} = A_k^i \mathbf{e}^{k*}, \tag{A4.5}$$

where $A_k^i M_j^k = \delta_j^i$, i.e. $A = (M^T)^{-1}$, and the components of the vectors $\mathbf{v} = v^i \mathbf{e}_i = v^{\prime i} \mathbf{e}'_i$ and of the covectors $\vartheta = \vartheta_i \mathbf{e}^{i*} = \vartheta'_i \mathbf{e}^{i'*}$ are transformed according to the following rules:

$$\begin{aligned} v^{\prime i} &= A_j^i v^j, \\ \vartheta'_i &= M_i^j \vartheta_j. \end{aligned} \tag{A4.6}$$

Because of this transformation property, the components $v^i$ of the vectors are called *contravariant* and the components $\vartheta_i$ of the covectors are called *covariant*. Indeed, they are transformed, respectively, through the matrix $A$, the (transposed) inverse of the change of basis, and the matrix $M$ of the change of basis.

DEFINITION A4.2  *An (algebraic) $k$-form is a map $\omega : V^k \to \mathbf{R}$, where $V^k = V \times \ldots \times V$ ($k$ times), multilinear and skew-symmetric: for any choice of $k$ vectors $(\mathbf{v}_1, \ldots, \mathbf{v}_k) \in V^k$, $\mathbf{v}'_1 \in V$ and two scalars $\lambda_1, \lambda_2 \in \mathbf{R}$ we have*

$$\omega(\lambda_1 \mathbf{v}_1 + \lambda_2 \mathbf{v}'_1, \mathbf{v}_2, \ldots, \mathbf{v}_k) = \lambda_1 \omega(\mathbf{v}_1, \ldots, \mathbf{v}_k) + \lambda_2 \omega(\mathbf{v}'_1, \ldots, \mathbf{v}_k) \tag{A4.7}$$

*and*

$$\omega(\mathbf{v}_{i_1}, \ldots, \mathbf{v}_{i_k}) = (-1)^\nu \omega(\mathbf{v}_1, \ldots, \mathbf{v}_k), \tag{A4.8}$$

*where $\nu = 0$ if the permutation $(i_1, \ldots, i_k)$ of $(1, \ldots, k)$ is even, and $\nu = 1$ if it is odd.* ∎

Recall that a permutation is even if it is obtained by an even number of exchanges of pairs of indices.

*Example* A4.1
The oriented area of the parallelogram in $\mathbf{R}^2$ with sides $\mathbf{v}_1, \mathbf{v}_2$ is given by

$$\omega(\mathbf{v}_1, \mathbf{v}_2) = \det \begin{pmatrix} v_1^1 & v_1^2 \\ v_2^1 & v_2^2 \end{pmatrix}.$$

This is clearly an algebraic 2-form. Similarly the oriented volume of the solid with parallel sides $\mathbf{v}_1, \ldots, \mathbf{v}_l$ in $\mathbf{R}^l$ is an algebraic $l$-form, while the oriented volume of the projection of such a solid onto $x_1, \ldots, x_k$ is a $k$-form. ∎

*Example* A4.2
A symplectic vector space $V$ is endowed with a skew-symmetric linear form $\omega$ which is clearly an example of a 2-form. ∎

## A4.1  Algebraic forms, differential forms, tensors

The set of all the $k$-forms is a vector space, if we introduce the operations of sum and product with a scalar $\lambda \in \mathbf{R}$:

$$(\omega_1 + \omega_2)(\mathbf{v}_1, \ldots, \mathbf{v}_k) = \omega_1(\mathbf{v}_1, \ldots, \mathbf{v}_k) + \omega_2(\mathbf{v}_1, \ldots, \mathbf{v}_k),$$
$$(\lambda\omega)(\mathbf{v}_1, \ldots, \mathbf{v}_k) = \lambda\omega(\mathbf{v}_1, \ldots, \mathbf{v}_k). \tag{A4.9}$$

We denote this space by $\Lambda^k(V)$.

DEFINITION A4.3  Let $\alpha \in \Lambda^r$, $\beta \in \Lambda^s$. The *exterior product of* $\alpha$ *and* $\beta$, denoted by $\alpha \wedge \beta$, is the $(r+s)$-form given by

$$(\alpha \wedge \beta)(\mathbf{v}_1, \ldots, \mathbf{v}_{r+s}) = \sum_{\sigma \in P} \nu(\sigma)\alpha(\mathbf{v}_{\sigma_1}, \ldots, \mathbf{v}_{\sigma_r})\beta(\mathbf{v}_{\sigma_{r+1}}, \ldots, \mathbf{v}_{\sigma_{r+s}}), \tag{A4.10}$$

where $\sigma = (\sigma_1, \ldots, \sigma_{r+s})$, $P$ denotes the set of all possible permutations of $(1, \ldots, r+s)$ and $\nu(\sigma) = \pm 1$ according to whether $\sigma$ is even or odd. ∎

It is not difficult to check that the exterior product satisfies the following properties: if $\alpha \in \Lambda^r$, $\beta \in \Lambda^s$ and $\gamma \in \Lambda^t$, we have

$$\alpha \wedge (\beta \wedge \gamma) = (\alpha \wedge \beta) \wedge \gamma,$$
$$\alpha \wedge (\beta + \gamma) = \alpha \wedge \beta + \alpha \wedge \gamma \quad (t = s), \tag{A4.11}$$
$$\alpha \wedge \beta = (-1)^{rs}\beta \wedge \alpha.$$

Hence it is associative, distributive and anticommutative.

*Example A4.3*
Let $V = \mathbf{R}^{2l}$, $\omega = \sum_{i=1}^{l} \mathbf{e}^{i*} \wedge \mathbf{e}^{(i+l)*}$, where $(\mathbf{e}_1, \ldots, \mathbf{e}_{2l})$ denotes the canonical basis of $\mathbf{R}^{2l}$. It is immediate to check that for every $k = 1, \ldots, l$, setting $\Omega^k = \omega \wedge \ldots \wedge \omega$ ($k$ times), we have

$$\Omega^k = (-1)^{k-1} k! \sum_{1 \le i_1 < \cdots < i_k \le l} \mathbf{e}^{i_1*} \wedge \ldots \wedge \mathbf{e}^{i_k*} \wedge \mathbf{e}^{(i_1+l)*} \wedge \ldots \wedge \mathbf{e}^{(i_k+l)*}. \tag{A4.12}$$

∎

*Example A4.4*
Let $\omega$ be a 2-form on $\mathbf{R}^3$. If $(\mathbf{e}_1, \mathbf{e}_2, \mathbf{e}_3)$ is a basis of $\mathbf{R}^3$ it can be checked that for every $\mathbf{v}, \mathbf{w} \in \mathbf{R}^3$ we have

$$\omega(\mathbf{v}, \mathbf{w}) = \omega(v^i \mathbf{e}_i, w^j \mathbf{e}_j) = (v^1 w^2 - v^2 w^1)\omega(\mathbf{e}_1, \mathbf{e}_2)$$
$$+ (v^2 w^3 - v^3 w^2)\omega(\mathbf{e}_2, \mathbf{e}_3) + (v^3 w^1 - v^1 w^3)\omega(\mathbf{e}_3, \mathbf{e}_1) \tag{A4.13}$$
$$= (\omega_{12} \mathbf{e}^{1*} \wedge \mathbf{e}^{2*} + \omega_{23} \mathbf{e}^{2*} \wedge \mathbf{e}^{3*} + \omega_{31} \mathbf{e}^{3*} \wedge \mathbf{e}^{1*})(\mathbf{v}, \mathbf{w}),$$

where clearly $\omega_{12} = \omega(\mathbf{e}_1, \mathbf{e}_2)$, $\omega_{23} = \omega(\mathbf{e}_2, \mathbf{e}_3)$ and $\omega_{31} = \omega(\mathbf{e}_3, \mathbf{e}_1)$. Therefore $\dim \Lambda^2(\mathbf{R}^3) = 3$. ∎

*Example* A4.5
Let $V = \mathbf{R}^3$. Because of the Euclidean space structure of $\mathbf{R}^3$, we can associate with each vector in $\mathbf{R}^3$, a 1-form $\vartheta_\mathbf{v}$ and a 2-form $\omega_\mathbf{v}$ by setting

$$\vartheta_\mathbf{v}(\mathbf{w}) = \mathbf{v} \cdot \mathbf{w}, \quad \omega_\mathbf{v}(\mathbf{w}_1, \mathbf{w}_2) = \mathbf{v} \cdot \mathbf{w}_1 \times \mathbf{w}_2,$$

where, as usual, $\mathbf{w}_1 \times \mathbf{w}_2$ denotes the vector product of $\mathbf{w}_1$ and $\mathbf{w}_2$. We can check then that, for a fixed orthonormal basis $(e_1, e_2, e_3)$ of $\mathbf{R}^3$, we have

$$\vartheta_\mathbf{v} = v^1 \mathbf{e}^{1*} + v^2 \mathbf{e}^{2*} + v^3 \mathbf{e}^{3*},$$
$$\omega_\mathbf{v} = v^1 \mathbf{e}^{2*} \wedge \mathbf{e}^{3*} + v^2 \mathbf{e}^{3*} \wedge \mathbf{e}^{1*} + v^3 \mathbf{e}^{1*} \wedge \mathbf{e}^{2*}.$$  ∎

THEOREM A4.1 *Let $(e_1, \ldots, e_l)$ be a basis of $V$. A basis of $\Lambda^k(V)$ is given by*

$$\left\{ \sum_{1 \leq i_1 < i_2 < \cdots < i_k \leq l} \mathbf{e}^{i_1*} \wedge \ldots \wedge \mathbf{e}^{i_k*} \right\}.$$

*Therefore* $\dim \Lambda^k(V) = \binom{l}{k}$ *and every k-form $\alpha$ can be uniquely expressed as follows:*

$$\alpha = \sum_{1 \leq i_1 < i_2 < \cdots < i_k \leq l} \alpha_{i_1 \ldots i_k} \mathbf{e}^{i_1*} \wedge \ldots \wedge \mathbf{e}^{i_k*}, \tag{A4.14}$$

*where*

$$\alpha_{i_1 \ldots i_k} = \alpha(e_{i_1}, \ldots, e_{i_k}). \tag{A4.15}$$  ∎

The proof of this theorem is a good exercise, that we leave to the reader.

Such $k$-forms have additional transformation properties under changes of basis, or under the action of a linear map. These properties generalise the properties of covectors. If $(e_1, \ldots, e_l)$ and $(e'_1, \ldots, e'_l)$ are two bases of $V$, $M$ is the matrix of the change of basis, and $A$ is given by (A4.5), for every $k$-form we have the representations $\omega = \sum \omega_{i_1 \ldots i_k} \mathbf{e}^{i_1*} \wedge \mathbf{e}^{i_k*} = \sum \omega'_{i_1 \ldots i_k} \mathbf{e}'^{i_1*} \wedge \mathbf{e}'^{i_k*}$, where

$$\omega'_{i_1 \ldots i_k} = M_{i_1}^{j_1} \ldots M_{i_k}^{j_k} \omega_{j_1 \ldots j_k}. \tag{A4.16}$$

Every linear map $f : V \to V$ induces a linear map $f^*$ on $\Lambda^k(V)$:

$$(f^*(\alpha))(\mathbf{v}_1, \ldots, \mathbf{v}_k) = \alpha(f(\mathbf{v}_1), \ldots, f(\mathbf{v}_k)). \tag{A4.17}$$

If $(f_i^j)$ is the matrix representing $f$, $f(\mathbf{v}) = f\left(\sum v^i e_i\right) = \sum v^i f_i^j e_j$, if $\alpha = \sum \alpha_{i_1, \ldots i_k} \mathbf{e}^{i_1*} \wedge \mathbf{e}^{i_k*}$, setting

$$(f^*\alpha) = \sum_{1 \leq i_1 < \ldots < i_k \leq l} (f^*\alpha)_{i_1 \ldots i_k} \mathbf{e}^{i_1*} \wedge \ldots \wedge \mathbf{e}^{i_k*}, \tag{A4.18}$$

we find

$$(f^*\alpha)_{i_1\ldots i_k} = f_{i_1}^{j_1} \ldots f_{i_k}^{j_k} \alpha_{j_1\ldots j_k}. \tag{A4.19}$$

Equation (A4.19) is immediately verified, once one shows that $f^*$ preserves the exterior product:

$$f^*(\alpha \wedge \beta) = (f^*\alpha) \wedge (f^*\beta). \tag{A4.20}$$

## A4.2  Differential forms

Let $M$ be a connected differentiable manifold of dimension $l$.

DEFINITION A4.4  *The dual space $T_P^*M$ of the tangent space $T_PM$ to $M$ in $P$ is called* the cotangent space *to $M$ in $P$. The elements $\vartheta \in T_P^*M$ are called* cotangent vectors *to $M$ in $P$.* ∎

It is possible to identify the tangent vectors with differentiations (along a curve), so that if $(x^1, \ldots, x^l)$ is a local parametrisation of $M$ a basis of $T_PM$ is given by $\partial/\partial x^1, \ldots, \partial/\partial x^l$. In the same way, every cotangent vector is identified with the differential of a function. Therefore a basis of $T_P^*M$ is given by $(dx^1, \ldots, dx^l)$ and every cotangent vector $\vartheta \in T_P^*M$ can be written as

$$\vartheta = \vartheta_i \, dx^i. \tag{A4.21}$$

It is immediate to check that, if $(x'^1, \ldots, x'^l)$ is a different local parametrisation of $M$ in $P$, setting

$$\vartheta = \vartheta'_i \, dx'^i,$$

we have

$$\vartheta'_i = \vartheta_j \frac{\partial x^j}{\partial x'^i}.$$

Hence the components of a cotangent vector are covariant.

DEFINITION A4.5  *We call the* cotangent bundle $T^*M$ *of the manifold $M$ the union of the cotangent spaces to $M$ at all of its points:*

$$T^*M = \bigcup_{P \in M} \{P\} \times T_P^*M. \tag{A4.22}$$

∎

*Remark* A4.1
The cotangent bundle $T^*M$ is naturally endowed with the structure of a differentiable manifold of dimension $2l$. If $(x^1, \ldots, x^l)$ is a local parametrisation of $M$, and $(\vartheta_1, \ldots, \vartheta_l)$ are the components of a covector with respect to the basis $(dx^1, \ldots, dx^l)$ of $T_P^*M$, a local parametrisation of $T^*M$ can be obtained by considering $(x^1, \ldots, x^l, \vartheta_1, \ldots, \vartheta_l)$. ∎

*Example* A4.6
If $M = \mathbf{R}^l$, $T^*M \simeq \mathbf{R}^{2l}$; if $M = \mathbf{T}^l$, $T^*M \simeq \mathbf{T}^l \times \mathbf{R}^l$. ∎

*Example* A4.7
Let $(M, (\mathrm{d}s)^2)$ be a Riemannian manifold and $V : M \to \mathbf{R}$ be a regular function. Consider the Lagrangian

$$L : TM \to \mathbf{R}, \quad L = \frac{1}{2}\left|\frac{\mathrm{d}s}{\mathrm{d}t}\right|^2 - V. \tag{A4.23}$$

If $(q^1, \ldots, q^l, \dot{q}^1, \ldots, \dot{q}^l)$ is a local parametrisation of $TM$ and

$$(\mathrm{d}s)^2 = g_{ij}(\mathbf{q})\,\mathrm{d}q^i\,\mathrm{d}q^j, \tag{A4.24}$$

we have

$$L(\mathbf{q}, \dot{\mathbf{q}}) = \frac{1}{2} g_{ij}(\mathbf{q}) \dot{q}^i \dot{q}^j - V(\mathbf{q}). \tag{A4.25}$$

The kinetic moments $p_1, \ldots, p_l$ conjugate to $(q^1, \ldots, q^l)$:

$$p_i = \frac{\partial L}{\partial \dot{q}^i} = g_{ij}(\mathbf{q})\dot{q}^j \tag{A4.26}$$

are covariant and can therefore be considered as the components of a cotangent vector to $M$ at the point with coordinates $(q^1, \ldots, q^l)$. The Hamiltonian of the system

$$H(\mathbf{p}, \mathbf{q}) = \frac{1}{2} g^{ij}(\mathbf{q}) p_i p_j + V(\mathbf{q}), \tag{A4.27}$$

where $g^{ij}(\mathbf{q}) g_{jk}(\mathbf{q}) = \delta^i_k$, is a regular function defined on the cotangent bundle of $M$:

$$H : T^*M \to \mathbf{R}. \tag{A4.28}$$

It follows that the *Hamiltonian phase space* of the system coincides with the cotangent bundle of $M$. ∎

The cotangent bundle $T^*M$ is endowed with a natural projection:

$$\begin{aligned}\pi : T^*M &\to M, \\ (P, \vartheta) &\mapsto P.\end{aligned} \tag{A4.29}$$

Note that $\pi^{-1}(P) = T^*_P M$.

DEFINITION A4.6  *The field of cotangent vectors (or differential 1-forms on $M$) of a manifold $M$ is a section of $T^*M$, i.e. a regular map*

$$\Theta : M \to T^*M \tag{A4.30}$$

such that

$$\pi \circ \Theta = \mathrm{id}_M. \tag{A4.31}$$

## A4.2 Algebraic forms, differential forms, tensors

If $(x^1, \ldots, x^l, \vartheta_1, \ldots, \vartheta_l)$ is a local parametrisation of $T^*M$, $\Theta$ can be written in the form

$$\Theta(\mathbf{x}) = \vartheta_i(\mathbf{x})\, \mathrm{d}x^i, \tag{A4.32}$$

where the functions $\vartheta_i$ are regular. ∎

**Remark A4.2**
A field of cotangent vectors $\Theta$ can be identified with a regular map $\widetilde{\Theta}$ on the tangent bundle $TM$ with values in $\mathbf{R}$, linear on each tangent space $T_PM$:

$$\begin{aligned}\widetilde{\Theta} &: TM \to \mathbf{R}, \\ (P, \mathbf{v}) &\mapsto \widetilde{\Theta}(P, \mathbf{v}) = \Theta(P)(\mathbf{v}).\end{aligned} \tag{A4.33}$$

If $(x^1, \ldots, x^l)$ is a local parametrisation of $M$, we have

$$\Theta(\mathbf{x}, \mathbf{v}) = \vartheta_i(\mathbf{x})v^i. \tag{A4.34}$$
∎

**Example A4.8**
Let $f: N \to \mathbf{R}$ be a regular function. The differential of $f$:

$$\mathrm{d}f(\mathbf{x}) = \frac{\partial f}{\partial x^i}(\mathbf{x})\, \mathrm{d}x^i \tag{A4.35}$$

defines a field of cotangent vectors. It is indeed immediate to verify the covariance of its components:

$$\vartheta'_i = \frac{\partial f}{\partial x'^i} = \frac{\partial f}{\partial x^j}\frac{\partial x^j}{\partial x'^i} = \frac{\partial x^j}{\partial x'^i}\vartheta_j.$$
∎

**DEFINITION A4.7** Let $P \in M$ and denote by $\Lambda^k_P(M)$ the vector space of algebraic $k$-forms on $T_PM$. We call a differential $k$-form on $M$ a regular map

$$\Omega: M \to \bigcup_{P \in M} \{P\} \times \Lambda^k_P(M), \tag{A4.36}$$

that associates with every point $P \in M$ a $k$-form on $T_PM$ with a regular dependence on $P$. The space of differential $k$-forms $M$ is denoted by $\Lambda^k(M)$. ∎

**Remark A4.3**
Note that $\bigcup_{P \in M} \{P\} \times \Lambda^1_P(M) = T^*M$. ∎

In local coordinates $(x^1, \ldots, x^l)$ we have

$$\Omega(x^1, \ldots, x^l) = \sum_{1 \leq i_1 < \cdots < i_k \leq l} \Omega_{i_1 \ldots i_k}(\mathbf{x})\, \mathrm{d}x^{i_1} \wedge \ldots \wedge \mathrm{d}x^{i_k}, \tag{A4.37}$$

and the $\binom{l}{k}$ functions $\Omega_{i_1 \ldots i_k}: M \to \mathbf{R}$ are regular.

Given a function $f : M \to \mathbf{R}$, which can be considered as a 'differential 0-form', its differential $df$ is a 1-form. This procedure can be generalised to an operation called 'exterior derivation' that transforms $k$-forms into $(k+1)$-forms.

**DEFINITION A4.8** Let $\Omega \in \Lambda^k(M)$. The exterior derivative $d\Omega \in \Lambda^{k+1}(M)$ is the $(k+1)$-form

$$d\Omega = \sum_{j=1}^{l} \sum_{1 \le i_1 < \ldots < i_k \le l} \frac{\partial \Omega_{i_1 \ldots i_k}}{\partial x^j} \, dx^j \wedge dx^{i_1} \wedge \ldots \wedge dx^{i_k}. \tag{A4.38}$$

∎

*Remark A4.4*
It is immediate to check that if $\Omega \in \Lambda^0(M)$, i.e. a function, $d\Omega$ is its differential.

∎

**THEOREM A4.2** (properties of exterior differentiation)
(1) If $\Omega, \tilde{\Omega} \in \Lambda^k(M)$, then $d(\Omega + \tilde{\Omega}) = d\Omega + d\tilde{\Omega}$.
(2) If $\Omega \in \Lambda^k(M)$, $\tilde{\Omega} \in \Lambda^j(M)$, then $d(\Omega \wedge \tilde{\Omega}) = d\Omega \wedge \tilde{\Omega} + (-1)^k \Omega \wedge d\tilde{\Omega}$.
(3) For any $\Omega \in \Lambda^k(M)$, we have $d(d\Omega) = 0$.

*Proof*
We leave (1) and (2) as an exercise, and we prove (3). Setting $\mathbf{i} = (i_1, \ldots, i_k)$, where $1 \le i_1 < \cdots < i_k \le l$, and $d\mathbf{x}^{\mathbf{i}} = dx^{i_1} \wedge \ldots \wedge dx^{i_k}$, by (A4.38) we have

$$d(d\Omega) = d \sum_{j=1}^{l} \sum_{\mathbf{i}} \frac{\partial \Omega_{\mathbf{i}}}{\partial x^j} \, dx^j \wedge d\mathbf{x}^{\mathbf{i}}$$

$$= \sum_{j,k=1}^{l} \sum_{\mathbf{i}} \frac{\partial^2 \Omega_{\mathbf{i}}}{\partial x^j \partial x^k} \, dx^k \wedge dx^j \wedge d\mathbf{x}^{\mathbf{i}}$$

$$= \sum_{\mathbf{i}} \sum_{j<k} \left( \frac{\partial^2 \Omega_{\mathbf{i}}}{\partial x^j \partial x^k} - \frac{\partial^2 \Omega_{\mathbf{i}}}{\partial x^k \partial x^j} \right) dx^k \wedge dx^j \wedge d\mathbf{x}^{\mathbf{i}} = 0.$$

∎

*Example A4.9:* vector calculus in $\mathbf{R}^3$
If $f : \mathbf{R}^3 \to \mathbf{R}$ is a regular function, its differential

$$df = \frac{\partial f}{\partial x^1} dx^1 + \frac{\partial f}{\partial x^2} dx^2 + \frac{\partial f}{\partial x^3} dx^3$$

is identified with the gradient vector field of $f$:

$$\nabla f = \frac{\partial f}{\partial x^1} \mathbf{e}_1 + \frac{\partial f}{\partial x^2} \mathbf{e}_2 + \frac{\partial f}{\partial x^3} \mathbf{e}_3,$$

where $\mathbf{e}_1, \mathbf{e}_2, \mathbf{e}_3$ are the unit vectors of the canonical of $\mathbf{R}^3$.

If $\Theta \in \Lambda^1(\mathbf{R}^3)$, $\Theta(\mathbf{x}) = \vartheta_1(\mathbf{x})\,dx^1 + \vartheta_2(\mathbf{x})\,dx^2 + \vartheta_3(\mathbf{x})\,dx^3$, by identifying $\Theta(\mathbf{x})$ with the vector field $\Theta = \vartheta_1(\mathbf{x})\mathbf{e}_1 + \vartheta_2(\mathbf{x})\mathbf{e}_2 + \vartheta_3(\mathbf{x})\mathbf{e}_3$, the exterior derivative $d\Theta$:

$$d\Theta = \left(\frac{\partial \vartheta_2}{\partial x^1} - \frac{\partial \vartheta_1}{\partial x^2}\right) dx^1 \wedge dx^2 + \left(\frac{\partial \vartheta_3}{\partial x^1} - \frac{\partial \vartheta_1}{\partial x^3}\right) dx^1 \wedge dx^3$$
$$+ \left(\frac{\partial \vartheta_3}{\partial x^2} - \frac{\partial \vartheta_2}{\partial x^3}\right) dx^2 \wedge dx^3$$

can be identified with the curl of the vector field $\Theta$:

$$\nabla \times \Theta = \left(\frac{\partial \vartheta_3}{\partial x^2} - \frac{\partial \vartheta_2}{\partial x^3}\right)\mathbf{e}_1 + \left(\frac{\partial \vartheta_1}{\partial x^3} - \frac{\partial \vartheta_3}{\partial x^1}\right)\mathbf{e}_2 + \left(\frac{\partial \vartheta_2}{\partial x^1} - \frac{\partial \vartheta_1}{\partial x^2}\right)\mathbf{e}_3,$$

since $dx^2 \wedge dx^3 = \mathbf{e}_2 \times \mathbf{e}_3 = \mathbf{e}_1$, $dx^1 \wedge dx^2 = \mathbf{e}_1 \times \mathbf{e}_2 = \mathbf{e}_3$ and $dx^1 \wedge dx^3 = \mathbf{e}_1 \times \mathbf{e}_3 = -\mathbf{e}_2$.

Finally, if $\Omega \in \Lambda^2(\mathbf{R}^3)$, $\Omega(\mathbf{x}) = \Omega_{12}(\mathbf{x})\,dx^1 \wedge dx^2 - \Omega_{31}(\mathbf{x})\,dx^1 \wedge dx^3 + \Omega_{23}(\mathbf{x})\,dx^2 \wedge dx^3$, identifying this with the vector field $\Omega = \Omega_{23}\,\mathbf{e}_1 + \Omega_{31}\,\mathbf{e}_2 + \Omega_{12}\,\mathbf{e}_3$, the exterior derivative

$$d\Omega = \left(\frac{\partial \Omega_{12}}{\partial x^3} + \frac{\partial \Omega_{23}}{\partial x^1} + \frac{\partial \Omega_{31}}{\partial x^2}\right) dx^1 \wedge dx^2 \wedge dx^3$$

can be identified with the divergence of $\Omega$:

$$\nabla \cdot \Omega = \frac{\partial \Omega_{12}}{\partial x^3} + \frac{\partial \Omega_{31}}{\partial x^2} + \frac{\partial \Omega_{23}}{\partial x^1}.$$

The vanishing of $d^2$ (the second exterior derivative), in the context of vector analysis in $\mathbf{R}^3$, summarises the two classical results $\nabla \times (\nabla f) = 0$ and $\nabla \cdot (\nabla \times \Theta) = 0$. ∎

DEFINITION A4.9 *A differential $k$-form $\Omega \in \Lambda^k(M)$ is* closed *if $d\Omega = 0$, and* exact *if there exists a $(k-1)$-form $\Theta$ such that $\Omega = d\Theta$.* ∎

Remark A4.5
By property (3) (Theorem A4.2) of exterior differentiation, *every exact form is closed*. The converse is in general false, except in open simply connected sets of $\mathbf{R}^l$. ∎

THEOREM A4.3 (Poincaré's lemma) *Let $A \subseteq \mathbf{R}^l$ be an open simply connected set, and $\Omega$ be a differential $k$-form on $A$. If $\Omega$ is closed, then $\Omega$ is exact.* ∎

For the proof, see any one of the books suggested in this appendix.

From Poincaré's lemma it immediately follows that a vector field $\mathbf{X}$ on $\mathbf{R}^l$ is a gradient vector field if and only if the Jacobian matrix $\partial X_i / \partial x^j$ of the field is symmetric.

We end this brief introduction to forms by studying their behaviour under the action of a diffeomorphism $f$ between two differentiable manifolds $M$ and $N$. Let $\Omega \in \Lambda^k(N)$, and let $f : M \to N$ be a diffeomorphism. Since the differential

$f_*(P) = \mathrm{d}f(P)$ is an isomorphism of $T_PM$ and $T_{f(P)}N$ (cf. Section 1.7) it is possible to associate with $\Omega$ a $k$-form on $\Lambda^k(M)$, denoted by $f^*\Omega$ (the *pull-back* of $\Omega$) defined as follows:

$$(f^*\Omega)(P)(\mathbf{v}_1,\ldots,\mathbf{v}_k) = \Omega(f(P))(f_*(P)\mathbf{v}_1,\ldots,f_*(P)\mathbf{v}_k), \qquad (A4.39)$$

where $P \in M$ and $\mathbf{v}_1,\ldots,\mathbf{v}_k \in T_PM$.

If $(y^1,\ldots,y^l)$ are local coordinates on $N$ and

$$\Omega(\mathbf{y}) = \sum_\mathbf{i} \Omega_\mathbf{i}(\mathbf{y})\,\mathrm{d}\mathbf{y}^\mathbf{i},$$

it is straightforward to check that (cf. (A4.19))

$$(f^*\Omega)(\mathbf{x}) = \sum_{\mathbf{i},\mathbf{j}} \Omega_\mathbf{i}(\mathbf{f}(\mathbf{y})) \frac{\partial f^{i_1}}{\partial x^{j_1}} \cdots \frac{\partial f^{i_k}}{\partial x^{j_k}}\,\mathrm{d}x^{j_1} \wedge \ldots \wedge \mathrm{d}x^{j_k}, \qquad (A4.40)$$

where $\mathbf{i} = (i_1,\ldots,i_k)$, $\mathbf{j} = (j_1,\ldots,j_k)$.

## A4.3 Stokes' theorem

The theory of integration of differential forms defined on a manifold is rather rich, and cannot be covered in this brief introduction. We can however devote a little space to Stokes' theorem, whose statement is very simple and which has useful consequences for a better understanding of some of the topics considered in this book (e.g. the Poincaré–Cartan form, see Section 10.3).

A *differentiable manifold with boundary* $M$ is a manifold whose atlas $(U_\alpha, \mathbf{x}_\alpha)_{\alpha \in A}$ contains the two following types of charts. If we denote by $\mathbf{H}^l = \{\mathbf{x} \in \mathbf{R}^l | x^l \geq 0\}$ the upper half-space $\mathbf{R}^l$, for some charts $\mathbf{x}_\alpha^{-1}(U_\alpha)$ is an open set of $\mathbf{R}^l$ homeomorphic to $\mathbf{R}^l$, and for others it is homeomorphic to $\mathbf{H}^l$. The *boundary* $\partial M$ of $M$ is then the set of points $P$ of $M$ whose images $\mathbf{x}^{-1}(P) \in \partial \mathbf{H}^l = \{\mathbf{x} \in \mathbf{R}^l | x^l = 0\} \simeq \mathbf{R}^{l-1}$.

Clearly the boundary $\partial M$ of $M$ is a smooth manifold of dimension $l-1$.

*Example* A4.10
The half-sphere

$$M = \{\mathbf{x} \in \mathbf{R}^{l+1} | (x^1)^2 + \cdots + (x^l)^2 + (x^{l+1})^2 = 1, x^{l+1} \geq 0\}$$

is a differentiable manifold with boundary

$$\partial M = \{\mathbf{x} \in \mathbf{R}^l | (x^1)^2 + \cdots + (x^l)^2 = 1, x^{l+1} = 0\} \simeq \mathbf{S}^{l-1}. \qquad \blacksquare$$

Let $M$ be an $l$-dimensional oriented manifold with boundary $\partial M$ (hence a manifold of dimension $l-1$ coherently oriented). It is not difficult to introduce the notion of the integral of an $l$-form on $M$, although a rigorous definition requires the use of a partition of unity (see Singer and Thorpe 1980). We simply note that if $(x^1, \ldots, x^l)$ is a local parametrisation of $M$, any $l$-form $\Omega$ can be written as

$$\Omega(x) = w(x) \, dx^1 \ldots dx^l,$$

and hence is identified with a function $w(x)$ that can be integrated on the corresponding chart $U$. Evidently the theorem of change of integration variable ensures that the result is independent of the parametrisation. Indeed, if $(y^1, \ldots, y^l)$ is a new parametrisation, and $V$ is the image of the open $U$ in the new local coordinates, from (A4.40) it follows that

$$\int_U w(\mathbf{x}) \, d^l\mathbf{x} = \int_V w(\mathbf{x}(\mathbf{y})) \det\left(\frac{\partial x_i}{\partial y_j}\right) d^l\mathbf{y}$$

(note that $\det(\partial x_i/\partial y_j) > 0$ since $M$ is oriented).

Hence the integral can be extended to $M$ (and we denote it by $\int_M \Omega$) or to a part of it.

THEOREM A4.4 (Stokes) *Let $M$ be an $l$-dimensional oriented manifold with boundary, and let $\Omega \in \Lambda^{l-1}(M)$; then*

$$\int_M d\Omega = \int_{\partial M} \Omega. \tag{A4.41}$$

∎

*Remark A4.6*
If $M = [a, b]$, $\Omega = f : [a, b] \to \mathbf{R}$, equation (A4.41) becomes

$$\int_a^b f'(x) \, dx = [f(x)]_a^b = f(b) - f(a),$$

and we recover the fundamental theorem of calculus. Similarly it is not difficult to check, using example A4.24, that Stokes' theorem summarises Green's theorem, the divergence theorem and the classical Stokes' theorem of vector calculus (see Giusti 1989).
∎

COROLLARY 4.1 (Stokes' lemma) *If $\omega$ is a non-singular 1-form (see Definition 10.9) in $\mathbf{R}^{2l+1}$ and $\gamma_1$, $\gamma_2$ are two closed curves enclosing the same tube of characteristics of $\omega$, then*

$$\int_{\gamma_1} \omega = \int_{\gamma_2} \omega.$$

*Proof*
By Remark 10.15, $d\omega = 0$ on a tube of characteristics. If $\sigma$ is the portion of the tube of characteristics having as boundary $\gamma_1 - \gamma_2$ (Fig. A4.1) we have

$$\int_{\gamma_1} \omega - \int_{\gamma_2} \omega = \int_{\partial \sigma} \omega = \int_\sigma d\omega = 0.$$

∎

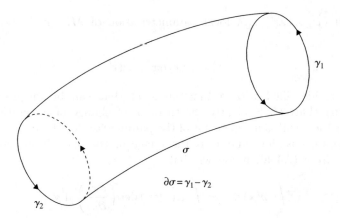

**Fig. 4.1** $\partial_\sigma = \gamma_1 - \gamma_2$.

## A4.4 Tensors

We only give the definition of a tensor field on a differentiable manifold, and refer the interested reader to Dubrovin *et al.* (1991a).

Let $M$ be a differentiable manifold of dimension $l$, $P \in M$.

DEFINITION A4.10 *An element of the vector space of multilinear forms*

$$\tau : \underbrace{T_P M \times \cdots \times T_P M}_{m \text{ times}} \times \underbrace{T_P^* M \times \cdots \times T_P^* M}_{n \text{ times}} \to \mathbf{R} \tag{A4.42}$$

*is an $m$ times covariant, $n$ times contravariant tensor (or of type $(m,n)$ and order $m+n$).*

If $(x^1, \ldots, x^l)$ is a local parametrisation in a neighbourhood of $P \in M$ and we denote by $(\mathbf{e}_1, \ldots, \mathbf{e}_l)$ the basis $(\partial/\partial x^1, \ldots, \partial/\partial x^l)$ of $T_P M$, and by $(\mathbf{e}^{1*}, \ldots, \mathbf{e}^{l*})$ the (dual) basis $(\mathrm{d}x^1, \ldots, \mathrm{d}x^l)$ of $T_P^* M$, a tensor of type $(m,n)$ can be expressed through its components:

$$\tau = \tau^{i_{m+1} \ldots i_{m+n}}_{i_1 \ldots i_m} \mathbf{e}^{i_1*} \otimes \cdots \otimes \mathbf{e}^{i_m*} \otimes \mathbf{e}_{i_{m+1}} \otimes \cdots \otimes \mathbf{e}_{i_{m+n}}, \tag{A4.43}$$

where $\otimes$ denotes the tensor product: if $\vartheta_1, \vartheta_2 \in T_P^* M$ then

$$\mathbf{e}_i \otimes \mathbf{e}_j(\vartheta_1, \vartheta_2) = (\vartheta_1)_i (\vartheta_2)_j, \tag{A4.44}$$

where $(\vartheta_1)_i$ denotes the $i$th component of $\vartheta_1 = \vartheta_{1i} \mathbf{e}^{i*}$. If, on the other hand, $\mathbf{v}_1$, $\mathbf{v}_2 \in T_P M$, we have

$$\mathbf{e}^{i*} \otimes \mathbf{e}^{j*}(\mathbf{v}_1, \mathbf{v}_2) = v_1^i v_2^j. \tag{A4.45}$$

∎

## A4.4      Algebraic forms, differential forms, tensors

The vector space of tensors of type $(m, n)$ therefore coincides with the tensor product

$$\bigotimes^{m} T_P^*M \otimes \bigotimes^{n} T_PM \tag{A4.46}$$

of $m$ cotangent spaces $T_P^*M$ with $n$ tangent spaces $T_PM$, and hence has dimension $l^{m+n}$.

*Remark A4.7*
It is necessary to take care of the fact that the tensor product is distributive with respect to addition, it is associative (as follows from (A4.44) and (A4.45)) but it is *not commutative*: the spaces $T_P^*M \otimes T_PM$ and $T_PM \otimes T_P^*M$ are both vector spaces of tensors once covariant and once contravariant, but $T_P^*M \otimes T_PM \neq T_PM \otimes T_P^*M$. ∎

Let $(y^1, \ldots, y^l)$ be a new choice of local coordinates in a neighbourhood of $P \in M$. Denote by $(\mathbf{e}'_1, \ldots, \mathbf{e}'_l) = (\partial/\partial y^1, \ldots, \partial/\partial y^l)$ the associated basis of $T_PM$ and by $(\mathbf{e}'^{1*}, \ldots, \mathbf{e}'^{l*}) = (dy^1, \ldots, dy^l)$ the dual basis. If we denote by

$$J^j_k = \frac{\partial x^j}{\partial y^k}, \quad (J^{-1})^j_k = \frac{\partial y^j}{\partial x^k} \tag{A4.47}$$

the elements of the Jacobian matrix of the transformation of the change of chart and of its inverse, a tensor (A4.43) is transformed into

$$(\tau')^{i_{m+1}\ldots i_{m+n}}_{i_1 \ldots i_m} \mathbf{e}'^{i_1*} \otimes \cdots \otimes \mathbf{e}'^{i_m*} \otimes \mathbf{e}'_{i_{m+1}} \otimes \cdots \otimes \mathbf{e}'_{i_{m+n}}, \tag{A4.48}$$

where the old and new components are related by

$$\tau^{i_{m+1}\ldots i_{m+n}}_{i_1 \ldots i_m} = (J^{-1})^{j_1}_{i_1} \cdots (J^{-1})^{j_m}_{i_m} J^{i_{m+1}}_{j_{m+1}} \cdots J^{i_{m+n}}_{j_{m+n}} \cdot (\tau')^{j_{m+1}\ldots j_{m+n}}_{j_1 \ldots j_m}. \tag{A4.49}$$

**DEFINITION A4.11**    A tensor field *of type $(m,n)$ on a manifold $M$ is a regular map*

$$\tau : M \to \bigcup_{P \in M} \{P\} \times \bigotimes^{m} T_P^*M \otimes \bigotimes^{n} T_PM \tag{A4.50}$$

*that associates with every point $P \in M$ a tensor of type $(m, n)$ belonging to $\bigotimes^{m} T_P^*M \otimes \bigotimes^{n} T_PM$ and depending regularly on $P$.*
In local coordinates $(x^1, \ldots, x^l)$ we have

$$\tau(x^1, \ldots, x^l) = \tau(\mathbf{x})^{i_{m+1}\ldots i_{m+n}}_{i_1 \ldots i_m} dx^{i_1} \ldots dx^{i_m} \frac{\partial}{\partial x^{i_{m+1}}} \cdots \frac{\partial}{\partial x^{i_{m+n}}}. \qquad ∎$$

Vector fields are examples of tensor fields of type $(0, 1)$. The fields of cotangent vectors are tensor fields of type $(1, 0)$. A differential $k$-form on $M$ is a tensor field of type $(k, 0)$ that is *totally skew-symmetric*: if we denote by $\tau_{i_1 \ldots i_k}(\mathbf{x})$ its components, then $\tau_{i'_1 \ldots i'_k} = \pm \tau_{i_1 \ldots i_k}$, where the sign is positive if $i'_1 \ldots i'_k$ is an even permutation of $i_1 \ldots i_k$, and negative otherwise (hence $\tau_{i_1 \ldots i_j i_j \ldots i_k} = 0$).

# APPENDIX 5: PHYSICAL REALISATION OF CONSTRAINTS

The notion of a system subject to an ideal bilateral frictionless holonomic constraint was introduced in Chapter 1 and further discussed in Chapters 2 and 4. Our discussion, however rigorous from the point of view of mathematical modelling, does not consider the more complex issue of the physical phenomenon responsible for the effect of the constraint.

Consider, as an example, a point particle of mass $m$ constrained to move on the surface of a sphere of radius $R$ and subject to gravity (spherical pendulum). If $(r, \vartheta, \varphi)$ indicate spherical coordinates in $\mathbf{R}^3$, the Lagrangian of the system is given by

$$L(\vartheta, \varphi, \dot\vartheta, \dot\varphi) = \frac{1}{2} mR^2(\dot\vartheta^2 + \sin^2\vartheta\, \dot\varphi^2) - mgR\cos\vartheta. \tag{A5.1}$$

From a physical point of view, we expect the constraint reactions to be due to the elastic reaction of the materials subject to a small deformation due to contact—in the example under consideration—between the particle and the sphere. Assume then that the constraint is realised physically through a spring of negligible mass, length at rest equal to $R$ and a very large elastic constant $k$. We then want to use a limiting procedure for $k \to \infty$, in analogy with what was done in Section 2.6 to study unilateral constraints.

Setting $\xi = r - R$, the Lagrangian of the system (that now has three degrees of freedom) can be written as

$$\hat{L} = \frac{1}{2} m\dot\xi^2 + \frac{1}{2} m(R+\xi)^2(\dot\vartheta^2 + \sin^2\vartheta\, \dot\varphi^2) - mg(R+\xi)\cos\vartheta - \frac{1}{2} k\xi^2. \tag{A5.2}$$

The problem is to compare the solutions of the system (A5.2) for very large values of $k$, corresponding to initial conditions belonging to the sphere of radius $R$ (i.e. $\xi = 0$, $\dot\xi = 0$) with the solutions of the system (A5.1). Note that, in spite of the fact that for $t = 0$ we have $\xi = 0$, $\dot\xi = 0$, for $t > 0$ in general $\xi \neq 0$, $\dot\xi \neq 0$, since the Lagrange equations associated with (A5.2) couple the variables $(\xi, \dot\xi)$ with $(\vartheta, \varphi, \dot\vartheta, \dot\varphi)$.

More generally, consider a point particle $P$ of mass $m$ and Cartesian coordinates $\mathbf{x} = (x_1, x_2, x_3)$ subject to a smooth holonomic constraint with equation

$$f(x_1, x_2, x_3) = 0, \tag{A5.3}$$

and under the action of a conservative force field with potential energy $V(x_1, x_2, x_3)$. If $f$ is of class $\mathcal{C}^2$ and $\nabla_x f \neq 0$, equation (A5.3) defines a surface $S \subset \mathbf{R}^3$. Assume that the constraint is realised through a recoiling elastic force proportional to $f^2$, vanishing on $S$ and with a very large constant $k$. The Lagrangian corresponding to this system is given by

$$\hat{L}(\mathbf{x}, \dot{\mathbf{x}}) = \frac{m}{2}(\dot{x}_1^2 + \dot{x}_2^2 + \dot{x}_3^2) - V(x_1, x_2, x_3) - \frac{1}{2}k(f(x_1, x_2, x_3))^2. \tag{A5.4}$$

Let $\mathbf{x}(t, k)$ be a solution of the Lagrange equations associated with (A5.4) with initial conditions

$$\mathbf{x}(0, k) = \mathbf{x}_0 \in S, \quad \dot{\mathbf{x}}(0, k) = \dot{\mathbf{x}}_0 \in T_{x_0} S, \tag{A5.5}$$

independent of $k$. If $\mathbf{x} = \mathbf{x}(q_1, q_2)$ is a local parametrisation of $S$ in a neighbourhood of $\mathbf{x}_0$, let $(q_1(0), q_2(0), \dot{q}_1(0), \dot{q}_2(0))$ be the local coordinates corresponding to the initial conditions (A5.5). To compare with (A5.4) we then consider the Lagrangian of the system subject to the constraint, given by

$$L(\mathbf{q}, \dot{\mathbf{q}}) = \frac{m}{2}(E(q_1, q_2)\dot{q}_1^2 + 2F(q_1, q_2)\dot{q}_1\dot{q}_2 + G(q_1, q_2)\dot{q}_2^2) - V(\mathbf{x}(q_1, q_2)) \tag{A5.6}$$

(where $E$, $F$, $G$ are the coefficients of the first fundamental form associated with $S$) and the solution $q_1(t)$, $q_2(t)$ of the Lagrange equations associated with (A5.6) and corresponding to initial conditions $(q_1(0), q_2(0), \dot{q}_1(0), \dot{q}_2(0))$. The Lagrangian (A5.6) is obtained from (A5.4) by imposing the restrictions $\mathbf{x} \in S$, $\dot{\mathbf{x}} \in T_\mathbf{x} S$ and introducing local coordinates.

Under the above assumptions the following important result holds.

THEOREM A5.1 (Rubin, Hungar, Takens) *Fix $T > 0$. Then for every $t \in [0, T]$ the limit*

$$\hat{\mathbf{x}}(t) = \lim_{k \to \infty} \mathbf{x}(t, k) \tag{A5.7}$$

*exists. The function $t \to \hat{\mathbf{x}}(t)$ is of class $C^2$, $\hat{\mathbf{x}}(t)$ belongs to $S$ for every $t \in [0, T]$ and hence can be written using the local parametrisation of $S$ as*

$$\hat{\mathbf{x}}(t) = \mathbf{x}(q_1(t), q_2(t)). \tag{A5.8}$$

*Here $(q_1(t), q_2(t))$ is a solution of the Lagrange equations associated with the Lagrangian (A5.6) and satisfying the initial conditions $(q_1(0), q_2(0), \dot{q}_1(0), \dot{q}_2(0))$ associated with (A5.5) by the parametrisation.* ∎

We shall not give the proof of this theorem (see Rubin and Hungar 1957, Takens 1970, or Gallavotti 1980), which guarantees that the limiting procedure used to build a physical model of a system subject to a constraint converges to the ideal notion discussed in Chapters 1, 2, and 4. We remark, however, that a fundamental step in the proof is the following observation. If $e$ denotes the total mechanical energy of the system with Lagrangian (A5.4), by the theorem of conservation of energy $\mathbf{x}(t, k)$ cannot have a distance larger than $e/\sqrt{k} \to 0$ from $S$, as $k \to \infty$.

The problem of the physical realisation of constraints is also of considerable interest in the study of classical statistical mechanics. In particular, it is at the centre of several research projects on the so-called problem of the 'ultraviolet catastrophe' predicted by classical statistical mechanics (see Born 1927), and hence the conflict between the predictions of the equipartition theorem and of some experimental evidence relating to the specific heat of polyatomic gases, or to the form of the spectrum of black body radiation.[1] This problem deserves a more detailed discussion, which however cannot find space in this brief introduction.

---

[1] In a remarkable series of recent papers, Benettin, Galgani and Giorgilli have studied in great detail the consequences of the modern canonical theory of perturbations for this problem, and deduced interesting physical outcomes: see the original papers cited in the bibliography (1985, 1987a, 1987b, 1989).

# APPENDIX 6: KEPLER'S PROBLEM, LINEAR OSCILLATORS AND GEODESIC FLOWS

In this appendix we briefly study the connections between some seemingly unrelated dynamical systems.

(1) *Kepler's problem* (cf. Section 5.2): a point particle of unit mass moves in the plane of coordinates $(x,y)$ under the action of the central field with potential energy $V(r) = -k/r$, with fixed $k > 0$.
(2) *Linear oscillators*: a point particle of unit mass moves in the $(x,y)$ plane under the action of a central field with potential energy $V(r) = \frac{1}{2}\alpha r^2 = \frac{1}{2}\alpha(x^2 + y^2)$, $\alpha > 0$ fixed.
(3a) *Geodesic flow on the sphere* (cf. Section 1.6): a point particle of unit mass moves freely on the sphere of radius $R$ immersed in three-dimensional Euclidean space.
(3b) *Geodesic flow on the Poincaré disc* (cf. Section 1.7): a point particle of unit mass moves freely on the Riemannian manifold $(\mathbf{D}, \mathrm{d}s_R^2)$ where $\mathbf{D} = \{w = x + iy \in \mathbf{C} | \, |w| < 1\}$ and

$$\mathrm{d}s_R^2 = 4R^2 \frac{\mathrm{d}w\, \mathrm{d}\bar{w}}{(1 - |w|^2)^2} = 4R^2 \frac{(\mathrm{d}x)^2 + (\mathrm{d}y)^2}{(1 - x^2 - y^2)^2}. \tag{A6.1}$$

*Remark A6.1*
The Gaussian curvature (see Appendix 3) of the sphere of radius $R$ is equal to $1/R^2$, while that of the Poincaré disc $(\mathbf{D}, \mathrm{d}s_R^2)$ is $-1/R^2$. ∎

The dynamical systems (1) and (2) are related by the theorem of Bertrand (Theorem 5.3). This is not by chance; in fact, by an appropriate coordinate and time transformation, it is possible to transform one into the other (see Levi-Civita 1920).

THEOREM A6.1 (Levi-Civita transformation) *Setting*

$$z = x + iy \in \mathbf{C}, \tag{A6.2}$$

*the change of (space and time) variables*

$$z = \frac{w^2}{2}, \quad w \in \mathbf{C},$$
$$\frac{\mathrm{d}t}{\mathrm{d}\tau} = |w|^2 = 2|z| \tag{A6.3}$$

transforms the equations of motion of the Kepler problem

$$\ddot{z} = -k\frac{z}{|z|^3}, \qquad (A6.4)$$

for a fixed value of the energy $E$:

$$E = \frac{1}{2}|\dot{z}|^2 - \frac{k}{|z|}, \qquad (A6.5)$$

into the equations of a linear oscillator:

$$w'' - 2Ew = 0, \qquad (A6.6)$$

where $w'' = \mathrm{d}^2 w/\mathrm{d}\tau^2$.

*Proof*
We check first of all that equation (A6.4) coincides with the equation of motion for Kepler's problem. Separating real and imaginary parts in (A6.4) we find

$$\ddot{x} = -\frac{k}{r^2}\frac{x}{r}, \qquad \ddot{y} = -\frac{k}{r^2}\frac{y}{r}, \qquad (A6.7)$$

and similarly we check that equation (A6.6) coincides with the equation of a linear oscillator (set $w = \xi + i\eta$, $\alpha = -2E$) and that (A6.5) is the energy of Kepler's problem.

From equations (A6.3) it follows that

$$\dot{z} = \frac{\mathrm{d}z}{\mathrm{d}t} = \frac{\mathrm{d}z}{\mathrm{d}w}\frac{\mathrm{d}w}{\mathrm{d}\tau}\frac{\mathrm{d}\tau}{\mathrm{d}t} = \frac{w}{|w|^2}w' \qquad (A6.8)$$

and $|z| = |w|^2/2$; hence in the new variables the energy is expressed as

$$E = \frac{1}{2|w|^2}|w'|^2 - \frac{2k}{|w|^2},$$

and multiplying by $|w|^2$, it follows that

$$\frac{1}{2}|w'|^2 = E|w|^2 + 2k. \qquad (A6.9)$$

Equation (A6.6) is obtained by computing

$$\ddot{z} = \frac{1}{|w|^2}\frac{\mathrm{d}}{\mathrm{d}\tau}\frac{ww'}{|w|^2} = \frac{1}{|w|^2}\frac{\mathrm{d}}{\mathrm{d}\tau}\frac{w'}{\overline{w}},$$

substituting this into (A6.4) and taking into account (A6.9). ∎

*Remark A6.2*
An analogous theorem holds for the three-dimensional Kepler's problem (the transformation then goes under the name of Kustaanheimo–Stiefel–Schiefele; see Schiefele and Stiefel 1971). ∎

*Remark A6.3*
It is interesting—and applicable to the numerical study of orbits—to note that equations (A6.6) are regular at $|w| = |z| = 0$. At the same point, equations (A6.4) are singular. ∎

*Remark A6.4*
Theorem A6.1 relates Keplerian motions with fixed energy $E$ and solutions of the linear oscillator with $\alpha = -2E$ (then with the harmonic oscillator of frequency $\omega^2 = -2E$, if $E < 0$, while for $E > 0$ the oscillator if hyperbolic). ∎

The relation between problem (1) (Kepler) (and, thanks to Theorem A6.1, linear oscillators) and the geodesic flows (3a) and (3b) is the object of the following theorem (see Moser 1970 and Alekseev 1981).

THEOREM A6.2  *The phase flow of the plane Kepler problem for a fixed value $E$ of the energy is equivalent, up to a time parametrisation, to the geodesic flow on a surface with constant Gaussian curvature $-2E/k^2$.*

*After the addition of a point, this surface is isometric to the sphere of radius $R = k/\sqrt{-2E}$ if $E < 0$, to the Euclidean plane if $E = 0$ or to the Poincaré disc $(\mathbf{D}, \mathrm{d}s_R^2)$, with $R = k/\sqrt{2E}$ if $E > 0$.* ∎

*Remark A6.5*
The meaning of the 'equivalence' of two phase flows will be made clear in the course of the proof. ∎

*Remark A6.6*
An analogous theorem holds for the Kepler problem in three-dimensional space. ∎

*Proof of Theorem A6.2*
Let $\mathbf{q} = (x, y)$, $\mathbf{p} = \dot{\mathbf{q}}$. The Hamiltonian of the plane Kepler problem is

$$H(\mathbf{p}, \mathbf{q}) = \frac{1}{2}|\mathbf{p}|^2 - \frac{k}{|\mathbf{q}|}. \tag{A6.10}$$

Consider a fixed value of the energy $E$ and the three-dimensional manifold of constant energy

$$M_E = \{(\mathbf{p}, \mathbf{q}) \in \mathbf{R}^2 \times \mathbf{R}^2 \setminus \{(0,0)\} | H(\mathbf{p}, \mathbf{q}) = E\}. \tag{A6.11}$$

Introduce the transformation

$$\frac{\mathrm{d}t}{\mathrm{d}\tau} = \frac{|\mathbf{q}|}{k}. \tag{A6.12}$$

We want to show that on the manifold $M_E$, Hamilton's equations for (A6.10) become

$$\frac{dp_i}{d\tau} = -\frac{\partial \widetilde{\mathcal{H}}}{\partial q_i},$$
$$\frac{dq_i}{d\tau} = \frac{\partial \widetilde{\mathcal{H}}}{\partial p_i},$$
(A6.13)

where $i = 1, 2,$

$$\widetilde{\mathcal{H}}(\mathbf{p},\mathbf{q}) = \frac{|\mathbf{q}|}{k}(H(\mathbf{p},\mathbf{q}) - E) = \frac{|\mathbf{q}|}{2k}(|\mathbf{p}|^2 - 2E) - 1 \quad (A6.14)$$

and the manifold $M_E$ coincides with

$$\widetilde{M}_E = \{(\mathbf{p},\mathbf{q}) \in \mathbf{R}^2 \times \mathbf{R}^2 \setminus \{(0,0)\} | \widetilde{\mathcal{H}}(\mathbf{p},\mathbf{q}) = 0\}. \quad (A6.15)$$

Indeed on $\widetilde{M}_E$ we have

$$\frac{\partial \widetilde{\mathcal{H}}}{\partial p_i} = \frac{|\mathbf{q}|}{k}\frac{\partial H}{\partial p_i} = \frac{dt}{d\tau}\frac{dq_i}{dt} = \frac{dq_i}{d\tau},$$

$$-\frac{\partial \widetilde{\mathcal{H}}}{\partial q_i} = -\frac{1}{k}\left(\frac{\partial}{\partial q_i}|\mathbf{q}|\right)[H(\mathbf{p},\mathbf{q}) - E]\bigg|_{M_E} - \frac{|\mathbf{q}|}{k}\frac{\partial H}{\partial q_i} = \frac{dt}{d\tau}\dot{p}_i = \frac{dp_i}{d\tau}.$$

In addition, note that $M_E = \widetilde{M}_E$ and the following identity holds:

$$\frac{|\mathbf{q}|}{2k}(|\mathbf{p}|^2 - 2E) = 1. \quad (A6.16)$$

This identity, applied to Hamilton's equations (A6.13), yields

$$\frac{dp_i}{d\tau} = -\frac{|\mathbf{q}|}{2k}(|\mathbf{p}|^2 - 2E)\frac{\partial}{\partial q_i}\left[\frac{|\mathbf{q}|}{2k}(|\mathbf{p}|^2 - 2E)\right] = -\frac{\partial \mathcal{H}}{\partial q_i},$$
$$\frac{dq_i}{d\tau} = \frac{|\mathbf{q}|}{2k}(|\mathbf{p}|^2 - 2E)\frac{\partial}{\partial p_i}\left[\frac{|\mathbf{q}|}{2k}(|\mathbf{p}|^2 - 2E)\right] = \frac{\partial \mathcal{H}}{\partial p_i},$$
(A6.17)

where we set

$$\mathcal{H}(\mathbf{p},\mathbf{q}) = \frac{|\mathbf{q}|^2(|\mathbf{p}|^2 - 2E)^2}{8k^2} \quad (A6.18)$$

and the manifolds $M_E = \widetilde{M}_E$ coincide with

$$\mathcal{M}_E = \left\{(\mathbf{p},\mathbf{q}) \in \mathbf{R}^2 \times \mathbf{R}^2 \setminus \{(0,0)\} | |\mathbf{p}|^2 > 2E, \mathcal{H}(\mathbf{p},\mathbf{q}) = \frac{1}{2}\right\} \quad (A6.19)$$

(the requirement that $|\mathbf{p}|^2 > 2E$ allows one to select the sign of $\mathcal{H}^{1/2}$).

From the first of equations (A6.17) we deduce $\mathbf{q}$ as a function of $\mathbf{p}$ and $d\mathbf{p}/d\tau$: indeed, taking into account (A6.18) we find

$$\frac{d\mathbf{p}}{d\tau} = -\nabla_\mathbf{q}\mathcal{H} = -\mathbf{q}\frac{(|\mathbf{p}|^2 - 2E)^2}{4k^2}, \qquad (A6.20)$$

from which it follows that

$$\mathbf{q} = -\frac{4k^2}{(|\mathbf{p}|^2 - 2E)^2}\frac{d\mathbf{p}}{d\tau}. \qquad (A6.21)$$

Differentiating the expression (A6.21) we have

$$\frac{d\mathbf{q}}{d\tau} = \frac{8k^2}{(|\mathbf{p}|^2 - 2E)^3} 2\left(\mathbf{p}\cdot\frac{d\mathbf{p}}{d\tau}\right)\frac{d\mathbf{p}}{d\tau} - \frac{4k^2}{(|\mathbf{p}|^2 - 2E)^2}\frac{d^2\mathbf{p}}{d\tau^2}, \qquad (A6.22)$$

to be compared with

$$\frac{d\mathbf{q}}{d\tau} = \nabla_\mathbf{p}\mathcal{H} = \frac{|\mathbf{q}|^2}{2k^2}(|\mathbf{p}|^2 - 2E)\mathbf{p} = \frac{8k^2}{(|\mathbf{p}|^2 - 2E)^3}\left|\frac{d\mathbf{p}}{d\tau}\right|^2 \mathbf{p} \qquad (A6.23)$$

(where we used (A6.21) in the last equality). Equating (A6.22) and (A6.23) we find

$$\frac{8k^2}{(|\mathbf{p}|^2 - 2E)^3} 2\left(\mathbf{p}\cdot\frac{d\mathbf{p}}{d\tau}\right)\frac{d\mathbf{p}}{d\tau} - \frac{8k^2}{(|\mathbf{p}|^2 - 2E)^3}\left|\frac{d\mathbf{p}}{d\tau}\right|^2 \mathbf{p} = \frac{4k^2}{(|\mathbf{p}|^2 - 2E)^2}\frac{d^2\mathbf{p}}{d\tau^2},$$

and hence the second-order equation

$$\frac{d^2\mathbf{p}}{d\tau^2} = \frac{2}{|\mathbf{p}|^2 - 2E}\left[2\left(\mathbf{p}\cdot\frac{d\mathbf{p}}{d\tau}\right)\frac{d\mathbf{p}}{d\tau} - \mathbf{p}\left|\frac{d\mathbf{p}}{d\tau}\right|^2\right]. \qquad (A6.24)$$

From (A6.20), (A6.16) it also follows that

$$\left|\frac{d\mathbf{p}}{d\tau}\right| = \frac{(|\mathbf{p}|^2 - 2E)^2}{4k^2}|\mathbf{q}| = \frac{|\mathbf{p}|^2 - 2E}{2k}. \qquad (A6.25)$$

Consider the projection

$$\pi : \mathcal{M}_E \to \mathbf{R}^2,$$

$$(\mathbf{p}, \mathbf{q}) \mapsto \pi(\mathbf{p}, \mathbf{q}) = \mathbf{p}.$$

It is immediate to check that:
(a) if $E < 0$, $\pi(\mathcal{M}_E) = \mathbf{R}^2$;
(b) if $E = 0$, $\pi(\mathcal{M}_E) = \mathbf{R}^2 \backslash \{(0,0)\}$;
(c) if $E > 0$, $\pi(\mathcal{M}_E) = \mathbf{R}^2 \backslash \overline{D}_{\sqrt{2E}} = \{(p_1, p_2) \in \mathbf{R}^2 | p_1^2 + p_2^2 > 2E\}$.

Let $A_E = \pi(\mathcal{M}_E)$ and introduce the Riemannian metric

$$(ds_E)^2 = \frac{4k^2}{(|\mathbf{p}|^2 - 2E)^2}[(dp_1)^2 + (dp_2)^2]. \qquad (A6.26)$$

We now check that equation (A6.24) is the geodesic equation on the two-dimensional Riemannian manifold $(A_E, (ds_E)^2)$, and that $\tau$ is the arc length parameter along the geodesic. By (A6.26) the components of the metric tensor are

$$g_{ij}(E) = \frac{4k^2 \delta_{ij}}{(|\mathbf{p}|^2 - 2E)^2}, \quad g^{ij}(E) = \frac{(|\mathbf{p}|^2 - 2E)^2}{4k^2}\delta_{ij}, \qquad (A6.27)$$

and hence it is immediate to compute the Christoffel symbols (see (1.69))

$$\Gamma^i_{jk}(E) = \frac{1}{2}\sum_{n=1}^{2} g^{ni}(E)\left(\frac{\partial g_{jn}(E)}{\partial p_k} + \frac{\partial g_{kn}(E)}{\partial p_j} - \frac{\partial g_{jk}(E)}{\partial p_n}\right)$$

$$= \frac{1}{2}\sum_{n=1}^{2}\frac{(|\mathbf{p}|^2 - 2E)^2}{4k^2}\delta^{ni}\left(-\frac{8k^2}{(|\mathbf{p}|^2 - 2E)^3}\right)[2p_k\delta_{jn} + 2p_j\delta_{kn} - 2p_n\delta_{jk}]$$

$$= -\frac{2}{(|\mathbf{p}|^2 - 2E)}\sum_{n=1}^{2}\delta^{ni}[p_k\delta_{jn} + p_j\delta_{kn} - p_n\delta_{jk}]$$

$$= -\frac{2}{(|\mathbf{p}|^2 - 2E)}(\delta_{ij}p_k + \delta_{ik}p_j - p_i\delta_{jk}). \qquad (A6.28)$$

The geodesic equation (see (1.68)) on $(A_E, (ds_E)^2)$ can then be written as

$$\frac{d^2 p_i}{ds^2} = -\sum_{j,k=1}^{2}\Gamma^i_{jk}(E)\frac{dp_j}{ds}\frac{dp_k}{ds}$$

$$= \frac{2}{(|\mathbf{p}|^2 - 2E)}\sum_{j,k=1}^{2}(\delta_{ij}p_k + \delta_{ik}p_j - \delta_{jk}p_i)\frac{dp_j}{ds}\frac{dp_k}{ds} \qquad (A6.29)$$

$$= \frac{2}{(|\mathbf{p}|^2 - 2E)}\left[2\left(\mathbf{p}\cdot\frac{d\mathbf{p}}{ds}\right)\frac{dp_i}{ds} - p_i\left|\frac{d\mathbf{p}}{ds}\right|^2\right],$$

which coincides with (A6.24) by setting $s = \tau$.

Finally, we consider the map

$$\psi : \mathcal{M}_E \to TA_E,$$
$$(\mathbf{p}, \mathbf{q}) \mapsto (\mathbf{p}, \mathbf{u}), \qquad (A6.30)$$

where $TA_E$ is the tangent bundle of $A_E$ and

$$\mathbf{u} = \frac{d\mathbf{p}}{d\tau}. \qquad (A6.31)$$

Since from (A6.25), (A6.26) it follows that

$$|\mathbf{u}|_{ds_E} = \frac{2k}{(|\mathbf{p}|^2 - 2E)} \left|\frac{d\mathbf{p}}{d\tau}\right| = 1, \qquad (A6.32)$$

and the projection $\pi : \mathcal{M}_E \to A_E$ is onto, the map $\psi$ is a diffeomorphism from $\mathcal{M}_E$ to $UA_E$, the bundle of tangent vectors of unit length:

$$UA_E = \{(\mathbf{p}, \mathbf{u}) \in TA_E | \, |\mathbf{u}|_{ds_E} = 1\}. \qquad (A6.33)$$

In addition, $\psi$ transforms the Hamiltonian flow $\mathcal{M}_E$ into the geodesic flow, since, as we have seen, $s = \tau$ and (A6.24) coincides with (A6.29) (recall that $\mathbf{q}$ is uniquely determined by the knowledge of $\mathbf{p}$ and of $d\mathbf{p}/d\tau$, see (A6.21)).

Due to Remark A3.2, it is immediate to compute the Gaussian curvature of the Riemannian manifold $(A_E, ds_E^2)$. Indeed, setting

$$g = \frac{4k^2}{(|\mathbf{p}|^2 - 2E)^2}, \qquad (A6.34)$$

the Gaussian curvature $K$ of $(A_E, ds_E^2)$ is given by

$$\begin{aligned} K &= -\frac{1}{g}\left[\frac{\partial}{\partial p_1}\frac{1}{\sqrt{g}}\frac{\partial}{\partial p_1}\sqrt{g} + \frac{\partial}{\partial p_2}\frac{1}{\sqrt{g}}\frac{\partial}{\partial p_2}\sqrt{g}\right] \\ &= \frac{(|\mathbf{p}|^2 - 2E)^2}{4k^2}\left[\frac{\partial}{\partial p_1}\left(\frac{2p_1}{|\mathbf{p}|^2 - 2E}\right) + \frac{\partial}{\partial p_2}\left(\frac{2p_2}{|\mathbf{p}|^2 - 2E}\right)\right] \\ &= -\frac{1}{4k^2}[4p_1^2 + 4p_2^2 - 4(|\mathbf{p}|^2 - 2E)] = -\frac{2E}{k^2}. \end{aligned} \qquad (A6.35)$$

If $E < 0$, the stereographic projection

$$p_1 = \frac{k}{\sqrt{2|E|}}\frac{x_1}{k/\sqrt{2|E|} - x_0}, \quad p_2 = \frac{k}{\sqrt{2|E|}}\frac{x_2}{k/\sqrt{2|E|} - x_0} \qquad (A6.36)$$

transforms $A_E$ into the sphere

$$x_0^2 + x_1^2 + x_2^2 = \frac{k^2}{2|E|} \qquad (A6.37)$$

without the point $x_0 = k/\sqrt{2|E|}$, $x_1 = x_2 = 0$. The sphere (A6.37) has radius $k/\sqrt{2|E|}$, and it is immediate to check that the metric (A6.26) is transformed into the first fundamental form (A6.37).

If $E = 0$, setting

$$\begin{aligned} z &= p_1 + ip_2 = Re^{i\vartheta}, \\ w &= x_1 + ix_2 = re^{i\varphi}, \end{aligned} \qquad (A6.38)$$

the inversion

$$z = \frac{2k}{w} \qquad (A6.39)$$

transforms $A_E$ into $\mathbf{R}^2\setminus\{(0,0)\}$ with the Euclidean metric $(dx_1)^2 + (dx_2)^2 = dw\,d\bar{w} = (dr)^2 + r^2(d\varphi)^2$.

Finally, for $E > 0$, with the same notation as in (A6.38), the inversion

$$z = \frac{\sqrt{2E}}{w}$$

transforms $A_E$ into $\mathbf{D}$ and (A6.26) into

$$(ds_E)^2 = \frac{4k^2}{(|z|^2 - 2E)^2}\,dz\,d\bar{z} = \frac{4k^2}{(2E/|w|^2 - 2E)^2}\,\frac{2E}{|w|^4}\,dw\,d\bar{w}$$

$$= \frac{4}{2E/k^2}\,\frac{dw\,d\bar{w}}{(1-|w|^2)^2}, \qquad (A6.40)$$

which coincides with (A6.1) by setting $R = k/\sqrt{2E}$. ∎

# APPENDIX 7: FOURIER SERIES EXPANSIONS

Fourier trigonometric series are a particular case of the more general concept of expansion of any element of a Hilbert space with respect to a complete orthonormal basis.

We briefly summarise the essential notions (for more details and proofs, see for example Rudin (1974)).

A real (respectively, complex) vector space $V$ is endowed with a *norm* if to all its elements $x$ we can associate a real number $\|x\|$ (its norm) with the following properties:

(i) $\|x\| \geq 0, \ \forall \ x \in V$;
(ii) $\|\lambda x\| = |\lambda| \|x\|, \ \forall \ x \in V, \ \forall \ \lambda \in \mathbf{R}$ (respectively, $\forall \ \lambda \in \mathbf{C}$);
(iii) $\|x + y\| \leq \|x\| + \|y\|, \ \forall \ x, y \in V$.

A sequence $\{x_n\}$ in a normed space $V$ *converges in the norm* to the limit $x \in V$ if $\lim_{n \to \infty} \|x_n - x\| = 0$.

A sequence $\{x_n\}$ in a normed space is a *Cauchy sequence* if $\|x_n - x_m\| \to 0$ for $n, m \to +\infty$.

A normed space is *complete* if all Cauchy sequences are convergent.

A complete normed space is called a *Banach space*.

A Banach space whose norm is generated by a *scalar product* is called a *Hilbert space*. If the space is real (respectively, complex) a scalar product $(x, y)$ is a symmetric (respectively, Hermitian) positive definite bilinear form.

It is easily seen that $(x, x)^{1/2}$ has the properties of a norm. In a Hilbert space we then set $\|x\|^2 = (x, x)$.

A *complete orthonormal basis* $\{e_i\}$, $i \in \mathbf{N}$, of a Hilbert space is a basis such that $(e_i, e_j) = \delta_{ij}$ and $(x, e_i) = 0, \ \forall \ i \ \Rightarrow \ x = 0$.

THEOREM A7.1  *Let $\{e_i\}$ be a complete orthonormal basis of a Hilbert space $H$. Then every element $x \in H$ can be uniquely represented by the series*

$$x = \sum_i \alpha_i e_i. \tag{A7.1}$$

*The coefficients $\alpha_i$ are called Fourier coefficients and are defined by*

$$\alpha_i = (x, e_i). \tag{A7.2}$$

*The Parseval identity*

$$\|x\|^2 = \sum_i \alpha_i^2 \tag{A7.3}$$

*holds.* ∎

THEOREM A7.2  *A Hilbert space has a complete orthonormal basis if and only if it is separable.* ∎

COROLLARY A7.1  *Every separable real (respectively, complex) Hilbert space (infinite dimensional) is isomorphic to the space $\ell^2$ of sequences $(\alpha_i)_{i\in \mathbf{N}}$ in $\mathbf{R}$ (respectively, in $\mathbf{C}$) for which the series of $\sum_i |\alpha_i|^2$ is convergent with respect to the scalar product $(x,y) = \sum_{i=1}^\infty x_i y_i$ (respectively $(x,y) = \sum_{i=1}^\infty x_i \bar{y}_i$).* ∎

Consider the Hilbert space $L^2((0, 2\pi), \mathbf{R})$ of measurable functions $v : (0, 2\pi) \to \mathbf{R}$ which are square integrable, with the scalar product $(u, v) = \int_0^{2\pi} u(x)v(x)\,\mathrm{d}x$. A complete orthonormal basis for it is given by the functions $1/\sqrt{2\pi}$, $(1/\sqrt{\pi})\sin nx$, $(1/\sqrt{\pi})\cos nx$, $n \in \mathbf{N}$, $n > 0$. Every measurable function $V : \mathbf{R} \to \mathbf{R}$, $2\pi$-periodic (hence such that $V(x + 2\pi) = V(x)$ for every $x \in \mathbf{R}$), such that its restriction $v = V|_{(0,2\pi)}$ is also square integrable, determines uniquely an element $v \in L^2((0, 2\pi), \mathbf{R})$ which can be written as

$$v(x) = a_0 + \sum_{n=1}^\infty [a_n \cos(nx) + b_n \sin(nx)], \tag{A7.4}$$

where

$$a_0 = \frac{1}{\sqrt{2\pi}} \int_0^{2\pi} v(x)\,\mathrm{d}x, \quad a_n = \frac{1}{\sqrt{\pi}} \int_0^{2\pi} v(x)\cos(nx)\,\mathrm{d}x, \quad b_n = \frac{1}{\sqrt{\pi}} \int_0^{2\pi} v(x)\sin(nx)\,\mathrm{d}x. \tag{A7.5}$$

Equations (A7.4) and (A7.5) take a particularly compact form if we express sine and cosine through complex exponentials:

$$v(x) = \sum_{k\in\mathbf{Z}} \hat{v}_k e^{ikx}, \quad \hat{v}_k = \frac{1}{\sqrt{2\pi}} \int_0^{2\pi} v(x)e^{-ikx}\,\mathrm{d}x. \tag{A7.6}$$

The relation between the coefficients $(\hat{v}_k)_{k\in\mathbf{Z}}$ and $(a_n, b_n)_{n\in\mathbf{N}}$ is given by

$$\hat{v}_0 = a_0, \quad \hat{v}_k = \frac{1}{2}(a_k - ib_k) \text{ if } k \geq 1, \quad \hat{v}_k = \frac{1}{2}(a_{-k} + ib_{-k}) = \hat{v}^*_{-k} \text{ if } k \leq -1. \tag{A7.7}$$

Now let $v : \mathbf{R}^l \to \mathbf{R}$ be a measurable function, periodic of period $2\pi$ in each of its arguments:

$$\begin{aligned} v(x_1 + 2\pi, \ldots, x_l) &= v(x_1, x_2 + 2\pi, \ldots, x_l) = \ldots \\ &= v(x_1, x_2, \ldots, x_l + 2\pi) \\ &= v(x_1, \ldots, x_l). \end{aligned} \tag{A7.8}$$

Assume also that its restriction $v$ to $(0, 2\pi)^l$ is square integrable.

The function $v$ is hence defined on the $l$-dimensional torus $\mathbf{T}^l = (\mathbf{R}/2\pi\mathbf{Z})^l$, and it is possible to expand it in Fourier series:

$$v(\mathbf{x}) = \sum_{\mathbf{m} \in \mathbf{Z}^l} \hat{v}_{\mathbf{m}} e^{i\mathbf{m}\cdot\mathbf{x}}. \tag{A7.9}$$

The coefficients $\hat{v}_{\mathbf{m}}$ of the series expansion (A7.9) are determined as follows:

$$\hat{v}_{\mathbf{m}} = \frac{1}{(2\pi)^l} \int_{\mathbf{T}^l} v(\mathbf{x}) e^{-i\mathbf{m}\cdot\mathbf{x}} \, d\mathbf{x}. \tag{A7.10}$$

If the function $v$ is regular, its Fourier coefficients belong to $\ell^2(\mathbf{Z}^l, \mathbf{C})$ and decay at infinity at a rate related to the degree of regularity of the function.

THEOREM A7.3  *If $v$ is of class $\mathcal{C}^r$, there exists a constant $M > 0$ such that for every $\mathbf{m} \in \mathbf{Z}^l$, $\mathbf{m} \neq \mathbf{0}$ we have*

$$|\hat{v}_{\mathbf{m}}| \leq M(|m_1| + \cdots + |m_l|)^{-r}. \tag{A7.11}$$

*If $v$ is of class $\mathcal{C}^\infty$, for every positive integer $r$ there exists a constant $M > 0$ such that for every $\mathbf{m} \in \mathbf{Z}^l$, $\mathbf{m} \neq \mathbf{0}$ we have*

$$|\hat{v}_{\mathbf{m}}| \leq M(|m_1| + \cdots + |m_l|)^{-r}. \tag{A7.12}$$

*If $v$ is analytic, there exist two constants $M > 0$ and $\delta > 0$ such that for every $\mathbf{m} \in \mathbf{Z}^l$ we have*

$$|\hat{v}_{\mathbf{m}}| \leq M e^{-\delta(|m_1| + \cdots + |m_l|)}. \tag{A7.13}$$

*Proof*
For simplicity we only prove the estimates (A7.11) and (A7.12). For the proof of (A7.13) see for example Sternberg (1969) and Rudin (1974).
We first remark that (A7.12) is an obvious consequence of (A7.11).
Let $l = 1$. The proof of (A7.11) depends on the identity

$$e^{-imx} = \frac{1}{-im}\frac{d}{dx}e^{-imx}, \qquad m \neq 0. \tag{A7.14}$$

From (A7.10) it follows that

$$\hat{v}_m = \frac{1}{2\pi}\int_0^{2\pi} v(x)e^{-imx}\,dx = \frac{1}{2\pi}\int_0^{2\pi} v(x)\frac{1}{-im}\frac{d}{dx}e^{-imx}\,dx,$$

and hence, integrating by parts, we find

$$\hat{v}_m = \frac{1}{2\pi}\left[\frac{1}{(-im)}v(x)e^{-imx}\right]_{x=0}^{x=2\pi} + \frac{1}{i2\pi m}\int_0^{2\pi} v'(x)e^{-imx}\,dx.$$

The first term is zero due to the periodicity of $v(x)e^{-imx}$. If $v$ is of class $\mathcal{C}^r$ we can iterate this procedure $r$ times to obtain

$$\hat{v}_m = \frac{1}{2\pi(im)^r} \int_0^{2\pi} v^{(r)}(x) e^{-imx} \, dx,$$

from which it immediately follows that

$$|\hat{v}_m| \le \frac{1}{2\pi|m|^r} \int_0^{2\pi} |v^{(r)}(x)| \, dx \le \frac{M}{|m|^r},$$

where $M = \max_{0 \le x \le 2\pi} |v^{(r)}(x)|$. The existence of $M$ is guaranteed by the assumption that $v$ is of class $\mathcal{C}^r$.

If $l \ge 1$, it is sufficient to observe the following. Since $\mathbf{m} \ne \mathbf{0}$, there exists at least one component $m_j \ne 0$. From (A7.10), integrating by parts, we find

$$\hat{v}_\mathbf{m} = \frac{1}{(2\pi)^l} \int_{\mathbf{T}^l} v(\mathbf{x}) \left( -\frac{1}{im_j} \frac{\partial}{\partial x_j} \right) e^{-i\mathbf{m}\cdot\mathbf{x}} d\mathbf{x}$$

$$= -\frac{1}{im_j} \frac{1}{(2\pi)^l} \int_{\mathbf{T}^l} \frac{\partial v}{\partial x_j}(\mathbf{x}) e^{-i\mathbf{m}\cdot\mathbf{x}} d\mathbf{x},$$

from which, iterating this procedure $r$ times, we obtain

$$|\hat{v}_\mathbf{m}| \le \frac{\tilde{M}}{(\max_{1 \le j \le l} |m_j|)^r}, \qquad (A7.15)$$

where $\tilde{M} = \max_{1 \le j \le l} \max_{\mathbf{x} \in \mathbf{T}^l} |\partial^r v / \partial x_j^r(\mathbf{x})|$.

On the other hand, $\max_{1 \le j \le l} |m_j| \le |m_1| + \cdots + |m_l| \le l \max_{1 \le j \le l} |m_j|$. Substituting this into (A7.15) yields the desired estimate (A7.11), with $M = l^r \tilde{M}$. ∎

# APPENDIX 8: MOMENTS OF THE GAUSSIAN DISTRIBUTION AND THE EULER Γ FUNCTION

The *moments of the Gaussian distribution* are the integrals of the type

$$\mu_n = \int_{-\infty}^{+\infty} x^n e^{-ax^2}\, dx, \tag{A8.1}$$

with $a$ being a positive constant and $n \in \mathbf{N}$.

PROPOSITION A8.1  *The moments $\mu_n$ of the Gaussian distribution are*

$$\mu_0 = \sqrt{\frac{\pi}{a}},$$

$$\mu_{2n+1} = 0, \qquad\qquad \text{for every } n \geq 0, \tag{A8.2}$$

$$\mu_{2n} = (2n-1)!!\sqrt{\frac{\pi}{a}}\,(2a)^{-n}, \quad \text{for every } n \geq 0.$$

*Proof*
The proof is a direct computation. First of all, consider $\mu_0^2$:

$$\mu_0^2 = \left(\int_{-\infty}^{+\infty} e^{-ax^2}\, dx\right)^2 = \int_{-\infty}^{+\infty} e^{-ax^2}\, dx \int_{-\infty}^{+\infty} e^{-ay^2}\, dy$$

$$= \int_{-\infty}^{+\infty} dx \int_{-\infty}^{+\infty} e^{-a(x^2+y^2)}\, dy = \int_0^{2\pi} d\varphi \int_0^{+\infty} r e^{-ar^2}\, dr = \frac{\pi}{a},$$

where we have used the substitution $x = r\cos\varphi$, $y = r\sin\varphi$.

Evidently $\mu_{2n+1} = 0$, because the integrand is an odd function of $x$. To compute $\mu_{2n}$ we first note that

$$x^{2n} e^{-ax^2} = (-1)^n \frac{\partial^n}{\partial a^n} e^{-ax^2},$$

From which it immediately follows that

$$\mu_{2n} = (-1)^n \frac{d^n}{da^n} \int_{-\infty}^{+\infty} e^{-ax^2}\, dx$$

$$= \sqrt{\pi}(-1)^n \left(-\frac{1}{2}\right)\left(-\frac{3}{2}\right)\cdots\left(-\frac{2n-1}{2}\right) a^{-(2n+1)/2}$$

$$= (2n-1)!!\sqrt{\frac{\pi}{a}}\,(2a)^{-n}. \qquad\blacksquare$$

It is often also useful to compute the integral between 0 and $\infty$. For integrals with even $n$ it is enough to divide the previous result by two. For those with odd $n$, we can check with a sequence of integration by parts that

$$\int_0^\infty x^{2n+1} e^{-x^2} \, dx = \frac{1}{2} n!, \tag{A8.3}$$

and hence

$$\int_0^\infty x^{2n+1} e^{-ax^2} \, dx = \frac{1}{2} a^{-(n+1)} n!. \tag{A8.4}$$

There exists an obvious relation between the moments of the Gaussian distribution and the *Euler $\Gamma$ function*

$$\Gamma(z) = \int_0^\infty t^{z-1} e^{-t} \, dt, \quad z > 0. \tag{A8.5}$$

Indeed, after the substitution $ax^2 = t$ we immediately find

$$\int_0^\infty x^n e^{-ax^2} \, dx = \frac{1}{2} a^{-(n+1)/2} \, \Gamma\left(\frac{n+1}{2}\right). \tag{A8.6}$$

For the function $\Gamma$ we have

$$\Gamma(n+1) = n!, \quad \forall \, n \in \mathbf{N}. \tag{A8.7}$$

This can be deduced by induction from equation (A8.10) below, while equation (A8.2) yields

$$\Gamma\left(\frac{1}{2}\right) = \sqrt{\pi}, \quad \Gamma\left(k + \frac{1}{2}\right) = 2^{-k}(2k-1)!\sqrt{\pi}. \tag{A8.8}$$

Another interesting formula is

$$\Gamma(z)\Gamma(1-z) = \frac{\pi}{\sin z\pi}, \quad 0 < z < 1, \tag{A8.9}$$

illustrating that $\Gamma(z)$ diverges for $z \to 0+$ as $1/z$.

Equation (A8.9) is important because, by using recursively the property

$$\Gamma(z+1) = z \Gamma(z), \tag{A8.10}$$

we can reduce to the computation of $\Gamma$ only for $z \in (0,1)$ and, due to (A8.9), for $z \in \left(0, \frac{1}{2}\right)$.

We can also see that $\Gamma(z)$ has a unique minimum for $z \simeq 1.4616\ldots$.

For $z \gg 1$ the famous *Stirling formula* holds:

$$\Gamma(z+1) = \sqrt{2\pi z} \, z^z \, e^{-z} \, e^{\alpha(z)/12z}, \quad \text{with } \alpha \in (0,1), \tag{A8.11}$$

A8    Moments of the Gaussian distribution and the Euler Γ function    747

and the approximation

$$\Gamma(z+1) \simeq \sqrt{2\pi z}\left(\frac{z}{e}\right)^z \qquad (A8.12)$$

can be used when $z(\log z - 1) \gg 1$ (e.g. for $z = 10$ we have $z(\log z - 1) \simeq 15.4$ and the relative error in (A8.12) is less than 1%, while for $z = 50$, $z(\log z - 1) \simeq 1680$ and the relative error is about 0.2%).

# Bibliography

## (a) Books

Abraham R., Marsden J.E. (1978). *Foundations of mechanics*, Benjamin Cummings, Reading, MA.

Agostinelli C., Pignedoli A. (1989). *Meccanica analitica*, Mucchi, Modena.

Amann H. (1990). *Ordinary differential equations*, de Gruyter Studies in Mathematics, Vol. 13, W. de Gruyter, Berlin.

Arnol'd V.I. (1978a). *Mathematical methods of classical mechanics*, Springer-Verlag, New York.

—— (1978b). *Ordinary differential equations*, MIT Press, Boston, MA.

—— (1983). *Geometric methods in the theory of ordinary differential equations*, Springer-Verlag, New York.

—— (1990). *Huygens and Barrow, Newton and Hooke*, Birkhäuser Verlag, Boston.

—— (1991). *Theory of singularities and its applications* (Lezioui Fermiane), Cambridge University Press.

Arnol'd V.I., Avez A. (1968). *Ergodic problems of classical mechanics*, Benjamin, New York.

Arnol'd V.I., Kozlov V.V., Neishtadt A.I. (1988). *Dynamical systems III*, Encyclopedia of Mathematical Sciences, Springer-Verlag, Berlin.

Arrowsmith D.K., Place C.M. (1990). *An introduction to dynamical systems*, Cambridge University Press.

Barrow-Green J. (1997). *Poincaré and the three body problem*, History of Mathematics **11**, American Mathematical Society and London Mathematical Society, Providence, RI.

Bedford T., Keane M., Series C. (1991). *Ergodic theory, symbolic dynamics and hyperbolic spaces*, Oxford University Press.

Beletski V. (1986). *Essais sur le mouvement des corps cosmiques*, Mir, Moscow.

Benettin G., Galgani L., Giorgilli A. (1991). *Appunti di meccanica razionale*, CUSL, Milano.

Bertin G. (2000). *Dynamics of galaxies*, Cambridge University Press.

Binney J., Tremaine S. (1987). *Galactic dynamics*, Princeton University Press.

Boltzmann L. (1912). *Vorlesungen über Gastheorie*, Ambrosius Barth, Leipzig.

Born M. (1927). *The mechanics of the atom*, Bell and Sons, London.

Burgatti P. (1919). *Lezioni di meccanica razionale*, Zanichelli, Bologna.

Carleman T. (1957). *Problémes mathématiques dans la théorie cinétique des gaz*, Publications Scientifiques de l'Institut Mittag-Leffler, Vol. 2, Uppsala.

Cercignani C. (1972). *Teoria e applicazioni delle serie di Fourier*, Tamburini, Bologna.

—— (1976a). *Vettori matrici geometria*, Zanichelli, Bologna.

—— (1976b). *Spazio tempo movimento. Introduzione alla meccanica razionale*, Zanichelli, Bologna.
Cercignani C. (1988). *The Boltzmann equation and its Applications*, Springer-Verlag, Berlin.
—— (1997). *Ludwig Boltzmann e la meccanica statistica*, Percorsi della Fisica, La Goliardica Pavese, Pavia.
Cercignani C., Jona-Lasinio G., Parisi G., Radicati di Bronzolo L.A. (eds) (1997). *Boltzmann's legacy 150 years After His birth*, Atti dei Convegni Lincei **131**, Accademia Nazionale dei Lincei, Roma.
Cornfeld I.P., Fomin S.V., Sinai YA.G. (1982). *Ergodic theory*, Springer-Verlag, Berlin.
Courant R., Hilbert D. (1953). *Methods of mathematical physics*, 2 vols., Interscience, New York.
Danby J.M.A. (1988). *Fundamentals of celestial mechanics*, Willmann-Bell, Richmond.
Dell'Antonio G. (1996). *Elementi di meccanica I: Meccanica classica*, Liguori, Napoli.
Dieudonné J. (1968). *Calcul infinitésimal*, Hermann, Paris.
—— (1978). *Abregé d'histoire des mathématiques*, Hermann, Paris.
Do Carmo M.P. (1994). *Riemannian geometry*, Birkhauser.
Dubrovin B., Fomenko A., Novikov S. (1985). *Modern geometry—methods and applications: Part II: The geometry and topology of manifolds*, Graduate Texts in Mathematics, Springer-Verlag.
—— (1991). *Modern geometry—methods and applications: Part I: The geometry of surfaces, transformation groups, and fields*, Graduate Texts in Mathematics, Springer-Verlag.
Fasano A., De Rienzo V., Messina A. (2001). *Corso di meccanica razionale*, Laterza, Bari.
Flanders H. (1963). *Differential forms with applications to the physical sciences*, Academic Press, New York.
Fletcher N.H., Rossing T.D. (1991). *The physics of musical instruments*, Springer-Verlag, New York.
Fox C. (1987). *An Introduction to the calculus of variations*, Dover, New York.
Gallavotti G. (1983). *The elements of mechanics*, Texts and Monographs in Physics, Springer-Verlag.
—— (1995). *Meccanica statistica. Trattatello.* Quaderni del GNF.M. n. 50, Roma.
—— (1999). *Statistical mechanics: A short treatise*, Springer-Verlag, Berlin.
Gallavotti G., Bonetto F., Gentile G. (2004). *Aspects of ergodic, qualitative and statistical theory of motion*, Texts and Monographs in Physics, Springer-Verlag.
Giaquinta M., Modica G. (2003). *Mathematical analysis: Functions in the variable*, Birkhäuser.
—— (1999). *Analisi matematica 2: Approssimazione e processi discreti*, Pitagora Editrice, Bologna.
—— (2000). *Analisi matematica 3: Strutture lineari e metriche, continuità*, Pitagora Editrice, Bologna.

Gibbs W. (1902). *Elementary principles of statistical mechanics*, Yale University Press, New Haven, CT.

Gilbar D., Trudinger N.S. (1977). *Elliptic partial differential equations of second order*, Springer-Verlag, Berlin.

Giorgilli A. (1990). *Appunti del corso di meccanica celeste*, dispense inedite.

Giusti E. (1987). *Analisi matematica 1*, Bollati Boringhieri, Torino.

—— (1989). *Analisi matematica 2*, Bollati Boringhieri, Torino.

Hirsch G., Smale S. (1974). *Differential equations, dynamical systems and linear algebra*, Academic Press, New York.

Hörmander L. (1994). *Notions of convexity*, Progress in Mathematics **127**, Birkhäuser Verlag, Boston, MA.

Huang K. (1987). *Statistical mechanics*, second edn, Wiley, New York.

Khinchin A.I. (1949). *Mathematical foundations of statistical mechanics*, Dover, New York.

—— (1957). *Mathematical foundations of information theory*, Dover, New York.

Krylov N.S. (1979). *Works on the foundations of statistical physics*, Princeton University Press.

Ladyzenskaya O.A., Ural'ceva N.N. (1968). *Equations aux derivées partielles de type elliptique*, Monogr. Univ. Math. 31, Dunod, Paris.

Landau L.D., Lifschitz E.M. (1982). *Course of theoretical physics: Mechanics*, Butterworth-Heinemann.

—— (1986). *The theory of elasticity*, Butterworth-Heinemann.

—— (1987). *Fluid mechanics*, Butterworth-Heinemann.

Lang S. (1970). *Algebra lineare*, Boringhieri, Torino.

—— (1975). *Complex analysis*, second edn, Springer-Verlag, Berlin.

La Salle J., Lefschetz F. (1961). *Stability by Lyapunov's direct method with applications*, Academic Press, New York.

Lasota A., Mackey M.C. (1985). *Probabilistic properties of deterministic systems*, Cambridge University Press.

Levi-Civita T., Amaldi U. (1927). *Lezioni di meccanica razionale. Volume secondo. Parte seconda. Dinamica dei sistemi con un numero finito di gradi di libertà*, Zanichelli, Bologna.

Mach E. (1915). *The science of mechanics: A critical and historical account of its development*, Open Court Publishing.

Mañe R. (1987) *Ergodic theory and differentiable dynamics*, Springer-Verlag, Berlin.

McKean H., Moll V. (1999). *Elliptic curves*, Cambridge University Press.

Meyer K.R., Hall G.R. (1992). *Introduction to Hamiltonian dynamical systems and the N-body problem*, Applied Mathematical Sciences 90, Springer-Verlag, Berlin.

Meyer Y. (1972). *Algebraic numbers and harmonic analysis*, North Holland Mathematical Library, Vol. 2, Amsterdam.

Moser J. (1973). *Stable and random motions in dynamical systems*, Annals of Mathematical Studies 77, Princeton University Press.

Percival I.C., Richards (1986). *An introduction to dynamics*, Cambridge University Press.

Piccinini L.C., Stampacchia G., Vidossich G. (1984). *Ordinary differential equations in $\mathbf{R}^n$*, Applied Mathematical Sciences, Vol. 39, Springer-Verlag, Berlin.

Poincaré H. (1892). *Les méthodes nouvelles de la mécanique céleste. Tome I*, Gauthier-Villars, Paris.

—— (1893). *Les méthodes nouvelles de la mécanique céleste. Tome II*, Gauthier-Villars, Paris.

—— (1899). *Les méthodes nouvelles de la mécanique céleste. Tome III*, Gauthier-Villars, Paris.

—— (1905). *Leçons de mécanique céleste*, Tome I, Gauthier-Villars, Paris.

Pollard H. (1966). *Mathematical introduction to celestial mechanics*, Prentice Hall, Englewood Cliffs, NJ.

—— (1976). *Celestial mechanics*, The Icarus Mathematical Monographs, Vol. 18, The Mathematical Association of America.

Rudin W. (1974). *Analisi reale e complessa*, Boringhieri, Torino.

Ruelle D. (1969). *Statistical mechanics: Rigorous results*, W.A. Benjamin, New York.

Schiefele G., Stiefel E. (1971). *Linear and regular celestial mechanics*, Springer-Verlag, Berlin.

Schmidt W.M. (1980). *Diophantine approximation*, Lecture Notes in Mathematics **785**, Springer-Verlag, Berlin.

—— (1991). *Diophantine approximations and diophantine equations*, Lecture Notes in Mathematics **1467**, Springer-Verlag, Berlin.

Sernesi E. (1989). *Geometria 1*, Bollati Boringhieri, Torino.

—— (1994). *Geometria 2*, Bollati Boringhieri, Torino.

Siegel C.L., Moser J. (1971). *Lectures on celestial mechanics*, Springer-Verlag, Berlin.

Sinai YA.G. (1982). *Theory of phase transitions: Rigorous results*, Pergamon, Oxford.

Singer I.M., Thorpe J.A. (1980). *Lezioni di topologia elementare e di geometria*, Boringhieri, Torino.

Sternberg S. (1969). *Celestial mechanics*, W.A. Benjamin, New York.

Struik D.J. (1988). *Lectures on classical differential geometry*, Dover, New York.

Tabachnikov S. (1995). *Billiards*, Panoramas et Synthéses **1**, Société Mathématique de France.

Thompson C.J. (1972). *Mathematical statistical mechanics*, Princeton University Press.

—— (1988). *Classical equilibrium statistical mechanics*, Clarendon Press, Oxford.

Thorpe J.A. (1978). *Elementary topics in differential geometry*, Springer-Verlag, Berlin.

Tricomi F. (1937). *Funzioni ellittiche*, Zanichelli, Bologna.

Truesdell C. (1968). *Essays in the history of mechanics*, Springer-Verlag, Berlin.

Uhlenbeck G.E., Ford G.W. (1963). *Lectures in statistical mechanics*, American Mathematical Society, Providence, RI.

Walters P. (1982). *An introduction to ergodic theory*, Graduate Texts in Mathematics **79**, Springer-Verlag, Berlin.

Watson G.N. (1980). *A treatise on the theory of Bessel functions*, Cambridge University Press.
Weeks J.R. (1985). *The shape of space*, Marcel Dekker, New York.
Whittaker E.T. (1936). *A treatise on the analytical dynamics of particles and rigid bodies*, Cambridge University Press.
Whittaker E.T., Watson G.N. (1927). *A course of modern analysis*, Cambridge University Press.
Wintner A. (1941). *The analytical foundations of celestial mechanics* Princeton University Press.

## (b) Articles

Albouy A. (2000). Lectures on the two-body problem, in *The Recife lectures in celestial mechanics*, F. Diacu, H. Cabral, eds, Princeton University Press.
Alekseev V.M. (1981). Quasirandom oscillations and qualitative questions in celestial mechanics, *Amer. Math. Soc. Transl.*, **116**, 97–169.
Anosov D.V. (1963). Ergodic properties of geodesic flows on closed Riemannian manifolds of negative curvature, *Sov. Math. Dokl.*, **4**, 1153–6.
——(1967). Geodesic flows on compact Riemannian manifolds of negative curvature, *Proc. Steklov Inst. Math.*, **90**, 1–209.
Arnol'd V.I. (1961). Small denominators I: on the mappings of a circle into itself, Translations of the A.M.S., 2nd series, **46**, 213.
——(1963a). Proof of A.N. Kolmogorov's theorem on the preservation of quasiperiodic motions under small perturbations of the Hamiltonian, *Russ. Math. Surv.*, **18**, 9.
——(1963b). Small denominators and problems of stability of motion in classical and celestial mechanics, *Russ. Math. Surv.*, **18**, 85.
——(1991). A Mathematical trivium, *Russ. Math. Surv.*, **46**, 271–8.
Benettin G., Ferrari G., Galgani L., Giorgilli A. (1982). An extension of the Poincaré–Fermi theorem on the nonexistence of invariant manifolds in nearly integrable Hamiltonian systems, *Nuovo Cimento*, **72**(B), 137.
Benettin G., Galgani L., Giorgilli A., Strelcyn J.M. (1984). A proof of Kolmogorov's theorem on invariant tori using canonical transformations defined by the Lie method, *Nuovo Cimento*, **79B**, 201.
Benettin G., Galgani L., Giorgilli A. (1985). Boltzmann's ultraviolet cutoff and Nekhoroshev's theorem on Arnol'd diffusion, *Nature*, **311**, 444.
——(1987a). Exponential law for the equipartition times among translational and vibrational degrees of freedom, *Phys. Lett. A*, **120**, 23.
——(1987b). Realization of holonomic constraints and freezing of high frequency degrees of freedom in the light of classical perturbation theory. Part I, *Commun. Math. Phys.*, **113**, 87–103.
——(1989). Realization of holonomic constraints and freezing of high frequency degrees of freedom in the light of classical perturbation theory. Part II, *Commun. Math. Phys.*, **121**, 557–601.

Bertrand J. (1873). Théorème relatif au mouvement d'un point attiré vers un centre fixe, *Comptes Rendus*, **77**, 849–53.

Birkhoff G.D. (1931). Proof of the ergodic theorem, *Proc. Nat. Acad. Sci. USA*, **17**, 656–60.

Bost J.B. (1986). Tores invariants des systèmes dynamiques hamiltoniens, Seminaire Bourbaki 639, Astérisque, **133–4**, 113–57.

Boutroux P. (1914). Lettre de M. Pierre Boutroux à M. Mittag-Leffler, *Acta Math.*, **38**, 197–201.

Brin M. Katok A. (1983). On local entropy, in *Geometric dynamics*, Springer Lecture Notes in Math. **1007**, 30–8.

Cayley A. (1861). Tables of the developments of functions in the theory of elliptic motion, *Mem. Roy. Astron. Soc.*, **29**, 191–306 (also in *Collected mathematical papers*, Vol. III, pp. 360–474, Cambridge (1890)).

Celletti A. (1990). Analysis of resonances in the spin–orbit problem in celestial mechanics, *J. Appl. Math. Phys. (ZAMP)*, **41**, 174–204, 453–79.

Cercignani C. (1988). Le radici fisiche e matematiche dell'irreversibilità temporale: vecchi problemi e nuovi risultati, Atti del Convegno in Onore del Prof. Antonio Pignedoli, Università di Bologna, CNR, 1–13.

Chenciner A. (1990). Séries de Lindstedt, Note S028, Bureau des Longitudes, Paris.

Cherry T.M. (1924a). On integrals developable about a singular point of a Hamiltonian system of differential equations, *Proc. Cambridge Phil. Soc.*, **22**, 325–49.

——(1924b). On integrals developable about a singular point of a Hamiltonian system of differential equations II, *Proc. Cambridge Phil. Soc.*, **22**, 510–33.

Chierchia L., Gallavotti G. (1982). Smooth prime integrals for quasi-integrable Hamiltonian systems, *Nuovo Cimento*, **67**(B), 277.

Di Perna R., Lions P.L. (1990). On the Cauchy problem for the Boltzmann equation: Global existence and weak stability results. *Annals of Math.*, **130**, 321–66.

Diana E., Galgani L., Giorgilli A., Scotti A. (1975). On the direct construction of formal integrals of a Hamiltonian system near an equilibrium point, *Boll. U.M.I.*, **11**, 84–9.

Dyson F.J., Lenard A. (1967). Stability of matter, I, *J. Math. Phys.*, **8**, 282.

Eckmann J.P., Ruelle D. (1985). Ergodic theory of chaos and strange attractors, *Rev. Mod. Phys.*, **57**, 617–56.

Escande D.F. (1985). Stochasticity in classical hamiltonian systems: Universal aspects, *Phys. Rep.*, **121**.

Fermi E. (1923a). Dimostrazione che in generale un sistema meccanico normale è quasi-ergodico, *Nuovo Cimento*, **25**, 267.

——(1923b). Generalizzazione del teorema di Poincaré sopra la non esistenza di integrali uniformi di un sistema di equazioni canoniche normali, *Nuovo Cimento*, **26**, 105–15.

——(1923c). Beweis dass ein mechanisches Normalsystem im allgemeinen quasi-ergodisch ist, *Phys. Z.*, **24**, 261–5.

——(1924). Über die Existenz quasi-ergodischer systeme, *Phys. Z.*, **25**, 166–7.

Fermi E., Pasta, J., Ulam S. (1954). Collected papers of E. Fermi, Vol. 2, University of Chicago Press, Chicago, 978.

Gallavotti G. (1984). Quasi integrable mechanical systems, in, *Phénomènes critiques, systèmes aléatoires, théories de jauge*, K. Osterwalder, R. Stora (eds), Elsevier Science Publishers, Amsterdam.

—— (1998). Chaotic hypothesis and universal large deviations properties, Documenta Mathematica Extra Volume ICM 1998, I, 205–33.

Gallavotti G., Cohen E.G.D. (1995). Dynamical ensembles in nonequilibrium, statistical mechanics, *Phys. Rev. Lett.*, **74**, 2694–7.

Gallavotti G., Ornstein D.S. (1974). Billiards and Bernoulli schemes, *Commun. Math. Phys.*, **38**, 83–101.

Gallavotti G., Ruelle D. (1997). SRB states and non-equilibrium statistical mechanics close to equilibrium, *Commun. Math. Phys.*, **190**, 279–85.

Giorgilli A., Delshams A., Fontich E., Galgani L., Simó C. (1989). Effective stability for a Hamiltonian system near an elliptic equilibrium point, with an application to the restricted three body problem, *J. Diff. Eqs*, **77**, 167–98.

Goldreich P., Peale S.J. (1966). Spin–orbit coupling in the solar system, *Astron. J.*, **71**, 425–38.

Golin S., Marmi S. et al. (1990). A class of systems with measurable Hannay angles, *Nonlinearity*, **3**, 507–18.

Golin S., Knauf A., Marmi S. (1989). The Hannay angles: geometry, adiabaticity and an example, *Commun. Math. Phys.*, **123**, 95–122.

Graffi S. (1993). Le radici della quantizzazione, Quaderni di Fisica Teorica, Università degli Studi di Pavia.

Hadamard, (1898). Les Surfaces à courbures opposées et leurs lignes géodésiques, *J. Math. Pures et Appl.*, **4**, 27.

Herman M.R. (1998). Some open problems in dynamical system, Documenta Mathematica Extra Volume ICM 1998, II, 797–808.

Ising E. (1925). Beitrag zur Theorie des Ferromagnetismus, *Z. Physik*, **31**, 253.

Katznelson Y. (1975). Ergodic automorphisms of $\mathbf{T}^n$ are Bernoulli shifts, *Israel J. Math.*, **10**, 186–95.

Kolmogorov A.N. (1954). Preservation of conditionally periodic movements with small change in the Hamiltonian function, *Dokl. Akad. Nauk SSSR*, **98**, 527–30 (in Russian; English translation, in G. Casati, J. Ford, Lecture Notes in Physics, Vol. 93, pp. 51–6, Springer-Verlag, Berlin).

Kozlov V.V. (1983). Integrability and non-integrability in Hamiltonian mechanics, *Russ. Math. Surv.*, **38**, 1–76.

Lanford III O. (1975). Time evolution of large classical systems, in, *Dynamical Systems, Theory and Applications* J. Moser (ed.), Lecture Notes in Physics, V. 35, Springer-Verlag, Berlin.

—— (1989a). Les variables de Poincaré et le développement de la fonction perturbatrice, Note S026, Bureau des Longitudes, Paris.

—— (1989b). A numerical experiment on the chaotic behaviour of the Solar System, *Nature*, **338**, 237–8.

——(1990). The chaotic behaviour of the Solar System: a numerical estimate of size of the chaotic zones, *Icarus*, **88**, 266–91.

Laskar J. (1992). La stabilité du système solaire, in, *Chaos et déterminisme*, A. Dahan Dalmedico, J.-L. Chabert e K. Chemla (eds), Éditions du Seuil, Paris.

Laskar J., Robutel P. (1993). The chaotic obliquity of the planets, *Nature*, **361**, 608–12.

Lebowitz J.L. (1993). Boltzmann's entropy and time's arrow, *Physics Today*, **46**, 32–8.

Lee T.D., Yang C.N. (1952a). Statistical theory of equations of state and phase transitions, I. Theory of condensation, *Phys. Rev.*, **87**, 404.

——(1952b). Statistical theory of equations of state and phase transitions, II. Lattice gas and Ising model, *Phys. Rev.*, **87**, 410.

Levi-Civita T. (1920). Sur la régularisation du problème des trois corps, *Acta Math.*, **42**, 99–144.

Littlewood J.E. (1959a). On the equilateral configuration in the restricted problem of three bodies, *Proc. London Math. Soc.*, **9**, 343–72.

——(1959b). The Lagrange configuration in celestial mechanics, *Proc. London Math. Soc.*, **9**, 525–43.

Lorentz H. (1905). The motion of electrons in metallic bodies, *Proc. Amsterdam Acad.*, **7**, 438, 585, 604.

Manacorda T. (1991). Origin and development of the concept of wave, *Meccanica*, **26**, 1–5.

Marmi S. (2000). Chaotic behaviour in the Solar System, Séminaire Bourbaki n. 854, *Astérisque*, **266**, 113–36.

Milnor J. (1983). On the geometry of the Kepler problem, *Amer. Math. Monthly*, **90**, 353–65.

Moser J. (1962). On the invariant curves of area-preserving mappings of an annulus, *Nachr. Akad. Wiss. Göttingen Math. Phys. Kl.*, **6**, 87–120.

——(1967). Convergent series expansion for quasi-periodic motion, *Math. Ann.*, **169**, 136–76.

——(1968). Lectures on Hamiltonian systems, *Mem. Am. Math. Soc.*, **81**, 1.

——(1970). Regularisation of Kepler's problem and the averaging method on a manifold, *Comm. Pure Appl. Math.*, **23**, 609–36.

——(1986). Recent developments in the theory of Hamiltonian systems, *SIAM Review*, **28**, 459–85.

Neishtadt A.I. (1976). Averaging in multifrequency systems II, *Sov. Phys. Dokl.*, **21**, 80–2.

Nekhoroshev N.N. (1972). Action-angle variables and their generalizations, *Trans. Moscow Math. Soc.*, **26**, 180–98.

——(1977). Exponential estimate of the stability time for near-integrable Hamiltonian systems, *Russ. Math. Surv.*, **32**, 1–65.

Ornstein D.S. (1970). Bernoulli shifts with the same entropy are isomorphic, *Adv. Math.*, **4**, 337–52.

Oseledec V.I. (1968). A multiplicative ergodic theorem: Lyapunov characteristic numbers for dynamical systems, *Trans. Moscow Math. Soc.*, **19**, 197–231.

Osserman R. (1990). Curvature in the eighties, *Amer. Math. Monthly*, **97**, 731–56.

Pesin Ya.B. (1977). Characteristic Lyapunov exponents and smooth ergodic theory, *Russ. Math. Surveys*, **32**, 55–114.

Pöschel J. (1982). Integrability of Hamiltonian systems on Cantor sets, *Commun. Pure Appl. Math.*, **35**, 653–96.

Rohlin V.A. (1964). Exact endomorphisms of a Lebesgue space, *Amer. Math. Soc. Translations Ser. 2*, **39**, 1–36.

Rubin H., Hungar P. (1957). Motion under a strong constraining force, *Commun. Pure Appl. Math.*, **10**, 65–87.

Ruelle D., (1978). An inequality of the entropy of differentiable maps, *Bol. Soc. Bra. Mat.*, **9**, 83–7.

—— (1996). Positivity of entropy production in non-equilibrium statistical mechanics, *J. Stat. Phys.*, **85**, 1–25.

—— (1997). Entropy production in non-equilibrium statistical mechanics, *Commun. Math. Phys.*, **189**, 365–71.

—— (1999). Smooth dynamics and new theoretical ideas in non-equilibrium statistical mechanics, *J. Stat. Phys.*, **95**, 393–468.

Saari D.G. (1990). A visit to the Newtonian $n$-body problem via elementary complex variables, *Amer. Math. Monthly*, **97**, 105–19.

Salomon D., Zehnder E. (1989). KAM theory in configuration space, *Comm. Math. Helvetici*, **64**, 84–132.

Siegel C.L. (1941). On the integrals of canonical systems, *Ann. Math.*, **42**, 806–22.

—— (1954). Über die Existenz einer Normalform analytischer Hamiltonscher Differentialglei- chungen in der Nähe einer Gleichgewichtslösung, *Math. Ann.*, **128**, 144–70.

Sinai Ya.G. (1970). Dynamical systems with elastic reflections, *Russ. Math. Surv.*, **25**, 137–89.

—— (1979). Development of Krylov's ideas, in *Works on the foundations of statistical physics*, Krylov N.S., Princeton University Press, pp. 239–81.

Smale S. (1967). Differentiable dynamical systems, *Bull. Amer. Math. Soc.*, **73**, 747.

—— (1970a). Topology and mechanics I, *Inventiones Math.*, **10**, 305–31.

—— (1970b). Topology and mechanics II. The planar $n$-body problem, *Inventiones Math.*, **11**, 45–64.

Sundman K.F. (1907). Recherches sur le probléme des trois corps, *Acta Soc. Sci. Fennicae*, **34**, 6.

Takens F. (1970). Motion under the influence of a strong constraining force, in *Global theory of Dynamical Systems*, Z. Nitecki, C. Robinson (eds), Lecture Notes in Mathematics, Vol. 819, Springer-Verlag, Berlin.

Ulam S.M., Von Neumann J. (1947). On combinations of stochastic and deterministic properties, *Bull. Amer. Math. Soc.*, **53**, 1120.

Urbanski N. (1924). *Phys. Z.*, **25**, 47.

Yoccoz J.-C. (1992). Travaux de Herman sur les tores invariants, Seminaire Bourbaki 754, *Astérisque*, **206**, 311–44.

Yoccoz J.-C. (1995). Introduction to hyperbolic dynamics, in *Real and complex dynamical systems*, B. Branner, P. Hjorth, (eds). NATO ASI Series **C464**, Kluwer, Dordrecht, pp. 265–91.

Young L.S. (1995). Ergodic theory of differentiable dynamical systems, in *Real and complex dynamical systems*, B. Branner, P. Hjorth, (eds), NATO ASI Series **C464**, Kluwer, Dordrecht, pp. 293–336.

# Index

absolute pressure and absolute temperature, in ideal monatomic gas 602–4
absolute rigid motion 225
acceleration 10
  Coriolis 223
  of a holonomic system 57–8
action and reaction principle 71
action-angle variables
  for the Kepler problem 466–71
  for systems with one degree of freedom 431–9
  for systems with several degrees of freedom 453–8
adiabatic invariants 529–34
  perpetual 530
algebraic forms 715–18
algebras 545, 546
analysis of motion due to a positional force 92–5
angle between two intersecting curves 26
angular momentum 242–3
angular velocity 217
apparent forces 227
Archimede's spiral 58
area of a surface 27
Arnol'd theorem 446–53
'Arnol'd's cat' 553
ascending node longitude 469
asymptotically stable equilibrium points 99
attitude equation (Poisson's formula) 217
attracting node 618
attractor 563
axis of motion, instantaneous 219–21

'baker's transformation' 552, 563
Banach spaces 741

barotropic fluids 673
base spaces, of tangent bundles 41
basin of attraction 99, 553
beats 107
Bernoulli schemes 571–5
  isomorphic 574
Bernoulli systems 575
Bernoulli trinomial 674
Bertrand theorem 190
Bessel functions 212
bidimensional torus, parametrisation 21
billiard, definition 575
billiards
  dispersive 575–8
  examples of plane *576*
binormal unit vectors 12
Biot-Savart field 76
Birkhoff series 516–22
Birkhoff's theorem 558
body reference frames 213–14
Boltzmann equation 592–6
Boltzmann 'H theorem' 605–9
Boltzmann, Ludwig 591
Borel $\sigma$-algebras 546
brachistochrone 305–8
bridges, suspended 689–90
Brin–Katok theorem 570, 581

cables, suspended 685–90
canonical, and completely canonical transformations 340–52
canonical elements 466–71
canonical formalism 331–411
  problems 399–404
  solved problems 405–11
canonical isomorphism 337
canonical partition function 638

canonical perturbation theory 487–544
  fundamental equation of 507–13
  introduction to 487–99
  problems 532–4
  solved problems 535–44
canonical sets 636–40
  definition 636
  and energy fluctuations 646–7
  Helmholtz free energy and orthodicity of 645–6
canonical transformations 340–52
  infinitesimal and near-to-identity 384–93
  preservation of canonical structure of Hamilton equations 345–6
Cantor sets 528
cardinal equations of dynamics 125–7
cartesian product 47–8
catenary curves 688
Cauchy problem 675
Cauchy theorem 672–3
Cauchy–Riemann equations 410
celestial mechanics 426
central configuration 210
central fields
  motion in 179–212
  orbits in 179–85
central force fields 76
centre 700
centrifugal force 227
centrifugal moments 236
Četaev theorem 162
Chasles theorem 230
chaos 570
chaotic behaviour, of the orbits of planets in the Solar System 582–4
chaotic motion, introduction to 545–90
characteristics, tubes of 353
chemical potential 648
Christoffel symbols 29, 45
circle, osculating 9
circular orbits 187
  Lagrange stability 188
Clairaut's theorem 32

classical mechanics, axioms of 69–71
classical perturbation theory, fundamental equation of 495
closed isolated systems 624–7
closed orbits, potentials admitting 187–93
closed systems, with prescribed temperature 636–40
coefficient of dynamic friction 81
collapse of a system of $n$ particles 204
collisions
  between molecules 592, 593
  elastic 287, 575
commutativity, measuring lack of 375
commutators
  Lie derivatives and 374–9
  of two vector fields 376
complete elliptic integrals 705–707
compound rigid motion 225
computation of entropy 571–5
cones 23
  dynamic friction 81
  Poinsot 229
  static friction 80
configuration manifolds 52
configuration spaces, rigid body 214
conformal parametrisation 27
conservation laws, symmetries and 147–50
conservative autonomous systems in one dimension 420
conservative fields, work and 75–6
conservative systems 138–41
constant electric or magnetic fields, motion of a charge in 144–6
constrained rigid bodies, dynamics of 245–50
constrained systems, and Lagrangian coordinates 49–52
constraint equations, validity of 53
constraint reactions 77
  determination of 136–8
constraints
  with friction 80–1, 136–8
  holonomic 53

physical realisation of 729–31
simple 80
smooth 77, 127–8
continuous functions, quasi-periodic 462–3
continuous models, passage from discrete to 676–8
continuum mechanics
  applications of Lagrangian formalism to 680–3
  Lagrangian formalism in 671–93
  Lagrangian formulation of 678–80
  summary of fundamental laws of 671–6
Coriolis acceleration 223
Coriolis force 86, 227
cosine amplitude 706
Cotes spiral 184
critical points 99
curvature of plane curves 7–11
curvature vectors 709
curves
  angle between two intersecting 26
  curvature of plane 7–11
  length of a curve and natural parametrisation 3–7
  in the plane 1–3
  in $R^3$ 12–15
  regular 1
  vector fields and integral 15–16
cylinders, geodesic curves 29–30

d'Alembert solution 675
d'Alembert's principle 228
damped oscillations 103–7
Darboux theorem 398, 399
deformations 524
degenerate quasi-integrable Hamiltonian systems 516
degree of non-commutativity of two flows 375
Delaunay elements 469
delta amplitude 707

determination of constraint reactions 136–8
diffeomorphism maps 41
differentiable maps 39
differentiable Riemannian manifolds 33–46
differential forms 719–24
  non-singular 352, 354
Dirichlet stability criterion 160
Dirichlet theorem 151–2
discrete models, passage to continuous 676–8
discrete subgroups 450
discrete systems, dynamics of 125–78
dispersive billiards 575–8
dissipative forces 159–60
distribution functions 591–2
divisors, problem of small 504
double constraints 80
drag velocity 55
$D_w u = v$ equation 502–7
dynamic friction, coefficient of 81
dynamic friction cones 81
dynamical systems
  isomorphism of 574, 575
  on manifolds 701–4
  measurable 550–4
dynamics
  cardinal equations of 125–7
  of constrained rigid bodies 245–50
  of discrete systems 125–78
  of free systems 244–5
  general laws and the dynamics of a point particle 69–90
  of a point constrained by smooth holonomic constraints 77–80
  relative 226–8
  of rigid bodies 126, 235–77
dynamics of rigid systems, relevant quantities in 242–4

eccentric anomaly 193, 194, 195–7
effective potential energy 180
Egregium theorem of Gauss 713

Ehrenfest model 609–10
eigenspaces 241, 692
eigenvalues 241, 692
eikonal 475
elastic collisions 287, 575
elastic potential 139
electric fields, motion of a charge in constant 144–6
ellipsoids
  parametrisation 20
  separability of Hamilton–Jacobi equation for the geodesic motion on 429–31
ellipsoids of inertia (polhodes) 236–9, 252, 256
ellipsoids of revolution 256
  polhodes for 255
elliptic coordinates 426–8
elliptic functions 707
elliptic integrals 705–6
elliptic paraboloids 21
energy
  canonical sets and fluctuations of 646–7
  equipartition of the 634–6, 640–5
  internal 603
  prescribed total 634–6
energy integral 75
entropy 565–70, 605–9
  characteristic exponents and 581–2
  computation of 571–5
equations
  Boltzmann 592–6
  cardinal 125–7
  Euler 674
  Euler-Lagrange 312
  geometrical optics 475
  Hamilton 284–5
  Hamilton-Jacobi 413–20
  Kepler 193–7, 199, 211
  Lagrange 128–36
  Laplace 410
  of motion 86
  perturbation theory 502–7
  Schroedinger 476

equilibrium
  of holonomic systems with smooth constraints 141–2
  phase plane and 98–103
  stability and small oscillations 150–9
equilibrium configuration, stable 87
equilibrium of continua, as a variational problem 685–90
equilibrium points 86, 87, 99
equipartition
  of the energy (prescribed temperature) 640–5
  of the energy (prescribed total energy) 634–6
equipartition theorem 635–6
ergodic hypothesis, averages and measurements of observable quantities 616–20
ergodic problem, and the existence of first integrals 621–4
ergodic theory and chaotic motion
  introduction to 545–90
  problems 584–6
  solved problems 586–9
ergodicity, and frequency of visits 554–62
Euler angles 213–16
Euler equations 674
  integration of 256–8
  for precessions 250–1
  for stationary functionals 302–11
Euler function 746
Euler–Lagrange equations 312
existence and uniqueness theorem 695
exponents and entropy, characteristic 581–2
external forces 125, 138

Fermi theorem 623
Fermi–Pasta–Ulam model 488
first form of the orbit equation 181
first integrals
  ergodic problems and the existence of 621–4
  symmetries and 393–5

first integrals of the motion 513–16
fixed plane curve 252
fixed ruled surface, definition 228
flat torus 49
fluids 673
focus 700
foliation of phase space 453
forced oscillations 103–7
forces
  apparent 227
  Coriolis 227
  dissipative 159–60
  external 125
  inertial 227
  internal 125
Fourier coefficients 741
Fourier Series expansions 741–4
free point, motion in the absence of forces 318
free point particles 417–18
free systems, dynamics of 244–5
Frenet's theorem 10, 13
frequency of visits, ergodicity and 554–62
friction, constraints with 80–1, 136–8
friction cones, dynamic 81
friction torque 250
fugacity 649
functions
  canonical partition 638
  distribution 591–2
  generating 364–71
  quasi-periodic continuous 462–3
fundamental equation of
  canonical perturbation theory 507–13
  classical perturbation theory 495
fundamental form of a surface
  first 25, 27
  second 709–14
fundamental formula, kinematics of rigid systems 216–19
fundamental Poisson brackets 371, 373

Galilean group 71
Galilean relativity principle and interaction forces 71–4
Galilean space 71
gas, hard spheres 575, *576*
Gauss, egregium theorem of 713
Gauss transformation 552
Gaussian distribution, moments of the 745
general laws and the dynamics of a point particle, solved problems 83–90
generalised potentials 142–4
generating functions 364–71
generators 450
  infinitesimal 394
geodesic curvature vectors 709
geodesic curves 28, 45
  reversal on a surface of revolution *32*
  on a surface of revolution *31*
geodesic flow
  on the Poincaré disc 733
  on the sphere 733
geodesic motion on an ellipsoid, separability of Hamilton–Jacobi equation for 429–31
geodesic motion on a surface of revolution, separability of Hamilton–Jacobi equation 428–9
geodesics 45
  Riemannian manifold 307
geometric and kinematic foundations of Lagrangian mechanics 1–68
  problems 58–61
  solved problems 62–8
geometrical optics approximation 476
geometric properties, euler angles 213–16
geometry, of masses 235–6
geometry and kinematics, rigid bodies 213–34
Gibbs' paradox 631–4

Gibbs sets 613–69
  problems 656–9
  solved problems 662–9
grand canonical partition function 649
grand canonical potential 652
grand canonical sets 647–51
  fluctuations in 651–3
gravitational potential 139
groups, actions of tori and 46–9
gyroscopic precessions 259–60


'H theorem' of Boltzmann 605–9
Hamilton characteristic function 415
Hamilton equations, canonical
    structure 342, 345
Hamilton–Jacobi equation 413–20
  complete integrals 414
  separability for the geodesic motion
    on an ellipsoid 429–31
  separability for the geodesic motion
    on a surface of revolution
    428–9
  separation of variables for the
    421–31
Hamilton–Jacobi theory and
    integrability 413–86
  problems 477–80
  solved problems 481–6
Hamiltonian 282–3
  of the harmonic oscillator 283
  of restricted three-body problem 489
Hamiltonian action 312
Hamiltonian density 684
Hamiltonian dynamical systems,
    symplectic manifolds and 397–9
Hamiltonian flow 347
  time-dependent canonical
    transformation 361
Hamiltonian form, Hamilton's
    variational principle 314–15
Hamiltonian formalism 279–300, 684–5
  problems 288–90
  solved problems 291–300
Hamiltonian function 282–3

Hamiltonian phase space, symplectic
    structure 331–40
Hamiltonian systems
  degenerate quasi-integrable 516
  linear 331
  quasi-integrable 487
Hamiltonian vector fields 338
Hamiltonians, non-degenerate 525
Hamilton's canonical equations 284
Hamilton's characteristic function 415
Hamilton's equations 284–5
Hamilton's principal function 414
Hamilton's principle 679
Hamilton's variational principle
  Hamiltonian form 314–15
  Lagrangian form 312–14
hard spheres gas 575, *576*
hard spheres model 596–9
harmonic functions 410
harmonic oscillator 313–14, 418–20
  equation of 190
  Hamiltonian of 283
  perturbations of 516–22
  symplectic rectification of 380
harmonic potential 182–4
heavy gyroscopes, precessions of 261–3
helical motion 220
Helmholtz free energy, and orthodicity
    of the canonical set 645–6
  for monoatomic ideal gases 646
herpolhodes 252
Hilbert space 741, 742
holonomic constraints 53
holonomic systems 52–4
  accelerations of 57–8
  with fixed constraints, kinetic energy
    129
  with smooth constraints 127–8
  equilibrium of 141–2
homography of inertia 239–42
Huygens' theorem 236
hyperbolic paraboloids 21
hyperboloids 20

ideal fluids 673
ideal monatomic gas, absolute pressure and absolute temperature in 602–4
inertia
  ellipsoid and principal axes of 236–9
  homography of 239–42
  moment of 236
  polar moment of 204
  precessions by 251–4
  product 236
  tensor of 241
inertial force 227
inertial mass 70
inertial observers 69
infinitesimal canonical transformations 384–93
infinitesimal coordinate transformations 384
infinitesimal generators 394
instantaneous axis of motion 219–21
integrability
  by quadratures 439–46
  measurable functions 548–50
integrable systems
  with one degree of freedom 431–9
  with several degrees of freedom 453–8
integral curves, vector fields and 15–16
integral invariants 395–7
integration, of Euler equations 256–8
interaction forces, Galilean relativity principle and 71–4
interaction pairs 138
interaction potentials 139
internal energy 603
internal forces 125, 139
invariant $l$-dimensional tori 446–53
invariant subsets, metrically indecomposable 619
invariants
  adiabatic 529–34
  perpetual adiabatic 530
inversion points 93
isochronous motion 432

isomorphic systems 574
isomorphism
  canonical 337
  of dynamical systems 571–5
isoperimetric problem 311

Jacobi elliptic functions 707
Jacobi identity 379
Jacobi metric 318–22
Jacobi theorem 414
Jordan node 699
Jukowski function 411

KAM theorem 522–9, 532, 535
Katznelson theorem 574
Kepler's equation 193–7, 199, 211
Kepler's first law 186
Kepler's problem 185–7, 733, 734
  action-angle variables for 466–71
Kepler's second law 179, 187, 195
Kepler's third law 187, 469
kinematic states 56
kinematics, relative 223–6
kinematics of rigid systems, fundamental formula 216–19
kinetic energy, holonomic systems with smooth constraints 129
kinetic momenta 130
kinetic theory 591–612
  problems 609–10
  solved problems 610–11
Kolmogorov–Arnol'd–Moser theorem (KAM) 522–9, 532, 535
Kolmogorov–Sinai theorem 571, 573
König theorem 243
Koopman's operator 563

Lagrange formula (series inversion) 197–200
Lagrange stability of a circular orbit 188
Lagrange theorem 198
Lagrange–Jacobi identity 203

Lagrange's equations 128–36
Lagrangian of an electric charge in an
    electromagnetic 142–4
Lagrangian coordinates, constrained
    systems and 49–52
Lagrangian density 678, 680
Lagrangian form, Hamilton's
    variational principle 312–14
Lagrangian formalism 125–78
  applications to continuum mechanics
    680–3
  in continuum mechanics 671–93
    problems 690–1
    solved problems 691–3
  in dynamics of discrete systems
    125–78
    problems 162–5
    solved problems 165–78
Lagrangian formulation, of continuum
    mechanics 678–80
Lagrangian function 138–41, 676–8
Lagrangian mechanics
  geometric and kinematic foundations
    1–68
    problems 58–61
    solved problems 62–8
Lagrangian systems, natural 133
Laplace's equation 410
Lebesgue measure 528, 547, 624
Legendre transformations 279–82
length of a curve, and natural
    parametrisation 3–7
Levi–Civita transformation 733
libration motion 431
Lie algebra structure 333
Lie condition 356–8, 361
Lie derivatives and commutators 374–9
Lie product (commutator) 333
Lie series 384–93
limiting polhodes 253
Lindstedt series 524
  proving convergence of 527
linear acoustics 674
linear automorphisms of tori 565
linear Hamiltonian system 331

linear oscillators 733, 734
Liouville's theorem 285–7, 439 46
Lobačevskij: see Četaev 43
local parametrisation 36
longitudinal vibrations, of a rod 676
Lorentz, H. 575
loxodrome 60
Lusin theorem 557
Lyapunov
  characteristic exponents and entropy
    581–2
  characteristic exponents of 578–81
Lyapunov functions 159–62
Lyapunov stable equilibrium point 99


magnetic field, motion of a charge in a
    constant 144–6
manifolds
  configuration 52
  differentiable Riemannian 33–46
  dynamical systems on 701–4
  Riemannian 33–46, 52, 307, 395,
    701–4
  symplectic 395
maps
  diffeomorphism 41
  differentiable 39
masses, geometry of 235–6
material orthogonal symmetry 237
Maupertuis' principle 316–18
  stationary action of 317
Maxwell–Boltzmann distribution
    599–601, 627–31
mean anomaly 193
mean free path 604–5
mean longitude 470
mean quadratic fluctuation 620
measurable dynamical systems 550–4
measurable functions, integrability
    548–50
measure, concept of 545–7
measure space 546, 547
mechanics, wave interpretation of
    471–6

mechanics of rigid bodies 235–77
　problems 265–6
　solved problems 266–77
mechanics variational problems,
　　introduction 301–2
meridian curves 30, *31*
metrically indecomposable invariant
　　subsets 619
metrics
　Jacobi 318–22
　Riemannian 43
microcanonical sets 624–7
　definition 625
　fluctuations in the 627–31
mixing of measurable dynamical
　　systems 563–5
Möbius strip 24, 25f
models, passage from discrete to
　　continuous 676–8
molecular chaos 593, 609
molecules, collisions between 592, 593
moment of inertia 236
moments of the Gaussian distribution
　　745
motion
　analysis due to a positional force
　　92–5
　analytic first integrals of 513
　equations of 86
　of a free point in the absence of
　　forces 318
　helical 220
　instantaneous axis of 219–21
　isochronous 432
　libration or oscillatory 431
　one-dimensional 91–123
　periodic 93, 94
　of a point on an equipotential surface
　　318
　quasi-periodic 524
　quasi-periodic functions and 458–66
　rigid 221, 225
motion in a central field 179–212
　problems 205–7
　solved problems 208–12

motion of a charge, in a constant
　　electric or magnetic field 144–6
motion of a point mass 89
　in a one-dimensional field 320–1
　under gravity mass 312–13
motion of a point particle 87–9
　on an equipotential surface 79
motion of a spaceship around a planet
　　426
multiplicative ergodic theorem 579

$n$-body problem 201–5, 207
natural Lagrangian systems 133
natural parametrisation, length of a
　　curve and 3–7
Newton, view on the Solar System 582
Newton's binomial formula 392
Noether's theorem 147–50, 181, 395
nodes
　attracting 698
　Jordan 699
　repulsive 699
　star 699
non-degenerate Hamiltonians 525
non-holonomic constraints 53
non-singular differentiable forms 352,
　　354
non-singular points 1, 3
normal curvature vectors 709

one-dimensional motion 91–123
　problems 108–12
　solved motion 113–23
one-dimensional uniform motion, time
　　periodic perturbations of
　　499–502
open systems, with fixed temperature
　　647–51
orbit equation
　first form of the 181
　second form of the 184
orbits, potentials admitting closed
　　187–93
orbits in a central field 179–85

orbits of the planets in the Solar
    System 469–71
  chaotic behaviour of 582–4
ordinary differential equations 695–704
  general results 695–7
oriented surfaces 24
Ornstein theorem 574
orthodic statistical sets 615
orthogonal parametrisation 27
oscillations, damped and forced 103–7
oscillators, linear 734
oscillatory motion 431
osculating circle 9
osculating plane 12
Oseledec theorem 578–81

$p$-adic transformations 552
parabolic coordinates 425–6
paraboloids 21
parallel curves *31*
parametrisation
  of ellipsoids 20
  length of a curve and natural 3–7
  of spheres 19
Parseval identity 741
partition functions
  canonical 638
  grand canonical 649
pendulum, simple 96–8
perfect fluids 673
perihelion argument 469
period, of oscillations of a heavy point
    particle 94–5
periodic motion 93, 94
permanent rotations 254–6
perpetual adiabatic invariants 530
perturbation methods 487
perturbation theory
  canonical 487–544
  fundamental equation of classical 495
perturbations, of harmonic oscillators
    516–22
Pesin's formula 582
phase flow 702

phase plane and equilibrium 98–103
phase space 54–6
  of precessions 221–3
phase transitions 654–6
physical realisation of constraints
    729–31
plane curves, curvature of 7–11
plane rigid motions 221
plane waves 674
planets in the Solar System
  chaotic behaviour of orbits of 582–4
  orbits of the 469–71
Poincaré 207, 526, 527, 545
Poincaré-Cartan differential form 363
Poincaré disc, geodesic flow on the 733
Poincaré recurrence theorem 287–8,
    551
Poincaré theorem 622–3
  on the non-existence of first integrals
    of the motion 513–16
Poincaré variables 466–71
Poincaré–Bendixon theorem 704
Poincaré–Cartan integral invariant
    352–64
Poincaré's lemma 723
Poinsot cones 229
Poinsot theorem 251
point mass
  motion in a one-dimensional field
    320–1
  motion under gravity mass
    312–13
point motion
  in the absence of forces 318
  on an equipotential surface 318
point particles
  general laws and the dynamics of
    69–90
  isolated 69
  motion on an equipotential surface
    79
  motion of 87–9
  subject to unilateral constraints
    81–3
point transformations 343, 344

points
  critical 99
  equilibrium 86, 87, 99
  non-singular 1, 3
Poisson brackets 371–4
  properties of 378
Poisson's formula (attitude equation) 217
polar moment of inertia 204
polhodes 236–9, 252
  classification of *253*
  for ellipsoid of revolution *255*
  limiting 253
potentials 138
  admitting closed orbits 187–93
  generalised 142–4
  interaction 139
power 75
precessions 221
  by inertia 251–4
  composition with the same pole 225
  Euler equations for 250–1
  gyroscopic 259–60
  of a heavy gyroscope 261–3
  phase space of 221–3
  of a spinning top 261–3
principal axes of inertia 236–9
principal curvatures 157, 712
principal normal vector 8
principal reference frame 237
principle of the stationary action 316–18
probability density 615
probability measure 547
probability space 547
problem of small divisors 504
problems
  canonical formalism 399–404
  canonical perturbation theory 532–5
  ergodic theory and chaotic motion 584–6
  geometric and kinematic foundations of Lagrangian mechanics 58–61
  Gibbs sets 656–9

Hamilton–Jacobi theory and integrability 477–80
Hamiltonian formalism 288–90
kinetic theory 609–10
Lagrangian formalism in continuum mechanics 690–1
Lagrangian formalism in dynamics of discrete systems 162–5
mechanics of rigid bodies 265–6
motion in a central field 205–7
one-dimensional motion 108–12
rigid bodies, geometry and kinematics 230–1
second fundamental form of a surface 713–14
variational principles 323–4
*see also* solved problems
product manifold 47
product measure 547
product space 547
progressive waves 674

quadratic dispersion 620
quadratic fluctuations, mean 620
quadratures, integrability by 439–46
quasi-integrable Hamiltonian systems 487, 522, 535
  degenerate 516
quasi-periodic continuous functions 462–3
quasi-periodic motions 524
  and functions 458–66

radius of curvature 7
radius of gyration 236
Radon–Nikodym theorem 549–50
Rayleigh dissipation function 138
regressive waves 675
regular curves 1
regular submanifolds, parametrising 35
regular surfaces 17
relative dynamics 226–8
relative kinematics 223–6
repulsive node 699

resonance 103–7
resonance frequency 106
resonance modules 460, 461
resonance multiplicity 460
restricted three-body problem,
    Hamiltonian of 489
revolution
    ellipsoids of 255, 256
    surfaces of 20
Riemannian geometry 61
Riemannian manifolds 133
    differentiable 33–46
    geodesics 307
Riemannian metrics, on differentiable
    manifolds 42
rigid bodies
    configuration space 214
    definition 213
    dynamics of constrained 245–50
    dynamics of 126
    geometry and kinematics 213–34
    mechanics of 235–77
    problems 230–1
    solved problems 231–4
rigid motions
    composition of 225f
    plane 221
    precessions 221
    rotations 221
    ruled surfaces in 228–30
rigid systems, kinematics of 216–19
rod, longitudinal vibrations of a 676
rotations 221, 263–5, 433
    permanent 254–6
Rubin, Hungar, Takens theorem 730
Ruelle's inequality 582
ruled surfaces, in a rigid motion 228–30

saddle 699
satellites, effect of solar radiation on
    426
scale transformations 342
Schrödinger equation of wave
    mechanics 476

second form of the orbit equation 184
second fundamental form of a surface
    709–14
    problems 713–14
separatrix curve 101
series inversion (Lagrange formula)
    197–200
sets, microcanonical 624–7
Shannon–Breiman–McMillan theorem
    569–70
Siegel theorem 520
simple constraints 80
simple pendulum 96–8
Sinai billiards 577
sine amplitude 706
small divisors, problem of 504
small oscillations, equilibrium, stability
    and 150–9
smooth constraints 77
    holonomic systems with 127–8
smooth holonomic constraints,
    dynamics of a point constrained
    by 77–80
solar radiation pressure, effect on
    satellites 426
Solar System
    chaotic behaviour of orbits of planets
        582–4
    orbits of the planets in the 469–70
solved problems
    canonical formalism 405–11
    canonical perturbation theory
        535–44
    dynamics of discrete systems,
        Lagrangian formalism 165–78
    ergodic theory and chaotic motion
        586–9
    general laws and the dynamics of a
        point particle 83–90
    geometric and kinematic foundations
        of Lagrangian mechanics 62–8
    Gibbs sets 662–9
    Hamilton–Jacobi theory and
        integrability 481–6
    Hamiltonian formalism 291–300

kinetic theory 610–11
Lagrangian formalism in continuum
    mechanics 691–3
mechanics of rigid bodies 266–77
motion in a central field 208–12
one-dimensional motion 113–23
rigid bodies, geometry and
    kinematics 231–4
variational principles 324–9
see also problems
spheres
  geodesic flow on 733
  parametrisation of 19
spherical coordinates 423–4
spin–orbit problem 490–3
spinning top, precessions of a 261–3
square billiards 557
stability and small oscillations,
    equilibrium 150–9
stable equilibrium configuration 87
static friction coefficients 80
static friction cones 80
stationary action, principle of the
    316–18
stationary action of Maupertuis'
    principle 317
stationary functionals, Euler equations
    for 302–11
stationary subgroups 450
stationary waves 675
statistical mechanics
  Gibbs sets 613–69
  kinetic theory 591–612
statistical sets 613–15
star node 699
Stirling's formula 197, 747
Stokes' lemma 353, 725
Stokes' theorem 724–5
stress symmetry 673
string, vibrating 675
subgroups
  discrete 450
  stationary 450
submanifolds 33
  tangent space to regular 34

subsets, metrically indecomposable
    invariant 619
Sundman inequality 204
Sundman theorem 204
surface of revolution 20
  geodesic curves on *31*
  reversal of geodesics *32*
  separability of Hamilton–Jacobi
    equation for geodesic motion
    428–9
surfaces 16–33
  area of 27
  computing area of 27
  first fundamental form 25
  oriented 24
  regular 17
  second fundamental form of
    709–14
suspended bridges 689–90
suspended cables 685–90
symmetries, and first integrals 393–5
symmetries and conservation laws
    147–50
symmetry, material orthogonal 237
symplectic diffeomorphism 396
symplectic manifolds 395
  and Hamiltonian dynamical systems
    397–9
symplectic matrix, determinant of
    405–6
symplectic rectification 380–4
systems
  closed isolated 624–7
  separable with respect to elliptic
    coordinates 426–8
  separable with respect to parabolic
    coordinates 425–6
  separable with respect to spherical
    coordinates 423–4
systems of equations, with constant
    coefficients 697–701

tangent bundles 41
  base spaces of 41

tangent spaces 334
  to differentiable manifold 39
  to regular submanifold 34
tangent vectors 8
temperature 602–4, 627
tensor of inertia 241
tensors 725–7
thermodynamical limits 651–3
time dependence 91
time periodic perturbations of one-dimensional uniform motions 499–502
topology 38
tori 21, 49
  actions of groups and 46–9
  invariant $l$-dimensional 446–53
  linear automorphisms of 565
torsion 13
trajectories in phase space 287
transformations
  baker's 552
  canonical and completely canonical 340–52
  Gauss 552
  $p$-adic 552
tubes of characteristics 353
two-body problem 200–1

unilateral constraints, point particle subject to 81–3
unit tangent vectors 8
unit vectors 8
  binormal 12
  principal normal 8
  tangent 8

variables
  action-angle 431–9

Delaunay and Poincaré 466–71
  separation for the Hamilton–Jacobi equation 421–31
variance 620
variational principles 301–29
  problems 323–4
  solved problems 324–9
variational problems, introduction to 301–2
vector fields
  commutator of two 376
  complete 16
  Hamiltonian 338
  and integral curves 15–16
velocity 10, 55
  angular 217
velocity field 216
vibrating string 675
virial theorem 635
virtual velocity 55
visits, ergodicity and frequency of 554–62

wave interpretation of mechanics 471–6
wave mechanics, Schrödinger equation 476
waves
  plane 674
  progressive 674
  regressive 675
  stationary 675
Weierstrass, K. 527
Weierstrass's theorem 464
work, and conservative fields 75–6

Young inequality 280
Young's modulus 676, 677

Non senza fatica si giunge al fine

(Girolamo Frascobaldi, Toccata IX del II libro, 1627)

(Not without effort is the end gained)